Lecture Notes in Computer Science 8874

Commenced Publication in 1973
Founding and Former Series Editors:
Gerhard Goos, Juris Hartmanis, and Jan van Leeuwen

T0215184

Palash Sarkar   Tetsu Iwata (Eds.)

Advances in Cryptology – ASIACRYPT 2014

20th International Conference on the Theory
and Application of Cryptology and Information Security
Kaoshiung, Taiwan, December 7-11, 2014
Proceedings, Part II

 Springer

Volume Editors

Palash Sarkar
Indian Statistical Institute
Applied Statistics Unit
203, B.T. Road, Kolkata 700108, India
E-mail: palash@isical.ac.in

Tetsu Iwata
Nagoya University
Department of Computer Science and Engineering
Furo-cho, Chikusa-ku, Nagoya 464-8603, Japan
E-mail: iwata@cse.nagoya-u.ac.jp

ISSN 0302-9743   e-ISSN 1611-3349
ISBN 978-3-662-45607-1   e-ISBN 978-3-662-45608-8
DOI 10.1007/978-3-662-45608-8
Springer Heidelberg New York Dordrecht London

Library of Congress Control Number: 2014954246

LNCS Sublibrary: SL 4 – Security and Cryptology

Typesetting: Camera-ready by author, data conversion by Scientific Publishing Services, Chennai, India

Printed on acid-free paper

Springer is part of Springer Science+Business Media (www.springer.com)

Preface

It is with great pleasure that we present the proceedings of Asiacrypt 2014 in two volumes of *Lecture Notes in Computer Science* published by Springer. The year 2014 marked the 20th edition of the International Conference on Theory and Application of Cryptology and Information Security held annually in Asia by the International Association for Cryptologic Research (IACR). The conference was sponsored by the IACR and was jointly organized by the following consortium of universities and government departments of the Republic of China (Taiwan): National Sun Yat-sen University; Academia Sinica; Ministry of Science and Technology; Ministry of Education; and Ministry of Economic Affairs. The conference was held in Kaohsiung, Republic of China (Taiwan), during December 7-11, 2014.

An international Program Committee (PC) consisting of 48 scientists was formed approximately one year earlier with the objective of determining the scientific content of the conference. As for previous editions, Asiacrypt 2014 also stimulated great interest among the scientific community of cryptologists. A total of 255 technical papers were submitted for possible presentations approximately six months prior to the conference. Authors of the submitted papers are spread all over the world. Each PC member could submit at most two co-authored papers or at most one single-authored paper, and the PC co-chairs did not submit any paper. All the submissions were screened by the PC members and 55 papers were finally selected for presentation at the conference. These proceedings contain the revised versions of the papers that were selected. The revisions were not checked and the responsibility of the papers rest with the authors and not the PC members.

The selection of papers for presentations was made through a double-blind review process. Each paper was assigned four reviewers and submissions by PC members were assigned five reviewers. Apart from the PC members, the selection process was assisted by a total of 397 external reviewers. The total number of reviews for all the papers was more than 1,000. In addition to the reviews, the selection process involved an extensive discussion phase. This phase allowed PC members to express opinion on all the submissions. The final selection of 55 papers was the result of this extensive and rigorous selection procedure.

The decision of the best paper award was based on a vote among the PC members, and it was conferred upon the paper "Solving LPN Using Covering Codes" authored by Qian Guo, Thomas Johansson, and Carl Löndahl. In addition to the best paper, three other papers were recommended for solicitations by the Editor-in-Chief of the *Journal of Cryptology* to submit expanded versions to the journal. These papers are "Secret-Sharing for NP" authored by Ilan Komargodski, Moni Naor, and Eylon Yogev; "Mersenne Factorization Factory" authored by Thorsten Kleinjung, Joppe W. Bos, and Arjen K. Lenstra; and

"Jacobian Coordinates on Genus 2 Curves" authored by Huseyin Hisil and Craig Costello.

In addition to the regular presentations, the conference featured two invited talks. The invited speakers were decided through an extensive multi-round discussion among the PC members. This resulted in very interesting talks on two different aspects of the subject. Kennth G. Paterson spoke on "Big Bias Hunting in Amazonia: Large-Scale Computation and Exploitation of RC4 Biases," a topic of importance to practical cryptography, while Helaine Leggat spoke on "The Legal Infrastructure Around Information Security in Asia," which had an appeal to a wide audience.

Along with the regular presentations and the invited talks, a rump session was organized. This session contained short presentations on latest research results, announcements of future events, and other topics of interest to the audience.

Many people contributed to Asiacrypt 2014. We would like to thank the authors of all papers for submitting their research works to the conference. Thanks are due to the PC members for their enthusiastic and continued participation for over a year in different aspects of selecting the technical program. The selection of the papers was made possible by the timely reviews from external reviewers, and thanks are due to them. A list of external reviewers is provided in these proceedings. We have tried to ensure that the list is complete. Any omission is inadvertent and if there is an omission, we apologize to that person.

Special thanks are due to D. J. Guan, the general chair of the conference, for working closely with us and ensuring that the PC co-chairs were insulated from the organizational work. This work was carried out by the Organizing Committee and they deserve thanks from all the participants for the wonderful experience. We thank Daniel J. Bernstein and Tanja Lange for expertly organizing and chairing the rump session.

We thank Shai Halevi for developing the IACR conference management software, which was used for the whole process of submission, reviewing, discussions, and preparing these proceedings. We thank Josh Benaloh, our IACR liaison, and San Ling, Asiacrypt Steering Committee Representative, for guidance and advice on several issues. Springer published the volumes and made these available before the conference. We thank Alfred Hofmann, Anna Kramer, Christine Reiss and their team for the professional and efficient handling of the production process.

December 2014   Palash Sarkar
                                                                       Tetsu Iwata

Asiacrypt 2014
The 20th Annual International Conference on Theory and Application of Cryptology and Information Security

Sponsored by the *International Association for Cryptologic Research (IACR)*

December 7–11, 2014, Kaohsiung, Taiwan (R.O.C.)

General Chair

D. J. Guan — National Sun Yat-sen University, Taiwan, and National Chung Hsing University, Taiwan

Program Co-chairs

Palash Sarkar — Indian Statistical Institute, India
Tetsu Iwata — Nagoya University, Japan

Program Committee

Masayuki Abe — NTT Secure Platform Laboratories, Japan
Elena Andreeva — K.U. Leuven, Belgium
Paulo S. L. M. Barreto — University of Sao Paulo, Brazil
Daniel J. Bernstein — University of Illinois at Chicago, USA, and Technische Universiteit Eindhoven, The Netherlands
Guido Bertoni — STMicroelectronics, Italy
Jean-Luc Beuchat — ELCA, Switzerland
Debrup Chakraborty — CINVESTAV-IPN, Mexico
Chen-Mou Cheng — National Taiwan University, Taiwan
Jung Hee Cheon — Seoul National University, Korea
Ashish Choudhury — IIIT Bangalore, India
Sherman S.M. Chow — Chinese University of Hong Kong, Hong Kong SAR
Kai-Min Chung — Academia Sinica, Taiwan
Carlos Cid — Royal Holloway, University of London, UK
Jean-Sébastien Coron — University of Luxembourg, Luxembourg

Additional Reviewers

Aurélie Bauer
Carsten Baum
Anja Becker
Amos Beimel
Rishiraj Bhattacharya
Begül Bilgin
Olivier Billet
Elia Bisi
Nir Bitansky
Olivier Blazy
Céline Blondeau
Andrej Bogdanov
Alexandra Boldyreva
Joppe W. Bos
Elette Boyle
Zvika Brakerski
Nicolas Bruneau
Christina Brzuska
Sébastien Canard
Anne Canteaut
Claude Carlet
Angelo De Caro
David Cash
Dario Catalano
André Chailloux
Donghoon Chang
Pascale Charpin
Sanjit Chatterjee
Jie Chen
Wei-Han Chen
Yu-Chi Chen
Ray Cheung
Céline Chevalier
Dong Pyo Chi
Ji-Jian Chin
Alessandro Chisea
Chongwon Cho
Kim-Kwang Raymond
  Choo
HeeWon Chung
Craig Costello
Giovanni Di Crescenzo
Dana Dachman-Soled
Ivan Damgård
Jean-Luc Danger

Bernardo David
Patrick Derbez
David Derler
Srinivas Devadas
Sandra Diaz-Santiago
Vassil Dimitrov
Ning Ding
Yi Ding
Christoph Dobraunig
Matthew Dodd
Nico Döttling
Rafael Dowsley
Frédéric Dupuis
Stefan Dziembowski
Maria Eichlseder
Martianus Frederic
  Ezerman
Liming Fang
Xiwen Fang
Pooya Farshim
Sebastian Faust
Omar Fawzi
Serge Fehr
Victoria Fehr
Matthieu Finiasz
Dario Fiore
Rob Fitzpatrick
Pierre-Alain Fouque
Tore Kasper Frederiksen
Georg Fuchsbauer
Eiichiro Fujisaki
Philippe Gaborit
Tommaso Gagliardoni
David Galindo
Wei Gao
Pierrick Gaudry
Peter Gaži
Laurie Genelle
Irene Giacomelli
Sergey Gorbunov
Dov Gordon
Samuel Dov Gordon
Robert Granger
Jens Groth
Felix Guenther

Nicolas Guillermin
Sylvain Guilley
Siyao Guo
Divya Gupta
Patrick Haddad
Nguyen Manh Ha
Iftach Haitner
Shai Halevi
Fabrice Ben Hamouda
Shuai Han
Christian Hanser
Mitsuhiro Hattori
Carmit Hazay
Qiongyi He
Brett Hemenway
Jens Hermans
Takato Hirano
Jeffrey Hoffstein
Dennis Hofheinz
Deukjo Hong
Hyunsook Hong
Wei-Chih Hong
Sebastiaan de Hoogh
Jialin Huang
Kyle Huang
Qiong Huang
Yan Huang
Yun Huang
Zhengan Huang
Andreas Hülsing
Michael Hutter
Jung Yeon Hwang
Malika Izabachene
Abhishek Jain
Dirmanto Jap
Stanislaw Jarecki
Eliane Jaulmes
Jérémy Jean
Mahabir Jhanwar
Guo Jian
Shaoquan Jiang
Pascal Junod
Chethan Kamath
Pierre Karpman
Aniket Kate

Jonathan Katz
Elif Bilge Kavun
Akinori Kawachi
Yutaka Kawai
Sriram Keelveedhi
Dakshita Khurana
Franziskus Kiefer
Eike Kiltz
Jihye Kim
Jinsu Kim
Minkyu Kim
Miran Kim
Myungsun Kim
Sungwook Kim
Taechan Kim
Mehmet Sabir Kiraz
Susumu Kiyoshima
Ilya Kizhvatov
Markulf Kohlweiss
Ilan Komargodski
Takeshi Koshiba
Simon Kramer
Ranjit Kumaresan
Po-Chun Kuo
Thijs Laarhoven
Fabien Laguillaumie
Russell W.F. Lai
Tanja Lange
Adeline Langlois
Martin M. Laurisden
Rasmus Winther
    Lauritsen
Changmin Lee
Hyung Tae Lee
Kwangsu Lee
Moon Sung Lee
Younho Lee
Wang Lei
Tancrède Lepoint
Gaëtan Leurent
Kevin Lewi
Allison Lewko
Liangze Li
Wen-Ding Li
Guanfeng Liang

Kaitai Liang
Benoît Libert
Changlu Lin
Huijia (Rachel) Lin
Tingting Lin
Yannis Linge
Helger Lipmaa
Feng-Hao Liu
Joseph Liu
Zhen Liu
Daniel Loebenberger
Victor Lomné
Yu Long
Patrick Longa
Cuauhtemoc
    Mancillas-López
Atul Luykx
Vadim Lyubashevsky
Houssem Maghrebi
Mohammad Mahmoody
Alex Malozemoff
Mark Manulis
Xianping Mao
Joana Treger Marim
Giorgia Azzurra Marson
Ben Martin
Daniel Martin
Takahiro Matsuda
Mitsuru Matsui
Ingo von Maurich
Filippo Melzani
Florian Mendel
Bart Mennink
Sihem Mesnager
Arno Mittelbach
Payman Mohassel
Amir Moradi
Tomoyuki Morimae
Kirill Morozov
Nicky Mouha
Pratyay Mukherjee
Gregory Neven
Khoa Nguyen
Phon Nguyen
Ivica Nikolić

Ventzislav Nikov
Svetla Nikova
Ryo Nishimaki
Adam O'Neill
Miyako Ohkubo
Tatsuaki Okamoto
Cristina Onete
Claudio Orlandi
David Oswald
Elisabeth Oswald
Khaled Ouafi
Carles Padro
Jiaxin Pan
Omer Paneth
Anat Paskin
Rafael Pass
Kenneth G. Paterson
Arpita Patra
Roel Peeters
Chris Peikert
Geovandro
    C. C. F. Pereira
Olivier Pereira
Ludovic Perret
Edoardo Persichetti
Krzysztof Pietrzak
Bertram Poettering
Geong-Sen Poh
David Pointcheval
Antigoni Polychroniadou
Raluca Ada Popa
Manoj Prabhakaran
Baodong Qin
Somindu C. Ramanna
Samuel Ranellucci
C. Pandu Rangan
Vanishree Rao
Jean-René Reinhard
Ling Ren
Oscar Reparaz
Alfredo Rial
Jefferson E. Ricardini
Silas Richelson
Ben Riva
Matthieu Rivain

Thomas Roche
Francisco
Rodríguez-Henríquez
Lil María
Rodríguez-Henríquez
Mike Rosulek
Arnab Roy
Hansol Ryu
Minoru Saeki
Amit Sahai
Yusuke Sakai
Olivier Sanders
Fabrizio De Santis
Yu Sasaki
Alessandra Scafuro
Christian Schaffner
John Schanck
Tobias Schneider
Peter Schwabe
Gil Segev
Nicolas Sendrier
Jae Hong Seo
Karn Seth
Yannick Seurin
Ronen Shaltiel
Elaine Shi
Koichi Shimizu
Ji Sun Shin
Naoyuki Shinohara
Joseph Silverman
Marcos A. Simplicio Jr
Boris Skoric
Daniel Slamanig
Nigel Smart
Fang Song
Douglas Stebila
Damien Stehlé
Rainer Steinwandt
Marc Stottinger

Mario Strefler
Takeshi Sugawara
Ruggero Susella
Koutarou Suzuki
Alan Szepieniec
Björn Tackmann
Katsuyuki Takashima
Syh-Yuan Tan
Xiao Tan
Qiang Tang
Christophe Tartary
Yannick Teglia
Sidharth Telang
Isamu Teranishi
Adrian Thillard
Aishwarya
Thiruvengadam
Enrico Thomae
Susan Thomson
Mehdi Tibouchi
Tyge Tiessen
Elmar Tischhauser
Arnaud Tisserand
Yosuke Todo
Jacques Traoré
Roberto Trifiletti
Viet Cuong Trinh
Raylin Tso
Toyohiro Tsurumaru
Hoang Viet Tung
Yu-Hsiu Tung
Dominique Unruh
Berkant Ustaoglu
Meilof Veeningen
Muthuramakrishnan
Venkitasubramaniam
Daniele Venturi
Frederik Vercauteren
Damien Vergnaud

Andrea Visconti
Ivan Visconti
Niels de Vreede
Mingqiang Wang
Wei Wang
Yanfeng Wang
Yuntao Wang
Hoeteck Wee
Puwen Wei
Qiaoyan Wen
Erich Wenger
Qianhong Wu
Keita Xagawa
Hong Xu
Weijia Xue
Takashi Yamakawa
Bo-Yin Yang
Guomin Yang
Wun-She Yap
Scott Yilek
Eylon Yogev
Kazuki Yoneyama
Ching-Hua Yu
Yu Yu
Tsz Hon Yuen
Aaram Yun
Mark Zhandry
Cong Zhang
Guoyan Zhang
Liang Feng Zhang
Tao Zhang
Wei Zhang
Ye Zhang
Yun Zhang
Zongyang Zhang
Yongjun Zhao
Yunlei Zhao
Vassilis Zikas

Organizing Committee

Advisors

Lynn Batten	Deakin University, Australia
Eiji Okamoto	Tsukuba University, Japan
San Ling	Nanyang Technological University, Singapore
Kwangjo Kim	Korea Advanced Institute of Science and Technology, Korea
Xuejia Lai	Shanghai Jiaotong University, China
Der-Tsai Lee	National Chung Hsing University, Taiwan, and Academia Sinica, Taiwan
Tzong-ChenWu	National Taiwan University of Science and Technology, Taiwan

Secretary

Chun-I Fan	National Sun Yat-sen University, Taiwan

Treasurer

Chia-Mei Chen	National Sun Yat-sen University, Taiwan

Local Committee Members

Shiuhpyng Shieh	National Chiao Tung University, Taiwan
Ching-Long Lei	National Taiwan University, Taiwan
Wen-Guey Tzeng	National Chiao Tung University, Taiwan
Hung-Min Sun	National Tsing Hua University, Taiwan
Chen-Mou Cheng	National Taiwan University, Taiwan
Bo-Yin Yang	Institute of Information Science, Academia Sinica, Taiwan

Sponsors

National Sun Yat-sen University
Academia Sinica
Ministry of Science and Technology
Ministry of Education
Ministry of Economic Affairs

The Legal Infrastructure around Information Security in Asia (Abstract)

Helaine Leggat
Information Legal, Australia
helaine@informationlegal.com.au
http://www.informationlegal.com.au

Abstract. If the history of the Internet can be said to have commenced with the development of electronic computers in the 1950s, it took almost fifty years for the world to embrace the need to facilitate and to regulate electronic commerce and communications. This was brought about initially, through the United Nations Commission on International Trade Law (UNCITRAL) Model Law on Electronic Commerce in 1996. It was followed by the UNCITRAL Model Law on Electronic Signatures in 2001 and the United Nations Convention on the Use of Electronic Communications in International Contracts adopted by the United Nations General Assembly in 2005.

The theory and application of cryptology and information security are directly connected to these model laws and conventions.

In this talk, we will look, at a high level, at these developments and then shift the focus, with more detail, to the development and status of electronic law in Asia, including but not limited to Taiwan, Hong Kong, Japan, Singapore, India, Australia and New Zealand.

We will look at the overarching and interpretive nature of electronic transactions laws on laws that pre-date the electronic age and at subsequent ICT-specific laws, concentrating on three major trends (i) access to (or freedom of) information, (ii) monitoring and surveillance and (iii) information privacy.

We will look at competing rights and limitations in the global and national contexts. Including, developments in the light of economic, geographic and political shifts in dominance from West to East, the diffusion of power from state to society and the importance of knowledge based capital in the form of intangible assets.

The intention of the talk is to empower persons with an interest in cryptology and information security to look more broadly at the regulatory context and changing paradigms that can inspire hot topic research arising from law as a social science in relation to cryptology.

keywords: Law, information, security, cryptology, Asia.

Table of Contents – Part II

Encryption Schemes

Concise Multi-challenge CCA-Secure Encryption and Signatures with
Almost Tight Security 1
    Benoît Libert, Marc Joye, Moti Yung, and Thomas Peters

Efficient Identity-Based Encryption over NTRU Lattices 22
    Léo Ducas, Vadim Lyubashevsky, and Thomas Prest

Order-Preserving Encryption Secure Beyond One-Wayness 42
    Isamu Teranishi, Moti Yung, and Tal Malkin

Outsourcing and Delegation

Statistically-secure ORAM with $\tilde{O}(\log^2 n)$ Overhead 62
    Kai-Min Chung, Zhenming Liu, and Rafael Pass

Adaptive Security of Constrained PRFs 82
    *Georg Fuchsbauer, Momchil Konstantinov, Krzysztof Pietrzak, and
Vanishree Rao*

Obfuscation

Poly-Many Hardcore Bits for Any One-Way Function and a Framework
for Differing-Inputs Obfuscation 102
    Mihir Bellare, Igors Stepanovs, and Stefano Tessaro

Using Indistinguishability Obfuscation via UCEs 122
    Christina Brzuska and Arno Mittelbach

Indistinguishability Obfuscation versus Multi-bit Point Obfuscation
with Auxiliary Input 142
    Christina Brzuska and Arno Mittelbach

Bootstrapping Obfuscators via Fast Pseudorandom Functions.......... 162
    Benny Applebaum

Homomorphic Cryptography

Homomorphic Authenticated Encryption Secure against
Chosen-Ciphertext Attack.. 173
    Chihong Joo and Aaram Yun

Authenticating Computation on Groups: New Homomorphic Primitives
and Applications 193
    Dario Catalano, Antonio Marcedone, and Orazio Puglisi

Compact VSS and Efficient Homomorphic UC Commitments 213
    Ivan Damgård, Bernardo David, Irene Giacomelli,
    and Jesper Buus Nielsen

Secret Sharing

Round-Optimal Password-Protected Secret Sharing and T-PAKE
in the Password-Only Model 233
    Stanislaw Jarecki, Aggelos Kiayias, and Hugo Krawczyk

Secret-Sharing for NP 254
    Ilan Komargodski, Moni Naor, and Eylon Yogev

Block Ciphers and Passwords

Tweaks and Keys for Block Ciphers: The TWEAKEY Framework 274
    Jérémy Jean, Ivica Nikolić, and Thomas Peyrin

Memory-Demanding Password Scrambling 289
    Christian Forler, Stefan Lucks, and Jakob Wenzel

Side Channel Analysis II

Side-Channel Analysis of Multiplications in $GF(2^{128})$:
Application to AES-GCM 306
    Sonia Belaïd, Pierre-Alain Fouque, and Benoît Gérard

Higher-Order Threshold Implementations 326
    Begül Bilgin, Benedikt Gierlichs, Svetla Nikova, Ventzislav Nikov,
    and Vincent Rijmen

Masks Will Fall Off: Higher-Order Optimal Distinguishers 344
    Nicolas Bruneau, Sylvain Guilley, Annelie Heuser, and Olivier Rioul

Black-Box Separation

Black-Box Separations for One-More (Static) CDH
and Its Generalization 366
    Jiang Zhang, Zhenfeng Zhang, Yu Chen, Yanfei Guo,
    and Zongyang Zhang

Black-Box Separations for Differentially Private Protocols 386
    Dakshita Khurana, Hemanta K. Maji, and Amit Sahai

Composability

Composable Security of Delegated Quantum Computation 406
Vedran Dunjko, Joseph F. Fitzsimons, Christopher Portmann, and Renato Renner

All-But-Many Encryption: A New Framework for Fully-Equipped UC
Commitments 426
Eiichiro Fujisaki

Multi-Party Computation

Multi-valued Byzantine Broadcast: The $t < n$ Case 448
Martin Hirt and Pavel Raykov

Fairness versus Guaranteed Output Delivery in Secure Multiparty
Computation 466
Ran Cohen and Yehuda Lindell

Actively Secure Private Function Evaluation........................ 486
Payman Mohassel, Saeed Sadeghian, and Nigel P. Smart

Efficient, Oblivious Data Structures for MPC 506
Marcel Keller and Peter Scholl

Author Index 527

Table of Contents – Part I

Cryptology and Coding Theory

Solving LPN Using Covering Codes 1
  Qian Guo, Thomas Johansson, and Carl Löndahl

Algebraic Attack against Variants of McEliece with Goppa Polynomial
of a Special Form 21
  Jean-Charles Faugère, Ludovic Perret, and Frédéric de Portzamparc

New Proposals

Bivariate Polynomials Modulo Composites and Their Applications 42
  Dan Boneh and Henry Corrigan-Gibbs

Cryptographic Schemes Based on the ASASA Structure: Black-box,
White-box, and Public-key (Extended Abstract) 63
  Alex Biryukov, Charles Bouillaguet, and Dmitry Khovratovich

Authenticated Encryption

Beyond $2^{c/2}$ Security in Sponge-Based Authenticated
Encryption Modes 85
  Philipp Jovanovic, Atul Luykx, and Bart Mennink

How to Securely Release Unverified Plaintext in Authenticated
Encryption 105
  *Elena Andreeva, Andrey Bogdanov, Atul Luykx, Bart Mennink,
  Nicky Mouha, and Kan Yasuda*

Forging Attacks on Two Authenticated Encryption Schemes COBRA
and POET 126
  Mridul Nandi

Symmetric Key Cryptanalysis

Low Probability Differentials and the Cryptanalysis of Full-Round
CLEFIA-128 141
  *Sareh Emami, San Ling, Ivica Nikolić, Josef Pieprzyk, and
  Huaxiong Wang*

Automatic Security Evaluation and (Related-key) Differential
Characteristic Search: Application to SIMON, PRESENT, LBlock,
DES(L) and Other Bit-Oriented Block Ciphers 158
    Siwei Sun, Lei Hu, Peng Wang, Kexin Qiao, Xiaoshuang Ma, and
    Ling Song

Scrutinizing and Improving Impossible Differential Attacks:
Applications to CLEFIA, Camellia, LBlock and SIMON 179
    Christina Boura, María Naya-Plasencia, and Valentin Suder

A Simplified Representation of AES 200
    Henri Gilbert

Side Channel Analysis I

Simulatable Leakage: Analysis, Pitfalls, and New Constructions 223
    Jake Longo, Daniel P. Martin, Elisabeth Oswald,
    Daniel Page, Martijin Stam, and Michael J. Tunstall

Multi-target DPA Attacks: Pushing DPA Beyond the Limits of a
Desktop Computer 243
    Luke Mather, Elisabeth Oswald, and Carolyn Whitnall

GLV/GLS Decomposition, Power Analysis, and Attacks on ECDSA
Signatures with Single-Bit Nonce Bias 262
    Diego F. Aranha, Pierre-Alain Fouque, Benoît Gérard,
    Jean-Gabriel Kammerer, Mehdi Tibouchi,
    and Jean-Christophe Zapalowicz

Soft Analytical Side-Channel Attacks 282
    Nicolas Veyrat-Charvillon, Benoît Gérard,
    and François-Xavier Standaert

Hyperelliptic Curve Cryptography

On the Enumeration of Double-Base Chains with Applications to
Elliptic Curve Cryptography 297
    Christophe Doche

Kummer Strikes Back: New DH Speed Records 317
    Daniel J. Bernstein, Chitchanok Chuengsatiansup, Tanja Lange,
    and Peter Schwabe

Jacobian Coordinates on Genus 2 Curves 338
    Huseyin Hisil and Craig Costello

Factoring and Discrete Log

Mersenne Factorization Factory 358
    Thorsten Kleinjung, Joppe W. Bos, and Arjen K. Lenstra

Improving the Polynomial time Precomputation of Frobenius
Representation Discrete Logarithm Algorithms: Simplified Setting for
Small Characteristic Finite Fields 378
    Antoine Joux and Cécile Pierrot

Invited Talk I

Big Bias Hunting in Amazonia: Large-Scale Computation and
Exploitation of RC4 Biases (Invited Paper) 398
    Kenneth G. Paterson, Bertram Poettering, and Jacob C.N. Schuldt

Cryptanalysis

Multi-user Collisions: Applications to Discrete Logarithm,
Even-Mansour and PRINCE 420
    Pierre-Alain Fouque, Antoine Joux, and Chrysanthi Mavromati

Cryptanalysis of Iterated Even-Mansour Schemes with Two Keys 439
    Itai Dinur, Orr Dunkelman, Nathan Keller, and Adi Shamir

Meet-in-the-Middle Attacks on Generic Feistel Constructions 458
    Jian Guo, Jérémy Jean, Ivica Nikolić, and Yu Sasaki

XLS is Not a Strong Pseudorandom Permutation 478
    Mridul Nandi

Signatures

Structure-Preserving Signatures on Equivalence Classes and Their
Application to Anonymous Credentials 491
    Christian Hanser and Daniel Slamanig

On Tight Security Proofs for Schnorr Signatures 512
    Nils Fleischhacker, Tibor Jager, and Dominique Schröder

Zero-Knowledge

Square Span Programs with Applications to Succinct NIZK
Arguments 532
    George Danezis, Cédric Fournet, Jens Groth, and Markulf Kohlweiss

Better Zero-Knowledge Proofs for Lattice Encryption and Their
Application to Group Signatures   551
    Fabrice Benhamouda, Jan Camenisch, Stephan Krenn,
    Vadim Lyubashevsky, and Gregory Neven

Author Index   573

Concise Multi-challenge CCA-Secure Encryption and Signatures with Almost Tight Security

Benoît Libert[1,*], Marc Joye[2], Moti Yung[3], and Thomas Peters[4,**]

[1] Ecole Normale Supérieure de Lyon,
Laboratoire de l'Informatique du Parallélisme, France
[2] Technicolor, USA
[3] Google Inc. and Columbia University, USA
[4] Université catholique de Louvain, Crypto Group, Belgium

Abstract. To gain strong confidence in the security of a public-key scheme, it is most desirable for the security proof to feature a *tight* reduction between the adversary and the algorithm solving the underlying hard problem. Recently, Chen and Wee (Crypto '13) described the first Identity-Based Encryption scheme with almost tight security under a standard assumption. Here, "almost tight" means that the security reduction only loses a factor $O(\lambda)$ —where λ is the security parameter— instead of a factor proportional to the number of adversarial queries. Chen and Wee also gave the shortest signatures whose security almost tightly relates to a simple assumption in the standard model. Also recently, Hofheinz and Jager (Crypto '12) constructed the first CCA-secure public-key encryption scheme in the multi-user setting with tight security. These constructions give schemes that are significantly less efficient in length (and thus, processing) when compared with the earlier schemes with loose reductions in their proof of security. Hofheinz and Jager's scheme has a ciphertext of a few hundreds of group elements, and they left open the problem of finding truly efficient constructions. Likewise, Chen and Wee's signatures and IBE schemes are somewhat less efficient than previous constructions with loose reductions from the same assumptions. In this paper, we consider space-efficient schemes with security almost tightly related to standard assumptions. We construct an efficient CCA-secure public-key encryption scheme whose chosen-ciphertext security in the multi-challenge, multi-user setting almost tightly relates to the DLIN assumption (in the standard model). Quite remarkably, the ciphertext size decreases to 69 group elements under the DLIN assumption whereas the best previous solution required about 400 group elements. Our scheme is obtained by taking advantage of a new almost tightly secure signature scheme (in the standard model) which is based on the recent concise proofs of linear subspace membership in the quasi-adaptive non-interactive zero-knowledge setting (QA-NIZK) defined by Jutla and Roy (Asiacrypt '13). Our signature scheme reduces the length

---

* Part of this work was done while this author was with Technicolor (France).
** This author was supported by the CAMUS Walloon Region Project.

P. Sarkar and T. Iwata (Eds.): ASIACRYPT 2014, PART II, LNCS 8874, pp. 1–21, 2014.

of the previous such signatures (by Chen and Wee) by 37% under the Decision Linear assumption, by almost 50% under the K-LIN assumption, and it becomes only 3 group elements long under the Symmetric eXternal Diffie-Hellman assumption. Our signatures are obtained by carefully combining the proof technique of Chen and Wee and the above mentioned QA-NIZK proofs.

Keywords: CCA-secure encryption, multi-user, multi-challenge, signature, IND-CCA2 security, QA-NIZK proofs, tight security, efficiency.

1   Introduction

Security of public-key cryptographic primitives is established by demonstrating that any successful probabilistic polynomial time (PPT) adversary $\mathcal{A}$ implies a PPT algorithm $\mathcal{B}$ solving a hard problem. In order to be convincing, such "reductionist" arguments should be as *tight* as possible. Ideally, algorithm $\mathcal{B}$'s probability of success should be about as large as the adversary's advantage. The results of Bellare and Rogaway [9] initiated an important body of work devoted to the design of primitives validated by tight security reductions in the random oracle model [22,23,38,20,21,10,24,48,1,37] and in the standard model [21,7,48].

Tight security proofs may be hard to achieve and are even known not to exist at all in some situations [23,37,33]. On the positive side, long-standing open problems have been resolved in the recent years. Hofheinz and Jager [31] showed the first public-key encryption scheme whose chosen-ciphertext security [45,46] in the multi-user setting tightly relates to a standard hardness assumption, which solved a problem left open by Bellare, Boldyreva and Micali [6] although their ciphertext is a few hundreds group elements long. Chen and Wee [19] answered an important open question raised by Waters [51] by avoiding the concrete security loss, proportional to the number of adversarial queries, that affected the security reductions of all prior identity-based encryption (IBE) [14,49] schemes based on simple assumptions, including those based on the dual system paradigm [52,39]. The results of [19] also implied the shortest signatures almost tightly related to simple assumptions[1] in the standard model. In the terminology of [19], "almost tight security" refers to reductions where the degradation factor only depends on the security parameter λ, and not on the number q of adversarial queries, which is potentially much larger as it is common to assume $\lambda = 128$ and $q \approx 2^{30}$.

The tighter security results of Chen and Wee [19] overcame an important barrier since, as pointed out in [19], all earlier short signatures based on standard assumptions in the standard model [51,34,32,53,12] suffered a $\Theta(q)$ loss in terms of exact security. On the other hand, the Chen-Wee schemes are less efficient than previous solutions based on similar assumptions [51,39,18,12]. Likewise,

---

[1] By "simple assumptions," we mean non-interactive (and thus falsifiable [43]) assumptions that can be described using a *constant* number of group elements. In particular, the number of input elements in the description of the assumption does not depend on the number of adversarial queries.

encryption schemes with tight multi-challenge chosen-ciphertext security [31,3] come at the expense of much longer ciphertexts than constructions (e.g., [25]) in the single-challenge setting.[2] In order to exploit concrete security improvements in the choice of parameters, it is desirable to keep schemes as efficient —from both computational and space viewpoints— as their counterparts backed by loose reductions. This paper aims at rendering the constructions and techniques of [31,19] truly competitive with existing signatures and encryption schemes based on simple assumptions in the standard model.

OUR CONTRIBUTIONS. In this paper, we construct a new public-key encryption scheme with almost tight chosen-ciphertext (IND-CCA2) security in the multi-user, multi-challenge setting [6] under the DLIN assumption. As in the setting of Chen and Wee, the underlying reduction is not as tight as those of [31,3] since we lose a factor of $O(\lambda)$. On the other hand, our construction provides much shorter ciphertexts than previous tightly IND-CCA2-secure systems [31,3] based on the same assumption. Moreover, our security bound does not depend on the number of users or on the number of challenges, so that our scheme can be safely instantiated in environments involving arbitrarily many users encrypting as many ciphertexts as they like.

As a tool for achieving our encryption scheme (and a result of independent interest), we devise a variant of the Chen-Wee signature scheme [19], which has been proved almost tightly secure under the DLIN assumption, with shorter signatures in prime-order groups. Under the DLIN assumption, each signature consists of 6 groups elements, instead of 8 in [19]. Under the K-linear assumption (which is believed weaker than DLIN when $K > 2$), we reduce the signature length of [19] from $4K$ to $2K + 2$ and thus save $\Theta(K)$ group elements.

By combining our technique and the recent non-interactive proof systems of Jutla and Roy [36], we can further shorten our signatures and obtain 5 group elements per signature under the DLIN assumption and $2K + 1$ elements under the K-linear assumption. Our DLIN-based (resp. K-linear-based) system thus improves upon the Chen-Wee constructions [19] by 37% (resp. nearly 50%) in terms of signature length. Under the Symmetric eXternal Diffie-Hellman assumption (namely, the hardness of DDH in $\mathbb{G}$ and $\hat{\mathbb{G}}$ for asymmetric pairings $e : \mathbb{G} \times \hat{\mathbb{G}} \to \mathbb{G}_T$), the same optimizations yield signatures comprised of only 3 group elements, which only exceeds the length of Waters signatures [51] by one group element. Since the SXDH-based signatures of [19] live in $\mathbb{G}^4$, we also shorten them by one element (or 25%) under the same assumption. Our SXDH-based scheme turns out to yield the shortest known signature with nearly tight security under a simple assumption.

While randomizable in their basic variant, our schemes can be made strongly unforgeable in a direct manner, without any increase of the signature length.

---

[2] Using a hybrid argument, Bellare, Boldyreva and Micali [6] showed that any CCA2-secure encryption scheme in the single-challenge setting remains secure if the adversary is given arbitrarily many challenge ciphertexts. However, the reduction is linearly affected by the number q of challenge ciphertexts.

In particular, we do not need generic transformations based on chameleon hash functions, such as the one of Boneh *el al.* [15], which tend to lengthen signatures by incorporating the random coins of the chameleon hashing algorithm. Using the SXDH assumption and asymmetric pairings, we thus obtain the same signature length as the CDH-based strongly unforgeable signatures of Boneh, Shen and Waters [15] with the benefit of a much better concrete security (albeit under a stronger assumption).

Then, our signature schemes can be applied to construct a new efficient public-key encryption scheme with almost tight chosen-ciphertext (IND-CCA) security in the multi-user, multi-challenge setting [6]. Indeed, the randomizable signatures described in this paper easily lend themselves to the construction of new unbounded simulation-sound proof systems (where the adversary remains unable to prove false statements after having seen polynomially many simulated proofs for possibly false statements) with almost tight security. In turn, this yields the most efficient constructions, to date, of IND-CCA-secure public-key encryption schemes in the multi-challenge setting. By following the approach of [29,31], we can obtain an almost tightly simulation-sound proof system by showing that either: *(i)* a set of pairing product equations is satisfiable; and *(ii)* committed group elements form a valid signature on the verification key of a one-time signature. In this case, our randomizable signatures are very interesting candidates since they reduce the number of signature components that must appear in committed form. In addition, the specific algebraic properties of our signature scheme make it possible to construct an optimized simulation-extractable proof system that allows proving knowledge of the plaintext using only 62 group elements, which reduces our ciphertexts to only 69 group elements under the DLIN assumption. This dramatically improves upon previous tightly secure constructions based on the same assumption [31,3] which require several hundreds of group elements per ciphertext. Moreover, unlike [3], our system can also be instantiated in asymmetric pairing configurations. We stress that, unlike [42] (which has a loose security reduction), our simulation-sound proof system does not provide constant-size proofs of linear subspace membership. Still, for the specific application of nearly tight CCA-security, our proof system suffices to obtain relatively concise ciphertexts.

Concurrent to our work, Blazy, Kiltz and Pan [11] independently gave different constructions of signature schemes with tight security under the SXDH, DLIN and other simple assumptions. Their technique extends to provide (hierarchical) identity-based encryption schemes. Under the DLIN and SXDH assumption, our optimized signatures are as short as theirs. Our approach bears similiarities with theirs in that each signature can be seen as a NIZK proof that a message authentication code is valid w.r.t. a committed key.

OUR TECHNIQUES. Underlying our results is a methodology of getting security proofs with a short chain of transitions from actual games to ideal ones. Our constructions build upon a signature scheme of Jutla and Roy [35, Section 5], which is itself inspired by [16, Appendix A.3]. In [35], each signature is a CCA2-secure encryption of the private key, where the message is included in the label [50] of the ciphertext. The signer also computes a non-interactive zero-knowledge proof that the encrypted value is the private key. The security proof uses the dual system encryption method [52,40,28] and proceeds with a sequence of hybrid games heading for a game where all signatures encrypt a random value while the NIZK proofs are simulated.

While Camenisch *el al.* [16] used Groth-Sahai proofs, Jutla and Roy obtained a better efficiency using *quasi-adaptive* NIZK (QA-NIZK) proofs, i.e., where the common reference string (CRS) may depend on the specific language for which proofs are being generated but a single CRS simulator works for the entire class of languages. For the common task of proving that a vector of n group elements belongs to a linear subspace of rank t, Jutla and Roy [35] gave computationally sound QA-NIZK proofs of length $\Theta(n-t)$ where the Groth-Sahai (GS) techniques entail $\Theta(n+t)$ group elements per proof. They subsequently refined their techniques, reducing the proof's length to a constant [36], regardless of the number of equations or the number of variables. Libert *el al.* [42] independently obtained similar improvements using different techniques.

Our signature schemes rely on the observation that the constant-size QA-NIZK proofs of [42,36] make it possible to encode the label (which contains the message) in a bit-by-bit manner without affecting the signature length. In turn, this allows applying the technique of Chen and Wee [19] so as to avoid the need for q transitions, where q is the number of signing queries. As in the security proof of [19], the signing oracle uses a semi-functional private key which is obtained by shifting a normal private key by a factor consisting of a random function that depends on increasingly many bits of the message in each transition. In the last game, the random function depends on all the message bits, so that the shifting factor is thus totally unpredictable by the adversary.

Our encryption of almost tightly CCA2-secure encryption scheme is based on a modification of the Naor-Yung [45] paradigm due to [26,3]. The latter consists in combining an IND-CPA encryption and a simulation-extractable proof of knowledge of the plaintext. In order to build an optimized simulation-extractable proof, we take advantage of the simple algebraic structure of our signature scheme and its randomizability properties. Our proof system is a simplification of the one in [3] and shows that either: (i) A commitment is an extractable commitment to a function of the encryption exponents; or (ii) Another commitment contained in the proof contains a valid signature on the verfication key of a one-time signature. Our signature scheme allows implementing this very efficiently. Specifically, a real proof used by the encryption algorithm involves a commitment to a pseudo-signature —which can be generated without the signing key— whereas a simulated proof uses a real signature instead of a pseudo-signature. The perfect witness indistinguishability of Groth-Sahai proofs on a NIWI CRS guarantees that the adversary will not be able to distinguish committed pseudo-signatures from real signatures.

2   Background and Definitions

2.1   Hardness Assumptions

We consider groups $(\mathbb{G}, \hat{\mathbb{G}}, \mathbb{G}_T)$ of prime-order p endowed with a bilinear map $e : \mathbb{G} \times \hat{\mathbb{G}} \to \mathbb{G}_T$. In this setting, we rely on the standard Decision Linear assumption, which is a special case of the K-linear assumption for $K = 2$.

Definition 1 ([13]). *The* Decision Linear Problem *(DLIN) in a group $\mathbb{G}$, is to distinguish between the distributions $(g^a, g^b, g^{ac}, g^{bd}, g^{c+d})$ and $(g^a, g^b, g^{ac}, g^{bd}, g^z)$, with $a, b, c, d \xleftarrow{R} \mathbb{Z}_p$, $z \xleftarrow{R} \mathbb{Z}_p$. The* Decision Linear *assumption asserts the intractability of DLIN for any PPT distinguisher.*

It will sometimes be convenient to use the following assumption, which is implied by DLIN, as observed in [17].

Definition 2. *The* Simultaneous Double Pairing problem *(SDP) in $(\mathbb{G}, \hat{\mathbb{G}}, \mathbb{G}_T)$ is, given a tuple of group elements $(\hat{g}_z, \hat{g}_r, \hat{h}_z, \hat{h}_u) \in \hat{\mathbb{G}}^4$, to find a non-trivial triple $(z, r, u) \in \mathbb{G}^3 \backslash \{(1_\mathbb{G}, 1_\mathbb{G}, 1_\mathbb{G})\}$ satisfying the equalities $e(z, \hat{g}_z) \cdot e(r, \hat{g}_r) = 1_{\mathbb{G}_T}$ and $e(z, \hat{h}_z) \cdot e(u, \hat{h}_u) = 1_{\mathbb{G}_T}$.*

2.2   One-Time Linearly Homomorphic Structure-Preserving Signatures

In structure-preserving signatures [5,4], messages and public keys all consist of elements of a group over which a bilinear map $e : \mathbb{G} \times \hat{\mathbb{G}} \to \mathbb{G}_T$ is efficiently computable. Constructions based on simple assumptions were put forth in [2,3].

Libert *el al.* [41] considered structure-preserving schemes with linear homomorphic properties. This section recalls the one-time linearly homomorphic structure-preserving signature (LHSPS) of [41].

Keygen(λ, n): Given a security parameter λ and the dimension $n \in \mathbb{N}$ of the subspace to be signed, choose bilinear group $(\mathbb{G}, \hat{\mathbb{G}}, \mathbb{G}_T)$ of prime order p. Then, choose $\hat{g}_z, \hat{g}_r, \hat{h}_z, \hat{h}_u \xleftarrow{R} \hat{\mathbb{G}}$. For $i = 1$ to n, pick $\chi_i, \gamma_i, \delta_i \xleftarrow{R} \mathbb{Z}_p$ and compute $\hat{g}_i = \hat{g}_z{}^{\chi_i} \hat{g}_r{}^{\gamma_i}$, $\hat{h}_i = \hat{h}_z{}^{\chi_i} \hat{h}_u{}^{\delta_i}$. The private key is $\mathsf{sk} = \{(\chi_i, \gamma_i, \delta_i)\}_{i=1}^n$ while the public key is $\mathsf{pk} = (\hat{g}_z, \hat{g}_r, \hat{h}_z, \hat{h}_u, \{(\hat{g}_i, \hat{h}_i)\}_{i=1}^n) \in \hat{\mathbb{G}}^{2n+4}$.

Sign($\mathsf{sk}, (M_1, \ldots, M_n)$): To sign a vector $(M_1, \ldots, M_n) \in \mathbb{G}^n$ using the key $\mathsf{sk} = \{(\chi_i, \gamma_i, \delta_i)\}_{i=1}^n$, output $\sigma = (z, r, u) \in \mathbb{G}^3$, where $z = \prod_{i=1}^n M_i^{-\chi_i}$, $r = \prod_{i=1}^n M_i^{-\gamma_i}$ and $u = \prod_{i=1}^n M_i^{-\delta_i}$.

SignDerive($\mathsf{pk}, \{(\omega_i, \sigma^{(i)})\}_{i=1}^\ell$): Given pk as well as ℓ tuples $(\omega_i, \sigma^{(i)})$, parse $\sigma^{(i)}$ as $\sigma^{(i)} = (z_i, r_i, u_i)$ for $i = 1$ to ℓ. Compute and return $\sigma = (z, r, u)$, where $z = \prod_{i=1}^\ell z_i^{\omega_i}$, $r = \prod_{i=1}^\ell r_i^{\omega_i}$, $u = \prod_{i=1}^\ell u_i^{\omega_i}$.

Verify($\mathsf{pk}, \sigma, (M_1, \ldots, M_n)$): Given a signature $\sigma = (z, r, u) \in \mathbb{G}^3$ and a vector $(M_1, \ldots, M_n)$, return 1 if and only if $(M_1, \ldots, M_n) \neq (1_\mathbb{G}, \ldots, 1_\mathbb{G})$ and (z, r, u) satisfy the relations $1_{\mathbb{G}_T} = e(z, \hat{g}_z) \cdot e(r, \hat{g}_r) \cdot \prod_{i=1}^n e(M_i, \hat{g}_i)$, and $1_{\mathbb{G}_T} = e(z, \hat{h}_z) \cdot e(u, \hat{h}_u) \cdot \prod_{i=1}^n e(M_i, \hat{h}_i)$.

The one-time security of the scheme (in the sense of [41]) was proved [41] under the SDP assumption under a tight reduction. In short, the security notion implies the infeasibility of deriving a signature on a vector outside the subspace spanned by the vectors authenticated by the signer. Here, "one-time" security means that a given public key allows signing only one subspace.

3 Shorter Signatures Almost Tightly Related to the DLIN Assumption

This section shows that LHSPS schemes and constant-size QA-NIZK proofs for linear subspaces can be used to construct shorter signatures with nearly optimal reductions under the DLIN assumption.

The scheme builds on ideas used in a signature scheme suggested by Jutla and Roy [35, Section 5], where each signature is a CCA2-secure encryption —using the message to be signed as a label— of the private key augmented with a QA-NIZK proof (as defined in [35]) that the encrypted value is a persistent hidden secret. As in [52,40,28], the security proof uses a sequence of games which gradually moves to a game where all signatures contain an encryption of a random value while the QA-NIZK proofs are simulated. At each step of the transition, increasingly many signatures are generated without using the private key and the CCA2-security of the encryption scheme ensures that this should not affect the adversary's probability to output a signature that does encrypt the private key. In the security proof of [35], the latter approach implies that: (i) the number of transitions depends on the number of signing queries; and (ii) a CCA2-secure encryption scheme is needed since, at each transition, the reduction has to decrypt the ciphertext contained in the forgery.

Here, our key observation is that, by using a QA-NIZK proof system where the proof length is independent of the dimension of the considered linear subspace, the approach of [35] can be combined with the proof technique of Chen and Wee [19] so as to reduce the number of game transitions while retaining short signatures. In addition, the techniques of [19] allow us to dispense with the need for a CCA2-secure encryption scheme. The security analysis actually departs from that of [35] and rather follows the one of Chen and Wee [19]. The techniques of [35,16,28] argue that, even if the adversary is given signatures where the private key is blinded by a semi-functional component, its forgery will retain the distribution of a normal signature unless some indistinguishability assumption is broken. Here, we follow [19] and blind the outputs of the signing oracle by a random function of increasingly many bits of the message. Instead of using the same argument as in [35], however, we argue that the adversary's forgery will always have the same distribution as the signatures produced by the signing oracle. In the last game of the hybrid sequence, we prove that the adversary cannot retain the same behavior as the signing oracle since the latter's outputs are blinded by a random function of *all* message bits. In order to come up with the same kind of signature as the signing oracle, the adversary would have to predict the value of the random function on the forgery message $M^\star$, which is information-theoretically infeasible.

As in [19], by guessing exactly one bit of the target message, the reduction can efficiently test whether the forgery has the same distribution as outputs of the signing oracle while remaining able to embed a DLIN instance in outputs of signing queries. For L-bit messages, by applying arguments similar to those of [44,19], we need L game transitions to reach a game where each signature encrypts a random —and thus unpredictable— function of the message. As a result, we obtain DLIN-based signatures comprised of only 6 group elements.

Keygen(λ): Choose bilinear groups $(\mathbb{G}, \hat{\mathbb{G}}, \mathbb{G}_T)$ of prime order p together with $f, g, h, u_1, u_2 \xleftarrow{R} \mathbb{G}$.

1. For $\ell = 1$ to L, choose $V_{\ell,0}, V_{\ell,1}, W_{\ell,0}, W_{\ell,1} \xleftarrow{R} \mathbb{G}$ to assemble row vectors

$$\boldsymbol{V} = (V_{1,0}, V_{1,1}, \ldots, V_{L,0}, V_{L,1}), \quad \boldsymbol{W} = (W_{1,0}, W_{1,1}, \ldots, W_{L,0}, W_{L,1}) \in \mathbb{G}^{2L} .$$

2. Define the matrix $\mathbf{M} = (M_{i,j})_{i,j} \in \mathbb{G}^{(4L+2) \times (4L+3)}$ given by

$$\mathbf{M} = \left(\begin{array}{c|c|c|c|c} \boldsymbol{V}^\top & \mathbf{Id}_{f,2L} & 1^{2L \times 2L} & 1^{2L \times 1} & 1^{2L \times 1} \\ \hline \boldsymbol{W}^\top & 1^{2L \times 2L} & \mathbf{Id}_{h,2L} & 1^{2L \times 1} & 1^{2L \times 1} \\ \hline g & 1^{1 \times 2L} & 1^{1 \times 2L} & u_1 & 1 \\ \hline g & 1^{1 \times 2L} & 1^{1 \times 2L} & 1 & u_2 \end{array} \right) \quad (1)$$

with $\mathbf{Id}_{f,2L} = f^{\mathbf{I}_{2L}} \in \mathbb{G}^{2L \times 2L}$, $\mathbf{Id}_{h,2L} = h^{\mathbf{I}_{2L}} \in \mathbb{G}^{2L \times 2L}$, and where $\mathbf{I}_{2L} \in \mathbb{Z}_p^{2L \times 2L}$ is the identity matrix.

3. Generate a key pair $(\mathsf{sk}_{hsps}, \mathsf{pk}_{hsps})$ for the one-time linearly homomorphic signature of Section 2.2 in order to sign vectors of dimension $n = 4L+3$. Let $\mathsf{sk}_{hsps} = \{(\chi_i, \gamma_i, \delta_i)\}_{i=1}^{4L+3}$ be the private key, of which the corresponding public key is $\mathsf{pk}_{hsps} = (\hat{g}_z, \hat{g}_r, \hat{h}_z, \hat{h}_u, \{(\hat{g}_i, \hat{h}_i)\}_{i=1}^{4L+3})$.

4. Using $\mathsf{sk}_{hsps} = \{\chi_i, \gamma_i, \delta_i\}_{i=1}^{4L+3}$, generate one-time homomorphic signatures $\{(Z_j, R_j, U_j)\}_{j=1}^{4L+2}$ on the rows $\boldsymbol{M}_j = (M_{j,1}, \ldots, M_{j,4L+3})$ of $\mathbf{M}$. For each $j \in \{1, \ldots, 4L+2\}$, these are obtained as

$$(Z_j, R_j, U_j) = \left(\prod_{i=1}^{4L+3} M_{j,i}^{-\chi_i}, \quad \prod_{i=1}^{4L+3} M_{j,i}^{-\gamma_i}, \quad \prod_{i=1}^{4L+3} M_{j,i}^{-\delta_i} \right),$$

and, as part of the common reference string for the QA-NIZK proof system of [42], they will be included in the public key.

5. Choose $\omega_1, \omega_2 \xleftarrow{R} \mathbb{Z}_p$ and compute $\Omega_1 = u_1^{\omega_1} \in \mathbb{G}$, $\Omega_2 = u_2^{\omega_2} \in \mathbb{G}$.

The private key consists of $SK = (\omega_1, \omega_2)$ and the public key is

$$PK = \Big(f, \ g, \ h, \ u_1, \ u_2, \ \Omega_1, \ \Omega_2, \ \boldsymbol{V}, \ \boldsymbol{W},$$

$$\mathsf{pk}_{hsps} = (\hat{g}_z, \ \hat{g}_r, \ \hat{h}_z, \ \hat{h}_u, \ \{(\hat{g}_i, \hat{h}_i)\}_{i=1}^{4L+3}), \{(Z_j, R_j, U_j)\}_{j=1}^{4L+2} \Big) .$$

Sign(SK, M): Given $M = M[1] \ldots M[L] \in \{0,1\}^L$ and $SK = (\omega_1, \omega_2)$:

1. Choose $r, s \xleftarrow{R} \mathbb{Z}_p$ and compute

$$\sigma_1 = g^{\omega_1 + \omega_2} \cdot H(\boldsymbol{V}, M)^r \cdot H(\boldsymbol{W}, M)^s, \quad \sigma_2 = f^r, \quad \sigma_3 = h^s, \quad (2)$$

where $H(\boldsymbol{V}, M) = \prod_{\ell=1}^{L} V_{\ell, M[\ell]}$ and $H(\boldsymbol{W}, M) = \prod_{\ell=1}^{L} W_{\ell, M[\ell]}$.

2. Using $\{(Z_j, R_j, U_j)\}_{j=1}^{4L+2}$, derive a one-time homomorphic signature (Z, R, U) which will serve as a non-interactive argument showing that the vector

$$(\sigma_1, \sigma_2^{1-M[1]}, \sigma_2^{M[1]}, \ldots, \sigma_2^{1-M[L]}, \sigma_2^{M[L]}, \sigma_3^{1-M[1]}, \sigma_3^{M[1]},$$
$$\ldots, \sigma_3^{1-M[L]}, \sigma_3^{M[L]}, \Omega_1, \Omega_2) \quad (3)$$

is in the row space of $\mathbf{M}$, which ensures that $(\sigma_1, \sigma_2, \sigma_3)$ is of the form (2). Namely, compute

$$\begin{cases} Z = Z_{4L+1}^{\omega_1} \cdot Z_{4L+2}^{\omega_2} \cdot \prod_{i=1}^{L} \left(Z_{2i-\overline{M[i]}}^r \cdot Z_{2L+2i-\overline{M[i]}}^s \right) \\ R = R_{4L+1}^{\omega_1} \cdot R_{4L+2}^{\omega_2} \cdot \prod_{i=1}^{L} \left(R_{2i-\overline{M[i]}}^r \cdot R_{2L+2i-\overline{M[i]}}^s \right) \\ U = U_{4L+1}^{\omega_1} \cdot U_{4L+2}^{\omega_2} \cdot \prod_{i=1}^{L} \left(U_{2i-\overline{M[i]}}^r \cdot U_{2L+2i-\overline{M[i]}}^s \right). \end{cases} \quad (4)$$

Return the signature $\sigma = (\sigma_1, \sigma_2, \sigma_3, Z, R, U) \in \mathbb{G}^6$.

Verify$(\boldsymbol{PK}, \boldsymbol{M}, \boldsymbol{\sigma})$: Parse σ as $(\sigma_1, \sigma_2, \sigma_3, Z, R, U) \in \mathbb{G}^6$ and return 1 iff

$$e(Z, \hat{g}_z) \cdot e(R, \hat{g}_r) = e(\sigma_1, \hat{g}_1)^{-1} \cdot e(\sigma_2, \prod_{i=1}^{L} \hat{g}_{2i+M[i]})^{-1}$$

$$\cdot e(\sigma_3, \prod_{i=1}^{L} \hat{g}_{2L+2i+M[i]})^{-1} \cdot e(\Omega_1, \hat{g}_{4L+2})^{-1} \cdot e(\Omega_2, \hat{g}_{4L+3})^{-1}$$

$$e(Z, \hat{h}_z) \cdot e(U, \hat{h}_u) = e(\sigma_1, \hat{h}_1)^{-1} \cdot e(\sigma_2, \prod_{i=1}^{L} \hat{h}_{2i+M[i]})^{-1}$$

$$\cdot e(\sigma_3, \prod_{i=1}^{L} \hat{h}_{2L+2i+M[i]})^{-1} \cdot e(\Omega_1, \hat{h}_{4L+2})^{-1} \cdot e(\Omega_2, \hat{h}_{4L+3})^{-1}.$$

Each signature consists of 6 elements of $\mathbb{G}$, which is as short as Lewko's DLIN-based signatures [39, Section 4.3] where the security proof incurs a security loss proportional to the number of signing queries. Under the same assumption, the Chen-Wee signatures [19] require 8 group elements.

We emphasize that our security proof allows using any QA-NIZK proof system for linear subspaces and not only the one of [42] (which we used in order to keep the description as simple and self-contained as possible). Our constructions can thus be optimized if we replace the QA-NIZK proof system of [42] —which entails $K + 1$ group elements under the K-LIN assumption— by those recently suggested by Jutla and Roy, where only K group elements per proof are needed.

Under the DLIN (resp. K-linear) assumption, each signature is only comprised of 5 (resp. $2K+1$) group elements. We thus shorten signatures by 37% under the DLIN assumption. Under the K-Linear assumption, our improvement is more dramatic since, when K increases, our signatures become almost 50% shorter as we reduce the signature length of [19] from $4K$ to $2K+1$.

Under the SXDH assumption (namely, the 1-linear assumption), a direct adaptation of the above scheme entails 4 elements of $\mathbb{G}$ per signature, which is as long as [19]. However, as explained in the full version of the paper, the QA-NIZK proof system of Jutla and Roy [36] can supersede the one of [42] since, under the SXDH assumption, it only requires one group element per proof, instead of two in [42]. The signature thus becomes a triple $(\sigma_1, \sigma_2, Z) = (u^\omega \cdot H(\boldsymbol{V}, M)^r, f^r, Z)$, where Z is a QA-NIZK proof of well-formedness for (σ_1, σ_2).

Theorem 1. *The above signature scheme provides existential unforgeability under chosen-message attacks if the DLIN assumption holds in $\mathbb{G}$ and $\hat{\mathbb{G}}$. For L-bit messages, for any adversary $\mathcal{A}$, there exist DLIN distinguishers $\mathcal{B}$ and $\mathcal{B}'$ in $\hat{\mathbb{G}}$ and $\mathbb{G}$ such that $\mathbf{Adv}_{\mathcal{A}}(\lambda) \leq \mathbf{Adv}_{\mathcal{B}}^{\mathrm{DLIN}}(\lambda) + 2 \cdot L \cdot \mathbf{Adv}_{\mathcal{B}'}^{\mathrm{DLIN}}(\lambda) + \frac{2}{p}$ and with running times $t_{\mathcal{B}}, t_{\mathcal{B}'} \leq t_{\mathcal{A}} + q \cdot \mathsf{poly}(\lambda, L)$.*

Proof. The proof considers several kinds of valid signatures.

Type A signatures are produced by the real signing algorithm. Namely, if $\boldsymbol{V} = f^{\boldsymbol{v}}$ and $\boldsymbol{W} = h^{\boldsymbol{w}}$ for vectors $\boldsymbol{v} = (v_{1,0}, v_{1,1}, \ldots, v_{L,0}, v_{L,1}) \in \mathbb{Z}_p^{2L}$, $\boldsymbol{w} = (w_{1,0}, w_{1,1}, \ldots, w_{L,0}, w_{L,1}) \in \mathbb{Z}_p^{2L}$ and if we define $F(\boldsymbol{v}, M) = \sum_{\ell=1}^L v_{\ell, M[\ell]}$ and $F(\boldsymbol{w}, M) = \sum_{\ell=1}^L w_{\ell, M[\ell]}$, these signatures are such that

$$g^{\omega_1 + \omega_2} = \sigma_1 \cdot \sigma_2^{-F(\boldsymbol{v}, M)} \cdot \sigma_3^{-F(\boldsymbol{w}, M)}$$

and (Z, R, U) is a valid linearly homomorphic signature on the vector (3).

Type B signatures are valid signatures that are not Type A signatures. These are of the form

$$\sigma_1 = g^{\omega_1 + \omega_2 + \tau} \cdot H(\boldsymbol{V}, M)^r \cdot H(\boldsymbol{W}, M)^s, \qquad \sigma_2 = f^r, \qquad \sigma_3 = h^s,$$

for some $\tau \in_R \mathbb{Z}_p$, $r, s \in_R \mathbb{Z}_p$, and

$$\begin{cases} Z = g^{-\tau \cdot \chi_1} \cdot Z_{4L+1}^{\omega_1} \cdot Z_{4L+2}^{\omega_2} \cdot \prod_{i=1}^L \left(Z_{2i - \overline{M[i]}}^r \cdot Z_{2L + 2i - \overline{M[i]}}^s \right) \\ R = g^{-\tau \cdot \gamma_1} \cdot R_{4L+1}^{\omega_1} \cdot R_{4L+2}^{\omega_2} \cdot \prod_{i=1}^L \left(R_{2i - \overline{M[i]}}^r \cdot R_{2L + 2i - \overline{M[i]}}^s \right) \\ U = g^{-\tau \cdot \delta_1} \cdot U_{4L+1}^{\omega_1} \cdot U_{4L+2}^{\omega_2} \cdot \prod_{i=1}^L \left(U_{2i - \overline{M[i]}}^r \cdot U_{2L + 2i - \overline{M[i]}}^s \right) \end{cases} .$$

Note that Type B signatures also satisfy the verification algorithm since (Z, R, U) is a valid homomorphic signature on the vector (3). The term g^τ will be henceforth called the *semi-functional component* of the signature. Type B signatures include the following sub-classes.

Type B-k signatures ($1 \leq k \leq L$) are generated by choosing $r, s \xleftarrow{R} \mathbb{Z}_p$ and setting

$$\sigma_1 = g^{\omega_1 + \omega_2} \cdot R_k(M_{|k}) \cdot H(\boldsymbol{V}, M)^r \cdot H(\boldsymbol{W}, M)^s, \quad \sigma_2 = f^r, \quad \sigma_3 = h^s,$$

with $H(\boldsymbol{V}, M) = \prod_{\ell=1}^{L} V_{\ell, M[\ell]}$ and where $H(\boldsymbol{W}, M) = \prod_{\ell=1}^{L} W_{\ell, M[\ell]}$ and $R_k \colon \{0,1\}^k \to \mathbb{G}, M_{|k} \mapsto R_k(M_{|k})$ is a random function that depends on the first k bits of M. The (Z, R, U) components are simulated QA-NIZK proofs of subspace membership. They are obtained using $\{(\chi_i, \gamma_i, \delta_i)\}_{i=1}^{4L+3}$ to generate a homomorphic signature on the vector (3) by computing

$$\begin{cases} Z = \sigma_1^{-\chi_1} \cdot \sigma_2^{-\sum_{i=1}^{L} \chi_{2i+M[i]}} \cdot \sigma_3^{-\sum_{i=1}^{L} \chi_{2L+2i+M[i]}} \cdot \Omega_1^{-\chi_{4L+2}} \cdot \Omega_2^{-\chi_{4L+3}} \\ R = \sigma_1^{-\gamma_1} \cdot \sigma_2^{-\sum_{i=1}^{L} \sigma_{2i+M[i]}} \cdot \sigma_3^{-\sum_{i=1}^{L} \gamma_{2L+2i+M[i]}} \cdot \Omega_1^{-\gamma_{4L+2}} \cdot \Omega_2^{-\gamma_{4L+3}} \\ U = \sigma_1^{-\delta_1} \cdot \sigma_2^{-\sum_{i=1}^{L} \delta_{2i+M[i]}} \cdot \sigma_3^{-\sum_{i=1}^{L} \delta_{2L+2i+M[i]}} \cdot \Omega_1^{-\delta_{4L+2}} \cdot \Omega_2^{-\delta_{4L+3}} \end{cases}.$$

To prove the result, we consider the following sequence of games. For each i, we call S_i the event that the adversary wins in Game i. We also define E_i to be the event that, in Game i, $\mathcal{A}$'s forgery has the same type as the signatures it observes. Namely, if $\mathcal{A}$ obtains a Type A (resp. Type B-k) signature at each query, it should output a Type A (resp. Type B-k) forgery.

Game 0: This game is the real game. Namely, the adversary obtains Type A signatures at each signing query. At the end of the game, however, the challenger $\mathcal{B}$ checks if $\mathcal{A}$'s forgery is a Type A signature and we define E_0 the event that the forgery $\sigma^\star$ is a Type A forgery. We obviously have $\Pr[S_0] = \Pr[S_0 \wedge E_0] + \Pr[S_0 \wedge \neg E_0]$. Lemma 1 shows that, if the DLIN assumption holds in $\hat{\mathbb{G}}$, the adversary can only output a Type B signature with negligible probability. We have $\Pr[S_0 \wedge \neg E_0] \leq \mathbf{Adv}_{\hat{\mathbb{G}}}^{\mathrm{DLIN}}(\lambda) + 1/p$. We are thus left with the task of bounding $\Pr[S_0 \wedge E_0]$. To this end, we proceed using a sequence of L games.

Game 1: This game is identical to Game 0 with the difference that, at each signing query, the signature components (Z, R, U) are obtained as simulated QA-NIZK proofs of linear subspace membership. Namely, instead of computing (Z, R, U) as per (4), the challenger uses $\{\chi_i, \gamma_i, \delta_i\}_{i=1}^{4L+3}$ to compute (Z, R, U) as a one-time linearly homomorphic signature on the vector (3). Clearly (Z, R, U) retains the same distribution as in Game 0, so that $\mathcal{A}$'s view remains unchanged. We have $\Pr[S_1 \wedge E_1] = \Pr[S_0 \wedge E_0]$, where E_1 is the counterpart of event E_0 in Game 1.

Game 2.k ($1 \leq k \leq L$): In Game 2.k, all signing queries are answered by returning Type B-k signatures. For each k, we call $E_{2.k}$ the event that $\mathcal{A}$ outputs a Type B-k forgery in Game 2.k. Lemma 2 provides evidence that Game 2.1 is computationally indistinguishable from Game 1 under the DLIN assumption in $\mathbb{G}$: we have $|\Pr[S_{2.1} \wedge E_{2.1}] - \Pr[S_1 \wedge E_1]| \leq 2 \cdot \mathbf{Adv}_{\mathbb{G}}^{\mathrm{DLIN}}(\lambda)$. In the full version of the paper, we show that, under the DLIN assumption in $\mathbb{G}$, the probability of $\mathcal{A}$'s forgery to be of the same type as the outputs of signing queries is about the same in Game 2.k and in Game 2.$(k-1)$. We thus have $|\Pr[S_{2.k} \wedge E_{2.k}] - \Pr[S_{2.(k-1)} \wedge E_{2.(k-1)}]| \leq 2 \cdot \mathbf{Adv}_{\mathbb{G}}^{\mathrm{DLIN}}(\lambda)$.

When we reach Game 2.L, we know that $|\Pr[S_{2.L} \wedge E_{2.L}] - \Pr[S_{2.0} \wedge E_{2.0}]| \leq 2 \cdot L \cdot$ $\mathbf{Adv}_{\mathbb{G}}^{\mathrm{DLIN}}(\lambda)$ by the triangle inequality. However, in Game 2.L, it is easy to prove that, even though $\mathcal{A}$ only obtains Type B-k signatures throughout the game, its probability to output a Type B-k forgery is negligible even with an unbounded computational power. Indeed, a legitimate adversary that outputs a forgery on a new message $M^\star$ has no information on $R_L(M^\star)$. Hence, it can only produce a Type B-k forgery by pure chance and we thus have $\Pr[S_{2.L} \wedge E_{2.L}] \leq 1/p$. □

Lemma 1. *In Game 0, any PPT adversary outputting a Type B forgery with non-negligible probability implies an algorithm breaking the DLIN assumption in $\hat{\mathbb{G}}$ with nearly the same advantage.* (The proof is in the full version of the paper).

Lemma 2. *If the DLIN assumption holds in $\mathbb{G}$, $\mathcal{A}$'s probability to output a Type B-1 signature in Game 2.1 is about the same as its probability to output a Type A signature in Game 1.*

Proof. Let us assume that events $S_{2.1} \wedge E_{2.1}$ and $S_1 \wedge E_1$ occur with noticeably different probabilities in Game 2.1 and Game 1, respectively. We construct a DLIN distinguisher $\mathcal{B}$ in $\mathbb{G}$. Our algorithm $\mathcal{B}$ takes as input (f, g, h, f^a, h^b, T) with the task of deciding if $T = g^{a+b}$ or $T \in_R \mathbb{G}$. Similarly to [19, Lemma 6], the reduction $\mathcal{B}$ uses the random self-reducibility of DLIN to build q tuples $(F_j = f^{a_j}, H_j = h^{b_j}, T_j)$ such that, for each $j \in \{1, \ldots, q\}$, we have

$$T_j = \begin{cases} g^{a_j + b_j} & \text{if } T = g^{a+b} \\ g^{a_j + b_j + \tau_0} & \text{if } T \in_R \mathbb{G} \end{cases}$$

for some $\tau_0 \in_R \mathbb{Z}_p$. This is done by picking $\rho_0 \xleftarrow{R} \mathbb{Z}_p$ and $\rho_{a_j}, \rho_{b_j} \xleftarrow{R} \mathbb{Z}_p$, for $j \in \{1, \ldots, q\}$, and setting

$$(F_j, H_j, T_j) = \left((f^a)^{\rho_0} \cdot f^{\rho_{a_j}}, (h^b)^{\rho_0} \cdot h^{\rho_{b_j}}, T^{\rho_0} \cdot g^{\rho_{a_j} + \rho_{b_j}} \right), \quad \forall j \in \{1, \ldots, q\} .$$

In addition, $\mathcal{B}$ generates an extra tuple $(u_1, u_2, \Omega_1, \Omega_2) \in \mathbb{G}^4$ by choosing random exponents $\alpha_{u,1}, \alpha_{u,2} \xleftarrow{R} \mathbb{Z}_p$ and setting

$$u_1 = f^{\alpha_{u,1}}, \qquad u_2 = h^{\alpha_{u,2}}, \qquad \Omega_1 = (f^a)^{\alpha_{u,1}}, \qquad \Omega_2 = (h^b)^{\alpha_{u,2}} .$$

Before generating the public key of the scheme, $\mathcal{B}$ flips a coin $b^\dagger \xleftarrow{R} \{0, 1\}$ hoping that the first bit of the target message $M^\star = M[1]^\star \ldots M[L]^\star \in \{0, 1\}^L$ will coincide with $b^\dagger$. To construct PK, $\mathcal{B}$ chooses $\boldsymbol{\alpha} = (\alpha_{1,0}, \alpha_{1,1}, \ldots, \alpha_{L,0}, \alpha_{L,1}) \xleftarrow{R} \mathbb{Z}_p^{2L}$, $\boldsymbol{\beta} = (\beta_{1,0}, \beta_{1,1}, \ldots, \beta_{L,0}, \beta_{L,1}) \xleftarrow{R} \mathbb{Z}_p^{2L}$ and $\zeta \xleftarrow{R} \mathbb{Z}_p$. It defines the vectors $\boldsymbol{V} = (V_{1,0}, V_{1,1}, \ldots, V_{L,0}, V_{L,1})$, $\boldsymbol{W} = (W_{1,0}, W_{1,1}, \ldots, W_{L,0}, W_{L,1})$ as

$$\begin{cases} (V_{\ell,0}, V_{\ell,1}) = (f^{\alpha_{\ell,0}}, f^{\alpha_{\ell,1}}), & (W_{\ell,0}, W_{\ell,1}) = (h^{\beta_{\ell,0}}, h^{\beta_{\ell,1}}) & \text{if } \ell \neq 1, \\ (V_{1,1-b^\dagger}, V_{1,b^\dagger}) = (f^{\alpha_{1,1-b^\dagger}} \cdot g^\zeta, f^{\alpha_{1,b^\dagger}}), & (W_{1,1-b^\dagger}, W_{1,b^\dagger}) = (h^{\beta_{1,1-b^\dagger}} \cdot g^\zeta, h^{\beta_{1,b^\dagger}}) . \end{cases}$$

The rest of PK, including $(\mathsf{sk}_{hsps}, \mathsf{pk}_{hsps})$ and $\{(Z_i, R_i, U_i)\}_{i=1}^{4L+2}$, is generated as in the real setup. The adversary $\mathcal{A}$ is run on input of

$$PK = \Big(f,\ g,\ h,\ u_1,\ u_2,\ \Omega_1 = u_1{}^a,\ \Omega_2 = u_2{}^b,\ \boldsymbol{V},\ \boldsymbol{W},$$

$$\mathsf{pk}_{hsps} = \big(\hat{g}_z,\ \hat{g}_r,\ \hat{h}_z,\ \hat{h}_u,\ \{(\hat{g}_i, \hat{h}_i)\}_{i=1}^{4L+3}\big),\ \{(\hat{Z}_j, \hat{R}_j, \hat{U}_j)\}_{j=1}^{4L+2}\Big)$$

and $\mathcal{B}$ keeps $(\{\chi_i, \gamma_i, \delta_i\}_{i=1}^{4L+3})$ to itself. Note that $a, b \in \mathbb{Z}_p$ are part of the original DLIN instance and are not available to $\mathcal{B}$. However, $\mathcal{B}$ will use the challenge value T —which is either g^{a+b} or a random element of $\mathbb{G}$— to answer signing queries.

During the game, signing queries are answered as follows. In order to handle the j-th signing query $M^j = M[1]^j \dots M[L]^j \in \{0,1\}^L$, the answer of $\mathcal{B}$ depends on the first bit $M[1]^j$ of M^j. Specifically, $\mathcal{B}$ considers the following cases.

- If $M[1]^j = b^\dagger$, $\mathcal{B}$ chooses $r, s \xleftarrow{R} \mathbb{Z}_p$ and sets

$$\sigma_1 = T \cdot H(\boldsymbol{V}, M)^r \cdot H(\boldsymbol{W}, M)^s, \qquad \sigma_2 = f^r, \qquad \sigma_3 = h^s,$$

where $H(\boldsymbol{V}, M) = \prod_{\ell=1}^L V_{\ell, M[\ell]}$ and $H(\boldsymbol{W}, M) = \prod_{\ell=1}^L W_{\ell, M[\ell]}$. The (Z, R, U) components of the private key are computed by generating a homomorphic structure-preserving signature on the vector

$$(\sigma_1, \sigma_2^{1-M[1]}, \sigma_2^{M[1]}, \dots, \sigma_2^{1-M[L]}, \sigma_2^{M[L]}, \sigma_3^{1-M[1]}, \sigma_3^{M[1]},$$
$$\dots, \sigma_3^{1-M[L]}, \sigma_3^{M[L]}, \Omega_1, \Omega_2),$$

by computing

$$\begin{cases} Z = \sigma_1^{-\chi_1} \cdot \sigma_2^{-\sum_{i=1}^L \chi_{2i+M[i]}} \cdot \sigma_3^{-\sum_{i=1}^L \chi_{2L+2i+M[i]}} \cdot \Omega_1^{-\chi_{4L+2}} \cdot \Omega_2^{-\chi_{4L+3}} \\ R = \sigma_1^{-\gamma_1} \cdot \sigma_2^{-\sum_{i=1}^L \gamma_{2i+M[i]}} \cdot \sigma_3^{-\sum_{i=1}^L \gamma_{2L+2i+M[i]}} \cdot \Omega_1^{-\gamma_{4L+2}} \cdot \Omega_2^{-\gamma_{4L+3}} \\ U = \sigma_1^{-\delta_1} \cdot \sigma_2^{-\sum_{i=1}^L \delta_{2i+M[i]}} \cdot \sigma_3^{-\sum_{i=1}^L \delta_{2L+2i+M[i]}} \cdot \Omega_1^{-\delta_{4L+2}} \cdot \Omega_2^{-\delta_{4L+3}} \end{cases} \quad (5)$$

Note that, if $T = g^{a+b+\tau}$ for some $\tau \in_R \mathbb{Z}_p$, (Z, R, U) can be written

$$\begin{cases} Z = g^{-\chi_1 \cdot \tau} \cdot Z_{4L+1}^a \cdot Z_{4L+2}^b \cdot \prod_{i=1}^L \big(Z_{2i-\overline{M[i]}}^r \cdot Z_{2L+2i-\overline{M[i]}}^s \big) \\ R = g^{-\gamma_1 \cdot \tau} \cdot R_{4L+1}^a \cdot R_{4L+2}^b \cdot \prod_{i=1}^L \big(R_{2i-\overline{M[i]}}^r \cdot R_{2L+2i-\overline{M[i]}}^s \big) \\ U = g^{-\delta_1 \cdot \tau} \cdot U_{4L+1}^a \cdot U_{4L+2}^b \cdot \prod_{i=1}^L \big(U_{2i-\overline{M[i]}}^r \cdot U_{2L+2i-\overline{M[i]}}^s \big) \end{cases}.$$

We observe that $(\sigma_1, \sigma_2, \sigma_3, Z, R, U)$ matches the distribution of signatures in both Game 2.1 if $\tau \neq 0$ and Game 1 if $\tau = 0$. Indeed, in the former case, we implicitly define the constant function $R_0(\varepsilon) = g^\tau$ and define the function R_1 so that $R_1(b^\dagger) = R_0(\varepsilon)$.

- If $M[1]^j = 1 - b^\dagger$, $\mathcal{B}$ implicitly defines

$$R_1(M_{|1}^j) = R_1(1 - b^\dagger) = \begin{cases} R_0(\varepsilon) \cdot g^{\zeta \cdot \tau_0} & \text{if } T \in_R \mathbb{G} \\ 1 & \text{if } T = g^{a+b} \end{cases}.$$

Namely, $\mathcal{B}$ uses the j-th tuple (F_j, H_j, T_j) to set

$$\sigma_1 = T \cdot F_j^{\sum_{\ell=1}^{L} \alpha_{\ell, M[\ell]}} \cdot H_j^{\sum_{\ell=1}^{L} \beta_{\ell, M[\ell]}} \cdot T_j^{\zeta} , \quad \sigma_2 = F_j = f^{a_j} , \quad \sigma_3 = H_j = h^{b_j} .$$

If $T = g^{a+b}$ (and thus $T_j = g^{a_j+b_j}$), this implicitly defines σ_1 as the product $\sigma_1 = g^{a+b} \cdot H(\boldsymbol{V}, M^j)^{a_j} \cdot H(\boldsymbol{W}, M^j)^{b_j}$, so that $(\sigma_1, \sigma_2, \sigma_3)$ has the same distribution as in Game 1. If $T = g^{a+b+\tau}$ (so that $T_j = g^{a_j+b_j+\tau}$), we have

$$\sigma_1 = g^{a+b} \cdot R_1(M_{|1}^{j}) \cdot H(\boldsymbol{V}, M^j)^{a_j} \cdot H(\boldsymbol{W}, M^j)^{b_j} ,$$

since $R_1(M_{|1}^{j}) = R_0(\varepsilon) \cdot g^{\zeta \cdot \tau_0}$, which is distributed as in Game 2.1. In either case, (Z, R, U) are computed using $\mathsf{sk}_{hsps} = \{(\chi_i, \gamma_i, \delta_i)\}_{i=1}^{4L+3}$ as in the previous case (i.e., as per (5)).

In the forgery stage, the adversary $\mathcal{A}$ outputs a new message $M^\star$ and a signature $\sigma^\star = (\sigma_1^\star, \sigma_2^\star, \sigma_3^\star, Z^\star, R^\star, U^\star)$. Our distinguisher $\mathcal{B}$ must determine if $\sigma^\star$ has the same type as the outputs of the simulated signing oracle. At this point, algorithm $\mathcal{B}$ halts and outputs a random bit if $M[1]^\star \neq b^\dagger$. Otherwise, $\mathcal{B}$ can compute $F(\boldsymbol{v}, M^\star) = \sum_{\ell=1}^{L} \alpha_{\ell, M[\ell]^\star}$ and $F(\boldsymbol{w}, M^\star) = \sum_{\ell=1}^{L} \beta_{\ell, M[\ell]^\star}$, which yields $\eta^\star = \sigma_1^\star \cdot \sigma_2^{\star - F(\boldsymbol{v}, M^\star)} \cdot \sigma_3^{\star - F(\boldsymbol{w}, M^\star)}$. If $\eta^\star = T$, $\mathcal{B}$ considers $(\sigma^\star, M^\star)$ as a forgery of the same type as outputs of the signing oracle and returns 1. Recall that $R_0(\varepsilon) = T/g^{a+b}$, so that $\sigma^\star$ matches the output distribution of the signing oracle in both Game 1 and Game 2.1. Otherwise, $\mathcal{B}$ decides that $\sigma^\star$ has a different distribution than signatures produced by the signing oracle and outputs 0. If the difference between $\mathcal{A}$'s probability to output the same kind of signatures as the signing oracle in Games 2.1 and 2.1 is ϵ, then $\mathcal{B}$'s advantage as a DLIN distinguisher is at least $\epsilon/2$ since $b^\dagger \in \{0,1\}$ is independent of $\mathcal{A}$'s view. $\square$

We remark that, while its signatures are randomizable, the system can be made strongly unforgeable in a simple manner and without increasing the signature length. In particular, we do not need a chameleon-hash-function-based transformation such as [15]. Using the QA-NIZK proofs of [36], we thus obtain strongly unforgeable signatures based on the SXDH assumption which are short as those of Boneh, Shen and Waters [15] with a nearly tight reduction. The details are given in the full version of the paper.

4 Almost Tightly CCA-Secure Encryption with Shorter Ciphertexts

Equipped with our signature scheme, we now present a public-key encryption scheme whose IND-CCA2 security in the multi-challenge-multi-user setting is almost tightly related to the DLIN assumption. Like [31], our scheme instantiates a variant of the Naor-Yung paradigm using Groth-Sahai proofs and the cryptosystem of Boneh, Boyen and Shacham (BBS) [13].

The construction can be seen as an instantiation of a technique suggested by Dodis *el al.* [26] as a modification of the Naor-Yung paradigm, where only one IND-CPA secure encryption suffices (instead of two in [45,47]) if it is accompanied with a NIZK proof of knowledge of the plaintext that is simulation-extractable (and not only simulation-sound). In [3], Abe *el al.* used a simulation-extractable proof system showing that either: (i) The IND-CPA encryption scheme encrypts the message containted in an extractable commitment; (ii) Another commitment included in the proof is a valid signature on the verification key VK of a one-time signature. Here, we show that, if this simulation-extractable proof system is combined with the BBS cryptosystem, it can be simplified by removing the commitment to the message and the proof that this commitment contains the encrypted plaintext. The reason is that, in each simulation-extractable proof, the commitments to the encryption exponents suffice to guarantee the extractability of the plaintext.

While our reduction is not quite as tight as in the results of [31,3] since we lose a factor of $\Theta(\lambda)$, our scheme is much more space-efficient as the ciphertext overhead reduces to 68 group elements. As a comparison, the most efficient solution of [3] incurs 398 group elements per ciphertext.

For simplicity, the description below uses symmetric pairings $e : \mathbb{G} \times \mathbb{G} \to \mathbb{G}_T$ (i.e., $\mathbb{G} = \hat{\mathbb{G}}$) but extensions to asymmetric pairings are possible.

Par-Gen(λ): Choose bilinear groups $(\mathbb{G}, \mathbb{G}_T)$ with generators $g, f, h \xleftarrow{R} \mathbb{G}$. Define common public parameters $\mathsf{par} = ((\mathbb{G}, \mathbb{G}_T), g, f, h)$.

Keygen(par): Parse par as $((\mathbb{G}, \mathbb{G}_T), g, f, h)$ and conduct the following steps.

1. Choose random exponents $x_1, y_1 \xleftarrow{R} \mathbb{Z}_p$ and set $f_1 = g^{x_1}$, $h_1 = g^{y_1}$.
2. Choose a strongly unforgeable one-time signature $\Sigma = (\mathcal{G}, \mathcal{S}, \mathcal{V})$ with verification keys of length $L \in \mathsf{poly}(\lambda)$.
3. For $\ell = 1$ to L, choose $V_{\ell,0}, V_{\ell,1}, W_{\ell,0}, W_{\ell,1} \xleftarrow{R} \mathbb{G}$ to assemble row vectors

$$\boldsymbol{V} = (V_{1,0}, V_{1,1}, \ldots, V_{L,0}, V_{L,1}),$$
$$\boldsymbol{W} = (W_{1,0}, W_{1,1}, \ldots, W_{L,0}, W_{L,1}) \in \mathbb{G}^{2L} .$$

4. Choose $\omega_1, \omega_2 \xleftarrow{R} \mathbb{Z}_p$, $u_1, u_2 \xleftarrow{R} \mathbb{G}$, and compute $\Omega_1 = u_1^{\omega_1}$, $\Omega_2 = u_2^{\omega_2}$.
5. Define the matrix $\mathbf{M} = (M_{i,j})_{i,j} \in \mathbb{G}^{(4L+2) \times (4L+3)}$ as

$$(M_{i,j})_{i,j} = \left(\begin{array}{c|c|c|c|c} \boldsymbol{V}^\top & \mathbf{Id}_{f,2L} & \mathbf{1}^{2L \times 2L} & \mathbf{1}^{2L \times 1} & \mathbf{1}^{2L \times 1} \\ \hline \boldsymbol{W}^\top & \mathbf{1}^{2L \times 2L} & \mathbf{Id}_{h,2L} & \mathbf{1}^{2L \times 1} & \mathbf{1}^{2L \times 1} \\ \hline g & \mathbf{1}^{1 \times 2L} & \mathbf{1}^{1 \times 2L} & u_1 & 1 \\ \hline g & \mathbf{1}^{1 \times 2L} & \mathbf{1}^{1 \times 2L} & 1 & u_2 \end{array} \right)$$

with $\mathbf{Id}_{f,2L} = f^{\mathbf{I}_{2L}} \in \mathbb{G}^{2L \times 2L}$, $\mathbf{Id}_{h,2L} = h^{\mathbf{I}_{2L}} \in \mathbb{G}^{2L \times 2L}$, where $\mathbf{I}_{2L} \in \mathbb{Z}_p^{2L \times 2L}$ is the identity matrix. Then, generate a key pair $(\mathsf{pk}_{hsps}, \mathsf{sk}_{hsps})$ for the one-time LHSPS scheme of Section 2.2 with $n = 4L + 3$. Let $\mathsf{pk}_{hsps} = (g_z, g_r, h_z, h_u, \{g_i, h_i\}_{i=1}^{4L+3})$ be the resulting public key and let $\mathsf{sk}_{hsps} = \{\chi_i, \gamma_i, \delta_i\}_{i=1}^{4L+3}$ be the underlying private key.

6. Generate one-time linearly homomorphic signatures $\{(z_j, r_j, u_j)\}_{j=1}^{4L+2}$ on the rows of $\mathbf{M}$.
7. Choose a perfectly WI Groth-Sahai CRS $\mathbf{g} = (\boldsymbol{G}_1, \boldsymbol{G}_2, \boldsymbol{G}_3)$ defined by vectors $\boldsymbol{G}_1 = (G_1, 1, G)$, $\boldsymbol{G}_2 = (1, G_2, G)$ and $\boldsymbol{G}_3 \in \mathbb{G}^3$, with $G, G_1, G_2 \xleftarrow{R} \mathbb{G}$ and $\boldsymbol{G}_3 \xleftarrow{R} \mathbb{G}^3$.
8. Define the private key as $SK = (x_1, y_1) \in \mathbb{Z}_p^2$. The public key is

$$PK = \big(g, f_1, \ h_1, \ \boldsymbol{V}, \ \boldsymbol{W}, \ u_1, \ u_2, \ \varOmega_1, \ \varOmega_2,$$
$$\mathsf{pk}_{hsps}, \ \{(z_j, r_j, u_j)\}_{j=1}^{4L+2}, \mathbf{g} = (\boldsymbol{G}_1, \boldsymbol{G}_2, \boldsymbol{G}_3), \ \varSigma\big),$$

whereas $\omega_1, \omega_2 \in \mathbb{Z}_p$ and sk_{hsps} are erased.

Encrypt(M, PK): To encrypt $M \in \mathbb{G}$, generate a one-time signature key pair $(\mathsf{SK}, \mathsf{VK}) \leftarrow \mathcal{G}(\lambda)$ and conduct the following steps:

1. Choose $\theta_1, \theta_2 \xleftarrow{R} \mathbb{Z}_p$ and compute $(C_0, C_1, C_2) = (M \cdot g^{\theta_1 + \theta_2}, f_1^{\theta_1}, h_1^{\theta_2})$.
2. Choose $r, s \xleftarrow{R} \mathbb{Z}_p$ and compute a pseudo-signature

$$\sigma_1 = H(\boldsymbol{V}, \mathsf{VK})^r \cdot H(\boldsymbol{W}, \mathsf{VK})^s, \qquad \sigma_2 = f^r, \qquad \sigma_3 = h^s,$$

where $H(\boldsymbol{V}, \mathsf{VK}) = \prod_{\ell=1}^{L} V_{\ell, \mathsf{VK}[\ell]}$ and $H(\boldsymbol{W}, \mathsf{VK}) = \prod_{\ell=1}^{L} W_{\ell, \mathsf{VK}[\ell]}$.
3. Define the variables $(W_1, W_2) = (g^{\theta_1}, g^{\theta_2})$ and compute Groth-Sahai commitments $\{\boldsymbol{C}_{W_i}\}_{i=1}^2$ to these.
4. Define the bit $b = 1$ and generate $\boldsymbol{C}_b = (1, 1, G^b) \cdot \boldsymbol{G}_1^{r_b} \cdot \boldsymbol{G}_2^{s_b} \cdot \boldsymbol{G}_3^{t_b}$, where $r_b, s_b, t_b \xleftarrow{R} \mathbb{Z}_p$, as a commitment to b. Then, compute a Groth-Sahai commitment $\boldsymbol{C}_{\sigma_1}$ to σ_1 and commitments $\boldsymbol{C}_{\Theta_1}, \boldsymbol{C}_{\Theta_2} \in \mathbb{G}^3$ and $\boldsymbol{C}_{\Gamma_g}$ to

$$\varTheta_1 = \varOmega_1^{1-b}, \quad \varTheta_2 = \varOmega_2^{1-b}, \quad \varGamma_g = g^b. \tag{6}$$

The vector

$$(\sigma_1, \sigma_2^{1-\mathsf{VK}[1]}, \sigma_2^{\mathsf{VK}[1]}, \dots, \sigma_2^{1-\mathsf{VK}[L]}, \sigma_2^{\mathsf{VK}[L]},$$
$$\sigma_3^{1-\mathsf{VK}[1]}, \sigma_3^{\mathsf{VK}[1]}, \dots, \sigma_3^{1-\mathsf{VK}[L]}, \sigma_3^{\mathsf{VK}[L]}, \varTheta_1, \varTheta_2) \in \mathbb{G}^{4L+3} \tag{7}$$

belongs to the subspace spanned by the first $4L$ rows of the matrix $\mathbf{M} \in \mathbb{G}^{(4L+2) \times (4L+3)}$. Hence, the algorithm can use $r, s \in \mathbb{Z}_p$ to derive a one-time linearly homomorphic signature $(Z, R, U) \in \mathbb{G}^3$ on the vector (7). Note that $(\sigma_1, \sigma_2, \sigma_3, Z, R, U)$ can be seen as a signature on VK, for the degenerated private key $(\omega_1, \omega_2) = (0, 0)$.
5. Generate commitments $\boldsymbol{C}_Z, \boldsymbol{C}_R, \boldsymbol{C}_U \in \mathbb{G}^3$. Then, compute a NIWI proof $\boldsymbol{\pi}_b \in \mathbb{G}^9$ that b satisfies $b^2 = b$ (which ensures that $b \in \{0, 1\}$) and NIWI proofs $\boldsymbol{\pi}_{\mathsf{PPE1}}, \boldsymbol{\pi}_{\mathsf{PPE2}} \in \mathbb{G}^3$ that variables $(\sigma_1, Z, R, U, \varTheta_1, \varTheta_2)$ satisfy

$$e(Z, g_z) \cdot e(R, g_r) = e(\sigma_1, g_1)^{-1} \cdot e(\sigma_2, \prod_{i=1}^{L} g_{2i+\mathsf{VK}[i]})^{-1}$$

$$\cdot e(\sigma_3, \prod_{i=1}^{L} g_{2L+2i+\mathsf{VK}[i]})^{-1} \cdot e(\Theta_1, g_{4L+2})^{-1} \cdot e(\Theta_2, g_{4L+3})^{-1},$$

$$e(Z, h_z) \cdot e(U, h_u) = e(\sigma_1, h_1)^{-1} \cdot e(\sigma_2, \prod_{i=1}^{L} h_{2i+\mathsf{VK}[i]})^{-1}$$

$$\cdot e(\sigma_3, \prod_{i=1}^{L} h_{2L+2i+\mathsf{VK}[i]})^{-1} \cdot e(\Theta_1, h_{4L+2})^{-1} \cdot e(\Theta_2, h_{4L+3})^{-1} .$$

6. Generate NIWI proofs π_g, $\{\pi_{\Theta_i}\}_{i=1}^{2}$ that (b, Θ_1, Θ_2), which are committed in $C_b, C_{\Theta_1}, C_{\Theta_2}$, satisfy (6). Each such proof lives in $\mathbb{G}^3$.
7. Generate a simulation-extractable proof that (W_1, W_2) satisfy

$$e(C_1, g) = e(f_1, W_1), \qquad\qquad e(C_2, g) = e(h_1, W_2) . \qquad (8)$$

To this end, prove that (W_1, W_2, Γ_g) satisfy

$$e(C_1, \Gamma_g) = e(f_1, W_1), \qquad\qquad e(C_2, \Gamma_g) = e(h_1, W_2) . \qquad (9)$$

This requires proofs π_1, π_2 of 3 group elements each.
8. Finally, compute a one-time signature $sig = \mathcal{S}(\mathsf{SK}, C_0, C_1, C_2, \pi)$ and output the ciphertext $C = (\mathsf{VK}, C_0, C_1, C_2, \pi, sig)$, where

$$\pi = (C_b, \pi_b, C_{\sigma_1}, \sigma_2, \sigma_3, \{C_{W_i}\}_{i=1}^{2}, C_Z, C_R, C_U, \{C_{\Theta_i}\}_{i=1}^{2}, C_{\Gamma_g},$$
$$\pi_g, \{\pi_{\Theta_i}\}_{i=1}^{2}, \pi_{\mathsf{PPE1}}, \pi_{\mathsf{PPE2}}, \pi_1, \pi_2) \qquad (10)$$

is a simulation-extractable proof of plaintext knowledge consisting of 62 elements of $\mathbb{G}$.

Decrypt(SK, C): Parse C as $C = (\mathsf{VK}, C_0, C_1, C_2, \pi, sig)$ and do the following.
1. Return $\bot$ if $\mathcal{V}(\mathsf{VK}, (C_0, C_1, C_2, \pi), sig) = 0$ or if π does not properly verify.
2. Using $SK = (x_1, y_1) \in \mathbb{Z}_p^2$, compute and return $M = C_0 \cdot C_1^{-1/x_1} \cdot C_2^{-1/y_1}$.

Note that π forms a proof that either $(\sigma_1, \sigma_2, \sigma_3)$ is a valid signature or $\{C_{W_i}\}_{i=1}^{2}$ are commitments to $(W_1, W_2) = (g^{\theta_1}, g^{\theta_2})$, where $\theta_1, \theta_2 \in \mathbb{Z}_p$ are the encryption exponents. A simulator holding the private key $(\omega_1, \omega_2) \in \mathbb{Z}_p^2$ of the signature scheme can simulate a proof π of plaintext knowledge by computing $(\sigma_1, \sigma_2, \sigma_3)$ as a real signature, by setting $b = 0$ at step 4 of the encryption algorithm and using the witnesses $(W_1, W_2) = (1_\mathbb{G}, 1_\mathbb{G})$ to prove relations (9).

We remark that each ciphertext must contain a proof comprised of 62 group elements. In an instantiation using the one-time signature of [31], the entire ciphertexts thus costs 69 group elements. The scheme can also be adapted to asymmetric pairings in a simple manner.

For simplicity, we follow [3] and only prove security in the single-user, multi-challenge case. However, as pointed out in [3], the single-user security results can always be simply extended to the scenario of multiple public keys by leveraging the random self-reducibility of the DLIN assumption in a standard manner. In the full version of the paper, we prove the following result.

Theorem 2. *The above scheme is $(1, q_e)$-IND-CCA secure provided: (i) Σ is a strongly unforgeable one-time signature; and (ii) the DLIN assumption holds in $\mathbb{G}$. For any adversary $\mathcal{A}$, there exist a one-time signature forger $\mathcal{B}'$ and a DLIN distinguisher $\mathcal{B}$ with running times $t_\mathcal{B}, t_{\mathcal{B}'} \leq t_\mathcal{A} + q_e \cdot \mathsf{poly}(\lambda, L)$ such that $\mathbf{Adv}_\mathcal{A}^{(1,q_e)\text{-cca}}(\lambda) \leq 2 \cdot \mathbf{Adv}_{\mathcal{B}'}^{n\text{-suf-ots}}(\lambda) + (4 \cdot L + 5) \cdot \mathbf{Adv}_\mathcal{B}^{\mathrm{DLIN}}(\lambda) + 5/p$, where L is the verification key length of Σ and q_e is the number of encryption queries.*

In order to extend the result to the multi-user setting, the main changes are that we need to rely on: *(i)* The random self-reducibility of DLIN, which is used as in [31]; *(ii)* The almost tight security of the signature scheme of Section 3 in the multi-user setting [27], which can also be proved using the random self-reducibility of DLIN. The latter proof notably relies on the tight security of the homomorphic signature of Section 2.2 in the multi-key setting, which is proved in the full version of the paper.

Acknowledgements. We thank the anonymous reviewers for very useful comments. In particular, we are grateful to one reviewer for suggesting a more efficient approach to the scheme of Section 4. The first author's work was supported in part by the ERC Starting Grant ERC-2013-StG-335086-LATTAC. This work has also been supported by the "Programme Avenir Lyon Saint-Etienne de l'Université de Lyon" in the framework of the programme "Inverstissements d'Avenir" (ANR-11-IDEX-0007).

References

1. Abdalla, M., Fouque, P.-A., Lyubashevsky, V., Tibouchi, M.: Tightly-secure signatures from lossy identification schemes. In: Pointcheval, D., Johansson, T. (eds.) EUROCRYPT 2012. LNCS, vol. 7237, pp. 572–590. Springer, Heidelberg (2012)
2. Abe, M., Chase, M., David, B., Kohlweiss, M., Nishimaki, R., Ohkubo, M.: Constant-size structure-preserving signatures: Generic constructions and simple assumptions. In: Wang, X., Sako, K. (eds.) ASIACRYPT 2012. LNCS, vol. 7658, pp. 4–24. Springer, Heidelberg (2012)
3. Abe, M., David, B., Kohlweiss, M., Nishimaki, R., Ohkubo, M.: Tagged one-time signatures: Tight security and optimal tag size. In: Kurosawa, K., Hanaoka, G. (eds.) PKC 2013. LNCS, vol. 7778, pp. 312–331. Springer, Heidelberg (2013)
4. Abe, M., Fuchsbauer, G., Groth, J., Haralambiev, K., Ohkubo, M.: Structure-preserving signatures and commitments to group elements. In: Rabin, T. (ed.) CRYPTO 2010. LNCS, vol. 6223, pp. 209–236. Springer, Heidelberg (2010)
5. Abe, M., Haralambiev, K., Ohkubo, M.: Signing on elements in bilinear groups for modular protocol design. In: Cryptology ePrint Archive: Report 2010/133 (2010)
6. Bellare, M., Boldyreva, A., Micali, S.: Public-key encryption in a multi-user setting: Security proofs and improvements. In: Preneel, B. (ed.) EUROCRYPT 2000. LNCS, vol. 1807, pp. 259–274. Springer, Heidelberg (2000)

7. Bellare, M., Ristenpart, T.: Simulation without the Artificial Abort: Simplified Proof and Improved Concrete Security for Waters' IBE Scheme. In: Joux, A. (ed.) EUROCRYPT 2009. LNCS, vol. 5479, pp. 407–424. Springer, Heidelberg (2009)
8. Bellare, M., Rogaway, P.: Random oracles are practical: A paradigm for designing efficient protocols. In: ACM CCS 1993, pp. 62–73. ACM Press (1993)
9. Bellare, M., Rogaway, P.: The Exact Security of Digital Signatures - How to Sign with RSA and Rabin. In: Maurer, U.M. (ed.) EUROCRYPT 1996. LNCS, vol. 1070, pp. 399–416. Springer, Heidelberg (1996)
10. Bernstein, D.J.: Proving Tight Security for Rabin-Williams Signatures. In: Smart, N.P. (ed.) EUROCRYPT 2008. LNCS, vol. 4965, pp. 70–87. Springer, Heidelberg (2008)
11. Blazy, O., Kiltz, E., Pan, J. (Hierarchical) identity-based encryption from affine message authentication. In: Garay, J.A., Gennaro, R. (eds.) CRYPTO 2014, Part I. LNCS, vol. 8616, pp. 408–425. Springer, Heidelberg (2014)
12. Böhl, F., Hofheinz, D., Jager, T., Koch, J., Seo, J.H., Striecks, C.: Practical signatures from standard assumptions. In: Johansson, T., Nguyen, P.Q. (eds.) EUROCRYPT 2013. LNCS, vol. 7881, pp. 461–485. Springer, Heidelberg (2013)
13. Boneh, D., Boyen, X., Shacham, H.: Short group signatures. In: Franklin, M. (ed.) CRYPTO 2004. LNCS, vol. 3152, pp. 41–55. Springer, Heidelberg (2004)
14. Boneh, D., Franklin, M.: Identity-based encryption from the Weil pairing. SIAM J. of Computing 32(3), 586–615 (2003)
15. Boneh, D., Shen, E., Waters, B.: Strongly unforgeable signatures based on computational diffie-hellman. In: Yung, M., Dodis, Y., Kiayias, A., Malkin, T. (eds.) PKC 2006. LNCS, vol. 3958, pp. 229–240. Springer, Heidelberg (2006)
16. Camenisch, J., Chandran, N., Shoup, V.: A public key encryption scheme secure against key dependent chosen plaintext and adaptive chosen ciphertext attacks. In: Joux, A. (ed.) EUROCRYPT 2009. LNCS, vol. 5479, pp. 351–368. Springer, Heidelberg (2009)
17. Cathalo, J., Libert, B., Yung, M.: Group encryption: Non-interactive realization in the standard model. In: Matsui, M. (ed.) ASIACRYPT 2009. LNCS, vol. 5912, pp. 179–196. Springer, Heidelberg (2009)
18. Chen, J., Lim, H.-W., Ling, S., Wang, H., Wee, H.: Shorter IBE and signatures via asymmetric pairings. In: Abdalla, M., Lange, T. (eds.) Pairing 2012. LNCS, vol. 7708, pp. 122–140. Springer, Heidelberg (2013)
19. Chen, J., Wee, H.: Fully (almost) tightly secure IBE and dual system groups. In: Canetti, R., Garay, J.A. (eds.) CRYPTO 2013, Part II. LNCS, vol. 8043, pp. 435–460. Springer, Heidelberg (2013)
20. Chevallier-Mames, B.: An efficient CDH-based signature scheme with a tight security reduction. In: Shoup, V. (ed.) CRYPTO 2005. LNCS, vol. 3621, pp. 511–526. Springer, Heidelberg (2005)
21. Chevallier-Mames, B., Joye, M.: A practical and tightly secure signature scheme without hash function. In: Abe, M. (ed.) CT-RSA 2007. LNCS, vol. 4377, pp. 339–356. Springer, Heidelberg (2006)
22. Coron, J.-S.: On the exact security of full domain hash. In: Bellare, M. (ed.) CRYPTO 2000. LNCS, vol. 1880, pp. 229–235. Springer, Heidelberg (2000)
23. Coron, J.-S.: Optimal security proofs for PSS and other signature schemes. In: Knudsen, L.R. (ed.) EUROCRYPT 2002. LNCS, vol. 2332, pp. 272–287. Springer, Heidelberg (2002)
24. Coron, J.-S.: A variant of Boneh-Franklin IBE with a tight reduction in the random oracle model. Designs, Codes & Cryptography 50(1), 115–133 (2009)

25. Cramer, R., Shoup, V.: A practical public key cryptosystem provably secure against adaptive chosen ciphertext attack. In: Krawczyk, H. (ed.) CRYPTO 1998. LNCS, vol. 1462, pp. 13–25. Springer, Heidelberg (1998)

26. Dodis, Y., Haralambiev, K., López-Alt, A., Wichs, D.: Efficient public-key cryptography in the presence of key leakage. In: Abe, M. (ed.) ASIACRYPT 2010. LNCS, vol. 6477, pp. 613–631. Springer, Heidelberg (2010)

27. Galbraith, S., Malone-Lee, J., Smart, N.: Public-key signatures in the multi-user setting. Information Processing Letters 83(5), 263–266 (2002)

28. Gerbush, M., Lewko, A., O'Neill, A., Waters, B.: Dual form signatures: An approach for proving security from static assumptions. In: Wang, X., Sako, K. (eds.) ASIACRYPT 2012. LNCS, vol. 7658, pp. 25–42. Springer, Heidelberg (2012)

29. Groth, J.: Simulation-sound NIZK proofs for a practical language and constant size group signatures. In: Lai, X., Chen, K. (eds.) ASIACRYPT 2006. LNCS, vol. 4284, pp. 444–459. Springer, Heidelberg (2006)

30. Groth, J., Sahai, A.: Efficient non-interactive proof systems for bilinear groups. In: Smart, N.P. (ed.) EUROCRYPT 2008. LNCS, vol. 4965, pp. 415–432. Springer, Heidelberg (2008)

31. Hofheinz, D., Jager, T.: Tightly secure signatures and public-key encryption. In: Safavi-Naini, R., Canetti, R. (eds.) CRYPTO 2012. LNCS, vol. 7417, pp. 590–607. Springer, Heidelberg (2012)

32. Hofheinz, D., Jager, T., Kiltz, E.: Short signatures from weaker assumptions. In: Lee, D.H., Wang, X. (eds.) ASIACRYPT 2011. LNCS, vol. 7073, pp. 647–666. Springer, Heidelberg (2011)

33. Hofheinz, D., Jager, T., Knapp, E.: Waters signatures with optimal security reduction. In: Fischlin, M., Buchmann, J., Manulis, M. (eds.) PKC 2012. LNCS, vol. 7293, pp. 66–83. Springer, Heidelberg (2012)

34. Hohenberger, S., Waters, B.: Short and stateless signatures from the RSA assumption. In: Halevi, S. (ed.) CRYPTO 2009. LNCS, vol. 5677, pp. 654–670. Springer, Heidelberg (2009)

35. Jutla, C., Roy, A.: Shorter quasi-adaptive NIZK proofs for linear subspaces. In: Sako, K., Sarkar, P. (eds.) ASIACRYPT 2013, Part I. LNCS, vol. 8269, pp. 1–20. Springer, Heidelberg (2013)

36. Jutla, C., Roy, A.: Switching lemma for bilinear tests and constant-size NIZK proofs for linear subspaces. In: Garay, J.A., Gennaro, R. (eds.) CRYPTO 2014, Part II. LNCS, vol. 8617, pp. 295–312. Springer, Heidelberg (2014)

37. Kakvi, S., Kiltz, E.: Optimal security proofs for full domain hash, revisited. In: Pointcheval, D., Johansson, T. (eds.) EUROCRYPT 2012. LNCS, vol. 7237, pp. 537–553. Springer, Heidelberg (2012)

38. Katz, J., Wang, N.: Efficiency improvements for signature schemes with tight security reductions. In: ACM-CCS 2003, pp. 155–164. ACM Press (2003)

39. Lewko, A.: Tools for simulating features of composite order bilinear groups in the prime order setting. In: Pointcheval, D., Johansson, T. (eds.) EUROCRYPT 2012. LNCS, vol. 7237, pp. 318–335. Springer, Heidelberg (2012)

40. Lewko, A., Waters, B.: New techniques for dual system encryption and fully secure HIBE with short ciphertexts. In: Micciancio, D. (ed.) TCC 2010. LNCS, vol. 5978, pp. 455–479. Springer, Heidelberg (2010)

41. Libert, B., Peters, T., Joye, M., Yung, M.: Linearly homomorphic structure-preserving signatures and their applications. In: Canetti, R., Garay, J.A. (eds.) CRYPTO 2013, Part II. LNCS, vol. 8043, pp. 289–307. Springer, Heidelberg (2013)

42. Libert, B., Peters, T., Joye, M., Yung, M.: Non-malleability from malleability: Simulation-sound quasi-adaptive NIZK proofs and CCA2-secure encryption from homomorphic signatures. In: Nguyen, P.Q., Oswald, E. (eds.) EUROCRYPT 2014. LNCS, vol. 8441, pp. 514–532. Springer, Heidelberg (2014)
43. Naor, M.: On cryptographic assumptions and challenges. In: Boneh, D. (ed.) CRYPTO 2003. LNCS, vol. 2729, pp. 96–109. Springer, Heidelberg (2003)
44. Naor, M., Reingold, O.: Number-theoretic constructions of efficient pseudo-random functions. In: FOCS 1997, pp. 458–467. IEEE Press (1997)
45. Naor, M., Yung, M.: Public-key cryptosystems provably secure against chosen ciphertext attacks. In: STOC 1990, ACM Press (1990)
46. Rackoff, C., Simon, D.: Non-interactive zero-knowledge proof of knowledge and chosen ciphertext attack. In: Feigenbaum, J. (ed.) CRYPTO 1991. LNCS, vol. 576, pp. 433–444. Springer, Heidelberg (1992)
47. Sahai, A.: Non-malleable non-interactive zero knowledge and adaptive chosen-ciphertext security. In: FOCS 1999, pp. 543–553. IEEE Press (1999)
48. Schäge, S.: Tight proofs for signature schemes without random oracles. In: Paterson, K.G. (ed.) EUROCRYPT 2011. LNCS, vol. 6632, pp. 189–206. Springer, Heidelberg (2011)
49. Shamir, A.: Identity-based cryptosystems and signature schemes. In: Blakely, G.R., Chaum, D. (eds.) CRYPTO 1984. LNCS, vol. 196, pp. 47–53. Springer, Heidelberg (1985)
50. Shoup, V.: A proposal for an ISO standard for public key encryption. Manuscript (December 20, 2001)
51. Waters, B.: Efficient identity-based encryption without random oracles. In: Cramer, R. (ed.) EUROCRYPT 2005. LNCS, vol. 3494, pp. 114–127. Springer, Heidelberg (2005)
52. Waters, B.: Dual system encryption: Realizing fully secure IBE and HIBE under simple assumptions. In: Halevi, S. (ed.) CRYPTO 2009. LNCS, vol. 5677, pp. 619–636. Springer, Heidelberg (2009)
53. Yamada, S., Hanaoka, G., Kunihiro, N.: Two-dimensional representation of cover free families and its applications: Short signatures and more. In: Dunkelman, O. (ed.) CT-RSA 2012. LNCS, vol. 7178, pp. 260–277. Springer, Heidelberg (2012)

Efficient Identity-Based Encryption over NTRU Lattices

Léo Ducas[1,*], Vadim Lyubashevsky[2,**], and Thomas Prest[3]

[1] University of California, San Diego
lducas@eng.ucsd.edu
[2] École Normale Supérieure, INRIA
vadim.lyubashevsky@inria.fr
[3] École Normale Supérieure, Thales Communications & Security
thomas.prest@ens.fr

Abstract. Efficient implementations of lattice-based cryptographic schemes have been limited to only the most basic primitives like encryption and digital signatures. The main reason for this limitation is that at the core of many advanced lattice primitives is a trapdoor sampling algorithm (Gentry, Peikert, Vaikuntanathan, STOC 2008) that produced outputs that were too long for practical applications. In this work, we show that using a particular distribution over NTRU lattices can make GPV-based schemes suitable for practice. More concretely, we present the first lattice-based IBE scheme with practical parameters – key and ciphertext sizes are between two and four kilobytes, and all encryption and decryption operations take approximately one millisecond on a moderately-powered laptop. As a by-product, we also obtain digital signature schemes which are shorter than the previously most-compact ones of Ducas, Durmus, Lepoint, and Lyubashevsky from Crypto 2013.

Keywords: Lattice Cryptography, Identity-Based Encryption, Digital Signatures, NTRU.

1 Introduction

Recent improvements in efficiency have firmly established lattice-based cryptography as one of the leading candidates to replace number-theoretic cryptography after the eventual coming of quantum computing. There are currently lattice-based encryption [HPS98, LPR13a, LPR13b], identification [Lyu12], and digital signature schemes [GLP12, DDLL13] that have run-times (both in software and in hardware), key sizes, and output lengths that are more or less on par with traditional number-theoretic schemes. But unfortunately, the extent of practical lattice-based cryptography stops here. While number-theoretic assumptions

---

* This research was supported in part by the DARPA PROCEED program and NSF grant CNS-1117936. Opinions, findings and conclusions or recommendations expressed in this material are those of the author(s) and do not necessarily reflect the views of DARPA or NSF.

** This research was partially supported by the ANR JCJC grant "CLE".

P. Sarkar and T. Iwata (Eds.): ASIACRYPT 2014, PART II, LNCS 8874, pp. 22–41, 2014.

allow for very efficient constructions of advanced schemes like identity-based encryption [BF01], group signatures [CS97, BBS04], etc. none of these schemes yet have practical lattice-based realizations.

One of the major breakthroughs in lattice cryptography was the work of Gentry, Peikert, and Vaikuntanathan [GPV08], that showed how to use a short trap-door basis to generate short lattice vectors *without revealing the trap-door*.[1] In [GPV08], this was used to give the first lattice-based construction of secure hash-and-sign digital signatures and identity-based encryption schemes. This vector-sampling algorithm has since become a key component in many other lattice constructions, ranging from hierarchical IBEs [CHKP12, ABB10] to the recent breakthrough in multi-linear map constructions [GGH13]. Unfortunately, even when using improved trap-doors [AP11, MP12] and instantiating with ideal lattices [LPR13a], signature schemes that used the GPV trap-door approach were far less practical (by about two orders of magnitude) than the Fiat-Shamir ones [GLP12, DDLL13], and identity-based encryption had ciphertexts that were even longer - having ciphertexts on the order of millions of bits.[2]

1.1 Our Results

Our main result is showing that the GPV sampling procedure can in fact be used as a basis for *practical* lattice cryptography. The two main insights in our work are that one can instantiate the GPV algorithm using a particular distribution of NTRU lattices that have nearly-optimal trapdoor lengths, and that a particular parameter in the GPV algorithm can be relaxed, which results in shorter vectors being output with no loss in security. As our main applications, we propose identity-based encryption schemes that have ciphertext (and key) sizes of two and four kilobytes (for approximately 80-bit and 192-bit security, respectively) and digital signatures that have outputs (and keys) of approximately 5120 bits for about 192-bits of security. We believe that this firmly places GPV-based cryptographic schemes into the realm of practicality. The IBE outputs are orders of magnitude smaller than previous instantiations and the signature sizes are smaller by about a factor of 1.3 than in the previously shortest lattice-based scheme based on the same assumption [DDLL13].

Our schemes, like all other practical lattice-based ones, work over the polynomial ring $\mathbb{Z}_q[x]/(x^N + 1)$, where N is a power of 2 and q is a prime congruent to 1 mod $2N$. For such a choice of q, the polynomial $x^N + 1$ splits into N linear factors over $\mathbb{Z}_q$, which greatly increases the efficiency of multiplication over the ring. Our hardness assumption is related to the hardness, in the random oracle model, of solving lattice problems over NTRU lattices. These assumptions underlie the NTRU encryption scheme [HPS98], the NTRU-based fully-homomorphic encryption scheme [LTV12], and the recent signature scheme BLISS [DDLL13].

---

[1] A very similar algorithm was earlier proposed by Klein [Kle00], but was utilized in a different context and was not fully analyzed.

[2] The only works that we are aware of that give actual parameters for candidate constructions that use trapdoor sampling are [MP12, RS10].

Table 1. Comparing our IBE (GPV) with a recent implementation [Gui13] of the Boneh-Franklin scheme (BF). Our implementation was done in C++, using the NTL and GnuMP libraries. Timings were performed on an Intel Core i5-3210M laptop with a 2.5GHz CPU and 6GB RAM. The complete implementation can be found on github.com/tprest/Lattice-IBE/.

Scheme	GPV-80	GPV-192	BF-128	BF-192
User Secret key size	11 kbits	27 kbits	0.25 kbits	0.62 kbits
Ciphertext size	13 kbits	30 kbits	3 kbits	15 kbits
User Key Generation	**8.6 ms**	**32.7 ms**	**0.55 ms**	**3.44 ms**
Encryption	**0.91 ms**	**1.87 ms**	**7.51 ms**	**40.3 ms**
Decryption	**0.62 ms**	**1.27 ms**	**5.05 ms**	**34.2 ms**

Table 2. IBE scheme parameters (see Section 5)

Security parameter λ	80	192
Root Hermite factor [GN08] γ	1.0075	1.0044
Polynomial degree N	512	1024
Modulus q	$\approx 2^{23}$	$\approx 2^{27}$
User Public key size	13 Kbits	30 Kbits
User Secret key size	11 Kbits	27 Kbits
Ciphertext size	13 Kbits	30 Kbits
Ciphertext expansion factor	26	30

And even though this assumption is not related to the hardness of worst-case lattice problems via some worst-case to average-case reduction[3], in the fifteen years that the assumption has been around, there were no serious cryptanalytic threats against it. The work of [DDLL13] also provided experimental evidence that the computational complexity of finding short vectors in these special NTRU lattices was consistent with the extensive experiments of Gama and Nguyen on more general classes of lattices [GN08], some of which are connected to the hardness of worst-case lattice problems.

We implemented our schemes in software (see Table 1), and most of the algorithms are very efficient. The slowest one is user key generation, but this procedure is not performed often. More important is the underlying encryption scheme, which in our case is the Ring-LWE scheme from [LPR13a, LPR13b], which already has rather fast hardware implementations [PG13]. And as can be seen from the tables, decryption and encryption are very fast in software as well and compare very favorably to state-of-the-art implementations of pairing-based constructions.

---

[3] The work of [SS11] showed a connection between problems on NTRU lattices and worst-case problems, but for choices of parameters that do not lead to practical instantiations.

1.2 Related Work

Following the seminal work of [GPV08], there were attempts to improve several aspects of the algorithm. There were improved trap-doors [AP11], more efficient trap-door sampling algorithms [Pei10, MP12], and an NTRU signature scheme proposal [SS13]. All these papers, however, only considered parameters that preserved a security proof to lattice problems that were known to have an average-case to worst-case connection. To the best of our knowledge, our work is the first that successfully utilizes GPV trapdoor sampling in practice.

1.3 Identity-Based Encryption Scheme

In a public-key IBE scheme, the public key of every user in the system is a combination of the *master authority's* public key along with an evaluation of a publicly-computable function on the user's name or i.d.. The secret key of each user is then derived by the master authority by using his master secret key. We now give a brief description of the IBE in this paper, which is built by using the GPV algorithm to derive the user's secret keys from an NTRU lattice [GPV08, SS13], and then using the Ring-LWE encryption scheme of [LPR13a, LPR13b] for the encryption scheme.

The master public key in the scheme will be a polynomial h and the secret key will consist of a "nice basis" for the $2N$-dimensional lattice generated by the rows of $\mathbf{A}_{h,q} = \begin{pmatrix} -\mathcal{A}(h) & I_N \\ qI_N & O_N \end{pmatrix}$, where $\mathcal{A}(h)$ is the anti-circulant matrix whose i^{th} row consists of the coefficients of the polynomial $hx^i \bmod x^N + 1$ (see Definition 1). A user with identity id will have a public key consisting of h as well as $t = H(id)$, where H is some publicly-known cryptographic hash function mapping into $\mathbb{Z}_q[x]/(x^N + 1)$. The user's secret key will consist of a small polynomial s_2 such that $s_1 + s_2h = t$, where s_1 is another small polynomial (how one generates these keys is explicited in Alg. 3 in Section 5). Encryption and decryption will proceed as in the Ring-LWE scheme of [LPR13a]. To encrypt a message $m \in \mathbb{Z}[x]/(x^N + 1)$ with binary coefficients, the sender chooses polynomials r, e_1, e_2 with small coefficients and sends the ciphertext

$$(u = rh + e_1, v = rt + e_2 + \lfloor q/2 \rfloor m).^4$$

To decrypt, the receiver computes $v - us_2 = rs_1 + e_2 + \lfloor q/2 \rfloor m - s_2e_1$. If the parameters are properly set, then the polynomial $rs_1 + e_2 - s_2e_1$ will have small coefficients (with respect to q), and so the coordinates in which m is 0 will be small, whereas the coordinates in which it is 1 will be close to $q/2$. Notice that for decryption, it is crucial for the polynomial $rs_1 + e_2 - s_2e_1$ to have small coefficients, which requires s_1 and s_2 to be as small as possible.

While the above follows the usual encryption techniques based on LWE, we need a little tweak to make the security proof based on KL-divergence work

---

[4] In fact, one can save almost a factor of 2 in the ciphertext length by only sending the three highest order bits of v, rather than all of v.

Table 3. Signature scheme parameters (see Section 5)

Security parameter λ	80	192
Root Hermite factor γ	1.0069	1.0042
Polynomial degree N	256	512
Modulus q	$\approx 2^{10}$	$\approx 2^{10}$
Public key size	2560 bits	5120 bits
Secret key size	1280 bits	2048 bits
Signature size	2560 bits	5120 bits
Verification time	0.62 ms	1.27 ms

(see Section 4), since this argument only applies to search problems (while CPA security is a decisional problem). To do so we use a key-encapsulation mechanism, that is we encrypt a random key k rather than m, and then use it as a one-time-pad to send $m \oplus H'(k)$ where H' is a hash function.

1.4 Interlude: A Hash-and-Sign Digital Signature Scheme

The first part of the above IBE is actually a hash-and-sign digital signature scheme. The public (verification) key corresponds to the master authority's public key, the secret (signing) key is the master secret key, messages correspond to user i.d.'s, and signatures are the user secret keys. To sign a message m, the signer uses his secret key to compute short polynomials s_1, s_2 such that $s_1 + s_2 h = H(m)$, and transmits s_2. The verifier simply checks that s_2 and $H(m) - hs_2$ are small polynomials. In the IBE, the modulus q is set deliberately large to avoid decryption errors, but this is not an issue in the signature scheme. By selecting a much smaller q, which allows one to sample from a tighter distribution, the signature size can be made more compact than the user secret key size in the IBE.

In Table 3, we present some possible parameters for such signature schemes. The size of the keys and signatures compare very favorably to those of the BLISS signature scheme [DDLL13]. For example, for the 192 bit security level, the signature size in BLISS is approximately 6500 bits, whereas signatures in this paper are approximately 5000 bits. In fact, further improvements to the signature size may be possible via similar techniques that were used for BLISS. The main drawback of the hash-and-sign signature scheme is that signing requires sampling a discrete Gaussian over a *lattice*, whereas the Fiat-Shamir based BLISS scheme only required Gaussian sampling over the integers. At this point, the signature scheme in this paper yields smaller signatures but BLISS is much faster. Since both BLISS and this current proposal are very new, we believe that there are still a lot of improvements left in both constructions.

1.5 Techniques and Paper Organization

The main obstacle in making the above schemes practical is outputting short s_1, s_2 such that $s_1 + s_2 h = t$ while hiding the trap-door that allows for this generation. [GPV08] provided an algorithm where the length of s_1, s_2 crucially depend on the length of the Gram-Schmidt orthogonalized vectors in the trapdoor basis of the public lattice. In Section 3 we show, by experimental evidence backed up by a heuristic argument, that there exist distributions of NTRU lattices that have trap-doors whose lengths are within a small factor of optimal. Once we have such short trap-doors (which correspond to the master secret key in the IBE), we can use the GPV algorithm to sample s_1, s_2 such that $s_1 + s_2 h = t$. In order for (s_1, s_2) to reveal nothing about the trap-door, it's important that s_1, s_2 come from a distribution such that seeing (h, s_1, s_2, t) does not reveal whether s_1, s_2 were generated first and then t was computed as $s_1 + hs_2 = t$, or whether t was first chosen at random and then s_1, s_2 were computed using the GPV sampler.

To prove this, [GPV08] showed that the distribution of s_1, s_2 produced by their sampler is *statistically-close* to some trapdoor-independent distribution. In Section 4, we show that the requirement of statistical closeness can be relaxed, and we can instead use Kullback-Leibler divergence to obtain shorter secret keys. The intuition behind using KL-divergence can be described by the following example. If $\mathcal{B}_c$ denotes a Bernoulli variable with probability c on 1, then trying to distinguish *with constant probability* $\mathcal{B}_{1/2+\epsilon/2}$ from $\mathcal{B}_{1/2}$ requires $O(1/\epsilon^2)$ samples. Therefore if there is no adversary who can forge in time less than t (for some $t > 1/\epsilon^2$) on a signature scheme where some parameter comes from the distribution $\mathcal{B}_{1/2}$, then we can conclude that no adversary can forge in time less than approximately $1/\epsilon^2$ if that same variable were distributed according to $\mathcal{B}_{1/2+\epsilon/2}$. This is because a successful forger is also clearly a distinguisher between the two distributions (since forgeries can be checked), but no distinguisher can work in time less than $1/\epsilon^2$. On the other hand, distinguishing $\mathcal{B}_\epsilon$ from $\mathcal{B}_0$ requires only $O(1/\epsilon)$ samples. And so if there is a time t forger against a scheme using $\mathcal{B}_0$, all one can say about a forger against the scheme using $\mathcal{B}_\epsilon$ is that he cannot succeed in time less than $1/\epsilon$. In both cases, however, we have statistical distance ϵ between the two distributions. In this regard, statistical distance based arguments are not tight; but the KL-divergence is finer grained and can give tighter proofs. Indeed, in the first case, we can set $1/\epsilon$ to be the square root of our security parameter, whereas in the second case, $1/\epsilon$ would have to be the security parameter. In Section 4, we show that the GPV algorithm produces samples in a way that allows us to work with parameters for which the inverse of the statistical distance is the square root of the security parameter, whereas previous work required it to be the security parameter itself.

1.6 Conclusions and Open Problems

Trapdoor sampling is at the heart of many "advanced" lattice constructions, yet it has not been previously considered to be viable in practice. In this paper, we showed that with a proper distribution on the trap-door as well as analyzing the outputs using KL divergence instead of statistical distance, one can have schemes that are rather efficient and have their security based on the hardness of lattice problems over NTRU lattices. We believe that this opens the door to further practical implementations of lattice primitives having the GPV trap-door sampling algorithm at their core.

Our work used a distribution over NTRU lattices that is somewhat new – rather than having very short vectors, our secret key has vectors with a small Gram-Schmidt maximum. It is unclear how this compares in terms of dificulty to the hardness of lattice problems under the "standard" NTRU distribution. On the one hand, the vectors in our secret key are longer, but on the other hand, our secret key is more "orthogonal". General lattice algorithms (such as BKZ) don't seem to exploit this feature, but it is an interesting open problem whether other techniques could be used for improved cryptanalysis of our schemes.

2   Preliminaries

2.1   The Ring $\mathbb{Z}[x]/(x^N + 1)$

For the rest of the paper, N will be a power-of-two integer. We will work in the ring $\mathcal{R} \triangleq \mathbb{Z}[x]/(x^N + 1)$ (and occasionally $\mathcal{R}' \triangleq \mathbb{Q}[x]/(x^N + 1)$). Among other useful properties, $x^N + 1$ is irreducible, so $\mathcal{R}'$ is a cyclotomic field.

We clarify a few notations. Let $f = \sum_{i=0}^{N-1} f_i x^i$ and $g = \sum_{i=0}^{N-1} g_i x^i$ be polynomials in $\mathbb{Q}[x]$.

- fg denotes polynomial multiplication in $\mathbb{Q}[x]$, while $f * g \triangleq fg \mod (x^N + 1)$.
- (f) is the vector whose coefficients are $f_0, ..., f_{N-1}$. $(f, g) \in \mathbb{R}^{2N}$ is the concatenation of (f) and (g).
- $\lfloor f \rceil$ is the coefficient-wise rounding of f. The same notation applies for vectors.

2.2   Anticirculant Matrices

Definition 1 (Anticirculant matrix). *An N-dimensional anticirculant matrix of f is the following Toeplitz matrix:*

$$
\mathcal{A}_N(f) = \begin{pmatrix} f_0 & f_1 & f_2 & \cdots & f_{N-1} \\ -f_{N-1} & f_0 & f_1 & \cdots & f_{N-2} \\ \ddots & \ddots & \ddots & \ddots & \ddots \\ -f_1 & -f_2 & \cdots & \cdots & f_0 \end{pmatrix} = \begin{pmatrix} (f) \\ (x * f) \\ \vdots \\ (x^{N-1} * f) \end{pmatrix}
$$

When it is clear from context, we will drop the subscript N, and just write $\mathcal{A}(f)$. Anticirculant matrices verify this useful property:

Lemma 1. *Let $f, g \in \mathcal{R}$. Then $\mathcal{A}_N(f) + \mathcal{A}_N(g) = \mathcal{A}_N(f + g)$, and $\mathcal{A}_N(f) \times \mathcal{A}_N(g) = \mathcal{A}_N(f * g)$.*

2.3   Gaussian Sampling

Gaussian sampling was introduced in [GPV08] as a technique to use a short basis as a trap-door without leaking any information about the short basis; in particular it provably prevents any attack in the lines of [NR06, DN12b] designed against the NTRUSign scheme. The discrete distribution is defined as follows.

Definition 2 (Discrete Gaussians). *The n-dimensional Gaussian function* $\rho_{\sigma,c} : \mathbb{R}^n \to (0, 1]$ *is defined by:*

$$\rho_{\sigma,c}(\boldsymbol{x}) \overset{\Delta}{=} \exp\left(-\frac{\|\boldsymbol{x} - \boldsymbol{c}\|^2}{2\sigma^2}\right)$$

For any lattice $\Lambda \subset \mathbb{R}^n$, $\rho_{\sigma,c}(\Lambda) \overset{\Delta}{=} \sum_{\boldsymbol{x} \in \Lambda} \rho_{\sigma,c}(\boldsymbol{x})$. *Normalizing* $\rho_{\sigma,c}(\boldsymbol{x})$ *by* $\rho_{\sigma,c}(\Lambda)$, *we obtain the probability mass function of the discrete Gaussian distribution* $\mathcal{D}_{\Lambda,\sigma,c}$.

Using an algorithm inspired by Klein [Kle00], Gentry *et al.* [GPV08] showed that it was possible to sample vectors according to this discrete Gaussian distribution using a short basis $\mathbf{B}$ of the lattice Λ. There is a requirement on the width of the Gaussian σ related to the so called smoothing parameter. In section 4 we detail this sampler and show, using KL-divergence that the condition on the width σ can be reduced by a factor $\sqrt{2}$.

2.4   Hardness Assumptions

We can base the hardness of our IBE scheme on two assumptions that have been previously used in the literature. The first assumption deals with NTRU lattices and states that if we take two random small polynomials $f, g \in \mathcal{R}_q$, their quotient g/f is indistinguishable from random in $\mathcal{R}_q$. This assumption was first formally stated in [LTV12], but it has been studied since the introduction of the NTRU cryptosystem [HPS98] in its computational form (i.e. recovering the polynomials f and g from h). Despite more than fifteen years of cryptanalytic effort, there has not been any significant algorithmic progress towards solving either the search or decision version of this problem. As a side note, Stehle and Steinfeld [SS11] showed that for large enough f and g generated from a discrete Gaussian distribution, the quotient g/f is actually uniform in $\mathcal{R}_q$. Thus if one were to use larger polynomials, the NTRU assumption would be unnecessary. Using smaller polynomials, however, results in much more efficient schemes.

The other assumption we will be using is the Ring-LWE assumption [LPR13a] stating that the distribution of $(h_i, h_i s + e_i)$, where h_i is random in $\mathcal{R}_q$ and s, e_i are small polynomials, is indistinguishable from uniform. When the number of such samples given is polynomial (with respect to the degree of s), the coefficients of e_i cannot be too small [AG11], however, if we only give one or two samples (as is done for Ring-LWE encryption), there have been no specific attacks found if the coefficients of s, e_1, e_2 are taken from a very small set like $\{-1, 0, 1\}$. In our work, we choose to sample them from such a small set, but the scheme can be changed to sample from any other slightly larger set at the expense of slightly increasing the size of the modulus. For the concrete parameter choices, we will be using the standard methods of Gama and Nguyen [GN08] based on the currently most efficient lattice reduction algorithms [CN11].

3   Optimizing the Master Key Generation

One of the most important parameters in the scheme is the Master Secret Key: its size impacts the speed of the computations and, more importantly, the size of the users' secret keys. The smaller these secret keys will be, the more secure and efficient the scheme is (with the additional advantage that these keys can be sent more easily). While our scheme can be instantiated in any ring lattice, we choose to work in the family of NTRU lattices because the Gram-Schmidt norm of some bases are both small and easy to compute. In the end, this is what will determine the size of the users' secret keys.

3.1   The NTRU Lattices

Definition 3 (NTRU lattices). *Let N be a power-of-two integer, q a positive integer, and $f, g \in \mathcal{R}$. Let $h = g * f^{-1} \mod q$. The NTRU lattice associated to h and q is*

$$\Lambda_{h,q} = \{(u, v) \in \mathcal{R}^2 | u + v * h = 0 \mod q\}$$

$\Lambda_{h,q}$ *is a full-rank lattice of $\mathbb{Z}^{2N}$ generated by the rows of* $\mathbf{A}_{h,q} = \begin{pmatrix} -\mathcal{A}_N(h) & I_N \\ qI_N & O_N \end{pmatrix}$.

This basis is storage-efficient since it is uniquely defined by a single polynomial $h \in \mathcal{R}_q$, however it proves to be a poor choice if one wants to perform standard lattice operations. Assuming h is uniformly distributed in $\mathcal{R}_q$, $\mathbf{A}_{h,q}$ has a very large orthogonal defect.

This makes this basis not very appropriate for solving usual lattice problems such as finding the closest lattice vector to a target point. However, as explained in [HHGP+03], another basis can be found by computing $F, G \in \mathcal{R}$ such that:

$$f * G - g * F = q \tag{1}$$

Finding such (F, G) can be achieved efficiently and we describe one way (which is not new) of doing it in Section 5. A short basis is then provided by the following proposition.

Proposition 1. *Let $f, g, F, G \in \mathcal{R}$ verifying (1) and $h = g * f^{-1} \mod q$. Then*
$\mathbf{A}_{h,q} = \begin{pmatrix} -\mathcal{A}(h) & I_N \\ qI_N & O_N \end{pmatrix}$ *and* $\mathbf{B}_{f,g} = \begin{pmatrix} \mathcal{A}(g) & -\mathcal{A}(f) \\ \mathcal{A}(G) & -\mathcal{A}(F) \end{pmatrix}$ *generate the same lattice.*

Proof. Consider $\mathbf{P} = \mathbf{A}_{h,q} \times \mathbf{B}_{f,g}^{-1}$ the change-of-basis matrix between $\mathbf{A}_{h,q}$ and $\mathbf{B}_{f,g}$. One can check that $q\mathbf{P} = O_{2N} \mod q$, so $\mathbf{P} \in \mathbb{Z}^{2N \times 2N}$. Also, $|\det(\mathbf{P})| = 1$ so $\mathbf{P}^{-1} \in \mathbb{Z}^{2N \times 2N}$. We can conclude that $\mathbf{A}_{h,q}$ and $\mathbf{B}_{f,g}$ both generate the same lattice. $\square$

Definition 4 (Gram-Schmidt norm [GPV08]). *Let* $\mathbf{B} = (\mathbf{b}_i)_{i \in I}$ *be a finite basis, and* $\tilde{\mathbf{B}} = (\tilde{\mathbf{b}}_i)_{i \in I}$ *be its Gram-Schmidt orthogonalization. The Gram-Schmidt norm of* $\mathbf{B}$ *is the value*

$$\|\tilde{\mathbf{B}}\| = \max_{i \in I} \|\tilde{\mathbf{b}}_i\|$$

An interesting property of NTRU lattices is related to the Gram-Schmidt norm of their bases: they can be small and can be computed quickly. These two facts and their benefits for our scheme are discussed in the following subsection.

3.2   Bounding the Gram-Schmidt Norm

The lattice over which we do Gaussian sampling is $\Lambda_{h,q}$, and the size of the secret keys we sample will be proportional to $\|\tilde{\mathbf{B}}\|$, where $\mathbf{B}$ is the basis of $\Lambda_{h,q}$ used in the Gaussian sampler. It is very important then that the Gram-Schmidt norm $\|\tilde{\mathbf{B}}\|$ of $\mathbf{B}$ is as small as possible.

Proposition 1 tells us that $\mathbf{B}_{f,g}$ is a basis of $\Lambda_{h,q}$. The second step is to compute its Gram-Schmidt norm $\|\tilde{\mathbf{B}}_{f,g}\|$. For general lattices, this is done by applying the Gram-Schmidt process to the basis and computing the maximum length of the resulting vectors. In the case of NTRU lattices, however, Lemmas 2 and 3 allow to compute $\|\tilde{\mathbf{B}}_{f,g}\|$ much faster, in time $\mathcal{O}(N \log(N))$ instead of $\mathcal{O}(N^3)$.

Lemma 2. *Let* $\mathbf{B}_{f,g}$ *be as defined in Proposition 1, and* $\mathbf{b}_1, ..., \mathbf{b}_{2N}$ *be the row vectors of* $\mathbf{B}_{f,g}$. *Then* $\|\tilde{\mathbf{B}}_{f,g}\| = max\{\|\tilde{\mathbf{b}}_1\|, \|\tilde{\mathbf{b}}_{N+1}\|\}$

Proof. For $\mathbf{V}$ a subspace of $\mathbb{R}^{2N}$ and $\mathbf{b} \in \mathbb{R}^{2N}$, let us denote $\mathbf{Proj_V}(\mathbf{b})$ the orthogonal projection of $\mathbf{b}$ over $\mathbf{V}$. We also call r the linear isometry $(f, g) \mapsto (x * f, x * g)$ (see the notations from Subsection 2.1), so that for any $i \leqslant N$, $\mathbf{b}_i = r^{i-1}(\mathbf{b}_1)$ and $\mathbf{b}_{N+i} = r^{i-1}(\mathbf{b}_{N+1})$. Let $\mathbf{V}_i = Span(\mathbf{b}_1, ..., \mathbf{b}_i)^{\perp}$. By definition of the Gram-Schmidt orthogonalization, for any $i \in [\![1, 2N]\!]$, $\tilde{\mathbf{b}}_i = \mathbf{Proj}_{\mathbf{V}_{i-1}}(\mathbf{b}_i)$.

Moreover, one can check the two following properties:
- $\|\mathbf{Proj_V}(\mathbf{b})\| \leqslant \|\mathbf{b}\|$
- $\mathbf{V} \subseteq \mathbf{W} \Rightarrow \|\mathbf{Proj_V}(\mathbf{b})\| \leqslant \|\mathbf{Proj_W}(\mathbf{b})\|$

From the first property comes the fact that for any $i \in [\![1, N]\!]$, $\|\tilde{\mathbf{b}}_i\| \leqslant \|\mathbf{b}_1\| = \|\tilde{\mathbf{b}}_1\|$. Proving $\|\tilde{\mathbf{b}}_{N+i}\| \leqslant \|\tilde{\mathbf{b}}_{N+1}\|$ is a bit trickier. Since $Span(\mathbf{b}_1, ...\mathbf{b}_N)$ is stable by r, so is $\mathbf{V}_N$. One can check that $\mathbf{Proj}_{\mathbf{V}_N}(\mathbf{b}_{N+i}) = r^{i-1}(\mathbf{Proj}_{\mathbf{V}_N}(\mathbf{b}_{N+1}))$. Now $\mathbf{V}_{N+i-1} \subseteq \mathbf{V}_N$, so :
$\|\tilde{\mathbf{b}}_{N+i}\| = \|\mathbf{Proj}_{\mathbf{V}_{N+i-1}}(\mathbf{b}_{N+i})\| \leqslant \|\mathbf{Proj}_{\mathbf{V}_N}(\mathbf{b}_{N+i})\| = \|\mathbf{Proj}_{\mathbf{V}_N}(\mathbf{b}_{N+1})\| = \|\tilde{\mathbf{b}}_{N+1}\|$
Which concludes this proof.   □

Figure 1 illustrates the result of Lemma 2. Before the reduction, all the vectors of each semi-basis are of the same size, but after the reduction, the largest vector of $\tilde{\mathbf{B}}_{f,g}$ is either $\tilde{\mathbf{b}}_1$ or $\tilde{\mathbf{b}}_{N+1}$.

Instead of computing $2N$ values $\|\tilde{\mathbf{b}}_1\|, ..., \|\tilde{\mathbf{b}}_{2N}\|$, there is now only two of them to compute. We already know that $\|\tilde{\mathbf{b}}_1\| = \|\mathbf{b}_1\|$, and we introduce a notation which will provide us an expression for $\|\tilde{\mathbf{b}}_{N+1}\|$.

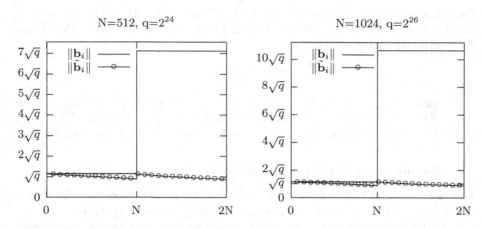

Fig. 1. Size of the vectors of $\tilde{\mathbf{B}}_{f,g}$ before and after Gram-Schmidt reduction

Definition 5. *Let $f \in \mathcal{R}'$. We denote $\bar{f}$ the unique polynomial in $\mathcal{R}'$ such that $\mathcal{A}(f)^t = \mathcal{A}(\bar{f})$. If $f(x) = \sum_{i=0}^{N-1} f_i x^i$, then $\bar{f}(x) = f_0 - \sum_{i=1}^{N-1} f_{N-i} x^i$*

This notation is only needed in the master key generation, to compute the Gram-Schmidt norm of the basis as well as reducing $(G, -F)$ modulo $(g, -f)$. The following lemma gives an exact expression for $\|\tilde{\mathbf{b}}_{N+1}\|$.

Lemma 3. $\|\tilde{\mathbf{b}}_{N+1}\| = \left\| \left(\frac{q\bar{f}}{f*\bar{f}+g*\bar{g}}, \frac{q\bar{g}}{f*\bar{f}+g*\bar{g}} \right) \right\|$

Proof. Let $k = \frac{\bar{f}*F+\bar{g}*G}{f*\bar{f}+g*\bar{g}} \mod (x^N + 1)$ and write $k(x) = \sum_{i=0}^{N-1} k_i x^i$. Then

$$c \overset{\triangle}{=} \mathbf{b}_{N+1} - \sum_{i=0}^{N-1} k_i \mathbf{b}_{i+1} = (G, -F) - (k*g, -k*f) = \left(\frac{q\bar{f}}{f*\bar{f}+g*\bar{g}}, \frac{q\bar{g}}{f*\bar{f}+g*\bar{g}} \right)$$

is orthogonal to $Span(\mathbf{b}_1, ..., \mathbf{b}_N)$. By the uniqueness of the Gram-Schmidt decomposition, $c = \tilde{\mathbf{b}}_{N+1}$ and the result follows. $\square$

This enables us to compute $\|\tilde{\mathbf{B}}_{f,g}\|$ only from (f, g), gaining some time when generating the Master Key. Knowing (f, g) is enough to know almost instantly whether the basis $\tilde{\mathbf{B}}_{f,g}$ will be a good one for Gaussian sampling. After deriving this formula for $\|\tilde{\mathbf{B}}_{f,g}\|$, we ran experiments to compute it for different values of N, q and initial vector $\mathbf{b}_1$.

For fixed values of $\|\mathbf{b}_1\|$ and q, experiments show no correlation between the dimension N and $\|\tilde{\mathbf{B}}_{f,g}\|$. Moreover, they suggest that $\|\tilde{\mathbf{b}}_{N+1}\|$ is actually pretty close to its lower bound $q/\|\mathbf{b}_1\|$ (see Lemma 4 for the proof of this bound). Both experiments and a heuristic indicate that the optimal choice for $\|\mathbf{b}_1\|$ is $\|\mathbf{b}_1\| \approx \sqrt{\frac{qe}{2}}$, since we then get $\|\tilde{\mathbf{b}}_{N+1}\| \approx \|\tilde{\mathbf{b}}_1\|$. And so in our experiments, we sample $\mathbf{b}_1$ with a norm slightly bigger than $\sqrt{\frac{qe}{2}} \approx 1.1658\sqrt{q}$. The heuristic can be found in the full version of this paper.

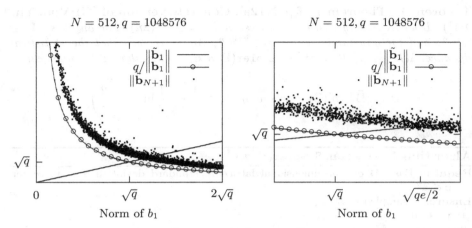

Fig. 2. Values of candidates $\|\tilde{\mathbf{b}}_{N+1}\|$ and $\|\tilde{\mathbf{b}}_1\|$ for $\|\tilde{\mathbf{B}}_{f,g}\|$, with $N = 512, q = 2^{20}$. $q/\|\mathbf{b}_1\|$ is the lower bound for $\|\tilde{\mathbf{b}}_{N+1}\|$ given in Lemma 4.

The following lemma provides a theoretical lower bound for $\|\tilde{\mathbf{b}}_{N+1}\|$, given q and $\|\mathbf{b}_1\|$. In our case, $\|\tilde{\mathbf{b}}_{N+1}\|$ is very close to its lover bound.

Lemma 4. *Let* $\mathbf{B} = (\mathbf{b}_i)_{1 \leqslant i \leqslant 2N}$ *be a NTRU basis.* $\|\tilde{\mathbf{b}}_{N+1}\|$ *admits the following lower bound:* $\|\tilde{\mathbf{b}}_{N+1}\| \geqslant q/\|\mathbf{b}_1\|$.

Proof. We have $|\det(\mathbf{B}_{f,g})| = |\det(\tilde{\mathbf{B}}_{f,g})| = \prod\limits_{i=1}^{N} \|\tilde{\mathbf{b}}_i\|$. We know that $|\det(\mathbf{B}_{f,g})| = q^N$ and that for any k in $[\![1; N]\!]$, $\|\tilde{\mathbf{b}}_i\| \leqslant \|\tilde{\mathbf{b}}_1\|$ and $\|\tilde{\mathbf{b}}_{N+i}\| \leqslant \|\tilde{\mathbf{b}}_{N+1}\|$. So $q^N \leqslant \|\mathbf{b}_1\|^N \|\tilde{\mathbf{b}}_{N+1}\|^N$. □

With the results of these experiments, we can now design an efficient Master Key Generator found in Section 5.

4   Optimizing the User Key Generation

Many arguments in lattice based cryptography are driven by a smoothing condition, that is that the parameter σ of some Gaussian is greater than the smoothing parameter. We recall that $D_{\Lambda,\sigma,\mathbf{c}}$ is defined in Definition 2.

Definition 6 (Smoothing parameter [MR07]). *For any n-dimensional lattice* Λ *and real* $\epsilon > 0$, *the smoothing parameter* $\eta_\epsilon(\Lambda)$ *is the smallest* $s > 0$ *such that* $\rho_{1/s\sqrt{2\pi},0}(\Lambda^* \setminus 0) \leqslant \epsilon$. *We also define a scaled version* $\eta'_\epsilon(\Lambda) = \frac{1}{\sqrt{2\pi}}\eta_\epsilon(\Lambda)$.

This is in particular the case for the correctness of the sampling algorithm of [GPV08]: it is correct up to negligible statistical distance for any choice of $s = \omega(\sqrt{\log n}) \cdot \|\tilde{\mathbf{B}}\|$. A concrete sufficient condition to ensure λ-bits of security was computed in [DN12a]

Theorem 1 (Theorem 1 of [DN12a], Concrete version of [GPV08, Th. 4.1]). *Let n, λ be any positive integers, and $\epsilon = 2^{-\lambda}/(2n)$. For any basis $\mathbf{B} \in \mathbb{Z}^{n \times n}$, and for any target vector $\mathbf{c} \in \mathbb{Z}^{1 \times n}$, Alg. 1 is such that the statistical distance $\Delta(D_{\Lambda(\mathbf{B}), \sigma, \mathbf{c}}, \mathtt{Gaussian\_Sampler}(\mathbf{B}, \sigma, \mathbf{c}))$ is less than $2^{-\lambda}$, provided:*

$$\sigma \geq \|\tilde{\mathbf{B}}\| \cdot \eta_\epsilon'(\mathbb{Z}) \quad where \quad \eta_\epsilon'(\mathbb{Z}) \approx \frac{1}{\pi} \cdot \sqrt{\frac{1}{2} \ln\left(2 + \frac{2}{\epsilon}\right)}.$$

Algorithm 1. $\mathtt{Gaussian\_Sampler}(\mathbf{B}, \sigma, \mathbf{c})$

Require: Basis $\mathbf{B}$ of a n-dimensional lattice Λ, standard deviation $\sigma > 0$, center $\mathbf{c} \in \mathbb{Z}^n$
Ensure: $\mathbf{v}$ sampled in $\mathcal{D}_{\Lambda, \sigma, \mathbf{c}}$
1: $\mathbf{v}_n \leftarrow \mathbf{0}$
2: $\mathbf{c}_n \leftarrow \mathbf{c}$
3: **for** $i \leftarrow n, ..., 1$ **do**
4:      $c_i' \leftarrow \langle \mathbf{c}_i, \tilde{\mathbf{b}}_i \rangle / \|\tilde{\mathbf{b}}_i\|^2$
5:      $\sigma_i' \leftarrow \sigma / \|\tilde{\mathbf{b}}_i\|$
6:      $z_i \leftarrow \mathtt{SampleZ}(\sigma_i', c_i')$
7:      $\mathbf{c}_{i-1} \leftarrow \mathbf{c}_i - z_i \mathbf{b}_i$ and $\mathbf{v}_{i-1} \leftarrow \mathbf{v}_i + z_i \mathbf{b}_i$
8: **end for**
9: **return** $\mathbf{v}_0$

The sub-algorithm $\mathtt{SampleZ}(\sigma', c')$ samples a 1-dimensional Gaussian $\mathcal{D}_{\mathbb{Z}, \sigma', c'}$. This can be achieved in various ways: rejection sampling, look-up tables, etc. For our implementation, we chose an hybrid method using the discrete Gaussian sampling from [DDLL13] and "standard" rejection sampling.

Fig. 3. Description of Klein-GPV Gaussian Sampler

In this section, we sketch why the condition $\epsilon = 2^{-\lambda}/(4N)$ can be relaxed to

$$\epsilon \leq 2^{-\lambda/2}/(4\sqrt{2}N);$$

asymptotically square-rooting the minimal value of ϵ; this impacts the value of $\eta_\epsilon'(\mathbb{Z})$ by a factor $\sqrt{2}$, that is one can use *the same algorithm* with a standard deviation σ shorter by a factor $\sqrt{2}$. To do so we rely on a finer grained measure of "distance"[5] between distributions, called Kullback-Leibler Divergence (or KL Divergence). Interestingly, this common notion from information theory, has to our knowledge been used in cryptography only in the context of symmetric key cryptanalysis [Vau03]. The use of KL Divergence recently found similar application in lattice based Cryptography, namely for an enhanced implementation [PDG14] of BLISS [DDLL13]. It is defined as follows:

---

[5] Technically it is not a distance: it is neither symmetric nor does it verifies the triangle inequality.

Definition 7 (Kullback-Leibler Divergence). *Let $\mathcal{P}$ and $\mathcal{Q}$ be two distributions over a common countable set Ω, and let $S \subset \Omega$ be the support of $\mathcal{P}$. The Kullback-Leibler Divergence, noted D_{KL} of $\mathcal{Q}$ from $\mathcal{P}$ is defined as:*

$$D_{KL}(\mathcal{P}\|\mathcal{Q}) = \sum_{i \in S} \ln\left(\frac{\mathcal{P}(i)}{\mathcal{Q}(i)}\right) \mathcal{P}(i)$$

with the convention that $\ln(x/0) = +\infty$ for any $x > 0$.

For complements the reader can refer to [CT91]. We only require two essential properties, additivity: $D_{\mathrm{KL}}(\mathcal{P}_0 \times \mathcal{P}_1 \| \mathcal{Q}_0 \times \mathcal{Q}_1) = D_{\mathrm{KL}}(\mathcal{P}_0\|\mathcal{Q}_0) + D_{\mathrm{KL}}(\mathcal{P}_1\|\mathcal{Q}_1)$, and the data processing inequality: $D_{\mathrm{KL}}(f(\mathcal{P})\|f(\mathcal{Q})) \leq D_{\mathrm{KL}}(\mathcal{P}\|\mathcal{Q}))$.

Lemma 5 (Bounding Success Probability Variations [PDG14]). *Let $\mathcal{E}^{\mathcal{P}}$ be an algorithm making at most q queries to an oracle sampling from a distribution $\mathcal{P}$ and outputting a bit. Let $\epsilon \geq 0$, $\mathcal{Q}$ be a distribution such that $D_{KL}(\mathcal{P}\|\mathcal{Q}) \leq \epsilon$, and x (resp. y) denote the probability that $\mathcal{E}^{\mathcal{P}}$ (resp. $\mathcal{E}^{\mathcal{Q}}$) outputs 1. Then,*

$$|x - y| \leq \frac{1}{\sqrt{2}}\sqrt{q\epsilon}.$$

Concrete Security. This lemma lets us conclude that if a scheme is λ-bit secure with access to a perfect oracle for distribution $\mathcal{P}$, then it is also about λ-bit secure with oracle access to $\mathcal{Q}$ if $D_{\mathrm{KL}}(\mathcal{P}\|\mathcal{Q}) \leq 2^{-\lambda}$.

To argue concrete security according to Lemma 5, consider a search problem $\mathcal{S}^{\mathcal{P}}$ using oracle access to a distribution $\mathcal{P}$, and assume it is not λ-bit hard; that is there exists an attacker $\mathcal{A}$ that solve $\mathcal{S}^{\mathcal{P}}$ with probability p and has running time less than $2^{\lambda}/p$; equivalently (repeating the attack until success) there exists an algorithm $\mathcal{A}'$ that solves $\mathcal{S}^{\mathcal{P}}$ in time $\approx 2^{\lambda}$ with probability at least $3/4$. Such algorithms make $q \leq 2^{\lambda}$ queries to $\mathcal{P}$. If $D_{\mathrm{KL}}(\mathcal{P}\|\mathcal{Q}) \leq 2^{-\lambda}$, Lemma 5 ensures us that the success of $\mathcal{A}'$ against $\mathcal{S}^{\mathcal{Q}}$ will be at least $1/4$; in other word if $\mathcal{S}^{\mathcal{Q}}$ is λ-bit secure, $\mathcal{S}^{\mathcal{P}}$ is also about λ-bit secure.

Note that this applies to search problems only, therefore, it is unclear if it could be directly applied to any CPA scheme: CPA security is a decisional problem, not a search problem. Yet our IBE design makes this argument valid: we designed encryption using key-encapsulation mechanism, the random key k being fed into a hash function H' to one-time-pad the message. Modeling H' as a random oracle, one easily proves that breaking CPA security with advantage p is as hard as recovering k with probability p; which is a search problem. $\square$

The point is that the KL Divergence is in some cases much smaller than statistical distance; and it will indeed be the case for Klein sampling as used in [GPV08] and described in Fig. 3.

Theorem 2 (KL Divergence of the Gaussian Sampler). *For any $\epsilon \in (0, 1/4n)$, if $\sigma \geq \eta'_\epsilon(\mathbb{Z}) \cdot \|\tilde{\mathbf{B}}\|$ then the KL Divergence between $D_{\Lambda(\mathbf{B}),\mathbf{c},\sigma}$ and the output of $\mathtt{Gaussian\_Sampler}(\mathbf{B}, \sigma, \mathbf{c})$ is bounded by $2\left(1 - \left(\frac{1+\epsilon}{1-\epsilon}\right)^n\right)^2 \approx 8n^2\epsilon^2$.*

Proof. The probability that Klein's algorithm outputs $\mathbf{x} = \tilde{\mathbf{x}}$ on inputs $\sigma, \mathbf{B}, \mathbf{c}$ is proportional to

$$\prod_{i=1}^{n} \frac{1}{\rho_{\sigma_i, c'_i}(\mathbb{Z})} \cdot \rho_{\sigma, \mathbf{c}}(\tilde{\mathbf{x}})$$

for $\sigma_i = \sigma/\|\tilde{\mathbf{b}}_i\|$ and some $c'_i \in \mathbb{R}$ that depends on $\mathbf{c}$ and $\mathbf{B}$. as detailed in [GPV08]. By assumption, $\sigma_i \geq \eta_\epsilon(\mathbb{Z})$, therefore $\rho_{\sigma_i, c'_i}(\mathbb{Z}) \in [\frac{1-\epsilon}{1+\epsilon}, 1] \cdot \rho_{\sigma_i}(\mathbb{Z})$ (see [MR07, Lemma 4.4]). The relative error to the desired distribution (proportional to $\rho_{\sigma, \mathbf{c}}(\tilde{\mathbf{x}})$) is therefore bounded by $1 - \left(\frac{1+\epsilon}{1-\epsilon}\right)^n$; we can conclude using Lemma 2 from [PDG14]. □

This last theorem implies that the condition $\epsilon \leq \frac{2^{-\lambda}}{4N}$ of [DN12a] can be relaxed to $\epsilon \leq \frac{2^{-\lambda/2}}{4\sqrt{2}N}$ which conclude this section.

5 The Schemes and Their Security Analysis

5.1 The IBE Scheme

We recall that an IBE scheme is composed of four algorithms: Master_Keygen, which generates the Master Keys, Extract, which uses the Master Secret Key to generate users' secret keys for any identity, Encrypt, which allows anybody to encrypt a message for an user given the Master Public Key and the user's identity, and Decrypt which enables an user to decrypt the messages intended to him with his secret key.

Algorithm 2. Master_Keygen(N, q)

Require: N, q
Ensure: Master Secret Key $\mathbf{B} \in \mathbb{Z}_q^{2N \times 2N}$ and Master Public Key $h \in \mathcal{R}_q$
1: $\sigma_f = 1.17\sqrt{\frac{q}{2N}}$ {σ_f chosen such that $\mathbb{E}[\|b_1\|] = 1.17\sqrt{q}$}
2: $f, g \leftarrow \mathcal{D}_{N, \sigma_f}$
3: Norm $\leftarrow \max\left(\|(g, -f)\|, \left\|\left(\frac{q\bar{f}}{f*\bar{f}+g*\bar{g}}, \frac{q\bar{g}}{f*\bar{f}+g*\bar{g}}\right)\right\|\right)$ {We compute $\|\tilde{\mathbf{B}}_{f,g}\|$}
4: if Norm$> 1.17\sqrt{q}$, go to step 2
5: Using extended euclidean algorithm, compute $\rho_f, \rho_g \in \mathcal{R}$ and $R_f, R_g \in \mathbb{Z}$ such that
    − $\rho_f \cdot f = R_f \mod (x^N + 1)$
    − $\rho_g \cdot g = R_g \mod (x^N + 1)$
6: if $GCD(R_f, R_g) \neq 1$ or $GCD(R_f, q) \neq 1$, go to step 2
7: Using extended euclidean algorithm, compute $u, v \in \mathbb{Z}$ such that $u \cdot R_f + v \cdot R_g = 1$
8: $F \leftarrow qv\rho_g, G \leftarrow -qu\rho_f$
9: $k = \left\lfloor \frac{F*\bar{f}+G*\bar{g}}{f*\bar{f}+g*\bar{g}} \right\rceil \in \mathcal{R}$
10: Reduce F and G: $F \leftarrow F - k*f, G \leftarrow G - k*g$
11: $h = g * f^{-1} \mod q$
12: $\mathbf{B} = \begin{pmatrix} \mathcal{A}(g) & -\mathcal{A}(f) \\ \mathcal{A}(G) & -\mathcal{A}(F) \end{pmatrix}$

Here, $\mathbf{B}$ is a short basis of $\Lambda_{h,q}$, making it a trapdoor for sampling short elements (s_1, s_2) such that $s_1 + s_2 * h = t$ for any t, without leaking any information about itself.

Algorithm 3. $\mathtt{Extract}(\mathbf{B}, id)$

Require: Master Secret Key $\mathbf{B} \in \mathbb{Z}_q^{2N \times 2N}$, hash function $H : \{0,1\}^* \to \mathbb{Z}_q^N$, user identity id
Ensure: User secret key $\mathbf{SK}_{id} \in \mathcal{R}_q$
 1: **if** $\mathbf{SK}_{id}$ is in local storage **then**
 2:     Output $\mathbf{SK}_{id}$ to user id
 3: **else**
 4:     $t \leftarrow H(id) \in \mathbb{Z}_q^N$
 5:     $(s_1, s_2) \leftarrow (t, 0) - \mathtt{Gaussian\_Sampler}(\mathbf{B}, \sigma, (t, 0))$ $\{s_1 + s_2 * h = t\}$
 6:     $\mathbf{SK}_{id} \leftarrow s_2$
 7:     Output $\mathbf{SK}_{id}$ to user id and keep it in local storage
 8: **end if**

Algorithm 3 stores the secret key for each user that has made a query. The reasons behind this choice, and alternatives are discussed in the full version of this paper.

Algorithm 4. $\mathtt{Encrypt}(id, m)$

Require: Hash functions $H : \{0,1\}^* \to \mathbb{Z}_q^N$ and $H' : \{0,1\}^N \to \{0,1\}^m$, message $m \in \{0,1\}^m$, Master Public Key $h \in \mathcal{R}_q$, identity id
Ensure: Encryption $(u, v, c) \in \mathcal{R}_q^2$ of m under the public key of id
 1: $r, e_1, e_2 \leftarrow \{-1, 0, 1\}^N$; $k \leftarrow \{0, 1\}^N$ (uniform)
 2: $t \leftarrow H(id)$
 3: $u \leftarrow r * h + e_1 \in \mathcal{R}_q$
 4: $v \leftarrow r * t + e_2 + \lfloor q/2 \rfloor \cdot k \in \mathcal{R}_q$
 5: Drop the least significant bits of v: $v \leftarrow 2^\ell \lfloor v/2^\ell \rceil$
 6: Output $(u, v, m \oplus H'(k))$

Note that encryption is designed using a key-encapsulation mechanism; the hash of the key k is used to one-time-pad the message. If H' is modeled as a random oracle, this makes the CPA security (a decisional problem) of the scheme as hard as finding the key k exactly (a search problem). Basing the security argument on a search problem is necessary for our KL Divergence-based security argument to hold, as explained in Section 4.

In order for an user to decrypt correctly, $y = r * s_1 + e_2 - e_1 * s_2$ must have all its coefficients in $(-\frac{q}{4}, \frac{q}{4})$, so we need to set q big enough. In practice, this gives $q \geqslant 5.1 \cdot 10^6$ for $\lambda = 80$, and $q \geqslant 5.6 \cdot 10^6$ for $\lambda = 192$.

Dropping bits can also lead to incorrect decryption. However, for $\ell \leqslant \lfloor \log_2 q \rfloor - 3$, it doesn't significantly affect the correct decryption rate of the scheme, so we take this value of ℓ as standard.

Algorithm 5. Decrypt($\mathbf{MS}_{id}, (u, v, c)$)

Require: User secret key $\mathbf{SK}_{id}$, encryption (u, v, c) of m
Ensure: Message $m \in \{0, 1\}^N$
  1: $w \leftarrow v - u * s_2$
  2: $k \leftarrow \left\lfloor \frac{w}{q/2} \right\rceil$
  3: Output $m \leftarrow c \oplus H'(k)$

The computations leading to these values of q and ℓ can be found in the full version of this paper.

The scheme described above is only CPA-secure. In practice, we would want to make it CCA-secure by using one of the standard transformations (e.g. [FO99]).

5.2   Security Analysis of the IBE Scheme

We now use the techniques in [GN08, CN11, DDLL13] to analyze the concrete security of the scheme. The way lattice schemes are analyzed is to determine the hardness of the underlying lattice problem, which is measured using the "root Hermite factor" introduced in [GN08]. If one is looking for a vector $\mathbf{v}$ in an n-dimensional lattice that is *larger* than the nth root of the determinant, then the associated root Hermite factor is

$$\frac{\|\mathbf{v}\|}{det(\Lambda)^{1/n}} = \gamma^n \qquad (2)$$

If one is looking for an unusually-short planted vector $\mathbf{v}$ in an NTRU lattice, then the associated root Hermite factor, according to the experiments in [DDLL13] is

$$\frac{\sqrt{n/(2\pi e)} \cdot det(\Lambda)^{1/n}}{\|\mathbf{v}\|} = .4\gamma^n. \qquad (3)$$

Based on the results in [GN08, CN11], one can get a very rough estimate of the hardness of the lattice problem based on the value of γ (unfortunately, there has not been enough lattice cryptanalysis literature to have anything more that just a rough estimate). For values of $\gamma \approx 1.007$, finding the vector is at least 2^{80}-hard. For values less that 1.004, the problem seems completely intractable and is approximated to be at least 192-bits hard.

The most vulnerable part of our IBE scheme will be the actual encryption. Still, we will first run through the best attacks on the master public key and user secret keys because these correspond exactly to attacks on the key and signature forgery, respectively, in the hahs-and-sign digital signature scheme. Our master public key polynomial h is not generated uniformly at random, but rather as $g * f^{-1}$. The best-known attack for distinguishing an NTRU polynomial from a random one is to find the polynomials f, g that are "abnormally short". This involves finding the short f and g such that $h * f - g = 0 \bmod q$. This is equivalent to finding the vector (f, g) in a $2N$-dimensional lattice with determinant

q^N. From Section 3, we know that the euclidean norm of the vector (f, g) is approximately $1.17\sqrt{q}$ and so calculating the value of γ using (3), we get

$$\frac{\sqrt{2N/(2\pi e)} \cdot \sqrt{q}}{1.17\sqrt{q}} = .4\gamma^{2N} \implies \gamma = (\sqrt{N}/1.368)^{1/2N},$$

which is 1.0054 for $N = 256$ and 1.0027 for $N = 512$, which is already beyond the realm of practical algorithms. The secret user keys (s_1, s_2) are generated with standard deviation of about $\sigma = 1.17\eta'_\epsilon(\mathbb{Z}) \cdot \|\tilde{\mathbf{B}}\|$, which gives $\sigma = 1.5110\sqrt{q}$ for $N = 256$ (resp. $\sigma = 2.2358\sqrt{q}$ for $N = 512$), and so the vector has length $\sigma\sqrt{2N}$, which by (2) results in a value of γ,

$$\frac{\sigma\sqrt{2N}}{\sqrt{q}} = \gamma^{2N} \implies \begin{cases} \gamma = (2.137\sqrt{N})^{1/2N} \text{ for } N = 256 \\ \gamma = (3.162\sqrt{N})^{1/2N} \text{ for } N = 512 \end{cases}$$

which is 1.0069 for $N = 256$ and 1.0042 for $N = 512$.

We now move on to the hardness of breaking the CPA-security of the scheme. Encryption (disregarding the message) consists of $(u = r * h + e_1, v = r * t + e_2)$, where the coefficients of r, e_1, e_2 have coefficients chosen from $\{-1, 0, 1\}$. In order to avoid decryption errors, the value of the modulus q has to be set fairly high (see Table 1). The best-known attack against the encryption scheme involves essentially recovering the errors e_1, e_2. From the ciphertext (u, v), we can set up the equation $(t * h^{-1}) * e_1 - e_2 = (t * h^{-1}) * u - v \bmod q$, which can be converted into the problem of finding the $2N + 1$-dimensional vector $(e_1, e_2, 1)$ in a $2N + 1$-dimensional lattice with determinant q^N. Using (3), we get

$$\frac{\sqrt{2N/(2\pi e)} \cdot \sqrt{q}}{\|(e_1, e_2, 1)\|} = .4\gamma^{2N} \implies \gamma = (.74\sqrt{q})^{1/2N},$$

which gives us $\gamma = 1.0075$ for $N = 512$ and $\gamma = 1.0044$ for $N = 1024$.

5.3 Analysis of the Signature Scheme

In our signature scheme, the Keygen is provided by Algorithm 2, Signature by Algorithm 3 and Verification by checking the norm of (s_1, s_2) as well as the equality $s_1 + s_2 * h = H(message)$. Since there is no encryption, we can discard the CPA-security analysis at the end of the previous section, as well as the issues regarding correctness of the encryption. This leads to much smaller values for N and q, which can be found in the Table 3 of Section 1.

We now analyze the bitsize of the secret key, public key and signature. The public key is $h \in \mathcal{R}_q$, as well as the signature s_1, so their bitsizes are $N\lceil \log_2 q \rceil$. The secret key is f such that $f * h = g \bmod q$. Given the procedure to generate $b_1 = (f, g)$, with high probability each coefficient of f has absolute value at most equal to $6\sigma_f$ (if it isn't the case, one just need to resample the coefficient). f can therefore be stored using $N(1 + \lceil \log_2(6\sigma_f) \rceil)$ bits, where $\sigma_f = 1.17\sqrt{\frac{q}{2N}}$.[6]

---

[6] Using Huffman coding, as in BLISS [DDLL13], may also be appropriate here for reducing the key length by several hundred bits.

Acknowledgments. The authors wish to thank David Xiao and Aurore Guillevic for helpful conversations, as well as the anonymous Asiacrypt'14 reviewers.

References

[ABB10]   Agrawal, S., Boneh, D., Boyen, X.: Lattice basis delegation in fixed dimension and shorter-ciphertext hierarchical IBE. In: Rabin, T. (ed.) CRYPTO 2010. LNCS, vol. 6223, pp. 98–115. Springer, Heidelberg (2010)

[AG11]   Arora, S., Ge, R.: New algorithms for learning in presence of errors. In: ICALP, vol. (1), pp. 403–415 (2011)

[AP11]   Alwen, J., Peikert, C.: Generating shorter bases for hard random lattices. Theory Comput. Syst. 48(3), 535–553 (2011)

[BBS04]   Boneh, D., Boyen, X., Shacham, H.: Short group signatures. In: Franklin, M. (ed.) CRYPTO 2004. LNCS, vol. 3152, pp. 41–55. Springer, Heidelberg (2004)

[BF01]   Boneh, D., Franklin, M.K.: Identity-based encryption from the weil pairing. In: Kilian, J. (ed.) CRYPTO 2001. LNCS, vol. 2139, pp. 213–229. Springer, Heidelberg (2001)

[CHKP12]   Cash, D., Hofheinz, D., Kiltz, E., Peikert, C.: Bonsai trees, or how to delegate a lattice basis. J. Cryptology 25(4), 601–639 (2012)

[CN11]   Chen, Y., Nguyen, P.Q.: BKZ 2.0: Better lattice security estimates. In: Lee, D.H., Wang, X. (eds.) ASIACRYPT 2011. LNCS, vol. 7073, pp. 1–20. Springer, Heidelberg (2011)

[CS97]   Camenisch, J.L., Stadler, M.A.: Efficient group signature schemes for large groups. In: Kaliski Jr., B.S. (ed.) CRYPTO 1997. LNCS, vol. 1294, pp. 410–424. Springer, Heidelberg (1997)

[CT91]   Cover, T.M., Thomas, J.: Elements of Information Theory. Wiley (1991)

[DDLL13]   Ducas, L., Durmus, A., Lepoint, T., Lyubashevsky, V.: Lattice signatures and bimodal gaussians. In: Canetti, R., Garay, J.A. (eds.) CRYPTO 2013, Part I. LNCS, vol. 8042, pp. 40–56. Springer, Heidelberg (2013)

[DN12a]   Ducas, L., Nguyen, P.Q.: Faster gaussian lattice sampling using lazy floating-point arithmetic. In: Wang, X., Sako, K. (eds.) ASIACRYPT 2012. LNCS, vol. 7658, pp. 415–432. Springer, Heidelberg (2012)

[DN12b]   Ducas, L., Nguyen, P.Q.: Learning a zonotope and more: Cryptanalysis of nTRUSign countermeasures. In: Wang, X., Sako, K. (eds.) ASIACRYPT 2012. LNCS, vol. 7658, pp. 433–450. Springer, Heidelberg (2012)

[FO99]   Fujisaki, E., Okamoto, T.: Secure integration of asymmetric and symmetric encryption schemes. In: Wiener, M. (ed.) CRYPTO 1999. LNCS, vol. 1666, pp. 537–554. Springer, Heidelberg (1999)

[GGH13]   Garg, S., Gentry, C., Halevi, S.: Candidate multilinear maps from ideal lattices. In: Johansson, T., Nguyen, P.Q. (eds.) EUROCRYPT 2013. LNCS, vol. 7881, pp. 1–17. Springer, Heidelberg (2013)

[GLP12]   Güneysu, T., Lyubashevsky, V., Pöppelmann, T.: Practical lattice-based cryptography: A signature scheme for embedded systems. In: Prouff, E., Schaumont, P. (eds.) CHES 2012. LNCS, vol. 7428, pp. 530–547. Springer, Heidelberg (2012)

[GN08]   Gama, N., Nguyen, P.Q.: Predicting lattice reduction. In: Smart, N.P. (ed.) EUROCRYPT 2008. LNCS, vol. 4965, pp. 31–51. Springer, Heidelberg (2008)

[GPV08]   Gentry, C., Peikert, C., Vaikuntanathan, V.: Trapdoors for hard lattices and new cryptographic constructions. In: STOC 2008, pp. 197–206. ACM, New York (2008)

[Gui13]   Guillevic, A.: Étude de l'arithmétique des couplages sur les courbes
             algébriques pour la cryptographie. These, Ecole Normale Supérieure de
             Paris - ENS Paris (December 2013)
[HHGP+03]   Hoffstein, J., Howgrave-Graham, N., Pipher, J., Silverman, J.H., Whyte,
             W.: Ntrusign: digital signatures using the ntru lattice. In: Joye, M. (ed.)
             CT-RSA 2003. LNCS, vol. 2612, pp. 122–140. Springer, Heidelberg (2003)
[HPS98]   Hoffstein, J., Pipher, J., Silverman, J.H.: Ntru: A ring-based public key
             cryptosystem. In: ANTS-III, pp. 267–288 (1998)
[Kle00]   Klein, P.: Finding the closest lattice vector when it's unusually close. In:
             SODA 2000, pp. 937–941 (2000)
[LPR13a]   Lyubashevsky, V., Peikert, C., Regev, O.: On ideal lattices and learning
             with errors over rings (Preliminary version appeared in EUROCRYPT
             2010). In: Gilbert, H. (ed.) EUROCRYPT 2010. LNCS, vol. 6110, pp.
             1–23. Springer, Heidelberg (2010)
[LPR13b]   Lyubashevsky, V., Peikert, C., Regev, O.: A toolkit for ring-LWE cryp-
             tography. In: Johansson, T., Nguyen, P.Q. (eds.) EUROCRYPT 2013.
             LNCS, vol. 7881, pp. 35–54. Springer, Heidelberg (2013)
[LTV12]   López-Alt, A., Tromer, E., Vaikuntanathan, V.: On-the-fly multiparty
             computation on the cloud via multikey fully homomorphic encryption.
             In: STOC, pp. 1219–1234 (2012)
[Lyu12]   Lyubashevsky, V.: Lattice signatures without trapdoors. In: Pointcheval,
             D., Johansson, T. (eds.) EUROCRYPT 2012. LNCS, vol. 7237, pp. 738–
             755. Springer, Heidelberg (2012)
[MP12]   Micciancio, D., Peikert, C.: Trapdoors for lattices: Simpler, tighter,
             faster, smaller. In: Pointcheval, D., Johansson, T. (eds.) EUROCRYPT
             2012. LNCS, vol. 7237, pp. 700–718. Springer, Heidelberg (2012)
[MR07]   Micciancio, D., Regev, O.: Worst-case to average-case reductions based
             on gaussian measures. SIAM J. Comput. 37(1), 267–302 (2007)
[NR06]   Nguyên, P.Q., Regev, O.: Learning a parallelepiped: Cryptanalysis of
             GGH and NTRU signatures. In: Vaudenay, S. (ed.) EUROCRYPT 2006.
             LNCS, vol. 4004, pp. 271–288. Springer, Heidelberg (2006)
[PDG14]   Pöppelmann, T., Ducas, L., Güneysu, T.: Enhanced lattice-based sig-
             natures on reconfigurable hardware. IACR Cryptology ePrint Archive
             (2014)
[Pei10]   Peikert, C.: An efficient and parallel gaussian sampler for lattices. In:
             Rabin, T. (ed.) CRYPTO 2010. LNCS, vol. 6223, pp. 80–97. Springer,
             Heidelberg (2010)
[PG13]   Pöppelmann, T., Güneysu, T.: Towards practical lattice-based public-
             key encryption on reconfigurable hardware. In: Lange, T., Lauter, K.,
             Lisoněk, P. (eds.) SAC 2013. LNCS, vol. 8282, pp. 68–85. Springer, Hei-
             delberg (2013)
[RS10]   Rückert, M., Schneider, M.: Estimating the security of lattice-based cryp-
             tosystems. IACR Cryptology ePrint Archive 2010, 137 (2010)
[SS11]   Stehlé, D., Steinfeld, R.: Making NTRU as secure as worst-case problems
             over ideal lattices. In: Paterson, K.G. (ed.) EUROCRYPT 2011. LNCS,
             vol. 6632, pp. 27–47. Springer, Heidelberg (2011)
[SS13]   Stehlé, D., Steinfeld, R.: Making ntruencrypt and ntrusign as secure
             as standard worst-case problems over ideal lattices. IACR Cryptology
             ePrint Archive 2013, 4 (2013)
[Vau03]   Vaudenay, S.: Decorrelation: A theory for block cipher security. J. Cryp-
             tology 16(4), 249–286 (2003)

Order-Preserving Encryption Secure Beyond One-Wayness

Isamu Teranishi[1], Moti Yung[2,3], and Tal Malkin[2]

[1] NEC
[2] Columbia University
[3] Google Inc.
teranisi@ah.jp.nec.com, {moti,tal}@cs.columbia.edu

Abstract. Semantic-security of individual plaintext bits given the corresponding ciphertext is a fundamental notion in modern cryptography. We initiate the study of this basic problem for Order-Preserving Encryption (OPE), asking "what plaintext information can be semantically hidden by OPE encryptions?" OPE has gained much attention in recent years due to its usefulness for secure databases, and has received a thorough formal treamtment with innovative and useful security notions. However, all previous notions are one-way based, and tell us nothing about partial-plaintext indistinguishability (semantic security).

In this paper, we propose the first indistinguishability-based security notion for OPE, which can ensure *secrecy of lower bits of a plaintext* (under essentially a random ciphertext probing setting). We then justify the definition, from the theoretical plausibility and practicality aspects. Finally, we propose a new scheme satisfying this security notion (the first one to do so). In order to be clear, we note that the earlier security notions, while innovative and surprising, nevertheless tell us nothing about the above partial- plaintext indistinguishability because they are limited to being one-way-based.

Keywords: Order-preserving encryption, secure encryption, security notions, indistinguishability, foundations.

1   Introduction

Securing cloud database with untrusted cloud servers needs to hide information from the database manager itself, and has resulted in new research areas.

Order-Preserving Encryption (OPE): This is, perhaps, the most promising new primitives in the area of encrypted database processing [1,17,3,7,8,28]. It is a symmetric encryption over the integers such that ciphertexts preserve the numerical orders of the corresponding plaintexts. That is, $\forall m, m'\{m < m' \Rightarrow \mathsf{Enc}_K(m) < \mathsf{Enc}_K(m')\}$. OPE was originally studied in an ad-hoc fashion in the database community by Agrawal, Kiernan, Ramakrishnan, Srikant and Xu [1], and seemed like a clever heuristics. However, its careful foundational study was initiated with surprising formal cryptographic models and proofs by

P. Sarkar and T. Iwata (Eds.): ASIACRYPT 2014, PART II, LNCS 8874, pp. 42–61, 2014.

Boldyreva, Chenette, Lee, and O'Neill [7,8]. Overall, it has received much recent attention in the cryptographic community [7,8,28], in the database community [1,17,3], as well as in other applied areas.

OPE is attractive since it allows one to simultaneously perform very efficiently over encrypted data numerous fundamental database operations: sorting, simple matching (i.e., finding m in a database), range queries (i.e., finding all messages m within a given range $\{i, \ldots, j\}$), and SQL operations [1,20,21,23]. Furthermore, OPE is more efficient than these other primitives. For instance, the simple matching operation realized by OPE only requires logarithmic time in the database size [1], while the same operation realized by, say, searchable encryption [9,22], needs linear time in the size, which is too costly for a database containing a few millions data items.

Security of OPE: Despite its importance, security of OPE is far from being understood at this time. Even the most fundamental problem: "what plaintext information can be semantically hidden" is open. This is important. Imagine the following "string embedding" problem: we concatenate numerical strings to get a larger number and we have degree of freedom in this concatenation, don't we want to hide the most crucial string by embedding it at a location within the large number which hides it better than otherwise? Hasn't this very issue (partial information security in a ciphertext) been at the heart of cryptographic formalisms of encryption technologies in the last 30 years or so? Indeed, a naturally defined indistinguishability notion for OPE, *indistinguishability under ordered CPA attack (IND-O-CPA)* [7], was not only unachievable but it was shown that **any** OPE under this notion is broken with **overwhelming** probability if the OPE scheme has a super-polynomial size message space. (And if the message space is only polynomial size, an OPE scheme completely loses its utility, of course.)

OPE Is an Inherently "Leaky" Method: The reason behind the above negative result is that an OPE scheme has to reveal something about plaintexts other than their order, i.e., information about the distance between the two plaintexts. By definition (as stated above), an OPE scheme's encryption function Enc_K has to satisfy the monotone increasing property, $m_0 < m_1 \Rightarrow \mathsf{Enc}_K(m_0) < \mathsf{Enc}_K(m_1)$. Hence, the difference $\mathsf{Enc}_K(m_1) - \mathsf{Enc}_K(m_0)$ of two ciphertexts has to become noticeably large if the difference $m_1 - m_0$ of the corresponding plaintexts becomes large. The negative result of [7] mentioned above is, in fact, proved using an attack based on this observation.

To date, no one can tell what exactly OPE *must* leak and what it can protect. Our motivation is the fact that the existing security notions are not really helpful in understanding this simple basic question. If we have started to take the formal approach to the problem, why should we stop short of answering such a question? Here are a few notions to date:

IND-O-CPA [7]: It is similar to the LOR-based indistinguishability notion [4] for symmetric key encryption, except that queries of the adversary have to satisfy some order-preserving property. This notion is natural but as we stated above, it is not achievable for schemes with a super-polynomial size message space [7].

POPF-CCA [7]: This is a very important notion which says that a CCA adversary cannot distinguish a pair of an encryption and decryption oracles from a pair of an order-preserving random oracle and its inverse. This notion is natural and therefore should be further studied. But currently, nothing is known about what partial information it can hide and what it cannot hide, as pointed out in [8].

$(r, q+1)$-**WOW [8] (Window One-Wayness):** It says that[1] no adversary, who gets $q + 1$ encryptions $C_*, C_1, \ldots, C_q$ of uniformly randomly selected unknown messages, can find an interval I of length $\leq r$ satisfying $\mathsf{Dec}_K(C^*) \in I$. This notion is important since it captures the following natural database setting: Randomly selected $q+1$ elements stored in a database system in their encrypted form and an adversary A who wants to know one of them breaches the database and gets all the ciphertexts in it. This notion, however, does not ensure anything about the secrecy of internal plaintext partial information, since it is "one-way-based" in nature.

$(r, q + 1)$-**WDOW [8]:** It is another one-way-based notion defined in [8]. Since it is one-way-based, it also does not tell us what partial information about the plaintexts is hidden.

1.1   Our Contributions

This paper presents the first attempt to give a new perspective to the fundamental open problem: "go beyond one-wayness security and investigate what internal plaintext partial information OPE can hide." Here (while respecting earlier important works on the subject) we propose the first achievable indistinguishability notion for OPE regarding partial plaintext information hiding. More specifically: we show that our notion can assure *secrecy of lower bits of a plaintext* in the same natural settings as WOW [8].

Our Security Notion — $(\mathcal{X}, \theta, q)$-indistinguishability: It is defined based on $(r, q + 1)$-WOW [8]. But since WOW is inherently one-way-based, our security notion is defined as a "hybrid" of WOW and indistinguishability as follows. Consider the same database setting as WOW, where an honest entity (not the adversary!) stores $q + 1$ his data elements $m^*, m_1, \ldots, m_q$ in their encrypted forms in a database and an adversary A, who wants to get knowledge of m^*, breaches the database system and gets all ciphertexts in it. Above, the messages $m_1, \ldots, m_q$ have been selected according to given distributions $\mathcal{X}_1, \ldots, \mathcal{X}_q$.

The difference from WOW is that m^* has been selected as follows: two messages m_0^* and m_1^* are generated using a polynomial time machine Mg called *message generator*, and set $m^* \leftarrow m_b^*$, where b is a random bit hidden from A.

For $\mathcal{X} = (\mathcal{X}_i)_{i=1,\ldots,q}$, an OPE scheme is called $(\mathcal{X}, \theta, q)$-*indistinguishable* if the advantage of A in the above game (guessing the bit b beyond probability $1/2$) is negligible for any A and for any Mg whose output satisfies

$$|m_1^* - m_0^*| \leq \theta. \tag{1.1}$$

---

[1] Here we adopt the simpler definition of the window one-wayness notion given in Appendix B of the full paper of [8], which can be reduced to the definition of Section 3 of that paper and and vise versa.

Restriction (1.1) enables us to avoid the known attack [7] since it applies only when the distance between m_0^* and m_1^* is large.

Our Results: We will show in Section 2 the following fact:

Fact 1 (informal). *If an OPE scheme satisfies $(\mathcal{X}, \theta, q)$-indistinguishability, the least significant $\lfloor \log_2 \theta \rfloor$ bits of a plaintext are hidden from the adversary in the above database setting.*

We then propose a new OPE scheme $\bar{\mathcal{E}}_{k,\theta}$ based on a pseudo-random function PRF and show the following facts in Section 4. Below, $\mathcal{X}_1, \ldots, \mathcal{X}_q$ are distributions on $[1..M]$ such that they are independent from one another and one can take a sample from $\mathcal{X}_i$ in time polynomial in λ.

Theorem 2 (informal). *Let β and t be constants satisfying $0 < t < \beta \le 1$. Suppose that the message space size M is super-polynomial in the security parameter λ. Then, for any $\mathcal{X} = (\mathcal{X}_i)_i$ satisfying $\forall i : H_\infty(\mathcal{X}_i) \ge \beta \log_2 M$, $\bar{\mathcal{E}}_{k,\theta}$ satisfies $(\mathcal{X}, M^t, q)$-indistinguishability under the condition that PRF is secure.*

Remarks: First, our security notion does not ensure the secrecy of higher bits of the plaintext, and, in fact, there is no known scheme which can ensure their secrecy, since the scheme of [7] also reveals its high order bits [8]. Second, since any distribution $\mathcal{Y}$ on $[1..M]$ satisfies $0 \le H_\infty(\mathcal{Y}) \le \log_2 M$, the condition $H_\infty(\mathcal{X}_i) \ge \beta \log_2 M$ means that the ratio of $H_\infty(\mathcal{X}_i)$ to the maximum $\log_2 M$ has to be more than β. Third, Theorem 2 requires that the message space size M is super-polynomial in λ: which is exactly the same condition assumed by Boldyreva et.al.[8] to get their results. Fourth, Theorem 2 shows stronger security when t is closer to β, though the advantage bound decrease is slower in this case.

Due to the above results, we can conclude the following crucial facts:

Knowledge of $\mathcal{X}$: Theorem 2 only requires $\mathcal{X}$ to satisfy the entropy bound. Hence, we can show $(\mathcal{X}, \theta, q)$-indistinguishability *even when we do know the tuple $\mathcal{X}$ of message distributions completely in advance.* This fact is very important because the complete knowledge of $\mathcal{X}$ is not realistic in a central application, like the secure database above, when, for instance, plaintexts are names of new students (with the lexicographic order) or scores of some examination.

Fraction $t < \beta$ of Lower Bits Are Hidden: Due to Fact 1, $(\mathcal{X}, M^t, q)$-indistinguishability implies secrecy of the least significant $\lfloor \log M^t \rfloor$ bits of a plaintext. Since the maximum bit length of a message in the message space $[1..M]$ is $\lfloor \log_2 M \rfloor + 1$, Theorem 2 shows that our scheme with the above parameters can ensure secrecy of the fraction

$$\frac{\lfloor \log M^t \rfloor}{\lfloor \log_2 M \rfloor + 1} \approx t < \beta \tag{1.2}$$

of the least significant bits of a plaintext. The above secrecy can be shown even when we do not have complete knowledge of plaintext distributions.

Any Fraction of Low-Order Bits Are Hidden in the Uniform Distribution Case: In the most significant case where plaintexts distribute uniformly at random, Theorem 2, in particular, shows that our scheme can ensure secrecy of any fraction of the least significant bits of the plaintext because the maximum $\log_2 M$ of the min-entropy is achieved by the uniform distribution and we therefore can set β to 1 in this case.

Allowing Decryption Queries: As in [8], we can naturally make our scheme secure even when we allow the adversary to make decryption queries at any time, using the "Encrypt-then-Mac" composition (adding MAC data) [5].

Open Problem: We can show that our scheme satisfies $\mathsf{Enc}_K(m+1) = \mathsf{Enc}_K(m) + 1$ with high probability. Hence, an adversary can break the scheme if she can get $\mathsf{Enc}_K(m_b^* + (\text{small value}))$, where (m_0^*, m_1^*) is a challenge query of her. (Our proof for Theorem 2 ensures that she can get it only with negligible probability.) Designing a scheme ensuring security for this case is an important open problem.

Finally, we give a note about the construction of our scheme. Since Boldyreva et.al. [7] already gave a natural security notion, POPF-CCA, one important approach to study indistinguishability of OPE is to show that POPF-CCA implies some indistinguishability notion, such as ours. However, we take a different approach in this paper because currently, we do not have much knowledge about the random order-preserving function used in the definition of POPF-CCA, which means that showing our security notion based on POPF-CCA seems to us to be hard. Rather, we define a specific scheme $\bar{\mathcal{E}}_{k,\theta}$ designed for showing our security notion. Showing some indistinguishability results for the more natural security notion, POPF-CCA, is, of course, of independent interest and we leave it as an important open problem.

1.2   Other Security Notions

We also introduce two more security notions for OPE.

(k, θ)-**FTG-O-nCPA:** This is an (artificial) variant of an indistinguishability notion. We will give the definition of it in Section 3 and show that this notion with suitable parameter implies $(\mathcal{X}, \theta, q)$-indistinguishability for any $\mathcal{X} = (\mathcal{X}_i)_i$ such that $H_\infty(\mathcal{X}_i)$ is larger than the predetermined constant. We then use this fact to show $(\mathcal{X}, \theta, q)$-indistinguishability of our proposed scheme.

WOWM — Stronger Variant of WOW [8]: Informally, $(r, q+1)$-WOWM says that no adversary given $\mathsf{Enc}_K(m^*)$, and $(m_i, \mathsf{Enc}_K(m_i))_{i=1,\ldots,q}$ can find an interval I of length $\leq r$ satisfying $m^* \in I$. This is stronger than $(r, q+1)$-WOW because it allows an adversary to watch $(m_i)_i$ while $(r, q+1)$-WOW prohibits her from doing this.

We will show in Section 5 the following facts. The $(\mathcal{X}, \theta, q)$-indistinguishability notion with suitable parameters implies $(r, q+1)$-WOWM. For any constant $0 \leq \rho < 1$, our scheme with suitable parameters satisfies $(M^\rho, q+1)$-WOWM (and therefore $(M^\rho, q+1)$-WOW).

1.3   Comparison with Known Results [7,8]

First, we clarify what our results owe to [8]: we consider the same natural "database as a service" setting of WOW as described in Section 1.1, and our results are shown under the same condition as WOW [8], that is, the message space size M is super-polynomial in λ. (Note that, technically, our proposed scheme owes the excellent "lazy sampling" of [7] as well.)

Next, we clarify the difference of them. The earlier results on OPE are indeed remarkable and opened the door to our investigation, but there are some crucial differences which we would like to point out explicitly.

About Our Security Notion: $(\mathcal{X}, \theta, q)$-indistinguishability of the scheme [7] is unknown, because our goal is newly defined. Moreover, we can prove that the known scheme [7] cannot satisfy $(\mathcal{U}^q, M^t, q)$-indistinguishability for $t > 1/2$. (See our full paper for the proof.)

Our scheme achieves $(\mathcal{U}^q, M^t, q)$-indistinguishability for any $0 \leq t < 1$, where $\mathcal{U}^q$ was the tuple of the uniform distributions on the message space. This means that it can hide (in the sense of semantic security) any fraction t of the least significant bits of a plaintext in our setting with uniformly randomly selections of plaintexts. Even when plaintext distributions are not the uniform ones, the scheme can hide fraction $t < \beta$ of lower bits of a plaintext. (β is determined depending on the min-entropy measure of other plaintexts).

About WOW [8]: The known best result [8] is $(1, q)$-WOW security of the scheme of [7]. But it is proved that this scheme cannot achieve $(M^\rho, q+1)$-WOW [8] for any $\rho > 1/2$. In contrast, for any constant $0 \leq \rho < 1$, our scheme with suitable parameters can satisfies $(M^\rho, q+1)$-WOWM (and therefore $(M^\rho, q+1)$-WOW, in particular).

Finally, we describe the POPF-CCA notion given in the seminal work [7].

About POPF-CCA [7]: POPF-CCA is very important notion which can ensure indistinguishability from an ideal object, while our security notion cannot ensure it. Hence, POPF-CCA, as a notion, is more natural and has much potential like other real-vs-ideal definitions and it can ensure security in lots of situations while ours can ensure it in the specific situation described before. E.g. our notion can ensure nothing when an adversary knows $\mathsf{Enc}_K(m)$ and $\mathsf{Enc}_K(m+1)$ while POPF-CCA can ensure something even in this situation. In particular, our notion does not imply POPF-CCA and therefore, POPF-CCA has independent interest.

But currently and unfortunately, nothing is known about what POPF-CCA can hide and what it cannot hide, as pointed out in [8]. This is the motivation behind our entire investigation. Our result is the first positive result in the sense of indistinguishability. Showing some indistinguishability for a more natural notion like, say, POPF-CCA, is an important open issue.

1.4   Other Related Works

Property preserving encryptions [18,2,10] was introduced by Pandey and Rouse-lakis [18] as a variants of the OPE. Although the security notions for this scheme can be the same as for OPE, almost the same attack as that of [7] can break any OPE scheme under these security notions when the scheme has a super-polynomial size message space. See our full paper for the details.

CEOE and *MOPE* schemes (introduced by Boldyreva, Chenette, Lee, and O'Neill [8]), *mOPE* and *stOPE* schemes (introduced by Popa, Li, and Zeldovich [19]), and *GOPE* schemes (introduced by Xiao and Yen [25]) achieve stronger security than OPE by sacrificing some of their functionalities, by allowing inter-actions, or by considering restrictive cases, respectively. *Comparable encryption* schemes (introduced by Furukawa [12,13]) consider an encrypted database where the database manager can search messages m satisfying $m > u$ on behalf of a user if a key K_u depending on u is given from the user as a query. These notions are of independent interests, some may require further formalizations, and are all beyond the scope of this work.

Yum, Kim, Kim, Lee and Hong [28] propose a more efficient method to com-pute the encryption and decryption functions of the known scheme [7]. Xiao, Yen, and Huynh [27] study OPE in a multi-user setting. Xiao and Yen [26] estimates the min-entropy of a plaintext encrypted by the known scheme [7].

2 $(\mathcal{X}, \theta, q)$-indistinguishability

We introduce notations and terminology and then define our security notion.

Intervals: For integers a and $b \geq a$, *interval* $[a..b]$ is the set $\{a, \ldots, b\}$. $[b]$, $(a..b]$, $[a..b)$, and $(a..b)$ denote $[1..b]$, $[a+1..b]$, $[a..b-1]$, and $[a+1..b-1]$, respectively.

Order-Preserving Encryption: An *OPE* scheme is a symmetric key encryp-tion $\mathcal{E} = (\mathsf{Kg}, \mathsf{Enc}, \mathsf{Dec})$ whose message space $\mathcal{M}$ and ciphertext space are inter-vals in $\mathbb{N}$ and which satisfies $m < m' \Rightarrow \mathsf{Enc}_K(m) < \mathsf{Enc}_K(m')$ for $\forall m, m' \in \mathcal{M}$ and $\forall K \leftarrow \mathsf{Kg}(1^\lambda)$. Here "$<$" represents the numerical order. Throughout this paper, we assume w.l.o.g. that $\mathcal{M}$ can be written as $[1..M]$.

Definition 3 $((\mathcal{X}, \theta, q)$-**indistinguishability**). Let λ, $\mathcal{E} = (\mathsf{Kg}, \mathsf{Enc}, \mathsf{Dec})$, $\theta = \theta(\lambda) > 0$, and $q = q(\lambda) > 0$ be a security parameter, an OPE scheme, a real number, and a polynomial respectively and $\mathcal{X} = (\mathcal{X}_i)_{i \in [1..q]}$ be a tuple of distri-butions on the message space of $\mathcal{E}$. $\mathcal{E}$ is said to be $(\mathcal{X}, \theta, q)$-*indistinguishable* if for any polynomial time machine Mg (called *message generator*) whose outputs $(m_0^*, m_1^*, \mathsf{info})$ satisfies

$$m_0^* < m_1^*, \qquad\qquad |m_1^* - m_0^*| \leq \theta \qquad\qquad (2.1)$$

and any polynomial time adversary A, $\mathsf{Adv}.\mathsf{Exp}_{\mathcal{E}}^{(\mathcal{X},\theta,q)\text{-indis.}}(\mathsf{Mg}, \mathsf{A}) = |\Pr[\mathsf{Exp}_{\mathcal{E}}^{(\mathcal{X},\theta,q)\text{-indis.}}(\mathsf{Mg}, \mathsf{A}, 1) = 1] - \Pr[\mathsf{Exp}_{\mathcal{E}}^{(\mathcal{X},\theta,q)\text{-indis.}}(\mathsf{Mg}, \mathsf{A}, 0) = 1]|$ is negligi-ble. Here $\mathsf{Exp}_{\mathcal{E}}^{(\mathcal{X},\theta,q)\text{-indis.}}(\mathsf{Mg}, \mathsf{A}, b)$ is defined as follows.

$$K \leftarrow \mathsf{Kg}(1^\lambda), (m_0^*, m_1^*, \mathsf{info}) \leftarrow \mathsf{Mg}(1^\lambda), m_1 \xleftarrow{\$} \mathcal{X}_1, \ldots, m_q \xleftarrow{\$} \mathcal{X}_q,$$
$$d \leftarrow \mathsf{A}(\mathsf{Enc}_K(m_b^*), (m_i, \mathsf{Enc}_K(m_i))_{i \in [1..q]}, \mathsf{info}), \text{Return } d.$$

Remarks: First, when we consider the above notion, the probability that $m_i \in [m_0^* .. m_1^*]$ has to be negligible because otherwise, an OPE scheme under the above notion is broken by an adversary simply by checking $\mathsf{Enc}_K(m_i^*) > \mathsf{Enc}_k(m_i)$. This condition will be automatically satisfied in our theorems due to the selection of parameters of our scheme. Second, due to the bit string info moving between the parties, we can re-interpret the above definition as Mg and A being the "guess and find stages" of a single adversary $(\mathsf{Mg}, \mathsf{A})$ where info is her state.

Low-Order Bits Can be Hidden: Our security notion ensures the secrecy of the least significant $\lfloor \log_2 \theta \rfloor$ bits of a plaintext, due to the following: Let $L = \lfloor \log_2 \theta \rfloor$ and take any (maximal) interval I satisfying the following theorem: for any two elements of I, all of their bits except the least significant L bits are the same. That is, I can be written as $I = \{2^L u + x \mid x \in [0..2^L - 1]\}$ for some u. By definition the length of I is not more than θ.

Then, our security notion, in particular, ensures that any element m_0^* of I is indistinguishable from that of a uniformly random element m_1^* of I, because our condition (1.1) is satisfied due to the definition of I. Since the least significant L bits of uniformly random element m_1^* of I distribute uniformly at random on the L-bit space $[0..2^L - 1]$, the indistinguishability of m_0^* and m_1^* can ensure secrecy of the least significant L bits of m_0^*.

3 (k, θ)-FTG-O-nCPA

In this section, we introduce a security notion, (k, θ)-FTG-O-nCPA, and using it, we give a sufficient condition for $(\mathcal{X}, \theta, q)$-indistinguishability.

(k, θ)-**FTG-O-nCPA:** It is Find-Then-Guess [4] type indistinguishability for nCPA adversary whose queries satisfy the conditions (3.1), ... (3.4) described later. Here $nCPA$ *(non-adaptive CPA)* [16,14,15] is a type of attack where the adversary is required to output encryption queries $m_1, \ldots, m_q$ and challenge query (m_0^*, m_1^*) together at the same time and gets their answers thereafter.

Definition 4 $((k, \theta)$-FTG-O-nCPA). For real numbers $k = k(\lambda) > 0$ and $\theta = \theta(\lambda) > 0$, an OPE $\mathcal{E}$ is said to be (k, θ)-*FTG-O-nCPA* secure if for any polynomial time adversary $\mathsf{A} = (\mathsf{A}_{\mathsf{find}}, \mathsf{A}_{\mathsf{guess}})$, the advantage $\mathsf{Adv.Exp}_{\mathcal{E}}^{(k,\theta)\text{-FTG-O-nCPA}}(\mathsf{A})$ $= |\Pr[\mathsf{Exp}_{\mathcal{E}}^{(k,\theta)\text{-FTG-O-nCPA}}(\mathsf{A}, 1) = 1] - \Pr[\mathsf{Exp}_{\mathcal{E}}^{(k,\theta)\text{-FTG-O-nCPA}}(\mathsf{A}, 0) = 1]|$ is negligible. Here $\mathsf{Exp}_{\mathcal{E}}^{(k,\theta)\text{-FTG-O-nCPA}}(\mathsf{A}, b)$ is defined as follows (below, q is an arbitrary number selected by A):

$$K \leftarrow \mathsf{Kg}(1^\lambda), \ ((m_0^*, m_1^*), (m_i)_{i \in [1..q]}, \mathsf{st}) \leftarrow \mathsf{A}_{\mathsf{find}}(1^\lambda),$$
$$d \leftarrow \mathsf{A}_{\mathsf{guess}}(\mathsf{Enc}_K(m_b^*), (\mathsf{Enc}_K(m_i))_{i \in [1..q]}, \mathsf{st}), \text{Return } d.$$

(m_0^*, m_1^*) and $m_1, \ldots, m_q$ are called a *challenge query* and *encryption queries* respectively. The output of A has to satisfy the following (3.1), (3.2), and (3.3). We also assume (3.4) throughout this paper w.l.o.g.

$$\forall i \ : \ m_i < m_0^* \ \Leftrightarrow \ m_i < m_1^*, \tag{3.1}$$

$$|m_0^* - m_1^*| \leq \theta, \tag{3.2}$$

$$\forall d \in \{0,1\}, \forall i \ : \ |m_d^* - m_i| \geq k\theta. \tag{3.3}$$

$$m_0^* < m_1^* \tag{3.4}$$

Above, (3.1) requires the order preserving property, (3.2) requires the same condition as $(\mathcal{X}, \theta, q)$-indistinguishability, and (3.3) requires the distance $|m_d^* - m_i|$ has to be bigger than the given constant $k\theta$ for any d and i. (3.3) is required because without it, an adversary can take (m_0^*, m_1^*) and m_1 such that $|m_1^* - m_1|$ is much larger than $|m_0^* - m_1|$ (when θ is big. Say, take any m_0^* and set $m_1^* \leftarrow m_0^* + \theta$ and $m_1 \leftarrow m_0^* - 1$). Then, since OPE reveals information about the distance between the two plaintexts, an adversary can know b by checking $|\mathsf{Enc}_K(m_b^*) - \mathsf{Enc}_K(m)|$.

Sufficient Condition: Using (k, θ)-FTG-O-nCPA, we can give the following sufficient condition for $(\mathcal{X}, \theta, q)$-indistinguishability. Below, λ, $\mathcal{E}$, $q = q(\lambda)$ are a security parameter, an OPE scheme on a message space $[1..M]$, and a polynomial respectively. $\mathcal{X}_1, \ldots, \mathcal{X}_q$ are distributions on $[1..M]$ such that they are independent from one another and one can take a sample from $\mathcal{X}_i$ in time polynomial in λ. (M and $\mathcal{X}$ can depend on λ.) A and Mg denote an adversary and a message generator for $(\mathcal{X}, \theta, q)$-indistinguishability respectively and B denotes an adversary for (k, θ)-FTG-O-nCPA.

Theorem 5 (Sufficient Condition for $(\mathcal{X}, \theta, q)$-indistinguishability). *Let* $\beta > 0$ *be any constant. For* $k = k(\lambda) > 0$, $\theta = \theta(\lambda) > 0$, *if*

$$\forall i \in [1..q] \ : \ H_\infty(\mathcal{X}_i) \geq \beta \log_2 M \tag{3.5}$$

holds for any λ, *then* $\forall \mathsf{Mg} \forall \mathsf{A} \exists \mathsf{B}$:

$$\mathsf{Adv.Exp}_{\mathcal{E}}^{(\mathcal{X}, \theta, q)\text{-indis.}}(\mathsf{Mg}, \mathsf{A}) \leq \mathsf{Adv.Exp}_{\mathcal{E}}^{(k, \theta)\text{-FTG-O-nCPA}}(\mathsf{B}) + O\left(\frac{qk\theta}{M^\beta}\right). \tag{3.6}$$

We next give two notes reg. Theorem 5. First, as in Theorem 2, condition (3.5) means that the ratio of $H_\infty(\mathcal{X}_i)$ to the maximum $\log_2 M$ has to be more than β. Second, the right hand side of (3.6) is negligible only when $k\theta/M^\beta$ is negligible. We will show that $k\theta/M^\beta$ is, in fact, negligible (for suitable parameters k and θ we will choose) in the proof of Theorem 7, which uses the above theorem.

Proof. For Mg and A for $(\mathcal{X}, \theta, q)$-indistinguishability, consider an adversary B for (k, θ)-FTG-O-nCPA which takes $(m_0^*, m_1^*, \mathsf{info}) \leftarrow \mathsf{Mg}(1^\lambda)$ and $m_1 \xleftarrow{\$} \mathcal{X}_1, \ldots, m_q \xleftarrow{\$} \mathcal{X}_q$, makes query $((m_0^*, m_1^*), m_1, \ldots, m_q)$, gives info and an answer to the query to A, and produces the output of A.

Let I be the interval $(m_0^* - k\theta..m_1^* + k\theta)$. The above B will violate constraint (3.3) if $m_i \in I$ holds for some i. But the probability that $m_i \in I$ holds for some i is $\sum_{i \in [1..q]} \Pr[m_i \leftarrow \mathcal{X}_i \ : \ m_i \in I] \leq (\text{length of } I) \cdot \sum_{i \in [1..q]} \max_{x \in I} \Pr[m_i \leftarrow \mathcal{X}_i \ : \ m_i = x] \leq \sum_{i \in [1..q]} \frac{(2k+1)\theta}{2^{H_\infty(\mathcal{X}_i)}} \leq O\left(\frac{qk\theta}{M^\beta}\right)$. When $m_i \notin I$ holds, (3.1) is also satisfied. Moreover (2.1) implies (3.2). Thus, Theorem 5 follows. $\square$

4   Our Scheme

4.1   Our Goal

This section is devoted to constructing our scheme $\bar{\mathcal{E}}_{k,\theta}$ satisfying the following theorem: Below, A and B are adversaries for (k,θ)-FTG-O-nCPA and PRF respectively, λ is a security parameter, and $\mathsf{neg}(\cdot)$ is some negligible function which is determined independently of (k,θ,A).

Theorem 6 $((k,\theta)$-FTG-O-nCPA of $\bar{\mathcal{E}}_{k,\theta})$. *For $k,\theta > 0$, $\forall\mathsf{A}\exists\mathsf{B}$:*

$$\mathsf{Adv.Exp}_{\bar{\mathcal{E}}_{k,\theta}}^{(k,\theta)\text{-FTG-O-nCPA}}(\mathsf{A}) \leq O\left(\frac{1}{\sqrt{k}}\right) + \mathsf{Adv.Exp}_{\mathsf{PRF}}(\mathsf{B}) + \mathsf{neg}(\lambda) \qquad (4.1)$$

holds when $k \to \infty$. (The value θ does not affect the advantage bound.)

   Moreover, the computational costs of algorithms of $\bar{\mathcal{E}}_{k,\theta}$ and the ciphertext length of it are within polynomial of $\log k$, $\log\theta$, $\log M$, and λ, where M is the size of the message space $[1..M]$.

Due to Theorem 5, our scheme satisfies the following theorem as well. Below, M is the size of message space $[1..M]$ of our scheme $\bar{\mathcal{E}}_{k,\theta}$, $q = q(\lambda)$ is a polynomial, and $\mathcal{X}_1,\ldots,\mathcal{X}_q$ are distributions on $[1..M]$ such that they are independent from one another and one can take a sample from $\mathcal{X}_i$ in time polynomial in λ, $\mathsf{neg}(\cdot)$ is some negligible function, A and Mg are an adversary and a message generator for $(\mathcal{X},M^t,q)$-indistinguishability, B is an adversary for PRF, and $\mathsf{Adv.Exp}_{\mathsf{PRF}}(\mathsf{B})$ is an advantage of B in the experiments of PRF.

Theorem 7 $((\mathcal{X},\theta,q)$-Indistinguishability of Our Scheme, Formal Version of Theorem 2). *Let $0 < \beta \leq 1$ be any constant. Suppose that $\mathcal{X} = (\mathcal{X}_1,\ldots,\mathcal{X}_q)$ satisfies*

$$\forall i \in [1..q] \quad : \quad H_\infty(\mathcal{X}_i) \geq \beta \log_2 M. \qquad (4.2)$$

Then, for any constant $0 < t < \beta(\leq 1)$, our scheme $\bar{\mathcal{E}}_{k,\theta}$ with suitable (k,θ) (depending on (M,β,t)) satisfies $\forall\mathsf{Mg}\forall\mathsf{A}\exists\mathsf{B}$:

$$\mathsf{Adv.Exp}_{\bar{\mathcal{E}}_{k,\theta}}^{(\mathcal{X},M^t,q)\text{-indis.}}(\mathsf{Mg},\mathsf{A}) \leq O\left(\frac{q}{M^{\frac{\beta-t}{3}}}\right) + \mathsf{Adv.Exp}_{\mathsf{PRF}}(\mathsf{B}) + \mathsf{neg}(\lambda). \qquad (4.3)$$

   Moreover, the computational costs of algorithms of $\bar{\mathcal{E}}_{k,\theta}$ and the ciphertext length of it are within polynomial of t, β, $\log M$, and λ.

The right hand sides of (4.3) becomes negligible under the condition that the message space size M is super-polynomial in λ.

Reduction from Theorem 7 to Theorem 5 and 6: Theorem 7 follows if we set parameters (k,θ) of our scheme $\bar{\mathcal{E}}_{k,\theta}$ as

$$(k,\theta) = (M^{2(\beta-t)/3}, M^t) \qquad (4.4)$$

because in this case, terms of (3.6) and (4.1) become $O\left(\frac{qk\theta}{M^\beta}\right) = O(\frac{qM^{2(\beta-t)/3}\cdot M^t}{M^\beta})$
$= O(\frac{q}{M^{(\beta-t)/3}})$ and $O(\frac{1}{\sqrt{k}}) = O(\frac{1}{M^{(\beta-t)/3}})$. They are negligible when $M \to \infty$
because the constants t and β satisfy the condition $0 < t < \beta \le 1$ of Theorem 7.
The computational costs of algorithms in our scheme and the ciphertext length
of it are polynomial in $\log M$ even when parameters are set as in (4.4), due to
the latter part of Theorem 6 and the condition $0 < t < \beta \le 1$.   □

4.2   Scheme $\mathcal{E}_{k,\theta}$ with Polysize Message Space

The goal of this section is designing an OPE scheme $\mathcal{E}_{k,\theta}$ whose advantage bound
regarding (k,θ)-FTG-O-nCPA is given in Theorem 6. But the message space
size M of $\mathcal{E}_{k,\theta}$ must be bounded by some polynomial in the security parameter
λ. (Hence, e.g. the upper bound (4.3) of an advantage for this scheme is not
negligible although the bound itself holds even for this scheme.) We stress that
$\mathcal{E}_{k,\theta}$ is *not* our proposed scheme.

The scheme $\mathcal{E}_{k,\theta}$ does not use PRF although Theorem 6 refers to it and the
discussion in this subsection is purely information theoretic ones. The PRF will
be used to design our proposed scheme $\bar{\mathcal{E}}_{k,\theta}$ in the next subsection.

Ideas behind Construction. The scheme $\mathcal{E}_{k,\theta}$ is constructed based mainly
on three ideas. Firstly, we write an OPE encryption $\mathsf{Enc}_K(m)$ on a message
space $[1..M]$ as $\mathsf{Enc}_K(m) = R + \sum_{i\in[2..m]} \delta_i$, where $R = \mathsf{Enc}_K(1)$ and $\delta_i = \mathsf{Enc}_K(i) - \mathsf{Enc}_K(i-1)$. Then, a design of an OPE encryption can be reduced to
the selections of R and (δ_i).

Secondly, we take some values j_0, j_1, ..., and set δ_{j_0}, δ_{j_1}, ... and/or R to
random values which are very large compare to other δ_i, so as to hide a (smaller)
secret value which the adversary wants to know. A naive way to apply this idea
is that we set R to a large random value, while setting all δ_i to 1. Then, the
large randomness R seems to hide the secret bit b of a challenge ciphertext
$\mathsf{Enc}_K(m_b^*) = R + \sum_{i\in[2..m_b^*]} \delta_i = m_b^* + R - 1$. But, in fact, the adversary can
recover b because she can cancel out R by computing $\mathsf{Enc}_K(m_b) - \mathsf{Enc}_K(m') = m_b - m'$, where m' and $\mathsf{Enc}_K(m')$ are her encryption query and its answer.

Therefore, we set some $\delta_{j_0}, \delta_{j_1}, \ldots$, to large random values as well and expect
that the set $\{j_0, j_1, \ldots\}$ of indices of them and queries of the adversary to satisfy
"good relation" in the sense that, for some j_s, the adversary cannot cancel out δ_{j_s}
even when she has encryption queries and their answers. (The precise meaning
of this "good relation" will be given later.)

But, the problem is that we cannot know her queries in advance. Therefore,
after we fix $j_0, j_1, \ldots$, she may choose her queries such that the queries and
$\{j_0, j_1, \ldots\}$ do not satisfy the good relation. So, thirdly, we solve the above
problem by introducing another key idea: changing the bit length of δ_i randomly.
Specifically, for each i, we flip a random coin ρ_i which becomes 0 with small
probability p and then samples δ_i randomly from some given large set if $\rho_i = 0$
and set $\delta_i \leftarrow 1$ otherwise. Then the set $I = \{j_0, j_1, \ldots, \}$ of indices of large δ_i
varies randomly and, (due to the definition of nCPA,) we can hide I from the
view of the adversary until she determines her queries. Hence, the adversary

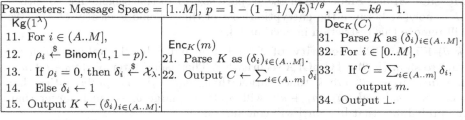

Parameters: Message Space $= [1..M]$, $p = 1 - (1 - 1/\sqrt{k})^{1/\theta}$, $A = -k\theta - 1$.		
$\mathsf{Kg}(1^\lambda)$		$\mathsf{Dec}_K(C)$
11. For $i \in (A..M]$,		31. Parse K as $(\delta_i)_{i \in (A..M]}$.
12. $\rho_i \xleftarrow{\$} \mathsf{Binom}(1, 1 - p)$.	$\mathsf{Enc}_K(m)$	32. For $i \in [0..M]$,
13. If $\rho_i = 0$, then $\delta_i \xleftarrow{\$} \mathcal{X}_\lambda$.	21. Parse K as $(\delta_i)_{i \in (A..M]}$.	33. If $C = \sum_{i \in (A..m]} \delta_i$,
14. Else $\delta_i \leftarrow 1$	22. Output $C \leftarrow \sum_{i \in (A..m]} \delta_i$	output m.
15. Output $K \leftarrow (\delta_i)_{i \in (A..M]}$.		34. Output $\perp$.

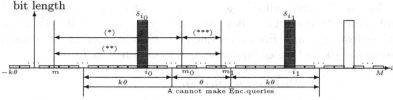

Fig. 1. The Scheme of Section 4.2 (upper) and the Intuition Behind Its Security (lower). In the lower figure, $\mathsf{Enc}_K(m_0) - \mathsf{Enc}_K(m)$, $\mathsf{Enc}_K(m_1) - \mathsf{Enc}_K(m)$, and the difference of them are the sum of δ_i in (*), (**), and (***) respectively. Since both (*) and (**) contain a large randomness δ_{i_0}, the difference (***), which is smaller, is hidden by this large randomness. $\mathsf{Enc}_K(m_0) - \mathsf{Enc}_K(m)$ and $\mathsf{Enc}_K(m_1) - \mathsf{Enc}_K(m)$ are therefore indistinguishable.

cannot arrange intentionally her queries such that the queries and I do not satisfy the good relation.

Note that this idea has resemblance to the partitioned technique [24] of Waters for an identity based encryption, where one takes some parameters (which determine a "partition") randomly and secretly and expects that queries of an adversary fall into some good places.

Scheme $\mathcal{E}_{k,\theta}$: The formal description of our scheme is given in Fig.1. Here k and θ be the values which we want to show (k, θ)-FTG-nCPA security for, p is a parameter which we will determine in (4.9), and $\mathsf{Binom}(n, p)$ is a binomial distribution.

We set in Fig.1 $\mathsf{Enc}_K(m) = \sum_{i \in (A..m]} \delta_i$ where $A = -k\theta - 1 < 0$ is a fixed value while in the idea described before, we set $\mathsf{Enc}_K(m) = R + \sum_{i \in [2..m]} \delta_i$. (That is, we set $R \leftarrow \sum_{i \in [A..1]} \delta_i$.) Due to this change, we can simplify the security proof for the case where an adversary take as a query a small value m, such as $m = 0$.

$\mathcal{X}_\lambda$ is a probability distribution such that a random variable taken from it can hide other values, specifically,

$$\exists \xi \; : \; \text{(negligible func.)}, \quad \forall \alpha, \beta \in [-\theta..\theta], \text{ for } \delta \xleftarrow{\$} \mathcal{X}_\lambda, \; \mathsf{SD}(\alpha + \delta, \beta + \delta) \leq \xi(\lambda),$$
$$(4.5)$$

where SD denotes statistical distance. We can use the uniform distribution on $[1..2^\lambda \theta]$ as $\mathcal{X}_\lambda$ for example. But the scheme in Section 4.3 will use another distribution due to a technical reason.

Message Space Size: The message space size M of this scheme has to satisfy $M = \mathsf{poly}(\lambda)$ because the encryption cost of this scheme is clearly $O(M)$. We will remove this restriction in Section 4.3.

(k, θ)**-FTG-nCPA Security of** $\mathcal{E}_{k,\theta}$: Let k and θ be the values which we want to show (k, θ)-FTG-nCPA security for. Then, since intervals $[m_0 - k\theta .. m_0]$ and $[m_1 .. m_1 + k\theta]$ are k times larger than $[m_0 .. m_1]$, the probabilities that $[m_0 - k\theta .. m_0]$ and $[m_1 .. m_1 + k\theta]$ will contain a large δ_i is much larger than the probability that $[m_0 .. m_1]$ will contain a large δ_i.

Therefore, if p is taken suitably, we can ensure the three properties below with high probability. (See Fig.1). Bellow, we call δ_i *large number* if it is taken from $[0 .. 2^\lambda M]$ and we say "$\delta_i = \mathsf{Enc}_K(i) - \mathsf{Enc}_K(i-1)$ is in interval I" to mean that both integers $i - 1$ and i used to define δ_i are contained in I.[2]

$$\text{All } \delta_i \text{ in } [m_0 .. m_1] \text{ are } 1, \tag{4.6}$$

$$\text{Some } \delta_{i_0} \text{ in } [m_0 - k\theta .. m_0] \text{ is large}, \tag{4.7}$$

$$\text{Some } \delta_{i_1} \text{ in } [m_1 .. m_1 + k\theta] \text{ is large}. \tag{4.8}$$

Note that the precise meaning of "good relation" given in "Ideas behind Construction" is that $(\delta_i)_{i \in (A .. M]}$ and queries of an adversary satisfy all of the above three properties.

Here we exploit constraints (3.3) and (3.4) of (k, θ)-FTG-nCPA. Due to them, encryption query m has to satisfy $m \leq m_0 - k\theta$ or $m \geq m_1 + k\theta$. In the former case, the difference $\mathsf{Enc}_K(m_b) - \mathsf{Enc}_K(m) = \sum_{i \in (m .. m_b]} \delta_i = \sum_{i \in (m .. m_0]} \delta_i + \sum_{i \in (m_0 .. m_b]} \delta_i$ contains the large dominant randomness δ_{i_0} as a summand. Since the term $\sum_{i \in (m_0 .. m_b]} \delta_i$ depending on b can be hidden by δ_{i_0}, an adversary cannot detect b from $\mathsf{Enc}_K(m_b) - \mathsf{Enc}_K(m)$.

In the latter case, similarly, the sum $\mathsf{Enc}_K(m) - \mathsf{Enc}_K(m_b) = \sum_{i \in (m_b .. m]} \delta_i$ contains the other large dominant randomness δ_{i_1}. An adversary therefore cannot detect b from $\mathsf{Enc}_K(m) - \mathsf{Enc}_K(m_b)$ due to a similar argument as above.

The above discussion shows that the secret bit b is hidden by "barriers" δ_{i_0} and δ_{i_1}. Based on the same idea, we can show, more generally, that the distribution of the secret bit b is independent from the view of an adversary even when she knows encryption queries and their answers, under the assumption that (4.6), (4.7), and (4.8) hold. (See the full paper for the formal proof.)

Upper Bound on Advantage: The rest of thing we have to do is to show the advantage bound of (4.1) by estimating the probabilities that (4.6), (4.7), and (4.8) hold. To this end, we set the parameter p of the scheme $\mathcal{E}_{k,\theta}$ as follows:

$$p = 1 - \left(1 - \frac{1}{\sqrt{k}}\right)^{\frac{1}{\theta}}. \tag{4.9}$$

---

[2] That is, δ_i is in $I = [a .. b]$ iff $i \in (a .. b]$. Seemingly asymmetry of the interval, which is a "left-open" one $(a .. b]$ but is not "right open" one $[a .. b)$, comes from how we number δ_i. If we set δ_i not to $\mathsf{Enc}_K(i) - \mathsf{Enc}_K(i-1)$ but to $\mathsf{Enc}_K(i+1) - \mathsf{Enc}_K(i)$, it becomes a right open one $[a .. b)$.

Then the advantage bound is calculated as follows. Let E_1, E_2, and E_3 be, respectively, the events that condition (4.6), (4.7), and (4.8) does *not* hold and Bad be $E_1 \vee E_2 \vee E_3$. Then, the previous discussion showed that the advantage of an adversary for our scheme is less than $\Pr[\mathsf{Bad}] + \mathsf{neg}(\lambda)$.

Recall that nCPA adversary has to make her challenge query (m_0, m_1) and encryption queries at the same time. Hence, she has to determine her challenge query (m_0, m_1) without knowing any information about ciphertexts, in particular, any information about δ_i. Therefore, the distributions of $(\delta_i)_i$ and (m_0, m_1) are independent. Since they are independent, E_1, E_2, E_3 are smaller than $1 - (1-p)^\theta = 1/\sqrt{k}$, $(1-p)^{k\theta} = (1-1/\sqrt{k})^k$, and $(1-p)^{k\theta} = (1-1/\sqrt{k})^k$, respectively. Due to the same reason, it follows that

$$
\Pr[\mathsf{Bad}] \leq \frac{1}{\sqrt{k}} + 2\left(1 - \frac{1}{\sqrt{k}}\right)^k = \frac{1}{\sqrt{k}} + 2\left\{\left(1 - \frac{1}{\sqrt{k}}\right)^{\sqrt{k}}\right\}^{\sqrt{k}}
$$
$$
= \frac{1}{\sqrt{k}} + O\left(e^{-\sqrt{k}}\right) = O\left(\frac{1}{\sqrt{k}}\right), \tag{4.10}
$$

which is the bound given in Theorem 6.

About CPA Security: The above proof crucially relies on the independence of the distributions of challenge query (m_0^*, m_1^*) and $(\delta_i)_i$, which is ensured in the nCPA setting. However, a CPA adversary can choose (m_0^*, m_1^*) in the region $(m_i..m_{i+1}]$ where $\mathsf{Enc}_K(m_{i+1}) - \mathsf{Enc}_K(m_i)$ is the smallest, where $m_1 < \ldots < m_q$ are the encryption queries and $(\mathsf{Enc}_K(m_i))_i$ are their answers. Then all δ_i contained in the sum $\mathsf{Enc}_K(m_{i+1}) - \mathsf{Enc}_K(m_i) = \sum_{i \in (m_i..m_{i+1}]} \delta_i$ must be small as well. This means that the probabilities that conditions (4.7) and (4.8) hold must be smaller than those of the case of nCPA. Hence, our proof does not work well in the CPA setting.

4.3   The Proposed Scheme

By improving the scheme $\mathcal{E}_{k,\theta}$ of Section 4.2, we achieve our proposed OPE scheme $\bar{\mathcal{E}}_{k,\theta}$. The encryption and decryption algorithms of it stay polynomial time in the logarithm in the message space M, which enables M to become a super-polynomial in the security parameter λ.

Idea of the Full Paper of [7]: The starting point of our improvement is the following excellent new "lazy sampling" [6] technique of Section 6 of the full paper of [7]: They construct a polynomial time algorithm[3] $\bar{G}$ which takes two pairs (u, C_u) and (v, C_v) of messages and their encryptions, and outputs a data whose distribution is the same as that of ciphertext C_w of w, where w is the "midpoint" $\lceil (u+v)/2 \rceil$ of u and v. Using $\bar{G}$, their improved encryption algorithm $\overline{\mathsf{Enc}}(m)$ computes a ciphertext C_m of m the following binary search recursion:

---

[3] To simplify, here we only consider the case where inputs of $\bar{G}$ are (u, C_u) and (v, C_v), although the full paper of [7] considers more general case due to some technical reasons.

It takes some initial values u, v such that $m \in (u..v]$ holds and C_u and C_v are known. (We denote by Init an algorithm which outputs the encryption C_u and C_v of the initial values.) $\overline{\mathsf{Enc}}(m)$ then computes C_w using $\bar{G}$, replaces interval $(u..v]$ with $(u..w]$ or $(w..v]$ depending on whether $m \le w$ or not, and recursively executes $\overline{\mathsf{Enc}}$ itself. The computational cost of $\overline{\mathsf{Enc}}$ is $O(\log M)$, where M is the message space size, because the binary search recursion is terminated in time $O(\log M)$. Their decryption algorithm $\overline{\mathsf{Dec}}$ is constructed in a similar fashion.

The Idea Behind Our Scheme: Our efficient encryption and decryption algorithms are constructed based on the above idea, but our innovation is that our algorithms $\bar{G}$ and Init are constructed based not on a ciphertext C_u itself but on I_u defined below. This is so, since our elaborated scheme of Section 4.2 does not allow construction of $\bar{G}$ to be based simply on C_u. Below, ρ_i, δ_i, and $A = -k\theta - 1$ are as defined in the scheme of Section 4.2.

$$I_u \leftarrow (C_u^{(0)}, C_u^{(1)}) \leftarrow \Big(\sum_{\substack{i \in (A..u] \\ \rho_i = 0}} \delta_i, \sum_{\substack{i \in (A..u] \\ \rho_i = 1}} \delta_i \Big). \tag{4.11}$$

We will construct Init and $\bar{G}$ satisfying the following properties:

Output Init is indistinguishable from (I_A, I_M). $\qquad(4.12)$

For any $u, v \in (A..M]$ and any I'_u and I'_v, the distribution of an output of $\bar{G}(u, v, I'_u, I'_v)$ is the same as the conditional distribution of I_w when $\quad(4.13)$ $(I_u, I_v) = (I'_u, I'_v)$ holds. Here $w = \lceil (u+v)/2 \rceil$.

Then our efficient encryption algorithm can get I_m in time logarithm $O(\log M)$ in the message space size M by executing a recursion based on Init and $\bar{G}$. It can get the ciphertext of m from $I_m = (C_m^{(0)}, C_m^{(1)})$ because an encryption $\mathsf{Enc}_K(m)$ of Section 4.2 is $\sum_{i \in (A..u]} \delta_i$, and therefore satisfies

$$\mathsf{Enc}_K(m) = C_m^{(0)} + C_m^{(1)}. \tag{4.14}$$

As in the case of [7], the efficient decryption algorithm is also constructed based on a similar idea.

Ideas Behind the Construction of Init and $\bar{G}$: The remaining issue to take care of is the construction of Init and $\bar{G}(I_u, I_v)$. To this end, we set $\mathcal{X}_\lambda$ of (4.5) to a binomial distribution

$$\mathcal{X}_\lambda = \mathcal{B}(2^\lambda \theta^2, 1/2) \tag{4.15}$$

with suitable parameters. Note that this $\mathcal{X}_\lambda$, in fact, satisfies (4.5), which is the property required to ensure the security of the scheme of Section 4.2. Formally, the following fact holds (See the full paper for the proof):

Proposition 8 (Binomial Satisfies (4.5)). *There exists a negligible function ξ such that for all $\alpha, \beta \in [-\theta..\theta]$, the statistical distance between the random variables $\alpha + \delta$ and $\beta + \zeta$ for $\delta, \zeta \xleftarrow{\$} \mathcal{B}(2^\lambda \theta^2, 1/2)$ is less than $\xi(\lambda)$.*

(4.15) allows us to write I_u by using two binomial distributions because (4.11) shows that I_u can be written as sums of δ_i, step 13 of Fig.1 and (4.15) show that δ_i is taken from a binomial distribution, and the sum of binomials is also binomial. Since $I_A = (0,0)$, this means that our algorithm Init satisfying (4.12) can be constructed by using two binomial distributions for generating I_M.

Moreover, it is also known that the conditional distributions of binomials can be written as hypergeometric distributions. Hence, our algorithm $\bar{G}$ satisfying (4.13) can be constructed by using hypergeometric distributions. Since the values which follow the binomial and hypergeometric distributions can be generated in polynomial time [11], our algorithms Init and $\bar{G}$ can terminate in polynomial time.

The description of our algorithms $\bar{G}$ and Init is given in Fig.2. Here $\mathsf{Binom}(n, p)$ and $\mathsf{HG}(a, b, c)$ are algorithms whose outputs follow binomial distribution and hypergeometric distribution. We can show that our algorithms Init and $\bar{G}$ in fact satisfy (4.12) and (4.13); see the full paper for the proof.

Proposition 9 (Init and $\bar{G}$ Work Well). *For constants A and M be given in Fig.1, tuples $(\delta_i)_{i \in (A..M]}$ and $(\rho_i)_{i \in (A..M]}$ generated as in $\mathsf{Kg}(1^\lambda)$ of Fig.1, and I_u defined as is (4.11), (4.12) and (4.13) hold.*

We denote the encryption function given in the above way by $\widetilde{\mathsf{Enc}}$. Then, from (4.12), (4.13), and the construction of $\widetilde{\mathsf{Enc}}$, the following proposition holds. (See the full paper for the formal proof.)

Proposition 10. *Take A, M, Kg, Enc as in Fig.1 and take $\overline{\mathsf{Kg}}$ as in Fig.2. Then for $\bar{K} \leftarrow \overline{\mathsf{Kg}}(1^\lambda)$ and $K \leftarrow \mathsf{Kg}(1^\lambda)$, the distributions of $(\widetilde{\mathsf{Enc}}_{\bar{K}}(i))_{i \in [A..M]}$ and $(\mathsf{Enc}_K(i))_{i \in [A..M]}$ are perfectly indistinguishable.*

Finally, we replace the randomness of $\widetilde{\mathsf{Enc}}$ with a pseudo-random value output by a pseudo-random function, so as to make it deterministic, as in [7]. Then our final encryption algorithm $\overline{\mathsf{Enc}}$ is obtained.

Formal Description of Our Scheme: It is given in Fig.2. Here k and θ are the values which we want to show (k, θ)-FTG-O-nCPA security for, M is the value such that the message space is $[1..M]$, and p and A are the same values used in the scheme of Section 1. Cph, in turn, is an algorithm which computes a ciphertext C_u from I_u based on (4.14). The notation $\bar{G}(u, v, I_u, I_v; cc)$ means that we compute $\bar{G}(u, v, I_u, I_v)$ using cc as the random tape. PRF is a pseudo-random function.

(k, θ)**-FTG-O-nCPA:** Theorem 6 follows from Proposition 8, 9, and 10, and the security of the scheme of Section 4.2. See the full paper for the formal proof of Proposition 8, 9, and 10 and Theorem 6.

Efficiency: The algorithms of our scheme can terminate within polynomial time in $\log M$, $\log k$, $\log \theta$, and security parameter λ due to our binary recursion search and polynomial time algorithms [11] of binomial and hypergeometric distributions. The ciphertext bit length is not more than $\lambda + 2\log_2 \theta + \log_2(M + k\theta + 1)$ because, due to Proposition 10, a ciphertext can be written as $\sum_{i \in [A..m)} \delta_i$

Parameters: · Message Space $= [1..M]$, · $p = 1 - (1 - 1/\sqrt{k})^{1/\theta}$, · $A = -k\theta - 1$.	$\mathsf{Cph}(I)$ 71. Parse I as $(C^{(0)}, C^{(1)})$. 72. Output $C^{(0)} + C^{(1)}$.
$\overline{\mathsf{Kg}}(1^\lambda)$ 41. Randomly take λ bit string K'. 42. $(I_A, I_M) \leftarrow \mathsf{Init}$. 43. Return $\bar{K} \leftarrow (K', A, M, I_A, I_M)$.	Init 81. $C_M^{(1)} \leftarrow \mathsf{Binom}(M - A, 1 - p)$, 82. $C_M^{(0)} \leftarrow \mathsf{Binom}(2^\lambda \theta^2 (M - A - C_M^{(1)}), 1/2)$, 83. $I_A \leftarrow (0, 0)$, $I_M \leftarrow (C_M^{(0)}, C_M^{(1)})$. 84. Output (I_A, I_M).

$\overline{\mathsf{Enc}}_{\bar{K}}(m)$	$\overline{\mathsf{Dec}}_{\bar{K}}(C)$
51. Parse $\bar{K}$ as (K', u, v, I_u, I_v).	61. Parse $\bar{K}$ as (K', u, v, I_u, I_v).

52. If $m = v$ holds,
    return $\mathsf{Cph}(I_v)$.
	62. If $C = \mathsf{Cph}(I_v)$ or $u = v$ holds,
	    return v or $\perp$ respectively.

53. $w \leftarrow \lceil (u+v)/2 \rceil$.	63. $w \leftarrow \lceil (u+v)/2 \rceil$.
54. $cc \leftarrow \mathsf{PRF}_{K'}(u, v)$	64. $cc \leftarrow \mathsf{PRF}_{K'}(u, v)$
55. $I_w \leftarrow \bar{G}(u, v, I_u, I_v; cc)$	65. $I_w \leftarrow \bar{G}(u, v, I_u, I_v; cc)$
56. $\bar{K} \leftarrow \begin{cases} (K', u, w, I_u, I_w) & \text{if } m \leq w \\ (K', w, v, I_w, I_v) & \text{otherwise} \end{cases}$	66. $\bar{K} \leftarrow \begin{cases} (K', u, w, I_u, I_w) & \text{if } C \leq \mathsf{Cph}(I_w) \\ (K', w, v, I_w, I_v) & \text{otherwise} \end{cases}$
57. Return $\overline{\mathsf{Enc}}_{\bar{K}}(m)$.	67. Return $\overline{\mathsf{Dec}}_{\bar{K}}(C)$

$\bar{G}(u, v, I_u, I_v)$
91. Parse I_u and I_v as $(C_u^{(0)}, C_u^{(1)})$ and $(C_v^{(0)}, C_v^{(1)})$. $w \leftarrow \lceil (u+v)/2 \rceil$.
92. $C_w^{(1)} \leftarrow C_u^{(1)} + \mathsf{HG}(v - u, C_v^{(1)} - C_u^{(1)}, w - u)$,
93. $C_w^{(0)} \leftarrow C_u^{(0)}$
    $+ \mathsf{HG}(2^\lambda \theta^2 ((v - u) - (C_v^{(1)} - C_u^{(1)})), C_v^{(0)} - C_u^{(0)}, 2^\lambda \theta^2 ((w - u) - (C_w^{(1)} - C_u^{(1)})))$,
94. Output $I_w \leftarrow (C_w^{(0)}, C_w^{(1)})$.

Fig. 2. The Schemes of Section 4.3, its Parameters, and its Subroutines

for some $m \in [1..M]$ and each δ_i is not more than $2^\lambda \theta^2$ due to (4.15). When we set $(k, \theta) = (M^{2(\beta-t)/3}, M^t)$ as in (4.4), the ciphertext bit length becomes $\lambda + 3 \log_2 M +$ (lower terms) due to $0 < t < \beta < 1$.

On the other hand, the known scheme [7] can ensure $(1, q+1)$-WOW if the ciphertext length is more than $(\log_2 M) + 1$ when M is super-polynomial of λ.

5   Stronger Window-OneWayness of Our Scheme

Finally, we study a stronger variant of (r, q)-WOW notion, called (r, q)-WOWM (studied in [8] intuitively as well). Our definition of WOWM is based on the simpler definition of WOW given in Appendix B of the full paper of [8] which can be reduced to the original WOW given in Section 3 of that paper and vise versa.

Definition 11 ((r, q)-WOWM). An OPE scheme $\mathcal{E}$ on the message space $[1..M]$ is said to be (r, q)-*WOWM (Window One-Way viewing Messages)* if for any polynomial time adversary A, $\mathsf{Succ.Exp}_{\mathcal{E}}^{(r,q)\text{-WOWM}}(\mathsf{A}) = \Pr[\mathsf{Exp}^{(r,q)\text{-WOWM}}_{\mathcal{E}}(\mathsf{A}) = 1]$ is negligible for the message space size M. Here, experiment

$\mathsf{Exp}_{\mathcal{E}}^{(r,q)\text{-WOWM}}(\mathsf{A})$ is defined as follows. Below, $\mathsf{Comb}_q(M)$ be the set of q-element subset of $[1..M]$.

$$K \leftarrow \mathsf{Kg}(1^\lambda), \boldsymbol{m} \xleftarrow{\$} \mathsf{Comb}_q(M), m_* \xleftarrow{\$} \boldsymbol{m},$$
$$(m_L, m_R) \leftarrow \mathsf{A}(\mathsf{Enc}_K(m_*), (m, \mathsf{Enc}_K(m))_{m \in \boldsymbol{m} \setminus \{m_*\}}),$$
$$\text{Return } 1 \text{ iff } m_* \in \mathcal{S}(m_L, m_R),$$

$$\text{where } \mathcal{S}(m_L, m_R) = \begin{cases} [m_L..m_R] & \text{if } m_L \leq m_R \\ [1..m_R] \cup [m_L..M] & \text{otherwise.}, \end{cases}$$

"$m_* \xleftarrow{\$} \boldsymbol{m}$" means that "choose a message m_* from the tuple $\boldsymbol{m}$ uniformly at random". The output (m_L, m_R) of A has to satisfy $\#\mathcal{S}(m_L, m_R) \leq r$.

The following property holds for WOWM and WOW of Appendix B of the full paper of [8] because they are the same except that A can view $\boldsymbol{m} \setminus \{m_*\}$.

$$\forall \mathsf{A} \quad : \quad \mathsf{Adv}.\mathsf{Exp}_{\mathcal{E}}^{(r,q)\text{-WOW}}(\mathsf{A}) \leq \mathsf{Adv}.\mathsf{Exp}_{\mathcal{E}}^{(r,q)\text{-WOWM}}(\mathsf{A}). \tag{5.1}$$

Lemma 12 (Relationship between $(\mathcal{U}^q, \theta, q)$-indis. and WOWM). *Let $q = q(\lambda)$ be a polynomial of security parameter λ, $\mathcal{E}$ be an OPE scheme with a message space $[1..M]$, $\mathcal{U}^q$ be the tuple of q uniform distributions on $[1..M]$, and $0 < t < 1$ be a constant. Suppose that $\mathcal{E}$ is $(\mathcal{U}^q, M^t, q)$-indistinguishable. Then for any constant ρ satisfying*

$$0 \leq \rho < t(< 1), \tag{5.2}$$

$\mathcal{E}$ is $(M^\rho, q+1)$-WOWM when M is super-polynomial of λ. Specifically,

$$\forall \mathsf{A} \exists \mathsf{Mg} \exists \mathsf{B} : \mathsf{Succ}.\mathsf{Exp}_{\mathcal{E}}^{(M^\rho, q+1)\text{-WOWM}}(\mathsf{A})$$
$$\leq \mathsf{Adv}.\mathsf{Exp}_{\mathcal{E}}^{(\mathcal{U}^q, M^t, q)\text{-indis}}(\mathsf{Mg}, \mathsf{B}) + O\left(\frac{1}{M^{t-\rho}}\right) + O\left(\frac{1}{M^{1-t}}\right) + O\left(\frac{q}{M}\right). \tag{5.3}$$

The right hand side of (5.3) is negligible when M is super-polynomial to λ because of (5.2). See the full paper for the formal proof of the above lemma. Lemma 12 and Theorem 7 show the following theorem.

Theorem 13 (WOWM of Our Scheme). *For a polynomial $q = q(\lambda)$ and for any constant*

$$0 \leq \rho < 1, \tag{5.4}$$

our scheme $\bar{\mathcal{E}}_{k,\theta}$ with suitable parameter (depending on (M, ρ)) is $(M^\rho, q+1)$-WOWM under security of PRF (although the advantage bound decreases slower when ρ becomes closer to 1). Specifically,

$$\forall \mathsf{A} \exists \mathsf{B} : \mathsf{Succ}.\mathsf{Exp}_{\bar{\mathcal{E}}_{k,\theta}}^{(M^\rho, q+1)\text{-WOWM}}(\mathsf{A}) \leq O\left(\frac{q}{M^{\frac{1-\rho}{4}}}\right) + \mathsf{Adv}.\mathsf{Exp}_{\mathsf{PRF}}(\mathsf{B}) + \mathsf{neg}(\lambda). \tag{5.5}$$

It achieve better ρ than [8]. See Section 1.3 for details.

Proof (Theorem 13). Take any ρ satisfying (5.4) and set

$$(\beta, t) = (1, (3\rho + 1)/4). \tag{5.6}$$

Let $\mathcal{U}$ be the tuple uniform distributions on the message space $[1..M]$ and let $\mathcal{X} = \mathcal{U}^q$. Then two conditions of Theorem 7, (4.2) and $\beta > t$, are satisfied due to $H_\infty(\mathcal{U}) = \log_2 M$, (5.6), and (5.4). Hence, our scheme $\bar{\mathcal{E}}_{k,\theta}$ with suitable parameter (k, θ) is $(\mathcal{U}^q, M^t, q)$-indistinguishable and satisfies (4.3). (Due to (4.4), the parameters are $(k, \theta) = (M^{2(\beta-t)/3}, M^t) = (M^{(1-\rho)/2}, M^{(3\rho+1)/4}))$. The condition (5.2) of Lemma 12 follows from (5.6) and (5.4). Hence, our scheme with the above parameters is $(M^\rho, q + 1)$-WOWM and satisfies (5.3). The bound (5.5) comes from (5.4) and (5.6) because in (4.3) and (5.3), $O(\frac{q}{M^{(\beta-t)/3}}) = O(\frac{q}{M^{\frac{1}{3} \cdot (1-(3\rho+1)/4)}}) = O(\frac{q}{M^{(1-\rho)/4}})$, $O(\frac{1}{M^{t-\rho}}) = O(\frac{1}{M^{(1-\rho)/4}}) \leq O(\frac{q}{M^{(1-\rho)/4}})$, $O(\frac{1}{M^{1-t}}) = O(\frac{1}{M^{3(1-\rho)/4}}) \leq O(\frac{1}{M^{(1-\rho)/4}})$, and $O(\frac{q}{M}) \leq O(\frac{q}{M^{(1-\rho)/4}})$. □

Acknowledgements. We thank the anonymous reviewers of ASIACRYPT 2014 for useful comments.

References

1. Agrawal, R., Kiernan, J., Srikant, R., Xu, Y.: Order-preserving encryption for numeric data. In: SIGMOD, pp. 563–574 (2004)
2. Agrawal, S., Agrawal, S., Badrinarayanan, S., Kumarasubramanian, A., Prabhakaran, M., Sahai, A.: Function private functional encryption and property preserving encryption: New definitions and positive results. Cryptology ePrint Archive, Report 2013/744 (2013)
3. Bebek, G.: Anti-Tamper Databases: Inference control techniques. Case Western Reserve University (2002)
4. Bellare, M., Desai, A., Jokipii, E., Rogaway, P.: A concrete security treatment of symmetric encryption. In: FOCS, pp. 394–403 (1997)
5. Bellare, M., Namprempre, C.: Authenticated encryption: Relations among notions and analysis of the generic composition paradigm. In: Okamoto, T. (ed.) ASIACRYPT 2000. LNCS, vol. 1976, p. 531. Springer, Heidelberg (2000)
6. Bellare, M., Rogaway, P.: The security of triple encryption and a framework for code-based game-playing proofs. In: Vaudenay, S. (ed.) EUROCRYPT 2006. LNCS, vol. 4004, pp. 409–426. Springer, Heidelberg (2006)
7. Boldyreva, A., Chenette, N., Lee, Y., O'Neill, A.: Order-preserving symmetric encryption. In: Joux, A. (ed.) EUROCRYPT 2009. LNCS, vol. 5479, pp. 224–241. Springer, Heidelberg (2009)
8. Boldyreva, A., Chenette, N., O'Neill, A.: Order-preserving encryption revisited: Improved security analysis and alternative solutions. In: Rogaway, P. (ed.) CRYPTO 2011. LNCS, vol. 6841, pp. 578–595. Springer, Heidelberg (2011)
9. Boneh, D., Waters, B.: Conjunctive, subset, and range queries on encrypted data. In: Vadhan, S.P. (ed.) TCC 2007. LNCS, vol. 4392, pp. 535–554. Springer, Heidelberg (2007)

10. Chatterjee, S., Das, M.P.L.: Property preserving symmetric encryption revisited. Cryptology ePrint Archive, Report 2013/830 (2013)
11. Devroye, L.: Non-Uniform Random Variate Generation. Springer (1986)
12. Furukawa, J.: Request-based comparable encryption. In: Crampton, J., Jajodia, S., Mayes, K. (eds.) ESORICS 2013. LNCS, vol. 8134, pp. 129–146. Springer, Heidelberg (2013)
13. Furukawa, J.: Short comparable encryption. In: CANS (2014)
14. Maurer, U.M., Oswald, Y.A., Pietrzak, K., Sjödin, J.: Luby-rackoff ciphers from weak round functions? In: Vaudenay, S. (ed.) EUROCRYPT 2006. LNCS, vol. 4004, pp. 391–408. Springer, Heidelberg (2006)
15. Maurer, U.M., Pietrzak, K., Renner, R.S.: Indistinguishability amplification. In: Menezes, A. (ed.) CRYPTO 2007. LNCS, vol. 4622, pp. 130–149. Springer, Heidelberg (2007)
16. Minematsu, K., Tsunoo, Y.: Hybrid symmetric encryption using known-plaintext attack-secure components. In: Won, D.H., Kim, S. (eds.) ICISC 2005. LNCS, vol. 3935, pp. 242–260. Springer, Heidelberg (2006)
17. Özsoyoglu, G., Singer, D.A., Chung, S.S.: Anti-tamper databases: Querying encrypted databases. In: De Capitani di Vimercati, S.I., Ray, I., Ray, I. (eds.) Data and Applications Security XVII. IFIP, vol. 142, pp. 133–146. Springer, Boston (2003)
18. Pandey, O., Rouselakis, Y.: Property preserving symmetric encryption. In: Pointcheval, D., Johansson, T. (eds.) EUROCRYPT 2012. LNCS, vol. 7237, pp. 375–391. Springer, Heidelberg (2012)
19. Popa, R.A., Li, F.H., Zeldovich, N.: An ideal-security protocol for order-preserving encoding. In: IEEE Symposium on S&P, pp. 463–477 (2013)
20. Popa, R.A., Redfield, C.M.S., Zeldovich, N., Balakrishnan, H.: CryptDB: protecting confidentiality with encrypted query processing. In: SOSP, pp. 85–100 (2011)
21. Popa, R.A., Redfield, C.M.S., Zeldovich, N., Balakrishnan, H.: CryptDB: processing queries on an encrypted database. Commun. ACM 55(9), 103–111 (2012)
22. Shi, E., Bethencourt, J., Chan, H.T.-H., Song, D.X., Perrig, A.: Multi-dimensional range query over encrypted data. In: IEEE Symposium on S&P, pp. 350–364 (2007)
23. Tu, S., Kaashoek, M.F., Madden, S., Zeldovich, N.: Processing analytical queries over encrypted data. VLDB Endowment 6(5), 289–300 (2013)
24. Waters, B.: Efficient identity-based encryption without random oracles. In: Cramer, R. (ed.) EUROCRYPT 2005. LNCS, vol. 3494, pp. 114–127. Springer, Heidelberg (2005)
25. Xiao, L., Yen, I.-L.: A note for the ideal order-preserving encryption object and generalized order-preserving encryption. IACR Cryptology ePrint Archive 2012, 350 (2012)
26. Xiao, L., Yen, I.-L.: Security analysis for order preserving encryption schemes. In: CISS, pp. 1–6 (2012)
27. Xiao, L., Yen, I.-L., Huynh, D.T.: Extending order preserving encryption for multi-user systems. IACR Cryptology ePrint Archive 2012, 192 (2012)
28. Yum, D.H., Kim, D.S., Kim, J.S., Lee, P.J., Hong, S.J.: Order-preserving encryption for non-uniformly distributed plaintexts. In: Jung, S., Yung, M. (eds.) WISA 2011. LNCS, vol. 7115, pp. 84–97. Springer, Heidelberg (2012)

Statistically-secure ORAM with $\tilde{O}(\log^2 n)$ Overhead

Kai-Min Chung[1], Zhenming Liu[2], and Rafael Pass[3,*]

[1] Institute of Information Science, Academia Sinica, Taiwan
kmchung@iis.sinica.edu.tw
[2] Department of Computer Science, Princeton University, Princeton, NJ, USA
lzhenming@post.harvard.edu
[3] Department of Computer Science, Cornell NYC Tech, Ithaca, NY, USA
rafael@cs.cornell.edu

Abstract. We demonstrate a simple, statistically secure, ORAM with computational overhead $\tilde{O}(\log^2 n)$; previous ORAM protocols achieve only computational security (under computational assumptions) or require $\tilde{\Omega}(\log^3 n)$ overheard. An additional benefit of our ORAM is its conceptual simplicity, which makes it easy to implement in both software and (commercially available) hardware.

Our construction is based on recent ORAM constructions due to Shi, Chan, Stefanov, and Li (Asiacrypt 2011) and Stefanov and Shi (ArXiv 2012), but with some crucial modifications in the algorithm that simplifies the ORAM and enable our analysis. A central component in our analysis is reducing the analysis of our algorithm to a "supermarket" problem; of independent interest (and of importance to our analysis,) we provide an upper bound on the rate of "upset" customers in the "supermarket" problem.

1 Introduction

In this paper we consider constructions of *Oblivious RAM (ORAM)* [10,11]. Roughly speaking, an ORAM enables executing a RAM program while hiding the access pattern to the memory. ORAM have several fundamental applications (see e.g. [11,24] for further discussion). Since the seminal works for Goldreich [10] and Goldreich and Ostrovksy [11], constructions of ORAM have been extensively studied. See, for example, [32,33,1,25,12,6,27,2,13,29,15] and references therein. While the original constructions only enjoyed "computational security" (under the the assumption that one-way functions exists) and required a computational overhead of $\tilde{O}(\log^3 n)$, more recent works have overcome both of these barriers, but only individually. State of the art ORAMs satisfy either of the following:

---

* Pass is supported in part by a Alfred P. Sloan Fellowship, Microsoft New Faculty Fellowship, NSF CAREER Award CCF-0746990, AFOSR YIP Award FA9550-10-1-0093, and DARPA and AFRL under contract FA8750-11-2-0211. The views and conclusions contained in this document are those of the authors and should not be interpreted as representing the official policies, either expressed or implied, of the Defense Advanced Research Projects Agency or the US government.

P. Sarkar and T. Iwata (Eds.): ASIACRYPT 2014, PART II, LNCS 8874, pp. 62–81, 2014.

- An overhead of $\tilde{O}(\log^2 n)$[1], but only satisfies computational security, assuming the existence of one-way functions. [25,12,15]
- Statistical security, but have an overhead of $O(\log^3 n)$. [1,6,27,8,5].

A natural question is whether both of these barriers can be simultaneously overcome; namely, does there exists a statistically secure ORAM with only $\tilde{O}(\log^2 n)$ overhead? In this work we answer this question in the affirmative, demonstrating the existence of such an ORAM.

Theorem 1. *There exists a statistically-secure ORAM with $\tilde{O}(\log^2(n))$ worst-case computational overhead, constant memory overhead, and CPU cache size* $\mathrm{poly}\log(n)$, *where n is the memory size.*

An additional benefit of our ORAM is its conceptual simplicity, which makes it easy to implement in both software and (commercially available) hardware. (A software implementation is available from the authors upon request.)

Our ORAM Construction. A conceptual breakthrough in the construction of ORAMs appeared in the recent work of Shi, Chan, Stefanov, and Li [27]. This work demonstrated a statistically secure ORAM with overhead $O(\log^3 n)$ using a new "tree-based" construction framework, which admits significantly simpler (and thus easier to implemented) ORAM constructions (see also [8,5] for instantiations of this framework which additionally enjoys an extremely simple proof of security).

On a high-level, each memory cell r accessed by the original RAM will be associated with a random leaf pos in a binary tree; the position is specified by a so-called "position map" Pos. Each node in the tree consists of a "bucket" which stores up to ℓ elements. The content of memory cell r will be found inside one of the buckets along the path from the root to the leaf pos; originally, it is put into the root, and later on, the content gets "pushed-down" through an eviction procedure—for instance, in the ORAM of [5] (upon which we rely), the eviction procedure consists of "flushing" down memory contents along a random path, while ensuring that each memory cell is still found on its appropriate path from the root to its assigned leaf. (Furthermore, each time the content of a memory cell is accessed, the content is removed from the tree, the memory cell is assigned to a new random leaf, and the content is put back into the root).

In the work of [27] and its follow-ups [8,5], for the analysis to go through, the bucket size ℓ is required to be $\omega(\log n)$. Stefanov and Shi [28] recently provided a different instantiation of this framework which only uses *constant size* buckets, but instead relies on a *single* $\mathrm{poly}\log n$ size "stash" into which potential "overflows" (of the buckets in the tree) are put;[2] Stefanov and Shi conjectured (but did not prove) security of such a construction (when appropriately evicting elements from the "stash" along the path traversed to access some memory cell).[3]

In this work, we follow the above-mentioned approaches, but with the following high-level modifications:

---

[1] The best protocol achieves $O(\log^2 n/\log\log n)$.

[2] We mention that the idea of using "stash" also appeared in the works [12,13,15,17].

[3] Although different, the "flush" mechanism in [5] is inspired by this eviction method.

– We consider a binary tree where the bucket size of all *internal* buckets is $O(\log \log n)$, but all the leaf nodes still have bucket size $\omega(\log n)$.
– As in [28], we use a "stash" to store potential "overflows" from the bucket. In our ORAM we refer to this as a "queue" as the main operation we require from it is to insert and "pop" elements (as we explain shortly, we additionally need to be able to find and remove any particular element from the queue; this can be easily achieved using a standard hash table). Additionally, instead of inserting memory cells directly into the tree, we insert them into the queue. When searching for a memory cell, we first check whether the memory cell is found in the queue (in which case it gets removed), and if not, we search for the memory cell in the binary tree along the path from the root to the position dictated by the position map.
– Rather than just "flushing" once (as in [5]), we repeat the following procedure "pop and random flush" procedure twice.
  • We "pop" an element from the queue into the root.
  • Next, we flush according to a *geometrically distributed* random variable with expectation 2.[4]

We demonstrate that such an ORAM construction is both (statistically) secure, and only has $\tilde{\Omega}(\log^2 n)$ overhead.

Our Analysis. The key element in our analysis is reducing the security of our ORAM to a "supermarket" problem. Supermarket problems were introduced by Mitzenmacher [20] and have seen been well-studied (see e.g., [20,31,23,26,21]). We here consider a simple version of a supermarket problem, but ask a new question: what is the rate of "upset" customers in a supermarket problem: There are D cashiers in the supermarket, all of which have empty queues in the beginning of the day. At each time step t: with probability $\alpha < 1/2$ a new customer arrives and chooses a random cashier[5] (and puts himself in that cashiers queue); otherwise (i.e., with probability $1 - \alpha$) a random cashier is chosen that "serves" the first customer in its queue (and the queue size is reduced by one). We say that a customer is *upset* if he chooses a queue whose size exceeds some bound φ. What is the rate of upset customers?[6]

We provide an upper bound on the rate of upset customers relying on Chernoff bounds for Markov chains [9,14,16,3]—more specifically, we develop a variant of traditional Chernoff bounds for Markov chains which apply also with "resets" (where at each step, with some small probability, the distribution is reset to the stationary distribution of the Markov chain), which may be of independent interest, and show how such a Chernoff bound can be used in a rather straight-forward way to provide a bound on the number of upset customers.

---

[4] Looking forward, our actual flush is a little bit different than the one in [5] in that we only pull down a *single* element between any two consecutive nodes along the path, whereas in [5] *all* elements that can be pulled down get flushed down.

[5] Typically, in supermarket problems the customer chooses d random cashiers and picks the one with the smallest queue; we here focus on the simple case when $d = 1$.

[6] Although we here consider a discrete-time version of the supermarket problem (since this is the most relevant for our application), as we remark in Remark 1, our results apply also to the more commonly studied continuous-time setting.

Intuitively, to reduce the security of our ORAM to the above-mentioned supermarket problem, each cashier corresponds to a bucket on some particular level k in the tree, and the bound φ corresponds to the bucket size, customers correspond to elements being placed in the buckets, and upset customers overflows. Note that for this translation to work it is important that the number of flushes in our ORAM is geometrically distributed—this ensures that we can view the sequence of operations (i.e., "flushes" that decrease bucket sizes, and "pops" that increase bucket sizes) as independently distributed as in the supermarket problem.

Independent Work. In a very recent independent work, Stefanov, van Dijk, Shi, Fletcher, Ren, Yu, and Devadas [30] prove security of the conjectured Path ORAM of [28]. This yields a ORAM with overhead $O(\log^2 n)$, whereas our ORAM has overhead $O(\log^2 n \log \log n))$. On the other hand, the data structure required to implement our queue is simpler than the one needed to implement the "stash" in the Path ORAM construction. More precisely, we simply need a standard queue and a standard hash table (both of which can be implemented using commodity hardware), whereas the "stash" in [28,30,18] requires using a data structure that additionally supports sorting, or "range queries" (thus a binary search tree is needed), which may make implementations less straightforward. We leave a more complete exploration of the benefits of these two independent approaches for future work.

In another concurrent work, Gentry, Goldman, Halevi, Jutla, Raykova, and Wichs optimize the ORAM of [27]. In particular, they improve the memory overhead from $O(\log n)$ to constant, but the time overhead remains $\tilde{O}(\log^3 n)$. We rely on their idea to achieve constant memory overhead.

2 Preliminaries

A Random Access Machine (RAM) with memory size n consists of a CPU with a small size cache (e.g., can store a constant or poly $\log(n)$ number of words) and an "external" memory of size n. To simplify notation, a word is either $\perp$ or a $\log n$ bit string.

The CPU executes a program Π (given n and some input x) that can access the memory by a $Read(r)$ and $Write(r, val)$ operations where $r \in [n]$ is an index to a memory location, and val is a word (of size $\log n$). The sequence of memory cell accesses by such read and write operations is referred to as the *memory access pattern* of $\Pi(n, x)$ and is denoted $\tilde{\Pi}(n, x)$. (The CPU may also execute "standard" operations on the registers, any may generate outputs).

Let us turn to defining an *Oblivious RAM Compiler*. This notion was first defined by Goldreich [10] and Goldreich and Ostrovksy [11]. We recall a more succinct variant of their definition due to [5].

Definition 1. *A polynomial-time algorithm C is an* Oblivious RAM (ORAM) compiler *with computational overhead $c(\cdot)$ and memory overhead $m(\cdot)$, if C given $n \in N$ and a deterministic RAM program Π with memory-size n outputs a program Π' with memory-size $m(n) \cdot n$ such that for any input x, the running-time of $\Pi'(n, x)$ is bounded by $c(n) \cdot T$ where T is the running-time of $\Pi(n, x)$, and there exists a negligible function μ such that the following properties hold:*

- **Correctness:** *For any $n \in N$ and any string $x \in \{0,1\}^*$, with probability at least $1 - \mu(n)$, $\Pi(n, x) = \Pi'(n, x)$.*
- **Obliviousness:** *For any two programs Π_1, Π_2, any $n \in N$ and any two inputs $x_1, x_2 \in \{0,1\}^*$ if $|\tilde{\Pi}_1(n, x_1)| = |\tilde{\Pi}_2(n, x_2)|$, then $\tilde{\Pi}'_1(n, x_1)$ is μ-close to $\tilde{\Pi}'_2(n, x_2)$ in statistical distance, where $\Pi'_1 = C(n, \Pi_1)$ and $\Pi'_2 = C(n, \Pi_2)$.*

Note that the above definition (just as the definition of [11]) only requires an oblivious compilation of *deterministic* programs Π. This is without loss of generality: we can always view a randomized program as a deterministic one that receives random coins as part of its input.

3   Our ORAM and Its Efficiency

This section presents the construction of our ORAM, followed by an analysis of its efficiency.

3.1   The Algorithm

Our ORAM data structure serves as a "big" memory table of size n and exposes the following two interfaces.

- READ(r): the algorithm returns the value of memory cell $r \in [n]$.
- WRITE(r, v): the algorithm writes value v to memory cell r, and returns the original value of r.

We start by assuming that the ORAM is executed on a CPU with cache size is $2n/\alpha + o(n)$ (in words) for a suitably large constant α (the reader may imagine $\alpha = 16$). Following the framework in [27], we can then reduce the cache size to $O(\text{poly} \log n)$ by recursively applying the ORAM construction; we provide further details on this transformation at the end of the section.

In what follows, we group each consecutive α memory cells in the RAM into a *block* and will thus have n/α blocks in total. We also index the blocks in the natural way, *i.e.* the block that contains the first α memory cells in the table has index 0 and in general the i-th block contains memory cells with addresses from αi to $\alpha(i + 1) - 1$.

Our algorithm will always operate at the block level, *i.e.* memory cells in the same block will always be read/written together. In addition to the content of its α memory cells, each block is associated with two extra pieces of information. First, it stores the index i of the block. Second, it stores a "position" p that specify its storage "destination" in the external memory, which we elaborate upon in the forthcoming paragraphs. In other words, a block is of the form (i, p, val), where val is the content of its α memory cells.

Our ORAM construction relies on the following three main components.

1. **A full binary tree at the in the external memory** that serves as the primary media to store the data.
2. **A position map in the internal cache** that helps us to search for items in the binary tree.
3. **A queue in the internal cache** that is the secondary venue to store the data.

We now walk through each of the building blocks in details.

The Full Binary Tree Tr. The depth of this full binary tree is set to be the smallest d so that the number of leaves $L = 2^d$ is at least $2(n/\alpha)/(\log n \log \log n)$ (*i.e.*, $L/2 < 2(n/\alpha)/(\log n \log \log n) \leq L$). (In [27,5] the number of leaves was set to n/α; here, we instead follow [8] and make the tree slightly smaller—this makes the memory overhead constant.) We index nodes in the tree by a binary strings of length at most d, where the root is indexed by the empty string λ, and each node indexed by γ has left and right children indexed $\gamma 0$ and $\gamma 1$, respectively. Each node is associated with a *bucket*. A bucket in an internal node can store up to ℓ blocks, and a bucket in a leaf can store up to ℓ' blocks, where ℓ and ℓ' are parameters to be determined later. The tree shall support the following two atomic operations:

- READ(Node: v): the tree will return all the blocks in the bucket associated with v to the cache.
- WRITE(Node: v, Blocks: b): the input is a node v and an array of blocks b (that will fit into the bucket in node v). This operation will replace the bucket in the node v by b.

The Position Map P. This data structure is an array that maps the indices of the blocks to leaves in the full binary tree. Specifically, it supports the following atomic operations:

- READ(i): this function returns the position $P[i] \in [L]$ that corresponds to the block with index $i \in [n/\alpha]$.
- WRITE(i, p): this function writes the position p to $P[i]$.

We assume that the position map is initialized with value $\bot$.

The Queue Q. This data structure stores a queue of blocks with maximum size q_{max}, a parameter to be determined later, and supports the following three atomic operations:

- INSERT(Block b): insert a block b into the queue.
- POPFRONT(): the first block in the queue is popped and returned.
- FIND(int: i, word: p): if there is a block b with index i and position p stored in the queue, then FIND returns b and deletes it from the queue; otherwise, it returns $\bot$.

Note that in addition to the usual INSERT and POPFRONT operations, we also require the queue to support a FIND operation that finds a given block, returns and deletes it from the queue. This operation can be supported using a standard hash table in conjunction with the queue. We mention that all three operations can be implemented in time less than $O(\log n \log \log n)$, and discuss the implementation details in Appendix A.

Our Construction. We now are ready to describe our ORAM construction, which relies the above atomic operations. Here, we shall focus on the read operation. The algorithm for the write operation is analogous.

For two nodes u and v in Tr, we use $\text{path}(u, v)$ to denote the (unique) path connecting u and v. Throughout the life cycle of our algorithm we maintain the following *block-path* invariance.

> **Block-Path Invariance:** *For any index $i \in [n/\alpha]$, either $P[i] = \bot$ and in this case both* Tr *and the queue do not contain any block with index i, or there exists a unique block b with index i that is located either in the queue, or in the bucket of one of the nodes on* $\text{path}(\lambda, P[i])$ *in* Tr

We proceed to describe our READ(r) algorithm. At a high-level, READ(r) consists of two sub-routines FETCH() and DEQUEUE(). READ(r) executes FETCH() and DEQUEUE() once in order. Additionally, at the end of every $\log n$ invocations of READ(r), *one extra* DEQUEUE() is executed. Roughly, FETCH() fetches the block b that contains the memory cell r from either path($\lambda, P[\lfloor r/\alpha \rfloor]$) in Tr or in Q, then returns the value of memory cell r, and finally inserts the block b to the queue Q. On the other hand, DEQUEUE() pops one block b from Q, inserts b to the root λ of Tr (provided there is a room), and performs a *random* number of "FLUSH" actions that gradually moves blocks in Tr down to the leaves.

Fetch: Let $i = \lfloor r/\alpha \rfloor$ be the index of the block b that contains the r-th memory cell, and $p = P[i]$ be the current position of b. If $P[i] = \perp$ (which means that the block is not initialized yet), let $P[i] \leftarrow [L]$ be a uniformly random leaf, create a block $b = (i, P[i], \perp)$, and insert b to the queue Q. Otherwise, FETCH performs the following actions in order.

   Fetch from Tree Tr **and Queue** Q**:** Search the block b with index i along path(λ, p) in Tr by reading all buckets in path(λ, p) once and writing them back. Also, search the block b with index i and position p in the queue Q by invoking FIND(i, p). By the block-path invariance, we must find the block b.

   Update Position Map P**.** Let $P[i] \leftarrow [L]$ be a uniformly random leaf, and update the position $p = P[i]$ of b.

   Insert to Queue Q**:** Insert the block b to Q.

Dequeue: This sub-routine consists of two actions PUT-BACK() and FLUSH(). It starts by executing PUT-BACK() once, and then performs a *random* number of FLUSH()es as follows: Let $C \in \{0, 1\}$ be a biased coin with $\Pr[C = 1] = 2/3$. It samples C, and if the outcome is 1, then it continues to perform one FLUSH() and sample another independent copy of C, until the outcome is 0. (In other words, the number of FLUSH() is a geometric random variable with parameter $2/3$.)

   Put-Back: This action moves a block from the queue, if any, to the root of Tr. Specifically, we first invoke a POPFRONT(). If POPFRONT() returns a block b then add b to λ .

   Flush: This procedure selects a random path (namely, the path connecting the root to a random leaf $p^* \leftarrow \{0, 1\}^d$) on the tree and tries to move the blocks along the path down subject to the condition that the block always finds themselves on the appropriate path from the root to their assigned leaf node (see the block-path invariance condition). Let $p_0(= \lambda)p_1...p_d$ be the nodes along path(λ, p^*). We traverse the path while carrying out the following operations for each node p_i we visit: in node p_i, find the block that can be "pulled-down" as far as possible along the path path(λ, p^*) (subject to the block-path invariance condition), and pull it down to p_{i+1}. For $i < d$, if there exists some $\eta \in \{0, 1\}$ such that p_i contains more than $\ell/2$ blocks that are assigned to leafs of the form $p_i||\eta||\cdot$ (see also Figure 1 in Appendix),[7] then select an arbitrary such block b, remove it from the bucket p_i and invoke an OVERFLOW(b) procedure, which re-samples a uniformly random

---

[7] Here, $a||b$ denotes the concatenation of string a and b.

position for the overflowed block b and inserts it back to the queue Q. (See the full version of the paper [4] for the pseudocode.)

Finally, the algorithm aborts and terminates if one of the following two events happen throughout the execution.

Abort-Queue: If the size of the queue Q reaches q_{max}, then the algorithm aborts and outputs ABORTQUEUE.

Abort-Leaf: If the size of any leaf bucket reaches ℓ' (i.e., it becomes full), then the algorithm aborts and outputs ABORTLEAF.

This completes the description of our READ(r) algorithm; the WRITE(r, v) algorithm is defined in essentially identically the same way, except that instead of inserting b into the queue Q (in the last step of FETCH), we insert a modified b' where the content of the memory cell r (inside b) has been updated to v.

It follows by inspection that the block-path invariance is preserved by our construction. Also, note that in the above algorithm, FETCH increases the size of the queue Q by 1 and PUT-BACK is executed twice which decreases the queue size by 2. On the other hand, the FLUSH action may cause a few OVERFLOW events, and when an OVERFLOW occurs, one block will be removed from Tr and inserted to Q. Therefore, the size of the queue changes by minus one plus the number of OVERFLOW for each READ operation. The crux of our analysis is to show that the number of OVERFLOW is sufficiently small in any given (short) period of time, except with negligible probability.

We remark that throughout this algorithm's life cycle, there will be at most $\ell - 2$ non-empty blocks in each internal node except when we invoke FLUSH($\cdot$), in which case some intermediate states will have $\ell - 1$ blocks in a bucket (which causes an invocation of OVERFLOW).

Reducing the Cache's Size. We now describe how the cache can be reduced to poly $\log(n)$ via recursion [27]. The key observation here is that the position map shares the same set of interfaces with our ORAM data structure. Thus, we may substitute the position map with a (smaller) ORAM of size $\lceil n/\alpha \rceil$. By recursively substituting the position map $O(\log n)$ times, the size of the position map will be reduced to $O(1)$.

A subtle issue here is that we need to update the position map when overflow occurs (in addition to the update for the fetched block), which results in an access to the recursive ORAM. This causes two problems. First, it reveals the time when overflow occurs, which kills obliviousness. Second, since we may make more than one recursive calls, the number of calls may blow up over $O(\log n)$ recursion levels.

To solve both problems, we instead defer the recursive calls for updating the position map to the time when we perform PUT-BACK operations. It is not hard to check that this does not hurt correctness. Recall that we do DEQUEUE once for each ORAM access, and additionally do an extra DEQUEUE for every $\log n$ ORAM accesses (to keep the cache size small). This is a deterministic pattern and hence restores obliviousness. Also note that this implies only $(\log n) + 1$ recursive calls are invoked for every $\log n$ ORAM accesses. Thus, intuitively, the blow-up rate is $(1 + (1/\log n))$ per level, and only results in a constant blow up over $O(\log n)$ levels. More precisely, consider a program execution with T ORAM access. It results in $T \cdot (1 + (1/\log n))$ access to the second ORAM, and $O(T)$ access to the final $O(1)$ size ORAM.

Now, we need to be slightly more careful to avoid the following problem. It might be possible that the one extra DEQUEUE occurs in multiple recursion levels simultaneous, resulting in unmanageable worst case running time. This problem can be avoided readily by schedule the extra DEQUEUE in different round among different recursion levels. Specifically, let $u = \log n$. For recursion level ℓ, the extra DEQUEUE is scheduled in the $(au + \ell)$-th (base-)ORAM access, for all positive integers a. Note that the extra DEQUEUE occurs in slightly slower rate in deeper recursion levels, but this will not change the asymptotic behavior of the system. As such, no two extra DEQUEUE's will be called in the same READ/WRITE operation.

On the other hand, recall that we also store the queue in the cache. We will set the queue size $q_{max} = \text{poly} \log(n)$ (specifically, we can set $q_{max} = O(\log^{2+\varepsilon} n)$ for an arbitrarily small constant ε). Since there are only $O(\log n)$ recursion levels, the total queue size is $\text{poly} \log(n)$.

3.2 Efficiency of Our ORAM

In this section, we discuss how to set the parameters of our ORAM and analyze its efficiency. We summarize the parameters of our ORAM and the setting of parameters as follows:

- ℓ: The bucket size (in terms of the number of blocks it stores) of the internal nodes of Tr. We set $\ell = \Theta(\log \log n)$.
- ℓ': The bucket size of the leaves of Tr. We set $\ell' = \Theta(\log n \log \log n)$.
- d: The depth of Tr. As mentioned, we set it to be the smallest d so that the number of leaves 2^d is at least $2(n/\alpha)/(\log n \log \log n)$.
- q_{max}: The queue size. As mentioned, we set $q_{max} = \Theta(\log^{2+\varepsilon} n)$ for an arbitrarily small constant ε.
- α: The number of memory cells in a block. As mentioned, we set α to be a constant, say 16.

We proceed to analyze the efficiency of our ORAM.

Memory Overhead: Constant. The external memory stores $O(\log n)$ copies of binary trees from $O(\log n)$ recursion levels. Let us first consider Tr of the top recursion level: there are $2^{d+1} - 1 = \Theta(n/\log n \log \log n)$ nodes, each of which has bucket of size at most $\ell' = \Theta(\log n \log \log n)$. The space complexity of Tr is $\Theta(n)$. As the size of Tr in each recursion level shrinks by a constant factor, one can see that the total memory overhead is constant.

Cache Size: $\text{poly} \log(n)$. As argued, the CPU cache stores the position map in the final recursion level, which has $O(1)$ size, and the queues from $O(\log n)$ recursion levels, each of which has at most $\Theta(\log^{2+\varepsilon} n)$ size. Thus, the total cache size is $O(\log^{3+\varepsilon} n)$. As we shall see below, $\text{poly} \log(n)$ queue size is required in our analysis to ensure that the queue overflows with negligible probability by concentration bounds. On the other hand, we mention that our simple simulation shows that the size of the queue in the top recursion level is often well below 50 for ORAM with reasonable size.

Worst-Case Computational Overhead: $\tilde{O}(\log^2 n)$. As above, we first consider the top recursion level. In the FETCH() sub-routine, we need to search from both Tr and the queue. Searching Tr requires us to traverse along a path from the root to a leaf. The time spent on each node is proportional to the size of the node's bucket. Thus, the cost here is $O(\log n \log \log n)$. One can also see searching the queue takes $O(\log n \log \log n)$ time. The total cost of FETCH() is $O(\log n \log \log n)$.

For the DEQUEUE() sub-routine, the PUT-BACK() action invokes (1) one POPFRONT(), which takes $O(\log n \log \log n)$ time, and (2) accesses the root node, which costs $O(\log \log n)$. It also writes to the position map and triggers recursive calls. Note that certain recursive levels may execute two consecutive DEQUEUE's after a READ/WRITE operation. But our construction ensures only one level will execute two DEQUEUE's for any READ/WRITE. Thus, the total cost here is $\tilde{O}(\log^2 n)$.

The FLUSH() sub-routine also traverses Tr along a path, and has cost $O(\log n \log \log n)$. However, since we do a *random* number of FLUSH() (according to a geometric random variable with parameter $2/3$), we only achieve *expected* $O(\log n \log \log n)$ runtime, as opposed to worst-case runtime.

To address this issue, recall that there are $O(\log n)$ recursion levels, and the total number of FLUSH() is the sum of $O(\log n)$ i.i.d. random variables. Thus, the probability of performing a total of more than $\omega(\log n)$ number of FLUSH()'s is negligible by standard concentration result. Thus, the total time complexity is upper bounded by $\omega(\log^2 n \log \log n)$ except with negligible probability. To formally get $\tilde{O}(\log^2 n)$ worst-case computational overhead, we can add an **Abort-Flush** condition that aborts when the total number of flush in one READ()/WRITE() operation exceeds some parameter $t \in \omega(\log n)$.

4   Security of Our ORAM

The following observation is central to the security of our ORAM construction (and an appropriate analogue of it was central already to the constructions of [27,5]):

> **Key Observation:** *Let X denote the sum of two independent geometric random variables with mean 2. Each Read and Write operation traverses the tree along $X + 1$ randomly chosen paths,* independent *of the history of operations so far.*

The key observation follows from the facts that (1) just as in the schemes of [27,5], each position in the position map is used exactly once in a traversal (and before this traversal, no information about the position is used in determining what nodes to traverse), and (2) we invokes the FLUSH action X times and the flushing, by definition, traverses a random path, independent of the history.

Armed with the key observation, the security of our construction reduces to show that our ORAM program does not abort except with negligible probability, which follows by the following two lemmas.

Lemma 1. *Given any program Π, let $\Pi'(n, x)$ be the compiled program using our ORAM construction. We have*

$$\Pr\left[\text{ABORTLEAF}\right] \leq \text{negl}(n).$$

Proof. The proof follows by a direct application of the (multiplicative) Chernoff bound. We show that the probability of overflow in any of the leaf nodes is small. Consider any leaf node γ and some time t. For there to be an overflow in γ at time t, there must be $\ell' + 1$ out of n/α elements in the position map that map to γ. Recall that all positions in the position map are uniformly and independently selected; thus, the expected number of elements mapping to γ is $\mu = \log n \log \log n$ and by a standard multiplicative version of Chernoff bound, the probability that $\ell'+1$ elements are mapped to γ is upper bounded by $2^{-\ell'}$ when $\ell' \geq 6\mu$ (see Theorem 4.4 in [19]). By a union bound, we have that the probability of *any* node ever overflowing is bounded by $2^{-(\ell')} \cdot (n/\alpha) \cdot T$

To analyze the full-fledged construction, we simply apply the union bound to the failure probabilities of the $\log_\alpha n$ different ORAM trees (due to the recursive calls). The final upper bound on the overflow probability is thus $2^{-(\ell')} \cdot (n/\alpha) \cdot T \cdot \log_\alpha n$, which is negligible as long as $\ell' = c \log n \log \log n$ for a suitably large constant c.   □

Lemma 2. *Given any program Π, let $\Pi'(n, x)$ be the compiled program using our ORAM construction. We have*

$$\Pr[\text{ABORTQUEUE}] \leq \text{negl}(n).$$

The proof of Lemma 2 is significantly more interesting. Towards proving it, in Section 5 we consider a simple variant of a "supermarket" problem (introduced by Mitzenmacher[20]) and show how to reduce Lemma 2 to an (in our eyes) basic and natural question that seems not to have been investigated before.

5   Proof of Lemma 2

We here prove Lemma 2: in Section 5.1 we consider a notion of "upset" customers in a supermarket problem [20,31,7]; in Section 5.2 we show how Lemma 2 reduced to obtaining a bound on the rate of upset customers, and in Section 5.3 we provide an upper bound on the rate of upset customers.

5.1   A Supermarket Problem

In a supermarket problem, there are D cashiers in the supermarket, all of which have empty queues in the beginning of the day. At each time step t,

- With probability $\alpha < 1/2$, an *arrival* event happens, where a new customer arrives. The new customer chooses d uniformly random cashiers and join the one with the shortest queue.
- Otherwise (*i.e.* with the remaining probability $1 - \alpha$), a *serving* event happens: a random cashier is chosen that "serves" the first customer in his queue and the queue size is reduced by one; if the queue is empty, then nothing happens.

We say that a customer is *upset* if he chooses a queue whose size exceeds some bound φ. We are interested in large deviation bounds on the number of upset customers for a given short time interval (say, of $O(D)$ or poly $\log(D)$ time steps).

Supermarket problems are traditionally considered in the continuous time setting [20,31,7]. But there exists a standard connection between the continuous model and its discrete time counterpart: conditioned on the number of events is known, the continuous time model behaves in the same way as the discrete time counterpart (with parameters appropriately rescaled).

Most of the existing works [20,31,7] study only the stationary behavior of the processes, such as the expected waiting time and the maximum load among the queues over the time. Here, we are interested in large deviation bounds on a statistics over a *short* time interval; the configurations of different cashiers across the time is highly correlated.

For our purpose, we analyze only the simple special case where the number of choice $d = 1$; *i.e.* each new customer is put in a random queue.

We provide a large deviation bound for the number of upset customers in this setting.[8]

Proposition 1. *For the (discrete-time) supermarket problem with D cashiers, one choice (i.e., $d = 1$), probability parameter $\alpha \in (0, 1/2)$, and upset threshold $\varphi \in \mathbb{N}$, for any T steps time interval $[t + 1, t + T]$, let F be the number of upset customers in this time interval. We have*

$$
\Pr\left[F \geq (1 + \delta)(\alpha/(1 - \alpha))^{\varphi}T\right] \leq \begin{cases} \exp\left\{-\Omega\left(\frac{\delta^2(\alpha/(1-\alpha))^{\varphi}T}{(1-\alpha)^2}\right)\right\} & \text{for } 0 \leq \delta \leq 1 \\ \exp\left\{-\Omega\left(\frac{\delta(\alpha/(1-\alpha))^{\varphi}T)}{(1-\alpha)^2}\right)\right\} & \text{for } \delta \geq 1 \end{cases}
$$

$$(1)$$

Note that Proposition 1 would trivially follow from the standard Chernoff bound if T is sufficiently large (i.e., $T \gg O(D)$) to guarantee that we *individually* get concentration on each of the D queue (and then relying on the union bound). What makes Proposition 1 interesting is that it applies also in a setting when T is poly log D.

The proof of Proposition 1 is found in Section 5.3 and relies on a new variant Chernoff bounds for Markov chains with "resets," which may be of independent interest.

Remark 1. *One can readily translate the above result to an analogous deviation bound on the number of upset customers for (not-too-short) time intervals in the continuous time model. This follows by noting that the number of events that happen in a time interval is highly concentrated (provided that the expected number of events is not too small), and applying the above proposition after conditioning on the number of events happen in the time interval (since conditioned on the number of events, the discrete-time and continous-time processes are identical).*

5.2   From ORAM to Supermarkets

This section shows how we may apply Proposition 1 to prove Lemma 2. Central to our analysis is a simple reduction from the execution of our ORAM algorithm at level k in Tr to a supermarket process with $D = 2^{k+1}$ cashiers. More precisely, we show there exists a coupling between two processes so that each bucket corresponds with two cashiers; the load in a bucket is always upper bounded by the total number of customers in the two cashiers it corresponds to.

To begin, we need the following Lemma.

---

[8] It is not hard to see that with D cashiers, probability parameter α, and "upset" threshold φ, the expected number of upset customers is at most $(\alpha/(1 - \alpha))^{\varphi} \cdot T$ for any T-step time interval.

Lemma 3. *Let* $\{a_i\}_{i \geq 1}$ *be the sequence of* PUT-BACK/FLUSH *operations defined by our algorithm, i.e. each* $a_i \in \{\text{PUT-BACK}, \text{FLUSH}\}$ *and between any consecutive* PUT-BACKs, *the number of* FLUSHes *is a geometric r.v. with parameter* $2/3$. *Then* $\{a_i\}_{i \geq 1}$ *is a sequence of i.i.d. random variables so that* $\Pr[a_i = \text{PUT-BACK}] = \frac{1}{3}$.[9]

To prove Lemma 3, we may view the generation of $\{a_i\}_{i \geq 1}$ as generating a sequence of i.i.d. Bernoulli r.v. $\{b_i\}_{i \geq 1}$ with parameter $\frac{2}{3}$. We set a_i be a FLUSH() if and only if $b_i = 1$. One can verify that the $\{a_i\}_{i \geq 1}$ generated in this way is the same as those generated by the algorithm.

We are now ready to describe our coupling between the original process and the supermarket process. At a high-level, a block corresponds to a customer, and 2^{k+1} sub-trees in level $k + 1$ of Tr corresponds to $D = 2^{k+1}$ cashiers. More specifically, we couple the configurations at the k-th level of Tr in the ORAM program with a supermarket process as follows.

- Initially, all cashiers have zero customers.
- For each PUT-BACK(), a corresponding arrival event occurs: if a ball b with position $p = (\gamma || \eta)$ (where $\gamma \in \{0, 1\}^{k+1}$) is moved to Tr, then a new customer arrives at the γ-th cashier; otherwise (*e.g.* when the queue is empty), a new customer arrives at a random cashier.
- For each FLUSH() along the path to leaf $p^* = (\gamma || \eta)$ (where $\gamma \in \{0, 1\}^{k+1}$), a serving event occurs at the γ-th cashier.
- For each FETCH(), nothing happens in the experiment of the supermarket problem. (Intuitively, FETCH() translates to extra "deletion" events of customers in the supermarket problem, but we ignore it in the coupling since it only decreases the number of blocks in buckets in Tr.)

Correctness of the Coupling. We shall verify the above way of placing and serving customers exactly gives us a supermarket process. First recall that both PUT-BACK and FLUSH actions are associated with uniformly random leaves. Thus, this corresponds to that at each timestep a random cashier will be chosen. Next by Lemma 3, the sequence of PUT-BACK and FLUSH actions in the execution of our ORAM algorithm is a sequence of i.i.d. variables with $\Pr[\text{PUT-BACK}] = \frac{1}{3}$. Therefore, when a queue is chosen at a new timestep, an (independent) biased coin is tossed to decide whether an arrival or a service event will occur.

Dominance. Now, we claim that at any timestep, for every $\gamma \in \{0, 1\}^{k+1}$, the number of customers at γ-th cashier is at least the number of blocks stored at or above level k in Tr with position $p = (\gamma || \cdot)$. This follows by observing that (i) whenever there is a block with position $p = (\gamma || \cdot)$ moved to Tr (from PUT-BACK()), a corresponding new customer arrives at the γ-th cashier, *i.e.* when the number of blocks increase by one, so does the number of customers, and (ii) for every FLUSH() along the path to $p^* = (\gamma || \cdot)$: if there is at least one block stored at or above level k in Tr with position $p = (\gamma || \cdot)$, then one such block will be flushed down below level k (since we flush the

---

[9] The first operation in our system is always a PUT-BACK. To avoid that $a_1 \equiv$ PUT-BACK, we can first execute a geometric number of FLUSHes before the system starts for the analysis purpose.

blocks that can be pulled down the furthest)—that is, when the number of customers decreases by one, so does the number of blocks (if possible). This in particular implies that throughout the coupled experiments, for every $\gamma \in \{0,1\}^k$ the number of blocks in the bucket at node γ is always upper bounded by the sum of the number of customers at cashier $\gamma 0$ and $\gamma 1$.

We summarize the above in the following lemma.

Lemma 4. *For every execution of our ORAM algorithm (i.e., any sequence of READ and WRITE operations), there is a coupled experiment of the supermarket problem such that throughout the coupled experiments, for every $\gamma \in \{0,1\}^k$ the number of blocks in the bucket at node γ is always upper bounded by the sum of the number of customers at cashier $\gamma 0$ and $\gamma 1$.*

From Lemma 4 and Proposition 1 to Lemma 2. Note that at any time step t, if the queue size is $\leq \frac{1}{2}\log^{2+\epsilon} n$, then by Proposition 1 with $\varphi = \ell/2 = O(\log\log n)$ and Lemma 4, except with negligible probability, at time step $t + \log^4 n$, there have been at most $\omega(\log n)$ overflows per level in the tree and thus at most $\frac{1}{2}\log^{2+\epsilon} n$ in total. Yet during this time "epoch", $\log^3 n$ element have been "popped" from the queue, so, except with negligible probability, the queue size cannot exceed $\frac{1}{2}\log^{2+\epsilon} n$.

It follows by a union bound over $\log^3 n$ length time "epochs", that except with negligible probability, the queue size never exceeds $\log^{2+\epsilon} n$.

5.3   Analysis of the Supermarket Problem

We now prove Proposition 1. We start with interpreting the dynamics in our process as evolutions of a Markov chain.

A Markov Chain Interpretation. In our problem, at each time step t, a random cashier is chosen and either an arrival or a serving event happens at that cashier (with probability α and $(1 - \alpha)$, respectively), which increases or decreases the queue size by one. Thus, the behavior of each queue is governed by a simple Markov chain M with state space being the size of the queue (which can also be viewed as a drifted random walk on a one dimensional finite-length lattice). More precisely, each state $i > 0$ of M transits to state $i + 1$ and $i - 1$ with probability α and $(1 - \alpha)$, respectively, and for state 0, it transits to state 1 and stays at state 0 with probability α and $(1 - \alpha)$, respectively. In other words, the supermarket process can be rephrased as having D copies of Markov chains M, each of which starts from state 0, and at each time step, one random chain is selected and takes a move.

We shall use Chernoff bounds for Markov chains [9,14,16,3] to derive a large deviation bound on the number of upset customers. Roughly speaking, Chernoff bounds for Markov chains assert that for a (sufficiently long) T-steps random walk on an ergodic finite state Markov chain M, the number of times that the walk visits a subset V of states is highly concentrated at its expected value $\pi(V) \cdot T$, provided that the chain M has spectral expansion[10] $\lambda(M)$ bounded away from 1. However, there are a few technical issues, which we address in turn below.

---

[10] For an ergodic reversible Markov chain M, the *spectral expansion* $\lambda(M)$ of M is simply the second largest eigenvalue (in absolute value) of the transition matrix of M. The quantity $1 - \lambda(M)$ is often referred to as the spectral gap of M.

Overcounting. The first issue is that counting the number of visits to a state set $V \subset S$ does not capture the number of upset customers exactly—the number of upset customers corresponds to the *number of transits* from state i to $i + 1$ with $i + 1 \geq \varphi$. Unfortunately, we are not aware of Chernoff bounds for counting the number of transits (or visits to an edge set). Nevertheless, for our purpose, we can set $V_\varphi = \{i : i \geq \varphi\}$ and the number of visits to V_φ provides an *upper bound* on the number of upset customers.

Truncating the Chain. The second (standard) issue is that the chain M for each queue of a cashier has infinite state space $\{0\} \cup \mathbb{N}$, whereas Chernoff bounds for Markov chains are only proven for finite-state Markov chains. However, since we are only interested in the supermarket process with finite time steps, we can simply truncate the chain M at a sufficiently large K (say, $K \gg t + T$) to obtain a chain M_K with finite states $S_K = \{0, 1, \ldots, K\}$; that is, M_K is identical to M, except that for state K, it stays at K with probability α and transits to $K - 1$ with probability $1 - \alpha$. Clearly, as we only consider $t + T$ time steps, the truncated chain M_K behaves identical to M. It's also not hard to show that M_K has stationary distribution π_K with $\pi_K(i) = (1 - \beta)\beta^i / (1 - \beta^{K+1})$, and spectral gap $1 - \lambda(M_K) \geq \Omega(1/(1 - \alpha)^2)$.[11]

Correlation over a Short Time Frame. The main challenge, however, is to establish large deviation bounds for a *short* time interval T (compared to the number D of chains). For example, $T = O(D)$ or even $\operatorname{poly} \log(D)$, and in these cases the expected number of steps each of the D chains take can be a small constant or even $o(1)$. Therefore, we cannot hope to obtain meaningful concentration bounds individually for each single chain. Finally, the D chains are not completely independent: only one chain is selected at each time step. This further introduces correlation among the chains.

We address this issue by relying on a new variant of Chernoff bounds for Markov chains with "resets," which allows us to "glue" walks on D separate chains together and yields a concentration bound that is as good as a T-step random walk on a single chain. We proceed in the following steps.

- Recall that we have D copies of truncated chains M_K starting from state 0. At each time step, a random chain is selected and we takes one step in this chain. We want to upper bound the total number of visits to V_φ during time steps $[t + 1, t + T]$.
- We first note that, as we are interested in upper bounds, we can assume that the chains start at the stationary distribution π_K instead of the 0 state (i.e., all queues have initial size drawn from π_K instead of being empty). This follows by noting that starting from π_K can only increase the queue size *throughout* the whole process for *all* of D queues, compared to starting from empty queues, and thus the number of visits to V_φ can only increase when starting from π_K in compared to starting from state 0 (this can be formalized using a standard coupling argument).
- Since walks from the stationary distribution remain at the stationary distribution, we can assume w.l.o.g. that the time interval is $[1, T]$. Now, as a thought experiment, we can decompose the process as follows. We first determine the number of steps each of the D chains take during time interval $[1, T]$; let c_j denote the number of steps taken in the j-th chain. Then we take c_j steps of random walk from the stationary

---

[11] One can see this by lower bounding the conductance of M_K and applying Cheeger's inequality.

distribution π_K for each copy of the chain M_K, and count the total number of visit to V_φ.

- Finally, we can view the process as taking a T-step random walk on M_K with "resets." Namely, we start from the stationary distribution π_K, take c_1 steps in M_K, "reset" the distribution to stationary distribution (by drawing an independent sample from π_K) and take c_2 more steps, and so on. At the end, we count the number of visits to V_φ, denoted by X, as an upper bound on the number of upset customers.

Intuitively, taking a random walk with resets injects additional randomness to the walk and thus we should expect at least as good concentration results. We formalize this intuition as the following Chernoff bound for Markov chains with "resets"—the proof of which follows relatively easy from recent Chernoff bounds for Markov chains [3] and is found Section 5.4—and use it to finish the proof of Proposition 1.

Theorem 2 (Chernoff Bounds for Markov Chains with Resets). *Let M be an ergodic finite Markov chain with state space S, stationary distribution π, and spectral expansion λ. Let $V \subset S$ and $\mu = \pi(V)$. Let $T, D \in \mathbb{N}$ and $1 = T_0 \leq T_1 \leq \cdots \leq T_D < T_{D+1} = T+1$. Let $(W_1, \ldots, W_T)$ denote a T-step random walk on M from stationary with resets at steps $T_1, \ldots, T_D$; that is, for every $j \in \{0, \ldots, D\}$, $W_{T_j} \leftarrow \pi$ and $W_{T_j+1}, \ldots, W_{T_{j+1}-1}$ are random walks from W_{T_j}. Let $X_i = 1$ iff $W_i \in V$ for every $i \in [T]$ and $X = \sum_{i=1}^{T} X_i$. We have*

$$\Pr\left[X \geq (1+\delta)\mu T\right] \leq \begin{cases} \exp\left\{-\Omega(\delta^2(1-\lambda)\mu T)\right\} & \text{for } 0 \leq \delta \leq 1 \\ \exp\left\{-\Omega(\delta(1-\lambda)\mu T)\right\} & \text{for } \delta \geq 1 \end{cases}$$

Now, recall that $1 - \lambda(M_K) = \Omega(1/(1-\alpha)^2)$ and $\pi_K(\varphi) = \beta^\varphi/(1-\beta^{K+1}) = (\alpha/1-\alpha)^\varphi/(1-\beta^{K+1})$. Theorem 2 says that for every possible $c_1, \ldots, c_D$ (corresponding to resetting time $T_j = \sum_{l=1}^{j} c_j + 1$),

$$\Pr\left[X \geq \frac{(1+\delta)(\alpha/1-\alpha)^\varphi T}{(1-\beta^{K+1})} \,\middle|\, c_1, \ldots, c_D\right] \leq \begin{cases} \exp\left\{-\Omega\left(\frac{\delta^2(\alpha/1-\alpha)^\varphi T}{(1-\alpha)^2(1-\beta^{K+1})}\right)\right\} & \text{for } 0 \leq \delta \leq 1 \\ \exp\left\{-\Omega\left(\frac{\delta(\alpha/1-\alpha)^\varphi T)}{(1-\alpha)^2(1-\beta^{K+1})}\right)\right\} & \text{for } \delta \geq 1 \end{cases}$$

Since X is an upper bound on the number of upset customers, and the above bound holds for every $c_1, \ldots, c_D$ and for every $K \geq t + T$, Proposition 1 follows by taking $K \rightarrow \infty$. □

5.4 Chernoff Bounds for Markov Chains with Reset

We now prove Theorem 2. The high level idea is simple—we simulate the resets by taking a sufficiently long "dummy" walk, where we "turn off" the counter on the number of visits to the state set V. However, formalizing this idea requires a more general version of Chernoff bounds that handles "time-dependent weight functions," which allows us to turn on/off the counter. Additionally, as we need to add long dummy walks, a multiplicative version (as opposed to an additive version) Chernoff bound is needed to derive meaningful bounds. We here rely on a recent generalized version of Chernoff bounds for Markov chains due to Chung, Lam, Liu and Mitzenmacher [3].

Theorem 3 ([3]). *Let M be an ergodic finite Markov chain with state space S, stationary distribution π, and spectral expansion λ. Let $\mathcal{W} = (W_1, \ldots, W_T)$ denote a T-step random walk on M starting from stationary distribution π, that is, $W_1 \leftarrow \pi$. For every $i \in [T]$, let $f_i : S \rightarrow [0, 1]$ be a weight function at step i with expected weight $\mathbb{E}_{v \leftarrow \pi}[f_i(v)] = \mu_i$. Let $\mu = \sum_i \mu_i$. Define the total weight of the walk $(W_1, \ldots, W_t)$ by $X \triangleq \sum_{i=1}^{t} f_i(W_i)$. Then*

$$\Pr\left[X \geq (1 + \delta)\mu\right] \leq \begin{cases} \exp\left\{-\Omega(\delta^2(1 - \lambda)\mu)\right\} & \text{for } 0 \leq \delta \leq 1 \\ \exp\left\{-\Omega(\delta(1 - \lambda)\mu)\right\} & \text{for } \delta > 1 \end{cases}$$

We now proceed to prove Theorem 2.

Proof of Theorem 2. We use Theorem 3 to prove the theorem. Let $f : S \rightarrow [0, 1]$ be an indicator function on $V \subset S$ (i.e., $f(s) = 1$ iff $s \in V$). The key component from Theorem 3 we need to leverage here is that the functions f_i can change over the time. Here, we shall design a very long walk $\mathcal{V}$ on M so that the marginal distribution of a specific collections of "subwalks" from $\mathcal{V}$ will be statistically close to $\mathcal{W}$. Furthermore, we design $\{f_i\}_{i \geq 0}$ in such a way that those "unused" subwalks will have little impact to the statistics we are interested in. In this way, we can translate a deviation bound on $\mathcal{V}$ to a deviation bound on $\mathcal{W}$. Specifically, let $T(\epsilon)$ be the mixing time for M (*i.e.* the number of steps needed for a walk to be ϵ-close to the stationary distribution in statistical distance). Here, we let $\epsilon \triangleq \exp(-DT)$ (ϵ is chosen in an arbitrary manner so long as it is sufficiently small). Given $1 = T_0 \leq T_1 \leq \cdots \leq T_D < T_{D+1} = T + 1$, we define $\mathcal{V}$ and f_i as follows: $\mathcal{V}$ will start from π and take $T_1 - 2$ steps of walk. In the mean time, we shall set $f_i = f$ for all $i < T_1$. Then we "turn off" the function f_i while letting $\mathcal{V}$ keep walking for $T(\epsilon)$ more steps, *i.e.* we let $f_i = 0$ for all $T_1 \leq i \leq T_1 + T(\epsilon) - 1$. Intuitively, this means we let $\mathcal{V}$ take a long walk until it becomes close to π again. During this time, f_i is turned off so that we do not keep track of any statistics. After that, we "turn on" the function f_i again for the next $T_2 - T_1$ steps (*i.e.* $f_i = f$ for all $T_1 + T(\epsilon) \leq i \leq T_2 + T(\epsilon) - 1$, followed by turning f_i off for another $T(\epsilon)$ steps. We continue this "on-and-off" process until we walk through all T_j's.

Let $\mathcal{V}'$ be the subwalks of $\mathcal{V}$ with non-zero f_i. One can see that the statistical distance between $\mathcal{V}'$ and $\mathcal{W}$ is $\text{poly}(D, T)\exp(-DT) \leq \exp(-T + o(T))$. Thus, for any θ we have

$$\begin{aligned} \Pr\left[\sum_{w \in \mathcal{W}} f(w) \geq \theta\right] &\leq \Pr\left[\sum_{v' \in \mathcal{V}'} f(v') \geq \theta\right] + \exp(-T + o(T)) \\ &= \Pr\left[\sum_{v \in \mathcal{V}} f(v) \geq \theta\right] + \exp(-T + o(T)). \end{aligned} \quad (2)$$

By letting $\theta = (1 + \delta)\mu T$ and using Theorem 3 to the right hand side of (2), we finish our proof. □

Acknowledgements. We are extremely grateful to an anonymous reviewer for pointing out a subtle missing implementation detail needed to make the recursion go through.

References

1. Ajtai, M.: Oblivious RAMs without cryptogrpahic assumptions. In: STOC, pp. 181–190 (2010)

2. Boneh, D., Mazieres, D., Popa, R.A.: Remote oblivious storage: Making oblivious RAM practical. CSAIL Technical Report: MIT-CSAIL-TR-2011-018 (2012)
3. Chung, K.M., Lam, H., Liu, Z., Mitzenmacher, M.: Chernoff-Hoeffding bounds for Markov chains: Generalized and simplified. In: Proceedings of the 29th International Symposium on Theoretical Aspects of Computer Science, STACS (2012)
4. Chung, K.-M., Liu, Z., Pass, R.: Statistically-secure ORAM with \tilde{O}(\log^2 n overhead. CoRR, abs/1307.3699 (2013)
5. Chung, K.-M., Pass, R.: A simple ORAM. Cryptology ePrint Archive, Report 2013/243 (2013)
6. Damgård, I., Meldgaard, S., Nielsen, J.B.: Perfectly secure oblivious RAM without random oracles. In: Ishai, Y. (ed.) TCC 2011. LNCS, vol. 6597, pp. 144–163. Springer, Heidelberg (2011)
7. Eager, D.L., Lazowska, E.D., Zahorjan, J.: Lazowska, and John Zahorjan. Adaptive load sharing in homogeneous distributed systems. IEEE Trans. Software Eng. 12(5), 662–675 (1986)
8. Gentry, C., Goldman, K.A., Halevi, S., Jutla, C.S., Raykova, M., Wichs, D.: Optimizing ORAM and using it efficiently for secure computation. In: Privacy Enhancing Technologies, pp. 1–18 (2013)
9. Gillman, D.: A Chernoff bound for random walks on expander graphs. SIAM Journal on Computing 27(4) (1997)
10. Goldreich, O.: Towards a theory of software protection and simulation by oblivious RAMs. In: STOC, pp. 182–194 (1987)
11. Goldreich, O., Ostrovsky, R.: Software protection and simulation on oblivious RAMs. J. ACM 43(3), 431–473 (1996)
12. Goodrich, M.T., Mitzenmacher, M.: Privacy-preserving access of outsourced data via oblivious RAM simulation. In: ICALP, vol. (2), pp. 576–587 (2011)
13. Goodrich, M.T., Mitzenmacher, M., Ohrimenko, O., Tamassia, R.: Privacy-preserving group data access via stateless oblivious RAM simulation. In: SODA, pp. 157–167 (2012)
14. Kahale, N.: Large deviation bounds for Markov chains. Combinatorics, Probability, and Computing 6(4) (1997)
15. Kushilevitz, E., Lu, S., Ostrovsky, R.: On the (in)security of hash-based oblivious RAM and a new balancing scheme. In: SODA, pp. 143–156 (2012)
16. Lezaud, P.: Chernoff-type bound for finite Markov chains. Annals of Applied Probability 8(3), 849–867 (1998)
17. Lu, S., Ostrovsky, R.: Distributed oblivious RAM for secure two-party computation. In: Sahai, A. (ed.) TCC 2013. LNCS, vol. 7785, pp. 377–396. Springer, Heidelberg (2013)
18. Maas, M., Love, E., Stefanov, E., Tiwari, M., Shi, E., Asanovic, K., Kubiatowicz, J., Song, D.: Phantom: Practical oblivious computation in a secure processor. In: CCS 2013, pp. 311–324. ACM (2013)
19. Mitzenmacher, M., Upfal, E.: Probability and Computing: Randomized Algorithms and Probabilistic Analysis. Cambridge University Press (2005)
20. Mitzenmacher, M.: The power of two choices in randomized load balancing. IEEE Trans. Parallel Distrib. Syst. 12(10), 1094–1104 (2001)
21. Mitzenmacher, M., Prabhakar, B., Shah, D.: Load balancing with memory. In: FOCS, pp. 799–808 (2002)
22. Mitzenmacher, M., Vadhan, S.: Why simple hash functions work: exploiting the entropy in a data stream. In: Proceedings of the Nineteenth Annual ACM-SIAM Symposium on Discrete Algorithms, SODA 2008, pp. 746–755 (2008)
23. Mitzenmacher, M., Vocking, B.: The asymptotics of selecting the shortest of two, improved. Proceedings of the Annual Allerton Conference on Communication Control and Computing 37, 326–327 (1999)

24. Ostrovsky, R., Shoup, V.: Private information storage (extended abstract). In: STOC, pp. 294–303 (1997)
25. Pinkas, B., Reinman, T.: Oblivious RAM revisited. In: Rabin, T. (ed.) CRYPTO 2010. LNCS, vol. 6223, pp. 502–519. Springer, Heidelberg (2010)
26. Shah, D., Prabhakar, B.: The use of memory in randomized load balancing. In: Proceedingsof the 2002 IEEE International Symposium on Information Theory, p. 125. IEEE (2002)
27. Shi, E., Chan, T.-H.H., Stefanov, E., Li, M.: Oblivious RAM with $O((\log N)^3)$ Worst-Case Cost. In: Lee, D.H., Wang, X. (eds.) ASIACRYPT 2011. LNCS, vol. 7073, pp. 197–214. Springer, Heidelberg (2011)
28. Stefanov, E., Shi, E.: Path O-RAM: An extremely simple oblivious RAM protocol. CoRR, abs/1202.5150v1 (2012)
29. Stefanov, E., Shi, E., Song, D.: Towards practical oblivious RAM. In: NDSS (2012)
30. Stefanov, E., Van Dijk, M., Shi, E., Fletcher, C., Ren, L., Yu, X., Devadas, S.: Path O-RAM: An extremely simple oblivious RAM protocol. In: CCS (2013)
31. Vvedenskaya, N.D., Dobrushin, R.L., Karpelevich, F.I.: Queueing system with selection of the shortest of two queues: An asymptotic approach. Problemy Peredachi Informatsii 32(1), 20–34 (1996)
32. Williams, P., Sion, R.: Usable PIR. In: NDSS (2008)
33. Williams, P., Sion, R., Carbunar, B.: Building castles out of mud: practical access pattern privacy and correctness on untrusted storage. In: ACM Conference on Computer and Communications Security, pp. 139–148 (2008)

A Implementation Details

This section discusses a number of implementation details in our algorithm.

The queue at the cache. We now describe how we may use a hash table and a standard queue (that could be encapsulated in commodity chips) to implement our queue with slightly non-standard behavior, which still suffices for our ORAM. Here, we only assume the hash table uses universal hash function and it resolves collisions by using a linked-list. To implement the INSERT(Block :b) procedure, we simply insert b to both the hash table and the queue. The key we use is b's value at the position map. Doing so we may make sure the maximum load of the hash table is $O(\log n)$ whp [22]. To implement FIND(int :i, word :p), we find the block b from the hash table. If it exists, return the block and delete it. However, for simplicity of implementation, we *do not* delete b at the queue. This introduces inconsistencies between the hash table and the queue, which we take care below in POPFRONT().

We now describe how we implement POPFRONT(). Here, we need to be careful with the inconsistencies. We first pop a block from the queue. Then we need to check whether the block is in hash table. If not, that means the block was already deleted earlier. In this case, POPFRONT() will not return anything (because we need a hard bound on the running time). Note that this does not effect the correctness of our analysis, since the queue size is indeed decreased by 1 for every PUT-BACK() action.

One can see that the above implementation relies only on standard hash table and queue, and INSERT() takes $O(1)$ time and the other two operations take $\omega(\log n)$ time (except with negligible probability).

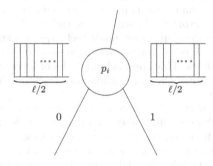

Fig. 1. In the FLUSH operation, we may imagine each bucket is splitted into two sub-arrays so that blocks that will travel to different subtrees are stored in different arrays. An overflow occurs when either sub-array's size reaches $\frac{\ell}{2}$.

Adaptive Security of Constrained PRFs[*]

Georg Fuchsbauer[1], Momchil Konstantinov[2],
Krzysztof Pietrzak[1], and Vanishree Rao[3]

[1] Institute of Science and Technology Austria
[2] London School of Geometry and Number Theory, UK
[3] UCLA, USA

Abstract. Constrained pseudorandom functions have recently been introduced independently by Boneh and Waters (Asiacrypt'13), Kiayias et al. (CCS'13), and Boyle et al. (PKC'14). In a standard pseudorandom function (PRF) a key k is used to evaluate the PRF on all inputs in the domain. Constrained PRFs additionally offer the functionality to delegate "constrained" keys k_S which allow to evaluate the PRF only on a subset S of the domain.

The three above-mentioned papers all show that the classical GGM construction (J.ACM'86) of a PRF from a pseudorandom generator (PRG) directly yields a constrained PRF where one can compute constrained keys to evaluate the PRF on all inputs with a given prefix. This constrained PRF has already found many interesting applications. Unfortunately, the existing security proofs only show selective security (by a reduction to the security of the underlying PRG). To achieve full security, one has to use complexity leveraging, which loses an exponential factor 2^N in security, where N is the input length.

The first contribution of this paper is a new reduction that only loses a quasipolynomial factor $q^{\log N}$, where q is the number of adversarial queries. For this we develop a new proof technique which constructs a distinguisher by interleaving simple guessing steps and hybrid arguments a small number of times. This approach might be of interest also in other contexts where currently the only technique to achieve full security is complexity leveraging.

Our second contribution is concerned with another constrained PRF, due to Boneh and Waters, which allows for constrained keys for the more general class of bit-fixing functions. Their security proof also suffers from a 2^N loss, which we show is inherent. We construct a meta-reduction which shows that any "simple" reduction of full security from a non-interactive hardness assumption must incur an exponential security loss.

Keywords: Constrained pseudorandom functions, full security, complexity leveraging, meta-reduction.

1 Introduction

PRFs. Pseudorandom functions (PRFs) were introduced by Goldreich, Goldwasser and Micali [GGM86]. A PRF is an efficiently computable keyed function

---

[*] Research supported by ERC starting grant (259668-PSPC).

P. Sarkar and T. Iwata (Eds.): ASIACRYPT 2014, PART II, LNCS 8874, pp. 82–101, 2014.

F: $\mathcal{K} \times \mathcal{X} \to \mathcal{Y}$, where $F(K, \cdot)$, instantiated with a random key $K \xleftarrow{\$} \mathcal{K}$, cannot be distinguished from a function randomly chosen from the set of all functions $\mathcal{X} \to \mathcal{Y}$ with non-negligible probability.

Constrained PRFs. The notion of constrained PRFs (CPRFs) was introduced independently by Boneh and Waters [BW13], Boyle, Goldwasser and Ivan [BGI14] and Kiayias, Papadopoulos, Triandopoulos and Zacharias [KPTZ13].[1]

A constrained PRF is defined with respect to a set system $\mathcal{S} \subseteq 2^{\mathcal{X}}$ and supports the functionality to "delegate" (short) keys that can only be used to evaluate the function $F: \mathcal{K} \times \mathcal{X} \to \mathcal{Y}$ on inputs specified by a subset $S \in \mathcal{S}$. Concretely, there is a "constrained" keyspace $\mathcal{K}_c$ and additional algorithms F.constrain: $\mathcal{K} \times \mathcal{S} \to \mathcal{K}_c$ and F.eval: $\mathcal{K}_c \times \mathcal{X} \to \mathcal{Y}$, which for all $k \in \mathcal{K}, S \in \mathcal{S}, x \in S$ and $k_S \leftarrow$ F.constrain(k, S), satisfy F.eval$(k_S, x) = F(k, x)$ if $x \in S$ and F.eval$(k_S, x) = \bot$ otherwise.

The GGM and the Boneh-Waters Construction. All the aforementioned papers [BW13, BGI14, KPTZ13] show that the classical GGM [GGM86] construction of the PRF GGM: $\{0,1\}^\lambda \times \{0,1\}^N \to \{0,1\}^\lambda$ from a length-doubling pseudorandom generator (PRG) G: $\{0,1\}^\lambda \to \{0,1\}^{2\lambda}$ directly gives a constrained PRF, where for any key K and input prefix $z \in \{0,1\}^{\leq N}$, one can generate a constrained key K_z that allows to evaluate GGM(K, x) for any x with prefix z. This simple constrained PRF has found many applications; apart from those discussed in [BW13, BGI14, KPTZ13], it can be used to construct so-called "punctured" PRFs, which are a key ingredient in almost all the recent proofs of indistinguishability obfuscation [SW14, BCPR13, HSW14].

Boneh and Waters [BW13] construct a constrained PRF for a much more general set of constraints, where one can delegate keys that fix any subset of bits of the input (not just the prefix, as in GGM). The construction is based on leveled multilinear maps [GGH13] and its security is proven under a generalization of the decisional Diffie-Hellman assumption.

Security of Constrained PRFs. The security definition for normal PRFs is quite intuitive. One considers two experiments: the "real" experiment and the "random" experiment, in both of which an adversary A gets access to an oracle $\mathcal{O}(\cdot)$ and then outputs a bit. In the real experiment $\mathcal{O}(\cdot)$ implements the PRF $F(K, \cdot)$ using a random key; in the random experiment $\mathcal{O}(\cdot)$ implements a random function. The PRF is secure if every efficient A outputs 1 in both experiments with (almost) the same probability.

Defining the security of constrained PRFs requires a bit more thought. We want to give an adversary access not only to $F(K, \cdot)$, but also to the constraining function F.constrain$(K, \cdot)$. But now we cannot expect the values $F(K, \cdot)$ to look random, as an adversary can always ask for a key $K_S \leftarrow$ F.constrain(K, S) and then for any $x \in S$ check whether $F(K, x) = $ F.eval(K_S, x).

---

[1] The name "constrained PRF" is from [BW13]; in [KPTZ13] and [BGI14] these objects are called "delegatable PRFs" and "functional PRFs", respectively. In this paper we follow the naming and notation from [BW13].

Instead, security is formalized by defining the experiments in two phases. In the first phase of both experiments the adversary gets access to the same oracles $F(K, \cdot)$ and $F.\text{constrain}(K, \cdot)$. The experiments differ in a second phase, where the adversary chooses some challenge query x^*. In the real experiment the adversary then obtains $F(K, x^*)$, whereas in the random experiment she gets a random value. Intuitively, when no efficient adversary can distinguish these two games, this means that the outputs of $F(K, \cdot)$ look random on all points that the adversary cannot compute by herself using the constrained keys she has received so far.

Selective vs. Full Security. In the above definition we let the adversary choose the challenge input x^* after she gets access to the oracles. This is the notion typically considered, and it is called "full security" or "adaptive security". One can also consider a weaker "selective security" notion, where the adversary must choose x^* before getting access to the oracles.

The reason to consider selective security notions, not only here, but also for other objects like identity-based encryption [BF01, BB04, AFL12] is that it is often much easier to achieve. Although there exists a simple generic technique called "complexity leveraging", which translates any selective security guarantee into a security bound for full security, this technique (which really just consists of guessing the challenge) typically loses an exponential factor (in the length of the challenge) in the quality of the reduction, often making the resulting security guarantee meaningless for practical parameters.

1.1   Our Contributions

All prior works [BW13, BGI14, KPTZ13] only show selective security of the GGM constrained PRF, and [BW13] also only give a selective-security proof for their bit-fixing constrained PRF. In this paper we investigate the full security of these two constructions. For GGM we achieve a positive result, giving a reduction that only loses a quasipolynomial factor. For the Boneh-Waters bit-fixing CPRF we give a negative result, showing that for a large class of reductions, an exponential loss is necessary. We now elaborate on these results.

A Quasipolynomial Reduction for GGM. To prove full security of GGM: $\{0,1\}^\lambda \times \{0,1\}^N \to \{0,1\}^\lambda$, the "standard" proof proceeds in two steps (we give a precise statement in Proposition 2).

1. A guessing step (a.k.a. complexity leveraging), which reduces full to selective security. This step loses an exponential factor 2^N in the input length N.
2. Now one applies a hybrid argument which loses a factor $2N$.

The above two steps transform an adversary A_f that breaks the full security of GGM with advantage ϵ into a new adversary that breaks the security of the underlying pseudorandom generator G (used to construct the GGM function) with advantage $\epsilon/(2N \cdot 2^N)$. As a consequence, even if one makes a strong exponential hardness assumption on the PRG G, one must use a PRG whose domain is $\Theta(N)$ bits in order to get any meaningful security guarantee.

The reason for the huge security loss is the first step, in which one guesses the challenge $x^* \in \{0,1\}^N$ the adversary will choose, which is correct with probability 2^{-N}. To avoid this exponential loss, one must avoid guessing the entire x^*. Our new proof also consists of a guessing step followed by a hybrid argument.

1. A guessing step, where (for some ℓ) we guess which of the adversary's queries will be the *first* one that agrees with x^* in the first ℓ positions.[2] This step loses a factor q, which denotes the number of queries made by the adversary.
2. A hybrid argument which loses a constant factor 3.

The above two steps only lose a factor $3q$. Unfortunately, after one iteration of this approach we do not get a distinguisher for G right away. At a high level, these two steps achieve the following: We start with two games which in some sense are at distance N from each other, and we end up with two games which are at distance $N/2$. We can iterate the above process $n := \log N$ times to end up with games at distance $N/2^n = 1$. Finally, from any distinguisher for games at distance 1 we can get a distinguisher for the PRG G with the same advantage. Thus, starting from an adversary against the full security of GGM with advantage ϵ, we get a distinguisher for the PRG with advantage $\epsilon/(3q)^{\log N}$.

We can optimize this by combining this approach with the original proof, and therby obtain a quasipolynomial loss of $2q \log q \cdot (3q)^{\log N - \log \log q}$. To give some numerical example, let the input length be $N = 2^{10} = 1024$ and the number of queries be $q = 2^{32}$. Then we get a loss of $2q \log q \cdot (3q)^{\log N - \log \log q} = 2 \cdot 2^{32} \cdot 32 \cdot (3 \cdot 2^{32})^{10-5} = 2^{198} \cdot 3^5 \leq 2^{206}$, whereas complexity leveraging loses $2N2^N = 2^{1035}$.

Although our proof is somewhat tailored to the GGM construction, the general "fine-grained" guessing approach outlined above might be useful to improve the bounds for other constructions (like CPRFs, and even IBE schemes) where currently the only proof technique that can be applied is complexity leveraging.

A Lower Bound for the Boneh-Waters CPRF and Hofheinz's Construction. We then turn our attention to the bit-fixing constrained PRF by Boneh and Waters [BW13]. For this construction too, complexity leveraging— losing an exponential factor—is the only known technique to prove full security. We give strong evidence that this is inherent (even when the construction is only used as a prefix-fixing CPRF).

Concretely, we prove that every "simple" reduction (which runs the adversary once without rewinding; see Sect. 5.2) of the full security of this scheme from any decisional (and thus also search) assumption must lose an exponential factor. Our proof is a so-called meta-reduction [BV98, Cor02, FS10], showing that any reduction that breaks the underlying assumption when given access to any adversary that breaks the CPRF, could be used to break the underlying assumption without the help of an adversary.

---

[2] This guessing is somewhat reminiscent of a proof technique from [HW09].

This impossibility result is similar to existing results, the closest one being a result of Lewko and Waters [LW14] ruling out security proofs without exponential loss for so-called "prefix-encryption" schemes (which satisfy some special properties). Other related results are those of Coron [Cor02] and Hofheinz et al. [HJK12], which show that security reductions for certain signature schemes must lose a factor polynomial in the number of signing queries.

The above impossibility proofs are for public-key objects, where a public key that uniquely determines the input/output distribution of the object. This property is crucially used in the proof, wherein one first gets the public key and then runs the reduction, rewinding the reduction multiple times to the point right after the public key has been received.

As we consider a secret-key primitive, the above approach seems inapplicable. We overcome this by observing that for the Boneh-Waters CPRF we can initially make some fixed "fingerprint" queries, which then uniquely determine the remaining outputs. We can therefore use the responses to these fingerprint queries instead of a public key as in [LW14].

Hofheinz [Hof14] has (independently and concurrently with us) investigated the adaptive security of bit-fixing constrained PRFs. He gives a new construction of such PRFs which is more sophisticated than the Boneh-Waters construction, and for which he can give a security reduction that only loses a polynomial factor. The main tool that allows Hofheinz to overcome our impossibility result is the use of a random oracle $H(\cdot)$. Very informally, instead of evaluating the PRF on an input X, it is evaluated on $H(X)$ which forces an attacker to make every query X explicit. Unfortunately, this idea does not work directly as it destroys the structure of the preimages, and thus makes the construction of short delegatable keys impossible. Hofheinz deals with this problem using several other ideas.

2 Preliminaries

For $a \in \mathbb{N}$, we let $[a] := \{1, 2, \ldots, a\}$ and $[a]_0 := \{0, 1, \ldots, a\}$. By $\{0,1\}^{\leq a} = \bigcup_{i \leq a} \{0,1\}^i$ we denote the set of bitstrings of length at most a, including the empty string $\emptyset$. By U_a we denote the random variable with uniform distribution over $\{0,1\}^a$. We denote sampling s uniformly from a set S by $s \xleftarrow{\cdot} S$. We let $x\|y$ denote the concatenation of the bitstrings x and y. For sets $\mathcal{X}, \mathcal{Y}$, we denote by $\mathcal{F}[\mathcal{X}, \mathcal{Y}]$ the set of all functions $\mathcal{X} \rightarrow \mathcal{Y}$; moreover, $\mathcal{F}[a, b]$ is short for $\mathcal{F}[\{0,1\}^a, \{0,1\}^b]$. For $x \in \{0,1\}^*$, we denote by x_i the i-th bit of x, and by $x[i \ldots j]$ the substring $x_i\|x_{i+1}\| \ldots \|x_j$.

Definition 1 (Indistinguishability). *Two distributions X and Y are (ϵ, s)-indistinguishable, denoted $X \sim_{(\epsilon,s)} Y$, if no circuit D of size at most s can distinguish them with advantage greater than ϵ, i.e.,*

$$X \sim_{(\epsilon,s)} Y \iff \forall \mathsf{D}, |\mathsf{D}| \leq s : \ \big| \Pr[\mathsf{D}(X) = 1] - \Pr[\mathsf{D}(Y) = 1]\big| \leq \epsilon \ .$$

$X \sim_\delta Y$ denotes that the statistical distance of X and Y is δ (i.e., $X \sim_{(\delta,\infty)} Y$), and $X \sim Y$ denotes that they have the same distribution.

Definition 2 (PRG). *An efficient function* $G: \{0,1\}^\lambda \to \{0,1\}^{2\lambda}$ *is an* (ϵ, s)-*secure (length-doubling)* **pseudorandom generator** *(PRG) if*

$$G(U_\lambda) \sim_{(\epsilon,s)} U_{2\lambda} \ .$$

Definition 3 (PRF). *A keyed function* $F: \mathcal{K} \times \mathcal{X} \to \mathcal{Y}$ *is an* (ϵ, s, q)-*secure* **pseudorandom function** *if for all adversaries* A *of size at most* s *making at most* q *oracle queries*

$$\left| \Pr_{K \xleftarrow{\$} \mathcal{K}}[A^{F(K,\cdot)} \to 1] - \Pr_{f \xleftarrow{\$} \mathcal{F}[\mathcal{X}, \mathcal{Y}]}[A^{f(\cdot)} \to 1] \right| \le \epsilon \ .$$

Constrained Pseudorandom Functions. Following [BW13], we say that a function $F: \mathcal{K} \times \mathcal{X} \to \mathcal{Y}$ is a *constrained PRF* for a set system $\mathcal{S} \subseteq 2^{\mathcal{X}}$, if there is a *constrained-key space* $\mathcal{K}_c$ and algorithms

$$F.\text{constrain}: \mathcal{K} \times \mathcal{S} \to \mathcal{K}_c \quad \text{and} \quad F.\text{eval}: \mathcal{K}_c \times \mathcal{X} \to \mathcal{Y} \ ,$$

which for all $k \in \mathcal{K}, S \in \mathcal{S}, x \in S$ and $k_S \leftarrow F.\text{constrain}(k, S)$ satisfy

$$F.\text{eval}(k_S, x) = \begin{cases} F(k, x) & \text{if } x \in S \\ \bot & \text{otherwise} \end{cases}$$

That is, $F.\text{constrain}(k, S)$ outputs a key k_S that allows evaluation of $F(k, \cdot)$ on all $x \in S$.

Informally, a constrained PRF F is secure, if no efficient adversary can distinguish $F(k, x^*)$ from random, even given access to $F(k, \cdot)$ and $F.\text{constrain}(k, \cdot)$ which he can query on all $x \ne x^*$ and $S \in \mathcal{S}$ where $x^* \notin S$, respectively. We will always assume that $\mathcal{S}$ contains all singletons, i.e., $\forall x \in \mathcal{X} : \{x\} \in \mathcal{S}$; this way we do not have to explicitly give access to $F(k, \cdot)$ to an adversary, as $F(k, x)$ can be learned by querying for $k_x \leftarrow F.\text{constrain}(k, \{x\})$ and computing $F.\text{eval}(k_x, x)$.

We distinguish between selective and full security. In the selective security game the adversary must choose the challenge x^* before querying the oracles. Both games are parametrized by the maximum number q of queries the adversary makes, of which the last query is the challenge query.

Exp$_{\text{CPRF}}^{\text{sel}}$(A, F, b, q)	**Exp**$_{\text{CPRF}}^{\text{full}}$(A, F, b, q)	**Oracle** $\mathcal{O}(S)$
$K \xleftarrow{\$} \mathcal{K}, \hat{S} := \emptyset, c := 0$	$K \xleftarrow{\$} \mathcal{K}, \hat{S} := \emptyset, c := 0$	if $c = q - 1$, return $\bot$
$x^* \leftarrow A$	$A^{\mathcal{O}(\cdot)}$	$c := c + 1$
$A^{\mathcal{O}(\cdot)}$	$x^* \leftarrow A$	$\hat{S} := \hat{S} \cup S$
$C_0 \xleftarrow{\$} \mathcal{Y}, C_1 := F(K, x^*)$	$C_0 \xleftarrow{\$} \mathcal{Y}, C_1 := F(K, x^*)$	$k_S \leftarrow F.\text{constrain}(K, S)$
A gets C_b	A gets C_b	return k_S
$\tilde{b} \leftarrow A$	$\tilde{b} \leftarrow A$	
if $x^* \in \hat{S}$, return 0	if $x^* \in \hat{S}$, return 0	
return $\tilde{b}$	return $\tilde{b}$	

For atk $\in \{\text{sel}, \text{full}\}$ we define A's advantage as

$$\text{Adv}_F^{\text{atk}}(A, q) = 2 \left| \Pr_{b \xleftarrow{\$} \{0,1\}}[\text{\textbf{Exp}}_{\text{CPRF}}^{\text{atk}}(A, F, b, q) = b] - \tfrac{1}{2} \right| \tag{1}$$

and denote with

$$\mathsf{Adv}_\mathsf{F}^\mathsf{atk}(s,q) = \max_{\mathsf{A},|\mathsf{A}|\leq s} \mathsf{Adv}_\mathsf{F}^\mathsf{atk}(\mathsf{A},q)$$

the advantage of the best q-query adversary of size at most s.

Definition 4 (Security of CPRFs). *A constrained PRF* F *is*

- **selectively** (ϵ, s, q)-**secure** *if* $\mathsf{Adv}_\mathsf{F}^\mathsf{sel}(s,q) \leq \epsilon$ *and*
- **fully** (ϵ, s, q)-**secure** *if* $\mathsf{Adv}_\mathsf{F}^\mathsf{full}(s,q) \leq \epsilon$.

Remark 1 (CCA1 vs. CCA2 security). In the selective and full security notion, we assume that the challenge query x^* is only made at the very end, when A has no longer access to the oracle (this is reminiscent of CCA1 security). All our positive results hold for stronger notions (reminiscent to CCA2 security) where A still has access to $\mathcal{O}(\cdot)$ after making the challenge query, but may not query on any S where $x^* \in S$.

Remark 2 (Multiple challenge queries). We only allow the adversary one challenge query. As observed in [BW13], this implies security against any $t > 1$ challenge queries, losing a factor of t in the distinguishing advantage, by a standard hybrid argument.

Using what is sometimes called "complexity leveraging", one can show that selective security implies full security: given an adversary A against full security, we construct a selective adversary B, which at the beginning *guesses* the challenge x^*, which it outputs, then runs the A and aborts if the challenge A eventually outputs is different from x^*. The distinguishing advantage drops thus by a factor of the domain size $|\mathcal{X}|$. We prove the following in the full version [FKPR14].

Lemma 1 (Complexity leveraging). *If a constrained PRF* $\mathsf{F}: \mathcal{K} \times \mathcal{X} \to \mathcal{Y}$ *is* (ϵ, s, q)-**selectively** *secure, then it is* $(\epsilon|\mathcal{X}|, s', q)$-**fully** *secure (where* $s' = s - O(\log|\mathcal{X}|)$), *i.e.,*

$$\mathsf{Adv}_\mathsf{F}^\mathsf{full}(s',q) \leq |\mathcal{X}| \cdot \mathsf{Adv}_\mathsf{F}^\mathsf{sel}(s,q) \ .$$

3   The GGM Construction

The GGM construction, named after its inventors Goldreich, Goldwasser and Micali [GGM86], is a keyed function $\mathsf{GGM}^\mathsf{G}: \{0,1\}^\lambda \times \{0,1\}^* \to \{0,1\}^\lambda$ from any length-doubling PRG $\mathsf{G}: \{0,1\}^\lambda \to \{0,1\}^{2\lambda}$, recursively defined as

$$\mathsf{GGM}(K_\emptyset, x) = K_x \ , \quad \text{where} \quad \forall z \in \{0,1\}^{\leq N-1}: K_{z\|0}\|K_{z\|1} = \mathsf{G}(K_z) \quad (2)$$

(cf. Fig. 1). In [GGM86] it is shown that when the inputs are restricted to $\{0,1\}^N$ then GGM^G is a secure PRF if G is a secure PRG. Their proof is one of the first applications of the so-called hybrid argument.[3] The proof loses a factor of $q \cdot N$ in distinguishing advantage, where q is the number of queries. We provide it in the full version [FKPR14].

---

[3] The first application is in the "probabilistic encryption" paper [GM84].

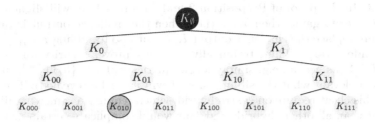

Fig. 1. Illustration of the GGM PRF. Every left child $K_{z\|0}$ of a node K_z is defined as the first half of $G(K_z)$, the right child $K_{z\|1}$ as the second half. The circled node corresponds to $GGM(K_\emptyset, 010)$.

Proposition 1 (GGM is a PRF [GGM86]). *If* $G: \{0,1\}^\lambda \to \{0,1\}^{2\lambda}$ *is an* (ϵ_G, s_G)*-secure PRG then (for any* N, q*)* $GGM^G: \{0,1\}^\lambda \times \{0,1\}^N \to \{0,1\}^\lambda$ *is an* (ϵ, s, q)*-secure PRF with*

$$\epsilon = \epsilon_G \cdot q \cdot N \quad and \quad s = s_G - O(q \cdot N \cdot |G|) \ .$$

3.1 GGM Is a Constrained PRF

As observed recently by three works independently [BW13, BGI14, KPTZ13], the GGM construction can be used as a constrained PRF for the set $\mathcal{S}_{pre}$ defined as

$$\mathcal{S}_{pre} = \{S_p \ : \ p \in \{0,1\}^{\leq N}\} \ , \quad \text{where} \quad S_p = \{p\|z \ : \ z \in \{0,1\}^{N-|p|}\} \ .$$

Thus, given a key K_p for the set S_p, one can evaluate $GGM^G(K, \cdot)$ on all inputs with prefix p. Formally, the constrained PRF with key $K = K_\emptyset$ is defined using (2) as follows:

$$GGM^G.\mathsf{constrain}(K_\emptyset, p) = GGM^G(K_\emptyset, p) = K_p$$
$$GGM^G.\mathsf{eval}(K_p, x = p\|z) = GGM^G(K_p, z) = K_x$$

Remark 3. When the domain is defined as $\mathcal{X} := \{0,1\}^*$ as for eq. (2) then the GGM construction is a secure *prefix-free* PRF, which means that none of the adversary's queries can be a prefix of another query (see [FKPR14]). One might be tempted to think that this fact together with the fact that constrained-key derivation is simply the GGM function itself, already implies that it is a secure constrained PRF. Unfortunately, this is not sufficient, as the (selective and full) security notions for CPRFs do allow queries that are prefixes of previous queries.

The *selective* security of this construction can be proven using a standard hybrid argument, losing only a factor of $2N$ in the distinguishing advantage. Proving full security seems much more challenging, and prior to our work it was only achieved by complexity leveraging (see Lemma 1), which loses an additional exponential factor 2^N in the distinguishing advantage, as stated in Proposition 2 below.

Remark 4. In the proof of Proposition 2 and Theorem 1 we will slightly cheat, as in the security game when $b = 0$ (i.e., when the challenge output is random) we not only replace the challenge output K_{x^*}, but also its sibling $K_{x^*[1...N-1]\overline{x}_N^*}$, with a random value. Thus, technically this only proves security for inputs of length $N - 1$ (as we can e.g. simply forbid queries $x\|0, x \in \{0,1\}^{N-1}$, in which case it is irrelevant what the sibling is, as it will never be revealed). The proofs without this cheat require one extra hybrid, which requires a somewhat different treatment than all others hybrids and thus would complicate certain proofs and definitions. Hence, we chose to not include it. The bounds stated in Proposition 2 and Theorem 1 are the bounds we get *without* this cheat.

Proposition 2. *If* $G: \{0,1\}^\lambda \to \{0,1\}^{2\lambda}$ *is an* (ϵ_G, s_G)-*secure PRG then (for any* N, q) $GGM^G: \{0,1\}^N \to \{0,1\}^\lambda$ *is a constrained PRF for* $\mathcal{S}_{pre}$ *which is*

1. **selectively** (ϵ, s, q)-*secure, with*

$$\epsilon = \epsilon_G \cdot 2N \quad and \quad s = s_G - O(q \cdot N \cdot |G|) \; ;$$

2. **fully** (ϵ, s, q)-*secure, with*

$$\epsilon = \epsilon_G \cdot 2^N 2N \quad and \quad s = s_G - O(q \cdot N \cdot |G|) \; .$$

Full security as stated in Item 2. of the proposition follows from selective security (Item 1.) by complexity leveraging as explained in Lemma 1. To prove selective security, we let H_0 be the real game for selective security and let H_{2N-1} be the random game, that is, where K_{x^*} is random. We then define intermediate hybrid games $H_1, \ldots, H_{2N-2}$ by embedding random values along the path to K_{x^*}. In particular, in hybrid H_i, for $1 \le i \le N$, the nodes $K_\emptyset, K_{x_1^*}, \ldots, K_{x^*[1...i]}$ are random and for $N + 1 \le i \le 2N - 1$ the nodes $K_\emptyset, K_{x_1^*}, \ldots, K_{x^*[1...2N-1-i]}$ and K_{x^*} are random. Thus two consecutive games H_i, H_{i+1} differ in one node that is real in one game and random in the other, and moreover the parent of that node is random, meaning we can embed a PRG challenge. From any distinguisher for two consecutive games we thus get a distinguisher for the PRG G with the same advantage. (A formal proof can be found in [FKPR14].)

This hybrid argument only loses a factor $2N$ in distinguishing advantage, but complexity leveraging loses a huge factor 2^N. In the next section we show how to prove full security avoiding such an exponential loss.

4   Full Security with Quasipolynomial Loss

Theorem 1. *If* $G: \{0,1\}^\lambda \to \{0,1\}^{2\lambda}$ *is an* (ϵ_G, s_G)-*secure PRG then (for any* N, q) $GGM^G: \{0,1\}^N \to \{0,1\}^\lambda$ *is a* **fully** (ϵ, s, q)-*secure constrained PRF for* $\mathcal{S}_{pre}$, *where*

$$\epsilon = \epsilon_G \cdot (3q)^{\log N} \quad and \quad s = s_G - O(q \cdot N \cdot |G|) \; .$$

At the end of this section we will sketch how to combine the proof of this theorem with the standard complexity leveraging proof from Proposition 2 to get a smaller loss of $\epsilon = \epsilon_G \cdot 2q \log q \cdot (3q)^{\log N - \log \log q}$.

Proof Idea. We can view the real and the random game for CPRF security as having distance N, in the sense that from the only node in which they differ (which is the challenge node K_{x^*}) we have to walk up N nodes until we reach a node that was chosen uniformly at random (which here is the root $K_\emptyset$).

As outlined in Sect. 1.1, our goal is to halve that distance. For this, we could define two intermediate hybrids which are defined as the real and the random games, except that the node half way down the path to x^*, i.e., $K_{x^*[1...N/2]}$, is a random node. This is illustrated in Fig. 2, where a row depicts the path from the root, labeled '0', to x^*, labeled '8' and where dark nodes correspond to random values. The path at the top of the figure is the real and the one at the bottom is the random game (ignore anything in the boxes for now), and the intermediate hybrids are the 2nd and the 3rd path. Of these 4 hybrids, each pair of consecutive hybrids has the following property: they differ in one node and its distance to the closest random node above is $N/2$.

There is a problem with this approach because the intermediate hybrid games we have just constructed are not even well-defined, as the value $x^*[1...N/2]$ is only known when the adversary makes his challenge query. This is also the case for x^* itself, but K_{x^*} only needs to be computed once x^* is queried; in contrast, $K_{x^*[1...N/2]}$ could have been computed earlier in the game, if the value of some constrained-key query is a descendant of it. In order to avoid possible inconsistencies, we do the following: we guess which of the adversary's queries will be the first one with a prefix $x^*[1...N/2]$. As there are at most q queries and there always exists a query with this property (at latest the challenge query itself), the probability of guessing correctly is $1/q$. If we guess correctly then the node $x^*[1...N/2]$ is known precisely when the value $K_{x^*[1...N/2]}$ is computed for the first time and we can correctly simulate the game. If our guess was wrong, we abort.

Assuming an attacker can distinguish between the real and the random game, there must be two consecutive hybrids of the 4 hybrids that it can distinguish with at least one third of his original advantage. Between these two hybrids, which differ in one node d, we can again embed two intermediate hybrids, which have a random value half way between d and the closest random node above (cf. the outer box in Fig. 2). We continue to do so until we reach two hybrids where there is a random node immediately above the differing node. A distinguisher between two such games can then be used to break the PRG.

Neighboring Sets with Low Weight. Before starting with the proof, we introduce some notation. It will be convenient to work with ternary numbers, which we represent as strings of digits from $\{0, 1, 2\}$ within angular brackets $\langle\ldots\rangle$. We denote repetition of digits as $0_n = 0\ldots0$ (n times). Addition will also be in ternary, e.g., $\langle 202\rangle + \langle 1\rangle = \langle 210\rangle$.

Let $N = 2^n$ be a power of 2. In the proof of Theorem 1 we will construct $3^n + 1$ subsets $\mathcal{S}_{\langle 0\rangle}, \ldots, \mathcal{S}_{\langle 10_n\rangle} \subset \{0, \ldots, N\}$. These sets will define the positions in the path to the challenge where we make random guesses in a particular hybrid. The following definition measures how "close" sets (that differ in one element) are and will be useful in defining neighboring hybrids.

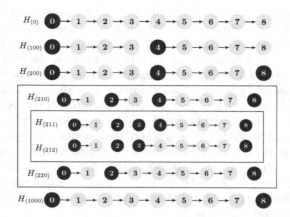

Fig. 2. Concrete example ($n = 3$) illustrating the iterative construction of hybrids in Theorem 1

Definition 5 (Neighboring sets). *For $k \in \mathbb{N}^+$, sets $\mathcal{S}, \mathcal{S}' \subset \mathbb{N}^0$, $\mathcal{S} \neq \mathcal{S}'$ are k-neighboring if*

1. $\mathcal{S} \bigtriangleup \mathcal{S}' := (\mathcal{S} \cup \mathcal{S}') \setminus (\mathcal{S} \cap \mathcal{S}') = \{d\}$ *for some $d \in \mathbb{N}^0$, i.e., they differ in exactly one element d.*
2. $d - k \in \mathcal{S}$.
3. $\forall i \in [k-1] : d - i \notin \mathcal{S}$.

We define the first set (with index $0 = \langle 0 \rangle$) and the and last set (with index $3^n = \langle 10_n \rangle$) as

$$\mathcal{S}_{\langle 0 \rangle} := \{0\} \quad \text{and} \quad \mathcal{S}_{\langle 10_n \rangle} := \{0, N\} . \quad (3)$$

(They will correspond to the real game, where only the root at depth '0' is random, and the random game, where the value for x^* at depth N is random too.) The remaining intermediate sets are defined recursively as follows. For $\ell = 0, \ldots, n$, we define the ℓ-th level of sets to be all the sets of the form $\mathcal{S}_{\langle ?0_{n-\ell} \rangle}$ (i.e., whose index in ternary ends with $(n - \ell)$ zeros). Thus, $\mathcal{S}_{\langle 0 \rangle}$ and $\mathcal{S}_{\langle 10_n \rangle}$ are the (only) level-0 sets.

Let $\mathcal{S}_I, \mathcal{S}_{I'}$ be two consecutive level-ℓ sets, by which we mean that $I' = I + \langle 10_{n-\ell} \rangle$. By construction, these sets will differ in exactly one element $\{d\}$ (i.e., $\mathcal{S}_I \neq \mathcal{S}_{I'}$; and $\mathcal{S}_I \cup \{d\} = \mathcal{S}_{I'}$ or $\mathcal{S}_{I'} \cup \{d\} = \mathcal{S}_I$). Then the two level-$(\ell + 1)$ sets between the level-ℓ sets $\mathcal{S}_I, \mathcal{S}_{I'}$ are defined as

$$\mathcal{S}_{I+\langle 10_{n-(\ell+1)} \rangle} := \mathcal{S}_I \cup \{d - \tfrac{N}{2^{\ell+1}}\} \quad \text{and} \quad \mathcal{S}_{I'-\langle 10_{n-(\ell+1)} \rangle} := \mathcal{S}_{I'} \cup \{d - \tfrac{N}{2^{\ell+1}}\} . \quad (4)$$

A concrete example for $N = 2^n = 2^3 = 8$ is illustrated in Fig. 2 (where the dark nodes of H_I correspond to $\mathcal{S}_I$).

An important fact we will use is that consecutive level-ℓ sets are $(N/2^\ell)$-neighboring (see Definition 5); in particular, consecutive level-n sets (the 4 lines in the box in Fig. 2 illustrate 4 consecutive sets) are thus 1-neighboring, i.e.,

$$\forall\, I \in \{\langle 0 \rangle, \ldots, \langle 2_n \rangle\}:\ \mathcal{S}_I \,\Delta\, \mathcal{S}_{I+\langle 1 \rangle} = \{d\} \quad \text{and} \quad d - 1 \in \mathcal{S}_I \ . \tag{5}$$

Proof of Theorem 1. Below we prove two lemmata (2 and 3) concerning the games defined in Fig. 3, from which the theorem follows quite immediately. As the games and the lemmata are rather technical, we first intuitively explain what is going on, going through a concrete example as illustrated in Fig. 2.

To prove the theorem, we assume that there exists an adversary A_f that breaks the *full* security of GGM^G with some advantage ϵ, and from this, we want to construct a distinguisher for G with advantage at least $\epsilon/(3q)^n$, where $n = \log N$. Like in the proof of Proposition 2, we can think of the two games that A_f distinguishes as the games where we let A_f query GGM^G, but along the path from the root $K_\emptyset$ down to the challenge K_{x^*} the nodes are either computed by G or they are random values. The position of the random values are defined by the set $\mathcal{S}_{\langle 0 \rangle} = \{0\}$ for the real game and by $\mathcal{S}_{\langle 10_n \rangle} = \{0, N\}$ for the random game: in both cases the root $K_\emptyset$ is random, and in the latter game the final output K_{x^*} is also random. We call these two games $H_{\langle 0 \rangle}^\emptyset$ and $H_{\langle 10_n \rangle}^\emptyset$, and they correspond to the games defined in Fig. 3 with $\mathcal{P} = \emptyset$, and $I = \langle 0 \rangle$ and $\langle 10_n \rangle$, respectively). As just explained, they satisfy

$$H_{\langle 0 \rangle}^\emptyset \sim \mathbf{Exp}_{\mathsf{CPRF}}^{\mathsf{full}}(A_f, GGM^G, 0, q) \quad \text{and} \quad H_{\langle 10_n \rangle}^\emptyset \sim \mathbf{Exp}_{\mathsf{CPRF}}^{\mathsf{full}}(A_f, GGM^G, 1, q) \ .$$

Thus, if A_f breaks the full security of GGM^G with advantage ϵ then

$$\left| \Pr[H_{\langle 0 \rangle}^\emptyset = 1] - \Pr[H_{\langle 10_n \rangle}^\emptyset = 1] \right| \geq \epsilon \ . \tag{6}$$

In the proof of Proposition 2 we were able to "connect" the real and random experiments H_0 and H_{2N-1} via intermediate hybrids $H_1, \ldots, H_{2N-2}$, such that from a distinguisher for any two consecutive hybrids we can build a distinguisher for G with the same advantage.

We did this by using random values (instead of applying G) in some steps along the path from the root $K_\emptyset$ to the challenge K_{x^*}. Here we cannot use the same approach to connect $H_{\langle 0 \rangle}^\emptyset$ and $H_{\langle 10_n \rangle}^\emptyset$, as these games consider full (and not selective) security, where we learn x^* only at the very end, and thus "the path to x^*" is not even defined until the adversary makes the challenge query.

We could of course reduce the problem from the full to the selective setting by guessing x^* at the beginning like in the proof of Lemma 1, but this would lose a factor 2^N, which is what we want to avoid.

Instead of guessing the entire x^*, we will guess something easier. During the experiment $H_{\langle 0 \rangle}$ we have to compute at most q children $K_{z\|0}\|K_{z\|1} = G(K_z)$ of nodes at level $N/2 - 1$, i.e., $z \in \{0,1\}^{N/2-1}$. One of these K_z satisfies $z = x^*[1 \ldots N/2 - 1]$, that is, it lies on the path from the root $K_\emptyset$ to the challenge K_{x^*} (potentially this happens only at the very last query $x_q = x^*$). We randomly guess $q_{N/2} \xleftarrow{} [q]$ for which invocation of G this will be the case *for the first time*. Note that we have to wait until A_f makes its last query $x_q = x^*$ before we know whether our guess was correct. If the guess was wrong, we output 0; otherwise we output A_f's output. We will denote the position of the node down to which

Experiment $H_I^{\mathcal{P}}$

    // $I \in \{\langle 0 \rangle, \ldots, \langle 10_n \rangle\}$
    // $\mathcal{P} = \{p_1, \ldots, p_t\} \subseteq [N-1]$
    // $\mathcal{S}_I \subseteq \mathcal{P} \cup \{0, N\}$,
    // $\mathcal{S}_I$ defined by eq. (3) and (4).
$\forall x \in \{0,1\}^{\leq N} :\ K_x := \bot$
$K_\emptyset \xleftarrow{\$} \{0,1\}^\lambda$
    // Initialize counters:
$\forall j = 1 \ldots N - 1:\ c_j = 0$
    // Make a random guess for each
    // element in $\mathcal{P} = \{p_1, \ldots, p_t\}$:
$\forall j \in [t]:\ q_{p_j} \xleftarrow{\$} [q]$
    // $\mathsf{A_f}$ can make exactly q distinct
    // oracle queries $x_1, \ldots, x_q$;
    // the last (challenge) query
    // $x_q = x^*$ must be in $\{0,1\}^N$:
$\mathsf{A_f}^{\mathcal{O}(\cdot)}$
$\tilde{b} \leftarrow \mathsf{A_f}$
    // Only if guesses $q_{p_1}, \ldots, q_{p_t}$
    // were correct, return $\tilde{b}$,
    // otherwise return 0:
if $\forall p \in \mathcal{P}:\ x^*[1 \ldots p-1] = z_{p-1}$
    then return $\tilde{b}$
else return 0 **fi**

$\mathcal{O}(x = x[1 \ldots \ell])$

    // Return K_x if it is already defined:
if $K_x \neq \bot$ **then return** K_x **fi**
    // Get parent of K_x recursively:
$K_{x[1\ldots\ell-1]} := \mathcal{O}(x[1 \ldots \ell - 1])$
    // Increase counter for level $\ell - 1$:
$c_{\ell-1} = c_{\ell-1} + 1$
    // Compute K_x and its sibling using G,
    // unless its parent $K_{x[1\ldots\ell-1]}$ is a node
    // which we guessed will be on the path
    // from $K_\emptyset$ and K_{x^*} and as $\ell \in \mathcal{P}$ we
    // must use a random value at this level;
    // OR this is the challenge query $x_q = x^*$
    // and $N \in \mathcal{S}_I$, which means the answer
    // to the challenge is random:
if $(\ell \in \mathcal{P}$ **and** $c_{\ell-1} = q_{\ell-1})$
        OR $(x = x_q$ **and** $N \in \mathcal{S}_I)$
    $K_{x[1\ldots\ell-1]\|0} \| K_{x[1\ldots\ell-1]\|1} \xleftarrow{\$} U_{2\lambda}$
    // Store this node to check if guess
    // was correct later:
    $z_{\ell-1} = x[1 \ldots \ell - 1]$
else
    $K_{x[1\ldots\ell-1]\|0} \| K_{x[1\ldots\ell-1]\|1} := \mathsf{G}(K_{x[1\ldots\ell-1]})$
fi
return K_x

Fig. 3. Definition of the hybrid games from the proof of Theorem 1. The sets $\mathcal{S}_I$ are as in Equations (3) and (4). The hybrid $H_I^{\mathcal{P}}$ is defined like the full security game of a q-query adversary $\mathsf{A_f}$ against the CPRF $\mathsf{GGM}^{\mathsf{G}}$, but where we "guess", for any value $p \in \mathcal{P}$, at which point in the experiment the node at depth p on the path from the root $K_\emptyset$ to the challenge K_{x^*} is computed. (Concretely, the guess is that it's the c_{p-1}-th time we compute the children of a node at level $p-1$, we define the p level node $K_{x^*[1\ldots\ell]}$ on the path.) At a subset of these points, namely $\mathcal{S}_I$, we embed random values. The final output is 0 unless all guesses were correct, in which case we forward $\mathsf{A_f}$'s output.

our guessed query should equal the path to x^* as superscript of the hybrid H. The experiment just described corresponds thus to hybrid $H_{\langle 0 \rangle}^{\{N/2\}}$, as defined in Fig. 3.

The games $H_{\langle 0 \rangle}^{\{N/2\}}$ and $H_{\langle 10_n \rangle}^{\{N/2\}}$ behave *exactly* like $H_{\langle 0 \rangle}^{\emptyset}$ and $H_{\langle 10_n \rangle}^{\emptyset}$, except for the final output, which in the former two hybrids is set to 0 with probability $1 - 1/q$, and left unchanged otherwise (namely, if our random guess $q_{N/2} \xleftarrow{\$} [q]$ turns out to be correct, which we know after learning x^*). This implies

$$\Pr[H_{\langle 0 \rangle}^{\{N/2\}} = 1] = \Pr[H_{\langle 0 \rangle}^{\emptyset} = 1] \cdot \frac{1}{q} \quad \text{and} \quad \Pr[H_{\langle 10_n \rangle}^{\{N/2\}} = 1] = \Pr[H_{\langle 10_n \rangle}^{\emptyset} = 1] \cdot \frac{1}{q} ,$$

and with (6)

$$\left| \Pr[H_{\langle 0 \rangle}^{\{N/2\}} = 1] - \Pr[H_{\langle 10_n \rangle}^{\{N/2\}} = 1] \right| \geq \epsilon/q . \tag{7}$$

What did we gain? We paid a factor q in the advantage for aborting when our guess $q_{N/2}$ was wrong. What we gained is that when we guess correctly we know $x^*[1 \ldots N/2]$, i.e., the node half way in between the root and the challenge.

We use this fact to define two new hybrids $H_{\langle 10_{n-1}\rangle}^{\{N/2\}}, H_{\langle 20_{n-1}\rangle}^{\{N/2\}}$ which are defined like $H_{\langle 0\rangle}^{\{N/2\}}, H_{\langle 10_n\rangle}^{\{N/2\}}$, respectively, but where the children of $K_{x^*[1\ldots N/2-1]}$ are uniformly random instead of being computed by applying G to $K_{x^*[1\ldots N/2-1]}$.

Fig. 2 (ignoring the boxes for now) illustrates the path from $K_\emptyset$ to K_{x^*} in the hybrids $H_{\langle 0\rangle}^{\{4\}}, H_{\langle 100\rangle}^{\{4\}}, H_{\langle 200\rangle}^{\{4\}}, H_{\langle 1000\rangle}^{\{4\}}$ assuming the guessing was correct (a node with label i corresponds to $K_{x^*[1\ldots i]}$, dark nodes are sampled at random, and green ones by applying G to the parent).

By (7) we can distinguish the first from the last hybrid with advantage ϵ/q, and thus there are two consecutive hybrids in the sequence $H_{\langle 0\rangle}^{\{N/2\}}, H_{\langle 10_{n-1}\rangle}^{\{N/2\}}$, $H_{\langle 20_{n-1}\rangle}^{\{N/2\}}, H_{\langle 10_n\rangle}^{\{N/2\}}$ that we can distinguish with advantage at least $\epsilon/(3q)$. For concreteness, let us fix parameters $N = 8 = 2^3 = 2^n$ as in Fig. 2 and assume that this is the case for the last two hybrids in the sequence, i.e.,

$$| \Pr[H_{\langle 200\rangle}^{\{4\}} = 1] - \Pr[H_{\langle 1000\rangle}^{\{4\}} = 1]| \geq \epsilon/(3q) \ . \tag{8}$$

The central observation here is that the above guessing step (losing a factor q) followed by a hybrid argument (losing a factor 3) transformed a distinguishing advantage ϵ for two hybrids $H_{\langle 0\rangle}^{\emptyset}, H_{\langle 1000\rangle}^{\emptyset}$ which have random values embedded along the path from $K_\emptyset$ to K_{x^*} on positions defined by N-neighboring sets $\mathcal{S}_{\langle 0\rangle}, \mathcal{S}_{\langle 1000\rangle}$, into a distinguishing advantage of $\epsilon/(3q)$ for two hybrids that correspond to $N/2$-neighboring sets, e.g. $\mathcal{S}_{\langle 200\rangle}$ and $\mathcal{S}_{\langle 1000\rangle}$.

We can now iterate this approach, in each iteration losing a factor $3q$ in distinguishing advantage, but getting hybrids that correspond to sets of half the neighboring distance. After $n = \log N$ iterations we end up with hybrids that correspond to 1-neighboring sets, and can be distinguished with advantage $\epsilon/(3q)^n$. We will make this formal in Lemma 3 below. From any distinguisher for hybrids corresponding to two 1-neighboring sets we can construct a distinguisher for G with the same advantage, as formally stated in Lemma 2 below. Let's continue illustrating the approach using the hybrids illustrated in Fig. 2.

Recall that we assumed that we can distinguish $H_{\langle 200\rangle}^{\{4\}}$ and $H_{\langle 1000\rangle}^{\{4\}}$ as stated in eq. (8). We now embed hybrids corresponding to the sets $\mathcal{S}_{\langle 210\rangle}, \mathcal{S}_{\langle 220\rangle}$ in between, illustrated in the outer box in Fig. 2 (ignore the inner box for now). Since $\mathcal{S}_{\langle 200\rangle} \, \Delta \, \mathcal{S}_{\langle 1000\rangle} = \{4\}$, by eq. (4) for $\ell = 1$, we construct $\mathcal{S}_{\langle 200\rangle + \langle 10\rangle} = \mathcal{S}_{\langle 200\rangle} \cup \{4 - \frac{8}{2^2} = 2\}$ and $\mathcal{S}_{\langle 1000\rangle - \langle 10\rangle} = \mathcal{S}_{\langle 1000\rangle} \cup \{2\}$. We add this new element $\{2\}$ to the "guessing set" $\{4\}$, at the price of losing a factor q in distinguishing advantage compared to eq. (8):

$$| \Pr[H_{\langle 200\rangle}^{\{2,4\}} = 1] - \Pr[H_{\langle 1000\rangle}^{\{2,4\}} = 1]| \geq \epsilon/(3q^2) \ . \tag{9}$$

We can now consider the sequence of hybrids $H_{\langle 200\rangle}^{\{2,4\}}, H_{\langle 210\rangle}^{\{2,4\}}, H_{\langle 220\rangle}^{\{2,4\}}, H_{\langle 1000\rangle}^{\{2,4\}}$. There must be two consecutive hybrids that can be distinguished with advantage

$\epsilon/(3^2 q^2)$. Let's assume this is the case for the middle two.

$$\left| \Pr[H^{\{2,4\}}_{\langle 210 \rangle} = 1] - \Pr[H^{\{2,4\}}_{\langle 220 \rangle} = 1] \right| \geq \epsilon/(3^2 q^2) \ . \tag{10}$$

Now $\mathcal{S}_{\langle 210 \rangle} \bigtriangleup \mathcal{S}_{\langle 220 \rangle} = \{4\}$, and $4 - 8/2^3 = 3$, so we add $\{3\}$ to the guessing set losing another factor q:

$$\left| \Pr[H^{\{2,3,4\}}_{\langle 210 \rangle} = 1] - \Pr[H^{\{2,3,4\}}_{\langle 220 \rangle} = 1] \right| \geq \epsilon/(3^2 q^3) \ , \tag{11}$$

and can now consider the games $H^{\{2,3,4\}}_{\langle 210 \rangle}, H^{\{2,3,4\}}_{\langle 211 \rangle}, H^{\{2,3,4\}}_{\langle 212 \rangle}, H^{\{2,3,4\}}_{\langle 220 \rangle}$ as shown inside the two boxes in Fig. 2. Two consecutive hybrids in this sequence must be distinguishable with advantage at least $1/3$ of the advantage we had for the first and last hybrid in this sequence; let's assume this is the case for the last two, then:

$$\left| \Pr[H^{\{2,3,4\}}_{\langle 212 \rangle} = 1] - \Pr[H^{\{2,3,4\}}_{\langle 220 \rangle} = 1] \right| \geq \epsilon/(3^3 q^3) \ . \tag{12}$$

We have thus shown the existence of two games $H^{\mathcal{P}}_I$ and $H^{\mathcal{P}}_{I+\langle 1 \rangle}$ (what $\mathcal{P}$ and I are exactly is irrelevant for the rest of the argument) that can be distinguished with advantage $\epsilon/(3q)^n$. Any two consecutive (i.e., 1-neighboring) hybrids have the following properties (cf. eq. 5). They only differ in one node on the path to x^* and its parent node is random. Moreover, the position of the differing node is in the guessing set $\mathcal{P}$, meaning we know its position in the tree. Together, this means we can use a distinguisher between $H^{\mathcal{P}}_I$ and $H^{\mathcal{P}}_{I+\langle 1 \rangle}$ to break G: Given a challenge for G we embed it as the value of the differing node and, depending whether it was real or random, simulate one hybrid or the other. We formalize this in the following lemma, which is proven in the full version [FKPR14].

Lemma 2. *For any* $I \in \{\langle 0 \rangle, \dots, \langle 2_n \rangle\}, \mathcal{P} \subset \{1, \dots, N-1\}$ *where* $\mathcal{S}_I \cup \mathcal{S}_{I+\langle 1 \rangle} \subseteq \mathcal{P} \cup \{0, N\}$ *(so the games* $H^{\mathcal{P}}_{I+\langle 1 \rangle}, H^{\mathcal{P}}_I$ *are defined) the following holds. If*

$$\left| \Pr[H^{\mathcal{P}}_I = 1] - \Pr[H^{\mathcal{P}}_{I+\langle 1 \rangle} = 1] \right| = \delta$$

then G *is not a* (δ, s)-*secure PRG for* $s = |\mathsf{A}_\mathsf{f}| - O(q \cdot N \cdot |\mathsf{G}|)$.

Lemma 3. *For* $\ell \in \{0, \dots, n-1\}$, *any consecutive level-*ℓ *sets* $\mathcal{S}_I, \mathcal{S}_{I'}$ *(i.e.,* I, I' *are of the form* $\langle ?0_{n-\ell} \rangle$ *and* $I' = I + \langle 10_{n-\ell} \rangle$) *and any* $\mathcal{P}$ *for which the hybrids* $H^{\mathcal{P}}_I, H^{\mathcal{P}}_{I'}$ *are defined (which is the case if* $\mathcal{S}_I \cup \mathcal{S}_{I'} \subseteq \mathcal{P} \cup \{0, N\}$), *the following holds. If*

$$\left| \Pr[H^{\mathcal{P}}_I = 1] - \Pr[H^{\mathcal{P}}_{I'} = 1] \right| = \delta \tag{13}$$

then for some consecutive level-$(\ell+1)$ *sets* $J \in \{I, I + \langle 10_{n-\ell-1} \rangle, I + \langle 20_{n-\ell-1} \rangle\}$ *and* $J' = J + \langle 10_{n-\ell-1} \rangle$ *and some* $\mathcal{P}'$:

$$\left| \Pr[H^{\mathcal{P}'}_J = 1] - \Pr[H^{\mathcal{P}'}_{J'} = 1] \right| = \delta/(3q) \ . $$

The proof of Lemma 3 is in [FKPR14]. The theorem now follows from Lemmata 2 and 3 as follows. Assume a q-query adversary A_f breaks the full security of $\mathsf{GGM}^{\mathsf{G}}$ for domain $\mathcal{X} = \{0,1\}^{2^n}$ with advantage ϵ, which, as explained in the paragraph before eq. (6), means that we can distinguish the two level-0 hybrids $H^{\emptyset}_{\langle 0 \rangle}$ and $H^{\emptyset}_{\langle 10_n \rangle}$ with advantage ϵ. Applying Lemma 3 n times, we get that there exist consecutive level-n hybrids $H^P_I, H^P_{I+\langle 1 \rangle}$ that can be distinguished with advantage $\epsilon/(3q)^n$, which by Lemma 2 implies that we can break the security of G with the same advantage $\epsilon/(3q)^n$. This concludes the proof of Theorem 1.

To reduce the loss to $2q \log q \cdot (3q)^{n - \log \log q}$ as stated below Theorem 1, we use the same proof as above, but stop after $n - \log \log q$ (instead of n) iterations. At this point, we have lost a factor $(3q)^{n - \log \log q}$, and have constructed games that are $(\log q)$-neighboring. We can now use a proof along the lines of the proof of Proposition 2, and guess the entire remaining path of length $\log q$ at once. This step loses a factor $2q \log q$ (a factor $2^{\log q} = q$ to guess the path, and another $2 \log q$ as we have a number of hybrids which is twice the length of the path).

5 Impossibility Result for the Boneh-Waters PRF

In this section we show that we cannot hope to prove full security without an exponential loss for another constrained PRF, namely the one due to Boneh and Waters [BW13].

5.1 The Boneh-Waters Constrained PRF

Leveled κ-linear Maps. The Boneh-Waters constrained PRF [BW13] is based on leveled multilinear maps [GGH13, CLT13], of which they use the following abstraction.

We assume a *group generator* $\mathcal{G}$ that takes as input a security parameter 1^{λ} and the number of levels $\kappa \in \mathbb{N}$ and outputs a sequence of groups $(\mathbb{G}_1, \ldots, \mathbb{G}_{\kappa})$, each of prime order $p > 2^{\lambda}$, generated by g_i, respectively, such that there exists a set of bilinear maps $\{e_{i,j} \colon \mathbb{G}_i \times \mathbb{G}_j \to \mathbb{G}_{i+j} \mid i,j \geq 1;\ i+j \leq \kappa\}$ with

$$\forall a,b \in \mathbb{Z}_p \colon\ e_{i,j}(g^a_i, g^b_j) = (g_{i+j})^{ab}\ .$$

(For simplicity we will omit the indices of e.) Security of the PRF is based on the following assumption.

The κ-*multilinear decisional Diffie-Hellman assumption* states that given the output of $\mathcal{G}(1^{\lambda}, \kappa)$ and $(g_1, g_1^{c_1}, \ldots, g_1^{c_{\kappa+1}})$ for random $(c_1, \ldots, c_{\kappa+1}) \xleftarrow{*} \mathbb{Z}_p^{\kappa+1}$, it is hard to distinguish $(g_{\kappa})^{\prod_{j \in [\kappa+1]} c_j}$ from a random element in $\mathbb{G}_{\kappa}$ with better than negligible advantage in λ.

The Boneh-Waters Bit-Fixing PRF. Boneh and Waters [BW13] define a PRF with domain $\mathcal{X} = \{0,1\}^N$ and range $\mathcal{Y} = \mathbb{G}_{\kappa}$, where $\kappa = N + 1$. The sets $S \subseteq \mathcal{X}$ for which constrained keys can be derived are subsets of $\mathcal{X}$ where certain bits are fixed; a set S is described by a vector $v \in \{0,1,?\}^N$ (where '?' acts as a wildcard) as $S_v := \{x \in \{0,1\}^N \mid \forall i \in [N] \colon (v_i = ?) \vee (x_i = v_i)\}$.

The PRF is set up for domain $\mathcal{X} = \{0,1\}^N$ by running $\mathcal{G}(1^\lambda, N+1)$ to generate a sequence of groups $(\mathbb{G}_1, \ldots, \mathbb{G}_{N+1})$. We let g denote the generator of $\mathbb{G}_1$. Secret keys are random elements from $\mathcal{K} := \mathbb{Z}_p^{2N+1}$:

$$k = (\alpha, d_{1,0}, d_{1,1}, \ldots, d_{N,0}, d_{N,1}) \ . \tag{14}$$

and the PRF is defined as

$$\mathsf{F}: \mathcal{K} \times \mathcal{X} \to \mathcal{Y} \ , \qquad (k, x) \mapsto (g_{N+1})^{\alpha \prod_{i \in [N]} d_{i,x_i}} \ .$$

$\mathsf{F}.\mathsf{constrain}(k,v)$: On input a key k as in (14) and $v \in \{0,1,?\}^N$ describing the constrained set, output the key $k_v := (v, K, \{D_{i,b}\}_{i \in [N] \setminus V, \, b \in \{0,1\}})$, where $V := \{i \in [N] \mid v_i \neq ?\}$ is the set of fixed indices,

$$K := (g_{|V|+1})^{\alpha \prod_{i \in V} d_{i,v_i}} \quad \text{and} \quad D_{i,b} := g^{d_{i,b}} \ , \quad \text{for } i \in [N] \setminus V, b \in \{0,1\} \ .$$

$\mathsf{F}.\mathsf{eval}(k_v, x)$: On input $k_v = (v, K, \{D_{i,b}\}_{i \in [N] \setminus V, \, b \in \{0,1\}})$ and $x \in \mathcal{X}$:
  - if for some $i \in V$: $x_i \neq v_i$, return $\perp$ (as x is not in S_v);
  - if $|V| = N$, output K (as $S_v = \{v\}$ and $K = \mathsf{F}(k,v)$);
  - else, compute $T := (g_{N-|V|})^{\prod_{i \in [N] \setminus V} d_{i,x_i}}$ via repeated application of the bilinear maps to the elements $D_{i,x_i} = g^{d_{i,x_i}}$ for $i \in [N] \setminus V$ and output $e(T, K) = (g_{N+1})^{\alpha \prod_{i \in [N]} d_{i,x_i}} = \mathsf{F}(k, x)$.

In [BW13] it is shown how to use an adversary breaking the constrained PRF for N-bit inputs with advantage $\epsilon(\lambda)$ to break the $(N+1)$-multilinear decisional Diffie-Hellman assumption with advantage $\frac{1}{2^N} \cdot \epsilon(\lambda)$. (The exponential factor comes from security leveraging.) In the next section we show that this is optimal in the sense that every simple reduction from a decisional problem must lose a factor that is exponential in the input length N.

We actually prove a stronger statement. First, this security loss is necessary even when the CPRF is only used as a *prefix-fixing* PRF, that is, constrained keys are only issued for sets $S_{(z, ?\ldots?)}$ with $z \in \{0,1\}^{\leq N}$. Second, the loss is necessary even when one only wants to prove *unpredictability* of the CPRF, where the adversary must *compute* $\mathsf{F}(k, x^*)$ instead of distinguishing it from random.

Definition 6 (Unpredictability). *Consider the following experiment for a constrained PRF* $(\mathsf{F}, \mathsf{F}.\mathsf{constrain}, \mathsf{F}.\mathsf{eval})$.

  - *The challenger chooses $k \xleftarrow{\hspace{0.3em}} \mathcal{K}$;*
  - *A can query $\mathsf{F}.\mathsf{constrain}$ for sets S_i;*
  - *A wins if it outputs $(x, \mathsf{F}(k, x))$ with $x \in \mathcal{X}$ and $x \notin S_i$ for all queried S_i.*

*The CPRF is (ϵ, t, q)-**unpredictable** if no A running in time at most t making at most q queries can win the above game with probability greater than ϵ.*

Since unpredictability follows from pseudorandomness without any security loss (we assume that the domain $\mathcal{X}$ is of superpolynomial size), our impossibility result holds a forteriori for pseudorandomness. In particular, this precludes security proofs for the Boneh-Waters CPRF using the technique from Sect. 4.

5.2   Adaptive Security of the Boneh-Waters CPRF

Hierarchical identity-based encryption (HIBE) [HL02] is a generalization of identity-based encryption where the identities are arranged in a hierarchy and from a key for an identity *id* one can derive keys for any identities that are below *id*. In the security game for HIBE the adversary receives the parameters and can query keys for any identity. He then outputs (id, m_0, m_1) and, provided that *id* is not below any identity for which he queried a key, receives the encryption for *id* of one of the two messages, and wins if he guesses which one it was.

Lewko and Waters [LW14], following earlier work [Cor02, HJK12], show that it is hard to prove full security of HIBE schemes if one can check whether secret keys and ciphertexts are correctly formed w.r.t. the public parameters. In particular, they show that a simple black-box reduction (that is, one that runs the attacker once without rewinding; see below) from a decisional assumption must lose a factor that is exponential in the depth of the hierarchy. We adapt their proof technique and show that a proof of full security of the Boneh-Waters PRF with constrained keys for prefix-fixing must lose a factor that is exponential in the length of the PRF inputs.

The proof idea in [LW14] is the following: Assume that there exists a reduction which breaks a challenge with some probability δ after interacting with an adversary that breaks the security of the HIBE with some probability ϵ. We define a concrete adversary A, which, after receiving the public parameters, guesses a random identity *id* at the lowest level of the hierarchy and then queries keys for all identities except *id*, checking whether they are consistent with the parameters. Finally it outputs a challenge query for *id*.

Given a challenge, we run the reduction and simulate this adversary until we have keys for all identities except *id*. We then rewind the reduction to the point right after it sent the parameters to A and simulate A again (choosing a fresh random identity id'; thus $id' \neq id$ with high probability). A now asks for a challenge for id' and can break it by using the key for id' it received in the first run. It is crucial that keys can be verified w.r.t. the parameters, as this guarantees that the reduction cannot detect that a key from the first run was used to win in the second run (the parameters being the same in both runs).

The reduction can thus be used to break the challenge without any adversary, as we can simulate the adversary ourselves. (The actual proof, as well as that of Theorem 2, is more complex, as we need to rewind more than once.) We formally define decisional problems and simple reductions, following [LW14].

Definition 7. *A non-interactive* **decisional problem** $\Pi = (C, \mathcal{D})$ *is described by a set of challenges C and a distribution $\mathcal{D}$ on C. Each $c \in C$ is associated with a bit $b(c)$, the solution for challenge c. An algorithm A (ϵ, t)-solves Π if A runs in time at most t and* $\Pr_{c \xleftarrow{\mathcal{D}} C} [b(c) \leftarrow A(c)] \geq \frac{1}{2} + \epsilon$.

Definition 8. *An algorithm $\mathcal{R}$ is a* **simple** $(t, \epsilon, q, \delta, t')$-**reduction** *from a decisional problem Π to breaking unpredictability of a CPRF if, when given black-box access to any adversary A that (t, ϵ, q)-breaks unpredictability, $\mathcal{R}$ (δ, t')-solves Π after simulating the unpredictability game once for A.*

We show that every simple reduction from a decisional problem to unpredictability of the Boneh-Waters CPRF must lose at least a factor exponential in N. Instead of checking validity of keys computed by the reduction w.r.t. the public parameters, as in [LW14], we show that after two concrete key queries, the secret key k used by the reduction is basically fixed; the two received constrained keys are thus a "fingerprint" of the secret key. Moreover, we show that, by using the multilinear map, correctness of any key can be verified w.r.t. to this fingerprint; which gives us the required checkability property. We define an adversary A that we can simulate by rewinding the reduction: After making the fingerprint queries, A chooses a random value $x^* \in \mathcal{X}$ and queries keys which allow it to evaluate all other domain points, checking every key is consistent with the fingerprint. (Note that keys for $(1 - x_1^*, ?, \ldots)$, $(x_1^*, 1 - x_2^*, ?, \ldots)$, $\ldots,(x_1^*, \ldots, x_{N-1}^*, 1 - x_N^*)$ allow evaluation of the PRF on $\mathcal{X} \setminus \{x^*\}$.)

By rewinding the reduction to the point after receiving the fingerprint and choosing a different x', we can break security by using one of the keys obtained in the first run to evaluate the function at x'. In [FKPR14] we prove the following.

Theorem 2. *Let $\Pi(\lambda)$ be a decisional problem for which no algorithm running in time $t = poly(\lambda)$ has an advantage non-negligible in λ. Let $\mathcal{R}$ be a simple $(t, \epsilon, q, \delta, t')$ reduction from Π to unpredictability of the Boneh-Waters prefix-constrained PRF with domain $\{0, 1\}^N$, with $t, t' = poly(\lambda)$, and $q \geq N - 1$. Then δ vanishes exponentially as a function of N (up to terms that are negligible in λ).*

The reason why our impossibility result does not apply to the GGM construction is that its constrained keys are not "checkable". This is why in the intermediate hybrids we can embed random nodes on the path to x^*, which lead to constrained keys that are not correctly computed.

References

[AFL12]   Abdalla, M., Fiore, D., Lyubashevsky, V.: From selective to full security: Semi-generic transformations in the standard model. In: Fischlin, M., Buchmann, J., Manulis, M. (eds.) PKC 2012. LNCS, vol. 7293, pp. 316–333. Springer, Heidelberg (2012)

[BB04]   Boneh, D., Boyen, X.: Efficient selective-ID secure identity-based encryption without random oracles. In: Cachin, C., Camenisch, J.L. (eds.) EUROCRYPT 2004. LNCS, vol. 3027, pp. 223–238. Springer, Heidelberg (2004)

[BCPR13]   Bitansky, N., Canetti, R., Paneth, O., Rosen, A.: Indistinguishability obfuscation vs. auxiliary-input extractable functions: One must fall. Cryptology ePrint Archive, Report 2013/641 (2013), http://eprint.iacr.org/

[BF01]   Boneh, D., Franklin, M.: Identity-based encryption from the weil pairing. In: Kilian, J. (ed.) CRYPTO 2001. LNCS, vol. 2139, p. 213. Springer, Heidelberg (2001)

[BGI14]   Boyle, E., Goldwasser, S., Ivan, I.: Functional signatures and pseudorandom functions. In: Krawczyk, H. (ed.) PKC 2014. LNCS, vol. 8383, pp. 501–519. Springer, Heidelberg (2014)

[BV98]   Boneh, D., Venkatesan, R.: Breaking RSA may not be equivalent to factoring. In: Nyberg, K. (ed.) EUROCRYPT 1998. LNCS, vol. 1403, pp. 59–71. Springer, Heidelberg (1998)

[BW13]   Boneh, D., Waters, B.: Constrained pseudorandom functions and their applications. In: Sako, K., Sarkar, P. (eds.) ASIACRYPT 2013, Part II. LNCS, vol. 8270, pp. 280–300. Springer, Heidelberg (2013)

[CLT13]   Coron, J.-S., Lepoint, T., Tibouchi, M.: Practical multilinear maps over the integers. In: Canetti, R., Garay, J.A. (eds.) CRYPTO 2013, Part I. LNCS, vol. 8042, pp. 476–493. Springer, Heidelberg (2013)

[Cor02]   Coron, J.-S.: Optimal security proofs for PSS and other signature schemes. In: Knudsen, L.R. (ed.) EUROCRYPT 2002. LNCS, vol. 2332, p. 272. Springer, Heidelberg (2002)

[FKPR14]   Fuchsbauer, G., Konstantinov, M., Pietrzak, K., Rao, V.: Adaptive security of constrained PRFs. Cryptology ePrint Archive, Report 2014/416 (2014), http://eprint.iacr.org/

[FS10]   Fischlin, M., Schröder, D.: On the impossibility of three-move blind signature schemes. In: Gilbert, H. (ed.) EUROCRYPT 2010. LNCS, vol. 6110, pp. 197–215. Springer, Heidelberg (2010)

[GGH13]   Garg, S., Gentry, C., Halevi, S.: Candidate multilinear maps from ideal lattices. In: Johansson, T., Nguyen, P.Q. (eds.) EUROCRYPT 2013. LNCS, vol. 7881, pp. 1–17. Springer, Heidelberg (2013)

[GGM86]   Goldreich, O., Goldwasser, S., Micali, S.: How to construct random functions. J. ACM 33(4), 792–807 (1986)

[GM84]   Goldwasser, S., Micali, S.: Probabilistic encryption. Journal of Computer and System Sciences 28(2), 270–299 (1984)

[HJK12]   Hofheinz, D., Jager, T., Knapp, E.: Waters signatures with optimal security reduction. In: Fischlin, M., Buchmann, J., Manulis, M. (eds.) PKC 2012. LNCS, vol. 7293, pp. 66–83. Springer, Heidelberg (2012)

[HL02]   Horwitz, J., Lynn, B.: Toward hierarchical identity-based encryption. In: Knudsen, L.R. (ed.) EUROCRYPT 2002. LNCS, vol. 2332, p. 466. Springer, Heidelberg (2002)

[Hof14]   Hofheinz, D.: Fully secure constrained pseudorandom functions using random oracles. IACR Cryptology ePrint Archive, Report 2014/372 (2014)

[HSW14]   Hohenberger, S., Sahai, A., Waters, B.: Replacing a Random Oracle: Full Domain Hash from Indistinguishability Obfuscation. In: Nguyen, P.Q., Oswald, E. (eds.) EUROCRYPT 2014. LNCS, vol. 8441, pp. 201–220. Springer, Heidelberg (2014)

[HW09]   Hohenberger, S., Waters, B.: Short and Stateless Signatures from the RSA Assumption. In: Halevi, S. (ed.) CRYPTO 2009. LNCS, vol. 5677, pp. 654–670. Springer, Heidelberg (2009)

[KPTZ13]   Kiayias, A., Papadopoulos, S., Triandopoulos, N., Zacharias, T.: Delegatable pseudorandom functions and applications. In: Sadeghi, A.-R., Gligor, V.D., Yung, M. (eds.) ACM CCS 2013, pp. 669–684. ACM Press (November 2013)

[LW14]   Lewko, A., Waters, B.: Why Proving HIBE Systems Secure Is Difficult. In: Nguyen, P.Q., Oswald, E. (eds.) EUROCRYPT 2014. LNCS, vol. 8441, pp. 58–76. Springer, Heidelberg (2014)

[SW14]   Sahai, A., Waters, B.: How to use indistinguishability obfuscation: deniable encryption, and more. In: 46th ACM STOC, pp. 475–484. ACM Press (2014)

Poly-Many Hardcore Bits for Any One-Way Function and a Framework for Differing-Inputs Obfuscation

Mihir Bellare[1], Igors Stepanovs[1], and Stefano Tessaro[2]

[1] Department of Computer Science and Engineering,
University of California San Diego, USA
http://cseweb.ucsd.edu/~mihir/
[2] Department of Computer Science, University of California Santa Barbara, USA
http://www.cs.ucsb.edu/~tessaro/

Abstract. We show how to extract an arbitrary polynomial number of simultaneously hardcore bits from any one-way function. In the case the one-way function is injective or has polynomially-bounded pre-image size, we assume the existence of indistinguishability obfuscation (iO). In the general case, we assume the existence of differing-input obfuscation (diO), but of a form weaker than full auxiliary-input diO. Our construction for injective one-way functions extends to extract hardcore bits on multiple, correlated inputs, yielding new D-PKE schemes. Of independent interest is a definitional framework for differing-inputs obfuscation in which security is parameterized by circuit-sampler classes.

1 Introduction

When RSA was invented [52], the understanding was that public-key encryption (PKE) would consist of applying the RSA one-way function f directly to the message. In advancing semantic security (unachieved by this plain RSA scheme) as the appropriate target for PKE, Goldwasser and Micali [37] created a new challenge. Even given RSA, how was one to achieve semantic security?

The answer was hardcore functions [17,58,37]. Let f be a one-way function, and h a function that has the same domain as f and returns strings of some length s. We say that h is hardcore for f if the distributions $(f, h, f(x), h(x))$ and $(f, h, f(x), r)$ are computationally indistinguishable when x is chosen at random from the domain of f and r is a random s-bit string.[1] We will refer to the output length s of h, which is the number of (simultaneously) hardcore bits produced by h, as the *span* of h, and say that f achieves a certain span if there exists a hardcore function h for f with the span in question. Hardcore predicates are hardcore functions with span one. Semantically-secure PKE can now be easily achieved: encrypt s-bit message m as $(f(x), h(x) \oplus m)$ for random x where trapdoor, injective one-way function f together with hardcore function

---

[1] In the formal definitions in Section 2, both one-way functions and hardcore functions are families. Think here of f, h as instances chosen at random from the respective families, their descriptions public.

P. Sarkar and T. Iwata (Eds.): ASIACRYPT 2014, PART II, LNCS 8874, pp. 102–121, 2014.

h constitute the public key. The span s determines the number of message bits that can be encrypted, so the larger it is, the better.

The quest for semantic security turned into a quest for hardcore functions [17,58,37,54,15,42,3,48,39,56,46,35,40]. It became quickly evident that finding hardcore predicates, let alone hardcore functions of span larger than one, was very challenging, and many sophisticated techniques were invented. With time, encryption methods avoiding hardcore functions did emerge [26]. But hardcore functions have never lost their position as a fundamental cryptographic primitive, as new usages, applications and constructions have arisen [28,25,1,2,27]. Hardcore functions have been particularly important in developing new forms of encryption, including lossy-TDF based encryption [51] and deterministic encryption [6,8,30].

GENERIC CONSTRUCTIONS. Across all this work and applications, something that stands out is the value and appeal of generic constructions, the representative result being that of Goldreich and Levin (GL) [35]. Prior to this, hardcore functions were obtained by dedicated and involved analyses that exploited the algebra underlying the one-way function itself. GL [35] showed that the inner product of x with a random string (the latter is part of the description of h) results in a hardcore predicate for any OWF. This result has had an enormous impact, as evidenced by over 900 citations to the paper to date. A generic construction such as that of GL allows theoreticians to develop constructions and proofs independently of specific algebraic assumptions. Furthermore, it allows us to immediately obtain hardcore functions, and thus encryption, from new, candidate one-way functions.

QUESTIONS AND ANSWERS. The GL construction however only provided a hardcore predicate, meaning span one, and by extension logarithmic span. The most desired goal was polynomial span.[2] Significant effort was been invested towards this goal, allowing it to be reached for specific algebraic OWFs [40,25,47,50,2] or ones satisfying extra properties [51]. But a generic construction providing polynomial span seemed out of reach.

This is the question resolved by our work. We present generic constructions of hardcore functions with polynomial span for both injective and non-injective one-way functions. The tools we use are indistinguishability obfuscation (iO) [5,31] and differing-inputs obfuscation (diO) [5,20,4]. See Fig. 1 for a summary of our results.

PRIOR WORK. In more detail, early results gave hardcore predicates (ie. span one) for specific one-way functions including discrete exponentiation modulo a prime, RSA and Rabin [17,58,54,15,42,3,48,39,28]. Eventually, as noted above, the impactful and influential work of Goldreich and Levin [35] gave a generic hardcore predicate for any one-way function. Extensions of these results are able to achieve logarithmic span [56,46,35,1,27].

---

[2] Once one can obtain a particular polynomial number of output bits, one can always expand to an arbitrary polynomial via a PRG, so we do not distinguish these cases.

OWF	Span	Assumption	Construction	See
injective	poly	iO	**HC1**	Cor. 5 of Th. 4
poly pre-image size	poly	iO	**HC2**	Cor. 8 of Th. 6
any	poly	diO$^-$	**HC2**	Cor. 7 of Th. 6

Fig. 1. Our results: We indicate the assumptions we make in order to construct hardcore functions with polynomial span. By diO$^-$ we mean a weakening of diO that we define.

Hardcore functions with polynomial span have been provided for specific, algebraic functions, usually under a stronger assumption than one-wayness of the function itself. These include the works by Håstad, Schrift and Shamir [40] for discrete exponentiation modulo a composite, by Patel and Sundaram [50] for discrete exponentiation modulo a safe prime, by Catalano, Gennaro and Howgrave-Graham [25] for the Paillier function [49], and by Akavia, Goldwasser and Vaikuntanathan [2] for certain LWE-based functions. Moreover, polynomial-span hardcore functions for RSA [45,55] and the Rabin trapdoor function [55] have been given under different assumptions. Peikert and Waters showed that lossy trapdoor functions [51] achieve polynomial span, yielding further examples of specific one-way functions with polynomial span [51,29].

The basic question that remains open was whether polynomial span is achievable for an arbitrary, given OWF, meaning whether there is a generic hardcore function with polynomial span. One answer was provided by [9], who showed that UCE-security (specifically, relative to split, computationally-unpredictable sources) of a function h with polynomial span suffices for h to be hardcore for any one-way function, but the assumption made is arguably close to the desired conclusion.

OUR RESULTS. Injective one-way functions are the most important case for aplications. (Most applications are related to some form of encryption.) Accordingly we begin by focusing on the case where f is injective. We give a construction of a hardcore function called **HC1** that provides polynomial span for *any* injective OWF f. Beyond one-wayness of f we assume iO [5,53].

We then provide the **HC2** construction of a hardcore function with polynomial span for any, even non-injective OWF f. In the case f has polynomially-bounded pre-image size, the extra assumption remains iO. In the general case, it is diO, but of a form that is weaker than full auxiliary-input diO as defined in [20,4].

As direct corollaries, we obtain hardcore functions of polynomial span for many specific OWFs f, in some cases providing the first hardcore function with polynomial span for this f, in others providing a construction under different and new assumptions compared to prior ones. Thus for the basic discrete exponentiation OWF over the integers modulo a prime or an elliptic curve group, we provide the first construction with polynomial span. (Results were known when exponents are short [33,50] and for discrete exponentiation modulo composites [40,36].)

Our results have the above-noted benefits of generic constructions. They yield conceptual simplifications for the development of theoretical cryptography, and also provide automatic ways to get hardcore functions for new, candidate one-way functions. We view our work as also been interesting from the point of view of showing the power and applicability of iO, in the wake of Sahai-Waters [53].

TECHNICAL APPROACH. We now take a closer look at our constructions and proofs to highlight the technical approach and novelties. Recall that the guarantee of iO [5,53] is that the obfuscations of two circuits are indistinguishable if the circuits themselves are equivalent, meaning return the same output on all inputs. Differing-inputs obfuscation (diO) [5,20,4] relaxes the equivalence condition, asking instead that it only be hard, given the (unobfuscated) circuits, to find an input where they differ. See Section 3 for formal definitions.

Our **HC1** construction is a natural one, namely to let h be an obfuscation of the circuit $G(gk, \cdot)$ where G is a PRF [34] and gk is a random key for G. Our proof assumes the PRF is punctured [19,43,21]. (This is not an extra assumption, as we assume the existence of one-way functions.) Moreover, the proof uses a weak form of diO shown by Boyle, Chung and Pass (BCP) [20] to be implied by iO. The proof considers an adversary $\mathcal{H}$ provided with $f, h, f(x^*), r^*$ (where f is injective) and we want to move from the real game, in which $r^* = G(gk, x^*)$, to the random game, in which r^* is random. We begin by using the SW technique [53] to move to a game in which r^* is random and h is an obfuscation of the circuit C^1 that embeds the target input x^*, a punctured PRF key, and a random point r^*, returning the latter when called on $x = x^*$ (the trigger) and otherwise returning $G(gk, x)$, computed via the punctured key. (This move relies on iO and punctured PRF security.) While this has made r^* random as desired, h is not what it should be in the random game, where it is in fact an obfuscation of the real circuit $G(gk, \cdot)$. The difficulty is to move h back to an obfuscation of this real circuit while leaving r^* random. We realize that such a move must exploit the one-wayness of f, which has not so far been used. A one-wayness adversary, given $f(x^*)$ and aiming to find x^*, needs to run $\mathcal{H}$. The problem is that $\mathcal{H}$ needs an obfuscation of the above-described circuit C^1 as input, and construction of C^1 requires knowing the very point x^* that the one-wayness adversary is trying to find. The difficulty is inherent rather than merely one of proof, for the forms of obfuscation being used give no guarantee that an obfuscation of C^1 does not reveal x^*. We get around this by changing the trigger check from $x = x^*$ to $f(x) = f(x^*)$, so that now the circuit can embed $f(x^*)$ rather than x^*, a quantity there is no harm in revealing. This is reminiscent of the technique in the security proof of the short signature scheme of Sahai and Waters [53]. The new check is equivalent to the old if f is injective, which is where we use this assumption. But we have still not arrived at the random game. We note that our modified circuit C^2 and the target circuit $G(gk, \cdot)$ of the random game are inherently non-equivalent, and iO would not apply. However, these circuits differ only at input x^*. We exploit the one-wayness of f to prove that it is hard to find this input even given the two circuits. The assumed diO-security of the obfuscator now implies that the obfuscations of these circuits are indistinguishable, allowing us to conclude. Finally, since we exploit diO only for circuits that differ at one (hard to find) point, BCP [20] says that iO in fact suffices, making iO the only assumption needed for the result beyond the necessary one-wayness of f.

The above argument makes crucial use of the assumption that f is injective. To handle an arbitrary one-way function, our **HC2** construction modifies the above so that h is an obfuscation of the circuit $G(gk, f(\cdot))$ where gk is a random key for punctured PRF G. Thus, $h(x) = G(gk, f(x))$. We refer to Section 5 for the proof but note that BCP [20] implies that iO suffices for our proof when the number of preimages of any output of f is polynomial, but in general it could be exponential. In this case, diO suffices, but in fact a weaker version of it, that we define, does as well.

In summary, in the important case of injective one-way functions, and even for one-way functions with polynomial pre-image size, we provide a hardcore function with polynomial span assuming only iO. For one-way functions with super-polynomial pre-image size, we provide a hardcore function with polynomial span assuming a weak form of diO that we denote by diO^- in Fig. 1.

EXTENSIONS AND APPLICATIONS. We show that our **HC1** hardcore function construction is able to extract random, independent bits even on inputs that are arbitrarily correlated, which is not true of most prior constructions and, combined with the fact that we get polynomially-many output bits for each input, yields new applications. In more detail, an injective function f is said to be one-way on a distribution $\mathcal{I}$ over vectors $\mathbf{x}$ if, given the result $f(\mathbf{x})$ of applying f to $\mathbf{x}$ component-wise, it is hard to recover any component of $\mathbf{x}$. We want a hardcore function such that the components of $h(\mathbf{x})$ look not only random but *independent* even given $f(\mathbf{x})$. If the components of $\mathbf{x}$ are independent, this is true for any hardcore function meeting the standard definition, and thus for ours, but if the components are correlated, standard hardcore functions give no such guarantee and may fail. As an example, many existing hardcore functions [17,58,54,15,42,3,48,39,28,56,46,35,1,27,40,25,49,2] return certain specific bits of their input and will thus fail to be hardcore relative to the distribution in which $\mathbf{x}[2]$ is the bitwise complement of $\mathbf{x}[1]$, even if f is one-way on this distribution. We show however that **HC1** remains hardcore for f on *any* efficiently sampleable distribution $\mathcal{I}$ over which f is one-way, even when the entries of the vectors produced by $\mathcal{I}$ are arbitrarily correlated. This answers open questions from [38,30].

Deterministic PKE (D-PKE) is useful for many applications, including efficient search on encrypted data [6] and providing resilience in the face of the low-quality randomness that pervades systems [7]. However, it cannot provide IND-CPA security. BBO [6] define what it means for a D-PKE scheme to provide PRIV-security over an input sampler $\mathcal{I}$, the latter returning vectors of *arbitrarily correlated* messages to be encrypted. We restrict attention to distributions that are admissible, meaning that there exists a family of injective, trapdoor functions that is one-way relative to $\mathcal{I}$. The basic question that emerges is, for which admissible distributions $\mathcal{I}$ does there exist a D-PKE scheme that is PRIV-secure over $\mathcal{I}$? We show that this is true for all admissible distributions that are efficiently sampleable, assuming only the existence of iO. We obtain this result using the security of **HC1** on correlated inputs and techniques from [30]. Previously, this was known only in the ROM [6], under the assumption that UCE-secure functions exist [9], for distributions with limited correlation between messages [8,18] or assuming lossy trapdoor functions [30]. See [12] for definitions, a precise statement of the result, and proofs.

PARAMETERIZED DIO. Of independent interest, we provide a definitional framework for diO in which security is parameterized by a class of circuit samplers. This allows us to unify and capture variant notions of iO and diO in the literature [5,53,20,4]. The framework makes it easy to define further variants that are weaker than full diO, yielding a language in which one can state assumptions that are closer to those actually used in the proof rather than being overkill, particularly with regard to the type of auxiliary inputs used [13,22,32]. This is useful in the light of work like [32] which indicates that diO with arbitrary auxiliary inputs may be implausible. The weaker notion of diO noted above and denoted diO⁻ in Fig. 1 is formally defined in our framework as diO security for "short" auxiliary inputs, which evades the negative results of [32]. See Section 3.

DISCUSSION AND RELATED WORK. Random oracles (ROs) are "ideal" hardcore functions, able to provide polynomial span for any one-way function [10]. Our results, akin to [44,9,41], can thus be seen as instantiating the RO in a natural ROM construction, in particular showing hardcore functions in the standard model that are just as good as those in the ROM. As a consequence, we are able to instantiate the RO in the BR93 PKE scheme [10] to obtain a standard-model IND-CPA scheme.

The hardcore function in our second construction is the reverse of the hash function used to instantiate FDH in HSW [41]: in our case, the circuit being obfuscated first applies a one-way function and then a punctured PRF, while in their case it first applies a punctured PRF and then a one-way function.

Our work adopts the standard definition of a one-way function in which any polynomial-time adversary must have negligible inversion advantage. Polynomial span is known to be achievable for any *exponentially* hard to invert function [35,27].

Given a one-way permutation f and a polynomial n it is possible to construct another one-way permutation g that has n hardcore bits. Namely let $g(x) = f^n(x)$ and let the hardcore function on x be the result of the Blum-Micali-Yao PRG [17,58] on seed x. A similar transform is provided in [16]. We provide hardcore functions directly for any given one-way function rather than for another function built from it.

GGHW [32] show that if what they call "special purpose" obfuscation is possible then full diO (diO for all circuits relative to arbitrary auxiliary inputs) is not possible. (Their result constructs a pathological auxiliary input that is itself a special-purpose obfuscated circuit.) Their result does not rule out the assumptions we use, namely iO or diO⁻. GGHW go on to show that if a special-purpose obfuscation of Turing Machines is possible then there is an artificial one-way function with exponential pre-image size (in particular it is not injective) for which there is no hardcore predicate that is output-dependent. (That is, $f(x_1) = f(x_2)$ implies $h(x_1) = h(x_2)$, a property possessed by our second construction.)

SUBSEQUENT WORK. Our proof for **HC1** avoids hardwiring the challenge input x^* in the circuit by hardwiring $y^* = f(x^*)$ and changing the test for $x = x^*$ to $f(x) = y^*$. Brzuska and Mittelbach [23] change this step in our proof to implement the test using auxiliary-input point function obfuscation (AIPO) [23,24,14]. As a result, under the additional assumption that AIPO is possible, they obtain a proof for our **HC1** construction which applies to arbitrary (not just injective) one-way functions. Zhandry [59] shows how to replace diO in our constructions with extractable witness PRFs.

2   Preliminaries

We recall definitions for one-way functions, hardcore predicates and punctured PRFs.

NOTATION. We denote by $\lambda \in \mathbb{N}$ the security parameter and by 1^λ its unary representation. We denote the size of a finite set X by $|X|$, and the length of a string $x \in \{0,1\}^*$ by $|x|$. We let ε denote the empty string. If C is a circuit then $|C|$ denotes its size, and if s is an integer then $\mathsf{Pad}_s(C)$ denotes C padded to have size s. If X is a finite set, we let $x \leftarrow_\$ X$ denote picking an element of X uniformly at random and assigning it to x. Algorithms may be randomized unless otherwise indicated. Running time is worst case. "PT" stands for "polynomial-time," whether for randomized algorithms or deterministic ones. If A is an algorithm, we let $y \leftarrow A(x_1,\ldots;r)$ denote running A with random coins r on inputs $x_1,\ldots$ and assigning the output to y. We let $y \leftarrow_\$ A(x_1,\ldots)$ be the result of picking r at random and letting $y \leftarrow A(x_1,\ldots;r)$. We let $[A(x_1,\ldots)]$ denote the set of all possible outputs of A when invoked with inputs $x_1,\ldots$. We say that $f : \mathbb{N} \to \mathbb{R}$ is negligible if for every positive polynomial p, there exists $n_p \in \mathbb{N}$ such that $f(n) < 1/p(n)$ for all $n > n_p$. We use the code based game playing framework of [11]. (See Fig. 2 for examples of games.) By $\mathrm{G}^{\mathcal{A}}(\lambda)$ we denote the event that the execution of game G with adversary $\mathcal{A}$ and security parameter λ results in the game returning true.

FUNCTION FAMILIES. A family of functions F specifies the following. PT key generation algorithm F.Kg takes 1^λ and possibly another input to return a key $fk \in \{0,1\}^{\mathsf{F.kl}(\lambda)}$, where F.kl: $\mathbb{N} \to \mathbb{N}$ is the key length function associated to F. Deterministic, PT evaluation algorithm F.Ev takes key fk and an input $x \in \{0,1\}^{\mathsf{F.il}(\lambda)}$ to return an output $\mathsf{F.Ev}(fk,x) \in \{0,1\}^{\mathsf{F.ol}(\lambda)}$, where F.il, F.ol: $\mathbb{N} \to \mathbb{N}$ are the input and output length functions associated to F, respectively. The pre-image size of F is the function $\mathrm{PREIMG}_\mathsf{F}$ defined for $\lambda \in \mathbb{N}$ by

$$\mathrm{PREIMG}_\mathsf{F}(\lambda) = \max_{fk,x^*} \left| \left\{ x \in \{0,1\}^{\mathsf{F.il}(\lambda)} \ : \ \mathsf{F.Ev}(fk,x) = \mathsf{F.Ev}(fk,x^*) \right\} \right|$$

Game $\mathrm{OW}_\mathsf{F}^{\mathcal{F}}(\lambda)$	Game $\mathrm{HC}_{\mathsf{F},\mathsf{H}}^{\mathcal{H}}(\lambda)$	Game $\mathrm{PPRF}_\mathsf{G}^{\mathcal{G}}(\lambda)$
$\mathit{fk} \leftarrow_\$ \mathsf{F}.\mathsf{Kg}(1^\lambda)$	$b \leftarrow_\$ \{0,1\}$	$b \leftarrow_\$ \{0,1\}$
$x^* \leftarrow_\$ \{0,1\}^{\mathsf{F}.\mathsf{il}(\lambda)}$	$\mathit{fk} \leftarrow_\$ \mathsf{F}.\mathsf{Kg}(1^\lambda)$	$\mathit{gk} \leftarrow_\$ \mathsf{G}.\mathsf{Kg}(1^\lambda)$
$y^* \leftarrow \mathsf{F}.\mathsf{Ev}(\mathit{fk}, x^*)$	$\mathit{hk} \leftarrow_\$ \mathsf{H}.\mathsf{Kg}(1^\lambda, \mathit{fk})$	$b' \leftarrow_\$ \mathcal{G}^{\mathrm{CH}}(1^\lambda)$
$x' \leftarrow_\$ \mathcal{F}(1^\lambda, \mathit{fk}, y^*)$	$x^* \leftarrow_\$ \{0,1\}^{\mathsf{F}.\mathsf{il}(\lambda)}$	Return $(b = b')$
Return $(y^* = \mathsf{F}.\mathsf{Ev}(\mathit{fk}, x'))$	$y^* \leftarrow \mathsf{F}.\mathsf{Ev}(\mathit{fk}, x^*)$	$\underline{\mathrm{CH}(x^*)}$
	If $b = 1$ then	$\mathit{gk}^* \leftarrow_\$ \mathsf{G}.\mathsf{PKg}(1^\lambda, \mathit{gk}, x^*)$
	$\quad r^* \leftarrow \mathsf{H}.\mathsf{Ev}(\mathit{hk}, x^*)$	If $b = 1$ then
	Else $r^* \leftarrow_\$ \{0,1\}^{\mathsf{H}.\mathsf{ol}(\lambda)}$	$\quad r^* \leftarrow \mathsf{G}.\mathsf{Ev}(\mathit{gk}, x^*)$
	$b' \leftarrow_\$ \mathcal{H}(1^\lambda, \mathit{fk}, \mathit{hk}, y^*, r^*)$	Else $r^* \leftarrow_\$ \{0,1\}^{\mathsf{G}.\mathsf{ol}(\lambda)}$
	Return $(b = b')$	Return (gk^*, r^*)

Game $\mathrm{DIFF}_\mathcal{S}^{\mathcal{D}}(\lambda)$	Game $\mathrm{IO}_{\mathsf{Obf},\mathcal{S}}^{\mathcal{O}}(\lambda)$
$(\mathrm{C}_0, \mathrm{C}_1, \mathit{aux}) \leftarrow_\$ \mathcal{S}(1^\lambda)$	$b \leftarrow_\$ \{0,1\}$; $(\mathrm{C}_0, \mathrm{C}_1, \mathit{aux}) \leftarrow_\$ \mathcal{S}(1^\lambda)$
$x \leftarrow_\$ \mathcal{D}(\mathrm{C}_0, \mathrm{C}_1, \mathit{aux})$	$\overline{\mathrm{C}} \leftarrow_\$ \mathsf{Obf}(1^\lambda, \mathrm{C}_b)$; $b' \leftarrow_\$ \mathcal{O}(1^\lambda, \overline{\mathrm{C}}, \mathit{aux})$
Return $(\mathrm{C}_0(x) \neq \mathrm{C}_1(x))$	Return $(b = b')$

Fig. 2. Games defining one-wayness of F, security of H as a hardcore function for F, punctured-PRF security of G, difference-security of circuit sampler $\mathcal{S}$ and iO-security of obfuscator Obf relative to circuit sampler $\mathcal{S}$

where the maximum is over all $x^* \in \{0,1\}^{\mathsf{F}.\mathsf{il}(\lambda)}$ and all $\mathit{fk} \in [\mathsf{F}.\mathsf{Kg}(1^\lambda)]$. We say that F is injective if $\mathrm{PREIMG}_\mathsf{F}(\lambda) = 1$ for all $\lambda \in \mathbb{N}$, meaning $\mathsf{F}.\mathsf{Ev}(\mathit{fk}, x_1) \neq \mathsf{F}.\mathsf{Ev}(\mathit{fk}, x_2)$ for all distinct $x_1, x_2 \in \{0,1\}^{\mathsf{F}.\mathsf{il}(\lambda)}$, all fk and all $\lambda \in \mathbb{N}$. We say that F has polynomial pre-image size if there is a polynomial p such that $\mathrm{PREIMG}_\mathsf{F}(\cdot) \leq p(\cdot)$.

ONE-WAYNESS AND HARDCORE FUNCTIONS. Function family F is one-way if $\mathsf{Adv}_{\mathsf{F},\mathcal{F}}^{\mathsf{ow}}(\cdot)$ is negligible for all PT adversaries $\mathcal{F}$, where $\mathsf{Adv}_{\mathsf{F},\mathcal{F}}^{\mathsf{ow}}(\lambda) = \Pr[\mathrm{OW}_\mathsf{F}^{\mathcal{F}}(\lambda)]$ and game $\mathrm{OW}_\mathsf{F}^{\mathcal{F}}(\lambda)$ is defined in Fig. 2. Let H be a family of functions with $\mathsf{H}.\mathsf{il} = \mathsf{F}.\mathsf{il}$. We say that H is hardcore for F if $\mathsf{Adv}_{\mathsf{F},\mathsf{H},\mathcal{H}}^{\mathsf{hc}}(\cdot)$ is negligible for all PT adversaries $\mathcal{H}$, where $\mathsf{Adv}_{\mathsf{F},\mathsf{H},\mathcal{H}}^{\mathsf{hc}}(\lambda) = 2\Pr[\mathrm{HC}_{\mathsf{F},\mathsf{H}}^{\mathcal{H}}(\lambda)] - 1$ and game $\mathrm{HC}_{\mathsf{F},\mathsf{H}}^{\mathcal{H}}(\lambda)$ is defined in Fig. 2.

PUNCTURED PRFs. A punctured function family G specifies (beyond the usual algorithms) additional PT algorithms $\mathsf{G}.\mathsf{PKg}, \mathsf{G}.\mathsf{PEv}$. On inputs 1^λ, a key $\mathit{gk} \in [\mathsf{G}.\mathsf{Kg}(1^\lambda)]$ and target input $x^* \in \{0,1\}^{\mathsf{G}.\mathsf{il}(\lambda)}$, algorithm $\mathsf{G}.\mathsf{PKg}$ returns a "punctured" key gk^* such that $\mathsf{G}.\mathsf{PEv}(\mathit{gk}^*, x) = \mathsf{G}.\mathsf{Ev}(\mathit{gk}, x)$ for all $x \in \{0,1\}^{\mathsf{G}.\mathsf{il}(\lambda)} \setminus \{x^*\}$. We say that G is a punctured PRF if $\mathsf{Adv}_{\mathsf{G},\mathcal{G}}^{\mathsf{pprf}}(\cdot)$ is negligible for all PT adversaries $\mathcal{G}$, where $\mathsf{Adv}_{\mathsf{G},\mathcal{G}}^{\mathsf{pprf}}(\lambda) = 2\Pr[\mathrm{PPRF}_\mathsf{G}^{\mathcal{G}}(\lambda)] - 1$ and game $\mathrm{PPRF}_\mathsf{G}^{\mathcal{G}}(\lambda)$ is defined in Fig. 2. Here $\mathcal{G}$ must make exactly one oracle query where it picks a target point x^* and gets back the corresponding punctured key together with a challenge for the value of $\mathsf{G}.\mathsf{Ev}$ on the target point.

The concept of punctured PRFs is due to [19,43,21] who note that they can be built via the GGM construction [34]. This however yields a family G with G.il = G.ol. For our purposes, we need a stronger result, namely a punctured PRF with arbitrary polynomial output length:

Proposition 1. *Let ι, ℓ be polynomials and assume one-way functions exist. Then there is a punctured PRF G with G.il = ι and G.ol = ℓ.*

The claimed punctured PRF G can be obtained by starting from a GGM-based punctured PRF $\overline{\mathsf{G}}$ with $\overline{\mathsf{G}}$.il = $\overline{\mathsf{G}}$.ol = ι and letting G.Ev(gk, x) = S.Ev($\overline{\mathsf{G}}$.Ev(gk, x)) where S is a PRG with input length ι and output length ℓ. We omit the details.

3   Parameterized diO Framework

We present a definitional framework for diO where security is parameterized by a class of circuit samplers. This is of conceptual value in enabling us to capture and unify existing forms of iO and diO. Further, it allows one to easily define new forms of diO that are *weaker* than the full auxiliary-input diO of [20,4] and thus obtain results under weaker assumptions. Our parameterized language leads to sharper and more fine-grained security claims in which we can state assumptions that are closer to what is actually used by the proof rather than being overkill, in particular with regard to what types of auxiliary input are used [13,22,32]. This allows us to circumvent the negative results of [32] which apply to diO with arbitrary auxiliary input. We note that previous definitions did parameterize the definition by a class of circuits but this is different and in particular will not capture differences related to auxiliary inputs.

CIRCUIT SAMPLERS. A circuit sampler is a PT algorithm $\mathcal{S}$ that on input 1^λ returns a triple $(\mathrm{C}_0, \mathrm{C}_1, aux)$ where $\mathrm{C}_0, \mathrm{C}_1$ are circuits which have the same size, number of inputs and number of outputs, and aux is a string. We say that a circuit sampler $\mathcal{S}$ is difference secure if $\mathsf{Adv}^{\mathrm{diff}}_{\mathcal{S},\mathcal{D}}(\cdot)$ is negligible for every PT adversary $\mathcal{D}$, where $\mathsf{Adv}^{\mathrm{diff}}_{\mathcal{S},\mathcal{D}}(\lambda) = \Pr[\mathrm{DIFF}^{\mathcal{D}}_{\mathcal{S}}(\lambda)]$ and game $\mathrm{DIFF}^{\mathcal{D}}_{\mathcal{S}}(\lambda)$ is defined in Fig. 2. Difference security of $\mathcal{S}$ means that given $\mathrm{C}_0, \mathrm{C}_1, aux$ it is hard to find an input on which the circuits differ.

IO-SECURITY. An obfuscator is a PT algorithm Obf that on input 1^λ and a circuit C returns a circuit $\overline{C}$ such that $\overline{C}(x) = C(x)$ for all x. If $\mathcal{S}$ is a circuit sampler and $\mathcal{O}$ is an adversary, we let $\mathsf{Adv}^{\mathrm{io}}_{\mathsf{Obf},\mathcal{S},\mathcal{O}}(\lambda) = 2\Pr[\mathrm{IO}^{\mathcal{O}}_{\mathsf{Obf},\mathcal{S}}(\lambda)] - 1$ where game $\mathrm{IO}^{\mathcal{O}}_{\mathsf{Obf},\mathcal{S}}(\lambda)$ is defined in Fig. 2. Now let $\boldsymbol{S}$ be a class (set) of circuit samplers. We say that Obf is $\boldsymbol{S}$-secure if $\mathsf{Adv}^{\mathrm{io}}_{\mathsf{Obf},\mathcal{S},\mathcal{O}}(\cdot)$ is negligible for every PT adversary $\mathcal{O}$ and every circuit sampler $\mathcal{S} \in \boldsymbol{S}$. This is our parameterized notion of security. The following obvious fact will often be useful:

Proposition 2. *Let $\boldsymbol{S}_1, \boldsymbol{S}_2$ be classes of circuit samplers and Obf an obfuscator. Suppose $\boldsymbol{S}_1 \subseteq \boldsymbol{S}_2$. Then if Obf is $\boldsymbol{S}_2$-secure it is also $\boldsymbol{S}_1$-secure.*

CAPTURING KNOWN NOTIONS. Different types of iO security in the literature can now be captured and unified by considering different classes of circuit samplers, as follows.

Let S_{diff} be the class of all difference-secure circuit samplers. Then Obf being S_{diff}-secure means it is a differing-inputs obfuscator as per [20,4].

Let $S^{\overline{\text{aux}}}$ be the class of circuit samplers S that do not have auxiliary inputs, meaning $aux = \varepsilon$ for all $\lambda \in \mathbb{N}$ and all $(C_0, C_1, aux) \in [S(1^\lambda)]$. Let $S^{\overline{\text{aux}}}_{\text{diff}} = S_{\text{diff}} \cap S^{\overline{\text{aux}}} \subseteq S_{\text{diff}}$ be the class of all difference-secure circuit samplers that do not have auxiliary inputs. Then Obf being $S^{\overline{\text{aux}}}_{\text{diff}}$-secure means it is a differing-inputs obfuscator as per [5].

We say that circuits C_0, C_1 are equivalent, written $C_0 \equiv C_1$, if they agree on all inputs. We say that circuit sampler S produces equivalent circuits if $C_0 \equiv C_1$ for all $\lambda \in \mathbb{N}$ and all $(C_0, C_1, aux) \in [S(1^\lambda)]$. Let S_{eq} be the class of all circuit samplers that produce equivalent circuits. Then Obf being S_{eq}-secure means it is an indistinguishability obfuscator as per [53]. Let $S^{\overline{\text{aux}}}_{\text{eq}} = S_{\text{eq}} \cap S^{\overline{\text{aux}}}$ be the class of circuit samplers without auxiliary inputs that produce equivalent circuits. Then Obf being $S^{\overline{\text{aux}}}_{\text{eq}}$-secure means it is an indistinguishability obfuscator as per [5].

If S produces equivalent circuits it is certainly difference-secure. This means that $S_{\text{eq}} \subseteq S_{\text{diff}}$ and $S^{\overline{\text{aux}}}_{\text{eq}} \subseteq S^{\overline{\text{aux}}}_{\text{diff}}$. Hence Proposition 2 says that any S_{diff}-secure obfuscator is a S_{eq}-secure obfuscator and any $S^{\overline{\text{aux}}}_{\text{diff}}$-secure obfuscator is a $S^{\overline{\text{aux}}}_{\text{eq}}$-secure obfuscator. That is, diO implies iO, both for the case with auxiliary input and the case without, a fact we will often use.

We say that circuit sampler S produces d-differing circuits, where $d\colon \mathbb{N} \to \mathbb{N}$, if C_0 and C_1 differ on at most $d(\lambda)$ inputs for all $\lambda \in \mathbb{N}$ and all $(C_0, C_1, aux) \in [S(1^\lambda)]$. Let $S_{\text{diff}}(d)$ be the class of all difference-secure circuit samplers that produce d-differing circuits, so that $S_{\text{eq}} \subseteq S_{\text{diff}}(d) \subseteq S_{\text{diff}}$. The interest of this definition is the following result of BCP [20]:

Proposition 3. *If d is a polynomial then any S_{eq}-secure obfuscator is also a $S_{\text{diff}}(d)$-secure obfuscator.*

We will exploit this to reduce our assumptions from S_{diff}-secure obfuscation to S_{eq}-secure obfuscation in some cases.

NEW CLASSES. Above, we have used our framework to express and capture existing variants of iO and diO. We now define a new variant, via a new class of samplers. Following the definition we will explain the motivation. We say that a circuit sampler S has short auxiliary inputs if $|aux| < |C_b|$ for all $b \in \{0,1\}$, all $\lambda \in \mathbb{N}$ and all $(C_0, C_1, aux) \in [S(1^\lambda)]$. We let S^{sh} be the class of all circuit samplers that have short auxiliary inputs. The assumption made in Theorem 6 is a S-secure obfuscator for a particular $S \subseteq S_{\text{diff}} \cap S^{\text{sh}}$, meaning diO is only required relative to circuits samplers with short auxiliary inputs. This is a potentially *weaker* assumption than a S_{diff}-secure obfuscator.

4 Poly-Many Hardcore Bits for Injective OWFs

In this section we consider the natural construction of a hardcore function with arbitrary span, namely an obfuscated PRF. We show that this works assuming the one-way function is injective and the obfuscation is diO-secure, yielding our first result, namely a hardcore function with arbitrary polynomial span for any injective one-way function.

CONSTRUCTION. Let G be a function family. Let Obf be an obfuscator and let $s: \mathbb{N} \to \mathbb{N}$. We define function family $\mathsf{H} = \mathbf{HC1}[\mathsf{G}, \mathsf{Obf}, s]$ as follows, with $\mathsf{H.il} = \mathsf{G.il}$ and $\mathsf{H.ol} = \mathsf{G.ol}$:

$$
\begin{array}{l|l}
\underline{\mathsf{H.Kg}(1^\lambda, fk)} & \underline{\mathsf{H.Ev}(hk, x)} \\
gk \leftarrow_\$ \mathsf{G.Kg}(1^\lambda) \; ; \; \mathsf{C} \leftarrow \mathsf{Pad}_{s(\lambda)}(\mathsf{G.Ev}(gk, \cdot)) & \overline{\mathsf{C}} \leftarrow hk \; ; \; r \leftarrow \overline{\mathsf{C}}(x) \\
\overline{\mathsf{C}} \leftarrow_\$ \mathsf{Obf}(1^\lambda, \mathsf{C}) & \text{Return } r \\
hk \leftarrow \overline{\mathsf{C}} \; ; \; \text{Return } hk &
\end{array}
$$

We give H.Kg two inputs because this is required by the syntax, but the second is ignored. Here $\mathsf{G.Ev}(gk, \cdot)$ represents the circuit that given x returns $\mathsf{G.Ev}(gk, x)$, and C is obtained by padding $\mathsf{G.Ev}(gk, \cdot)$ to size $s(\lambda)$. The padding length function s is a parameter of the construction that will depend on the one-way function F for which H will be hardcore. The description hk of the hardcore function is an obfuscation of circuit C. The hardcore function output is the result of this obfuscated circuit on x, which is simply $\mathsf{G.Ev}(gk, x)$, the result of the original circuit on x. The output of the hardcore function is thus the output of a PRF, the key for the latter embedded in an obfuscated circuit to prevent its being revealed.

RESULTS. Recall that a $\mathbf{S}_{\mathrm{diff}}(1)$-secure obfuscator is weaker than a full $\mathbf{S}_{\mathrm{diff}}$-secure obfuscator (see Section 3 for definitions) since it is only required to work on circuits that differ on at most one (hard to find) input. We have:

Theorem 4. *Let* F *be an injective one-way function family. Let* G *be a punctured PRF with* $\mathsf{G.il} = \mathsf{F.il}$. *Then there is a polynomial* s *such that the following is true. Let* Obf *be any* $\mathbf{S}_{\mathrm{diff}}(1)$*-secure obfuscator. Then the function family* $\mathsf{H} = \mathbf{HC1}[\mathsf{G}, \mathsf{Obf}, s]$ *defined above is hardcore for* F.

Proposition 1 yields punctured PRFs with as long an output as desired, so that the above allows extraction of an arbitrary polynomial number of hardcore bits. The one-way function assumption made in Proposition 1 is already implied by the assumption that F is one way. Theorem 4 assumes a differing-inputs obfuscator only for circuits that differ on at most one input, which by Proposition 3 can be obtained from an indistinguishability obfuscator, making the latter the only assumption beyond one-wayness of F needed to extract polynomially-many hardcore bits. Formally, we have:

Corollary 5. *Let* ℓ *be any polynomial. Let* F *be any injective one-way function. If there exists a* $\mathbf{S}_{\mathrm{eq}}$*-secure obfuscator then there exists a hardcore function* H *for* F *with* $\mathsf{H.ol} = \ell$.

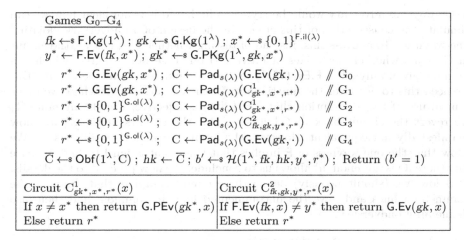

Fig. 3. Games for proof of Theorem 4

Proof (Theorem 4). We define s as follows: For any $\lambda \in \mathbb{N}$ let $s(\lambda)$ be a polynomial upper bound on $\max(|\mathsf{G.Ev}(gk, \cdot)|, |\mathrm{C}^1_{gk^*, x^*, r^*}|, |\mathrm{C}^2_{fk, gk, y^*, r^*}|)$ where the last two circuits are in Fig. 3 and the maximum is over all $gk \in [\mathsf{G.Kg}(1^\lambda)]$, $x^* \in \{0,1\}^{\mathsf{G.il}(\lambda)}$, $gk^* \in [\mathsf{G.PKg}(1^\lambda, gk, x^*)]$, $fk \in [\mathsf{F.Kg}(1^\lambda)]$, $y^* \in \{0,1\}^{\mathsf{F.ol}(\lambda)}$ and $r^* \in \{0,1\}^{\mathsf{G.ol}(\lambda)}$. Now let $\mathcal{H}$ be a PT adversary. Consider the games and associated circuits of Fig. 3. Lines not annotated with comments are common to all five games. The games begin by picking keys fk, gk for the one-way function F and the punctured-PRF G, respectively. They pick the challenge input x^* and then create a punctured PRF key gk^* for it. Then the games differ in how they define r^* and the circuit C to be obuscated. These defined, the games re-unite to obfuscate the circuit and run $\mathcal{H}$.

Game G_0 does not use the punctured keys, and is equivalent to the $b = 1$ case of $\mathrm{HC}^{\mathcal{H}}_{\mathsf{F},\mathcal{H}}(\lambda)$ while G_4, similarly, is its $b = 0$ case, so

$$\mathsf{Adv}^{\mathsf{hc}}_{\mathsf{F},\mathsf{H},\mathcal{H}}(\lambda) = \Pr[\mathrm{G}_0] - \Pr[\mathrm{G}_4] . \tag{1}$$

We now show that $\Pr[\mathrm{G}_{i-1}] - \Pr[\mathrm{G}_i]$ is negligible for $i = 1, 2, 3, 4$, which by Equation (1) implies that $\mathsf{Adv}^{\mathsf{hc}}_{\mathsf{F},\mathsf{H},\mathcal{H}}(\cdot)$ is negligible and proves the theorem. We begin with some intuition. In game G_1, we switch the circuit being obfuscated to one that uses the punctured key when $x \neq x^*$ and returns an embedded $r^* = \mathsf{G.Ev}(gk, x^*)$ otherwise. This circuit is equivalent to $\mathsf{G.Ev}(gk, \cdot)$ so iO-security, implied by diO, will tell us that the adversary $\mathcal{H}$ will hardly notice. This switch puts us in position to use the security of the punctured-PRF, based on which G_2 replaces r^* with a random value. These steps are direct applications of the Sahai-Waters technique [53], but now things get more difficult. Having made r^* random is not enough because we must now revert the circuit being obfuscated back to $\mathsf{G.Ev}(gk, \cdot)$. We also realize that we have not yet used the one-wayness of F, so this reversion must rely on it. But the circuit in G_2 embeds x^*, the point a one-wayness adversary would be trying to find, and it is not clear how we can simulate the construction of this circuit in the design of an adversary violating one-wayness. To address this, instead of testing whether x equals x^*, the circuit in G_3 tests whether the values of $F.Ev(fk, \cdot)$ on these points agree, which can be done given only $y^* = F.Ev(fk, x^*)$ and avoids putting x^* in the circuit. But we need this to not alter the functionality of the circuit. This is where we make crucial use of the assumption that $F.Ev(gk, \cdot)$ is injective. Finally, in game G_4 we revert the circuit back to $G.Ev(gk, \cdot)$. But the circuits in G_3, G_4 are now manifestedly *not* equivalent. However, the input on which they differ is x^*. We show that the one-wayness of F implies that this point is hard to find, whence diO (here iO is not enough) allows us to conclude. We now proceed to the details.

Below, we (simultaneously) define three circuit samplers that differ at the commented lines and have the uncommented lines in common, and then also define an iO-adversary:

Circuit Samplers $\mathcal{S}_1(1^\lambda)$, $\mathcal{S}_3(1^\lambda)$, $\mathcal{S}_4(1^\lambda)$

$fk \leftarrow_\$ F.Kg(1^\lambda)$; $gk \leftarrow_\$ G.Kg(1^\lambda)$
$x^* \leftarrow_\$ \{0,1\}^{F.il(\lambda)}$; $y^* \leftarrow F.Ev(fk, x^*)$; $gk^* \leftarrow_\$ G.PKg(1^\lambda, gk, x^*)$
$r^* \leftarrow G.Ev(gk, x^*)$; $C_1 \leftarrow Pad_{s(\lambda)}(G.Ev(gk, \cdot))$; $C_0 \leftarrow Pad_{s(\lambda)}(C^1_{gk^*, x^*, r^*})$ // $\mathcal{S}_1$
$r^* \leftarrow_\$ \{0,1\}^{G.ol(\lambda)}$; $C_1 \leftarrow Pad_{s(\lambda)}(C^1_{gk^*, x^*, r^*})$; $C_0 \leftarrow Pad_{s(\lambda)}(C^2_{fk, gk, y^*, r^*})$ // $\mathcal{S}_3$
$r^* \leftarrow_\$ \{0,1\}^{G.ol(\lambda)}$; $C_1 \leftarrow Pad_{s(\lambda)}(C^2_{fk, gk, y^*, r^*})$; $C_0 \leftarrow Pad_{s(\lambda)}(G.Ev(gk, \cdot))$ // $\mathcal{S}_4$
$aux \leftarrow (fk, y^*, r^*)$; Return (C_0, C_1, aux)

Adversary $\mathcal{O}(1^\lambda, \overline{C}, aux)$

$(fk, y^*, r^*) \leftarrow aux$; $hk \leftarrow \overline{C}$; $b' \leftarrow_\$ \mathcal{H}(1^\lambda, fk, hk, y^*, r^*)$
Return b'

Now we have

$$\Pr[G_{i-1}] - \Pr[G_i] = Adv^{io}_{Obf, \mathcal{S}_i, \mathcal{O}}(\lambda) \qquad \text{for } i \in \{1, 3, 4\} . \qquad (2)$$

We now make three claims: (1) $\mathcal{S}_1 \in \mathbf{S}_{eq}$ (2) $\mathcal{S}_3 \in \mathbf{S}_{eq}$ (3) $\mathcal{S}_4 \in \mathbf{S}_{diff}(1)$. Since $\mathbf{S}_{eq} \subseteq \mathbf{S}_{diff}(1)$ and Obf is assumed $\mathbf{S}_{diff}(1)$-secure, the RHS of Equation (2) is negligible in all three cases.

We now establish claim (1). If $x \neq x^*$ then $C^1_{gk^*, x^*, r^*}(x) = G.PEv(gk^*, x) = G.Ev(gk, x)$. If $x = x^*$ then $C^1_{gk^*, x^*, r^*}(x) = r^*$, but $\mathcal{S}_1$ sets $r^* = G.Ev(gk, x^*)$. This means that $\mathcal{S}_1$ produces equivalent circuits, and hence $\mathcal{S}_1 \in \mathbf{S}_{eq}$.

Next we establish claim (2). The assumed injectivity of F implies that circuits $C^1_{gk^*, x^*, r^*}$ and C^2_{fk, gk, y^*, r^*} are equivalent when $y^* = F.Ev(fk, x^*)$, and hence $\mathcal{S}_3 \in \mathbf{S}_{eq}$.

To establish claim (3), given any PT difference adversary $\mathcal{D}$ for $\mathcal{S}_4$, we build one-wayness adversary $\mathcal{F}$ via

---

Adversary $\mathcal{F}(1^\lambda, fk, y^*)$

$gk \leftarrow_\$ G.Kg(1^\lambda)$; $r^* \leftarrow_\$ \{0,1\}^{G.ol(\lambda)}$
$C_1 \leftarrow Pad_{s(\lambda)}(C^2_{fk, gk, y^*, r^*})$; $C_0 \leftarrow Pad_{s(\lambda)}(G.Ev(gk, \cdot))$
$aux \leftarrow (fk, y^*, r^*)$; $x \leftarrow_\$ \mathcal{D}(C_0, C_1, aux)$; Return x

If $C_1(x) \neq C_0(x)$ then it must be that $\mathsf{F}.\mathsf{Ev}(fk, x) = y^*$. Thus $\mathsf{Adv}^{\mathrm{diff}}_{\mathcal{S}_4, \mathcal{D}}(\cdot) \leq \mathsf{Adv}^{\mathrm{ow}}_{\mathsf{F}, \mathcal{F}}(\cdot)$. The assumed one-wayness of F thus means that $\mathcal{S}_4$ is difference-secure. But we also observe that, due to the injectivity of F, circuits C_0, C_1 differ on only one input, namely x^*. So $\mathcal{S}_4 \in \mathbf{S}_{\mathrm{diff}}(1)$.

One transition remains, namely that from G_1 to G_2. Here we have

$$\Pr[G_1] - \Pr[G_2] = \mathsf{Adv}^{\mathrm{pprf}}_{\mathsf{G}, \mathcal{G}}(\lambda) \tag{3}$$

where adversary $\mathcal{G}$ is defined via

Adversary $\mathcal{G}^{\mathrm{CH}}(1^\lambda)$

$fk \leftarrow_\$ \mathsf{F}.\mathsf{Kg}(1^\lambda)$; $x^* \leftarrow_\$ \{0, 1\}^{\mathsf{F}.\mathrm{il}(\lambda)}$; $y^* \leftarrow \mathsf{F}.\mathsf{Ev}(fk, x^*)$; $(gk^*, r^*) \leftarrow_\$ \mathrm{CH}(x^*)$
$C \leftarrow \mathsf{Pad}_{s(\lambda)}(C^1_{gk^*, x^*, r^*})$; $hk \leftarrow_\$ \mathsf{Obf}(1^\lambda, C)$; $b' \leftarrow_\$ \mathcal{H}(1^\lambda, fk, hk, y^*, r^*)$
Return b'

The RHS of Equation (3) is negligible by the assumption that G is a punctured PRF. This concludes the proof.

5 Poly-Many Hardcore Bits for Any OWF

The proof of Theorem 4 makes crucial use of the assumed injectivity of F. To remove this assumption, we modify the construction so that the obfuscated PRF is applied, not to x, but to the result of the one-way function on x.

CONSTRUCTION. Let F, G be function families with $\mathsf{G}.\mathrm{il} = \mathsf{F}.\mathrm{ol}$. Let Obf be an obfuscator and let $s \colon \mathbb{N} \to \mathbb{N}$. We define function family $\mathsf{H} = \mathbf{HC2}[\mathsf{F}, \mathsf{G}, \mathsf{Obf}, s]$ as follows, with $\mathsf{H}.\mathrm{il} = \mathsf{F}.\mathrm{il}$ and $\mathsf{H}.\mathrm{ol} = \mathsf{G}.\mathrm{ol}$:

$\mathsf{H}.\mathsf{Kg}(1^\lambda, fk)$	$\mathsf{H}.\mathsf{Ev}(hk, x)$
$gk \leftarrow_\$ \mathsf{G}.\mathsf{Kg}(1^\lambda)$; $C \leftarrow \mathsf{Pad}_{s(\lambda)}(\mathsf{G}.\mathsf{Ev}(gk, \mathsf{F}.\mathsf{Ev}(fk, \cdot)))$	$\overline{C} \leftarrow hk$; $r \leftarrow \overline{C}(x)$
$\overline{C} \leftarrow_\$ \mathsf{Obf}(1^\lambda, C)$	Return r
$hk \leftarrow \overline{C}$; Return hk	

This time the result of the hardcore function output on input x is $\mathsf{G}.\mathsf{Ev}(gk, y)$ where $y = \mathsf{F}.\mathsf{Ev}(fk, x)$.

RESULTS. The following says that $\mathbf{S}_{\mathrm{diff}}$-security of the obfuscator suffices for the security of the above hardcore function, but that in fact the assumption is weaker, the circuit samplers for which security is required being further restricted to have short auxiliary input as captured and to produce circuits that differ at a number of points bounded by the pre-image size d of the one-way function, formally $(\mathbf{S}_{\mathrm{diff}}(d) \cap \mathbf{S}^{\mathrm{sh}})$-security, where the classes were defined in Section 3. The proof is in [12]:

Theorem 6. *Let F be a one-way function family. Let G be a punctured PRF with $\mathsf{G}.\mathrm{il} = \mathsf{F}.\mathrm{ol}$. Let $d = \mathrm{PREIMG}_\mathsf{F}$. Then there is a polynomial s such that the following is true. Let $\mathbf{S} = \mathbf{S}_{\mathrm{diff}}(d) \cap \mathbf{S}^{\mathrm{sh}}$. Let Obf be any $\mathbf{S}$-secure obfuscator. Then the function family $\mathsf{H} = \mathbf{HC2}[\mathsf{F}, \mathsf{G}, \mathsf{Obf}, s]$ defined above is hardcore for F.*

This means that in the most general case we have:

Corollary 7. *Let ℓ be any polynomial. Let* F *be any one-way function. If there exists a* $(\boldsymbol{S}_{\mathrm{diff}} \cap \boldsymbol{S}^{\mathrm{sh}})$*-secure obfuscator then there exists a hardcore function* H *for* F *with* H.ol $= \ell$.

When the pre-image size of F is polynomial, however, we can again exploit Proposition 3 to obtain our conclusion assuming nothing beyond iO:

Corollary 8. *Let ℓ be any polynomial. Let* F *be any one-way function with polynomially-bounded pre-image size. If there exists a* $\boldsymbol{S}_{\mathrm{eq}}$*-secure obfuscator then there exists a hardcore function* H *for* F *with* H.ol $= \ell$.

6   Hardcore Functions for Correlated Inputs

We show that our hardcore functions are able to extract random bits even on sequences of inputs that are arbitrarily correlated. Somewhat more precisely, draw a vector **x** from an arbitrary distribution, in particuclar allowing its components to be arbitrarily correlated. Then applying our hardcore function componentwise to **x** results in a vector whose components look random *and independent* even given the result of applying f componentwise to **x**, making only the necessary assumption that f remains one-way on the distribution from which **x** was selected. This is an unusual property, not possessed by all constructions of hardcore functions. The ability to extract polynomially-many bits on correlated inputs leads to new constructions of deterministic PKE schemes.

NOTATION. We denote vectors by boldface lowercase letters, for example **x**. If **x** is a vector then $|\mathbf{x}|$ denotes the number of components of **x** and $\mathbf{x}[i]$ denotes the i-th component of **x**, for any $1 \leq i \leq |\mathbf{x}|$. We write $x \in \mathbf{x}$ to mean that $x = \mathbf{x}[i]$ for some $1 \leq i \leq |\mathbf{x}|$. If F is a family of functions, **x** is a vector over $\{0,1\}^{\mathsf{F}.\mathsf{il}(\lambda)}$ and $fk \in \{0,1\}^{\mathsf{F}.\mathsf{kl}(\lambda)}$, then we let $\mathsf{F}.\mathsf{Ev}(fk, \mathbf{x}) = (\mathsf{F}.\mathsf{Ev}(fk, \mathbf{x}[1]), \ldots, \mathsf{F}.\mathsf{Ev}(fk, \mathbf{x}[|\mathbf{x}|]))$. Let **Rnd** denote the algorithm that on input a vector **x** and an integer ℓ returns a vector **r** of the same length as **x** whose entries are random ℓ-bit strings except that if two entries of **x** are the same, the same is true of the corresponding entries of **r**. In detail, $\mathbf{Rnd}(\mathbf{x}, \ell)$ creates table T via: For $i = 1, \ldots, |\mathbf{x}|$ do: If not $T[\mathbf{x}[i]]$ then $T[\mathbf{x}[i]] \leftarrow_\$ \{0,1\}^\ell$. Then it sets $\mathbf{r}[i] \leftarrow T[\mathbf{x}[i]]$ for $i = 1, \ldots, |\mathbf{x}|$ and returns the vector **r**.

DEFINITIONS. An input sampler is an algorithm $\mathcal{I}$ that on input 1^λ returns a $\mathcal{I}.\mathsf{vl}(\lambda)$-vector of strings over $\{0,1\}^{\mathcal{I}.\mathsf{il}(\lambda)}$, where the vector-length $\mathcal{I}.\mathsf{vl}: \mathbb{N} \to \mathbb{N}$ and the input length $\mathcal{I}.\mathsf{il}: \mathbb{N} \to \mathbb{N}$ are polynomials associated to $\mathcal{I}$. We say that a function family F is one-way with respect to input sampler $\mathcal{I}$ if $\mathsf{F}.\mathsf{il} = \mathcal{I}.\mathsf{il}$ and $\mathsf{Adv}^{\mathrm{ow}}_{\mathsf{F},\mathcal{I},\mathcal{F}}(\cdot)$ is negligible for all PT adversaries $\mathcal{F}$, where $\mathsf{Adv}^{\mathrm{ow}}_{\mathsf{F},\mathcal{I},\mathcal{F}}(\lambda) = \Pr[\mathrm{OW}^{\mathcal{F}}_{\mathsf{F},\mathcal{I}}(\lambda) = 1]$ and game $\mathrm{OW}^{\mathcal{F}}_{\mathsf{F},\mathcal{I}}(\lambda)$ is defined in Fig. 4. We stress that for $\mathcal{F}$ to win in this game it needs to find the inverse under $\mathsf{F}.\mathsf{Ev}(fk, \cdot)$ of *some* component of $\mathbf{y}^*$, not all components. Let H be a family of functions with $\mathsf{H}.\mathsf{il} = \mathsf{F}.\mathsf{il}$. We say that H is hardcore for F with respect to input sampler $\mathcal{I}$ if $\mathsf{Adv}^{\mathrm{hc}}_{\mathsf{F},\mathsf{H},\mathcal{I},\mathcal{H}}(\cdot)$ is negligible for all PT adversaries $\mathcal{H}$, where $\mathsf{Adv}^{\mathrm{hc}}_{\mathsf{F},\mathsf{H},\mathcal{I},\mathcal{H}}(\lambda) = 2\Pr[\mathrm{HC}^{\mathcal{H}}_{\mathsf{F},\mathsf{H},\mathcal{I}}(\lambda)] - 1$ and game $\mathrm{HC}^{\mathcal{H}}_{\mathsf{F},\mathsf{H},\mathcal{I}}(\lambda)$ is defined in Fig. 4.

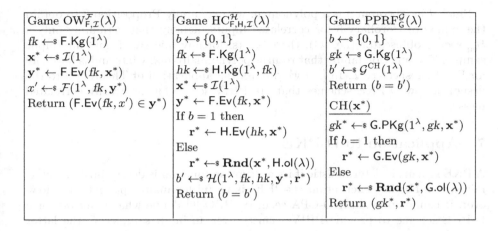

Fig. 4. Games defining one-wayness of F with respect to an input sampler $\mathcal{I}$, security of H as a hardcore function for F with respect to $\mathcal{I}$ and punctured-PRF security of G on multiple inputs

We extend punctured PRFs as defined in Section 2 to allow puncturing at multiple points. In a punctured function family G, algorithm G.PKg now takes 1^λ, a key $gk \in [\text{G.Kg}(1^\lambda)]$ and a vector $\mathbf{x}^*$ over $\{0,1\}^{\text{G.il}(\lambda)}$ (the target inputs) to return a "punctured" key gk^* such that $\text{G.PEv}(gk^*, x) = \text{G.Ev}(gk, x)$ for all $x \in \{0,1\}^{\text{G.il}(\lambda)}$ such that $x \notin \mathbf{x}^*$. G is a punctured PRF if $\text{Adv}^{\text{pprf}}_{\text{G},\mathcal{G}}(\cdot)$ is negligible for all PT adversaries $\mathcal{G}$, where $\text{Adv}^{\text{pprf}}_{\text{G}}(\lambda) = 2\Pr[\text{PPRF}^{\mathcal{G}}_{\text{G}}(\lambda)] - 1$ and game $\text{PPRF}^{\mathcal{G}}_{\text{G}}(\lambda)$ is defined in Fig. 4. Here $\mathcal{G}$ must make exactly one oracle query consisting of a vector over $\{0,1\}^{\text{G.il}(\lambda)}$. Proposition 1 extends, and we exploit this below.

RESULTS. The following says that our construction for injective functions F extracts hardcore bits for any PT input sampler $\mathcal{I}$ with respect to which F is one way, meaning even for arbitrarily correlated inputs. The proof is in [12]:

Theorem 9. *Let* F *be an injective function family. Let* G *be a punctured PRF with* G.il = F.il. *Let* $d\colon \mathbb{N} \to \mathbb{N}$ *be a polynomial. Then there is a polynomial* s *such that the following is true. Let* $\mathcal{I}$ *be any PT input sampler with* $\mathcal{I}.\text{vl} = d$ *and* $\mathcal{I}.\text{il} = \text{F.il}$. *Let* Obf *be any* $\mathbf{S}_{\text{diff}}(d)$-*secure obfuscator. Assume* F *is one-way with respect to* $\mathcal{I}$. *Then the function family* $H = \mathbf{HC1}[\text{G}, \text{Obf}, s]$ *defined in Section 4 is hardcore for* F *with respect to* $\mathcal{I}$.

A subtle point here is that s depends on $d = \mathcal{I}.\text{vl}$ in addition to F and G, which means that the size of the key hk describing the hardcore function grows with the number of correlated inputs d on which we want the function to be hardcore. This is expected and due to [57] may not be avoidable under falsifiable assumptions. Importantly for our applications, H does not depend on $\mathcal{I}$ beyond depending on $d = \mathcal{I}.\text{vl}$ and $\text{F.il} = \mathcal{I}.\text{il}$.

Since d in Theorem 9 is a polynomial, we may apply Proposition 3 to obtain the analog of Corollary 5 for correlated inputs, namely that, assuming only a S_{eq}-secure obfuscator (i.e. iO), there exists, for any injective F and any input sampler $\mathcal{I}$, a function family that returns polynomially-many bits and is hardcore for F with respect to $\mathcal{I}$. In [12] we discuss the application of Theorem 9 to the design of new D-PKE schemes that are PRIV-secure for arbitrarily correlated messages.

7   Application to D-PKE

A PKE scheme is deterministic if its encryption function is deterministic. Deterministic public-key encryption (D-PKE) is useful for many applications. However, it cannot provide IND-CPA security. BBO [6] define what it means for a D-PKE scheme to provide PRIV-security over an input sampler $\mathcal{I}$, the latter returning vectors of *arbitrarily correlated* messages to be encrypted. We restrict attention to distributions that are admissible, meaning that there exists a family of injective, trapdoor functions that is one-way relative to $\mathcal{I}$. The basic question that emerges is, for which admissible distributions $\mathcal{I}$ does there exist a D-PKE scheme that is PRIV-secure over $\mathcal{I}$? We show that this is true for all admissible distributions that are efficiently sampleable, assuming only the existence of iO. We obtain this result by combining Theorem 9 with techniques from [30]. Previously, this was known only in the ROM [6], under the assumption that UCE-secure functions exist [9], for distributions with limited correlation between messages [8,18] or assuming lossy trapdoor functions [30]. See [12] for details.

Acknowledgments. We thank Daniel Wichs and Elette Boyle for comments and discussions. Bellare and Stepanovs were supported in part by grants NSF CNS-1116800 and NSF CNS-1228890. Tessaro was supported in part by grant NSF CNS-1423566.

References

1. Akavia, A., Goldwasser, S., Safra, S.: Proving hard-core predicates using list decoding. In: 44th FOCS, pp. 146–159. IEEE Computer Society Press (October 2003)
2. Akavia, A., Goldwasser, S., Vaikuntanathan, V.: Simultaneous hardcore bits and cryptography against memory attacks. In: Reingold, O. (ed.) TCC 2009. LNCS, vol. 5444, pp. 474–495. Springer, Heidelberg (2009)
3. Alexi, W., Chor, B., Goldreich, O., Schnorr, C.P.: RSA and Rabin functions: Certain parts are as hard as the whole. SIAM Journal on Computing 17(2), 194–209 (1988)
4. Ananth, P., Boneh, D., Garg, S., Sahai, A., Zhandry, M.: Differing-inputs obfuscation and applications. Cryptology ePrint Archive, Report 2013/689, Version 20131031:060024 (October 31, 2013)
5. Barak, B., Goldreich, O., Impagliazzo, R., Rudich, S., Sahai, A., Vadhan, S.P., Yang, K.: On the (im)possibility of obfuscating programs. In: Kilian, J. (ed.) CRYPTO 2001. LNCS, vol. 2139, pp. 1–18. Springer, Heidelberg (2001)

6. Bellare, M., Boldyreva, A., O'Neill, A.: Deterministic and efficiently searchable encryption. In: Menezes, A. (ed.) CRYPTO 2007. LNCS, vol. 4622, pp. 535–552. Springer, Heidelberg (2007)
7. Bellare, M., Brakerski, Z., Naor, M., Ristenpart, T., Segev, G., Shacham, H., Yilek, S.: Hedged public-key encryption: How to protect against bad randomness. In: Matsui, M. (ed.) ASIACRYPT 2009. LNCS, vol. 5912, pp. 232–249. Springer, Heidelberg (2009)
8. Bellare, M., Fischlin, M., O'Neill, A., Ristenpart, T.: Deterministic encryption: Definitional equivalences and constructions without random oracles. In: Wagner, D. (ed.) CRYPTO 2008. LNCS, vol. 5157, pp. 360–378. Springer, Heidelberg (2008)
9. Bellare, M., Hoang, V.T., Keelveedhi, S.: Instantiating random oracles via UCEs. Cryptology ePrint Archive, Report 2013/424. Preliminary version in CRYPTO (2013)
10. Bellare, M., Rogaway, P.: Random oracles are practical: A paradigm for designing efficient protocols. In: Ashby, V. (ed.) ACM CCS 1993, pp. 62–73. ACM Press (November 1993)
11. Bellare, M., Rogaway, P.: The security of triple encryption and a framework for code-based game-playing proofs. In: Vaudenay, S. (ed.) EUROCRYPT 2006. LNCS, vol. 4004, pp. 409–426. Springer, Heidelberg (2006)
12. Bellare, M., Stepanovs, I., Tessaro, S.: Poly-many hardcore bits for any one-way function and a framework for differing-inputs obfuscation. Cryptology ePrint Archive, Report 2013/873 (2013), http://eprint.iacr.org/2013/873
13. Bitansky, N., Canetti, R., Paneth, O., Rosen, A.: On the existence of extractable one-way functions. In: Shmoys, D.B. (ed.) 46th ACM STOC, pp. 505–514. ACM Press (May/June 2014)
14. Bitansky, N., Paneth, O.: Point obfuscation and 3-round zero-knowledge. In: Cramer, R. (ed.) TCC 2012. LNCS, vol. 7194, pp. 190–208. Springer, Heidelberg (2012)
15. Blum, L., Blum, M., Shub, M.: A simple unpredictable pseudo-random number generator. SIAM Journal on Computing 15(2), 364–383 (1986)
16. Blum, M., Goldwasser, S.: An Efficient probabilistic public key encryption scheme which hides all partial information. In: Blakely, G.R., Chaum, D. (eds.) CRYPTO 1984. LNCS, vol. 196, pp. 289–299. Springer, Heidelberg (1985)
17. Blum, M., Micali, S.: How to generate cryptographically strong sequences of pseudorandom bits. SIAM Journal on Computing 13(4), 850–864 (1984)
18. Boldyreva, A., Fehr, S., O'Neill, A.: On notions of security for deterministic encryption, and efficient constructions without random oracles. In: Wagner, D. (ed.) CRYPTO 2008. LNCS, vol. 5157, pp. 335–359. Springer, Heidelberg (2008)
19. Boneh, D., Waters, B.: Constrained pseudorandom functions and their applications. In: Sako, K., Sarkar, P. (eds.) ASIACRYPT 2013, Part II. LNCS, vol. 8270, pp. 280–300. Springer, Heidelberg (2013)
20. Boyle, E., Chung, K.-M., Pass, R.: On extractability obfuscation. In: Lindell, Y. (ed.) TCC 2014. LNCS, vol. 8349, pp. 52–73. Springer, Heidelberg (2014)
21. Boyle, E., Goldwasser, S., Ivan, I.: Functional signatures and pseudorandom functions. In: Krawczyk, H. (ed.) PKC 2014. LNCS, vol. 8383, pp. 501–519. Springer, Heidelberg (2014)
22. Boyle, E., Pass, R.: Limits of extractability assumptions with distributional auxiliary input. Cryptology ePrint Archive, Report 2013/703 (2013)
23. Brzuska, C., Mittelbach, A.: Using indistinguishability obfuscation via UCEs. Cryptology ePrint Archive, Report 2014/381 (2014), http://eprint.iacr.org/2014/381

24. Canetti, R.: Towards realizing random oracles: Hash functions that hide all partial information. In: Kaliski Jr., B.S. (ed.) CRYPTO 1997. LNCS, vol. 1294, pp. 455–469. Springer, Heidelberg (1997)

25. Catalano, D., Gennaro, R., Howgrave-Graham, N.: The bit security of paillier's encryption scheme and its applications. In: Pfitzmann, B. (ed.) EUROCRYPT 2001. LNCS, vol. 2045, pp. 229–243. Springer, Heidelberg (2001)

26. Cramer, R., Shoup, V.: A practical public key cryptosystem provably secure against adaptive chosen ciphertext attack. In: Krawczyk, H. (ed.) CRYPTO 1998. LNCS, vol. 1462, pp. 13–25. Springer, Heidelberg (1998)

27. Dodis, Y., Goldwasser, S., Tauman Kalai, Y., Peikert, C., Vaikuntanathan, V.: Public-key encryption schemes with auxiliary inputs. In: Micciancio, D. (ed.) TCC 2010. LNCS, vol. 5978, pp. 361–381. Springer, Heidelberg (2010)

28. Fischlin, R., Schnorr, C.-P.: Stronger security proofs for RSA and Rabin bits. Journal of Cryptology 13(2), 221–244 (2000)

29. Freeman, D.M., Goldreich, O., Kiltz, E., Rosen, A., Segev, G.: More constructions of lossy and correlation-secure trapdoor functions. Journal of Cryptology 26(1), 39–74 (2013)

30. Fuller, B., O'Neill, A., Reyzin, L.: A unified approach to deterministic encryption: New constructions and a connection to computational entropy. In: Cramer, R. (ed.) TCC 2012. LNCS, vol. 7194, pp. 582–599. Springer, Heidelberg (2012)

31. Garg, S., Gentry, C., Halevi, S., Raykova, M., Sahai, A., Waters, B.: Candidate indistinguishability obfuscation and functional encryption for all circuits. In: 54th FOCS, pp. 40–49. IEEE Computer Society Press (October 2013)

32. Garg, S., Gentry, C., Halevi, S., Wichs, D.: On the implausibility of differing-inputs obfuscation and extractable witness encryption with auxiliary input. In: Garay, J.A., Gennaro, R. (eds.) CRYPTO 2014, Part I. LNCS, vol. 8616, pp. 518–535. Springer, Heidelberg (2014)

33. Gennaro, R.: An improved pseudo-random generator based on the discrete logarithm problem. Journal of Cryptology 18(2), 91–110 (2005)

34. Goldreich, O., Goldwasser, S., Micali, S.: How to construct random functions. Journal of the ACM 33, 792–807 (1986)

35. Goldreich, O., Levin, L.A.: A hard-core predicate for all one-way functions. In: 21st ACM STOC, pp. 25–32. ACM Press (May 1989)

36. Goldreich, O., Rosen, V.: On the security of modular exponentiation with application to the construction of pseudorandom generators. Journal of Cryptology 16(2), 71–93 (2003)

37. Goldwasser, S., Micali, S.: Probabilistic encryption. Journal of Computer and System Sciences 28(2), 270–299 (1984)

38. Goyal, V., O'Neill, A., Rao, V.: Correlated-input secure hash functions. In: Ishai, Y. (ed.) TCC 2011. LNCS, vol. 6597, pp. 182–200. Springer, Heidelberg (2011)

39. Håstad, J., Näslund, M.: The security of individual RSA bits. In: 39th FOCS, pp. 510–521. IEEE Computer Society Press (November 1998)

40. Håstad, J., Schrift, A., Shamir, A.: The discrete logarithm modulo a composite hides $O(n)$ bits. Journal of Computer and System Sciences 47(3), 376–404 (1993)

41. Hohenberger, S., Sahai, A., Waters, B.: Replacing a random oracle: Full domain hash from indistinguishability obfuscation. In: Nguyen, P.Q., Oswald, E. (eds.) EUROCRYPT 2014. LNCS, vol. 8441, pp. 201–220. Springer, Heidelberg (2014)

42. Kaliski Jr., B.S.: A pseudo-random bit generator based on elliptic logarithms. In: Odlyzko, A.M. (ed.) CRYPTO 1986. LNCS, vol. 263, pp. 84–103. Springer, Heidelberg (1987)

43. Kiayias, A., Papadopoulos, S., Triandopoulos, N., Zacharias, T.: Delegatable pseudorandom functions and applications. In: Sadeghi, A.-R., Gligor, V.D., Yung, M. (eds.) ACM CCS 2013, pp. 669–684. ACM Press (November 2013)

44. Kiltz, E., O'Neill, A., Smith, A.: Instantiability of RSA-OAEP under chosen-plaintext attack. In: Rabin, T. (ed.) CRYPTO 2010. LNCS, vol. 6223, pp. 295–313. Springer, Heidelberg (2010)

45. Lewko, M., O'Neill, A., Smith, A.: Regularity of lossy RSA on subdomains and its applications. In: Johansson, T., Nguyen, P.Q. (eds.) EUROCRYPT 2013. LNCS, vol. 7881, pp. 55–75. Springer, Heidelberg (2013)

46. Long, D.L., Wigderson, A.: The discrete logarithm hides $O(\log n)$ bits. SIAM Journal on Computing 17(2), 363–372 (1988)

47. MacKenzie, P.D., Patel, S.: Hard bits of the discrete log with applications to password authentication. In: Menezes, A. (ed.) CT-RSA 2005. LNCS, vol. 3376, pp. 209–226. Springer, Heidelberg (2005)

48. Näslund, M.: All bits in $ax + b$ mod p are hard. In: Koblitz, N. (ed.) CRYPTO 1996. LNCS, vol. 1109, pp. 114–128. Springer, Heidelberg (1996)

49. Paillier, P.: Public-key cryptosystems based on composite degree residuosity classes. In: Stern, J. (ed.) EUROCRYPT 1999. LNCS, vol. 1592, pp. 223–238. Springer, Heidelberg (1999)

50. Patel, S., Sundaram, G.S.: An efficient discrete log pseudo random generator. In: Krawczyk, H. (ed.) CRYPTO 1998. LNCS, vol. 1462, pp. 304–317. Springer, Heidelberg (1998)

51. Peikert, C., Waters, B.: Lossy trapdoor functions and their applications. In: Ladner, R.E., Dwork, C. (eds.) 40th ACM STOC, pp. 187–196. ACM Press (May 2008)

52. Rivest, R.L., Shamir, A., Adleman, L.M.: A method for obtaining digital signature and public-key cryptosystems. Communications of the Association for Computing Machinery 21(2), 120–126 (1978)

53. Sahai, A., Waters, B.: How to use indistinguishability obfuscation: deniable encryption, and more. In: Shmoys, D.B. (ed.) 46th ACM STOC, pp. 475–484. ACM Press (May/June 2014)

54. Alexi, W., Schnorr, C.-P.: RSA-bits are $0.5 + \epsilon$ secure. In: Beth, T., Cot, N., Ingemarsson, I. (eds.) EUROCRYPT 1984. LNCS, vol. 209, pp. 113–126. Springer, Heidelberg (1985)

55. Steinfeld, R., Pieprzyk, J., Wang, H.: On the provable security of an efficient RSA-based pseudorandom generator. In: Lai, X., Chen, K. (eds.) ASIACRYPT 2006. LNCS, vol. 4284, pp. 194–209. Springer, Heidelberg (2006)

56. Vazirani, U.V., Vazirani, V.V.: Efficient and secure pseudo-random number generation. In: Blakely, G.R., Chaum, D. (eds.) CRYPTO 1984. LNCS, vol. 196, pp. 193–202. Springer, Heidelberg (1985)

57. Wichs, D.: Barriers in cryptography with weak, correlated and leaky sources. In: Kleinberg, R.D. (ed.) ITCS 2013, pp. 111–126. ACM (January 2013)

58. Yao, A.C.-C.: Theory and applications of trapdoor functions (extended abstract). In: 23rd FOCS, pp. 80–91. IEEE Computer Society Press (November 1982)

59. Zhandry, M.: How to avoid obfuscation using witness PRFs. Cryptology ePrint Archive, Report 2014/301 (2014), http://eprint.iacr.org/2014/301

Using Indistinguishability Obfuscation via UCEs

Christina Brzuska[1] and Arno Mittelbach[2]

[1] Tel Aviv University, Israel
[2] Darmstadt University of Technology, Germany
brzuska@post.tau.ac.il, arno.mittelbach@cased.de

Abstract. We provide the first standard model construction for a powerful class of Universal Computational Extractors (UCEs; Bellare et al. Crypto 2013) based on indistinguishability obfuscation. Our construction suffices to instantiate q-query correlation-secure hash functions and to extract polynomially many hardcore bits from any one-way function.

For many cryptographic primitives and in particular for correlation-secure hash functions all known constructions are in the random-oracle model. Indeed, recent negative results by Wichs (ITCS 2013) rule out a large class of techniques to prove the security of correlation-secure hash functions in the standard model. Our construction is based on puncturable PRFs (Sahai und Waters; STOC 2014) and indistinguishability obfuscation. However, our proof also relies on point obfuscation under auxiliary inputs (AIPO). This is crucial in light of Wichs' impossibility result. Namely, Wichs proves that it is often hard to reduce two-stage games (such as UCEs) to a "one-stage assumption" such as DDH. In contrast, AIPOs and their underlying assumptions are inherently two-stage and, thus, allow us to circumvent Wichs' impossibility result.

Our positive result is also noteworthy insofar as Brzuska, Farshim and Mittelbach (Crypto 2014) have shown recently, that iO and some variants of UCEs are mutually exclusive. Our results, hence, validate some of the new UCE notions that emerged as a response to the iO-attack.

Keywords: Correlation-secure hash functions, hardcore functions, indistinguishability obfuscation, differing-inputs obfuscation, point-function obfuscation, auxiliary-input obfuscation, universal computational extractors (UCEs).

1 Introduction

For many cryptographic primitives, it is easy to construct a secure scheme in the random oracle model, but it is hard to give a construction in the standard model. For example, correlated-input hash functions (CIH) which were introduced by Goyal, O'Neill, and Rao [31], are easy to construct in the random oracle model, because the random oracle itself is secure under correlated inputs. However, up to now, no standard-model construction is known, and indeed, a recent black-box separation by Wichs [40] explains why it is so hard to construct them. Namely, the security definition of a CIH involves a pair of adversaries $(\mathcal{A}_1, \mathcal{A}_2)$ and is thus

P. Sarkar and T. Iwata (Eds.): ASIACRYPT 2014, PART II, LNCS 8874, pp. 122–141, 2014.

a two-stage game (i.e., the adversary is not a single algorithm but consists of two separate algorithms). The first adversary samples correlated inputs $(x_1, ..., x_t)$. Then a hash key hk is generated and the second adversary with access to hk needs to distinguish between getting a tuple of random strings and getting the tuple $(H(\text{hk}, x_1), ..., H(\text{hk}, x_t))$. Now, Wichs employs a meta reduction to show that it is unlikely to have a black-box reduction $\mathcal{R}$ from CIH to a (one-stage) cryptographic assumption such as the decisional Diffie–Hellman assumption (DDH). Namely, he shows that if such a reduction to DDH exists, then the DDH assumption is wrong. In his proof, he substantially exploits that the CIH game is a two-stage game. For a black-box reduction $\mathcal{R}$ it must hold that if the reduction $\mathcal{R}$ gets access to a pair of oracles $(\mathcal{A}_1, \mathcal{A}_2)$ that break CIH, then $\mathcal{R}^{\mathcal{A}_1, \mathcal{A}_2}$ must also break DDH. Wichs constructs a pair of inefficient adversaries $(\mathcal{A}_1, \mathcal{A}_2)$ which, however, can be efficiently emulated using a stateful simulator Sim. That is, the simulator simulates both adversaries together while sharing state between them. As the reduction cannot distinguish between the two settings $\mathcal{R}^{\mathcal{A}_1, \mathcal{A}_2}$ and $\mathcal{R}^{\text{Sim}}$ this breaks DDH, and hence, if we believe that DDH is a hard problem, then such an $\mathcal{R}$ cannot exist. Note that Wichs' proof is not specific to DDH, but rather applies to any one-stage assumption and presents a substantial barrier to prove security. Moreover, Wichs' impossibility result extends to a range of security notions that are specified by two-stage games.

In this paper, we use cryptographic obfuscation techniques to circumvent Wichs' impossibility result and achieve security notions that are based on two-stage assumptions. Towards this goal, the key idea is to combine point-function obfuscation and indistinguishability obfuscation.

Point and Indistinguishability Obfuscation. A point function p_x is a function that returns 1 on input x and $\bot$ on all other values. A point function obfuscator under auxiliary input AIPO returns a point function $p \leftarrow_\$ \text{AIPO}(x)$ that hides the point x even in case the adversary receives some side-channel information z about x. More formally, the security of AIPO is defined as security for all computationally unpredictable distributions $\mathcal{D}$, that is, $\mathcal{D}$ outputs a pair (x, z), where x is a point and z is some leakage that hides x computationally. AIPO is secure, if for all computationally unpredictable $\mathcal{D}$, $(\text{AIPO}(x), z)$ is indistinguishable from $(\text{AIPO}(u), z)$, where $(x, z) \leftarrow \mathcal{D}$ and u is a uniformly random point. Such AIPO schemes have been constructed in [20, 11].

While point function obfuscators are obfuscation schemes for a very specific class of functionalities (namely point functions) Garg et al. [26] have recently revived the study of general obfuscation schemes with their candidate construction of indistinguishability obfuscation. The notion of indistinguishability obfuscation is weaker than VBB-obfuscation—thereby circumventing the impossibility results of Barak et al. [3, 2]—and says intuitively that, for any two circuits that compute the same function, their obfuscations are indistinguishable. The publication of the candidate for indistinguishability obfuscation by Garg et al. inspired simultaneous breakthroughs for hard problems in various sub-areas of cryptography [39, 15, 1, 25, 33, 14, 9, 30] including functional and deniable encryption, two-round secure multi-party computation, full-domain hash, poly-many hard-core bits from any one-way function, multi-input functional encryption and more.

Correlated-Input Hash-Functions. In this paper, we give the first standard-model construction for q-query CIHs. Our CIH is not only one-way under correlated inputs, but also outputs elements that are indistinguishable from random. We will compare our notion of q-query CIH with other notions of CIHs shortly.

On a high-level, our construction is a de facto instantiation of a random oracle. As the behavior of a PRF is similar to that of a random function, we instantiate the random oracle by securely delegating a PRF, that is, we obfuscate a PRF with a hard coded key. Indeed, our hash-function construction only consists of a (puncturable) PRF that is obfuscated via an indistinguishability obfuscator (iO):

$$\text{\textbf{Hash Construction:} } \mathrm{iO}(\mathrm{PRF}(k,.)) \ .$$

Bellare, Stepanovs, and Tessaro (BST; [9]) already used this natural construction in the direct construction of hardcore functions for injective one-way functions from indistinguishability obfuscation. We will discuss BST and the relation to our our work shortly.

Note that before obfuscating the PRF we need to pad the circuit to a specific length. This is needed when using indistinguishability obfuscation to move from one circuit to another one in the security proof and thus the construction must be padded to the length of the biggest circuit needed within the security proof. Jumping ahead, we note that although our construction and that of BST look identical on the outside the padding is different. For BST, the construction needs to be padded differently depending on the size of the one-way function. In turn, our padding is universal and thus we yield a universal hardcore function that works for any one-way function.

Circumventing Wichs' Impossibility Result. Although the construction is natural, proving its security is non-trivial, as the security guarantees of iO do not even allow us to show easily that it is hard to extract the PRF key. Towards proving the security of our construction, we build on the puncturable PRF technique by Waters and Sahai [39] and combine it with point function obfuscators secure under auxiliary input (AIPO).

Using AIPOs is crucial to circumvent the impossibility result by Wichs [40], because the security of AIPOs is defined via a two-stage security game. The first AIPO adversary samples a point, and the second adversary tries to break the obfuscation of the point function. In a sense, the impossibility result of Wichs tells us that using a two-stage assumption such as AIPO in the proof is, indeed, necessary. In particular, iO and PRFs are both one-stage assumptions. Note that, as AIPOs are only used in the proof and not in the construction, it might be possible that the same construction can be proven secure without making use of AIPOs possibly through some other two-stage assumption.

Universal Hardcore Functions for Any One-Way Function. Bellare, Stepanovs, and Tessaro (BST; [9]) recently established that the same construction (with a different amount of padding) also yields a hardcore function for any injective one-way function, assuming a puncturable PRG and iO.

For general one-way functions, BST gave a second, different construction of a hardcore function and proved it based on so-called differing-inputs obfuscation. Differing-inputs obfuscation is a stronger assumption than iO and has been shown conditionally impossible by Garg et al. [27] assuming special-purpose obfuscators. Therefore, in the current version of their paper, Bellare et al. [9] use a weaker variant of diO that is not affected by the results of Garg et al. [27].

In an updated version of their paper, Garg et al. [28] show that, assuming a special-purpose obfuscator and indistinguishability obfuscation for Turing Machines, there are one-way functions for which the second construction of BST cannot be a secure hardcore function, because their hardcore function has "output-only dependence". This means that hardcore bits $h(x)$ are completely determined by $f(x)$, or in other words, for any inputs x and x' such that $f(x) = f(x')$ it holds that $h(x) = h(x')$. We note that the only candidate for iO for Turing machines is currently based on full diO.

The conditional negative result for output-only dependent hardcore functions does not apply to the construction $iO(PRF(k, .))$ which is the construction that we use throughout this paper and which BST—with a different amount of padding—prove to be a hardcore function for injective one-way functions. In turn, assuming AIPO in addition to iO allows us to prove this construction secure for all one-way functions, even those that have many pre-images. Another difference with the BST result is that we yield a universal hardcore function for any one-way function while their padding depends on the one-way function.

Our proof builds on ideas by BST, and we will come back to their result in the context of presenting our proof techniques. We note that for our security proof, we assume AIPO in addition to iO and thereby are able to avoid diO variants altogether. The assumption of point obfuscators is currently incomparable to the assumption of differing-inputs obfuscation as well as to more restricted versions that were used by BST. It is an interesting question to explore the relationship between these assumptions.

Modularizing Proofs via UCEs. We could prove the security of our construction directly, but instead, we split our proof into two parts. First, we show that our construction enjoys some useful, abstract properties. Then we use results by Bellare et al. [6] that show that these abstract properties suffice for the application at hand. This way, we provide a means of using iO in a black-box way. Our abstraction is a version of UCE security [6] that we discuss next.

The UCE Framework by Bellare, Hoang, and Keelveedhi (BHK; [6]) introduces assumptions that allow us to instantiate random oracles in a wide range of applications. Loosely speaking, UCEs are PRF-like assumptions that split the distinguisher into two parts: a first adversary S that gets access to a keyed hash function or a random oracle (and which is called the *source*), and a second adversary D that gets the hash key hk (and which is called the *distinguisher*). The two algorithms together try to guess whether the source was given access to a keyed hash function (under a randomly chosen key) or to a random oracle.

Concretely, the UCE notions are defined via a two-stage UCE game (we depict the communication flow in Figure 1 and the pseudocode in Figure 2). First, the source S is run with oracle access to HASH (which either implements a random oracle or the hash function with a randomly chosen key hk) to output some leakage L. Subsequently, distinguisher D is run on the leakage L and hash key hk but without access to oracle HASH. Distinguisher D outputs a single bit b indicating whether oracle HASH implements a random oracle or hash function H with key hk.

Without any restrictions, (S, D) can easily win the UCE game. For example, say, source S makes a random query x to receive $y \leftarrow \text{HASH}(x)$ and outputs (x, y) as leakage. As distinguisher D knows the hash key hk as well as the leakage (x, y), it can recompute the hash value and check whether $y = \text{H}(\text{hk}, x)$.

BHK present several possible restrictions on the source which give rise to various UCE notions. It turns out to be particularly useful to restrict sources to be computationally unpredictable, that is, the leakage created by the source S—when interacting with a random oracle—should not reveal (computationally) any of the source's queries to HASH. This notion is denoted by UCE[$\mathcal{S}^{\text{cup}}$], where $\mathcal{S}^{\text{cup}}$ denotes the class of computationally unpredictable sources [7]. BHK show that UCE[$\mathcal{S}^{\text{cup}}$]-secure hash functions can safely replace a random oracle in a large number of interesting applications such as hardcore functions or deterministic public-key encryption [6]. In a recent work Brzuska, Farshim and Mittelbach (BFM; [17]) show that UCE security with respect to computational unpredictability cannot be achieved in the standard model assuming indistinguishability obfuscation exists. Several refinements have been proposed since, including a statistical notion of unpredictability denoted by $\mathcal{S}^{\text{sup}}$ as well as source classes containing sources that are structurally required to produce output in a special way as well as sources which are restricted to only a fixed number of queries [7, 17, 36].

Our notion of UCE security strengthens the notion of unpredictability to what we call strong unpredictability and we denote the corresponding class of sources by $\mathcal{S}^{\text{s-cup}}$ for the computational variant and by $\mathcal{S}^{\text{s-sup}}$ for its statistical version. Namely, we demand that the leakage be computationally/statistically unpredictable even if the predictor additionally gets the answers to the queries that the source received from the oracle. We give the pseudo-code for strong unpredictability in Figure 3.

It turns out that UCEs for strongly computationally unpredictable sources that can only make a single query (denoted by UCE[$\mathcal{S}^{\text{s-cup}} \cap \mathcal{S}^{\text{1-query}}$]) already imply hardcore functions for any one-way function. Furthermore, UCEs for strongly statistically unpredictable sources that can only make q queries (denoted UCE[$\mathcal{S}^{\text{s-sup}} \cap \mathcal{S}^{\text{q-query}}$]) imply q-query correlation-secure hash functions. We note that strongly unpredictable sources can be regarded as a generalization of so-called split sources [7] which were introduced by BHK after the BFM impossibility results. We will discuss the exact relationship later.

So far UCEs have only been constructed in idealized models. BHK showed that a random oracle is UCE-secure in the strongest proposed settings and conjectured that HMAC is UCE-secure if the underlying compression function is modeled as an ideal function. This conjecture has recently been confirmed by Mittelbach [37] who shows that HMAC and various Merkle-Damgård variants are UCE-secure in the ideal compression function model. We note that so far, no standard model instantiation of any (non-trivial) UCE variant has been proposed and, hence, we present the first standard model construction of UCEs.[1]

Techniques. Our construction is based on indistinguishability obfuscation and similar to many other recent constructions from iO [39, 9, 33, 14] our construction also makes use of puncturable PRFs [39] which admit the generation of keys that allow to evaluate the PRF on all points except for points in a small target set (often containing just a single point). Our security reduction, however, differs from existing techniques. That is, we make use of point function obfuscations which allows us to hide the punctured points within our constructed circuits. Hiding the punctured points was also the key problem in the earlier discussed work by Bellare, Stepanovs and Tessaro [9] who construct hardcore-functions for one-way functions. They solve the problem elegantly by using the one-way function from the security game to blind the punctured point by embedding the image under the one-way function. However, when testing whether a given point is equivalent to the punctured point this test is ambiguous which is why they need to assume differing-inputs obfuscators for one-way functions that map more than polynomially many points to the same image value. This is where point function obfuscation comes into the picture which allows us to bypass any assumptions related to differing-input obfuscation variants. Yet, of course, point obfuscators are as far as is currently known an assumption incomparable to differing-inputs obfuscation.

Point Obfuscation and iO. In a recent and independent work, Hofheinz uses point obfuscation in a similar way to construct fully secure constrained pseudorandom functions [32] in the random oracle model. A constrained PRF is a generalized form of a puncturable PRF which allows for the generation of keys that enable the holder to evaluate the PRF on a set of points but not on all points. In contrast to previous constructions [13, 16, 34] Hofheinz uses point obfuscation and an extension he introduces as *extensible testers* in conjunction with indistinguishability obfuscation to hide which points a given key allows to honestly evaluate. This allows him to achieve full security without relying on complexity leveraging which was used in previous constructions entailing a superpolynomial loss of security in the adaptive setting. We note that unlike this work Hofheinz relies on the simpler assumption of plain point obfuscation (that is, obfuscation without auxiliary inputs) and he shows how to build extensible testers based on the DDH-based point obfuscator by Canetti [20].

---

[1] The UCE Framework is very flexible and it is, for example, possible to define a UCE restriction that corresponds to PRF security.

Brzuska and Mittelbach study the connection between point obfuscation with multi-bit output secure in the presence of auxiliary inputs and indistinguishability obfuscation [18]. They show that indistinguishability obfuscation and a strong form of multi-bit point obfuscation are mutually exclusive. Their results do not carry over to the setting of statistically hard-to-invert auxiliary information (which we rely on for CIHs) and it is not clear if their results can be extended to cover plain AIPO, that is point functions with single-bit outputs.

Our Results. We next discuss the specific UCE assumptions that our construction will meet and the relation to the specific point obfuscation schemes used. In Section 3 we will show that our construction is UCE[$\mathcal{S}^{\text{s-cup}} \cap \mathcal{S}^{\text{1-query}}$]-secure assuming iO, puncturable PRFs and the existence of AIPO. That is, we consider UCE-secure for computationally strongly unpredictable sources that make a single query. In Section 3.3, we prove that our construction is also UCE[$\mathcal{S}^{\text{s-sup}} \cap \mathcal{S}^{\text{q-query}}$]-secure, that is, secure against statistically unpredictable sources that make at most q queries.

As explained, we base the security of our construction on the existence of a different (incomparable) notion of point obfuscation. We consider a notion of AIPO which only needs to be secure against statistically unpredictable distributions but, in turn, we require it to be q-composable [21, 10]. Intuitively, q-composability says that an obfuscation remains secure even if an adversary sees q many (possibly related) obfuscations. The reason that we need q-composable AIPO is that now, the source is a allowed to make q queries and hence, we need to hide q points in the proof. q-composable AIPO implies multi-bit point function obfuscation [21] and thus does not exist, if iO exists [18].

However, as we here only consider sources in $\mathcal{S}^{\text{s-sup}}$, that is, sources which are only statistically strongly unpredictable, it suffices that our AIPO-notion is secure against statistically unpredictable samplers which weakens the notion of AIPO. Matsuda and Hanoka [35] have recently shown that q-composable AIPO secure against statistically unpredictable samplers is implied by composable VGB-AI point obfuscators, a notion that Bitansky and Canetti constructed under the q-Strong Vector Decision Diffie Hellman assumption [10]. Note that, for the proof to work, we need to let the circuit size of our construction grow, artificially, with the number of queries q. Towards this goal, we use some padding that does not have any functionality.

In summary we get the following results:

1. Our construction is UCE[$\mathcal{S}^{\text{s-cup}} \cap \mathcal{S}^{\text{1-query}}$]-secure assuming indistinguishability obfuscation for all circuits in $\mathcal{P}/\text{poly}$ and AIPO secure with respect to *computationally* hard-to-invert auxiliary information exist.
2. Our construction is UCE[$\mathcal{S}^{\text{s-sup}} \cap \mathcal{S}^{\text{q-query}}$]-secure assuming indistinguishability obfuscation for all circuits in $\mathcal{P}/\text{poly}$ and q-composable AIPO with respect to *statistically* hard-to-invert auxiliary information exist.

On the Feasibility of Our AIPO Assumptions. Standard AIPO secure against computationally unpredictable samplers has been constructed by Canetti in [20] under (non-standard) variants of the DDH assumption and by Bitansky and Paneth in [11] under (non-standard) assumptions on pseudorandom permutations. We discuss the constructions and the underlying assumptions in the full version of this work [19]. One might hope that AIPO is naturally composable. However, Canetti et al. show that this is generally not the case [21, 10]. On the other hand, Bitansky and Canetti [10] show that under the *t-Strong Vector Decision Diffie Hellman assumption* the original point obfuscation scheme of Canetti [20] composes in the so-called virtual grey-box (VGB) setting. The VGB setting was introduced by Bitansky and Canetti [10] and is a relaxation of the strongest obfuscation setting the virtual black-box (VBB) setting [3, 2]. Similarly to VBB obfuscation, VGB obfuscation is in general not achievable, yet for the class of point functions it seems in reach [10]. The VGB setting is particularly interesting because "plain" VGB and VGB with auxiliary information are equivalent [10]. This result stands in contrast to the VBB setting where allowing auxiliary information results in a stronger notion. Furthermore, we currently have no candidate constructions for composable point obfuscation schemes in this stronger setting. We note that for our purpose composable obfuscation in the VGB setting is sufficient for our purpose as Matsuda and Hanaoka [35] show that this setting already implies q-composable AIPO with respect to statistically unpredictable samplers which form the basis for our q-query correlation-secure hash functions.

In a very recent work Brzuska and Mittelbach (BM) investigate the connection between indistinguishability obfuscation and multi-bit output point obfuscation secure in the presence of auxiliary input (MB-AIPO) [18]. A multi-bit point function $p_{x,m}$ is zero everywhere except on x where it outputs m. BM show that various strong notions of MB-AIPO and indistinguishability obfuscation are mutually exclusive. However, their results do not seem to carry over to plain AIPO, that is to AIPO for plain point functions as needed in our constructions. We refer to [18] for a discussion on MB-AIPO and discuss the implications of an extension of the results of BM to plain AIPO shortly when talking about the feasibility of our UCE notions.

On the Feasibility of Our UCE Notions. In a recent work, Brzuska, Farshim, and Mittelbach (BFM; [17]) show that, assuming indistinguishability obfuscation exists, no standard model hash construction can be UCE-secure with respect to computationally unpredictable sources. Our construction achieves a weaker yet related notion of security, namely UCE-security with respect to strongly computationally unpredictable sources which raises the question whether the BFM result can be extended to this setting.

The BFM result crucially hinges on the possibility of extending the output-length of the studied hash construction such that it is significantly larger than the key size. For example, this can be achieved by using multiple queries to the hash construction or via extending the output size by applying a pseudo-random generator [17, 8]. Both approaches fail with our construction: the size of our hash key grows with the number of allowed queries and since we consider strong unpredictability it seems implausible to prove the construction $\mathsf{PRG}(\mathsf{H}(\cdot, \cdot))$-secure under the assumption that H is UCE-secure with respect to strongly computationally unpredictable sources. Thus, we think that extending the BFM attack is implausible. Furthermore, if it can be extended this would immediately imply that indistinguishability obfuscation implies the non-existence of AIPO, which would be a surprising result. We discuss the BFM result in greater detail in the full version [19] and note that, even if an extension of the BFM result were to break AIPOs with computational unpredictability, then the second construction would not be affected, as it only considers AIPOs secure with respect to statistically hard-to-invert auxiliary information.

Notions of Correlation-Secure Hash-Functions. We now compare our notion of q-query CIHs to different notions of correlated-input security. Note that q-query CIH means that the size of the hash-key can depend on the number of inputs q. However, and that is a crucial difference to previous works, each input value is hashed using the same hash-key. In turn, Freeman et al. [23] as well as Rosen and Segev [38] use a fresh hash-key for every input. Notably, the correlation-secure functions that they construct also have a trapdoor. Note that the correlated-input variant[2] of the IND security game for deterministic public-key encryption [5, 4, 12] and the CIH game are almost identical if it is required that the CIH has a trapdoor. We can then view the computation of the CIH as an encryption operation and the CIH game becomes a slightly stronger version of the IND security game (that is, a real-or-random rather than a left-or-right game). Hence, a CIH function which has a trapdoor is also a deterministic public-key encryption scheme.

As in the schemes of [23, 38] a new key needs to be generated for every new message, the constructions are not a deterministic public-key encryption scheme. In turn, if our q-query CIH were a trapdoor function, then by definition, it would also be a q-query deterministic public-key encryption scheme. Unfortunately, our construction of a q-query CIH does not come with a trapdoor, and we do not know whether this is possible.

Another related notion of CIH are *statistically* secure q-query CIHs. Here, as for our notion of q-query CIH, the key size may grow with the number of queries and one uses the same hash key for each query. In contrast to our security notion one here requires that the output is statistically close to random given the hash key. As we are concerned with statistical security, this notion is only achievable for distributions that come with a notable amount of entropy, that is, the q pre-images need to have entropy that is at least q times the output length. In turn, for the notion of entropy that we consider, the entropy of the pre-images does not need to grow with q and can also be less than the length of the output.

Hence, this notion of statistically secure CIH only applies to a substantially smaller class of distributions. In turn, while our construction relies on the strong assumption of indistinguishability obfuscation, statistically secure CIH can be achieved without any assumptions. That is, if the pre-images carry enough (true) entropy, then one can extract q uniformly random image values by using a q-wise independent hash-functions [24].

---

[2] Here, we refer to the variant where each message needs to have high entropy on its own, but might have low entropy conditioned on the other messages.

Finally, Goyal, O'Neill, and Rao [31] construct CIHs that are secure under polynomially related inputs and introduce a hierarchy of CIH notions: *One-wayness* under correlated inputs, *unpredictability* under correlated inputs and *pseudorandomness* under correlated inputs. These notions describe a hierarchy of security notions when we consider CIHs with superlogarithmic output length. We note that we achieve the strongest of these notions, namely *pseudorandomness* under correlated inputs.

Full Version. Due to space restrictions, this version should be regarded as an extended abstract as we defer many details and all proofs to the full version [19]. In the remainder of this extended abstract we present our main results and give some intuition for the underlying proofs.

2   Preliminaries

2.1   Obfuscation

Indistinguishability Obfuscation. While the strongest obfuscation notion, that is, virtual black-box obfuscation provably does not exist in general for all circuits [3], weaker notions such as *indistinguishability obfuscation* may well exist. VBB requires the existence of a simulator. On the other hand, an indistinguishability obfuscation (iO) scheme only ensures that the obfuscations of any two functionally equivalent circuits are computationally indistinguishable. Indistinguishability obfuscation was originally proposed by Barak et al. [3] as a potential weakening of virtual-black-box obfuscation. We recall the definition from [26].

Definition 1. *A* PPT *algorithm* iO *is called an* indistinguishability obfuscator *for a circuit ensemble* $\mathcal{C} = \{\mathcal{C}_\lambda\}_{\lambda \in \mathbb{N}}$ *if the following conditions are satisfied:*

- **Correctness.** *For all security parameters* $\lambda \in \mathbb{N}$, *for all* $C \in \mathcal{C}_\lambda$, *and for all inputs* x *we have that* $\Pr\left[C'(x) = C(x) : C' \leftarrow_\$ \mathsf{iO}(1^\lambda, C) \right] = 1$.
- **Security.** *For any* PPT *distinguisher* $\mathcal{D}$, *for all pairs of circuits* $C_0, C_1 \in \mathcal{C}_\lambda$ *such that* $C_0(x) = C_1(x)$ *on all inputs* x *the following distinguishing advantage is negligible:*

$$\left| \Pr\left[\mathcal{D}(1^\lambda, \mathsf{iO}(1^\lambda, C_1)) = 1 \right] - \Pr\left[\mathcal{D}(1^\lambda, \mathsf{iO}(1^\lambda, C_0)) = 1 \right] \right| \leq \mathsf{negl}(\lambda).$$

Closely related to indistinguishability obfuscation is the notion of *differing-inputs obfuscation* (diO) which also goes back to the seminal paper of Barak et al. [3]. Building on a theorem by Boyle, Chung and Pass [15], we are able to avoid diO as an assumption and only use it as an intermediary concept in our proof. We defer the details to the full version [19].

Point Obfuscation. While indistinguishability, as well as differing-inputs, obfuscation are obfuscation schemes for general circuits one can also study obfuscation schemes for particular function classes such as point functions. A point function p_x for some value $x \in \{0,1\}^*$ maps every input to $\perp$ except for x which is mapped to 1. We consider a variant of point function obfuscators under auxiliary input which was first formalized by Canetti [20], although in a slightly different context. We here give the definition from [11] presented in a game based formulation. The first definition formalizes unpredictable distributions which are in turn used to define obfuscators for point functions.

Definition 2 (Unpredictable distribution). *A distribution ensemble* $\mathcal{D} = \{D_\lambda = (Z_\lambda, X_\lambda)\}_{\lambda \in \mathbb{N}}$, *on pairs of strings is unpredictable if no poly-size (non-uniform) circuit can predict* X_λ *from* Z_λ. *That is, for every poly-size circuit sequence* $\{C_\lambda\}_{\lambda \in \mathbb{N}}$ *and for all large enough* λ:

$$\Pr_{(z,x) \leftarrow_\$ D_n} [C_\lambda(z) = x] \leq \mathsf{negl}(\lambda)$$

Definition 3 (Auxiliary input point obfuscation for unpredictable distributions (AIPO)). *A* PPT *algorithm* AIPO *is a point obfuscator for unpredictable distributions if it satisfies the functionality and polynomial slowdown requirements as VBB obfuscation [3, 2], and the following secrecy property: for any (efficiently sampleable) unpredictable distribution* $\mathcal{B}_1$ *over* $\{0,1\}^{\mathsf{poly}(\lambda)} \times \{0,1\}^\lambda$ *it holds for any* PPT *algorithm* $\mathcal{B}_2$ *that the probability that the following experiment outputs true for* $(\mathcal{B}_1, \mathcal{B}_2)$ *is negligibly close to* $\frac{1}{2}$:

$$
\begin{aligned}
&b \leftarrow_\$ \{0,1\} \\
&(z, x_0) \leftarrow_\$ \mathcal{B}_1(1^\lambda) \\
&x_1 \leftarrow_\$ \{0,1\}^\lambda \\
&p \leftarrow_\$ \mathsf{AIPO}(x_b) \\
&b' \leftarrow_\$ \mathcal{B}_2(1^\lambda, p, z) \\
&\mathbf{return}\ b = b'
\end{aligned}
$$

The probability is over the coins of adversary $(\mathcal{B}_1, \mathcal{B}_2)$, *the coins of* AIPO *and the choices of* x_1 *and* b.

2.2   Universal Computational Extractors (UCE)

The UCE Framework by Bellare, Hoang, and Keelveedhi (BHK; [6]) introduces assumptions that allow us to instantiate random oracles in a wide range of applications and which are not succeptible to the impossibility result by Canetti, Goldreich and Halevi [22]. Loosely speaking, UCEs are PRF-like assumptions that split the distinguisher into two parts: a first adversary S that gets access to a keyed hash function or a random oracle (and which is called the *source*), and a second adversary D that gets the hash key hk (and which is called the *distinguisher*). The two algorithms together try to guess whether the source was given access to a keyed hash function or to a random oracle.

Concretely, the UCE notions are defined via a two-stage UCE game (we depict the communication flow in Figure 1 and the pseudocode in Figure 2). First, the source S is run with oracle access to HASH to output some leakage L. Subsequently, distinguisher D is run on the leakage L and hash key hk but without

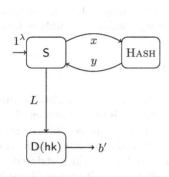

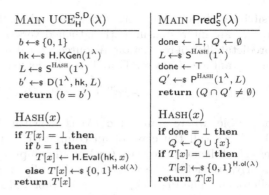

MAIN $\text{UCE}_H^{S,D}(\lambda)$	MAIN $\text{Pred}_S^P(\lambda)$
$b \leftarrow_\$ \{0, 1\}$	done $\leftarrow \bot$; $Q \leftarrow \emptyset$
$hk \leftarrow_\$ H.KGen(1^\lambda)$	$L \leftarrow_\$ S^{\text{HASH}}(1^\lambda)$
$L \leftarrow_\$ S^{\text{HASH}}(1^\lambda)$	done $\leftarrow \top$
$b' \leftarrow_\$ D(1^\lambda, hk, L)$	$Q' \leftarrow_\$ P^{\text{HASH}}(1^\lambda, L)$
return $(b = b')$	return $(Q \cap Q' \neq \emptyset)$
$\text{HASH}(x)$	$\text{HASH}(x)$
if $T[x] = \bot$ then	if done $= \bot$ then
if $b = 1$ then	$Q \leftarrow Q \cup \{x\}$
$T[x] \leftarrow H.\text{Eval}(hk, x)$	if $T[x] = \bot$ then
else $T[x] \leftarrow_\$ \{0, 1\}^{H.\text{ol}(\lambda)}$	$T[x] \leftarrow_\$ \{0, 1\}^{H.\text{ol}(\lambda)}$
return $T[x]$	return $T[x]$

Fig. 1. Schematic of the UCE game

Fig. 2. The UCE game together with the unpredictability game. In the UCE game source S has access to HASH, which returns real or ideal hash values, and leaks L to distinguisher D. The latter additionally gets the hash key and outputs a bit b'. On the right we give the unpredictability game.

access to oracle HASH. Distinguisher D outputs a single bit b indicating whether oracle HASH implements a random oracle or hash function H with key hk.

Without any restrictions, (S, D) can easily win the UCE game. For example, say, source S makes a random query x to receive $y \leftarrow \text{HASH}(x)$ and outputs (x, y) as leakage. As distinguisher D knows the hash key hk as well as the leakage (x, y), it can recompute the hash value and check whether $y = H(hk, x)$. BHK present several possible restrictions on the source which give rise to various UCE notions.

Formal UCE Definition. In line with [9] we consider families of functions F consisting of algorithms $F.\text{KGen}$, $F.\text{kl}$, $F.\text{Eval}$, $F.\text{il}$ and $F.\text{ol}$. Algorithm $F.\text{KGen}$ is a PPT algorithm taking the security parameter 1^λ and outputting a key $k \in \{0, 1\}^{F.\text{kl}(\lambda)}$ where $F.\text{kl} : \mathbb{N} \to \mathbb{N}$ denotes the key length. Functions $F.\text{il} : \mathbb{N} \to \mathbb{N}$ and $F.\text{ol} : \mathbb{N} \to \mathbb{N}$ denote the input and output length functions associated to F and for any $x \in \{0, 1\}^{F.\text{il}(\lambda)}$ and $k \leftarrow_\$ F.\text{KGen}(1^\lambda)$ we have that $F.\text{Eval}(k, x) \in \{0, 1\}^{F.\text{ol}(\lambda)}$, where the PPT algorithm $F.\text{Eval}$ denotes the "evaluation" function associated to F.

We denote hash functions by H. Let $H = (H.\text{KGen}, H.\text{Eval}, H.\text{kl}, H.\text{il}, H.\text{ol})$ be a hash-function family and let (S, D) be a pair of PPT algorithms. We define the UCE advantage of a pair (S, D) against H through

$$\text{Adv}_{H,S,D}^{\text{uce}}(\lambda) := 2 \cdot \Pr\left[\text{UCE}_H^{S,D}(\lambda) \right] - 1 ,$$

where game $\text{UCE}_H^{S,D}(\lambda)$ is shown in Figure 2 on the left (in Figure 1 we give a schematic overview of the communication within the game).

Unpredictability. Without any further restrictions there are PPT pairs (S, D) that achieve an advantage in the $\text{UCE}_H^{S,D}(\lambda)$ game close to 1. BHK define several possible restrictions for sources yielding various flavors of UCE assumptions [6]. Here, we are interested in a strengthened version of the original *computational* unpredictability [6] restriction. A source S is called *computationally unpredictable* if the advantage of any PPT predictor P, defined by

$$\mathsf{Adv}_{\mathsf{S},\mathsf{P}}^{\mathsf{pred}}(\lambda) := \Pr\left[\,\mathsf{Pred}_{\mathsf{S}}^{\mathsf{P}}(\lambda)\,\right],$$

is negligible, where game $\mathsf{Pred}_{\mathsf{S}}^{\mathsf{P}}(\lambda)$ is shown in Figure 2 on the right. In line with [7], we call the class of all computationally unpredictable sources $\mathcal{S}^{\mathrm{cup}}$, where $\mathcal{S}^{\mathrm{cup}}$ denotes the class (set) of all computationally unpredictable sources. Similarly, we define the class of statistically unpredictable sources where the predictor in game $\mathsf{Pred}_{\mathsf{S}}^{\mathsf{P}}(\lambda)$ can run in unbounded time but is still restricted to only polynomially many oracle queries. The class of statistically unpredictable sources is denoted by $\mathcal{S}^{\mathrm{sup}}$.

UCE *Security.* We say a hash function H is UCE secure for sources $\mathsf{S} \in \mathcal{S}$ denoted by UCE[$\mathcal{S}$], if for all PPT sources $\mathsf{S} \in \mathcal{S}$ and all PPT distinguishers D the advantage $\mathsf{Adv}_{\mathsf{H},\mathsf{S},\mathsf{D}}^{\mathsf{uce}}(\lambda)$ is negligible. In that way we get the UCE assumptions UCE[$\mathcal{S}^{\mathrm{cup}}$] and UCE[$\mathcal{S}^{\mathrm{sup}}$], that is, UCE with respect to computationally (resp. statistically) unpredictable sources.[3]

2.3   Puncturable PRFs

Besides point function obfuscation schemes, our main ingredient in the upcoming proofs are so-called puncturable pseudorandom functions (PRF) [39]. A family of puncturable PRFs $G :=(G.\mathsf{KGen}, G.\mathsf{Puncture}, G.\mathsf{kl}, G.\mathsf{Eval}, G.\mathsf{il}, G.\mathsf{ol})$ consists of functions that specify input length, output length and key length as well as a key generation algorithm $k \leftarrow G.\mathsf{KGen}$, a deterministic evaluation algorithm $G.\mathsf{Eval}(k,x)$ that takes a key k, an input x of length $G.\mathsf{il}(1^\lambda)$ and outputs a value y of length $G.\mathsf{ol}(1^\lambda)$. Additionally, there is a PPT puncturing algorithm $G.\mathsf{Puncture}$ which on input a polynomial-size set $S \subseteq \{0,1\}^{G.\mathsf{il}(\lambda)}$, outputs a special key k_S. A family of functions is called puncturable PRF if the following two properties are observed

- **Functionality Preserved under Puncturing.** For every PPT adversary $\mathcal{A}$ such that $\mathcal{A}(1^\lambda)$ outputs a polynomial-size set $S \subseteq \{0,1\}^{G.\mathsf{il}(\lambda)}$, it holds for all $x \in \{0,1\}^{G.\mathsf{il}(\lambda)}$ where $x \notin S$ that:

$$\Pr\left[\,G.\mathsf{Eval}(k,x) = G.\mathsf{Eval}(k_S,x) : k \leftarrow_{\$} G.\mathsf{KGen}(1^\lambda), k_S \leftarrow_{\$} G.\mathsf{Puncture}(k,S)\,\right] = 1$$

- **Pseudorandom at Punctured Points.** For every PPT adversary $(\mathcal{A}_1, \mathcal{A}_2)$ such that $\mathcal{A}_1(1^\lambda)$ outputs a set $S \subseteq \{0,1\}^{G.\mathsf{il}(\lambda)}$ and state st, consider an experiment where $k \leftarrow G.\mathsf{KGen}(1^\lambda)$ and $k_S = G.\mathsf{Puncture}(k,S)$. Then we have

$$\left|\Pr\left[\,\mathcal{A}_2(\mathsf{st}, k_S, S, G.\mathsf{Eval}(k,S)) = 1\,\right] - \Pr\left[\,\mathcal{A}_2(\mathsf{st}, k_S, S, U_{G.\mathsf{ol}(\lambda)\cdot|S|}) = 1\,\right]\right| \le \mathsf{negl}(\lambda)$$

---

[3] The notion UCE[$\mathcal{S}^{\mathrm{cup}}$] was originally named UCE1 and later changed to UCE[$\mathcal{S}^{\mathrm{cup}}$] [6, 7]. The notion of statistical unpredictability was introduced in [17, 7].

where $\mathsf{Eval}(k, S)$ denotes the concatenation of $\mathsf{Eval}(k, x_1), \ldots, \mathsf{Eval}(k, x_k)$ where $S = \{x_1, \ldots, x_k\}$ is the enumeration of the elements of S in lexicographic order, negl is a negligible function, and U_ℓ denotes the uniform distribution over $\{0, 1\}^\ell$.

As observed by [13, 16, 34] puncturable PRFs can, for example, be constructed from pseudorandom generators via the GGM tree-based construction [29]. As AIPO implies one-way functions [19] AIPO also implies puncturable PRFs.

3   UCEs from iO and Point Obfuscation

In this section we present our constructions of UCEs from iO and AIPO. We first define the precise UCE notions that our constructions achieve and introduce the UCE restriction of *strong unpredictability*. We will then in Section 3.2 present a construction of a UCE-secure function with respect to sources which are strongly computationally-unpredictable and which make exactly one oracle query. In Section 3.3 we will show how to extend the construction to allow for an a-priory fixed number of queries by switching to a statistical version of strong unpredictability.

Interestingly, our two constructions are basically the same modulo circuit padding. That is, our constructions depend on an obfuscation of a circuit, which in both cases is the same but padded to a different length. A larger but functionally equivalent circuit seems to be necessary to allow for multiple source queries.

We discuss applications of our constructions in the full version of this work [19]. Due to space limitations we also defer to the full version [19] for a discussion on why our construction does not (seem to) fall pray to the BFM attacks on computationally unpredictable sources [17].

3.1   Strongly Unpredictable and q-Query Sources

We now introduce the precise source restrictions for our upcoming UCE constructions. We define a new restriction that we call *strong unpredictability* and which can be seen as either a stronger form of unpredictability or a relaxed version of split sources. Secondly, we consider sources that make only a bounded number of oracle queries.

Strong Unpredictability. We consider sources which are strongly unpredictable both in the computational and in the statistical sense. We denote by $\mathcal{S}^{\text{s-cup}}$ the class of sources which are strongly, computationally unpredictable and by $\mathcal{S}^{\text{s-sup}}$ the class of strongly, statistically unpredictable sources. Strong unpredictability is a stronger requirement than unpredictability and we require that the leakage hides queries to HASH even if the predictor is given the query results. We say that a source S is called *strongly computationally unpredictable* if the advantage of any PPT predictor P, defined by

$$\mathsf{Adv}_{\mathsf{S},\mathsf{P}}^{\text{stpred}}(\lambda) := \Pr\left[\mathsf{stPred}_{\mathsf{S}}^{\mathsf{P}}(\lambda)\right],$$

MAIN stPred$_S^P(\lambda)$

$X^*, Y^* \leftarrow \emptyset$
$b \leftarrow_\$ \{0,1\}$
$L \leftarrow_\$ S^{\text{HASH}}(1^\lambda)$
$x' \leftarrow_\$ P^{\text{HASH}}(1^\lambda, L, Y^*)$
return $(x' \in X^*)$

HASH(x)

$X^* \leftarrow X^* \cup \{x\}$
$y \leftarrow_\$ \{0,1\}^{\text{H.ol}(\lambda)}$
$Y^* \leftarrow Y^* \cup \{y\}$
return y

Splt SOURCE S$^{\text{HASH}}(1^\lambda)$

$(L_0, \mathbf{x}) \leftarrow_\$ S_0(1^\lambda)$
for $i = 1, \ldots, |\mathbf{x}|$ **do** $\mathbf{y}[i] \leftarrow_\$ \text{HASH}(\mathbf{x}[i])$
$L_1 \leftarrow_\$ S_1(1^\lambda, \mathbf{y}); L \leftarrow (L_0, L_1)$
return L

Fig. 3. On the left: the strong unpredictability game where the predictor, in addition to the leakage is also given the result of the HASH queries. On the right: the definition of split sources [7]. A split source $S = \text{Splt}[S_0, S_1]$ consists of two parts S_0 and S_1 that jointly generate leakage L and neither part gets direct oracle access to HASH.

is negligible, where game stPred$_S^P(\lambda)$ is shown in Figure 3 on the left. For the case of strongly statistically unpredictable sources ($\mathcal{S}^{\text{s-sup}}$) we allow the predictor to be unbounded in its running time, but restrict the number of oracle queries to be bounded polynomially.

In order to circumvent the BFM attacks on computationally unpredictable sources BHK introduce the notion of split sources [7, 17]. A source S is called split source, denoted by $S \in \mathcal{S}^{\text{splt}}$ if it can be decomposed into two algorithms S_0 and S_1 such that neither part gets direct access to oracle HASH. We give the pseudocode of split sources in Figure 3 on the right. In a first step algorithm S_0 outputs a leakage string L_0 together with a vector $\mathbf{x}$. Then, each of the entries in $\mathbf{x}$ is queried to HASH and the results stored in vector $\mathbf{y}$. Finally, the second algorithm S_1 is run on vector $\mathbf{y}$ to produce the second part of the leakage L_1.

One can prove that split sources are a (strict) subclass of strongly unpredictable sources, that is, $\mathcal{S}^{\text{splt}} \cap \mathcal{S}^{\text{cup}} \subsetneq \mathcal{S}^{\text{s-cup}}$ (and similarly in the statistical case $\mathcal{S}^{\text{splt}} \cap \mathcal{S}^{\text{sup}} \subsetneq \mathcal{S}^{\text{s-sup}}$). For further information on the implications see the full version of this work [19].

q-*Query UCE.* Our first construction only admits sources which make exactly one query. We call such sources single-query sources and denote the corresponding source class by $\mathcal{S}^{1\text{-query}}$. We also consider a relaxed notion to allow for polynomially bounded number of queries for some polynomial $q := q(\lambda)$. We call the corresponding sources q-query sources and denote their class by $\mathcal{S}^{q\text{-query}}$. We note that sources restricted to a constant number of queries are discussed in [7].

3.2   A UCE Construction Secure against Sources in $\mathcal{S}^{\text{s-cup}} \cap \mathcal{S}^{1\text{-query}}$

We will now present our construction which depending on different assumptions on the existence of point obfuscators will achieve UCE[$\mathcal{S}^{\text{s-cup}} \cap \mathcal{S}^{1\text{-query}}$]-security or UCE[$\mathcal{S}^{\text{s-sup}} \cap \mathcal{S}^{q\text{-query}}$]-security. Note that depending on the number of supported queries the construction needs to pad the circuit before obfuscating it.

Construction 1. *Let $s : \mathbb{N} \to \mathbb{N}$, let G be a puncturable PRF and let iO be an indistinguishability obfuscator for all circuits in $\mathcal{P}/\text{poly}$. We define our hash function family H as*

H.*KGen*(1^λ)	H.Eval(hk, x)
$k \leftarrow_\$ G.KGen(1^\lambda)$	$\overline{C} \leftarrow$ hk
hk $\leftarrow_\$$ iO(PAD($s(\lambda), G.$Eval($k, \cdot$)))	***return*** $\overline{C}(x)$
return hk	

where PAD $: \mathbb{N} \times \{0,1\}^* \longrightarrow \{0,1\}^*$ *denotes a deterministic padding algorithm that takes as input an integer and a circuit and outputs a functionally equivalent circuit padded to length $s(\lambda)$.*[4]

In other words, the key generation algorithm H.KGen(1^λ) runs $k \leftarrow G.$KGen(1^λ) and returns iO($G.$Eval($k, \cdot$)), i.e., an obfuscation of the evaluation circuit of PRF G with key k hardwired into it. Function H.Eval is basically a universal Turing machine which runs input x on the obfuscated circuit hk.

Theorem 2. *If G is a secure puncturable PRF, if iO is a secure indistinguishability obfuscator and if AIPO exists, then the hash function family H defined in Construction 1 is* UCE[$\mathcal{S}^{\text{s-cup}} \cap \mathcal{S}^{1\text{-query}}$]*-secure.*

We prove the theorem via a sequence of 5 games (depicted in Figure 4 on page 138) where game Game_1 denotes the original UCE[$\mathcal{S}^{\text{s-cup}} \cap \mathcal{S}^{1\text{-query}}$] game with hidden bit b fixed to 1. We present the proof in the full version of this work [19].

3.3 A UCE Construction Secure against Sources in $\mathcal{S}^{\text{s-sup}} \cap \mathcal{S}^{q\text{-query}}$

In this section we show that our construction is also UCE-secure with respect to sources which are strongly unpredictable in a statistical sense and which allow the source to make q-many queries for any polynomial $q := q(\lambda)$. That is, we consider sources in class $\mathcal{S}^{\text{s-sup}} \cap \mathcal{S}^{q\text{-query}}$.

In case we allow the source to make q many queries, the first observation is that we need to choose the size of our obfuscated circuit such we can puncture at q many points. For each point, we will encode a random string into the circuit and thus, the circuit size grows with the number of points we need to puncture out. Besides this, the construction is identical to the one before with the exception that we need a different (incomparable) security property of our point function obfuscation scheme. That is, we require the point obfuscator to be a q-composable VGB obfuscator secure in the presence of statistically unpredictable auxiliary information which implies an AIPO obfuscator with statistically unpredictable auxiliary information. We refer to the full version for further details [19].

---

[4] Function s needs to be chosen in accordance with the puncturable PRF to allow for the required number of puncturings.

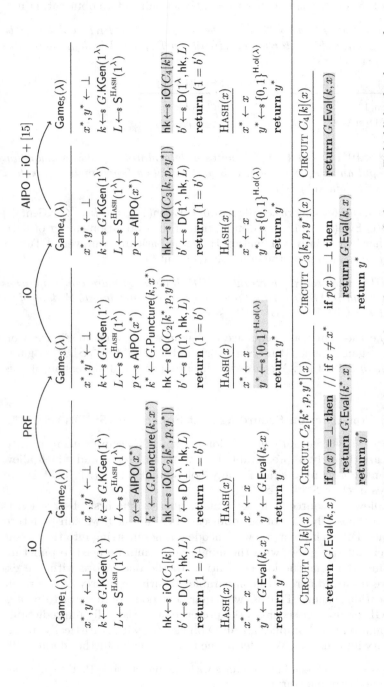

Fig. 4. The games used in the proof of Theorem 2 on the top and the used circuits on the bottom. To highlight the changes from game to game we have marked the changed lines with a light gray background color. By $C[k](x)$ we denote that circuit C depends on k (during construction time) and takes x as input. The arrows above the games indicate the security reduction to get from Game_i to Game_{i+1}.

Theorem 3. *Let q be a polynomial. If G is a secure puncturable PRF, if iO is a secure indistinguishability obfuscator and if there exist a q-composable VGB point obfuscator for statistically unpredicatable auxiliary input, then the hash function family H defined in Construction 1 is $\text{UCE}[\mathcal{S}^{\text{s-sup}} \cap \mathcal{S}^{q\text{-query}}]$-secure.*

The proof follows analogously to the proof of Theorem 2, except for puncturing at several points instead of a single point and therefore, we reduce to q-composable VGB point obfuscation. We defer the proof to the full version [19].

Acknowledgments. We thank the Asiacrypt 2014 reviewers for the many constructive comments. We especially thank Paul Baecher, Mihir Bellare, Pooya Farshim, Victoria Fehr, Giorgia Azzurra Marson, Adam O'Neill and Daniel Wichs for many helpful comments and discussions throughout the various stages of this work. Christina Brzuska was supported by the Israel Science Foundation (grant 1076/11 and 1155/11), the Israel Ministry of Science and Technology grant 3-9094), and the German-Israeli Foundation for Scientific Research and Development (grant 1152/2011). Arno Mittelbach was supported by CASED (www.cased.de) and the German Research Foundation (DFG) SPP 1736.

References

1. Ananth, P., Boneh, D., Garg, S., Sahai, A., Zhandry, M.: Differing-inputs obfuscation and applications. Cryptology ePrint Archive, Report 2013/689 (2013), http://eprint.iacr.org/2013/689
2. Barak, B., Goldreich, O., Impagliazzo, R., Rudich, S., Sahai, A., Vadhan, S., Yang, K.: On the (im)possibility of obfuscating programs. J. ACM 59(2), 6:1–6:48 (2012), http://doi.acm.org/10.1145/2160158.2160159
3. Barak, B., Goldreich, O., Impagliazzo, R., Rudich, S., Sahai, A., Vadhan, S.P., Yang, K.: On the (im)possibility of obfuscating programs. In: Kilian, J. (ed.) CRYPTO 2001. LNCS, vol. 2139, pp. 1–18. Springer, Heidelberg (2001)
4. Bellare, M., Boldyreva, A., O'Neill, A.: Deterministic and efficiently searchable encryption. In: Menezes, A. (ed.) CRYPTO 2007. LNCS, vol. 4622, pp. 535–552. Springer, Heidelberg (2007)
5. Bellare, M., Fischlin, M., O'Neill, A., Ristenpart, T.: Deterministic encryption: Definitional equivalences and constructions without random oracles. In: Wagner, D. (ed.) CRYPTO 2008. LNCS, vol. 5157, pp. 360–378. Springer, Heidelberg (2008)
6. Bellare, M., Hoang, V.T., Keelveedhi, S.: Instantiating random oracles via uCEs. In: Canetti, R., Garay, J.A. (eds.) CRYPTO 2013, Part II. LNCS, vol. 8043, pp. 398–415. Springer, Heidelberg (2013)
7. Bellare, M., Hoang, V.T., Keelveedhi, S.: Instantiating random oracles via UCEs. Cryptology ePrint Archive, Report 2013/424 (2013), http://eprint.iacr.org/2013/424
8. Bellare, M., Hoang, V.T., Keelveedhi, S.: Personal communication (Semptember 2013)
9. Bellare, M., Stepanovs, I., Tessaro, S.: Poly-many hardcore bits for any one-way function and a framework for differing-inputs obfuscation. In: Sarkar, P., Iwata, T. (eds.) ASIACRYPT 2014. LNCS, vol. 8874, Springer, Heidelberg (2014)

10. Bitansky, N., Canetti, R.: On strong simulation and composable point obfuscation. In: Rabin, T. (ed.) CRYPTO 2010. LNCS, vol. 6223, pp. 520–537. Springer, Heidelberg (2010)

11. Bitansky, N., Paneth, O.: Point obfuscation and 3-round zero-knowledge. In: Cramer, R. (ed.) TCC 2012. LNCS, vol. 7194, pp. 190–208. Springer, Heidelberg (2012)

12. Boldyreva, A., Fehr, S., O'Neill, A.: On notions of security for deterministic encryption, and efficient constructions without random oracles. In: Wagner, D. (ed.) CRYPTO 2008. LNCS, vol. 5157, pp. 335–359. Springer, Heidelberg (2008)

13. Boneh, D., Waters, B.: Constrained pseudorandom functions and their applications. In: Sako, K., Sarkar, P. (eds.) ASIACRYPT 2013, Part II. LNCS, vol. 8270, pp. 280–300. Springer, Heidelberg (2013)

14. Boneh, D., Zhandry, M.: Multiparty key exchange, efficient traitor tracing, and more from indistinguishability obfuscation. In: Garay, J.A., Gennaro, R. (eds.) CRYPTO 2014, Part I. LNCS, vol. 8616, pp. 480–499. Springer, Heidelberg (2014)

15. Boyle, E., Chung, K.M., Pass, R.: On extractability obfuscation. In: Lindell, Y. (ed.) TCC 2014. LNCS, vol. 8349, pp. 52–73. Springer, Heidelberg (2014)

16. Boyle, E., Goldwasser, S., Ivan, I.: Functional signatures and pseudorandom functions. In: Krawczyk, H. (ed.) PKC 2014. LNCS, vol. 8383, pp. 501–519. Springer (Mar 2014)

17. Brzuska, C., Farshim, P., Mittelbach, A.: Indistinguishability obfuscation and uCEs: The case of computationally unpredictable sources. In: Garay, J.A., Gennaro, R. (eds.) CRYPTO 2014, Part I. LNCS, vol. 8616, pp. 188–205. Springer, Heidelberg (2014)

18. Brzuska, C., Mittelbach, A.: Indistinguishability obfuscation versus multi-bit point obfuscation with auxiliary input. In: Sarkar, P., Iwata, T. (eds.) ASIACRYPT 2014. LNCS, vol. 8874, Springer, Heidelberg (2014)

19. Brzuska, C., Mittelbach, A.: Using indistinguishability obfuscation via uces. Cryptology ePrint Archive, Report 2014/381 (2014), http://eprint.iacr.org/

20. Canetti, R.: Towards realizing random oracles: Hash functions that hide all partial information. In: Kaliski Jr., B.S. (ed.) CRYPTO 1997. LNCS, vol. 1294, pp. 455–469. Springer, Heidelberg (1997)

21. Canetti, R., Dakdouk, R.R.: Obfuscating point functions with multibit output. In: Smart, N.P. (ed.) EUROCRYPT 2008. LNCS, vol. 4965, pp. 489–508. Springer, Heidelberg (2008)

22. Canetti, R., Goldreich, O., Halevi, S.: The random oracle methodology, revisited (preliminary version). In: 30th ACM STOC, pp. 209–218. ACM Press (May 1998)

23. Freeman, D.M., Goldreich, O., Kiltz, E., Rosen, A., Segev, G.: More constructions of lossy and correlation-secure trapdoor functions. Journal of Cryptology 26(1), 39–74 (2013)

24. Fuller, B., O'Neill, A., Reyzin, L.: A unified approach to deterministic encryption: New constructions and a connection to computational entropy. In: Cramer, R. (ed.) TCC 2012. LNCS, vol. 7194, pp. 582–599. Springer, Heidelberg (2012)

25. Garg, S., Gentry, C., Halevi, S., Raykova, M.: Two-round secure MPC from indistinguishability obfuscation. In: Lindell, Y. (ed.) TCC 2014. LNCS, vol. 8349, pp. 74–94. Springer, Heidelberg (2014)

26. Garg, S., Gentry, C., Halevi, S., Raykova, M., Sahai, A., Waters, B.: Candidate indistinguishability obfuscation and functional encryption for all circuits. In: 54th FOCS. IEEE Computer Society Press (October 2013)

27. Garg, S., Gentry, C., Halevi, S., Wichs, D.: On the implausibility of differing-inputs obfuscation and extractable witness encryption with auxiliary input. In: Garay, J.A., Gennaro, R. (eds.) CRYPTO 2014, Part I. LNCS, vol. 8616, pp. 518–535. Springer, Heidelberg (2014)
28. Garg, S., Gentry, C., Halevi, S., Wichs, D.: On the implausibility of differing-inputs obfuscation and extractable witness encryption with auxiliary input. Cryptology ePrint Archive, Report 2013/860 (June 13, 2014), http://eprint.iacr.org/
29. Goldreich, O., Goldwasser, S., Micali, S.: How to construct random functions (extended abstract). In: 25th FOCS, pp. 464–479. IEEE Computer Society Press (October 1984)
30. Goldwasser, S., Gordon, S.D., Goyal, V., Jain, A., Katz, J., Liu, F.H., Sahai, A., Shi, E., Zhou, H.S.: Multi-input functional encryption. In: Nguyen, P.Q., Oswald, E. (eds.) EUROCRYPT 2014. LNCS, vol. 8441, pp. 578–602. Springer, Heidelberg (2014)
31. Goyal, V., O'Neill, A., Rao, V.: Correlated-input secure hash functions. In: Ishai, Y. (ed.) TCC 2011. LNCS, vol. 6597, pp. 182–200. Springer, Heidelberg (2011)
32. Hofheinz, D.: Fully secure constrained pseudorandom functions using random oracles. Cryptology ePrint Archive, Report 2014/372 (2014), http://eprint.iacr.org/
33. Hohenberger, S., Sahai, A., Waters, B.: Replacing a random oracle: Full domain hash from indistinguishability obfuscation. In: Nguyen, P.Q., Oswald, E. (eds.) EUROCRYPT 2014. LNCS, vol. 8441, pp. 201–220. Springer, Heidelberg (2014)
34. Kiayias, A., Papadopoulos, S., Triandopoulos, N., Zacharias, T.: Delegatable pseudorandom functions and applications. In: Sadeghi, A.R., Gligor, V.D., Yung, M. (eds.) ACM CCS 2013, pp. 669–684. ACM Press (November 2013)
35. Matsuda, T., Hanaoka, G.: Chosen ciphertext security via point obfuscation. In: Lindell, Y. (ed.) TCC 2014. LNCS, vol. 8349, pp. 95–120. Springer, Heidelberg (2014)
36. Matsuda, T., Hanaoka, G.: Chosen ciphertext security via UCE. In: Krawczyk, H. (ed.) PKC 2014. LNCS, vol. 8383, pp. 56–76. Springer, Heidelberg (2014)
37. Mittelbach, A.: Salvaging indifferentiability in a multi-stage setting. In: Nguyen, P.Q., Oswald, E. (eds.) EUROCRYPT 2014. LNCS, vol. 8441, pp. 603–621. Springer, Heidelberg (2014)
38. Rosen, A., Segev, G.: Chosen-ciphertext security via correlated products. SIAM Journal on Computing 39(7), 3058–3088 (2010)
39. Sahai, A., Waters, B.: How to use indistinguishability obfuscation: deniable encryption, and more. In: Shmoys, D.B. (ed.) 46th ACM STOC, pp. 475–484. ACM Press (May/June 2014)
40. Wichs, D.: Barriers in cryptography with weak, correlated and leaky sources. In: Kleinberg, R.D. (ed.) ITCS 2013, pp. 111–126. ACM (January 2013)

Indistinguishability Obfuscation versus Multi-bit Point Obfuscation with Auxiliary Input

Christina Brzuska[1] and Arno Mittelbach[2]

[1] Tel Aviv University, Israel
[2] Darmstadt University of Technology, Germany
brzuska@post.tau.ac.il, arno.mittelbach@cased.de

Abstract. In a recent celebrated breakthrough, Garg et al. (FOCS 2013) gave the first candidate for so-called indistinguishability obfuscation (iO) thereby reviving the interest in obfuscation for a general purpose. Since then, iO has been used to advance numerous sub-areas of cryptography. While indistinguishability obfuscation is a general purpose obfuscation scheme, several obfuscators for specific functionalities have been considered. In particular, special attention has been given to the obfuscation of so-called *point functions* that return zero everywhere, except for a single point x. A strong variant is point obfuscation with auxiliary input (AIPO), which allows an adversary to learn some non-trivial auxiliary information about the obfuscated point x (Goldwasser, Tauman-Kalai; FOCS, 2005).

Multi-bit point functions are a strengthening of point functions, where on x, the point function returns a string m instead of 1. Multi-bit point functions with auxiliary input (MB-AIPO) have been constructed from composable AIPO by Canetti and Dakdouk (Eurocrypt 2008) and have been used by Matsuda and Hanaoka (TCC 2014) to construct CCA-secure public-key encryption schemes and by Bitansky and Paneth (TCC 2012) to construct three-round weak zero-knowledge protocols for NP.

In this paper we present both positive and negative results. We show that if indistinguishability obfuscation exists, then MB-AIPO does not. Towards this goal, we build on techniques by Brzuska, Farshim and Mittelbach (Crypto 2014) who use indistinguishability obfuscation as a mean to attack a large class of assumptions from the Universal Computational Extractor framework (Bellare, Hoang and Keelveedhi; Crypto 2013). On the positive side we introduce a weak version of MB-AIPO which we deem to be outside the reach of our impossibility result. We build this weak version of MB-AIPO based on iO and AIPO and prove that it suffices to construct a public-key encryption scheme that is secure even if the adversary can learn an arbitrary leakage function of the secret key, as long as the secret key remains computationally hidden. Thereby, we strengthen a result by Canetti et al. (TCC 2010) that showed a similar connection in the symmetric-key setting.

Keywords: Indistinguishability obfuscation, differing-inputs obfuscation, point function obfuscation, multi-bit point function obfuscation, auxiliary input obfuscation, leakage resilient PKE.

P. Sarkar and T. Iwata (Eds.): ASIACRYPT 2014, PART II, LNCS 8874, pp. 142–161, 2014.

1 Introduction

The obfuscation of a program should hide its inner workings while preserving the functionality of the program. Inspired by heuristic code-obfuscation techniques [28], obfuscation turned into a major research area of cryptography due to its manifold applications. The formal definition of Virtual Black-Box Obfuscation (VBB) demands that an obfuscated program is as good as a black-box that provides the same input-output behaviour as the program. Since the seminal paper of Barak et al. [5, 4], we know that this strong notion of obfuscation is generally not achievable.

Hence, research focused on special-purpose obfuscators and, in particular, there are various positive results for obfuscating so-called *point functions* p_x, that map all strings to 0, except for a single string x that they map to 1 [23, 27, 30, 52, 40, 24, 29, 26, 9, 13]. Other positive examples include obfuscating re-encryption [41] and encrypted signatures [38].

Point Functions vs. Point Functions with Multi-bit Output. When considering point function obfuscation, we need to make a clear distinction between plain point functions such as p_x which map every input to 0 except for the single input x that is mapped to 1 and point functions with multi-bit output (MBPF) such as $p_{x,m}$ where input x is mapped to string m. Obfuscators for plain point functions are constructed in [23, 52, 40, 29].

Another important distinction is, whether the adversary is given some "leakage" about x, so-called *auxiliary information*, as introduced by Goldwasser and Tauman-Kalai [36]. We note that the obfuscator by Canetti [23] also allow for auxiliary information about the point x to leak and the obfuscator by Dodis et al. [29] allows for auxiliary information that hides the point statistically.

Although very similar, obfuscation schemes for MBPFs seem to be harder to construct than obfuscation schemes for plain point functions. Indeed, Canetti and Dakdouk initiated the study of obfuscation for MBPFs and showed that such obfuscation schemes are closely related to *composable* obfuscation schemes for plain point functions [24]. They show that obfuscators for MBPFs exist if composable obfuscators for plain point functions exist. Moreover, they show that composability is a non-trivial property. Both of these results carry over to obfuscation in the presence of auxiliary information, as long as the auxiliary information does not allow to recover the point. We refer to this type of auxiliary information as hard-to-invert or more specifically to computationally hard-to-invert.

Bitansky and Paneth [13] provide a clean treatment of auxiliary inputs and introduce the notion of *point obfuscation with auxiliary input* secure against unpredictable distributions (AIPO). Assuming composable AIPO they construct a three-round weak zero-knowledge protocol for $\mathcal{NP}$. Matsuda and Hanaoka [49] extend the notion of AIPO to the multi-bit point function case (MB-AIPO) and show how to use it to build CCA-secure public-key encryption. We adopt the notions AIPO and MB-AIPO in this paper.

Indistinguishability Obfuscation. Simultaneously to constructing task-specific obfuscation schemes, the quest for general obfuscators continued, and in a celebrated breakthrough [32], Garg, Gentry, Halevi, Raykova, Sahai and Waters presented a candidate construction for indistinguishability obfuscation (iO). The notion of indistinguishability obfuscation is weaker than VBB-obfuscation and assures that, for any two circuits that compute the same function, their obfuscations are indistinguishable. As Goldwasser and Rothblum [37] establish, this seemingly weak notion of obfuscation is actually the *best possible* notion of obfuscation. And indeed, the work by Garg et al. [32] inspired simultaneous breakthroughs for hard problems in several sub-areas of cryptography [51, 16, 1, 31, 42, 15, 8, 22] such as functional encryption, deniable encryption, two-round secure multi-party computation, full-domain hash, poly-many hardcore bits for any one-way function and more.

Contribution. In this paper we give both positive and negative results. We show that the existence of indistinguishability obfuscation contradicts the existence of multi-bit point function obfuscation in the presence of computationally hard-to-invert auxiliary information (MB-AIPO), a notion which was built upon in [13, 49]. That is, if indistinguishability obfuscation exists, then MB-AIPO does not exist and some of the results in [13, 49] are based on a false assumption. (We discuss the precise implications shortly.) Or, equivalently, if MB-AIPO exists, then indistinguishability obfuscation does not exist and all candidate assumptions are false [32, 50, 34]. However, we do not have a candidate construction for MB-AIPO[1], but we do have a candidate construction for iO. Therefore, given the current advancements in the understanding of indistinguishability obfuscation—for example, Gentry et al. [34] show in a very recent work that iO can be based on the Multilinear Subgroup Elimination Assumption thereby giving the first construction based on an instance-independent assumption—we consider the existence of iO to be more likely.

In summary, we derive the following negative results.

Theorem (informal). *If indistinguishability obfuscation exists, then MB-AIPO and hence composable AIPO do not exist.*

Our proof is inspired by the result by Barak et al. [5, 4]. Technically, they show that multi-bit output point functions cannot be VBB-obfuscated when "coupled" with a particularly chosen second function. Let $p_{x,m}$ be a multi-bit output point function that maps all strings to 0, except for the single point x which the function maps to the string m. Now, the second function is a *test function* $\mathcal{T}_{x,m}$ that takes as input a circuit C and tests whether $C(x)$ is equal to m. Now, if an adversary is given access to two oracles that compute $p_{x,m}$ and $\mathcal{T}_{x',m'}$ then it cannot check whether the two functions "match", i.e., whether $(x', m') = (x, m)$.

---

[1] Note that the construction by Canetti and Dakdouk [24] is from composable AIPO for which we do not have a candidate construction. The construction by Bitansky and Canetti [9, 10] achieves composable point obfuscation in the virtual grey-box setting (VGB) which implies MB-AIPO, but only for statistically hard-to-invert leakage [49].

In turn, when given a circuit C that computes $p_{x,m}$, the adversary can run $\mathcal{T}_{x,m}$ on C and simply check whether $\mathcal{T}_{x,m}(C)$ returns 1. Hence, the obfuscation of $p_{x,m}$ and the obfuscation of $\mathcal{T}_{x,m}$ leak more information than two oracles for $p_{x,m}$ and $\mathcal{T}_{x,m}$ thus establishing a counterexample for VBB obfuscation.

Although the starting point of Barak et al.'s result is a point function $p_{x,m}$, they actually construct an unobfuscateble function that is a combination of the point function $p_{x,m}$ together with test function $\mathcal{T}_{x,m}$ and thus their result is an impossibility result for general VBB obfuscation rather than an impossibility result for point function obfuscation.

In order to obtain a result for point function obfuscation based on the above idea, we proceed in two steps. Firstly, we think of the test circuit $\mathcal{T}_{x,m}$ as "auxiliary information" [36] about the point function $p_{x,m}$. Secondly, we do not use the "plain" test function $\mathcal{T}_{x,m}$ but rather, based on indistinguishability obfuscation, we construct an obfuscated circuit that approximates the behaviour of $\mathcal{T}_{x,m}$.

Matsuda and Hanoaka [49] introduce MB-AIPO as follows. A first stage of the adversary $\mathcal{B}_1$ defines a distribution over a point address x, a message m and auxiliary input z—we sometimes refer to the auxiliary input as "leakage".

Now, a second stage of the adversary $\mathcal{B}_2$ gets the leakage z as well as an obfuscation of the point function $p_{x,m}$ or an obfuscation of the point function $p_{x,m'}$, where m' is drawn at random. The distinguisher $\mathcal{B}_2$ tries to guess which of the two it received.

A multi-bit point function obfuscator is called secure, if for all efficiently computable distributions[2] $\mathcal{B}_1$, for the second stage of the adversary $\mathcal{B}_2$, given z, obfuscations of $p_{x,m}$ and obfuscations of $p_{x,m'}$ are indistinguishable.

As such, the definition is not satisfiable, because $\mathcal{B}_1$ can leak the pair (x, m) so that $\mathcal{B}_2$ can check whether this pair "matches" the point function that $\mathcal{B}_1$ received. Hence, we additionally require that $\mathcal{B}_1$ be computationally *unpredictable*, that is, for all efficient predictors Pred, it holds that with high probability over $(z, x, m) \leftarrow_{\$} \mathcal{B}_1$, given z, the algorithm Pred outputs x at most with negligible probability.

To recap, $\mathcal{B}_1$ outputs a point address x, a point value m and some leakage z such that z hides the value x. Then, the second stage of the adversary $\mathcal{B}_2$ receives z as well as an obfuscation of $p_{x,m}$ or an obfuscation of $p_{x,m'}$ and needs to distinguish between the two. See Definition 5 for a formal definition.

Hence, to attack MB-AIPO, we need to define an adversarial distribution $\mathcal{B}_1$ that is unpredictable and that returns some leakage z that allows $\mathcal{B}_2$ to distinguish between obfuscations of $p_{x,m}$ and obfuscations of $p_{x,m'}$. Our adversarial distribution $\mathcal{B}_1$ draws a random value x and a random value m. Moreover, as auxiliary information z, it will output a specially devised obfuscation that approximates the behaviour of the test function $T_{x,m}$.

Given the circuit z and a multi-bit point function p, the second stage of the adversary $\mathcal{B}_2$ outputs whatever the circuit z outputs when run on p. It distinguishes successfully between an obfuscation p of the "matching" multi-bit

---

[2] We add the condition of unpredictability in the next paragraph.

point function $p_{x,m}$ and the obfuscation p of a non-matching multi-bit point function $p_{x,m'}$.

We now explain how adversary $\mathcal{B}_1$ constructs z. The hardness resides in constructing an obfuscation of the test function $\mathcal{T}_{x,m}$ such that indeed, x is unpredictable given the description of the obfuscated test function. Towards this goal, we build on techniques developed by Brzuska, Farshim and Mittelbach [20] who show a similar 1-out-of-2 result, namely that indistinguishability obfuscation and a large class of assumptions of the Universal Computational Extractor framework (UCE) [6] are mutually exclusive. We obfuscate the test function via indistinguishability obfuscation and prove that it is indistinguishable from an obfuscation of the zero circuit $\mathbf{0}$, the circuit that returns 0 on all inputs. As the zero circuit does not contain any information about x, indistinguishability obfuscation guarantees that likewise, an obfuscation of the test function $\mathcal{T}_{x,m}$ hides x computationally.

In detail, let y be the output of a pseudo-random generator G when applied to m. The circuit z is an indistinguishability obfuscation of the following circuit $C[x,y]$ with parameters x and y hard-coded. Circuit $C[x,y]$ gets as input a circuit p, runs p on x and checks whether $\mathsf{G}(p(x))$ is equal to y. If yes, it outputs 1. Else, it outputs 0.

For simplicity, let us assume that the G is injective. Then, $C[x,y]$ behaves exactly like the test function $\mathcal{T}_{x,m}$. Interestingly, and that is the key idea, we do not actually use m to compute the circuit $C[x,y]$, we only need $y = \mathsf{G}(m)$. In particular, as G is a one-way function, y does not leak m. Moreover, as G is a pseudo-random generator, y does not even leak whether a pre-image m exists.

We will now use the PRG property to argue that an indistinguishability obfuscation of $C[x,y]$ does not leak anything about x. Namely, if y is in the image of the PRG, then $C[x,y]$ is equal to the test function $\mathcal{T}_{x,m}$. In turn, when y is not in the image of the PRG, then $C[x,y]$ is the all-zero function. Due to the PRG security, these two distributions—$C[x,y]$ when y is drawn as an output from the G and $C[x,y]$ when y is drawn at random—are computationally indistinguishable. Moreover, when the PRG has enough stretch, then with overwhelming probability, a random y is not in the image of the PRG, and hence, with overwhelming probability over a random y, the circuit $C[x,y]$ is the all-zero circuit $\mathbf{0}$. For the two functionally equivalent circuits $C[x,y]$ and $\mathbf{0}$, it holds that $\mathrm{iO}(C[x,y])$ is computationally indistinguishable from $\mathrm{iO}(\mathbf{0})$. As $\mathbf{0}$ leaks nothing about x, we can argue that also $\mathrm{iO}(C[x,y])$ leaks nothing about x and hence, x is unpredictable from the leakage of $\mathcal{B}_1$ as required by the definition of MB-AIPO.

We note that our usage of the PRG is somewhat similar to the use by Sahai and Waters in their construction of a CCA-secure PKE scheme from iO [51] as well as the range-extension of Matsuda and Hanaoka [49] of a multi-bit point function to obtain shorter point values and the range-extension of a UCE1-secure hash-function by Bellare et al. [7] used to strengthen the impossibility result by Brzuska et al. [20].

To recap, we use the pseudo-random generator to hide m, and we use the indistinguishability obfuscation to hide x. Note that unpredictability in the MB-AIPO definition only requires that x is unpredictable from the leakage. Therefore, hiding

m might seem unnecessary. Interestingly, it turns out that this is not merely an artefact of our proof. Namely, we define a strong notion of unpredictability where x needs to be unpredictable from the pair (z, m), and we show that MB-AIPO can achieved under this definition, assuming plain AIPO in conjunction with iO.

Indeed, our negative results do not carry over to the setting of obfuscating plain point functions in the presence of auxiliary information, that is, to plain AIPO (assuming they are not composable[3]). This is due the fact that we cannot apply the PRG to a function that only outputs a single bit.

Analogously, it looks unlikely that the result of Barak et al. [5, 4] carries over to plain point functions, because it seems crucial that the point function $p_{x,m}$ has a multi-bit output m. Imagine that $\mathcal{T}_x$ takes the circuit C as input and returns 1 if and only if $C(x) = 1$. Then, an adversary can perform binary search and recover x, even when only given access to $\mathcal{T}_x$ and p_x as oracles.[4] Hence, also their result does not carry over to standard point functions.

On the positive side, as hinted above, we show ways to work around our impossibility result. Firstly, note that Canetti et al. [26] introduce weaker versions of MB-AIPO that are not affected by our negative results. In particular, they use these weaker notions to build a symmetric-key encryption scheme that is secure in the presence of hard-to-invert leakage about the key. We strengthen their result insofar, as we present a notion that lies between their weaker versions of MB-AIPO and full MB-AIPO.

Our weak notion of MB-AIPO requires that the auxiliary information L computationally hides the point x even when given the corresponding point value m for some multi-bit point function $p_{x,m}$.

This definition circumvents our impossibility result because we cannot use the security of the PRG anymore. In the proof of the impossibility result, we used that the circuit $C[x, y]$ does not need m as a parameter and only needs $y = \mathsf{G}(m)$. In the presence of the value m, the reduction to the PRG-security does not carry through.

This argument merely shows that our proof fails. However, we provide positive evidence for the new security notion. Assuming AIPO and iO, we give a construction that achieves MB-AIPO for strongly unpredictable distributions. We show that this weaker notion of MB-AIPO is useful for applications. Based on our weak MB-AIPO construction, we build a public-key encryption scheme which is leakage resilient in the presence of hard-to-invert leakage of the key. Previously, such a result was only known for symmetric-key encryption [26]. We next discuss existing notions of multi-bit point obfuscation.

Notions of Multi-bit Point-Obfuscation. Lynn et al. [47] initiate the study of obfuscators for point functions with multi-bit output (MBPF) in the idealized random oracle model (ROM) and give a construction of a VBB obfuscator in the ROM. Though they do not explicitly introduce auxiliary information, it is easily seen that their construction allows for computationally hard-to-invert auxiliary

---

[3] Canetti and Dakdouk [24] show that composable AIPO already implies MB-AIPO.

[4] Access to the testing function $\mathcal{T}_x$ suffices to recover x, even when not given access to p_x neither as a circuit nor as an oracle.

information. Canetti and Dakdouk [24] initiated the study of MBPF-obfuscators in the standard model and showed that these exist if so-called t-composable obfuscators exist for plain point functions. Building on these results Canetti and Bitansky [9, 10] show that the point obfuscator by Canetti [23] meets the requirements of a t-composable point function obfuscator down to a strong variant of the decisional Diffie–Hellman assumption (DDH), namely the t-strong vector DDH assumption. Note that the notion they achieve is the so-called notion of Virtual Grey-Box obfuscation (VGB)—the virtual grey box notion was introduced by Bitansky and Canetti [9, 10] and allows the simulator to run in unbounded time—and not the stronger notion of VBB obfuscation. In [26] Canetti et al. show that obfuscators for MBPFs are closely related to symmetric encryption and that obfuscators for MBPFs secure in the presence of (certain types of) auxiliary inputs imply the existence of (certain types of) leakage resilient symmetric encryption schemes. Bitansky and Paneth [13] introduce a clean treatment of a form of auxiliary information which hides the obfuscated point computationally (AIPO) and Matsuda and Hanaoka [49] extend their notion to multi-bit output functions which is also the notion considered in this paper (MB-AIPO). Using composable AIPOs Bitansky and Paneth construct a three-round weak zero-knowledge protocol for $\mathcal{NP}$ based on composable AIPO [13] thereby circumventing a black-box impossibility result [35]. Matsuda and Hanaoka (MH, [49]) introduce also an average case variant of MB-AIPO and a more restricted version of MB-AIPO which requires the auxiliary input to statistically hide the obfuscated point. They further study the relation between these average case MB-AIPO notions and the worst-case notions of point obfuscation, that is, virtual black-box and virtual grey-box. MH show how to construct CCA secure public-key encryption schemes from an IND-CPA secure encryption scheme using MB-AIPO with computationally hard-to-invert auxiliary information, as well as, how to achieve CCA security starting from a CPA-secure lossy encryption scheme and using MB-AIPO with statistically hard-to-invert auxiliary information. In a very recent work, Canetti et al. [25] show how to build fuzzy extractors using t-composable point obfuscation secure in the presence of auxiliary information in the virtual grey-box setting. MH show that this form of point obfuscation implies MB-AIPO with respect to statistically hard-to-invert auxiliary information [49], it is, however, not known if it can be shown to also imply MB-AIPO with computationally hard-to-invert auxiliary information.

Our negative result shows that if indistinguishability obfuscation exists that MB-AIPO with computationally hard-to-invert auxiliary information does not exist. This applies to the first of the two constructions of CCA secure PKE schemes by Matsuda and Hanaoka [49] as well as to the construction of a three-round weak zero-knowledge protocol for $\mathcal{NP}$ by Bitansky and Paneth [13].[5] We

---

[5] Bitanski and Paneth actually consider the stronger notion of composable AIPO which implies MB-AIPO. We also note that the construction of 3-message witness-hiding protocols from AIPO [13] as well as the construction of a CCA secure PKE scheme from a lossy encryption scheme and MB-AIPO with statistically hard-to-invert information [49] are not affected by our result.

leave as open problems, whether our negative results can be strengthened to encompass further uses of MBPF obfuscation or, whether the above constructions can be based on weaker notions of MBPF obfuscation not ruled out by our result.

Finally, we note that our result can be regarded as a random oracle uninstantiability result. One can show that the VBB obfuscator given by Lynn et al. [47] is a secure MB-AIPO in the random oracle model, even if hard-to-invert leakage is allowed. Our results shows that, if indistinguishability obfuscation exists, then there is no hash-function that instantiates the random oracle securely according to this notion of security.

Point Obfuscation and Indistinguishability Obfuscation. For our positive result, a construction of weak MB-AIPO and subsequently a construction of a leakage resilient PKE scheme, we combine AIPOs and indistinguishability obfuscation. In a recent work Brzuska and Mittelbach (BM, [22]) show that combining these techniques allows to build powerful primitives and they give the first construction of a standard model hash function which is UCE secure for a non-trivial UCE notion which implies universal hardcore-functions and q-query correlated input secure hash functions. Furthermore, we note that our notion of weak MB-AIPO is inspired by the UCE notion introduced by BM: UCE security with respect to strongly unpredictable sources.

In a recent and independent work, Hofheinz constructs fully secure constrained pseudorandom functions [39] in the random oracle model. A constrained PRF allows for the generation of keys that enable the holder to evaluate the PRF on a set of points but not on all points, and various forms have been suggested [14, 17, 44]. In contrast to previous works Hofheinz uses point obfuscation and an extension he calls *extensible testers*—an extensible tester can be regarded as an obfuscation of a set of points Z which can be combined with a known set Z' into a tester for set $(Z \cup Z')$—in conjunction with indistinguishability obfuscation to hide which points a given key allows to honestly evaluate. This allows him to achieve full security without relying on complexity leveraging which was used in previous constructions entailing a superpolynomial loss of security in the adaptive setting. We note that unlike this work (and the work by BM) Hofheinz relies on the simpler assumption of plain point obfuscation (that is, obfuscation without auxiliary inputs) and shows how to build extensible testers based on the DDH-based point obfuscator by Canetti [23].

Further 1-Out-of-2 Results. Indistinguishability obfuscation has led to many surprising breakthroughs in a number of sub-areas of cryptography [51, 16, 1, 31, 42, 15, 8, 22]. Interestingly, the existence of indistinguishability obfuscation collides with the existence of other desirable primitives. If indistinguishability obfuscation exists, then it draws a fine line between what is possible and what is impossible, e.g., MB-AIPO and iO are mutually exclusive, but weak MB-AIPO can be build from iO (and AIPO).

If indistinguishability obfuscations does not exist, then 1-out-of-2 results are a promising way to prove such an impossibility result. In particular, it would be highly interesting to show a 1-out-of-2 result for iO and some other primitive for which we have a candidate construction, e.g., AIPO. Whether indistinguishability obfuscation exists or not, 1-out-of-2 results for iO help us explore the boundaries of what is possible. Either, they increase our understanding of iO, or they increase our understanding of other primitives.

Before our result, several 1-out-of-2 results have been established for iO. We already discussed the result by Brzuska et al. [20] who show that iO is mutually exclusive with a large class of assumptions from the UCE framework [6].

Interestingly, several notions of obfuscation are mutually exclusive with iO. Bitansky et al. [11] show that iO implies the non-existence of average-case virtual black-box obfuscation with auxiliary input (AI-VBB) for circuit families with super-polynomial pseudo-entropy. In particular, AI-VBB obfuscation is impossible for all pseudo-random function families. Moreover, they show that indistinguishability obfuscation implies the non-existence of average-case virtual black-box obfuscation with a universal simulator for circuit families with a superpolynomial amount of pseudo-entropy. Bitansky et al. [12] show that if indistinguishability obfuscation exists, then for every extractable one-way function family there is an (unbounded polynomial-length) auxiliary input distribution $\mathcal{L}$ and an adversary $\mathcal{A}$ such that all extractors fail for $\mathcal{A}$. Similar to our result for MB-AIPO, they embed an attack circuit into the auxiliary input. Boyle and Pass [18] strengthen this result under the assumption of differing-input obfuscation (diO). If diO exists, then the quantifiers can be reversed so that $\mathcal{L}$ does not depend on the one-way function family.

Moreover, Bitansky et al. [12] show how to construct extractable one-way functions with *bounded* auxiliary input under relatively standard assumptions. Finally, Marcedone et al. [48], as well as Koppula et al. [46] show that if indistinguishability obfuscation exists, then IND-CPA-security of an encryption scheme does not imply its circular security, even if the cycles are of arbitrary polynomial-length.

On the Plausibility of iO. Barak et al. [5, 4] introduce Indistinguishability Obfuscation as a notion of obfuscation that is not ruled out by their impossibility result for virtual black-box obfuscation. The amount and quality of positive results based on iO as well as the number of 1-out-of-2 results indicate that indeed, indistinguishability obfusaction is a strong assumption and Komargodski et al. [45] show that (even imperfect) indistinguishability obfuscation does not exist in Pessiland [43], a world where NP is hard but one-way functions do not exist. Their result does not carry over to a world where one-way functions exist.

Garg et al. [33] show that differing-inputs obfuscation—a stronger form of indistinguishably obfuscation that was also introduced in the seminal paper by Barak et al. [5, 4]—is mutually exclusive with some special-purpose obfuscator. As the particular special-purpose obfuscator that they consider seems to be a relatively mild assumption, we interpret their result as a conditional impossibility result for differing-inputs obfuscation. However, their result does not apply to indistinguishability obfuscation. In particular, recent results show how to improve the assumptions that underly indistinguishability obfuscation [50, 19, 3, 2, 34] supporting its plausibility.

Auxiliary Input. Auxiliary input (AI) has been introduced by Goldwasser and Tauman-Kalai [36] and the specifics of how AI is modeled are very important when it comes to the (im)possibility of notions of obfuscation. Notably, for extractable one-way functions, the aforementioned results by Bitansky et al. [12] show that, assuming iO, this notion of security is impossible under unbounded AI, but possible when the length of the AI is bounded by a fixed polynomial that is known a priori. Potentially, bounded AI—for example, if the amount of AI is restricted to be less than the size of an MB-AIPO—could also be used to circumvent our iO-based impossibility result while preserving a reasonably wide range of applications.

Moreover, one can consider independent AI rather than dependent AI, which would also help to circumvent our impossibility result. However, AI is usually useful for composition where partial information about the obfuscated circuit/point is leaked to the outside and thus, dependent AI is often quite powerful in applications. However, even security under independent AI is non-trivial to achieve. Assuming iO, Bitansky et al. [11] show that a large class of functions cannot be VBB-obfuscated in the presence of independent AI.

A further possibility to circumvent our impossibility result is to consider a statistical notion of unpredictability rather than computational unpredictability. Statistical unpredictability has already proved useful for the construction of q-query secure correlation-secure hash functions [22] and CCA secure PKE schemes [49].

While in the VBB-setting AI is a strong notion that corresponds to the existence of a universal simulator [11], in the VGB-setting AI is trivial. That is, it is equivalent whether one considers VGB security with AI or without AI. The reason is that the VGB simulator is unbounded and hence able to compute the best AI itself [9]. Secure AIPOs in the VGB-setting imply AIPOs with statistically hard-to-invert leakage [49]. Our result does not rule out composable AIPOs in the VGB-setting, and indeed, this assumption has been used very recently by Canetti et al. [25] to build computationally secure fuzzy extractors that work for classes of sources that have more errors than entropy.

In light of the subtle modeling of AI, it remains to investigate whether those results in [13] and [49] that use an assumption which is mutually exclusive with iO can be based on an alternative assumption that is compatible with iO. Towards this goal, one might consider our weakened notion of MB-AIPO or model AI in a way that circumvents our result. Finally, it would then be interesting to come up with candidate assumptions for such a notion of security.

Conclusion and Future Work. We show that indistinguishability obfuscation and MB-AIPO—that is, MB-AIPO as used in [13, 49] and with computationally hard-to-invert auxiliary information—are mutually exclusive. It remains to investigate whether the positive results in [13, 49] can be salvaged through weaker notions of MB-AIPO or, perhaps, when combining AIPO and iO in a similar way as we do in the full version [21] to receive our positive result for weak MB-AIPO. We note, however, that, at a first glance, it is not straightforward to base the applications in [13, 49] on our weakened notion of MB-AIPO.[6]

On the other hand, one might ask whether our negative result can be extended to showing that AIPO and iO are mutually exclusive. Currently, we do not know whether this is possible. We consider such a result to be a highly interesting finding and suspect that it would require different techniques than the ones we use. Our result implies directly that differing-inputs obfuscation (diO) and MB-AIPO are mutually exclusive. Perhaps, using different techniques, one might be able to first show that diO and AIPO are mutually exclusive, for example, by showing that we can instantiate the special-purpose obfuscator by Garg et al. [33] using AIPO.

We hope that our work sparks further interest in studying the connections between iO/diO on the one hand and notions of (multi-bit) point obfuscation on the other hand. More generally, we believe that it is an interesting question to identify notions of security that collide with indistinguishability obfuscation and we expect more results of that flavor in the future.

Full Version. Due to space restrictions, this version should be regarded as an extended abstract as we defer many details and all proofs to the full version [21]. In the remainder of this extended abstract we present our main impossibility result and give some intuition for the underlying proof. For details as well as for our positive results (our weak MB-AIPO notion and the construction of a leakage resilient PKE scheme) we refer to the full version [21].

2   Preliminaries

Indistinguishability Obfuscation. Virtual black-box (VBB) obfuscation [5, 36, 4] requires that for any PPT adversary given the code of some functionality (and some auxiliary input) there exists a PPT simulator that given only black-box access to the functionality (and as input the same auxiliary input) produces a computationally indistinguishable distribution. While VBB obfuscation provably does not exist for all circuits [5, 4], weaker notions such as *indistinguishability obfuscation* may well do. An indistinguishability obfuscation (iO) scheme, on the other hand, only ensures that the obfuscations of any two functionally equivalent circuits are computationally indistinguishable. Indistinguishability obfuscation

---

[6] Note that [13] use composable point functions which is a stronger security notion than MB-AIPO for showing the existence of 3-round protocols that are weakly zero-knowledge. Also note, that their second result, a 3-round witness-hiding protocol, is not affected by our result. Likewise, our result only affects the CCA-encryption scheme in [49] that is based on CPA-security and MB-AIPO. They also build a CCA-secure encryption scheme based on *lossy* IND-CPA secure encryption and MB-AIPO with statistically hard-to-invert auxiliary input. The latter result is not affected by our result.

was originally proposed by Barak et al. [5] as a potential weakening of virtual-black-box obfuscation. We recall the definition from [32].

Definition 1. *A* PPT *algorithm* iO *is called an* indistinguishability obfuscator *for a circuit ensemble* $C = \{C_\lambda\}_{\lambda \in \mathbb{N}}$ *if the following conditions are satisfied:*

– **Correctness.** *For all security parameters* $\lambda \in \mathbb{N}$, *for all* $C \in C_\lambda$, *and for all inputs* x *we have that* $\Pr\left[C'(x) = C(x) : C' \leftarrow_\$ iO(1^\lambda, C)\right] = 1$.
– **Security.** *For any* PPT *distinguisher* $\mathcal{D}$, *for all pairs of circuits* $C_0, C_1 \in C_\lambda$ *such that* $C_0(x) = C_1(x)$ *on all inputs* x *the following distinguishing advantage is negligible:*

$$\left|\Pr\left[\mathcal{D}(1^\lambda, iO(1^\lambda, C_1)) = 1\right] - \Pr\left[\mathcal{D}(1^\lambda, iO(1^\lambda, C_0)) = 1\right]\right| \leq \mathsf{negl}(\lambda).$$

Differing-Inputs Obfuscation. Differing-inputs obfuscation is closely related to indistinguishability obfuscation and also goes back to the seminal paper of Barak et al. [5, 4]. Building on a theorem by Boyle, Chung and Pass [16], we are able to avoid diO as an assumption and only use it as an intermediary concept in our proof. We refer for details to the full version of this work [21].

Point Obfuscation. Besides the general purpose indistinguishability obfuscator we consider obfuscators for the specific class of so-called point functions. A point function p_x for some value $x \in \{0, 1\}^*$ is defined as outputting $\perp$ on all inputs except for x where it outputs 1. In this paper, we consider a variant of point function obfuscators under auxiliary input which was first formalized by Canetti [23]. We here give the definition from [13] presented in a game based formulation. The first definition formalizes unpredictable distributions which are in turn used to define obfuscators for point functions.

Definition 2 (Unpredictable distribution). *A distribution ensemble* $\mathcal{D} = \{D_\lambda = (Z_\lambda, X_\lambda)\}_{\lambda \in \mathbb{N}}$, *on pairs of strings is unpredictable if no poly-size (non-uniform) circuit can predict* X_λ *from* Z_λ. *That is, for every poly-size circuit sequence* $\{C_\lambda\}_{\lambda \in \mathbb{N}}$ *and for all large enough* λ:

$$\Pr_{(z,x) \leftarrow_\$ D_\lambda}\left[C_\lambda(z) = x\right] \leq \mathsf{negl}(\lambda)$$

Definition 3 (Auxiliary input point obfuscation for unpredictable distributions (AIPO)). *A* PPT *algorithm* AIPO *is a point obfuscator for unpredictable distributions if it satisfies the functionality and polynomial slowdown requirements as in VBB-obfuscation [5, 4], and the following secrecy property: for any (efficiently sampleable) unpredictable distribution* $\mathcal{B}_1$ *over* $\{0,1\}^{\mathsf{poly}(\lambda)} \times \{0,1\}^\lambda$ *it holds for any* PPT *algorithm* $\mathcal{B}_2$ *that the probability that the following experiment outputs true for* $(\mathcal{B}_1, \mathcal{B}_2)$ *is negligibly close to* $\frac{1}{2}$:

$$b \leftarrow_\$ \{0, 1\}$$
$$(z, x_0) \leftarrow_\$ \mathcal{B}_1(1^\lambda)$$
$$x_1 \leftarrow_\$ \{0, 1\}^\lambda$$
$$p \leftarrow_\$ \mathsf{AIPO}(x_b)$$
$$b' \leftarrow_\$ \mathcal{B}_2(1^\lambda, p, z)$$
$$\mathbf{return}\ b = b'$$

The probability is over the coins of adversary $(\mathcal{B}_1, \mathcal{B}_2)$*, the coins of* AIPO *and the choices of* x_1 *and* b.

Obfuscation for Point Functions with Multi-bit Output. While point functions only return a single bit, a point function with multi-bit output (MBPF) $p_{x,m}$ for values $x, m \in \{0,1\}^*$ is defined as $\perp$ on any input except for input x which is mapped to m. For an MBPF $p_{x,m}$ we call x the point address and m the point value. Similar to AIPO we can define MB-AIPO via an unpredictable distribution—the notion was introduced by Matsuda and Hanaoka [49] in an average case formulation called AIND-δ-cPUAI—where the distribution outputs a tuple (x,m) (defining a point function $p_{x,m}$) together with auxiliary information z. We require that it be computationally infeasible to recover the point address x given auxiliary information z. Thus, in the MBPF setting we define the unpredictable distribution as $\mathcal{D} = \{D_\lambda = (Z_\lambda, X_\lambda, M_\lambda)\}_{\lambda \in \mathbb{N}}$ but still require that the point address x remains hidden given the auxiliary input. An MB-AIPO assures that the obfuscation of $p_{x,m}$ is indistinguishable from an obfuscation with a changed point value m' that is chosen uniformly at random, which captures that the obfuscation does not reveal any information about the point value m.

Definition 4 (Unpredictable distribution). *A distribution ensemble* $\mathcal{D} = \{D_\lambda = (Z_\lambda, X_\lambda, M_\lambda)\}_{\lambda \in \mathbb{N}}$*, on triples of strings is unpredictable if no poly-size (non-uniform) circuit can predict* X_λ *from* Z_λ*. That is, for every poly-size circuit sequence* $\{C_\lambda\}_{\lambda \in \mathbb{N}}$ *and for all large enough* λ:

$$\Pr_{(z,x,m) \twoheadleftarrow\$ D_\lambda}[C_\lambda(z) = x] \leq \mathsf{negl}(\lambda)$$

Definition 5 (Auxiliary input point obfuscation for unpredictable distributions (MB-AIPO)). *A* PPT *algorithm* MB-AIPO *is a* multi-bit point obfuscator for unpredictable distributions *if it satisfies the functionality and polynomial slowdown requirements as in VBB-obfuscation [5, 4], and the following secrecy property: for any (efficiently sampleable) unpredictable distribution* $\mathcal{B}_1$ *over* $\{0,1\}^{\mathsf{poly}(\lambda)} \times \{0,1\}^\lambda \times \{0,1\}^{\mathsf{poly}(\lambda)}$ *it holds for any* PPT *algorithm* $\mathcal{B}_2$ *that the probability that the following experiment outputs true for* $(\mathcal{B}_1, \mathcal{B}_2)$ *is negligibly close to* $\frac{1}{2}$:

$$b \twoheadleftarrow\$ \{0,1\}$$
$$(z, x, m_0) \twoheadleftarrow\$ \mathcal{B}_1(1^\lambda)$$
$$m_1 \twoheadleftarrow\$ \{0,1\}^\lambda$$
$$p \twoheadleftarrow\$ \mathsf{MB\text{-}AIPO}(x, m_b)$$
$$b' \twoheadleftarrow\$ \mathcal{B}_2(1^\lambda, p, z)$$
$$\textbf{return } b = b'$$

The probability is over the coins of adversary $(\mathcal{B}_1, \mathcal{B}_2)$*, the coins of* AIPO *and the choices of* x*,* m_0*,* m_1 *and* b.

We note that also different definitional choices are possible and we discuss various choices in the full version of this work [21]

Average-Case Point Obfuscation and Statistical Unpredictability. The above notions for point obfuscation are for arbitrary high-entropy distributions over the point address. Instead, we can consider a weaker variant where the point address is sampled according to the uniform distribution. Indeed, Matsuda and Hanaoka [49] recently presented constructions of CCA-secure public-key encryption schemes based on this version of point obfuscation. They call AIPO with arbitrary high-entropy samplers a worst-case notion, and AIPO with the uniform distribution an average-case notion and denote it by AIND-δ-cPUAI. Our impossibility result also applies to AIND-δ-cPUAI which we refer to as *average case MB-AIPO.*

A second avenue to weaken the security requirements of point obfuscators is to require that the auxiliary input needs to hide the point address statistically. We call unpredictable distributions for which this is the case *statistically unpredictable.* Our impossibility result does not carry over to this notion.

3   IO Implies the Impossibility of MB-AIPO

In the following we present our negative result, namely that indistinguishability obfuscation and multi-bit point function obfuscation in the presence of auxiliary information (MB-AIPO) are mutually exclusive. This holds for MB-AIPO as defined in Definition 5 as well as for the two alternative definitions discussed below the definition. We discuss implications of our result in Section 3.2.

3.1   IO and MB-AIPO Are Mutually Exclusive

Multi-bit point obfuscation with auxiliary inputs is a powerful primitive and has, for example, been used to construct CCA-secure encryption schemes [49] and to circumvent black-box impossibility results for three-round weak zero-knowledge protocols for $\mathcal{NP}$ [13]. Our following result says that, if indistinguishability obfuscation and pseudo-random generators exist, then MB-AIPOs (as defined in Definition 5) cannot exist. The result remains valid even if we consider average case MB-AIPOs (where point address x is chosen uniformly at random). Technically our result builds on techniques used by Brzuska, Farshim and Mittelbach (BFM; [20]). BFM show a similar 1-out-of-2 result, namely that if indistinguishability obfuscation exists, then certain kinds of UCE-secure hash functions—a hash function security notion recently introduced in [6]—cannot exist [20]. In the UCE-framework, a hash function H gets a hash key hk and an input x and outputs y. BFM obfuscate the circuit $(H(\cdot, x) = y)$, that given a hash-key hk checks whether hk "matches" the pair (x, y), that is, whether $H(hk, .)$ maps x to y. They show that, if $|hk| < 2|y|$, then it is likely (in the corresponding experiment) that the circuit is the 0-circuit that outputs 0 on all inputs and hence, the indistinguishability obfuscation of this circuit does not leak x.

We will use a similar technique to hide the point address. In order to break AIPO with indistinguishability obfuscation, we need to show that, given the auxiliary input, it is hard to recover the point address, but that, given the auxiliary input and the point function, one can distinguish. Similarly, for UCEs, one needs to show that, given some leakage about x and y, it is hard to recover x, but that, given the leakage and the hash-key hk, one can distinguish whether y was generated by applying $H(hk, .)$ to x or whether y was drawn at random.

Showing that an indistinguishability obfuscation hides a certain value is usually the crux in proofs involving iO. For this, we construct a new technique which may be of independent interest and which we discuss in further detail in the full version [21].

Theorem 1. *If indistinguishability obfuscation exists for all circuits in $\mathcal{P}/\text{poly}$, then average-case obfuscation for multi-bit point functions secure under auxiliary input (MB-AIPO) does not exist.*

This theorem applies to the average-case version where the point address is sampled uniformly, because our adversary samples both, x and m uniformly at random. It also applies to other variants of the MB-AIPO definition which we discuss in the full version of this work [21].

To prove Theorem 1 we use indistinguishability obfuscation to construct an unpredictable distribution $\mathcal{B}_1$ together with an adversary $\mathcal{B}_2$ that, given leakage from the unpredictable distribution can distinguish between point obfuscations from $\mathcal{B}_1$ and point obfuscations from the uniform distribution.

We first give the unpredictable distribution $\mathcal{B}_1$ which takes as input the security parameter 1^λ and outputs two values x, m together with some auxiliary information (resp. leakage) z. Here leakage z will be the indistinguishability obfuscation of a predicate circuit that takes as input a description of a circuit C, evaluates the circuit on a hard-coded value x, runs the result through a pseudo-random generator G and finally compares this result with some hard-coded value y. That is, we consider the circuit

$$C[x, y](\cdot) := \text{iO}\left(G(uC(\cdot, x)) = y\right),$$

where uC denotes a universal circuit taking as input a circuit description C of a fixed length and a value x and which outputs $C(x)$. This use of a PRG allows us later to argue that if value y is chosen uniformly at random that with high probability it falls outside the image of the PRG and thus the circuit is 0 on all inputs, that is, it implements the zero-circuit $\mathbf{0}$.

We next formally define the unpredictable distribution. For this let n and ℓ be two polynomials and let $G : \{0,1\}^{n(\lambda)} \to \{0,1\}^{2n(\lambda)}$ be a pseudo-random generator with stretch 2. Note that we do not need to additionally assume the existence of PRGs as AIPOs (and in particular MB-AIPOs) already imply one-way functions.[7] Let, furthermore, $uC(\cdot, x)$ be a universal circuit that on input a description of a circuit C and value x outputs $C(x)$. Adversary $\mathcal{B}_1$ computes an unpredictable distribution over (z, x, m) as follows:

---

[7] Canetti et al. [26] show that multi-bit point function obfuscation is tightly related to symmetric encryption and that MB-AIPO implies the existence of (leakage-resilient) IND-CPA symmetric encryption schemes.

$$\overbrace{}^{\text{PRG}} \qquad \overbrace{}^{\text{Bad}} \qquad \overbrace{}^{\text{iO}}$$

$\mathsf{Game}_1(\lambda)$	$\mathsf{Game}_2(\lambda)$	$\mathsf{Game}_3(\lambda)$	$\mathsf{Game}_4(\lambda)$
$m \leftarrow\!\!\text{\tiny\$}\, \{0,1\}^{n(\lambda)}$	$m \leftarrow\!\!\text{\tiny\$}\, \{0,1\}^{n(\lambda)}$	$m \leftarrow\!\!\text{\tiny\$}\, \{0,1\}^{n(\lambda)}$	$m \leftarrow\!\!\text{\tiny\$}\, \{0,1\}^{n(\lambda)}$
$x \leftarrow\!\!\text{\tiny\$}\, \{0,1\}^{\ell(\lambda)}$	$x \leftarrow\!\!\text{\tiny\$}\, \{0,1\}^{\ell(\lambda)}$	$x \leftarrow\!\!\text{\tiny\$}\, \{0,1\}^{\ell(\lambda)}$	$x \leftarrow\!\!\text{\tiny\$}\, \{0,1\}^{\ell(\lambda)}$
$y \leftarrow \mathsf{G}(m)$	$y \leftarrow\!\!\text{\tiny\$}\, \{0,1\}^{\|\mathsf{G}(m)\|}$	$y \leftarrow\!\!\text{\tiny\$}\, \{0,1\}^{\|\mathsf{G}(m)\|}$	$y \leftarrow\!\!\text{\tiny\$}\, \{0,1\}^{\|\mathsf{G}(m)\|}$
		abort if $(x,y) \in \mathsf{Bad}(\lambda)$	abort if $(x,y) \in \mathsf{Bad}(\lambda)$
$C_{x,y} \leftarrow (\mathsf{G}(\mathsf{uC}(\cdot, x)) = y)$	$C_{x,y} \leftarrow (\mathsf{G}(\mathsf{uC}(\cdot, x)) = y)$	$C_{x,y} \leftarrow (\mathsf{G}(\mathsf{uC}(\cdot, x)) = y)$	
$\tilde{C} = \leftarrow\!\!\text{\tiny\$}\, \mathsf{iO}(C_{x,y})$	$\tilde{C} = \leftarrow\!\!\text{\tiny\$}\, \mathsf{iO}(C_{x,y})$	$\tilde{C} = \leftarrow\!\!\text{\tiny\$}\, \mathsf{iO}(C_{x,y})$	$\tilde{C} = \leftarrow\!\!\text{\tiny\$}\, \mathsf{iO}(\mathbf{0})$
$b' \leftarrow\!\!\text{\tiny\$}\, \mathsf{Dist}(1^\lambda, \tilde{C})$	$b' \leftarrow\!\!\text{\tiny\$}\, \mathsf{Dist}(1^\lambda, \tilde{C})$	$b' \leftarrow\!\!\text{\tiny\$}\, \mathsf{Dist}(1^\lambda, \tilde{C})$	$b' \leftarrow\!\!\text{\tiny\$}\, \mathsf{Dist}(1^\lambda, \tilde{C})$
return $(1 = b')$	**return** $(1 = b')$	**return** $(1 = b')$	**return** $(1 = b')$

Fig. 1. The hybrids for the proof of Theorem 1. We have highlighted the changes between the games with a light-grey background.

$$m \leftarrow\!\!\text{\tiny\$}\, \{0,1\}^{n(\lambda)}$$
$$y \leftarrow \mathsf{G}(m)$$
$$x \leftarrow\!\!\text{\tiny\$}\, \{0,1\}^{\ell(\lambda)}$$
$$z \leftarrow\!\!\text{\tiny\$}\, \mathsf{iO}\left(\mathsf{G}(\mathsf{uC}(\cdot, x)) = y\right)$$
$$\textbf{output: } (z, x, m)$$

We now present the adversary $\mathcal{B}_2$ that, given the leakage z from $\mathcal{B}_1$, breaks the security of the multi-bit point obfuscator. We then argue that $\mathcal{B}_1$, indeed, implements an unpredictable distribution. Adversary $\mathcal{B}_2$ gets values p and z as input, where p is either a point obfuscation of $p_{x,m}$ sampled according to $\mathcal{B}_1$ or an obfuscation for $p_{x,u}$ for a uniformly random value u. Adversary $\mathcal{B}_2$ computes $z(p)$ and outputs the result. If p is an obfuscation of $p_{x,m}$, then $\mathcal{B}_2$ computes the predicate function

$$\mathsf{G}(p_{x,m}(x)) = y$$

where y is computed as $\mathsf{G}(m)$ and outputs 1. In turn, if p is an obfuscation of $p_{x,u}$, then with overwhelming probability over the choice of u, adversary $\mathcal{B}_2$ returns 0. Thus, $(\mathcal{B}_1, \mathcal{B}_2)$ is a successful pair of adversaries. To prove that $(\mathcal{B}_1, \mathcal{B}_2)$ is also a valid pair of adversaries, we need to show that $\mathcal{B}_1$ is an unpredictable distribution. Under the assumption of indistinguishability obfuscation, the leakage computed by $\mathcal{B}_1$ is indistinguishable from an obfuscated zero circuit $\mathbf{0}$, the circuit that returns 0 on all inputs and which is padded to the same length as the (unobfuscated) leaked circuit, that is, the circuit $(\mathsf{G}(\mathsf{uC}(\cdot, x)) = y)$. As the zero circuit does not leak any information about y, the leakage is unpredictable. Formally we prove the unpredictably of $\mathcal{B}_1$ via a sequence of four hybrids depicted in Figure 1. We defer a formal proof to the full version [21].

3.2 Implications

Average case MB-AIPO is a relaxed notion of virtual-black-box point obfuscation in the presence of auxiliary input and in particular implied by it [49].

Consequently our impossibility result also shows that VBB obfuscation of multi-bit point functions secure in the presence of auxiliary input cannot exist if indistinguishability obfuscation exist:

Corollary 1. *If indistinguishability obfuscation exists, then VBB multi-bit point obfuscation secure with auxiliary input does not exist.*

We note that VBB multi-bit point obfuscation is also often referred to as *Digital Lockers*. Canetti and Dakdouk [24] study the composition of point function obfuscation and show that composable AIPO implies the existence of composable MB-AIPO. And hence, applying our result we get the following corollary.

Corollary 2. *If indistinguishability obfuscation exists, then composable AIPO does not exist.*

Several results have been based on the existence of MB-AIPO (or composable AIPO). Matsuda and Hanaoka give a CCA secure public-key encryption scheme based on MB-AIPO [49] and Bitansky and Paneth give a three-round weak zero-knowledge protocol for $\mathcal{NP}$ based on composable AIPO [13].[8] In the full version [21] we present a weakened notion of MB-AIPO that we deem to fall outside our impossibility result. It is not clear whether this weaker notion suffices for the applications in [13, 49] and such a proof is not straightforward, so it remains to study whether one could use other weak variants of MB-AIPO.

A Random Oracle Uninstantiability. Lynn et al. [47] construct VBB obfuscators for multi-bit point functions in the idealized random oracle model and their result can easily be seen to encompass auxiliary information. Thus, assuming iO exists our result rules out the existence of a standard model hash function that can instantiate the random oracle in their construction.

Corollary 3. *If indistinguishability obfuscation exists, then the multi-bit output point function obfuscator by Lynn et al. [47] cannot be instantiated in the standard model so that it achieves VBB security with auxiliary input.*

Acknowledgments. We thank the Asiacrypt 2014 reviewers for the many constructive comments. We especially thank Paul Baecher, Mihir Bellare, Nir Bitansky, Victoria Fehr, Peter Gazi, Dennis Hofheinz, Giorgia Azzurra Marson and Alon Rosen for many helpful comments and discussions throughout the various stages of this work. Christina Brzuska was supported by the Israel Science Foundation (grant 1076/11 and 1155/11), the Israel Ministry of Science and Technology grant 3-9094), and the German-Israeli Foundation for Scientific Research and Development (grant 1152/2011). Arno Mittelbach was supported by CASED (www.cased.de) and the German Research Foundation (DFG) SPP 1736.

---

[8] We note that the construction of 3-message witness-hiding protocols from AIPO [13] as well as the construction of a CCA secure PKE scheme from lossy encryption schemes and MB-AIPO with statistically hard-to-invert information [49] are not affected by our result.

References

1. Ananth, P., Boneh, D., Garg, S., Sahai, A., Zhandry, M.: Differing-inputs obfuscation and applications. Cryptology ePrint Archive, Report 2013/689 (2013), http://eprint.iacr.org/2013/689
2. Ananth, P., Gupta, D., Ishai, Y., Sahai, A.: Optimizing obfuscation: Avoiding barrington's theorem. Cryptology ePrint Archive, Report 2014/222 (2014), http://eprint.iacr.org/
3. Barak, B., Garg, S., Kalai, Y.T., Paneth, O., Sahai, A.: Protecting obfuscation against algebraic attacks. In: Nguyen, P.Q., Oswald, E. (eds.) EUROCRYPT 2014. LNCS, vol. 8441, pp. 221–238. Springer, Heidelberg (2014)
4. Barak, B., Goldreich, O., Impagliazzo, R., Rudich, S., Sahai, A., Vadhan, S., Yang, K.: On the (im)possibility of obfuscating programs. J. ACM 59(2), 1–6 (2012), http://doi.acm.org/10.1145/2160158.2160159
5. Barak, B., Goldreich, O., Impagliazzo, R., Rudich, S., Sahai, A., Vadhan, S.P., Yang, K.: On the (im)possibility of obfuscating programs. In: Kilian, J. (ed.) CRYPTO 2001. LNCS, vol. 2139, pp. 1–18. Springer, Heidelberg (2001)
6. Bellare, M., Hoang, V.T., Keelveedhi, S.: Instantiating random oracles via uCEs. In: Canetti, R., Garay, J.A. (eds.) CRYPTO 2013, Part II. LNCS, vol. 8043, pp. 398–415. Springer, Heidelberg (2013)
7. Bellare, M., Hoang, V.T., Keelveedhi, S.: Personal communication (September 2013)
8. Bellare, M., Stepanovs, I., Tessaro, S.: Poly-many hardcore bits for any one-way function. In: Sarkar, P., Iwata, T. (eds.) ASIACRYPT 2014, vol. 8874, Springer, Berlin (2014)
9. Bitansky, N., Canetti, R.: On strong simulation and composable point obfuscation. In: Rabin, T. (ed.) CRYPTO 2010. LNCS, vol. 6223, pp. 520–537. Springer, Heidelberg (2010)
10. Bitansky, N., Canetti, R.: On strong simulation and composable point obfuscation. J. Cryptology 27(2), 317–357 (2014)
11. Bitansky, N., Canetti, R., Cohn, H., Goldwasser, S., Kalai, Y.T., Paneth, O., Rosen, A.: The impossibility of obfuscation with auxiliary input or a universal simulator. In: Garay, J.A., Gennaro, R. (eds.) CRYPTO 2014, Part II. LNCS, vol. 8617, pp. 71–89. Springer, Heidelberg (2014)
12. Bitansky, N., Canetti, R., Paneth, O., Rosen, A.: On the existence of extractable one-way functions. In: Shmoys, D.B. (ed.) 46th ACM STOC, pp. 505–514. ACM Press (May/June 2014)
13. Bitansky, N., Paneth, O.: Point obfuscation and 3-round zero-knowledge. In: Cramer, R. (ed.) TCC 2012. LNCS, vol. 7194, pp. 190–208. Springer, Heidelberg (2012)
14. Boneh, D., Waters, B.: Constrained pseudorandom functions and their applications. In: Sako, K., Sarkar, P. (eds.) ASIACRYPT 2013, Part II. LNCS, vol. 8270, pp. 280–300. Springer, Heidelberg (2013)
15. Boneh, D., Zhandry, M.: Multiparty key exchange, efficient traitor tracing, and more from indistinguishability obfuscation. In: Garay, J.A., Gennaro, R. (eds.) CRYPTO 2014, Part I. LNCS, vol. 8616, pp. 480–499. Springer, Heidelberg (2014)
16. Boyle, E., Chung, K.M., Pass, R.: On extractability obfuscation. In: Lindell, Y. (ed.) TCC 2014. LNCS, vol. 8349, pp. 52–73. Springer, Heidelberg (2014)
17. Boyle, E., Goldwasser, S., Ivan, I.: Functional signatures and pseudorandom functions. In: Krawczyk, H. (ed.) PKC 2014. LNCS, vol. 8383, pp. 501–519. Springer, Heidelberg (2014)

18. Boyle, E., Pass, R.: Limits of extractability assumptions with distributional auxiliary input. Cryptology ePrint Archive, Report 2013/703 (2013), http://eprint.iacr.org/2013/703
19. Brakerski, Z., Rothblum, G.N.: Virtual black-box obfuscation for all circuits via generic graded encoding. In: Lindell, Y. (ed.) TCC 2014. LNCS, vol. 8349, pp. 1–25. Springer, Heidelberg (2014)
20. Brzuska, C., Farshim, P., Mittelbach, A.: Indistinguishability obfuscation and uCEs: The case of computationally unpredictable sources. In: Garay, J.A., Gennaro, R. (eds.) CRYPTO 2014, Part I. LNCS, vol. 8616, pp. 188–205. Springer, Heidelberg (2014)
21. Brzuska, C., Mittelbach, A.: Indistinguishability obfuscation versus multi-bit point obfuscation with auxiliary input. Cryptology ePrint Archive, Report 2014/405 (2014), http://eprint.iacr.org/
22. Brzuska, C., Mittelbach, A.: Using indistinguishability obfuscation via uces. In: Sarkar, P., Iwata, T. (eds.) ASIACRYPT 2014, December 7–11, vol. 8874, Springer, Berlin (2014)
23. Canetti, R.: Towards realizing random oracles: Hash functions that hide all partial information. In: Kaliski Jr., B.S. (ed.) CRYPTO 1997. LNCS, vol. 1294, pp. 455–469. Springer, Heidelberg (1997)
24. Canetti, R., Dakdouk, R.R.: Obfuscating point functions with multibit output. In: Smart, N.P. (ed.) EUROCRYPT 2008. LNCS, vol. 4965, pp. 489–508. Springer, Heidelberg (2008)
25. Canetti, R., Fuller, B., Paneth, O., Reyzin, L.: Key derivation from noisy sources with more errors than entropy. Cryptology ePrint Archive, Report 2014/243 (2014), http://eprint.iacr.org/
26. Canetti, R., Tauman Kalai, Y., Varia, M., Wichs, D.: On symmetric encryption and point obfuscation. In: Micciancio, D. (ed.) TCC 2010. LNCS, vol. 5978, pp. 52–71. Springer, Heidelberg (2010)
27. Canetti, R., Micciancio, D., Reingold, O.: Perfectly one-way probabilistic hash functions (preliminary version). In: 30th ACM STOC, pp. 131–140. ACM Press (May 1998)
28. Collberg, C., Thomborson, C., Low, D.: A taxonomy of obfuscating transformations. Technical Report 148, Department of Computer Science, University of Auckland (Jul 1997), http://citeseer.ist.psu.edu/collberg97taxonomy.html
29. Dodis, Y., Kalai, Y.T., Lovett, S.: On cryptography with auxiliary input. In: Mitzenmacher, M. (ed.) 41st ACM STOC, pp. 621–630. ACM Press (May/June 2009)
30. Fischlin, M.: Pseudorandom function tribe ensembles based on one-way permutations: Improvements and applications. In: Stern, J. (ed.) EUROCRYPT 1999. LNCS, vol. 1592, pp. 432–445. Springer, Heidelberg (1999)
31. Garg, S., Gentry, C., Halevi, S., Raykova, M.: Two-round secure MPC from indistinguishability obfuscation. In: Lindell, Y. (ed.) TCC 2014. LNCS, vol. 8349, pp. 74–94. Springer, Heidelberg (2014)
32. Garg, S., Gentry, C., Halevi, S., Raykova, M., Sahai, A., Waters, B.: Candidate indistinguishability obfuscation and functional encryption for all circuits. In: 54th FOCS, pp. 40–49. IEEE Computer Society Press (October 2013)
33. Garg, S., Gentry, C., Halevi, S., Wichs, D.: On the implausibility of differing-inputs obfuscation and extractable witness encryption with auxiliary input. In: Garay, J.A., Gennaro, R. (eds.) CRYPTO 2014, Part I. LNCS, vol. 8616, pp. 518–535. Springer, Heidelberg (2014)
34. Gentry, C., Lewko, A., Sahai, A., Waters, B.: Indistinguishability obfuscation from the multilinear subgroup elimination assumption. Cryptology ePrint Archive, Report 2014/309 (2014), http://eprint.iacr.org/2014/309

35. Goldreich, O., Krawczyk, H.: On the composition of zero-knowledge proof systems. SIAM J. Comput. 25(1), 169–192 (1996), http://dx.doi.org/10.1137/S0097539791220688
36. Goldwasser, S., Kalai, Y.T.: On the impossibility of obfuscation with auxiliary input. In: 46th FOCS, pp. 553–562. IEEE Computer Society Press (October 2005)
37. Goldwasser, S., Rothblum, G.N.: On best-possible obfuscation. In: Vadhan, S.P. (ed.) TCC 2007. LNCS, vol. 4392, pp. 194–213. Springer, Heidelberg (2007)
38. Hada, S.: Secure obfuscation for encrypted signatures. In: Gilbert, H. (ed.) EUROCRYPT 2010. LNCS, vol. 6110, pp. 92–112. Springer, Heidelberg (2010)
39. Hofheinz, D.: Fully secure constrained pseudorandom functions using random oracles. Cryptology ePrint Archive, Report 2014/372 (2014), http://eprint.iacr.org/
40. Hofheinz, D., Malone-Lee, J., Stam, M.: Obfuscation for cryptographic purposes. In: Vadhan, S.P. (ed.) TCC 2007. LNCS, vol. 4392, pp. 214–232. Springer, Heidelberg (2007)
41. Hohenberger, S., Rothblum, G.N., Shelat, A., Vaikuntanathan, V.: Securely Obfuscating Re-encryption. In: Vadhan, S.P. (ed.) TCC 2007. LNCS, vol. 4392, pp. 233–252. Springer, Heidelberg (2007)
42. Hohenberger, S., Sahai, A., Waters, B.: Replacing a random oracle: Full domain hash from indistinguishability obfuscation. In: Nguyen, P.Q., Oswald, E. (eds.) EUROCRYPT 2014. LNCS, vol. 8441, pp. 201–220. Springer, Heidelberg (2014)
43. Impagliazzo, R.: A personal view of average-case complexity. In: Proceedings of the 10th Annual Structure in Complexity Theory Conference, SCT 1995, p. 134. IEEE Computer Society, Washington, DC (1995), http://dl.acm.org/citation.cfm?id=829497.829786
44. Kiayias, A., Papadopoulos, S., Triandopoulos, N., Zacharias, T.: Delegatable pseudorandom functions and applications. In: Sadeghi, A.R., Gligor, V.D., Yung, M. (eds.) ACM CCS 2013, pp. 669–684. ACM Press (November 2013)
45. Komargodski, I., Moran, T., Naor, M., Pass, R., Rosen, A., Yogev, E.: One-way functions and (im)perfect obfuscation. Cryptology ePrint Archive, Report 2014/347 (2014), http://eprint.iacr.org/
46. Koppula, V., Ramchen, K., Waters, B.: Separations in circular security for arbitrary length key cycles. Cryptology ePrint Archive, Report 2013/683 (2013), http://eprint.iacr.org/2013/683
47. Lynn, B., Prabhakaran, M., Sahai, A.: Positive results and techniques for obfuscation. In: Cachin, C., Camenisch, J.L. (eds.) EUROCRYPT 2004. LNCS, vol. 3027, pp. 20–39. Springer, Heidelberg (2004)
48. Marcedone, A., Orlandi, C.: Obfuscation $= = >$ (ind-cpa security $= / = >$ circular security) (2014)
49. Matsuda, T., Hanaoka, G.: Chosen ciphertext security via point obfuscation. In: Lindell, Y. (ed.) TCC 2014. LNCS, vol. 8349, pp. 95–120. Springer, Heidelberg (2014)
50. Pass, R., Seth, K., Telang, S.: Indistinguishability obfuscation from semantically-secure multilinear encodings. In: Garay, J.A., Gennaro, R. (eds.) CRYPTO 2014, Part I. LNCS, vol. 8616, pp. 500–517. Springer, Heidelberg (2014)
51. Sahai, A., Waters, B.: How to use indistinguishability obfuscation: deniable encryption, and more. In: Shmoys, D.B. (ed.) 46th ACM STOC, pp. 475–484. ACM Press (May/June 2014)
52. Wee, H.: On obfuscating point functions. In: Gabow, H.N., Fagin, R. (eds.) 37th ACM STOC, pp. 523–532. ACM Press (May 2005)

Bootstrapping Obfuscators
via
Fast Pseudorandom Functions*

Benny Applebaum

School of Electrical Engineering, Tel-Aviv University
benny.applebaum@gmail.com

Abstract. We show that it is possible to upgrade an obfuscator for a weak complexity class **WEAK** into an obfuscator for arbitrary polynomial size circuits, assuming that the class **WEAK** can compute pseudorandom functions. Specifically, under standard intractability assumptions (e.g., hardness of factoring, Decisional Diffie-Hellman, or Learning with Errors), the existence of obfuscators for $\mathbf{NC}^1$ or even $\mathbf{TC}^0$ implies the existence of general-purpose obfuscators for **P**. Previously, such a bootstrapping procedure was known to exist under the assumption that there exists a fully-homomorphic encryption whose decryption algorithm can be computed in **WEAK**. Our reduction works with respect to virtual black-box obfuscators and relativizes to ideal models.

1 Introduction

General-purpose program obfuscation allows us to transform an arbitrary computer program into an "unintelligible" form while preserving its functionality. The latter property is formalized via the notion of *Virtual Black-Box* which asserts that the code of the obfuscated program reveals nothing more than what can be learned via oracle access to its input-output behavior. The seminal result of [7] shows that general purpose obfuscation is impossible in the standard model. Nevertheless, in a sequence of recent exciting works [10,8,6], it was shown that general-purpose obfuscation can be achieved in idealized models such as the Generic Colored Matrix Model, or the Generic Graded Encoding model.

All these works share a similar outline. First it is shown how to use the idealized model to obfuscate a weak complexity class such as $\mathbf{NC}^1$, and then the weak obfuscator is bootstrapped into a general-purpose obfuscator for arbitrary polynomial-size circuits.[1] The bootstrapping step in all these works employs a fully homomorphic encryption [11] whose decryption algorithm can be implemented in $\mathbf{NC}^1$. While recent constructions of FHE (e.g., [9]) make the latter

---

* Supported by Alon Fellowship, ISF grant 1155/11, Israel Ministry of Science and Technology (grant 3-9094), GIF grant 1152/2011, and the Check Point Institute for Information Security.

[1] The class $\mathbf{NC}^1$ is the class of polynomial-size circuits with logarithmic depth and bounded fan-in gates.

P. Sarkar and T. Iwata (Eds.): ASIACRYPT 2014, PART II, LNCS 8874, pp. 162–172, 2014.

assumption reasonable, the existence of FHE is still a strong public-key assumption and it is natural to ask whether it can be relaxed. Indeed, $\mathbf{NC}^1$ obfuscation already seems to put us at the "Heights of Cryptomania" [12], and so one may suspect whether the extra power of FHE is really needed for bootstrapping.

1.1 Our Results

In this note we show that bootstrapping can be based on a "Minicrypt" type assumption.

Theorem 1 (main theorem – informal). *Assume that the complexity class* **WEAK** *can compute a pseudorandom function (PRF). Then an obfuscator for* **WEAK** *(in some idealized model) can be bootstrapped into an obfuscator for every polynomial-size circuit family.*

(See Theorem 5 for a formal statement.) Since practical and theoretical implementations of PRFs tend to be highly efficient, we can instantiate the theorem with relatively low complexity classes. For example, by relying on PRFs constructions from [18,5], we derive the following corollary.

Corollary 1. *Assuming the hardness of factoring, Decisional Diffie-Hellman, or Learning with Errors, the following holds. If* $\mathbf{TC}^0$ *can be obfuscated (in some idealized model), then every polynomial-size circuit family can be obfuscated as well.*[2]

Tightness. One may ask whether the PRF assumption in Theorem 1 can be further relaxed and replaced with the existence of a weaker primitive such as a one-way function or a pseudorandom generator. We note that the answer seems to be negative as such primitives can be computed in $\mathbf{NC}^0$ (under standard assumptions [4]), a class which is learnable and therefore trivially obfuscatable in the standard model. Therefore, such a strengthening of Theorem 1 would contradict the impossibility results of [7].

In fact, we do not expect to prove a statement like Corollary 1 for classes lower than $\mathbf{TC}^0$, as $\mathbf{TC}^0$ is the lowest complexity class for which the impossibility results of [7] apply. More generally, [7] essentially show that if a complexity class **WEAK** contains a PRF then it cannot be obfuscated in the standard model. Our results complement the picture by showing that if such a class can be obfuscated in some idealized model, then everything can be obfuscated. This suggests the existence of a zero-one law: if an idealized model admits non-trivial obfuscation (i.e., for some class which is non-obfuscatable in the standard model) then it admits general purpose obfuscation for all (polynomial-size) circuits.

---

[2] The class $\mathbf{TC}^0$ is the class of all Boolean circuits with constant depth and polynomial size, containing only unbounded-fan in AND gates, OR gates, and majority gates. This class is a subclass of $\mathbf{NC}^1$.

Virtual Black-Box vs. Indistinguishability Obfuscation. Our results hold with respect to the strongest security notion of *Virtual Black-Box* (VBB) obfuscation [7] relative to some ideal model. An alternative (weaker) notion of security is *indistinguishability Obfuscation.* The latter notion is highly attractive as it may be achievable in the standard model (no impossibility results are known), and, quite surprisingly, it suffices for a wide range of applications [19]. In this work we focus on VBB obfuscators, as we believe that when *constructing* obfuscators it is best to strive for the strongest form of security. (See a detailed discussion in [6].) We do not know whether our results apply in the indistinguishability setting, and leave this question for future research. Interestingly, the FHE-based bootstrapping works in both settings [10,8].

1.2   Techniques

Our main technical tool is randomized encoding (RE) of functions [17,4]. Intuitively, a function $f(x)$ is encoded by a randomized function $\hat{f}(x; r)$ if the distribution $\hat{f}(x)$, induced by a random choice of r, reveals nothing but $f(x)$. Formally, a sample from $\hat{f}(x)$ can be *decoded* to $f(x)$, and vice versa, given $f(x)$ one can efficiently *simulate* the distribution $\hat{f}(x)$ – so the functions are essentially "equivalent". REs become non-trivial (and useful) when their complexity is smaller than the complexity of f. This "equivalence" between a complicated function f to a simpler encoding $\hat{f}$, was exploited in various applications to reduce a complex task to a simpler one. (See the surveys [1] and [16].) We can adopt a similar approach in our context as well.

In order to obfuscate a function f taken from a family $\mathcal{F}$ let us obfuscate its low-complexity encoding $\hat{f}$ and release the latter obfuscated program composed with the decoder algorithm. For this to work, let us assume the existence of universal decoder and universal simulator that work uniformly for all functions in $\mathcal{F}$. (This can be guaranteed by encoding the evaluation function of the collection $\mathcal{F}$ which maps a circuit of f and an input x to the value $f(x)$.)

Unfortunately, this approach is somewhat problematic as the encoding $\hat{f}$ employs internal randomness. One potential solution is to treat the randomness as an additional input and let the user of the obfuscated program choose it. While decoding still succeeds, this solution fails to be secure. Once the randomness r is revealed, the function f can be fully recovered. (Technically, universal simulation cannot be achieved anymore.) Another (flawed) solution, is to fix some secret randomness r and obfuscate the mapping $x \mapsto \hat{f}(x; r)$. Unfortunately, the privacy of the encoding holds only when fresh randomness is being used, and one can easily recover the circuit of f when r is fixed.

The problem is solved via the extra use of a PRF. Specifically, we choose a PRF $h \xleftarrow{R} \mathcal{H}$ and obfuscate the function $\hat{f}(x; h(x))$. The security of the PRF ensures that this function behaves essentially as a standard encoding whose internal randomness is freshly chosen in each invocation, and so simulation succeeds. By using the low-complexity encoding from [3], the complexity of $\hat{f}$ is dominated by the complexity of the PRF $\mathcal{H} \in \mathbf{WEAK}$ (assuming that the class $\mathbf{WEAK}$ satisfies some basic closure properties), and one can prove Theorem 1.

We note that a similar usage of PRF-derandomized randomized encoding was made by [14] in the context of Functional Encryption.

2   Preliminaries

We say that a function $\varepsilon : \mathbb{N} \rightarrow \mathbb{R}$ is *negligible* if for every constant $c > 0$ there exists an integer $n_0 \in \mathbb{N}$ such that $\varepsilon(n) < n^{-c}$ for every $n > n_0$. We will sometimes use $\mathrm{neg}(\cdot)$ to denote an unspecified negligible function.

One-Way Functions. An efficiently computable function $g : \{0,1\}^* \rightarrow \{0,1\}^*$ is *one-way* if for every (non-uniform) efficient adversary $\mathcal{A}$ we have that

$$\Pr_{x \xleftarrow{R} \{0,1\}^n} [\mathcal{A}(1^n, f(x)) \in f^{-1}(f(x))] < \mathrm{neg}(n). \tag{1}$$

Circuit Families. A family of polynomial-size boolean circuits $\mathcal{F}$ is an infinite sequence of circuit families $\{\mathcal{F}_n\}_{n \in \mathbb{N}}$ where for every $n \in \mathbb{N}$ the family $\mathcal{F}_n$ consists of boolean circuits with n inputs, $m(n)$ outputs, and circuit size $\ell(n)$ where m, ℓ are polynomials in n. For such a family there is always a universal efficient evaluator F such that for every length parameter n, circuit $f \in \mathcal{F}_n$ and input $x \in \{0,1\}^n$, we have that $F(f, x) = f(x)$.

Pseudorandom Functions [13]. Let $\mathcal{H} = \{\mathcal{H}_n\}_{n \in \mathbb{N}}$ be a family of polynomial-size boolean circuits and let $\mathcal{K}$ be a PPT sampling algorithm that on input 1^n samples a circuit in $\mathcal{H}_n$. (The probability distribution induced by $\mathcal{K}$ is not necessarily uniform.) We say that $\mathcal{H}$ is a *pseudorandom function family* (PRF) if for every (non-uniform) efficient oracle-aided adversary $\mathcal{A}$ we have that

$$\left| \Pr_{h \xleftarrow{R} \mathcal{K}(1^n)} [\mathcal{A}^h(1^n) = 1] - \Pr_{R_n}[\mathcal{A}^{R_n}(1^n) = 1] \right| \leq \mathrm{neg}(n), \tag{2}$$

where R_n is a uniformly chosen function with the same input and output lengths as the functions in $\mathcal{H}_n$. To simplify notation, we will typically make the sampler implicit and write $h \xleftarrow{R} \mathcal{H}_n$ with the understanding that the distribution is induced by some efficient sampler.

2.1   Randomized Encoding of Functions

Let F be a polynomial-time computable function that maps n bits to $m(n)$ bits. Intuitively, a randomized function $\hat{F}$ is an "encoding" of F if for every input x the distribution $\hat{F}(x)$ reveals the value of $F(x)$ but no other additional information. We formalize this via the notion of *computationally private randomized encoding* from [3].

Definition 1 (Computational randomized encoding). *Let* $F : \{0,1\}^n \to \{0,1\}^{m(n)}$ *be an efficiently computable function and let* $\hat{F}$ *be an efficiently computable randomized function. We say that* $\hat{F}$ *is a* computational randomized encoding *of* F *(or encoding for short), if there exist an efficient decoder algorithm* D *and an efficient probabilistic simulator algorithm* Sim *that satisfy the following:*

- **Perfect Correctness.** *For every* n *and every input* $x \in \{0,1\}^n$,

$$\mathsf{D}(1^n, \hat{F}(x)) = F(x).$$

- **Computational Privacy.** *For every non-uniform efficient oracle-aided adversary* $\mathcal{A}$ *we have*

$$\left| \Pr[\mathcal{A}^{\hat{F}_n}(1^n) = 1] - \Pr[\mathcal{A}^{\mathsf{Sim}(1^n, F_n(\cdot))}(1^n) = 1] \right| \le \mathrm{neg}(n), \tag{3}$$

where $\hat{F}_n$ *and* F_n *are the restrictions of* $\hat{F}$ *and* F *to* n-*bit inputs, and both oracles are probabilistic functions (and fresh randomness is used in each invocation).*

Encoding Collections. We encode a family of polynomial-size boolean circuits $\mathcal{F}$ by encoding its evaluation algorithm $F(f,x)$. Specifically, for every $f \in \mathcal{F}$ we define the randomized function $\hat{f}(x;r) = \hat{F}(f,x;r)$ where $\hat{F}$ is the encoder of F. By definition, the decoder D and the simulator Sim of F apply universally for all $f \in \mathcal{F}$. Formally, for every n and $f \in \mathcal{F}_n$:

$$\mathsf{D}(1^{|x|}, \hat{f}(x)) = f(x) \quad \forall x \in \{0,1\}^n.$$

Also for every non-uniform efficient oracle-aided adversary $\mathcal{A}$ and every sequence of functions $\{f_n\}$ and their encodings $\{\hat{f}_n\}$ we have that

$$\left| \Pr[\mathcal{A}^{\hat{f}_n}(1^n) = 1] - \Pr[\mathcal{A}^{\mathsf{Sim}(1^n, f_n(\cdot))}(1^n) = 1] \right| \le \mathrm{neg}(n). \tag{4}$$

(The oracles in (4) are simply the restriction of the oracles in (3) to inputs of the form $(f_n, \cdot)$ and so (4) follows immediately from (3).)

2.2   Obfuscation

Definition 2 (Virtual Black-Box Obfuscator [7]). *Let* $\mathcal{F} = \{\mathcal{F}_n\}_{n \in \mathbb{N}}$ *be a family of polynomial-size boolean circuits. An obfuscator* $\mathcal{O}$ *for* $\mathcal{F}$ *is a PPT algorithm which maps a circuit* $f \in \mathcal{F}_n$ *to a new circuit* $[f]$ *(not necessarily in* $\mathcal{F}$*) such that the following properties hold:*

1. ***Preserving Functionality.*** *For every* $n \in \mathbb{N}$, *every* $f \in \mathcal{F}_n$ *and every input* $x \in \{0,1\}^n$

$$\Pr_{[f] \overset{R}{\leftarrow} \mathcal{O}(f)} [[f](x) \ne f(x)] \le \mathrm{neg}(n).$$

2. **Polynomial Slowdown.** There exists a polynomial p such that for every $n \in \mathbb{N}$ and $f \in \mathcal{F}_n$ the circuit $\mathcal{O}(f)$ is of size at most $p(|f|)$.

3. **Virtual Black-Box.** For every (non-uniform) efficient adversary $\mathcal{A}$ there exists a (non-uniform) efficient simulator Sim such that for every n and every $f \in \mathcal{F}_n$:

$$\left| \Pr[\mathcal{A}(\mathcal{O}(f)) = 1] - \Pr[\mathsf{Sim}^f(1^{|f|}, 1^n) = 1] \right| \leq \mathrm{neg}(n). \tag{5}$$

A complexity class $\mathbf{C}$ is obfuscatable if there exists an efficiently computable mapping that maps every efficiently computable function family $\mathcal{F} \in \mathbf{C}$ (represented by its evaluator F) to an obfuscator $\mathcal{O}$ for $\mathcal{F}$.

Obfuscation in an Idealized Model. An idealized model is captured by a sequence of probabilistic stateful oracles $\mathcal{M} = \{\mathcal{M}_n\}_{n \in \mathbb{N}}$ indexed by a security parameter n. We consider obfuscators which are implementable relative to $\mathcal{M}$. This means that the obfuscator $\mathcal{O}$ is allowed to make oracle queries to $\mathcal{M}_n$ and that Eq. 5 should hold even when $\mathcal{A}$ is allowed to query $\mathcal{M}_n$. (The circuit f cannot have oracle gates to $\mathcal{M}$.) In this case, Properties (1) and (2) should hold for every possible coins of $\mathcal{M}$.

3 Our Reduction

Let $\mathcal{F}$ be a family of polynomial-size boolean circuits with an evaluator F. We will construct an obfuscator $\mathcal{O}$ for $\mathcal{F}$ based on the following ingredients. (1) An encoding $\hat{F}$ of F; (2) a pseudorandom function family $\mathcal{H}$ where the output length of functions in $\mathcal{H}_n$ equals to the length of the random input of $\hat{F}_n$; and (3) a weak obfuscator weak$\mathcal{O}$ for the circuit family $\mathcal{G} = \{\mathcal{G}_n\}$ where $\mathcal{G}_n$ contain all circuits of the form

$$g_{f,h} : x \mapsto \hat{F}(f, x; h(x)), \qquad \forall f \in \mathcal{F}_n, h \in \mathcal{H}_n.$$

We allow the weak obfuscator to be implementable in some idealized model $\mathcal{M}$, but assume that the PRF and the randomized encoding implemented in the standard model and make no calls to $\mathcal{M}$.

Construction 2. *Given a circuit $f \in \mathcal{F}_n$ the obfuscator $\mathcal{O}^{\mathcal{M}_n}$ does the following:*

- *Sample a random $h \xleftarrow{R} \mathcal{H}$ and obfuscate $g_{f,h}$ by $[g] := \mathsf{weak}\mathcal{O}^{\mathcal{M}_n}(g_{f,h})$.*
- *Output the circuit $[f] = \mathsf{D} \circ [g]$ where D is the RE decoder and $\circ$ denotes function composition.*

Note that the construction is syntactically well defined as $g_{f,h}$ makes no calls to the oracle $\mathcal{M}$. (For this purpose, we had to assume that the PRF and the randomized encoding do not use $\mathcal{M}$.)

It is easy to verify that $[f]$ preserves the functionality of f.

Lemma 1. *The obfuscator $\mathcal{O}$ is functionality preserving.*

Proof. Fix some x and $h \in \mathcal{H}$, and let us condition on the event that the weak obfuscator preserves the functionality, namely, $[g](x) = g_{f,h}(x)$. Then, by the correctness of the encoding, we have

$$[f](x) = \mathsf{D}([g](x)) = \mathsf{D}(g_{f,h}(x)) = \mathsf{D}(\hat{F}(f, x; h(x))) = f(x).$$

Since the weak obfuscator is correct with all but negligible probability the claim follows.   □

In the next section, we will prove that the obfuscator is secure.

Lemma 2. *The obfuscator $\mathcal{O}$ satisfies the Virtual Black-Box property relative to $\mathcal{M}$.*

3.1   Security (Proof of Lemma 2)

Let $\mathcal{A}$ be an efficient adversary for the new obfuscator $\mathcal{O}$. Our goal is to construct a simulator Sim that simulates $\mathcal{A}$. For this aim, let us first define an oracle-aided adversary $\mathcal{B}$ for the weak obfuscator weak$\mathcal{O}$ as follows. Given an obfuscated circuit $[f]$, the adversary $\mathcal{B}$ applies $\mathcal{A}$ to the circuit $\mathsf{D} \circ [f]$ where D is the (universal) decoder of the encoding. If $\mathcal{A}$ makes oracle queries to the oracle $\mathcal{M}_n$ then $\mathcal{B}$ answers them using his own oracle $\mathcal{M}_n$. Let weakSim be the simulator of weak$\mathcal{O}$ which simulates the adversary $\mathcal{B}$, and let reSim be the (universal) RE simulator.

We define the simulator $\mathsf{Sim}^f(1^n)$ for the adversary $\mathcal{A}$ as follows. Invoke the oracle-aided weak simulator weakSim$^g(1^n)$, and whenever weakSim makes a new oracle query $x \in \{0,1\}^n$, answer it with $\mathsf{reSim}(1^n, f(x))$, where the latter is computed via the help of the oracle f. If x was previously queried, respond with the same answer as before.

Fix some $f \in \mathcal{F}_n$. We will prove that

$$\left| \Pr[\mathcal{A}^{\mathcal{M}_n}(\mathcal{O}(f)) = 1] - \Pr[\mathsf{Sim}^f(1^n) = 1] \right| \leq \varepsilon_1(n) + \varepsilon_2(n) + \varepsilon_3(n) \qquad (6)$$

where ε_1 (resp., $\varepsilon_2, \varepsilon_3$) upper-bounds the distinguishing advantage of efficient adversaries against the weak obfuscator (resp., against the PRF, against the encoding). Through the proof, we simplify notation by omitting the unary input 1^n; also for a binary random variable Y we write $\Pr[Y]$ to denote $\Pr[Y = 1]$.

Let $\hat{f}(x; r)$ be the circuit that computes the encoding of $f(x)$, i.e., $\hat{f}(x; r) = \hat{F}(f, x; r)$. For a function $h \in \mathcal{H}$, let $\hat{f}_h$ denote the circuit that computes $\hat{f}(x; h(x))$ and let $\mathcal{O}(f; h)$ denote the output of the obfuscator $\mathcal{O}$ with input f and PRF h. Then, by definition, for every h, we have that

$$\Pr[\mathcal{A}^{\mathcal{M}_n}(\mathcal{O}^{\mathcal{M}_n}(f; h))] = \Pr[\mathcal{B}^{\mathcal{M}_n}(\mathsf{weak}\mathcal{O}^{\mathcal{M}_n}(\hat{f}_h))], \qquad (7)$$

Also, by the VBB property of the weak obfuscator, for every h we have that

$$\left| \Pr[\mathcal{B}^{\mathcal{M}_n}(\mathsf{weak}\mathcal{O}^{\mathcal{M}_n}(\hat{f}_h))] - \Pr[\mathsf{weakSim}^{\hat{f}_h}] \right| \leq \varepsilon_1(n). \qquad (8)$$

By relying on the security of the PRF we will prove the following claim.

Claim 3

$$\left| \Pr_{h \xleftarrow{R} \mathcal{H}_n} [\mathsf{weakSim}^{\hat{f}_h}] - \Pr[\mathsf{weakSim}^{\hat{f}}] \right| \leq \varepsilon_2(n),$$

where $\hat{f}(\cdot)$ is viewed as a randomized function and repeated queries to this function are answered consistently based on the first answer.

Proof. If the claim does not hold, we can break the PRF as follows. Let T^R be an oracle aided adversary which calls weakSim and whenever weakSim makes a query x to its oracle, T answers the query with $\hat{f}(x; R(x))$. Observe that if R is a truly random function then T accepts with probability exactly $\Pr[\mathsf{weakSim}^{\hat{f}}]$. On the other hand, if the oracle is $R \xleftarrow{R} \mathcal{H}_n$ then the acceptance probability is exactly $\Pr_{h \xleftarrow{R} \mathcal{H}_n} [\mathsf{weakSim}^{\hat{f}_h}]$. It follows that T breaks the security of the PRF, and the claim follows. $\qquad\square$

Finally, we rely on the privacy of the randomized encoding to prove the following claim.

Claim 4

$$\left| \Pr[\mathsf{weakSim}^{\hat{f}}] - \Pr[\mathsf{Sim}^f] \right| \leq \varepsilon_3(n),$$

where $\hat{f}(\cdot)$ is viewed as a randomized function and repeated queries to this function are answered consistently based on the first answer.

Proof. Recall that $\mathsf{Sim}^{f(\cdot)}$ is simply $\mathsf{weakSim}^{\mathsf{reSim}(f(\cdot))}$, where repeated queries are answered consistently. Therefore, if the claim does not hold, we can use weakSim to distinguish between the randomized functions $\mathsf{reSim}(f(\cdot))$ and $\hat{f}(\cdot)$, in contradiction to the privacy of the RE (Eq. 4). (The distinguisher will simply invoke weakSim and will answer repeated queries based on the first answer.) $\quad\square$

The lemma (i.e., Eq. 6) now follows from Eqs. 7 and 8 and Claims 3 and 4. The "furthermore" part follows by noting that the proof relativizes. $\qquad\square$

4   Main Result

We say that a circuit complexity class **WEAK** is *admissible* if it satisfies the following basic properties:

1. The class **WEAK** contains the class $\mathbf{NC^0}$;
2. (Closure under concatenation) If each output bit of a multi-output function f is computable in **WEAK** then so is f;
3. (Closure under composition) if $f \in$ **WEAK** and $g \in$ **WEAK** then $g \circ f$ is in **WEAK**, where $\circ$ denotes function composition.

We can now prove our main theorem.

Theorem 5. *Let* **WEAK** *be an admissible complexity class, and let* $\mathcal{M}$ *be some probabilistic oracle (an idealized model). Assume that there exists a one-way function and a pseudorandom function* $\mathcal{H}$ *in* **WEAK**. *Then, if* **WEAK** *can be obfuscated relative to* $\mathcal{M}$, *every family of polynomial-size circuits can be obfuscated relative to* $\mathcal{M}$.

Proof. First, we claim that, under the above assumption, any efficiently computable function F has an encoding $\hat{F}$ computable in **WEAK**. In [3] it was shown, based on Yao's garbled circuit technique, that $F(x)$ can be encoded by an encoding $\hat{F}(x;r)$ which is reducible to a minimal-stretch pseudorandom generator $G : \{0,1\}^{\kappa} \to \{0,1\}^{\kappa+1}$ via a non-adaptive **NC⁰** reduction. Namely, $\hat{F}(x;r)$ can be written as $g(x,r,G(r_1),\ldots,G(r_t))$ where g is an **NC⁰** function and the r_i's are sub-blocks of the random string r. Such a PRG is also reducible to a one-way function via a (non-adaptive) **NC⁰** reduction by [4,15] (see also [2, Remark 4.5]). Therefore, $\hat{F}(x;r)$ can be written as $g'(x,r,G'(r_1),\ldots,G'(r_{t'}))$ where g' is in **NC⁰** and G' is a one-way function. We can now instantiate G' with a one-way function in **WEAK** whose existence is promised by the theorem's hypothesis. Since **WEAK** is closed under concatenation, we can view the t' copies $G'(r_1),\ldots,G'(r_{t'})$ as a single function in **WEAK**. Now, the resulting encoding can be written as a composition of an **NC⁰** function with a function in **WEAK** which results in a **WEAK** function.[3]

Next, we observe that, since **WEAK** is closed under concatenation, the assumption implies the existence of a PRF in **WEAK** whose output length is equal to the randomness complexity of the encoding. (By using direct product, one can transform a **WEAK** PRF with a single output bit into a new **WEAK** PRF with arbitrary polynomial number of output bits.) It follows that the circuit family $\mathcal{G}$ (from Construction 2) can be written as a composition of two **WEAK** functions, and the theorem follows from Lemmas 1 and 2.   □

Remark. We assume nothing on the complexity of the *sampler* of the PRF, and therefore the sampler can preprocess the function $h \in \mathcal{H}$. (This preprocessing is exploited in fast implementations of PRFs [18,5].) As a result, the additional one-wayness assumption does not seem to follow from the fact that $\mathcal{H}$ is in **WEAK**.[4]

Under standard intractability assumptions, one can instantiate **WEAK** with **NC¹** or even **TC⁰**. Indeed, the existence of PRFs in **TC⁰** can be based on the hardness of factoring, the DDH assumption or the intractability of lattice/learning problems [18,5], and the existence of one-way functions in **TC⁰** (or even in **NC⁰**) follow from these assumptions as well [4]. Therefore Corollary 1 follows from Theorem 5.

---

[3] We note that the construction of the RE makes a non black-box use of the code of the one-way function. This does not affect the overall argument as none of these primitives makes calls to the oracle $\mathcal{M}$.

[4] The one-wayness assumption becomes redundant if the mapping $(k,x) \mapsto h_k(x)$ (where $h_k = \mathcal{K}(1^n;k)$) is computable in **WEAK**.

Acknowledgement. We thank Alon Rosen and Zvika Brakerski for useful discussions.

References

1. Applebaum, B.: Randomly encoding functions: A new cryptographic paradigm. In: Fehr, S. (ed.) ICITS 2011. LNCS, vol. 6673, pp. 25–31. Springer, Heidelberg (2011)
2. Applebaum, B.: Cryptography in Constant Parallel Time. Springer Publishing Company, Incorporated (2014)
3. Applebaum, B., Ishai, Y., Kushilevitz, E.: Computationally private randomizing polynomials and their applications. Compuational Complexity, 15(2), 115–162 (2006)
4. Applebaum, B., Ishai, Y., Kushilevitz, E.: Cryptography in NC^0. SIAM J. Comput. 36(4), 845–888 (2006)
5. Banerjee, A., Peikert, C., Rosen, A.: Pseudorandom functions and lattices. In: Pointcheval, D., Johansson, T. (eds.) EUROCRYPT 2012. LNCS, vol. 7237, pp. 719–737. Springer, Heidelberg (2012)
6. Barak, B., Garg, S., Kalai, Y.T., Paneth, O., Sahai, A.: Protecting obfuscation against algebraic attacks. In: Nguyen, P.Q., Oswald, E. (eds.) EUROCRYPT 2014. LNCS, vol. 8441, pp. 221–238. Springer, Heidelberg (2014)
7. Barak, B., Goldreich, O., Impagliazzo, R., Rudich, S., Sahai, A., Vadhan, S.P., Yang, K.: On the (im)possibility of obfuscating programs. J. of the ACM 59(2), 6 (2012)
8. Brakerski, Z., Rothblum, G.N.: Virtual black-box obfuscation for all circuits via generic graded encoding. In: Lindell, Y. (ed.) TCC 2014. LNCS, vol. 8349, pp. 1–25. Springer, Heidelberg (2014)
9. Brakerski, Z., Vaikuntanathan, V.: Efficient fully homomorphic encryption from (standard) LWE. In: IEEE 52nd Annual Symposium on Foundations of Computer Science, FOCS (2011)
10. Garg, S., Gentry, C., Halevi, S., Raykova, M., Sahai, A., Waters, B.: Candidate indistinguishability obfuscation and functional encryption for all circuits. In: IEEE 54th Annual Symposium on Foundations of Computer Science, FOCS, pp. 40–49 (2013)
11. Gentry, C.: Fully homomorphic encryption using ideal lattices. In: Proceedings of the 41st Annual ACM Symposium on Theory of Computing, STOC, pp. 169–178 (2009)
12. Gentry, C.: Encrypted messages from the heights of cryptomania. In: Sahai, A. (ed.) TCC 2013. LNCS, vol. 7785, pp. 120–121. Springer, Heidelberg (2013)
13. Goldreich, O., Goldwasser, S., Micali, S.: How to construct random functions. J. of the ACM 33, 792–807 (1986)
14. Gorbunov, S., Vaikuntanathan, V., Wee, H.: Functional encryption with bounded collusions via multi-party computation. In: Safavi-Naini, R., Canetti, R. (eds.) CRYPTO 2012. LNCS, vol. 7417, pp. 162–179. Springer, Heidelberg (2012)
15. Haitner, I., Reingold, O., Vadhan, S.P.: Efficiency improvements in constructing pseudorandom generators from one-way functions. SIAM J. Comput 42(3), 1405–1430 (2013)
16. Ishai, Y.: Randomization techniques for secure computation. In: Prabhakaran, M., Sahai, A. (eds.) Secure Multi-Party Computation. Cryptology and Information Security Series, vol. 10, pp. 222–248. IOS press, Amsterdam (2012)

17. Ishai, Y., Kushilevitz, E.: Randomizing polynomials: A new representation with applications to round-efficient secure computation. In: IEEE 41st Annual Symposium on Foundations of Computer Science, FOCS, pp. 294–304 (2000)
18. Naor, M., Reingold, O.: Number-theoretic constructions of efficient pseudo-random functions. J. of the ACM 51(2), 231–262 (2004)
19. Sahai, A., Waters, B.: How to use indistinguishability obfuscation: Deniable encryption, and more. In: Proceedings of the 46th Annual ACM Symposium on Theory of Computing, STOC, pp. 475–484 (2014)

Homomorphic Authenticated Encryption Secure against Chosen-Ciphertext Attack

Chihong Joo and Aaram Yun

Ulsan National Institute of Science and Technology (UNIST)
Republic of Korea
{chihongjoo,aaramyun}@unist.ac.kr

Abstract. We study *homomorphic authenticated encryption*, where privacy and authenticity of data are protected simultaneously. We define homomorphic versions of various security notions for privacy and authenticity, and investigate relations between them. In particular, we show that it is possible to give a natural definition of IND-CCA for homomorphic authenticated encryption, unlike the case of homomorphic encryption. Also, we construct a simple homomorphic authenticated encryption scheme supporting arithmetic circuits, which is chosen-ciphertext secure both for privacy and authenticity. Our scheme is based on the error-free approximate GCD assumption.

Keywords: homomorphic authenticated encryption, homomorphic MAC, homomorphic encryption.

1 Introduction

Homomorphic cryptography allows processing of cryptographically protected data. For example, homomorphic encryption lets a third party which does not have the secret key to evaluate functions implicitly using only ciphertexts so that the computed ciphertext decrypts to the correct function value. Similarly, homomorphic signature allows a third party who is not the signer to derive a signature to the output of a function, given signatures of the inputs. This possibility for secure delegation of computation could potentially be used for many applications including cloud computing, and so it makes homomorphic cryptography a very interesting area, which was recently attracting many focused research activities, especially since Gentry's first construction [20] of fully homomorphic encryption (FHE) in 2009. While existing FHE schemes are still many orders slower than ordinary encryption schemes to be truly practical, many progresses are already being made in improving the efficiency of FHE [17,27,9,10,21,15,16,6,8,22,23,13,7,14]. Eventually, a truly practical FHE could be used to build secure cloud computing services where even the cloud provider cannot violate the privacy of the data stored and processed by the cloud.

But, if such data is important enough to protect its privacy, in many scenarios the authenticity of the data would also be worth protecting simultaneously. Indeed, in symmetric-key cryptography, the authenticated encryption [26,4,18,25]

P. Sarkar and T. Iwata (Eds.): ASIACRYPT 2014, PART II, LNCS 8874, pp. 173–192, 2014.
© International Association for Cryptologic Research 2014

is exactly such a primitive protecting both. Therefore, we would like to study *homomorphic authenticated encryption* (henceforth abbreviated as HAE), which is a natural analogue of the authenticated encryption for homomorphic cryptography. A HAE is a symmetric-key primitive which allows public evaluation of functions using only corresponding ciphertexts.

Just as in the case of homomorphic encryption, one important goal in this area would be to design a *fully homomorphic* authenticated encryption. Since there are several known FHE constructions, we may construct a fully homomorphic authenticated encryption scheme by generic composition [4] with a fully homomorphic signature, or even a fully homomorphic MAC. But until very recently, the homomorphic signature scheme closest to being fully homomorphic was [5], where only low-degree polynomial functions are supported. A fully homomorphic MAC is proposed by Gennaro and Wichs [19], but it supports only a limited number of verification queries, so that the solution is in a sense incomplete. So far, the problem of constructing a fully homomorphic authenticated encryption is still not solved completely satisfyingly.[1]

Our contribution in this paper is twofold. First, we define various security notions for HAE and study relations among them. For privacy, we define homomorphic versions of IND-CPA and IND-CCA. While IND-CCA is not achievable for homomorphic encryption due to malleability, nevertheless we may define a version of IND-CCA for HAE. It is because that for HAE, encryption of a plaintext is done with respect to a 'label', and similarly decryption of a ciphertext is done with respect to a 'labeled program'. So, while the ciphertext is still malleable by function evaluation, a decryption query should essentially declare how the ciphertext was produced. This allows a homomorphic version of IND-CCA to be defined naturally.

For authenticity, we define UF-CPA, the homomorphic version of the unforgeability when the adversary has access to the encryption oracle. We also consider UF-CCA, where the adversary has both the encryption oracle and the decryption oracle. Moreover, we consider strong unforgeability flavors of authenticity, and define homomorphic versions accordingly: SUF-CPA and SUF-CCA. We investigate relationship between these notions, and, for example, show that SUF-CPA implies SUF-CCA. And, we show that IND-CPA and SUF-CPA imply IND-CCA. Together, this shows that a HAE scheme with IND-CPA and SUF-CPA security is in fact IND-CCA and SUF-CCA.

The second contribution is that we propose a HAE scheme supporting arithmetic circuits. This scheme is somewhat homomorphic and not fully homomorphic, but we show that our scheme is secure and satisfies both IND-CCA and SUF-CCA. Another appeal of our scheme is that it is a straightforward construction based on the error-free approximate GCD (EF-AGCD) assumption. EF-AGCD assumption was used before [17,15,16,13,14] in constructing fully

---

[1] Very recently, some constructions of leveled fully homomorphic signature schemes are proposed [28,24], after the current paper has been submitted to Asiacrypt. So, at least the fully homomorphic authenticated encryption via generic composition would be possible now.

homomorphic encryption schemes supporting boolean circuits, but here we use it to construct a HAE scheme supporting arithmetic circuits on $\mathbb{Z}_Q$ for $Q \in \mathbb{Z}^+$.

2   Related Work

After this paper has been submitted to Asiacrypt, there were many progresses in the area of homomorphic signatures. In CRYPTO 2014, Catalano, Fiore, and Warinschi constructed homomorphic signature schemes for polynomial functions [12]. Compared with the scheme of Boneh and Freeman [5], their construction is in the standard model, and allows efficient verification.

Also, more relevantly, constructions of (leveled) fully homomorphic signature schemes are proposed by Wichs [28] and also by Gorbunov and Vaikuntanathan [24]. Therefore, the fully homomorphic authenticated encryption scheme via generic composition would be now possible, using these. However, currently known FHEs require large amount of ciphertext expansion, and that would become worse by generic composition. Designing more efficient fully homomorphic authenticated encryption would be an interesting problem.

Gennaro and Wichs [19] proposed the first construction of the fully homomorphic MAC. Their construction uses FHE, and exploits randomness in the encryption to hide data necessary for authentication. In fact, since their scheme encrypts plaintexts using FHE, it is already a fully homomorphic authenticated encryption. But, their construction essentially does not allow verification queries, so it satisfies only weaker security notions: IND-CPA and UF-CPA, according to our definition.

Catalano and Fiore [11] proposed two somewhat homomorphic MACs supporting arithmetic circuits on $\mathbb{Z}_p$ for prime modulus p. In their construction, a MAC for a message m is a polynomial $\sigma(X)$ such that its constant term $\sigma(0)$ is equal to the message m, and its value $\sigma(\alpha)$ on a secret random point α is equal to randomness determined by the label τ of the message m. While their construction is very simple and practical, it does not protect privacy of data, and it seems that this cannot be changed by simple modifications, for example by choosing a secret random value β as the value satisfying $\sigma(\beta) = m$. Also, the size of the modulus p is related to the security of the scheme, so it cannot be chosen arbitrarily.

Our scheme is not as efficient as the schemes of Catalano and Fiore, but certainly more efficient than the generically composed HAE of a FHE scheme and the Catalano-Fiore homomorphic MAC. And our scheme is also relatively straightforward. Moreover, in our construction, the security does not depend on the modulus Q so that it can be chosen arbitrarily depending on the application.

Our scheme can also be compared with a homomorphic encryption scheme called IDGHV presented in [13]. It supports encryption of a plaintext vector $(m_1, \ldots, m_\ell)$ where each m_i is an element in $\mathbb{Z}_{Q_i}$. Like our scheme, IDGHV also uses the Chinese remainder theorem, and indeed our construction can be seen as a special-case, symmetric-key variant of IDGHV where $\ell = 1$, and where encryption randomness is pseudorandomly generated from the label. We intentionally omitted encryption of multiple plaintexts for simplicity of exposition, but our construction can naturally be extended in this way.

Security notions of the authenticated encryption was studied before. Bellare and Namprempre [4] studied both privacy and authenticity of authenticated encryption schemes, and the authenticity notions are later studied further by Bellare, Goldreich and Mityagin [2]. Our UF-CPA and SUF-CPA can be considered as homomorphic versions of INT-PTXT-1 and INT-CTXT-1 of [2], respectively. Our UF-CCA and SUF-CCA are comparable to homomorphic versions of INT-PTXT-M and INT-CTXT-M, respectively, but in our (S)UF-CCA, the adversary has access to the decryption oracle, while in INT-PTXT-M and INT-CTXT-M, the adversary has access to the verification oracle.

3   Preliminary

3.1   Notations

For any $a \in \mathbb{R}$, the *nearest integer* $\lfloor a \rceil$ of a is defined as the unique integer in $[a - \frac{1}{2}, a + \frac{1}{2})$. The ring $\mathbb{Z}_n$ of integers modulo n is represented as the set $\mathbb{Z} \cap (-\frac{n}{2}, \frac{n}{2}]$. That is, $x \bmod n = x - \lfloor x/n \rceil \cdot n$ for any $x \in \mathbb{Z}$. For example, we have $\mathbb{Z}_2 = \{0, 1\}$, $\mathbb{Z}_3 = \{-1, 0, 1\}$.

For any mutually prime $n, m \in \mathbb{Z}^+$, $\mathrm{CRT}_{(n,m)}$ is the isomorphism $\mathbb{Z}_n \times \mathbb{Z}_m \to \mathbb{Z}_{nm}$, such that for any $(a, b) \in \mathbb{Z}_n \times \mathbb{Z}_m$, we have

$$(\mathrm{CRT}_{(n,m)}(a, b) \bmod n, \mathrm{CRT}_{(n,m)}(a, b) \bmod m) = (a, b) .$$

In this paper, the security parameter is always denoted as λ, and the expression $f(\lambda) = negl(\lambda)$ means that $f(\lambda)$ is a negligible function, that is, for any $c > 0$, we have $|f(\lambda)| \leq \lambda^{-c}$ for all $\lambda \in \mathbb{Z}^+$ large enough.

Also, lg means the logarithm to base 2. And $\Delta(D_1, D_2)$ denotes the statistical distance between two distributions D_1 and D_2.

A notation like $(\tau, \cdot) \in S$ is an abbreviation for $\exists x \, (\tau, x) \in S$. Naturally, $(\tau, \cdot) \notin S$ is its negation, $\forall x \, (\tau, x) \notin S$. This notation can also be generalized to n-tuples for $n > 2$, for example $(\tau, \cdot, \cdot) \in S$.

3.2   Security Assumptions

Here we define security assumptions we use in this paper. First, let us define some distributions. For $p, q_0, \rho \in \mathbb{Z}^+$, we define the distribution $\mathcal{D}(p, q_0, \rho)$ as

$$\mathcal{D}(p, q_0, \rho) := \{\text{choose } q \xleftarrow{\$} \mathbb{Z} \cap [0, q_0), \, r \xleftarrow{\$} \mathbb{Z} \cap (-2^\rho, 2^\rho) : \text{ output } pq + r\} .$$

Clearly, we can efficiently sample from the above distribution. In this paper, when a distribution is given as an input to an algorithm, it means that a sampling oracle for the distribution is given.

Let PRIME be the set of all prime numbers, and ROUGH(x) the set of all 'x-rough integers', that is, integers having no prime factors less than x.

In the following, the parameters ρ, η, γ are polynomially bounded functions of λ, and we assume they can be efficiently computed, given λ.

Definition 1 (Error-Free Approximate GCD Assumption). *The (computational) (ρ, η, γ)-EF-AGCD assumption is that, for any PPT adversary A, we have*

$$\Pr\left[A(1^\lambda, y_0, \mathcal{D}(p, q_0, \rho)) = p\right] = negl(\lambda) \ ,$$

where $p \xleftarrow{\$} [2^{\eta-1}, 2^\eta) \cap \mathrm{PRIME}$, $q_0 \xleftarrow{\$} [0, 2^\gamma/p) \cap \mathrm{ROUGH}(2^{\lambda^2})$, and $y_0 = pq_0$.

The EF-AGCD assumption is suggested by Coron et al. [15] to prove the security of their variant of the DGHV scheme [17]. There is also a decisional version, suggested by Cheon at al. [13]:

Definition 2 (Decisional Error-Free Approximate GCD Assumption). *The decisional (ρ, η, γ)-EF-AGCD assumption is that, for any PPT distinguisher D, the following value is negligible:*

$$\left| \Pr\left[D(1^\lambda, y_0, \mathcal{D}(p, q_0, \rho), z) = 1 \mid z \leftarrow \mathcal{D}(p, q_0, \rho)\right] \right.$$

$$\left. - \Pr\left[D(1^\lambda, y_0, \mathcal{D}(p, q_0, \rho), z) = 1 \mid z \xleftarrow{\$} \mathbb{Z}_{y_0}\right] \right| \ ,$$

where $p \xleftarrow{\$} [2^{\eta-1}, 2^\eta) \cap \mathrm{PRIME}$, $q_0 \xleftarrow{\$} [0, 2^\gamma/p) \cap \mathrm{ROUGH}(2^{\lambda^2})$, and $y_0 = pq_0$.

Recently Coron et al. proved the equivalence of the EF-AGCD assumption and the decisional EF-AGCD assumption [14]. In Theorem 6, we show that our scheme is IND-CPA under the decisional EF-AGCD assumption. Hence, our scheme's security is in fact based on the (computational) EF-AGCD assumption, due to the equivalence.

4 Homomorphic Authenticated Encryption

Here we define the homomorphic authenticated encryption and its security. In the following, $\mathcal{M}$, $\mathcal{C}$, $\mathcal{L}$, $\mathcal{F}$ are the *plaintext space*, the *ciphertext space*, the *label space*, and the *admissible function space*, respectively.

4.1 Syntax

Labeled Programs. First, let us define labeled programs, a concept first introduced in [19].

For each HAE, a set of admissible functions $\mathcal{F}$ is associated. In reality, an element f of $\mathcal{F}$ is a concrete representation of a function which can be evaluated in polynomial time. It is required that any $f \in \mathcal{F}$ should represent a function of form $f : \mathcal{M}^l \to \mathcal{M}$ for some $l \in \mathbb{Z}^+$ which depends on f. We will simply call an element $f \in \mathcal{F}$ an *admissible function*. The number l is the *arity* of f.

A HAE encrypts a plaintext $m \in \mathcal{M}$ under a 'label' $\tau \in \mathcal{L}$, and a labeled program is an admissible function together with information which plaintexts should be used as inputs. Formally, a *labeled program* is a tuple $P = (f, \tau_1, \ldots, \tau_l)$, where f is an arity-l admissible function, and $\tau_i \in \mathcal{L}$ are labels for $i = 1, \ldots, l$ for each input of f. The idea is that, if m_i are plaintexts encrypted under the label τ_i, respectively, then the evaluation of the labeled program $P = (f, \tau_1, \ldots, \tau_l)$ is $f(m_1, \ldots, m_l)$.

We also define the *identity labeled program* with label $\tau \in \mathcal{L}$, which is $I_\tau = (\mathrm{id}, \tau)$, where $\mathrm{id} : \mathcal{M} \to \mathcal{M}$ is the identity function.

Homomorphic Authenticated Encryption. A HAE Π consists of the following four PPT algorithms.

- Gen(1^λ): given a security parameter λ, Gen outputs a key pair (ek, sk), with a public evaluation key ek and a secret key sk.
- Enc(sk, τ, m): given a secret key sk, a label τ and a plaintext m, Enc outputs a ciphertext c.
- Eval($ek, f, c_1, \cdots, c_l$): given an evaluation key ek, an admissible function $f : \mathcal{M}^l \to \mathcal{M}$ and l ciphertexts $c_1, \cdots, c_l \in \mathcal{C}$, the deterministic algorithm Eval outputs a ciphertext $\tilde{c} \in \mathcal{C}$.
- Dec($sk, (f, \tau_1, \cdots, \tau_l), \hat{c}$): when given a secret key sk, a labeled program $(f, \tau_1, \cdots, \tau_l)$ and a ciphertext $\hat{c} \in \mathcal{C}$, the deterministic algorithm Dec outputs a message $m \in \mathcal{M}$ or $\bot$.

We assume that ek implicitly contains the information about $\mathcal{M}, \mathcal{C}, \mathcal{L}$, and $\mathcal{F}$. As mentioned above, we assume both Eval and Dec are deterministic algorithms.

Compactness. In order to exclude trivial constructions, we require that the output size of Eval($ek, \ldots$) and Enc($sk, \cdot, \cdot$) should be bounded by a polynomial of λ for any choice of their input, when $(ek, sk) \leftarrow$ Gen(1^λ). That is, the ciphertext size is independent of the choice of the admissible function f.

Correctness. A HAE scheme must satisfy the following correctness properties:

- $m = $ Dec($sk, I_\tau, $ Enc(sk, τ, m)), for any $\lambda \in \mathbb{Z}^+$, $\tau \in \mathcal{L}$ and $m \in \mathcal{M}$, when $(ek, sk) \leftarrow$ Gen(1^λ).
- $f(m_1, \ldots, m_l) = $ Dec($sk, (f, \tau_1, \ldots, \tau_l), c$), for any $\lambda \in \mathbb{Z}^+$, any $f \in \mathcal{F}$, any $\tau_i \in \mathcal{L}$, $m_i \in \mathcal{M}$ for $i = 1, \ldots, l$, when $(ek, sk) \leftarrow$ Gen(1^λ), $c_i \leftarrow$ Enc(sk, τ_i, m_i) for $i = 1, \ldots, l$, and $c \leftarrow$ Eval($ek, f, c_1, \ldots, c_l$).

In addition, we require that a HAE should satisfy a property we call ciphertext constant testability, which will be explained in Sect. 4.3.

4.2 Legal Encryption

As in the case of homomorphic MAC [19], it is required that a label τ used in an encryption Enc(sk, τ, m) should *not* be reused. In practice, this should be enforced by policy between valid users of HAE.

In the security model of HAE, this is expressed by legality of an encryption query of an adversary: the adversary makes adaptive encryption queries (τ, m), and this query is answered by $c \leftarrow \mathsf{Enc}(sk, \tau, m)$. Let us keep an encryption history S of all tuples (τ, m, c) occurring as the result of such an encryption query. We say that an encryption query (τ, m) is *illegal*, if $(\tau, \cdot, \cdot) \in S$. Also, we say that τ is *new* if $(\tau, \cdot, \cdot) \notin S$, and τ is *used* if $(\tau, \cdot, \cdot) \in S$. We say that an adversary is *legal*, if it does not make any illegal queries, including illegal encryption queries.

In this paper, we only consider legal adversaries, that is, we always *exclude* illegal adversaries, in the sense of the paper by Bellare, Hofheinz, and Kiltz [3]. Different security notions may have additional definitions of illegal queries (for example, illegal decryption queries), but any encryption query involving a used label will always be considered as illegal.

4.3 Constant Testability

For later use, we need to be able to check efficiently whether certain functions are constant or not. For example, we need this for the homomorphic evaluation of any admissible function f, regarded as a function mapping a ciphertext tuple to a ciphertext: $(c_1, \ldots, c_l) \mapsto \mathsf{Eval}(ek, f, c_1, \ldots, c_l)$. In fact, we need to consider slightly more general functions.

Fix a HAE $\varPi$ and an evaluation key ek of $\varPi$. Given an arity-l admissible function f, a subset I of the index set $\{1, \cdots, l\}$, plaintexts $(m_i)_{i \in I} \in \mathcal{M}^{|I|}$, and ciphertexts $(c_i)_{i \in I} \in \mathcal{C}^{|I|}$, we make the following definition:

Definition 3. *A* partial application *of f w.r.t. plaintexts $(m_i)_{i \in I}$ is the function $\tilde{f} : \mathcal{M}^{l-|I|} \to \mathcal{M}$ defined by $\tilde{f}((m_j)_{j \notin I}) := f(m_1, \ldots, m_l)$. We denote this $\tilde{f}$ by* $\mathsf{App}(f, (m_i)_{i \in I})$.

Definition 4. *A* partial homomorphic evaluation *of f w.r.t. ciphertexts $(c_i)_{i \in I}$ is the function $\tilde{e} : \mathcal{C}^{l-|I|} \to \mathcal{C}$ defined by $\tilde{e}((c_j)_{j \notin I}) := \mathsf{Eval}(ek, f, c_1, \ldots, c_l)$. We denote this $\tilde{e}$ by* $\mathsf{Eval}(f, (c_i)_{i \in I})$.

So, $\tilde{f} = \mathsf{App}(f, (m_i)_{i \in I})$ is the admissible function f with some inputs m_i for $i \in I$ already 'filled in', and $\tilde{f}$ becomes a function of remaining plaintext inputs. Similarly, $\tilde{e} = \mathsf{Eval}(f, (c_i)_{i \in I})$ is the homomorphic evaluation $\mathsf{Eval}(ek, f, c_1, \ldots, c_l)$ with some inputs c_i for $i \in I$ already filled in, and $\tilde{e}$ becomes a function of remaining ciphertext inputs. In particular, $\mathsf{Eval}(f, (c_i)_{i \in I})$ is a constant function if $I = \{1, \ldots, l\}$.

Informally, if a HAE $\varPi$ satisfies *ciphertext constant testability* (CCT), then there is an efficient algorithm which can determine whether such $\mathsf{Eval}(f, (c_i)_{i \in I})$ is constant or not with overwhelming probability.

In fact, we need a computational version of this property. Therefore, we formally define CCT using the following security game $\mathrm{CCT}_{\varPi, A, D}$ involving an adversary A and a 'constant tester' D:

$\mathrm{CCT}_{\varPi, A, D}(1^\lambda)$:

Initialization. A key pair $(ek, sk) \leftarrow \mathsf{Gen}(1^\lambda)$ is generated, a set S is initialized as the empty set $\emptyset$. Then ek is given to A.

Queries. A may make legal encryption queries adaptively. For each encryption query (τ, m) of A, the answer $c \leftarrow \mathsf{Enc}(sk, \tau, m)$ is returned to A, and S is updated as $S \leftarrow S \cup \{(\tau, m, c)\}$.

Challenge. A outputs a labeled program $(f, \tau_1, \ldots, \tau_l)$. Let I be the set of indices $i = 1, \ldots, l$ such that $(\tau_i, m_i, c_i) \in S$ for some[2] m_i, c_i. Then D outputs a bit $b \leftarrow D(ek, (f, \tau_1, \ldots, \tau_l), I, (c_i)_{i \in I})$.

Finalization. The game outputs 1 if $\tilde{e} := \mathsf{Eval}(f, (c_i)_{i \in I})$ is constant and $b = 0$, or $\tilde{e}$ is nonconstant and $b = 1$. The game outputs 0 otherwise.

The output bit b of the tester D is 1 iff D 'thinks' that $\mathsf{Eval}(f, (c_i)_{i \in I})$ is constant. Therefore, the event $\mathrm{CCT}_{\Pi, A, D}(1^\lambda) = 1$ happens when the tester D is wrong.

The advantage of an adversary A in the game CCT against D is defined as

$$\mathbf{Adv}_{\Pi, A, D}^{\mathrm{CCT}}(\lambda) := \Pr[\mathrm{CCT}_{\Pi, A, D}(1^\lambda) = 1] \ .$$

We say that a HAE Π satisfies the *ciphertext constant testability (CCT)*, if there exists a PPT constant tester D such that $\mathbf{Adv}_{\Pi, A, D}^{\mathrm{CCT}}(\lambda)$ is negligible for any *legal* PPT adversary A.

Similarly, we may define *plaintext constant testability* (PCT): informally, Π satisfies PCT if testing whether a partial application of an admissible function is constant or not can be done efficiently.

When the set of admissible functions of a HAE is simple, both PCT and CCT may be satisfied. But, PCT might be a difficult property to be satisfied in general; if a HAE supports boolean circuits and is fully homomorphic, then since satisfiability of a boolean circuit can be efficiently determined if constant testing is efficient, if such HAE satisfies PCT, we may use it to invert any one-way function.

On the other hand, we claim that a HAE to satisfy CCT is a relatively mild requirement: unlike the plaintext space $\mathcal{M}$, often the ciphertext space $\mathcal{C}$ might be a large ring, and $\mathsf{Eval}(f, (c_i)_{i \in I})$ is a polynomial on the ring $\mathcal{C}$, in which case we may use the Schwartz-Zippel lemma to perform polynomial identity testing. This applies to our HAE scheme to be presented in this paper, as shown in Theorem 5.

Moreover, we show that if Π is a HAE which does not necessarily satisfy CCT, then there is a simple generic transformation which turns it into another HAE Π' which satisfies CCT, while preserving original security properties satisfied by Π. This will be explained in Sect. 4.7.

Therefore, without (much) loss of generality, we assume the CCT property to be an additional requirement for a HAE to satisfy.

4.4   Privacy

Indistinguishability under Chosen-Plaintext Attack. First we define a homomorphic version of the IND-CPA security based on the left-or-right indistinguishability [1]. We use the following security game IND-CPA$_{\Pi, A}$ for an adversary A:

---

[2] For any $i \in I$, such m_i, c_i are necessarily unique.

IND-CPA$_{\Pi,A}(1^\lambda)$:

Initialization. A key pair $(ek, sk) \leftarrow \text{Gen}(1^\lambda)$ is generated, a set S is initialized as the empty set $\emptyset$. And a coin $b \xleftarrow{\$} \{0, 1\}$ is flipped. Then ek is given to A.

Queries. A may make encryption queries adaptively. For each encryption query (τ, m_0, m_1) of A, the answer $c \leftarrow \text{Enc}(sk, \tau, m_b)$ is returned to A, and S is updated as $S \leftarrow S \cup \{(\tau, (m_0, m_1), c)\}$.

Finalization. A outputs a bit b', and then the challenger returns 1 if $b = b'$, and 0 otherwise.

As usual, an encryption query is considered illegal if it involves a used label.[3] The advantage of A in the game IND-CPA for the scheme Π is defined by

$$\mathbf{Adv}_{\Pi,A}^{\text{IND-CPA}}(\lambda) := \left| \Pr[\text{IND-CPA}_{\Pi,A}(1^\lambda) = 1] - \frac{1}{2} \right| .$$

We say that Π satisfies IND-CPA, if $\mathbf{Adv}_{\Pi,A}^{\text{IND-CPA}}(\lambda)$ is negligible for any *legal* PPT adversary A.

Indistinguishability under Chosen-Ciphertext Attack. Though the usual IND-CCA security is not achievable for homomorphic encryption due to malleability, nevertheless we may define a version of IND-CCA for HAE. It is because that for HAE, a ciphertext is decrypted with respect to a labeled program; while the ciphertext is still malleable by function evaluation, a decryption query should essentially declare how it was produced. This allows a homomorphic version of IND-CCA to be defined naturally as follows.

The most important difference of our definition is on the legality of a decryption query. In our case, any decryption query for a ciphertext produced by function evaluation which may nontrivially depend on the bit b should be considered illegal, since decryption of that ciphertext might reveal the bit b. To formalize:

Let S be the encryption history as before. Then, we say that a decryption query $((f, \tau_1, \cdots, \tau_l), \hat{c})$ is *illegal*, if $\tilde{f}_0$ and $\tilde{f}_1$ are not equal, where

$$I := \{i \in \{1, \cdots, l\} \mid (\tau_i, (m_{i,0}, m_{i,1}), \cdot) \in S \text{ for some } (m_{i,0}, m_{i,1}) \in \mathcal{M} \times \mathcal{M}\} ,$$
$$\tilde{f}_b := \text{App}(f, (m_{i,b})_{i \in I}) \text{ for } b = 0, 1 .$$

Homomorphic IND-CCA for a HAE $\Pi = (\text{Gen}, \text{Enc}, \text{Eval}, \text{Dec})$ is defined via the security game IND-CCA$_{\Pi,A}$ whose formal description we omit here due to the space constraints. It is very similar to IND-CPA$_{\Pi,A}$, except that the adversary A may make both encryption and decryption queries at any time. In the full version of this paper, the formal description of the security game IND-CCA$_{\Pi,A}$ will be given, as well as other security games.

---

[3] As before, a label τ is used if $(\tau, \cdot, \cdot) \in S$.

The advantage of A in the game IND-CCA for the scheme Π, $\mathbf{Adv}_{\Pi,A}^{\text{IND-CCA}}(\lambda)$, is defined similarly. And we say that Π satisfies IND-CCA, if $\mathbf{Adv}_{\Pi,A}^{\text{IND-CCA}}(\lambda)$ is negligible for any *legal* PPT adversary A.

Remark 1. As we have discussed while defining constant testability in Sect. 4.3, in general it may not be feasible to check whether $\tilde{f}_0 = \tilde{f}_1$ or not, especially when the HAE in question is fully homomorphic and may process arbitrarily large boolean circuits. Therefore, in general, it may not be feasible to efficiently decide whether a decryption query is legal or not. Hence we use the exclusion-style definition, rather than the penalty-style, according to the classification of Bellare, Hofheinz, and Kiltz [3]: we regard only legal adversaries, which does not make any illegal queries. While in other cases the two styles of definitions are mostly compatible, in this case it is not.

Also, later in Sect. 4.7, we show that by using a secure PRF and a collision-resistant hash function (or a hash tree), we may transform a HAE scheme into another HAE which satisfies CCT. After we apply the transformation, if $Dec(sk, (f, \tau_1, \ldots, \tau_l), \hat{c}) \neq \bot$, then with overwhelming probability we should have $I = \{1, \ldots, l\}$. Therefore, any decryption query will either output $\bot$, or both $\tilde{f}_0$ and $\tilde{f}_1$ are constant, which makes deciding if a verification query is illegal trivial. This transform can be used if an application requires ability to efficiently decide whether a verification query is illegal or not.

4.5   Authenticity

Unforgeability under Chosen-Plaintext Attack. Our authenticity definition for HAE is an adaptation of the definition given by Catalano and Fiore [11] for homomorphic MACs.

First, we define the forgery of an adversary. Let $((f, \tau_1, \cdots, \tau_l), \hat{c})$ be a forgery attempt, and let S be the encryption history. We say that it is a *forgery*, if the following holds:

1. It is *valid*, that is, $\bot \neq Dec(sk, (f, \tau_1, \cdots, \tau_l), \hat{c})$ and,
2. One of the following holds:
   - Type 1 forgery: $\mathsf{App}(f, (m_i)_{i \in I})$ is not constant, or,
   - Type 2 forgery: $\mathsf{App}(f, (m_i)_{i \in I})$ is constantly equal to some $\tilde{m}$, but $\tilde{m} \neq Dec(sk, (f, \tau_1, \cdots, \tau_l), \hat{c})$,

   where I is the set of $i \in \{1, \ldots, l\}$ such that $(\tau_i, m_i, \cdot) \in S$ for some (unique) $m_i \in \mathcal{M}$.

We define the *unforgeability under chosen-plaintext attack* (UF-CPA) of a HAE Π using the security game UF-CPA$_{\Pi,A}$. In the game, the adversary A is given the evaluation key ek and the encryption oracle. Finally, A outputs $((f, \tau_1, \cdots, \tau_l), \hat{c})$. The game outputs 1 if it is a successful forgery, and 0 otherwise.

The advantage of A in the game UF-CPA for the scheme Π is defined as

$$\mathbf{Adv}_{\Pi,A}^{\text{UF-CPA}}(\lambda) := \Pr[\, \text{UF-CPA}_{\Pi,A}(1^\lambda) = 1\,] \ .$$

We say that Π satisfies UF-CPA, if $\mathbf{Adv}_{\Pi,A}^{\text{UF-CPA}}(\lambda)$ is negligible for any *legal* PPT adversary A.

Unforgeability under Chosen-Ciphertext Attack. It is also natural to consider a stronger variant of unforgeability, in which an adversary is allowed to make decryption queries as well as encryption queries. We call this variant UF-CCA. The only difference of UF-CCA from UF-CPA is that the adversary A can also make any decryption query $((f, \tau_1, \cdots, \tau_l), \hat{c})$, which is answered with $\text{Dec}(sk, (f, \tau_1, \cdots, \tau_l), \hat{c})$.

The advantage of A in the game UF-CCA for the scheme Π, $\mathbf{Adv}_{\Pi,A}^{\text{UF-CCA}}(\lambda)$, is defined similarly. And we say that Π satisfies UF-CCA, if $\mathbf{Adv}_{\Pi,A}^{\text{UF-CCA}}(\lambda)$ is negligible for any *legal* PPT adversary A.

Strong Unforgeability under Chosen-Plaintext Attack. Sometimes it is useful to consider stronger definition of authenticity. So let us define *strong unforgeability* for HAE. Let S be the encryption history. Then we say that a forgery attempt $((f, \tau_1, \cdots, \tau_l), \hat{c})$ is a *strong forgery*, if the following holds:

1. It is valid, that is, $\perp \neq \text{Dec}(sk, (f, \tau_1, \cdots, \tau_l), \hat{c})$ and,
2. One of the following holds:
   - Type 1 strong forgery: $\text{Eval}(f, (c_i)_{i \in I})$ is not constant, or,
   - Type 2 strong forgery: $\text{Eval}(f, (c_i)_{i \in I})$ is constantly equal to some $\tilde{c}$, but $\tilde{c} \neq \hat{c}$, where I is the set of $i \in \{1, \ldots, l\}$ such that $(\tau_i, \cdot, c_i) \in S$ for some (unique) $c_i \in \mathcal{C}$.

We define the *strong unforgeability under chosen-plaintext attack* (SUF-CPA) of a HAE Π using the game SUF-CPA$_{\Pi,A}$. In the game, the adversary A is given the evaluation key ek and the encryption oracle. Finally, A outputs $((f, \tau_1, \cdots, \tau_l), \hat{c})$. The game outputs 1 iff it is a successful strong forgery.

The advantage of A in the game SUF-CPA for the scheme Π, $\mathbf{Adv}_{\Pi,A}^{\text{SUF-CPA}}(\lambda)$, is defined similarly. And we say that Π satisfies SUF-CPA, if $\mathbf{Adv}_{\Pi,A}^{\text{SUF-CPA}}(\lambda)$ is negligible for any *legal* PPT adversary A.

Strong Unforgeability under Chosen-Ciphertext Attack. Also for strong unforgeability, we consider security against chosen-ciphertext attacks, which we call SUF-CCA. Again, the only difference of SUF-CCA from SUF-CPA is that the adversary A is also given the decryption oracle.

The advantage of A in the game SUF-CCA for the scheme Π, $\mathbf{Adv}_{\Pi,A}^{\text{SUF-CCA}}(\lambda)$, is defined similarly. And we say that Π satisfies SUF-CCA, if $\mathbf{Adv}_{\Pi,A}^{\text{SUF-CCA}}(\lambda)$ is negligible for any *legal* PPT adversary A.

4.6 Relations on Security Notions

In this section, we investigate relations between the six security notions defined in the previous section. First, we have trivial implications from CCA security to CPA security.

Theorem 1. *UF-CCA implies UF-CPA, SUF-CCA implies SUF-CPA, and also IND-CCA implies IND-CPA.*

The following theorem says that the strong unforgeability implies unforgeability.

Theorem 2. *SUF-CCA implies UF-CCA. And SUF-CPA implies UF-CPA.*

Proof. It is enough to show that a successful forgery is also a successful strong forgery. Let $((f, \tau_1, \cdots, \tau_l), \hat{c})$ be a forgery. If it is a type 1 forgery, then $\tilde{f} := \mathsf{App}(f, (m_i)_{i \in I})$ is not constant. That is, there exist two tuples $(m_j^1)_{j \notin I}$ and $(m_j^2)_{j \notin I}$ such that $\tilde{f}(m_j^1)_{j \notin I} \neq \tilde{f}(m_j^2)_{j \notin I}$. Then there exist two tuples $(c_j^1)_{j \notin I}$ and $(c_j^2)_{j \notin I}$ such that $m_j^1 = \mathsf{Dec}(sk, I_{\tau_j}, c_j^1)$ and $m_j^2 = \mathsf{Dec}(sk, I_{\tau_j}, c_j^2)$ for each $j \notin I$. Then we can show that $\tilde{e} := \mathsf{Eval}(f, (c_i)_{i \in I})$ is nonconstant; since we have $\mathsf{Dec}(sk, (f, \tau_1, \ldots, \tau_l), \tilde{e}(c_j^b)_{j \notin I}) = \tilde{f}(m_j^b)_{j \notin I}$ for $b = 1, 2$ by correctness, we see that $\tilde{e}(c_j^1)_{j \notin I} \neq \tilde{e}(c_j^2)_{j \notin I}$. So it is a type 1 strong forgery.

If it is a type 2 forgery but not a type 1 strong forgery, then both $\tilde{f}$ and $\tilde{e}$ are constants. Let the constant value of $\tilde{f}$ be $\tilde{m} \in \mathcal{M}$, and the constant value of $\tilde{e}$ be $\tilde{c} \in \mathcal{C}$. We have $\tilde{m} \neq \mathsf{Dec}(sk, (f, \tau_1, \cdots, \tau_l), \hat{c})$. But $\tilde{m} = \mathsf{Dec}(sk, (f, \tau_1, \cdots, \tau_l), \tilde{c})$, again by correctness. So $\hat{c} \neq \tilde{c}$, and thus it is a type 2 strong forgery.

Bellare et al. [2] showed that, in case of MAC, strong unforgeability implies strong unforgeability even when the adversary has access to the verification oracle, and in case of AE, integrity of ciphertexts implies integrity of ciphertexts even when the adversary has access to the verification oracle. The following can be considered as a homomorphic analogue to the result.

Theorem 3. *SUF-CPA implies SUF-CCA.*

The basic intuition of the proof of Theorem 3 is as follows: if a HAE scheme Π satisfies SUF-CPA, since it is infeasible to produce any strong forgery, essentially any decryption query $((f, \tau_1, \cdots, \tau_l), \hat{c})$ which should be answered with anything other than $\bot$ must be the output of the Eval algorithm with correct ciphertexts from encryption queries as inputs. Therefore, even if the decryption oracle is given to the adversary A, it would not give any useful information. The actual proof, which will be on the full version of this paper due to the page constraints, uses a hybrid argument where the decryption queries are in the end handled by a decryption simulation.

Theorem 4. *IND-CPA and SUF-CPA together imply IND-CCA.*

Proof of this theorem is similar to that of Theorem 3. Again, we use a hybrid argument to transform the IND-CCA game into another game that is essentially same as the IND-CPA game: the strong unforgeability allows us to simulate decryption oracle. Again, the complete proof will be given in the full version of this paper.

In conclusion, we see that IND-CPA and SUF-CPA together imply the strongest security notions, IND-CCA and SUF-CCA.

4.7   Generic Transformation for Ciphertext Constant Testability

Suppose that Π is a HAE which does *not* necessarily satisfy CCT. We describe a generic construction that transforms a HAE Π into another HAE Π' satisfying CCT, while preserving IND-CPA or SUF-CPA. The construction uses a PRF $F_k : \{0,1\}^\lambda \to \{0,1\}^\lambda$ and a family $\mathcal{H}$ of collision-resistant hash functions $H : \{0,1\}^* \to \{0,1\}^\lambda$.

Scheme $\Pi' = (\text{Gen}', \text{Enc}', \text{Eval}', \text{Dec}')$:

- $\text{Gen}'(1^\lambda)$: Generate keys $(ek, sk) \leftarrow \text{Gen}(1^\lambda)$, $k \leftarrow \{0,1\}^\lambda$ and $H \leftarrow \mathcal{H}$. Return (ek', sk') where $ek' = (ek, H)$ and $sk' = (sk, k)$.
- $\text{Enc}'(sk', \tau, m)$: Let $h = F_k(\tau)$ and $c \leftarrow \text{Enc}(sk, \tau, m)$. Return $c' = (h, c)$.
- $\text{Eval}'(ek', f, c_1', \cdots, c_l')$: Parse $c_i' = (h_i, c_i)$ for $i = 1, \cdots, l$. Let $\tilde{h} = H(h_1, \cdots, h_l)$ and $\tilde{c} \leftarrow \text{Eval}(ek, f, c_1, \cdots, c_l)$. Return $\tilde{c}' = (\tilde{h}, \tilde{c})$.
- $\text{Dec}'(sk', (f, \tau_1, \cdots, \tau_l), \tilde{c}')$: Parse $\tilde{c}' = (\tilde{h}, \tilde{c})$. For each $i = 1, \cdots, l$, let $h_i = F_k(\tau_i)$. If $\tilde{h} = H(h_1, \cdots, h_l)$, then return $\text{Dec}(sk, (f, \tau_1, \cdots, \tau_l), \tilde{c})$. Otherwise, return $\bot$.

It is clear that Π' satisfies correctness properties, as long as Π also does.

We claim that in addition Π' satisfies CCT. The constant tester D for Π' is simple: given ek', $(f, \tau_1, \ldots, \tau_l)$, I, and $(c_i')_{i \in I}$, the tester D outputs 1 if $I = \{1, \ldots, l\}$, and outputs 0 otherwise. Suppose there exists an adversary A with non-negligible advantage in the game CCT against this tester. Observe that D errs only when $I \neq \{1, \ldots, l\}$ and the function $\text{Eval}(f, (c_i')_{i \in I})$ is constant. Therefore, we may use A to construct a hash collision finding algorithm B as follows: B receives $H \leftarrow \mathcal{H}$, and simulates the CCT game. Since B itself generates (ek, sk) and k, it may answer any queries made by A. Eventually A outputs a labeled program $(f, \tau_1, \ldots, \tau_l)$. If all τ_i are used, then B aborts. But there is a non-negligible probability that $I \neq \{1, \ldots, l\}$ and $\text{Eval}(f, (c_i')_{i \in I})$ is constant, and this means that $\tilde{h} = H(h_1, \ldots, h_l)$ as a function of $(h_j)_{j \notin I}$ is also constant. Therefore, B may output a collision pair with non-negligible probability, because $\{1, \ldots, l\} \setminus I \neq \emptyset$ and B may arbitrarily choose $h_j \neq h_j'$ for $j \notin I$.

Also, if Π satisfies IND-CPA then so does Π': informally, in the ciphertext $c' = (h, c)$ the h-part $F_k(\tau)$ has no information about the plaintext m, and any information about the plaintext m in the c-part $\text{Enc}(sk, \tau, m)$ is computationally hidden since Π is IND-CPA.

And, if Π satisfies SUF-CPA, then so does Π': since any strong forgery $((f, \tau_1, \ldots, \tau_l), \hat{c}' = (\hat{h}, \hat{c}))$ of Π' has to be valid, it is easy to see that with negligible exception, all τ_i are used and $\text{Eval}(f, (c_i')_{i \in I})$ is constant. Then we may show that in fact $((f, \tau_1, \ldots, \tau_l), \hat{c})$ should be a type 2 strong forgery for Π.

We will provide proofs for all of the above in the full version of this paper. Note that one disadvantage of this transform is that Π' does not support composition of admissible functions; if Π supports boolean circuits, is fully homomorphic, and admissible functions are composable, then, we may want the same for Π'. For this, we may adopt the Merkle hash tree construction used by

Gennaro and Wichs [19] for their fully homomorphic MAC. This hash tree based transformation will also be given in the full version.

5   Construction

Here we describe our HAE Π and show that it satisfies correctness and CCT.

Parameters ρ, η, γ, $\bar{d}$ are polynomially bounded functions of the security parameter λ, and the modulus parameter Q is a function of λ satisfying $2 \leq Q \leq 2^\lambda$. We assume that all these parameters can be efficiently computed, given λ. Constraints on these parameters are given after the description of the scheme.

We use a PRF F in our construction. We may assume that $F_k : \{0,1\}^\lambda \to \mathbb{Z}_{q_0}$ for each $k \in \{0,1\}^\lambda$. The message space and the ciphertext space of our scheme is $\mathbb{Z}_Q$ and $\mathbb{Z}_{y_0}$, resp., and the label space is $\{0,1\}^\lambda$. To represent admissible functions we use arithmetic circuits, that is, circuits consisting of $+$ gates and $\times$ gates. Such a circuit f of arity l determines a polynomial $f : \mathbb{Z}^l \to \mathbb{Z}$ with integral coefficients. We use such a circuit to compute function values of plaintext inputs in $\mathbb{Z}_Q$, and also to homomorphically evaluate ciphertexts in $\mathbb{Z}_{y_0}$. The precise description of the admissible function space will be given after the scheme description, together with discussions on the correctness property.

Scheme Π = (Gen, Enc, Eval, Dec):

- Gen(1^λ): Choose $p \xleftarrow{\$} [2^{\eta-1}, 2^\eta) \cap \text{PRIME}$, $q_0 \xleftarrow{\$} [0, 2^\gamma/p) \cap \text{ROUGH}(2^{\lambda^2})$, and $k \leftarrow \{0,1\}^\lambda$. Let $y_0 = pq_0$. Return the key pair (ek, sk), where $ek = (1^\lambda, y_0)$, and $sk = (1^\lambda, p, q_0, k)$.
- Enc(sk, τ, m): Given the secret key sk, a label $\tau \in \{0,1\}^\lambda$ and a plaintext $m \in \mathbb{Z}_Q$, choose $r \xleftarrow{\$} \mathbb{Z} \cap (-2^\rho, 2^\rho)$. Let $a = rQ + m$ and $b = F_k(\tau)$. Return $c = \text{CRT}_{(p,q_0)}(a, b)$.
- Eval($ek, f, c_1, \cdots, c_l$): Given ek, an arithmetic circuit f of arity l and ciphertexts $c_1, \cdots, c_l$, return $\tilde{c} := f(c_1, \cdots, c_l) \bmod y_0$
- Dec($sk, (f, \tau_1, \cdots, \tau_l), \hat{c}$): For $i = 1$ to l, compute $b_i \leftarrow F_k(\tau_i)$ and $b = f(b_1, \cdots, b_l) \bmod q_0$. Return $m = (\hat{c} \bmod p) \bmod Q$, if $b = \hat{c} \bmod q_0$. Otherwise, return $\perp$.

5.1   Correctness

Here we determine when our HAE scheme is correct. Let $(ek, sk) \leftarrow \text{Gen}(1^\lambda)$. Let $c_i \leftarrow \text{Enc}(sk, \tau_i, m_i)$ for each $i = 1, \cdots, l$. And let $\tilde{c} \leftarrow \text{Eval}(ek, f, c_1, \cdots, c_l)$, for any arity-$l$ arithmetic circuit f of degree d. Then

$$\tilde{c} \bmod p = (f(c_1, \cdots, c_l) \bmod y_0) \bmod p = f(c_1, \cdots, c_l) \bmod p$$
$$= f(c_1 \bmod p, \cdots, c_l \bmod p) \bmod p$$
$$= f(r_1 Q + m_1, \cdots, r_l Q + m_l) \bmod p$$
$$= f(r_1 Q + m_1, \cdots, r_l Q + m_l)$$

The last equality in the above holds if $|f(r_1Q + m_1, \cdots, r_lQ + m_l)| \leq p/2$. And so, in this case,

$$(\tilde{c} \bmod p) \bmod Q = f(r_1Q + m_1, \cdots, r_lQ + m_l) \bmod Q$$
$$= f(m_1, \cdots, m_l) \bmod Q .$$

Since $|f(r_1Q + m_1, \cdots, r_lQ + m_l)| \leq \|f\|_1 \cdot 2^{d(\rho+\lambda)}$ and $2^{\eta-2} \leq p/2$, the correctness is guaranteed if $\|f\|_1 \cdot 2^{d(\rho+\lambda)} \leq 2^{\eta-2}$, where $\|f\|_1$ is the ℓ_1-norm of the coefficient vector of f.

So, we can see that if $\|f\|_1 \leq 2^{\eta/2}$ and $\eta \geq 2d(\rho + \lambda) + 4$, then the correct decryption is guaranteed for $\tilde{c}$. Let $\bar{d}$ be the parameter representing the maximum degree for our admissible functions. Then, as long as the condition $\eta \geq 2\bar{d}(\rho + \lambda) + 4$ is met, we may define an admissible function as an arithmetic circuit f such that $\deg f \leq \bar{d}$ and $\|f\|_1 \leq 2^{\eta/2}$ as a polynomial.

5.2   Constraints of the Parameters

In our scheme, the parameters must satisfy the following constraints:

- $\rho = \omega(\lg \lambda)$: to resist the brute force attack on the EF-AGCD problem.
- $\eta \geq 2\bar{d}(\rho + \lambda) + 4$: for the correctness.
- $\eta \geq \Omega(\lambda^2)$: to resist the factoring attack using the elliptic curve method (ECM). In fact, we also want $\eta \geq \lambda^2 + 1$ to make y_0 a 2^{λ^2}-rough integer.
- $\gamma = \eta^2\omega(\lg \lambda)$: to resist known attacks on the EF-AGCD problem as explained in [17,13].
- $2 \leq Q \leq 2^\lambda$: to ensure that $\gcd(Q, y_0) = 1$.

Assuming $\bar{d} = \Theta(\lambda)$, one possible choice of parameters which satisfies all of above is $\rho = \Theta(\lambda)$, $\eta = \Theta(\lambda^2)$, and $\gamma = \Theta(\lambda^5)$.

5.3   Ciphertext Constant Testability

Theorem 5. *The scheme Π satisfies CCT.*

Proof. Let ek be an evaluation key generated by $\mathsf{Gen}(1^\lambda)$, f be any admissible arity-l arithmetic circuit, and $(c_i)_{i \in I}$ be any element in $\mathbb{Z}_{y_0}^{|I|}$ for a subset I of $\{1, \cdots, l\}$.

The constant tester D for our scheme Π determines if $\tilde{e} := \mathsf{Eval}(f, (c_i)_{i \in I})$ is constant or not with overwhelming probability, as follows: given ek, $(f, \tau_1, \ldots, \tau_l)$, I, and $(c_i)_{i \in I}$, the tester D outputs 1 if $I = \{1, \ldots, l\}$. Otherwise, it samples two tuples of ciphertexts $(c_j^0)_{j \notin I}, (c_j^1)_{j \notin I} \xleftarrow{\$} (\mathbb{Z}_{y_0})^{l-|I|}$. Finally, if $\tilde{e}(c_j^0)_{j \notin I} \equiv \tilde{e}(c_j^1)_{j \notin I}$ (mod y_0), then D outputs 1, and otherwise D outputs 0.

The tester D is essentially doing the usual polynomial identity testing. In the scheme Π, $\tilde{e}$ can be considered as an $(l - |I|)$-variate polynomial over $\mathbb{Z}_{y_0}$ of degree $\leq \deg f$. We have

$$\tilde{e}(c_j)_{j \notin I} = f((c_i)_{i \in I}, (c_j)_{j \notin I}) = f(c_1, \ldots, c_l) \bmod y_0 .$$

When $I = \{1, \cdots, l\}$, $\tilde{e}$ is clearly constant and D outputs 1 correctly. In case $I \neq \{1, \cdots, l\}$, consider the function $\tilde{e}' := \tilde{e} - \tilde{e}(c_j^0)_{j \notin I} \bmod y_0$ for $(c_j^0)_{j \notin I} \in (\mathbb{Z}_{y_0})^{l-|I|}$. If $\tilde{e}$ is constant, then $\tilde{e}'$ is constantly zero and $\tilde{e}'(c_j^1)_{j \notin I} = \tilde{e}(c_j^1)_{j \notin I} - \tilde{e}(c_j^0)_{j \notin I} \equiv 0 \pmod{y_0}$ for any $(c_j^1)_{j \notin I} \in (\mathbb{Z}_{y_0})^{l-|I|}$. So, $\tilde{e}(c_j^0)_{j \notin I} \equiv \tilde{e}(c_j^1)_{j \notin I} \pmod{y_0}$ and D outputs 1 correctly. If $\tilde{e}$ is not constant, then $\tilde{e}'$ is not constantly zero and D outputs the incorrect answer 1 when $\tilde{e}(c_j^0)_{j \notin I} \equiv \tilde{e}(c_j^1)_{j \notin I} \pmod{y_0}$, that is, $\tilde{e}'(c_j^1)_{j \notin I} \equiv 0 \pmod{y_0}$. This is the only case when D is incorrect. So the error probability of the tester D is

$$\Pr\left[\tilde{e}'(c_j^1)_{j \notin I} \equiv 0 \bmod y_0 \mid (c_j^1)_{j \notin I} \xleftarrow{\$} (\mathbb{Z}_{y_0})^{l-|I|}\right] ,$$

when $\tilde{e}'$ is not constantly zero.

Since y_0 is chosen as a 2^{λ^2}-rough random integer, with negligible exception, y_0 is square-free and $\tilde{e}'$ is not constantly zero modulo a prime factor $p' \geq 2^{\lambda^2}$ of y_0. Then, using Schwartz-Zippel lemma,

$$\Pr\left[\tilde{e}'(c_j^1)_{j \notin I} \equiv 0 \bmod y_0 \mid (c_j^1)_{j \notin I} \xleftarrow{\$} (\mathbb{Z}_{y_0})^{l-|I|}\right]$$
$$\leq \Pr\left[\tilde{e}'(c_j^1)_{j \notin I} \equiv 0 \bmod p' \mid (c_j^1)_{j \notin I} \xleftarrow{\$} (\mathbb{Z}_{p'})^{l-|I|}\right]$$
$$\leq \frac{\deg f}{p'} \leq \frac{\bar{d}}{2^{\lambda^2}} = negl(\lambda) .$$

Therefore, the error probability of the tester D is negligible.

6   Security

In this section, we prove our HAE scheme satisfies both IND-CPA and SUF-CPA. From this, we conclude that Π is IND-CCA and SUF-CCA by Theorem 3 and Theorem 4. For simplicity, we consider the scheme Π as an ideal scheme obtained by replacing the PRF F with a real random function from $\{0,1\}^\lambda$ into $\mathbb{Z}_{q_0}$. If F is secure, then the real scheme is secure if the ideal scheme is.

6.1   Privacy

As mentioned earlier, Coron et al. proved the equivalence of the EF-AGCD and the decisional EF-AGCD in [14]. So, Theorem 6 actually says that Π is IND-CPA under the EF-AGCD assumption.

Theorem 6. *The scheme Π is IND-CPA under the decisional (ρ, η, γ)-EF-AGCD assumption.*

Proof. We prove this theorem by a hybrid argument to transform the game IND-CPA into another game that is infeasible to break.

Let A be a PPT adversary engaging in the game IND-CPA. Without loss of generality, we assume that A makes exactly $q = q(\lambda)$ encryption queries. For each $i \in \{0, \ldots, q\}$, define IND-CPA^i to be the game that is equal to IND-CPA except that the first i encryption queries are answered by a sample from the uniform distribution over the ciphertext space $\mathbb{Z}_{y_0}$.

By definition, IND-CPA^0 = IND-CPA. So,

$$\mathbf{Adv}_{\Pi,A}^{\text{IND-CPA}^0}(\lambda) = \mathbf{Adv}_{\Pi,A}^{\text{IND-CPA}}(\lambda) ,$$

And the game IND-CPA^q does not reveal any information about the randomly chosen bit b. So,

$$\mathbf{Adv}_{\Pi,A}^{\text{IND-CPA}^q}(\lambda) = 0 .$$

Now consider the difference of each consecutive two games. We want to show that for each $i \in \{1, \ldots, q\}$, the difference between $\mathbf{Adv}_{\Pi,A}^{\text{IND-CPA}^{i-1}}(\lambda)$ and $\mathbf{Adv}_{\Pi,A}^{\text{IND-CPA}^i}(\lambda)$ is not greater than the advantage for the the decisional (ρ, η, γ)-EF-AGCD problem. For this purpose, we construct a PPT distinguisher $D(1^\lambda, y_0, \mathcal{D}(p, q_0, \rho), z)$ for the decisional (ρ, η, γ)-EF-AGCD problem as follows: D starts the simulation of the game IND-CPA$_{\Pi,A}^{i-1}$ or IND-CPA$_{\Pi,A}^i$ giving y_0 as an evaluation key to A. And $b \xleftarrow{\$} \{0,1\}$. Let $(\tau, m_0, m_1) \in \{0,1\}^\lambda \times \mathbb{Z}_Q \times \mathbb{Z}_Q$ be the j-th encryption query of A. Then D replies A with $c := (xQ + m_b) \bmod y_0$, where x is chosen as below.

$$j \leq i - 1 \implies x \xleftarrow{\$} \mathbb{Z}_{y_0} ,$$
$$j = i \implies x = z ,$$
$$j \geq i + 1 \implies x \leftarrow \mathcal{D}(p, q_0, \rho) .$$

Finally, D returns b', which is the output of A.

Note that $\gcd(y_0, Q) = 1$ since y_0 is 2^{λ^2}-rough and $Q \leq 2^\lambda$. Consider the answer $c = (xQ + m_b) \bmod y_0$ produced by D for an encryption query. If $x \xleftarrow{\$} \mathbb{Z}_{y_0}$, then c is also uniformly distributed over y_0. And if $x \leftarrow \mathcal{D}(p, q_0, \rho)$, then the distribution of c is identical to the distribution of $\text{Enc}(sk, \tau, m_b)$. Therefore, if $z \xleftarrow{\$} \mathbb{Z}_{y_0}$, then D simulates the game IND-CPA^i. And if $z \leftarrow \mathcal{D}(p, q_0, \rho)$, then D simulates the game IND-CPA^{i-1}. So, the difference between $\mathbf{Adv}_{\Pi,A}^{\text{IND-CPA}^{i-1}}(\lambda)$ and $\mathbf{Adv}_{\Pi,A}^{\text{IND-CPA}^i}(\lambda)$ is not greater than the advantage of D for the the decisional (ρ, η, γ)-EF-AGCD problem, which is negligible by the decisional (ρ, η, γ)-EF-AGCD assumption. That is,

$$\left| \mathbf{Adv}_{\Pi,A}^{\text{IND-CPA}^{i-1}}(\lambda) - \mathbf{Adv}_{\Pi,A}^{\text{IND-CPA}^i}(\lambda) \right| = negl(\lambda) ,$$

for any $i \in \{1, \cdots, q\}$.

Hence,

$$\mathbf{Adv}_{\Pi,A}^{\text{IND-CPA}}(\lambda) \leq \mathbf{Adv}_{\Pi,A}^{\text{IND-CPA}^q}(\lambda) + \sum_{i=1}^{q} \left| \mathbf{Adv}_{\Pi,A}^{\text{IND-CPA}^{i-1}}(\lambda) - \mathbf{Adv}_{\Pi,A}^{\text{IND-CPA}^i}(\lambda) \right|$$

$$\leq 0 + q \cdot negl(\lambda)$$

$$= negl(\lambda) .$$

Consequently, $\mathbf{Adv}_{\Pi,A}^{\text{IND-CPA}}(\lambda)$ is negiligible for any PPT adversary A and Π is IND-CPA.

6.2   Authenticity

Theorem 7. *If the (ρ, η, γ)-EF-AGCD assumption holds, then the scheme Π is SUF-CPA.*

Proof. Suppose there exists a PPT adversary A for the game SUF-CPA such that
$$\Pr\left[\text{SUF-CPA}_{\Pi,A}(1^\lambda) = 1\right] \geq \epsilon(\lambda) \ ,$$
for some non-negligible function $\epsilon > 0$.

Then, we construct a PPT solver $B(1^\lambda, y_0, \mathcal{D}(p, q_0, \rho))$ for the (ρ, η, γ)-EF-AGCD problem as follows: B starts the simulation of the game SUF-CPA$_{\Pi,A}$ giving y_0 as an evaluation key to A. For an encryption query $(\tau, m) \in \{0,1\}^\lambda \times \mathbb{Z}_Q$ of A, B replies A with $c := (xQ + m) \bmod y_0$, where $x \leftarrow \mathcal{D}(p, q_0, \rho)$. Eventually, A outputs a forgery attempt $((f, \tau_1, \cdots, \tau_l), \hat{c})$. Let I be the set of $i \in \{1, \ldots, l\}$ where τ_i is used, and for each $i = 1, \ldots, l$, choose $c_i \xleftarrow{\$} \mathbb{Z}_{y_0}$ if $i \notin I$, and let c_i be the unique ciphertext returned by the encryption query involving τ_i if $i \in I$. Now B computes $\tilde{c} = f(c_1, \cdots, c_l) \bmod y_0$, and outputs $y_0/\gcd(y_0, \tilde{c} - \hat{c})$.

For the similar reason as in Theorem 6, the simulation of the encryption oracle by B is exact.

Consider the forgery attempt $((f, \tau_1, \cdots, \tau_l), \hat{c})$ made by A. If it is a type 1 strong forgery, then $\tilde{e} := \text{Eval}(f, (c_i)_{i \in I})$ is not constant. Since y_0 is chosen as a 2^{λ^2}-rough random integer, with negligible exception, y_0 is square-free and $\tilde{e}$ is not constantly $\hat{c}$ modulo a prime factor $p' \geq 2^{\lambda^2}$ of y_0. So, using Schwartz-Zippel lemma,

$$\Pr\left[\tilde{e}(c_j)_{j \notin I} \equiv \hat{c} \bmod y_0 \,|(c_j)_{j \notin I} \xleftarrow{\$} (\mathbb{Z}_{y_0})^{l-|I|}\right]$$
$$\leq \Pr\left[\tilde{e}(c_j)_{j \notin I} \equiv \hat{c} \bmod p' \,|\, (c_j)_{j \notin I} \xleftarrow{\$} (\mathbb{Z}_{p'})^{l-|I|}\right]$$
$$\leq \frac{\deg f}{p'} \leq \frac{\bar{d}}{2^{\lambda^2}} = negl(\lambda) \ .$$

This means that $\tilde{c} = \tilde{e}(c_j)_{j \notin I} \not\equiv \hat{c} \bmod y_0$ with overwhelming probability. If $((f, \tau_1, \cdots, \tau_l), \hat{c})$ is a type 2 strong forgery, then again we have $\tilde{e}(c_j)_{j \notin I} = \tilde{c} \not\equiv \hat{c} \bmod y_0$.

Hence in both cases, we have $\tilde{c} \not\equiv \hat{c} \bmod y_0$, but also $\tilde{c} \equiv \hat{c} \bmod q_0$, since any strong forgery is valid. Therefore, $\gcd(y_0, \hat{c} - \tilde{c}) = q_0$ and the output of B is exactly p with overwhelming probability if the forgery attempt of A is a successful strong forgery. Since A succeeds with non-negligible probability, B outputs the correct answer p with non-negligible probability.

Acknowledgments. We thank the anonymous reviewers for their careful reading of our paper and helpful comments.

This research was supported by Basic Science Research Program through the National Research Foundation of Korea (NRF) funded by the Ministry of Education (No. 2011-0025127) and was also supported by the year of 2010 Research Fund of the UNIST (Ulsan National Institute of Science and Technology).

References

1. Bellare, M., Desai, A., Jokipii, E., Rogaway, P.: A concrete security treatment of symmetric encryption. In: Proceedings of the 38th Annual Symposium on Foundations of Computer Science, pp. 394–403 (1997)
2. Bellare, M., Goldreich, O., Mityagin, A.: The power of verification queries in message authentication and authenticated encryption. Cryptology ePrint Archive, Report 2004/309 (2004)
3. Bellare, M., Hofheinz, D., Kiltz, E.: Subtleties in the definition of IND-CCA: When and how should challenge decryption be disallowed? J. Cryptol (2013)
4. Bellare, M., Namprempre, C.: Authenticated encryption: Relations among notions and analysis of the generic composition paradigm. J. Cryptol. 21(4), 469–491 (2008)
5. Boneh, D., Freeman, D.M.: Homomorphic signatures for polynomial functions. In: Paterson, K.G. (ed.) EUROCRYPT 2011. LNCS, vol. 6632, pp. 149–168. Springer, Heidelberg (2011)
6. Brakerski, Z.: Fully homomorphic encryption without modulus switching from classical GapSVP. In: Safavi-Naini, R., Canetti, R. (eds.) CRYPTO 2012. LNCS, vol. 7417, pp. 868–886. Springer, Heidelberg (2012)
7. Brakerski, Z., Gentry, C., Halevi, S.: Packed ciphertexts in LWE-based homomorphic encryption. In: Kurosawa, K., Hanaoka, G. (eds.) PKC 2013. LNCS, vol. 7778, pp. 1–13. Springer, Heidelberg (2013)
8. Brakerski, Z., Gentry, C., Vaikuntanathan, V. (Leveled) fully homomorphic encryption without bootstrapping. In: Proceedings of the 3rd Innovations in Theoretical Computer Science Conference, ITCS 2012, pp. 309–325. ACM (2012)
9. Brakerski, Z., Vaikuntanathan, V.: Efficient fully homomorphic encryption from (standard) LWE. In: 2011 IEEE 52nd Annual Symposium on Foundations of Computer Science (FOCS), pp. 97–106. IEEE Computer Society CPS (2011)
10. Brakerski, Z., Vaikuntanathan, V.: Fully homomorphic encryption from ring-LWE and security for key dependent messages. In: Rogaway, P. (ed.) CRYPTO 2011. LNCS, vol. 6841, pp. 505–524. Springer, Heidelberg (2011)
11. Catalano, D., Fiore, D.: Practical homomorphic MACs for arithmetic circuits. In: Johansson, T., Nguyen, P.Q. (eds.) EUROCRYPT 2013. LNCS, vol. 7881, pp. 336–352. Springer, Heidelberg (2013)
12. Catalano, D., Fiore, D., Warinschi, B.: Homomorphic signatures with efficient verification for polynomial functions. In: Garay, J.A., Gennaro, R. (eds.) CRYPTO 2014, Part I. LNCS, vol. 8616, pp. 371–389. Springer, Heidelberg (2014)
13. Cheon, J.H., Coron, J.S., Kim, J., Lee, M.S., Lepoint, T., Tibouchi, M., Yun, A.: Batch fully homomorphic encryption over the integers. In: Johansson, T., Nguyen, P.Q. (eds.) EUROCRYPT 2013. LNCS, vol. 7881, pp. 315–335. Springer, Heidelberg (2013)
14. Coron, J.S., Lepoint, T., Tibouchi, M.: Scale-invariant fully homomorphic encryption over the integers. In: Krawczyk, H. (ed.) PKC 2014. LNCS, vol. 8383, pp. 311–328. Springer, Heidelberg (2014)
15. Coron, J.S., Mandal, A., Naccache, D., Tibouchi, M.: Fully homomorphic encryption over the integers with shorter public keys. In: Rogaway, P. (ed.) CRYPTO 2011. LNCS, vol. 6841, pp. 487–504. Springer, Heidelberg (2011)
16. Coron, J.S., Naccache, D., Tibouchi, M.: Public key compression and modulus switching for fully homomorphic encryption over the integers. In: Pointcheval, D., Johansson, T. (eds.) EUROCRYPT 2012. LNCS, vol. 7237, pp. 446–464. Springer, Heidelberg (2012)

17. van Dijk, M., Gentry, C., Halevi, S., Vaikuntanathan, V.: Fully homomorphic encryption over the integers. In: Gilbert, H. (ed.) EUROCRYPT 2010. LNCS, vol. 6110, pp. 24–43. Springer, Heidelberg (2010)

18. Dworkin, M.: Recommendation for block cipher modes of operation: Galois/Counter Mode (GCM) for confidentiality and authentication. Special Publication 800-38D, NIST, pp. 800–838 (2007)

19. Gennaro, R., Wichs, D.: Fully homomorphic message authenticators. In: Sako, K., Sarkar, P. (eds.) ASIACRYPT 2013, Part II. LNCS, vol. 8270, pp. 301–320. Springer, Heidelberg (2013)

20. Gentry, C.: Fully homomorphic encryption using ideal lattices. In: Proceedings of the 41st Annual ACM Symposium on Theory of Computing, STOC 2009, pp. 169–178. ACM (2009)

21. Gentry, C., Halevi, S.: Implementing gentry's fully-homomorphic encryption scheme. In: Paterson, K.G. (ed.) EUROCRYPT 2011. LNCS, vol. 6632, pp. 129–148. Springer, Heidelberg (2011)

22. Gentry, C., Halevi, S., Smart, N.P.: Fully homomorphic encryption with polylog overhead. In: Pointcheval, D., Johansson, T. (eds.) EUROCRYPT 2012. LNCS, vol. 7237, pp. 465–482. Springer, Heidelberg (2012)

23. Gentry, C., Halevi, S., Smart, N.P.: Homomorphic evaluation of the AES circuit. In: Safavi-Naini, R., Canetti, R. (eds.) CRYPTO 2012. LNCS, vol. 7417, pp. 850–867. Springer, Heidelberg (2012)

24. Gorbunov, S., Vaikuntanathan, V.: (Leveled) fully homomorphic signatures from lattices. Cryptology ePrint Archive, Report 2014/463 (2014)

25. Krovetz, T., Rogaway, P.: The software performance of authenticated-encryption modes. In: Joux, A. (ed.) FSE 2011. LNCS, vol. 6733, pp. 306–327. Springer, Heidelberg (2011)

26. Rogaway, P., Bellare, M., Black, J.: OCB: A block-cipher mode of operation for efficient authenticated encryption. ACM Trans. Inf. Syst. Secur. 6(3), 365–403 (2003)

27. Smart, N.P., Vercauteren, F.: Fully homomorphic encryption with relatively small key and ciphertext sizes. In: Nguyen, P.Q., Pointcheval, D. (eds.) PKC 2010. LNCS, vol. 6056, pp. 420–443. Springer, Heidelberg (2010)

28. Wichs, D.: Leveled fully homomorphic signatures from standard lattices. Cryptology ePrint Archive, Report 2014/451 (2014)

Authenticating Computation on Groups: New Homomorphic Primitives and Applications[*]

Dario Catalano[1], Antonio Marcedone[2,**], and Orazio Puglisi[1]

[1] Dipartimento di Matematica ed Informatica, Università di Catania, Italy
{catalano,opuglisi}@dmi.unict.it
[2] Cornell University, USA
and
Scuola Superiore di Catania, Università di Catania, Italy
marcedone@cs.cornell.edu

Abstract. In this paper we introduce new primitives to authenticate computation on data expressed as elements in (cryptographic) groups. As for the case of homomorphic authenticators, our primitives allow to verify the correctness of the computation *without* having to know of the original data set. More precisely, our contributions are two-fold.

First, we introduce the notion of *linearly homomorphic authenticated encryption with public verifiability* and show how to instantiate this primitive (in the random oracle model) to support Paillier's ciphertexts. This immediately yields a very simple and efficient (publicly) verifiable computation mechanism for encrypted (outsourced) data based on Paillier's cryptosystem.

As a second result, we show how to construct linearly homomorphic signature schemes to sign elements in bilinear groups (LHSG for short). Such type of signatures are very similar to (linearly homomorphic) structure preserving ones, but they allow for more flexibility, as the signature is explicitly allowed to contain components which *are not* group elements. In this sense our contributions are as follows. First we show a very simple construction of LHSG that is secure against weak random message attack (RMA). Next we give evidence that RMA secure LHSG are interesting on their own right by showing applications in the context of on-line/off-line homomorphic and network coding signatures. This notably provides what seems to be the first instantiations of homomorphic signatures achieving on-line/off-line efficiency trade-offs. Finally, we present a generic transform that converts RMA-secure LHSG into ones that achieve full security guarantees.

1 Introduction

Homomorphic signatures allow to validate computation over authenticated data. More precisely, a signer holding a dataset $\{m_i\}_{i=1,...,t}$ can produce corresponding

---

[*] Extended Abstract. The full version of this paper is available at https://eprint.iacr.org/2013/801.pdf

[**] Part of the work done while visiting Aarhus University.

P. Sarkar and T. Iwata (Eds.): ASIACRYPT 2014, PART II, LNCS 8874, pp. 193–212, 2014.

signatures $\sigma_i = \mathbf{Sign}(\text{sk}, m_i)$ and store the signed dataset on a remote server. Later the server can (publicly) compute $m = f(m_1, \ldots, m_t)$ together with a (succinct) valid signature σ on it. A keynote feature of homomorphic signature is that the validity of σ can be verified *without* needing to know the original messages $m_1, \ldots, m_t$. Because of this flexibility homomorphic signatures have been investigated in several settings and flavors. Examples include homomorphic signatures for linear and polynomial functions [9,8], redactable signatures [26], transitive signatures and more [32,36]. In spite of this popularity, very few realizations of the primitive encompass the very natural case where the computation one wants to authenticate involves elements belonging to typical cryptographic groups (such as, for instance, groups of points over certain classes of elliptic curves, or groups of residues modulo a composite integer).

Our Contribution. In this paper we put forward new tools that allow to authenticate computation on elements in (cryptographic) groups. In this sense our contributions are two-fold. First, we define a new primitive that we call *Linearly Homomorphic Authenticated Encryption with Public Verifiability* (LAEPuV for short). Informally, this primitive allows to authenticate computation on (outsourced) encrypted data, with the additional benefit that the correctness of the computation can be publicly verified. The natural application of this primitive is the increasingly relevant scenario where a user wants to store (encrypted) data on the cloud in a way such that she can later delegate the cloud to perform computation on this data. As a motivating example, imagine that a teacher wants to use the cloud to store the grades of the homeworks of her students. To do so she can create a file identifier fid for each class (e.g. *Cryptography - Spring 2014*), sign each record tied with the corresponding fid and store everything offline. There are two problems with this solution. First, if the teacher wants to compute statistics (e.g. average grades) on these records she has to download all the data locally. Second, since data is stored in clear, outsourcing it offline might violate the privacy of students. *LAEPuV* solves both these issues, as it allows to delegate (basic) computations (i.e. linear functions) on encrypted data in a reliable and efficient way. In particular, it allows to verify the correctness of the computation *without* needing to download the original ciphertexts locally. Moreover, as for the case of homomorphic signatures, correctness of the computation can be publicly verified via a succinct tag whose size is *independent* of the size of the outsourced dataset.

We show an (efficient) realization of the primitive (in the random oracle model) supporting Paillier's ciphertexts. At an intuitive level our construction works by combining Paillier's encryption scheme with some appropriate additively homomorphic signature scheme. Slightly more in detail, the idea is as follows. One first decrypts a "masking" of the ciphertext C and then signs the masked plaintext using the linearly homomorphic signature. Thus we use the homomorphic signature to authenticate computations on ciphertexts by basically authenticating (similar) computations on the masked plaintexts. The additional advantage of this approach is that it allows to authenticate computation on Paillier's ciphertexts while preserving the possibility to re-randomize the ciphertexts.

This means, in particular, that our scheme allows to authenticate computation also on randomized versions of the original ciphertexts[1].

This result allows to implement a very simple and efficient (publicly) verifiable computation mechanism for encrypted (outsourced) data based on Paillier's cryptosystem [34]. Previous (efficient) solutions for this problem rely on linearly homomorphic structure preserving signatures (LHSPS, for short) [30] and, as such, only supported cryptosystems defined over pairing-friendly groups. Since, no (linearly homomorphic) encryption scheme supporting exponentially large message spaces is known to exist in such groups, our construction appears to be the first one achieving this level of flexibility.

Beyond this efficiency gain, we stress that our approach departs from the LHSPS-based one also from a methodological point of view. Indeed, the latter authenticates computation by signing outsourced ciphertexts, whereas we sign (masked versions of) the corresponding plaintexts. This is essentially what buy us the possibility of relying on basic linearly homomorphic signatures, rather than on, seemingly more complicate, structure preserving ones. On the negative side, our solutions require the random oracle, whereas the only known LHSPS-based construction works in the standard model.

As additional byproduct of this gained flexibility, we show how to generalize our results to encompass larger classes of encryption primitives. In particular, we show that our techniques can be adapted to work using any encryption scheme, with some well defined homomorphic properties, as underlying encryption primitive. Interestingly, this includes many well known linearly homomorphic encryption schemes such as [25,33,29].

Signing Elements in Bilinear Groups. As a second main contribution of this paper, we show how to construct a very simple linearly homomorphic signature scheme to sign elements in bilinear groups (LHSG for short). Such type of signatures are very similar to (linearly homomorphic) structure preserving ones, but they allow for more flexibility, as the signature is explicitly allowed to contain components which *are not* group elements (and thus signatures are not necessarily required to comply with the Groth-Sahai famework). More in detail, our scheme is proven secure against random message attack (RMA)[2] under a variant of the Computational Diffie-Hellman assumption introduced by Kunz-Jacques and Pointcheval in [28]. In this sense, our construction is less general (but also conceptually simpler) than the linearly homomorphic structure preserving signature recently given in [30]. Also, the scheme from [30] allows to sign vectors of arbitrary dimension, while ours supports vectors composed by one single component only.

---

[1] We stress however that this does not buy us privacy with respect to the functionality, i.e. the derived (authenticated) ciphertexts are not necessarily indistinguishable from freshly generated (authenticated) ones.

[2] Specifically, by random message security here we mean that the unforgeability guarantee holds only with respect to adversaries that are allowed to see signatures corresponding to messages randomly chosen by the signer.

Interestingly, we show that this simple tool has useful applications in the context of *on-line/off-line* (homomorphic) signatures. Very informally, on-line/off-line signatures allow to split the cost of signing in two phases. An (expensive) offline phase that can be carried out *without* needing to know the message m to be signed and a much more efficient on-line phase that is done once m becomes available. In this sense, on-line/off-line homomorphic signature could bring similar efficiency benefits to protocols relying on homomorphic signatures. For instance, they could be used to improve the overall efficiency of linear network coding routing mechanisms employing homomorphic signatures to fight pollution attacks[3].

We show that RMA-secure LHSG naturally fit this more demanding on-line/off-line scenario. Specifically, we prove that if one combines a RMA-secure LHSG with (vector) Σ protocols with some specific homomorphic properties, one gets a fully fledged linearly homomorphic signature achieving a very efficient on-line phase. Moreover, since the resulting signature scheme supports vectors of arbitrary dimensions as underlying message space, our results readily generalize to the case of network coding signatures [7]. More concretely, by combining our RMA-secure scheme together with (a variant of) Schnorr's identification protocol we get what seem to be the first constructions of secure homomorphic and network coding signatures offering online/offline efficiency tradeoffs both for the message and the file identifier.

To complete the picture, we provide an efficient and *generic* methodology to convert RMA-secure LHSG into ones that achieve full security We stress that while similar transforms were known for structure preserving signatures (e.g. [17]), to our knowledge this is the first such transform for the case of *linearly homomorphic* signatures in general.

Other Related Work. Authenticated Encryption (AE) allows to simultaneously achieve privacy and authentication. In fact AE is considered to be the standard for symmetric encryption, and many useful applications are based on this primitive. Formal definitions for (basic) AE where provided by Bellare and Namprempre in [6]. More closely related to our setting is the notion of homomorphic authenticated encryption recently proposed by Joo and Yun in [27]. With respect to ours, their definitions encompass a wider class of functionalities, but do not consider public verifiability.

COMPUTATIONALLY SOUND PROOFS. In the random oracle model, the problem of computing reliably on (outsourced) encrypted data can be solved in principle using Computationally Sound (CS) proofs [31]. The advantage of this solution, with respect to ours, is that it supports arbitrary functionalities. On the other hand, it is much less efficient as it requires the full machinery of the PCP theorem. Moreover, composition in CS proofs is quite complicate to achieve, whereas

---

[3] This is because the sender could preprocess many off-line computations at night or when the network traffic is low and then use the efficient online signing procedure to perform better when the traffic is high.

it comes for free in our solution, as the outputs of previous computations can always be used as inputs for new ones.

LINEARLY HOMOMORPHIC SIGNATURES. The concept of homomorphic signature scheme was originally introduced in 1992 by Desmedt [18], and then refined by Johnson, Molnar, Song, Wagner in 2002 [26]. Linearly homomorphic signatures were introduced in 2009 by Boneh *et al.* [7] as a way to prevent pollution attacks in network coding. Following [7] many other works further explored the notion of homomorphic signatures by proposing new frameworks and realizations [23,3,9,8,14,4,15,21,5,13,16]. In the symmetric setting constructions of homomorphic message authentication codes have been proposed by [7,24,11,12].

Recently Libert *et al.* [30] introduced and realized the notion of *Linearly Homomorphic Structure Preserving signatures* (LHSPS for short). Informally LHSPS are like ordinary SPS but they come equipped with a linearly homomorphic property that makes them interesting even beyond their usage within the Groth Sahai framework. In particular Libert *et al.* showed that LHSPS can be used to enable simple verifiable computation mechanisms on encrypted data. More surprisingly, they observed that linearly homomorphic SPS (generically) yield efficient simulation sound trapdoor commitment schemes [22], which in turn imply non malleable trapdoor commitments [19] to group elements.

ON-LINE/OFF-LINE SIGNATURES. On-Line/Off-Line digital signature were introduced by Even, Goldreich and Micali in [20]. In such schemes the signature process consists of two parts: a computationally intensive one that can be done Off-Line (i.e. when the message to be signed is not known) and a much more efficient online phase that is done once the message becomes available. There are two general ways to construct on-line/off-line signatures: using one time signatures [20] or using chameleon hash [35].

In [10] Catalano *et al.*, unified the two approaches by showing that they can be seen as different instantiations of the same paradigm.

2   Preliminaries and Notation

We denote with $\mathbb{Z}$ the set of integers, with $\mathbb{Z}_p$ the set of integers modulo p. An algorithm $\mathcal{A}$ is said to be PPT if it's modelled as a probabilistic Turing machine that runs in polynomial time in its inputs. If S is a set, then $x \xleftarrow{\$} S$ denotes the process of selecting one element x from S uniformly at random. A function f is said to be negligible if for all polynomial p there exists $n_0 \in \mathbb{N}$ such that for each $n > n_0$, $|f(n)| < \frac{1}{p(n)}$.

Computational Assumptions. We start by recalling a couple of relevant computational assumptions. Let $\mathbb{G}$ be a finite (multiplicative) group of prime order p. The 2-out-of-3 Computational Diffie-Hellman assumption was introduced by Kunz-Jacques and Pointcheval in [28] as a relaxation of the standard CDH assumption. It is defined as follows.

Definition 1 (2-3CDH). *We say that the 2-out-of-3 Computational Diffie-Hellmann assumption holds in* $\mathbb{G}$ *if, given a random generator* $g \in \mathbb{G}$, *there exists no PPT* $\mathcal{A}$ *that on input* (g, g^a, g^b) *(for random* $a, b \xleftarrow{\$} \mathbb{Z}_p$*) outputs* h, h^{ab} $(h \neq 1)$ *with more than negligible probability.*

Also, we recall the Decisional Composite Residuosity Assumption, introduced by Paillier in [34].

Definition 2 (DCRA). *We say that the Decisional composite residuosity assumption (DCRA) holds if there exists no PPT* $\mathcal{A}$ *that can distinguish between a random element from* $\mathbb{Z}_{N^2}^*$ *and one from the set* $\{z^N | z \in \mathbb{Z}_{N^2}^*\}$ *(i.e. the set of the* N*-th residues modulo* N^2*), when* N *is the product of two random primes of proper size.*

3   (Publicly) Verifiable Delegation of Computation on Outsourced Ciphertext

In this section, we introduce a new primitive that we call Linearly Homomorphic Authenticated Encryption with Public Verifiability (LAEPuV). Informally, this notion is inspired by the concept of homomorphic authenticated encryption, introduced by Joo and Yun [27]. Important differences are that our definition[4] focuses on linear functions and explicitly requires public verifiability.

Next, we provide an instantiation of this primitive supporting Paillier's scheme as the underlying encryption mechanism.

Additionally, in this and the following sections, we adopt the following conventions

- The set $\mathcal{F}$ of supported functionalities, is the set of linear combinations of elements of the group. Thus each function $f \in \mathcal{F}$ can be uniquely expressed as $f(m_1, \ldots, m_k) = \prod_{i=1}^{k} m_i^{\alpha_i}$, and therefore can be identified by a proper vector $(\alpha_1, \ldots, \alpha_k) \in \mathbb{Z}^k$.
- We identify each dataset by a string fid $\in \{0, 1\}^*$, and use an additional argument $i \in \{1, \ldots, n\}$ for the signing/encryption algorithm to specify that the signed/encrypted message can be used only as the i-th argument for each function $f \in \mathcal{F}$.

Definition 3 (LAEPuV). *A LAEPuV scheme is a tuple of 5 PPT algorithms* (**AKeyGen, AEncrypt, ADecrypt, AVerify, AEval**) *such that:*

- **AKeyGen**$(1^\lambda, k)$ *takes as input the security parameter* λ, *and an upper bound* k *for the number of messages encrypted in each dataset. It outputs a secret key sk and a public key vk (used for function evaluation and verification); the public key implicitly defines a message space* $\mathcal{M}$ *which is also a group, a file identifier space* $\mathcal{D}$ *and a ciphertext space* $\mathcal{C}$.

---

[4] For lack of space the formal definition is provided in the full version of this paper.

- **AEncrypt**(sk, fid, i, m) *is a probabilistic algorithm which takes as input the secret key, an element* $m \in \mathcal{M}$, *a dataset identifier* fid, *an index* $i \in \{1, \ldots, k\}$ *and outputs a ciphertext* c.
- **AVerify**(vk, fid, c, f) *takes as input the public key* vk, *a ciphertext* $c \in \mathcal{C}$, *an identifier* fid $\in \mathcal{D}$ *and* $f \in \mathcal{F}$. *It return 1 (accepts) or 0 (rejects).*
- **ADecrypt**(sk, fid, c, f) *takes as input the secret key* sk, *a ciphertext* $c \in \mathcal{C}$, *an identifier* fid $\in \mathcal{D}$ *and* $f \in \mathcal{F}$ *and outputs* $m \in \mathcal{M}$ *or* $\perp$ *(if* c *is not considered valid).*
- **AEval**$(vk, f, fid, \{c_i\}_{i=1\ldots k})$ *takes as input the public key* vk, *an admissible function* f *in its vector form* $(\alpha_1, \ldots, \alpha_k)$, *an identifier* fid, *a set of* k *ciphertexts* $\{c_i\}_{i=1\ldots k}$ *and outputs a ciphertext* $c \in \mathcal{C}$. *Note that this algorithm should also work if less than* k *ciphertexts are provided, as long as their respective coefficients in the function* f *are 0, but we don't explicitly account this to avoid heavy notation.*

The correctness conditions are the following:

- For any $(sk, vk) \leftarrow \mathbf{AKeyGen}(1^\lambda, k)$ honestly generated keypair, any $m \in \mathcal{M}$, any dataset identifier fid and any $i \in \{1, \ldots, k\}$, with overwhelming probability

$$\mathbf{ADecrypt}(sk, fid, \mathbf{AEncrypt}(sk, fid, i, m), e_i) = m$$

where e_i is the i-th vector of the standard basis of $\mathbb{Z}^k$.
- For any $(sk, vk) \leftarrow \mathbf{AKeyGen}(1^\lambda, k)$ honestly generated keypair, any $c \in C$

$$\mathbf{AVerify}(vk, fid, c, f) = 1 \iff \exists m \in \mathcal{M} : \mathbf{ADecrypt}(sk, fid, c, f) = m$$

- Let $(sk, vk) \leftarrow \mathbf{AKeyGen}(1^\lambda, k)$ be an honestly generated keypair, fid any dataset identifier, $c_1, \ldots, c_k \in \mathcal{C}$ any tuple of ciphertexts such that $m_i = \mathbf{ADecrypt}(sk, fid, c_i, f_i)$. Then, for any admissible function $f = (\alpha_1, \ldots, \alpha_k) \in \mathbb{Z}^k$, with overwhelming probability

$$\mathbf{ADecrypt}(sk, fid, \mathbf{AEval}(vk, f, fid, \{c_i\}_{i=1\ldots k}), \sum_{i=0}^{k} \alpha_i f_i) = f(m_1, \ldots, m_k)$$

Security definitions for LAEPuV are easy to derive, so, for lack of space, are omitted. We refer the reader to the full version of this paper.

3.1   An Instantiation Supporting Paillier's Encryption

Let (**HKeyGen, HSign, HVerify, HEval**) be a secure[5] linearly homomorphic signature scheme whose message space is $\mathbb{Z}_N$ (where N is the product of two distinct (safe) primes). Moreover, let $\mathcal{H}$ be a family of collision resistant hash functions (whose images can be interpreted as elements of $\mathbb{Z}^*_{N^2}$). Then we can construct a LAEPuV scheme as follows.

---

[5] Again, for lack of space, security definition for linearly homomorphic signatures is provided in the full version of the paper.

AKeyGen$(1^\lambda, k)$: Choose two primes p, q of size $\lambda/2$, set $N \leftarrow pq$ and choose a random element $g \in \mathbb{Z}^*_{N^2}$ of order N. Run[6] **HKeyGen**$(1^\lambda, k, N)$ to obtain a signing key sk$'$ and a verification key vk$'$. Pick a hash function $H \leftarrow \mathcal{H}$. Return vk $\leftarrow$ (vk$'$, g, N, H) as the public verification key and sk $=$ (sk$'$, p, q) as the secret signing key.

AEncrypt$(\text{sk}, m, \text{fid}, i)$: Choose random $\beta \leftarrow \mathbb{Z}^*_{N^2}$, compute $C \leftarrow g^m \beta^N$ mod N^2. Set $R \leftarrow H(\text{fid}\|i)$, and use the factorization of N to compute $(a, b) \in \mathbb{Z}_N \times \mathbb{Z}^*_N$ such that $g^a b^N = RC$ mod N^2. Compute $\sigma \leftarrow$ **HSign**(sk', fid, i, a) and return $c = (C, a, b, \sigma)$.

AVerify$(\text{vk}, \text{fid}, c, f)$: Parse $c = (C, a, b, \sigma)$ and vk $\leftarrow$ (vk$'$, g, N, H), then check that:

$$\mathbf{HVerify}(\text{vk}', \text{fid}, a, f, \sigma) = 1$$

$$g^a b^N = C \prod_{i=1}^{k} H(\text{fid}\|i)^{f_i} \mod N^2$$

If both the above equations hold output 1, else output 0.

ADecrypt$(\text{sk}, \text{fid}, c, f)$: If **AVerify**$(\text{vk}, \text{fid}, c, f) = 0$, return $\bot$. Otherwise, use the factorization of N to compute (m, β) such that $g^m \beta^N = C$ mod N^2 and return m.

AEval$(\text{vk}, \alpha, \text{fid}, c_1, \ldots, c_k)$: Parse $\alpha = (\alpha_1, \ldots, \alpha_k)$ and $c_i = (C_i, a_i, b_i, \sigma_i)$, set

$$C \leftarrow \prod_{i=i}^{k} C_i^{\alpha_i} \mod N^2, \quad a \leftarrow \sum_{i=i}^{k} a_i \alpha_i \mod N,$$

$$b \leftarrow \prod_{i=i}^{k} b_i^{\alpha_i} \mod N^2, \quad \sigma \leftarrow \mathbf{HEval}(vk', \text{fid}, f, \{\sigma_i\}_{i=1,\ldots,k})$$

and return $c = (C, a, b, \sigma)$.

Remark 1. (**Supporting Datasets of Arbitrary Size**). In the construction above the number k of ciphertexts supported by each dataset needs to be fixed once and for all at setup time. This might be annoying in practical scenarios where more flexibility is preferable. We remark, that in the random oracle model, the scheme can be straightforwardly modified in order to remove this limitation. The idea would be to use the random oracle *also* in the underlying (homomorphic) signature scheme (see the full version of this paper for details) More precisely, rather than publishing the h_i as part of the public key, one computes different h_i's on the fly for each dataset by setting $h_i = H'(fid, i)$(where H' is some appropriate random oracle). Slightly more in detail, the elements from dataset fid are then authenticated by replacing the h_i with $h_{\text{fid},i} = H'(\text{fid}, i)$.

---

[6] Notice that the signature scheme must support $\mathbb{Z}_N$ as underlying message space. This is why we give N to the **HKeyGen** algorithm as additional parameter. Note that, this means that, in general, the signature algorithm cannot not use the factorization of N as part of its private key.

Using this simple trick brings the additional benefit that the public key can be reduced to constant size.

The security of the scheme is provided by the following theorems (whose proofs appear in the full version).

Theorem 1. *Assuming that the DCRA holds, if* (**HKeyGen, HSign, HVerify, HEval**) *is a secure linearly homomorphic signature scheme for messages in* $\mathbb{Z}_N$ *and H is a random oracle, the scheme described above is LH-IND-CCA secure.*

Theorem 2. *If* Σ = (**HKeyGen, HSign, HVerify, HEval**) *is a secure linearly homomorphic signature scheme for messages in* $\mathbb{Z}_N$ *then the scheme described above is LH-Uf-CCA secure.*

Remark 2. (**Instantiating the underlying signature scheme**). As a concrete instantiation of the linearly homomorphic signature scheme (**HKeyGen, HSign, HVerify, HEval**),one can use use a simple variant of the (Strong) RSA based scheme from [15] adapted to use $\mathbb{Z}_N$ as underlying message space, see the full version for details.

3.2   A General Result

In this section we show how to generalize our results to support arbitrary encryption schemes satisfying some well defined homomorphic properties.

In such schemes, the message, randomness and ciphertext spaces are assumed to be finite groups, respectively denoted with $\mathcal{M}, \mathcal{R}, \mathcal{C}$ (the key spaces are treated implicitly). To adhere with the notation used in the previous section, we will denote the operation over $\mathcal{M}$ additively and the ones over $\mathcal{R}$ and $\mathcal{C}$ multiplicatively. We assume $\mathcal{T}$ to be an IND-CPA secure public key encryption scheme satisfying the following additional properties:

- We require the group operation and the inverse of an element to be efficiently computable over all groups, as well as efficient sampling of random elements. The integer linear combinations are thus defined and computed by repeatedly applying these operations.
- For any $m_1, m_2 \in \mathcal{M}, r_1, r_2 \in \mathcal{R}$, any valid public key pk it holds

$$\mathbf{Enc}_{\mathrm{pk}}(m_1, r_1) \cdot \mathbf{Enc}_{\mathrm{pk}}(m_2, r_2) = \mathbf{Enc}_{\mathrm{pk}}(m_1 + m_2, r_1 \cdot r_2)$$

- For any honest key pair (pk, sk) and any $c \in \mathcal{C}$ there exists $m \in \mathcal{M}$ and $r \in \mathcal{R}$ such that $\mathbf{Enc}_{\mathrm{pk}}(m, r) = c$ (i.e. the encryption function is surjective over the group $\mathcal{C}$). Moreover, we assume that such m and r are efficiently computable given the secret key.

Now, let (**HKeyGen, HSign, HVerify, HEval**) be a secure linearly homomorphic signature scheme for elements in $\mathcal{M}$, let $\mathcal{H}$ be a family of collision resistant hash functions $H_K : \{0,1\}^* \to \mathcal{C}$ and let $\mathcal{T} = \{\mathbf{Gen, Enc, Dec}\}$ be an encryption scheme as above. We construct a LAEPuV scheme as follows:

AKeyGen$(1^\lambda, k)$: Run **HKeyGen**$(1^\lambda, k)$ to obtain a signing key sk$'$ and a verification key vk$'$ and **Gen**(1^λ) to obtain a public key $\overline{\text{pk}}$ and a secret key $\overline{\text{sk}}$. Pick a hash function $H \leftarrow \mathcal{H}$. Return vk $\leftarrow$ (vk$'$, $\overline{\text{pk}}$, H) as the public verification key and sk $=$ (sk$'$, $\overline{\text{sk}}$) as the secret key.

AEncrypt$(\text{sk}, m, \text{fid}, i)$: Choose random $r \leftarrow \mathcal{R}$, compute $C \leftarrow \textbf{Enc}_{\overline{\text{pk}}}(m, r)$ and compute, using the secret key sk, $\overline{m}$ and $\overline{r}$ such that $\textbf{Enc}_{\overline{\text{pk}}}(\overline{m}, \overline{r}) = H(\text{fid}||i)$. Compute $\sigma \leftarrow \textbf{HSign}(sk', \text{fid}, i, m + \overline{m})$ and return $c = (C, m + \overline{m}, r \cdot \overline{r}, \sigma)$.

AVerify$(\text{vk}, \text{fid}, c, f)$: Parse $c = (C, a, b, \sigma)$ and vk $\leftarrow$ (vk$'$, $\overline{\text{pk}}$), then check that:

$$\textbf{HVerify}(\text{vk}', \text{fid}, a, f, \sigma) = 1$$

$$\textbf{Enc}_{\overline{\text{pk}}}(a, b) = C \prod_{i=1}^{k} H(\text{fid}||i)^{f_i}$$

If both the above equations hold output 1, else output 0.

ADecrypt$(\text{sk}, \text{fid}, c, f)$: Parse $c = (C, a, b, \sigma)$. If **AVerify**$(\text{vk}, \text{fid}, c, f) = 0$, return $\perp$. Otherwise, use the secret key $\overline{\text{sk}}$ to compute $m \leftarrow \textbf{Dec}_{\overline{\text{sk}}}(C)$

AEval$(\text{vk}, \alpha, \text{fid}, c_1, \ldots, c_k)$: Parse $\alpha = (\alpha_1, \ldots, \alpha_k)$ and $c_i = (C_i, a_i, b_i, \sigma_i)$, set

$$C \leftarrow \prod_{i=i}^{k} C_i^{\alpha_i}, \quad a \leftarrow \sum_{i=i}^{k} a_i \alpha_i,$$

$$b \leftarrow \prod_{i=i}^{k} b_i^{\alpha_i}, \quad \sigma \leftarrow \textbf{HEval}(vk', \text{fid}, f, \{\sigma_i\}_{i=1,\ldots,k})$$

and return $c = (C, a, b, \sigma)$.

Theorem 3. *Assuming $\mathcal{T}$ is an is IND-CPA secure public key encryption scheme satisfying the conditions detailed above,* (**HKeyGen, HSign, HVerify, HEval**) *is a secure linearly homomorphic signature scheme supporting $\mathcal{M}$ as underlying message space and H is a random oracle, then the scheme described above has indistinguishable encryptions.*

Theorem 4. *If $\Sigma = $ (**HKeyGen, HSign, HVerify, HEval**) *is a secure linearly homomorphic signature scheme for messages in $\mathcal{M}$ then the scheme described above is unforgeable.*

Proofs of theorems 3 and 4 are almost identical to the (corresponding) proofs of theorems 1 and 2 and are thus omitted.

4   Linearly Homomorphic Signature Scheme to Sign Elements in Bilinear Groups

Here we introduce the notion of linearly homomorphic signature scheme to sign elements in bilinear groups. This essentially adapts the definition from [21] to support a bilinear group as underlying message space. The formal definition is given in the full version of the paper.

4.1   A Random Message Secure Construction

Let $\mathbb{G}, \mathbb{G}_T$ be groups of prime order p such that $e : \mathbb{G} \times \mathbb{G} \to \mathbb{G}_T$ is a bilinear map and $\mathcal{S} = (\textbf{KeyGen}, \textbf{Sign}, \textbf{Verify})$ a standard signature with message space $\mathcal{M}$. The scheme works as follows:

HKeyGen$(1^\lambda, 1, k)$: Choose a random generator $g \in \mathbb{G}$ and run **KeyGen**(1^λ)
   to obtain a signing key sk_1 and a verification key vk_1. Pick random $w \xleftarrow{\$} \mathbb{Z}_p$
   and set $W \leftarrow g^w$. Select random group elements $h_1, \dots, h_k, \xleftarrow{\$} \mathbb{G}$.
   Set vk $\leftarrow (\text{vk}_1, g, W, h_1, \dots, h_k)$ as the public verification key and sk $=$
   (sk_1, w) as the secret signing key.

HSign$(\text{sk}, m, \text{fid}, i)$: This algorithm stores a list $\mathcal{L}$ of all previously returned
   dataset identifiers fid (together with the related secret information r and
   public information σ, τ defined below) and works as follows

   If fid $\notin \mathcal{L}$, then choose $r \xleftarrow{\$} \mathbb{Z}_p$, set $\sigma \leftarrow g^r$, $\tau \leftarrow \textbf{Sign}(\text{sk}, \text{fid}, \sigma)$

   else if fid $\in \mathcal{L}$, then retrieve the associated r, σ, τ from memory.
   Then set $M \leftarrow m^w, V \leftarrow (h_i M)^r$ (if a signature for the same fid and the
   same index i was already issued, then abort). Finally output $\pi \leftarrow (\sigma, \tau, V, M)$
   as a signature for m w.r.t. the function e_i (where e_i is the i-th vector of the
   canonical basis of $\mathbb{Z}^n$).

HVerify$(\text{vk}, \pi, m, \text{fid}, f)$: Parse the signature π as (σ, τ, V, M) and f as
   $(f_1, \dots, f_k)$. Then check that:

$$\textbf{Verify}(\text{vk}, \tau, (\text{fid}, \sigma)) = 1 \qquad e(M, g) = e(m, W) \qquad e(V, g) = e(\prod_{i=1}^{k} h_i^{f_i} M, \sigma)$$

   If all the above equations hold output 1, else output 0.

HEval $(\text{vk}, \alpha, \pi_1, \dots, \pi_k)$: Parse α as $(\alpha_1, \dots, \alpha_k)$ and π_i as $(\sigma_i, \tau_i, V_i, M_i)$,
   $\forall i = 1, \dots, k$. Then, compute $V \leftarrow \prod_{i=1}^{k} V_i^{\alpha_i}$, $M \leftarrow \prod_{i=1}^{k} M_i^{\alpha_i}$ and output
   $\pi = (\sigma_1, \tau_1, V, M)$ (or $\bot$ if the σ_i are not all equal).

The security of the scheme follows from the following theorem (whose proof is deferred to the full version of this paper).

Theorem 5. *If the 2-3CDH assumption holds and $\mathcal{S}$ is a signature scheme unforgeable under adaptive chosen message attack then the scheme described above is a LHSG scheme secure against a random message attack .*

Remark 3. If the application considered allows the fid to be a group element and not simply a string, we can replace the signature $\mathcal{S}$ with a Structure preserving Signature satisfying the same hypothesis of theorem 5. This allows to obtain the first example of a linearly homomorphic structure preserving signature scheme (LHSPS) where all parts of the signature are actually elements of the group. This is in contrast with the construction from [30], where the fid is inherently used as a bit string. In addition, if the identifier can be chosen at random by the signer and not by the adversary, we can even define σ to be the identifier itself and thus further improve efficiency. In practical instantiation it's possible to use the SPS of [1].

5   Applications to On-Line/Off-Line Homomorphic Signatures

In this section, we show a general construction to build (efficient) on-line/off-line homomorphic (and network coding) signature schemes by combining a LHSG unforgeable against a random message attack (like the one described in section 4.1) with a certain class of sigma protocols. The intuitive idea is that in order to sign a certain message m, one can choose a Σ-Protocol whose challenge space contains m, then sign the first message of the Σ-Protocol with a standard signature (this can be done off-line) and use knowledge of the witness of the protocol to later compute the response (third message) of the protocol associated to the challenge m. This is secure because, if an adversary could produce a second signature with respect to the same first message, by the special soundness of the Σ-Protocol, he would be able to recover the witness itself. We show how, if both the signature scheme and the Σ-Protocol have specific homomorphic properties, this construction can be extended to build (linearly) homomorphic signatures as well.

Informally the properties we require from the underlying sigma protocol are: (1) it is linearly homomorphic, (2) its challenge space can be seen as a vector space and (3) the third message of the protocol can be computed in a very efficient way (as it is used in the online phase of the resulting scheme). In what follows, we first adapt the definition of linearly homomorphic signature (LHSG) to the On-line/Off-line case. Then, we formally define the properties required by the sigma protocol, and we describe (and prove secure) our construction.

Linearly Homomorphic On-Line/Off-Line Signatures. First, we remark that the only difference between a LHSG and a LHOOS is in the signing algorithm. When signing m the latter can use some data prepared in advance (by running a dedicated algorithm **OffSign**) to speed up the signature process. The definitions of unforgeability are therefore analogous to the ones of traditional LHSG schemes and are omitted to avoid repetition[7].

Definition 4 (LHOOS). *A Linearly Homomorphic On-line/Off-line signature scheme is a tuple of PPT algorithms (**KeyGen**, **OffSign**, **OnSign**, **Verify**, **Eval**) such that:*

- **KeyGen**$(1^\lambda, n, k)$ *takes as input the security parameter λ, an integer n denoting the length of vectors to be signed and an upper bound k for the number of messages signed in each dataset. It outputs a secret signing key sk and a public verification key vk; the public key implicitly defines a message space that can be seen as a vector space of the form $\mathcal{M} = \mathbb{F}^n$ (where $\mathbb{F}$ is a field), a file identifier space $\mathcal{D}$ and a signature space Σ.*

---

[7] We stress, however, that those definitions are stronger than the ones traditionally introduced for network coding (i.e. the adversary is more powerful and there are more types of forgeries), and therefore our efficient instantiation perfectly integrates in that framework.

- **OffSign**(sk) *takes as input the secret key and outputs some information I.*
- **OnSign**($sk, fid, I, \mathbf{m}, i$) *takes as input the secret key, an element $\mathbf{m} \in \mathcal{M}$, an index $i \in \{1, \ldots, k\}$, a dataset identifier fid and an instance of I output by **OffSign**. This algorithm must ensure that all the signatures issued for the same fid are computed using the same information I (i.e. by associating each fid with one specific I and storing these couples on a table). It outputs a signature σ.*
- **Verify** ($vk, \sigma, \mathbf{m}, fid, f$) *takes as input the public key vk, a signature $\sigma \in \Sigma$, a message $\mathbf{m} \in \mathcal{M}$, a dataset identifier $fid \in \mathcal{D}$ and a function $f \in \mathbb{Z}^k$; it outputs 1 (accept) or 0 (reject).*
- **Eval**($vk, fid, f, \{\sigma_i\}_{i=1\ldots k}$) *takes as input the public key vk, a dataset identifier fid, an admissible function f in its vector form $(\alpha_1, \ldots, \alpha_k)$, a set of k signatures $\{\sigma_i\}_{i=1\ldots k}$ and outputs a signature $\sigma \in \Sigma$. Note that this algorithm should also work if less than k signatures are provided, as long as their respective coefficients in the function f are 0, but we don't to explicitly account this to avoid heavy notation.*

The correctness conditions of our scheme are the following:

- Let $(sk, vk) \leftarrow \mathbf{KeyGen}(1^\lambda, n, k)$ be an honestly generated keypair, $\mathbf{m} \in \mathcal{M}$, fid any dataset identifier and $i \in 1, \ldots, k$. If $\sigma \leftarrow \mathbf{Sign}(sk, fid, \mathbf{OffSign}(sk), \mathbf{m}, i)$, then with overwhelming probability

$$\mathbf{Verify}(vk, \sigma, \mathbf{m}, fid, e_i) = 1,$$

where e_i is the ith vector of the standard basis of $\mathbb{Z}^k$.
- Let $(sk, vk) \leftarrow \mathbf{KeyGen}(1^\lambda, n, k)$ be an honestly generated keypair, $\mathbf{m}_1, \ldots,$ $\mathbf{m}_k \in \mathcal{M}$ any tuple of messages signed (or derived from messages originally signed) w.r.t the same fid (and therefore using the same offline information I), and let $\sigma_1, \ldots, \sigma_k \in \Sigma$, $f_1, \ldots, f_k \in \mathcal{F}$ such that for all $i \in \{1, \ldots, k\}$, $\mathbf{Verify}(vk, \sigma_i, \mathbf{m}_i, fid, f_i) = 1$. Then, for any admissible function $f = (\alpha_1, \ldots, \alpha_k) \in \mathbb{Z}^k$, with overwhelming probability

$$\mathbf{Verify}(vk, \mathbf{Eval}(vk, fid, f, \{\sigma_i\}_{i=1\ldots k}), f(\mathbf{m}_1, \ldots, \mathbf{m}_k), fid, \sum_{i=0}^{k} \alpha_i f_i) = 1$$

Vector and Homomorphic Σ-Protocols. Informally, a Σ-Protocol can be described as a tuple of four algorithms (Σ-**Setup**, Σ-**Com**, Σ-**Resp**, Σ-**Verify**), where the first one generates a statement/witness couple, Σ-**Com** and Σ-**Resp** generate the first and third message of the protocol, and Σ-**Verify** is used by the verifier to decide on the validity of the proof (a more formal and detailed description is given in the full version of this paper). This notion can be extended to the vector case[8]. For this purpose we adapt the notion of Homomorphic Identification Protocol originally introduced in [2] to the Sigma protocol framework.

---

[8] The intuition is that it should be more efficient to run a vector Σ-Protocol once than a standard Σ-Protocol multiple times in parallel).

Given a language L and an integer $n \in \mathbb{N}$, we can consider the language $L^n = \{(x_1, \ldots, x_n) \mid x_i \in L \ \forall i = 1, \ldots, n\}$. A natural witness for a tuple (vector) in this language is the tuple of the witnesses of each of its components for the language L. As before we can consider the relation $\mathcal{R}^n$ associated to L^n, where $(\boldsymbol{x}, \boldsymbol{w}) = (x_1, \ldots, x_n, w_1, \ldots, w_n) \in \mathcal{R}^n$ if $(x_1, \ldots, x_n)$ is part of L^n and w_i is a witness for x_i. A *vector Σ-Protocol* for $\mathcal{R}^n$ is a three round protocol defined similarly as above with the relaxation that the special soundness property is required to hold in a weaker form. Namely, we require the existence of an efficient extractor algorithm $\boldsymbol{\Sigma_n}$-**Ext** such that $\forall \boldsymbol{x} \in L^n$, $\forall R, \boldsymbol{c}, \boldsymbol{s}, \boldsymbol{c'}, \boldsymbol{s'}$ such that $(c, s) \neq (c', s')$, $\boldsymbol{\Sigma_n}$-**Verify**$(\boldsymbol{x}, R, \boldsymbol{c}, \boldsymbol{s}) = 1$ and $\boldsymbol{\Sigma_n}$-**Verify**$(\boldsymbol{x}, R, \boldsymbol{c'}, \boldsymbol{s'}) = 1$, outputs $(x, w) \leftarrow \boldsymbol{\Sigma_n}$-**Ext**$(\boldsymbol{x}, R, \boldsymbol{c}, \boldsymbol{s}, \boldsymbol{c'}, \boldsymbol{s'})$ where x is one of the components of $\boldsymbol{x}$ and $(x, w) \in \mathcal{R}$.

Another important requirement for our construction to work is the following property.

Definition 5. *A Σ-Protocol $\Sigma = (\Sigma\text{-}\mathbf{Setup}, \Sigma\text{-}\mathbf{Com}, \Sigma\text{-}\mathbf{Resp}, \Sigma\text{-}\mathbf{Verify})$ for a relation $\mathcal{R}$ is called* group homomorphic *if*

- *The outputs of the Σ-**Com** algorithm and the challenge space of the protocol can be seen as elements of two groups $(\mathbb{G}_1, \odot)$ and $(\mathbb{G}_2, \otimes)$ respectively*
- *There exists a PPT algorithm* **Combine** *such that, for all $(x, w) \in \mathcal{R}$ and all $\alpha \in \mathbb{Z}^n$, if transcripts $\{(R_i, c_i, s_i)\}_{i=1,\ldots,n}$ are such that $\Sigma\text{-}\mathbf{Verify}(x, R_i, c_i, s_i) = 1$ for all i, then*

$$\Sigma\text{-}\mathbf{Verify}\left(x, \bigodot_{i=1}^{n} R_i^{\alpha_i}, \bigotimes_{i=1}^{n} c_i^{\alpha_i}, \mathbf{Combine}(x, \alpha, \{(R_i, c_i, s_i)\}_{i=1,\ldots,n})\right) = 1$$

Although it is given for the standard case, this property can easily be extended to vector Σ-Protocols: in particular, the group $\mathbb{G}_2$ can be seen as the group of vectors of elements taken from another group $\mathbb{G}$. To sum up, we define a class of vector Σ-Protocols having all the properties required by our construction:

Definition 6 (1-n (vector) Σ-Protocol). *Let $(\mathbb{G}_1, \odot)$, $(\mathbb{G}_2, \otimes)$ be two computational groups. A 1-n vector sigma protocol consists of four PPT algorithm $\Sigma_n = (\Sigma_n\text{-}\mathbf{Setup}, \Sigma_n\text{-}\mathbf{Com}, \Sigma_n\text{-}\mathbf{Resp}, \Sigma_n\text{-}\mathbf{Verify})$ defined as follows:*

$\boldsymbol{\Sigma_n}$-**Setup**$(1^\lambda, n, \mathcal{R}^n) \to (\mathbf{x}, \mathbf{w})$. *It takes as input a security parameter λ, a vector size n and a relation $\mathcal{R}^n$ over a language L^n. It returns a vector of statements and witnesses $(x_1, \ldots, x_n, w_1, \ldots, w_n)$. The challenge space is required to be* **ChSp**$\subseteq \mathbb{G}_2^n$.

$\boldsymbol{\Sigma_n}$-**Com**$(\mathbf{x}) \to (R, r)$. *It's a PPT algorithm run by the prover to get the first message R to send to the verifier and some private state to be stored. We require that $R \in \mathbb{G}_1$.*

$\boldsymbol{\Sigma_n}$-**Resp**$(\mathbf{x}, \mathbf{w}, r, \mathbf{c}) \to s$. *It's a deterministic algorithm run by the prover to compute the last message of the protocol. It takes as input the statements and witnesses $(\mathbf{x}, \mathbf{w})$ the challenge string $\mathbf{c} \in$ **ChSp** (sent as second message of the protocol) and some state information r. It outputs the third message of the protocol, s.*

Σ_n-**Verify**$(\mathbf{x}, R, \mathbf{c}, s) \to \{0, 1\}$. *It's the verification algorithm that on input the message R, the challenger $\mathbf{c} \in \mathbf{ChSp}$ and a response s, outputs 1 (accept) or 0 (reject).*

We require this protocol to be group homomorphic and to satisfy the completeness and special honest verifier zero knowledge properties. Moreover, the protocol must guarantee either the vector special soundness outlined above or a stronger soundness property that we define below.

Roughly speaking, this property requires that the extractor, upon receiving the witnesses for all but one statements of the vector $\boldsymbol{x}$, has to come up with a witness for the remaining one.

Definition 7 (Strong (Vector) Special Soundness). *Let $\Sigma = (\Sigma$-Setup, Σ-Com, Σ-Resp, Σ-Verify) be a 1-n Σ-Protocol for a relation $\mathcal{R}^n$. We say that Σ has the Strong Special Soundness property if there exists an efficient extractor algorithm Σ_n-Ext such that $\forall \, \boldsymbol{x} \in L^n, \forall j^* \in \{1, \ldots, n\}, \forall \, R, \mathbf{c}, \boldsymbol{s}, \mathbf{c}', \boldsymbol{s}'$ such that $c_{j^*} \neq c'_{j^*}$, Σ_n-Verify$(\boldsymbol{x}, R, \mathbf{c}, \boldsymbol{s}) = 1$ and Σ_n-Verify$(\boldsymbol{x}, R, \mathbf{c}', \boldsymbol{s}') = 1$, outputs $w_{j^*} \leftarrow \Sigma_n$-Ext$(\boldsymbol{x}, R, \mathbf{c}, \boldsymbol{s}, \mathbf{c}', \boldsymbol{s}', \{w_j\}_{j \neq j^*})$ such that $(x_{j^*}, w_{j^*}) \in \mathcal{R}$.*

In the full version of this paper we show that a simple variant of the well known identification protocol by Schnorr is a 1-n Σ-Protocol (with Strong Vector Special Soundness).

5.1   A Linearly Homomorphic On-Line/Off-Line Signature

Suppose $\mathcal{S} = ($**KeyGen, Sign, Verify, Eval**$)$ is a randomly secure LHSG (even one that only allows to sign scalars), $\Sigma_n = (\Sigma_n$-**Setup** ,Σ_n-**Com** ,Σ_n-**Resp** ,Σ_n-**Verify**) is a 1-n Σ-Protocol and $\mathcal{H} = ($**CHGen, CHEval, CHFindColl**$)$ defines a family of chameleon hash functions. Moreover, suppose that the LHSG's message space is the same as the group $\mathbb{G}_1$ of the outputs of Σ_n-**Com**. Our generic construction uses the challenge space of the Σ-Protocol as a message space and works as follows:

ON/OFFKeyGen $(1^\lambda, k, n)$: It runs $(vk_1, sk_1) \leftarrow$ **KeyGen**$(1^\lambda, 1, k)$, $(\mathbf{x}, \mathbf{w}) \leftarrow \Sigma_n$-**Setup** $(1^\lambda, n, \mathcal{R}^n)$ and $(hk, ck) \leftarrow$ **CHGen**(1^λ). It outputs $vk \leftarrow (vk_1, \mathbf{x}, hk)$, $sk \leftarrow (sk_1, \mathbf{w}, ck)$.

OFFSign (sk): This algorithm runs the Σ_n-**Com** algorithm k times to obtain $(R_i, r_i) \leftarrow \Sigma_n$-**Com** $(\mathbf{x})$, chooses a random string fid' from the dataset identifiers' space and randomness ρ' and sets $\overline{fid} \leftarrow$ **CHEval**(hk, fid', ρ'). Then it signs each R_i using the LHSG signing algorithm $\overline{\sigma_i} \leftarrow$ **Sign**$(sk_1, R_i, \overline{fid}, i)$ and outputs $I_{fid'} = \{(i, r_i, R_i, \overline{\sigma_i}, fid', \rho')\}_{i=1,\ldots,k}$.

ONSign $(vk, sk, \mathbf{m}, fid, I_{fid'}, i)$: It parses $I_{fid'}$ as $\{(i, r_i, R_i, \overline{\sigma_i}, fid', \rho')\}_{i=1,\ldots,k}$, computes $s \leftarrow \Sigma_n$-**Resp** $(\mathbf{x}, \mathbf{w}, r_i, \mathbf{m})$, $\rho \leftarrow$ **CHFindColl**(ck, fid', ρ', fid) and outputs $\sigma \leftarrow (R_i, \overline{\sigma_i}, s, \rho)$. As explained in the definition, this algorithm must ensure that all the messages signed with respect to the same fid are computed from the same information $I_{fid'}$

ON/OFFVerify $(\mathrm{vk}, \sigma, \mathbf{m}, \mathrm{fid}, f)$: It parses σ as $(R, \overline{\sigma}, s, \rho)$ and vk as $(\mathrm{vk}_1, \mathbf{x}, \mathrm{hk})$. Then it checks that

$$\mathbf{Verify}(\mathrm{vk}_1, \overline{\sigma}, R, \mathbf{CHEval}(\mathrm{fid}, \rho), f) = 1 \quad \text{and} \quad \Sigma_{\mathbf{n}}\text{-}\mathbf{Verify}(\mathbf{x}, R, \mathbf{m}, s) = 1.$$

If both the above equations hold it returns 1, else it returns 0.

ON/OFFEval $(\mathrm{vk}, \alpha, \sigma_1, \ldots, \sigma_k)$: it parses σ_i as $(R_i, \overline{\sigma}_i, s_i, \rho)$ for each $i = 1, \ldots, k$ and vk as $(\mathrm{vk}_1, \mathbf{x})$. Then it computes:

$$R \leftarrow R_1^{\alpha_1} \odot \cdots \odot R_k^{\alpha_k}, \quad \overline{\sigma} \leftarrow \mathbf{Eval}(\mathrm{vk}_1, \alpha, \overline{\sigma}_1, \ldots, \overline{\sigma}_k),$$

$$s \leftarrow \mathbf{Combine}\left(\mathbf{x}, \alpha, \{(R_i, c_i, s_i)\}_{i=1,\ldots,k}\right).$$

Finally it returns $(R, \overline{\sigma}, s, \rho)$ (as a signature for the message $\mathbf{m}_1^{\alpha_1} \otimes \cdots \otimes \mathbf{m}_k^{\alpha_k}$).

Remark 4. The construction presented above applies to any LHSG. However, if the LHGS itself is obtained as described in section 4.1, the use of the chameleon hash function could be avoided by substituting the signature scheme $\mathcal{S}$ used for the fid with an on-line/off-line one. This improves efficiency.

Theorem 6. *If $\mathcal{S} = (\mathbf{KeyGen}, \mathbf{Sign}, \mathbf{Verify}, \mathbf{Eval})$ is a random message secure LHSG, $\Sigma_n = (\Sigma_{\mathbf{n}}\text{-}\mathbf{Setup}, \Sigma_{\mathbf{n}}\text{-}\mathbf{Com}, \Sigma_{\mathbf{n}}\text{-}\mathbf{Resp}, \Sigma_{\mathbf{n}}\text{-}\mathbf{Verify})$ is a 1-n Σ-Protocol for a non trivial relation $\mathcal{R}^n$, and $\mathcal{H}$ implements a family of chameleon hash functions then the LHOOS described above is secure against a chosen message attack .*

For lack of space, again, this proof is omitted. The security obtained by this construction can be strengthened by assuming additional properties on the underlying LHSG scheme: if $\mathcal{S}$ is strongly secure against a random message attack, then we can prove that the resulting construction is strongly secure (against a CMA) as well.

Theorem 7. *If $\mathcal{S} = (\mathbf{KeyGen}, \mathbf{Sign}, \mathbf{Verify}, \mathbf{Eval})$ is a LHSG scheme strongly unforgeable against a random message attack and $\Sigma_n = (\Sigma_{\mathbf{n}}\text{-}\mathbf{Setup}, \Sigma_{\mathbf{n}}\text{-}\mathbf{Com}, \Sigma_{\mathbf{n}}\text{-}\mathbf{Resp}, \Sigma_{\mathbf{n}}\text{-}\mathbf{Verify})$ is a 1-n Σ-Protocol for a non trivial relation $\mathcal{R}^n$, then the on-line/off-line scheme described above is **strongly** unforgeable against chosen message attacks.*

The proof is straightforward and similar to the previous one and is omitted.

6 From Random Message Security to Chosen Message Security

In this section we present a general transform to construct an LHSG secure against chosen message attack from one secure under random message attack. Our transformation requires the RMA secure scheme to satisfy some additional, but reasonable, requirements (a slightly more generic transformation is given in the full version of this paper). In particular we require it to be *almost deterministic*. Informally, this means that given a file identifier fid $\in \mathcal{D}$ and a signature on a message m with respect to fid, the signature of any other $m' \in \mathcal{M}$ w.r.t. to any admissible function $f \in \mathcal{F}$ and the same fid is uniquely determined.

Remark 5. We stress that while we present our theorems in the context of linearly homomorphic signatures (LHSG), if they are applied to linearly homomorphic structure preserving signatures, the structure preserving property is preserved.

Let $\mathcal{S} = (\textbf{HKeyGen}, \textbf{HSign}, \textbf{HVerify}, \textbf{HEval})$ be a LHSG which is RMA-secure and almost deterministic. The transformation below shows how to produce a new LHSG $\mathcal{T} = (\textbf{TKeyGen}, \textbf{TSign}, \textbf{TVerify}, \textbf{TEval})$ which is secure under CMA.

- **TKeyGen**$(1^\lambda, n, k)$ takes as input the security parameter λ, the vector size n and an upper bound k for the number of messages signed in each dataset. It runs two times the **HKeyGen** algorithm to obtain $(\text{sk}_1, \text{vk}_1) \leftarrow$ **HKeyGen**$(1^\lambda, n, k)$ and $(\text{sk}_2, \text{vk}_2) \leftarrow$ **HKeyGen**$(1^\lambda, n, k)$.
  It outputs $\text{sk} = (\text{sk}_1, \text{sk}_2)$ as the secret signing key and $\text{vk} = (\text{vk}_1, \text{vk}_2)$ as the public verification key. The message space $\mathcal{M}$ is the same of $\mathcal{S}$.
- **TSign**$(\text{sk}, \textbf{m}, \text{fid}, i)$ It chooses random $\textbf{m}_1 = (m_{1,1}, \ldots, m_{1,n}) \xleftarrow{\$} \mathcal{M}$ and computes $\textbf{m}_2 \leftarrow \left(\frac{m_1}{m_{1,1}}, \ldots, \frac{m_n}{m_{1,n}}\right)$ (where $\textbf{m} = (m_1, \ldots, m_n)$).
  Then it computes $\sigma_1 \leftarrow$ **HSign**$(\text{sk}_1, \textbf{m}_1, i, \text{fid})$, $\sigma_2 \leftarrow$ **HSign**$(\text{sk}_2, \textbf{m}_2, i, \text{fid})$ and outputs $\sigma = (\text{fid}, \textbf{m}_1, \sigma_1, \sigma_2)$.
- **TVerify**$(\text{vk}, \sigma, \textbf{m}, \text{fid}, f)$ parses σ as $(\text{fid}, \textbf{m}_1, \sigma_1, \sigma_2)$, computes
  $\textbf{m}_2 \leftarrow \left(\frac{m_1}{m_{1,1}}, \ldots, \frac{m_n}{m_{1,n}}\right)$ and checks that the following equations hold:

$$\textbf{HVerify}(\text{vk}_i, m_i, \sigma_i, \text{fid}, f) = 1 \quad \text{for} \quad i = 1, 2.$$

- **Eval**$(\text{vk}, \text{fid}, f, \{\sigma^{(i)}\}_{i=1\ldots k})$ parses $\sigma^{(i)}$ as $(\text{fid}^{(i)}, \textbf{m}_1^{(i)}, \sigma_1^{(i)}, \sigma_2^{(i)})$ and f as $(\alpha_1, \ldots, \alpha_k)$, then checks that $\text{fid} = \text{fid}^{(i)}$ for all i and, if not, aborts. Finally it sets

$$\sigma_1 \leftarrow \textbf{HEval}(\text{vk}_1, \text{fid}, \{\sigma_1^{(i)}\}_{i=1\ldots k}, f),$$

$$\sigma_2 \leftarrow \textbf{HEval}(\text{vk}_2, \text{fid}, \{\sigma_2^{(i)}\}_{i=1\ldots k}, f),$$

$$\textbf{m}_1 = \left(\prod_{i=1}^{k}(m_{1,1}^{(i)})^{\alpha_i}, \ldots, \prod_{i=1}^{k}(m_{1,n}^{(i)})^{\alpha_i}\right)$$

and returns

$$\sigma \leftarrow (\text{fid}, \textbf{m}_1, \sigma_1, \sigma_2)$$

Theorem 8. *Suppose $\mathcal{S}$ is a LHSG secure against a random message attack with almost deterministic signatures. Moreover assume that the underlying message space is a group where one can efficiently solve systems of group equations. Then the scheme $\mathcal{T}$ described above is a LHSG secure against a chosen message attack.*

Again, the proof of the above theorem is deferred to the full version of this paper.

Acknowledgements. We would like to thank Dario Fiore for helpful discussions and one of the anonymous reviewers of Crypto 2014 for suggesting us the generalized scheme presented in section 3.2.

References

1. Abe, M., Groth, J., Haralambiev, K., Ohkubo, M.: Optimal structure-preserving signatures in asymmetric bilinear groups. In: Rogaway, P. (ed.) CRYPTO 2011. LNCS, vol. 6841, pp. 649–666. Springer, Heidelberg (2011)
2. Ateniese, G., Kamara, S., Katz, J.: Proofs of storage from homomorphic identification protocols. In: Matsui, M. (ed.) ASIACRYPT 2009. LNCS, vol. 5912, pp. 319–333. Springer, Heidelberg (2009)
3. Attrapadung, N., Libert, B.: Homomorphic network coding signatures in the standard model. In: Catalano, D., Fazio, N., Gennaro, R., Nicolosi, A. (eds.) PKC 2011. LNCS, vol. 6571, pp. 17–34. Springer, Heidelberg (2011)
4. Attrapadung, N., Libert, B., Peters, T.: Computing on authenticated data: New privacy definitions and constructions. In: Wang, X., Sako, K. (eds.) ASIACRYPT 2012. LNCS, vol. 7658, pp. 367–385. Springer, Heidelberg (2012)
5. Attrapadung, N., Libert, B., Peters, T.: Efficient completely context-hiding quotable and linearly homomorphic signatures. In: Kurosawa, K., Hanaoka, G. (eds.) PKC 2013. LNCS, vol. 7778, pp. 386–404. Springer, Heidelberg (2013)
6. Bellare, M., Namprempre, C.: Authenticated encryption: Relations among notions and analysis of the generic composition paradigm. Journal of Cryptology 21(4), 469–491 (2008)
7. Boneh, D., Freeman, D., Katz, J., Waters, B.: Signing a linear subspace: Signature schemes for network coding. In: Jarecki, S., Tsudik, G. (eds.) PKC 2009. LNCS, vol. 5443, pp. 68–87. Springer, Heidelberg (2009)
8. Boneh, D., Freeman, D.M.: Homomorphic signatures for polynomial functions. In: Paterson, K.G. (ed.) EUROCRYPT 2011. LNCS, vol. 6632, pp. 149–168. Springer, Heidelberg (2011)
9. Boneh, D., Freeman, D.M.: Linearly homomorphic signatures over binary fields and new tools for lattice-based signatures. In: Catalano, D., Fazio, N., Gennaro, R., Nicolosi, A. (eds.) PKC 2011. LNCS, vol. 6571, pp. 1–16. Springer, Heidelberg (2011)
10. Catalano, D., Di Raimondo, M., Fiore, D., Gennaro, R.: Off-line/On-line signatures: Theoretical aspects and experimental results. In: Cramer, R. (ed.) PKC 2008. LNCS, vol. 4939, pp. 101–120. Springer, Heidelberg (2008)
11. Catalano, D., Fiore, D.: Practical homomorphic mACs for arithmetic circuits. In: Johansson, T., Nguyen, P.Q. (eds.) EUROCRYPT 2013. LNCS, vol. 7881, pp. 336–352. Springer, Heidelberg (2013)
12. Catalano, D., Fiore, D., Gennaro, R., Nizzardo, L.: Generalizing homomorphic mACs for arithmetic circuits. In: Krawczyk, H. (ed.) PKC 2014. LNCS, vol. 8383, pp. 538–555. Springer, Heidelberg (2014)
13. Catalano, D., Fiore, D., Gennaro, R., Vamvourellis, K.: Algebraic (Trapdoor) one-way functions and their applications. In: Sahai, A. (ed.) TCC 2013. LNCS, vol. 7785, pp. 680–699. Springer, Heidelberg (2013)
14. Catalano, D., Fiore, D., Warinschi, B.: Adaptive pseudo-free groups and applications. In: Paterson, K.G. (ed.) EUROCRYPT 2011. LNCS, vol. 6632, pp. 207–223. Springer, Heidelberg (2011)

15. Catalano, D., Fiore, D., Warinschi, B.: Efficient network coding signatures in the standard model. In: Fischlin, M., Buchmann, J., Manulis, M. (eds.) PKC 2012. LNCS, vol. 7293, pp. 680–696. Springer, Heidelberg (2012)
16. Catalano, D., Fiore, D., Warinschi, B.: Homomorphic signatures with efficient verification for polynomial functions. In: Garay, J.A., Gennaro, R. (eds.) CRYPTO 2014, Part I. LNCS, vol. 8616, pp. 371–389. Springer, Heidelberg (2014)
17. Chase, M., Kohlweiss, M., Lysyanskaya, A., Meiklejohn, S.: Malleable proof systems and applications. In: Pointcheval, D., Johansson, T. (eds.) EUROCRYPT 2012. LNCS, vol. 7237, pp. 281–300. Springer, Heidelberg (2012)
18. Desmedt, Y.: Computer security by redefining what a computer is. NSPW (1993)
19. Dolev, D., Dwork, C., Naor, M.: Non-malleable cryptography. In: 23rd ACM STOC Annual ACM Symposium on Theory of Computing, pp. 542–552. ACM Press (May 1991)
20. Even, S., Goldreich, O., Micali, S.: On-line/off-line digital signatures. Journal of Cryptology 9(1), 35–67 (1996)
21. Freeman, D.M.: Improved security for linearly homomorphic signatures: A generic framework. In: Fischlin, M., Buchmann, J., Manulis, M. (eds.) PKC 2012. LNCS, vol. 7293, pp. 697–714. Springer, Heidelberg (2012)
22. Garay, J.A., MacKenzie, P.D., Yang, K.: Strengthening zero-knowledge protocols using signatures. In: Biham, E. (ed.) EUROCRYPT 2003. LNCS, vol. 2656, Springer, Heidelberg (2003)
23. Gennaro, R., Katz, J., Krawczyk, H., Rabin, T.: Secure network coding over the integers. In: Nguyen, P.Q., Pointcheval, D. (eds.) PKC 2010. LNCS, vol. 6056, pp. 142–160. Springer, Heidelberg (2010)
24. Gennaro, R., Wichs, D.: Fully homomorphic message authenticators. In: Sako, K., Sarkar, P. (eds.) ASIACRYPT 2013, Part II. LNCS, vol. 8270, pp. 301–320. Springer, Heidelberg (2013)
25. Goldwasser, S., Micali, S.: Shafi Goldwasser and Silvio Micali. Probabilistic encryption. Journal of Computer and System Sciences 28(2), 270–299 (1984)
26. Johnson, R., Molnar, D., Song, D., Wagner, D.: Homomorphic signature schemes. In: Preneel, B. (ed.) CT-RSA 2002. LNCS, vol. 2271, p. 244. Springer, Heidelberg (2002)
27. Joo, C., Yun, A.: Homomorphic authenticated encryption secure against chosen-ciphertext attack. Cryptology ePrint Archive, Report 2013/726 (2013), http://eprint.iacr.org/
28. Kunz-Jacques, S., Pointcheval, D.: About the security of MTI/C0 and MQV. In: De Prisco, R., Yung, M. (eds.) SCN 2006. LNCS, vol. 4116, pp. 156–172. Springer, Heidelberg (2006)
29. Joye, M., Libert, B.: Efficient cryptosystems from 2^k-th power residue symbols. In: Johansson, T., Nguyen, P.Q. (eds.) EUROCRYPT 2013. LNCS, vol. 7881, pp. 76–92. Springer, Heidelberg (2013)
30. Libert, B., Peters, T., Joye, M., Yung, M.: Linearly homomorphic structure-preserving signatures and their applications. In: Canetti, R., Garay, J.A. (eds.) CRYPTO 2013, Part II. LNCS, vol. 8043, pp. 289–307. Springer, Heidelberg (2013)
31. Micali, S.: Cs proofs. In: FOCS, pp. 436–453. IEEE Computer Society (1994)
32. Micali, S., Rivest, R.L.: Transitive signature schemes. In: Preneel, B. (ed.) CT-RSA 2002. LNCS, vol. 2271, p. 236. Springer, Heidelberg (2002)
33. Naccache, D., Stern, J.: A new public key cryptosystem based on higher residues. In: ACM CCS 1998 Conference on Computer and Communications Security, pp. 59–66. ACM Press (November 1998)

34. Paillier, P.: Public-key cryptosystems based on composite degree residuosity classes. In: Stern, J. (ed.) EUROCRYPT 1999. LNCS, vol. 1592, p. 223. Springer, Heidelberg (1999)
35. Shamir, A., Tauman, Y.: Improved online/Offline signature schemes. In: Kilian, J. (ed.) CRYPTO 2001. LNCS, vol. 2139, p. 355. Springer, Heidelberg (2001)
36. Yi, X.: Directed transitive signature scheme. In: Abe, M. (ed.) CT-RSA 2007. LNCS, vol. 4377, pp. 129–144. Springer, Heidelberg (2006)

Compact VSS and Efficient Homomorphic UC Commitments[*]

Ivan Damgård, Bernardo David, Irene Giacomelli, and Jesper Buus Nielsen

Dept. of Computer Science, Aarhus University

Abstract. We present a new compact verifiable secret sharing scheme, based on this we present the first construction of a homomorphic UC commitment scheme that requires only cheap symmetric cryptography, except for a small number of seed OTs. To commit to a k-bit string, the amortized communication cost is $O(k)$ bits. Assuming a sufficiently efficient pseudorandom generator, the computational complexity is $O(k)$ for the verifier and $O(k^{1+\epsilon})$ for the committer (where $\epsilon < 1$ is a constant). In an alternative variant of the construction, all complexities are $O(k \cdot polylog(k))$. Our commitment scheme extends to vectors over any finite field and is additively homomorphic. By sending one extra message, the prover can allow the verifier to also check multiplicative relations on committed strings, as well as verifying that committed vectors $\boldsymbol{a}, \boldsymbol{b}$ satisfy $\boldsymbol{a} = \varphi(\boldsymbol{b})$ for a linear function φ. These properties allow us to non-interactively implement any one-sided functionality where only one party has input (this includes UC secure zero-knowledge proofs of knowledge). We also present a perfectly secure implementation of any multiparty functionality, based directly on our VSS. The communication required is proportional to a circuit implementing the functionality, up to a logarithmic factor. For a large natural class of circuits the overhead is even constant. We also improve earlier results by Ranellucci et al. on the amount of correlated randomness required for string commitments with individual opening of bits.

1 Introduction

A commitment scheme is perhaps the most basic primitive in cryptographic protocol theory, but is nevertheless very powerful and important both in theory and practice. Intuitively, a commitment scheme is a digital equivalent of a secure

[*] The authors acknowledge support from the Danish National Research Foundation and The National Science Foundation of China (under the grant 61061130540) for the Sino-Danish Center for the Theory of Interactive Computation and also from the CFEM research center (supported by the Danish Strategic Research Council), within which part of this work was performed. Partially supported by Danish Council for Independent Research via DFF Starting Grant 10-081612. Partially supported by the European Research Commission Starting Grant 279447.

P. Sarkar and T. Iwata (Eds.): ASIACRYPT 2014, PART II, LNCS 8874, pp. 213–232, 2014.

box: it allows a prover P to commit to a secret s by putting it into a locked box and give it to a verifier V. Since the box is locked, V does not learn s at commitment time, we say the commitment is *hiding*. Nevertheless, P can later choose to give V the key to the box to let V learn s. Since P gave away the box, he cannot change his mind about s after commitment time, we say the commitment is *binding*.

Commitment schemes with stand-alone security (i.e., they only have the binding and hiding properties) can be constructed from any one-way function, and already this most basic form of commitments implies zero-knowledge proofs for all NP languages. Commitments with stand-alone security can be very efficient as they can be constructed from cheap symmetric cryptography such as pseudo-random generators [Nao91].

However, in many cases one would like a commitment scheme that composes well with other primitives, so that it can be used as a secure module that will work no matter which context it is used in. The strongest form of security we can ask for here is UC security [Can01]. UC commitments cannot be constructed without set-up assumptions such as a common reference string [CF01]. On the other hand, a construction of UC commitment in such models implies public-key cryptography [DG03] and even multiparty computation [CLOS02] (but see [DNO10] for a construction based only on 1-way functions, under a stronger set-up assumption).

With this in mind, it is not surprising that constructions of UC commitments are significantly less efficient than stand-alone secure commitments. The most efficient UC commitment schemes known so far are based on the DDH assumption and requires several exponentiations in a large group [Lin11,BCPV13]. This means that the computational complexity for committing to k-bit strings is typically $\Omega(k^3)$.

Our Contribution. We first observe that even if we cannot build practical UC commitments without using public-key technology, we might still confine the use of it to a small once-and-for-all set-up phase. This is exactly what we achieve: given initial access to a small number of oblivious transfers, we show a UC secure commitment scheme where the only computation required is pseudorandom bit generation and a few elementary operations in a finite field. The number of oblivious transfers we need does not depend on the number of commitments we make later. The main observation we make is that we can reach our goal by combining the oblivious transfers with a "sufficiently compact" Verifiable Secret Sharing Scheme (VSS) that we then construct. The VSS has applications on its own as we detail below.

To commit to a k-bit string, the amortized communication cost is $O(k)$ bits. The computational complexity is $O(k)$ for the verifier and $O(k^{1+\epsilon})$ for the committer (where $\epsilon < 1$ is a constant). This assumes a pseudorandom generator with linear overhead per generated bit.[1] In an alternative variant of the construction, all complexities are $O(k \cdot polylog(k))$. After the set-up phase is done, the prover can commit by sending a single string. Our construction extends to commitment to strings over any finite field and is additively homomorphic, meaning that given commitments to strings a, b, the verifier can on his own compute a commitment to $a + b$, and the prover can open it while revealing nothing beyond $a + b$. Moreover, if the prover sends one extra string, the verifier can also check that committed vectors a, b, c satisfy $c = a * b$, the component-wise product. Finally, again by sending one extra string and allowing one extra opening, the verifier can compute a commitment to $\varphi(a)$, given the commitment to a, for any linear function φ. These extra strings have the same size as a commitment, up to a constant factor.

On the technical side, we take the work from [FJN+13] as our point of departure. As part of their protocol for secure 2-party computation, they construct an imperfect scheme (which is not binding for all commitments). While this is good enough for their application, we show how to combine their scheme with an efficient VSS that is compact in the sense that it allows to share several values from the underlying field, while shares only consist of a single field element. This is also known as packed secret sharing [FY92].

Our construction generalises the VSS from [CDM00] to the case of packed secret sharing. We obtain a VSS where the communication needed is only a constant factor larger than the size of the secret. Privacy for a VSS usually just says that the secret remains unknown to an unqualified subset of players until the entire secret is reconstructed. We show an extended form of privacy that may be of independent interest: the secret in our VSS is a set of ℓ vectors $s_1, ..., s_\ell$, each of length ℓ. We show that any linear combination of $s_1, ..., s_\ell$ can be (verifiably) opened and players will learn nothing beyond that linear combination. We also build two new VSS protocols, both of which are non-trivial extensions. The first allows the dealer to generate several sharings of the vector 0^ℓ. For an honest dealer, the shares distributed are random even given the extra verification information an adversary would see during the VSS. This turns out to be crucial in achieving secure multiplication of secret-shared or committed values. The second new protocol allows us to share two sets of vectors $s_1, ..., s_\ell$ and $\tilde{s}_1, ..., \tilde{s}_\ell$ such that it can be verified that $\varphi(s_1) = \tilde{s}_1, ..., \varphi(s_\ell) = \tilde{s}_\ell$ for a linear function φ. In the commitment scheme, this is what allows us to verify that two shared or committed vectors satisfy a similar linear relation.

Before we discuss applications, a note on an alternative way to view our commitment scheme: A VSS is essentially a multiparty commitment scheme.

---

[1] This seems a very plausible assumption as a number of different sufficient conditions for such PRG's are known. In [IKOS08] it is observed that such PRGs follow Alekhnovich's variant of the Learning Parity with Noise assumption. Applebaum [App13] shows that such PRGs can be obtained from the assumption that a natural variant of Goldreich's candidate for a one-way function in NC0 is indeed one-way. The improved HILL-style result of Vadhan and Zheng [VZ12] implies that such PRGs can be obtained from any exponentially strong OWF that can be computed by a linear-size circuit.

Therefore, given our observation that VSS and OT gives us efficient UC commitment, it is natural to ask whether our construction could be obtained using "MPC-in-the-head" techniques. Specifically, the IPS compiler [IPS08] is a general tool that transforms a multiparty protocol into a 2-party protocol implementing the same functionality in the OT hybrid model. Indeed, applying IPS to our VSS does result in a UC commitment protocol. However, while this protocol is somewhat similar to ours, it is more complicated and less efficient.

Applications. One easily derived application of our commitment scheme is an implementation of any two-party functionality where only one party has input, we call this a *one-sided* functionality. This obviously includes UC secure zero-knowledge proofs of knowledge for any NP relation. Our implementation is based on a Boolean circuit C computing the desired output.

We will focus on circuits that are not too "oddly shaped". Concretely, we assume that every layer of the circuit is $\Omega(\ell)$ gates wide, except perhaps for a constant number of layers. Here one may think of ℓ as a statistical security parameter, as well as the number of bits one of our commitments contains. Second, we want that the number of bits that are output from layer i in the circuit and used in layer j is either 0 or $\Omega(\ell)$ for all $i < j$. We call such circuits *well-formed*. In a nutshell, well-formed circuit are those that allow a modest amount of parallelization, namely a RAM program computing the circuit can always execute $\Omega(\ell)$ bit operations in parallel and when storing bits for later use or retrieving, it can always address $\Omega(\ell)$ bits at a time. In practice, since we can treat ℓ as a statistical security parameter, its value can be quite small(e.g., 80), in particular very small compared to the circuit size, and hence a requirement that the circuit be well-formed seems rather modest. Using the parallelisation technique from [DIK10], we can evaluate a well-formed circuit using only parallel operations on ℓ-bit blocks, and a small number of different permutations of bits inside blocks. This comes at the cost of a log-factor overhead.

Some circuits satisfy an even nicer condition: if we split the bits coming into a layer of C into ℓ-bit blocks, then each such block can be computed as a linear function of blocks from previous layers, where the function is determined by the routing of wires in the circuit. Such a function is called a block function. If each block function depends only on a constant number of previous blocks and if each distinct block function occurs at least ℓ times, then C is called *regular* (we can allow that a constant number of block functions do not satisfy the condition). For instance, block ciphers and hash functions do not spread the bits around much in one round, but repeat the same operations over many rounds and hence tend to have regular circuits. Also many circuits for arithmetic problems have a simple repetitive structure and are therefore regular.

Theorem 1. *For any one-sided two-party functionality F that can be computed by Boolean circuit C, there exists a UC secure non-interactive implementation of F in the OT hybrid model. Assuming C is well-formed and that there exists a linear overhead PRG, the communication as well as the receiver's computation is in $O(\log(|C|)|C|)$. If C is regular, the complexities are $O(|C|)$.*

We stress that the protocol we build works for any circuit, it will just be less efficient if C is not well-formed.[2] We can also apply our VSS directly to implement multiparty computation in the model where there are clients who have inputs and get outputs and servers who help doing the computation.

Theorem 2. *For any functionality F, there exists a UC perfectly secure implementation of F in the client/server model assuming at most a constant fraction of the servers and all but one of the clients may be corrupted. If C is well-formed, the* total communication complexity *is in $O(\log(|C|)|C|)$. If C is regular, the complexity is $O(|C|)$.*

We are not aware of any other approach that would allow us to get perfect security and "constant rate" for regular circuits.[3]

A final application comes from the fact that our commitment protocol can be interpreted as an unconditionally secure protocol in the model where correlated randomness is given. In this model, it was shown in [RTWW11] that any unconditionally secure protocol that allows commitment to N bits where each bit can be *individually* opened, must use $\Omega(Nk)$ bits of correlated randomness, where k the security parameter. They also show a positive result that partially circumvents this lower bound by considering a functionality $F_{com}^{N,r}$ that allows commitment to N bits where only $r < N$ bits can be selectively and individually opened. When r is $O(1)$, they implement this functionality at constant rate, i.e., the protocol requires only $O(1)$ bits of correlated randomness per bit committed to. We can improve this as follows:

Theorem 3. *There exists a constant-rate statistically secure implementation of $F_{com}^{N,r}$ in the correlated randomness model, where $r \in O(N^{1-\epsilon})$ for any $\epsilon > 0$.*

We find it quite surprising that r can be "almost" N, and still the lower bound for individual opening does not apply. What the actual cut-off point is remains an intriguing open question.

Related Work. In [DIK+08], a VSS was constructed that is also based on packed secret sharing (using Shamir as the underlying scheme). This construction relies crucially on hyper invertible matrices which requires the field to grow with the number of players. Our construction works for any field, including $\mathbb{F}_2$. This would not be so important if we only wanted to commit and reveal bits: we could use [DIK+08] with an extension field, pack more bits into a field element

---

[2] It is possible to use MPC-in-the-head techniques to prove results that have some (but not all) of the properties of Theorem 1. Essentially one applies the IPS compiler to a multiparty protocol, either a variant of [DI06] (described in [IKOS09]), or the protocol from [DIK10]. In the first case, the verifier's computation will be asymptotically larger than in our protocol, in the second case, one cannot obtain the result for regular circuits since [DIK10] has at least logarithmic overhead for any circuit since it cannot be based on fields of constant size.

[3] Using [DIK10] would give at least logarithmic overhead for any circuit, using variants of [DI06] would at best give statistical security.

and still get constant communication overhead, but we want to do (Boolean) operations on committed bits, and then "bit-packing" will not work. It therefore seems necessary to construct a more compact VSS in order to get our results. In [BBDK00], techniques for computing functions of shared secrets using both broadcast channels and private interactive evaluation are introduced. However, their constructions are based specifically on Shamir's LSSS and do not allow verification of share validity.

In recent independent work [GIKW14], Garay *et al.* also construct UC commitments using OT, VSS and pseudorandom generators as the main ingredients. While the basic approach is closely related to ours, the concrete constructions are somewhat different, leading to incomparable results. In [GIKW14] optimal rate is achieved, as well as a negative result on extension of UC commitments. On the other hand, we focus more on computational complexity and achieve homomorphic properties as well as non-interactive verification of linear relations inside committed vectors.[4]

2   Preliminaries

In this section we introduce the basic definitions and notation that will be used throughout the paper. We denote sampling a value r from a distribution $\mathcal{D}$ as $r \leftarrow \mathcal{D}$. We say that a function ϵ is negligible if there exists a constant c such that $\epsilon(n) < \frac{1}{p(n)}$ for every polynomial p and $n > c$. Two sequences $X = \{X_\kappa\}_{\kappa \in \mathbb{N}}$ and $Y = \{Y_\kappa\}_{\kappa \in \mathbb{N}}$ of random variables are said to be *computationally indistinguishable*, denoted by $X \stackrel{c}{\approx} Y$, if for every non-uniform probabilistic polynomial-time (*PPT*) distinguisher D there exists a negligible function $\epsilon(\cdot)$ such that for every $\kappa \in \mathbb{N}$, $| \, Pr[D(X_\kappa) = 1] - Pr[D(Y_\kappa) = 1] \, | < \epsilon(\kappa)$. Similarly two sequences X and Y of random variables are said to be *statistically indistinguishable*, denoted by $X \stackrel{s}{\approx} Y$, if the same relation holds for unbounded non-uniform distinguishers.

2.1   Universal Composability

The results presented in this paper are proven secure in the Universal Composability (UC) framework introduced by Canetti in [Can01]. We consider security against static adversaries, *i.e.* all corruptions take place before the execution of the protocol. We consider active adversaries who may deviate from the protocol in any arbitrary way. It is known that UC commitments cannot be obtained in the plain model [CF01]. In order to overcome this impossibility, our protocol is proven secure in the $\mathcal{F}_{OT}$-hybrid model in, where all parties are assumed to have access to an ideal 1-out-of-2 OT functionality. In fact, our protocol is constructed in the $\mathcal{F}_{OT}^{t,n}$-hybrid model (*i.e.* assuming access to t-out-of-n OT), which can be subsequently reduced to the $\mathcal{F}_{OT}$-hybrid model via standard techniques for obtaining $\mathcal{F}_{OT}^{t,n}$ from $\mathcal{F}_{OT}$ [Nao91,BCR86,NP99]. We denote by $\mathcal{F}_{OT}^{t,n}(\lambda)$ an instance

---

[4] Our work has been recognised by the authors of [GIKW14] as being independent.

Functionality $\mathcal{F}_{\text{HCOM}}$

$\mathcal{F}_{\text{HCOM}}$ proceeds as follows, running with parties $P_1, \ldots, P_n$ and an adversary S:

- **Commit Phase**: Upon receiving a message (commit, $sid, ssid, P_s, P_r, m$) from P_s, where $m \in \{0,1\}^\lambda$, record the tuple $(ssid, P_s, P_r, m)$ and send the message (receipt, $sid, ssid, P_s, P_r$) to P_r and S. (The lengths of the strings λ is fixed and known to all parties.) Ignore any future commit messages with the same $ssid$ from P_s to P_r. If a message (abort, $sid, ssid$) is received from S, the functionality halts.

- **Reveal Phase**: Upon receiving a message (reveal, $sid, ssid$) from P_s: If a tuple $(ssid, P_s, P_r, m)$ was previously recorded, then send the message (reveal, $sid, ssid, P_s, P_r, m$) to P_r and S. Otherwise, ignore.

- **Addition**: Upon receiving a message (add, $sid, ssid, P_s, ssid_1, ssid_2, ssid_3$) from P_r: If tuples $(ssid_1, P_s, P_r, m_1)$, $(ssid_2, P_s, P_r, m_2)$ were previously recorded and $ssid_3$ is unused, record $(ssid_3, P_s, P_r, m_1 + m_2)$ and send the message (add, $sid, ssid, P_s, ssid_1, ssid_2, ssid_3, \text{success}$) to P_s, P_r and S.

- **Multiplication**: Upon receiving a message (mult, $sid, ssid, P_s, ssid_1, ssid_2, ssid_3$) from P_r: If tuples $(ssid_1, P_s, P_r, m_1)$, $(ssid_2, P_s, P_r, m_2)$ and $(ssid_3, P_s, P_r, m_3)$ were previously recorded, and if $m_3 = m_1 * m_2$, send the message (mult, $sid, ssid, P_s, ssid_1, ssid_2, ssid_3, \text{success}$) to P_s, P_r and S. Otherwise, send message (mult, $sid, ssid, P_s, ssid_1, ssid_2, ssid_3, \text{fail}$) to P_s, P_r and S.

- **Linear Function Evaluation**: Upon receiving a message (linear, $sid, ssid, P_s, \varphi, ssid_1, ssid_2$), where φ is a linear function, from P_s: If the tuple $(ssid_1, P_s, P_r, m_1)$ was previously recorded and $ssid_2$ is unused, store $(ssid_2, P_s, P_r, \varphi(m_1))$ and send (linear, $sid, ssid, P_s, ssid_1, ssid_2, \text{success}$) to P_s, P_r and S.

Fig. 1. Functionality $\mathcal{F}_{\text{HCOM}}$

Functionality $\mathcal{F}_{OT}^{t,n}$

$\mathcal{F}_{OT}^{t,n}$ interacts with a sender P_s, a receiver P_r and an adversary S.

- Upon receiving a message (sender, $sid, ssid, x_0, \ldots, x_n$) from P_s, where each $x_i \in \{0,1\}^\lambda$, store the tuple $(ssid, x_0, \ldots, x_n)$ (The lengths of the strings λ is fixed and known to all parties). Ignore further messages from P_s to P_r with the same $ssid$.

- Upon receiving a message (receiver, $sid, ssid, c_1, \ldots, c_t$) from P_r, check if a tuple $(ssid, x_0, \ldots, x_n)$ was recorded. If yes, send (received, $sid, ssid, x_{c_1}, \ldots, x_{c_t}$) to P_r and (received, $sid, ssid$) to P_s and halt. If not, send nothing to P_r (but continue running).

Fig. 2. Functionality $\mathcal{F}_{OT}^{t,n}$

of the functionality that takes as input from the sender messages in $\{0,1\}^\lambda$. Notice that $\mathcal{F}_{OT}$ can be efficiently UC-realized by the protocols in [PVW08], which can be used to instantiate our commitment protocol. We define our commitment functionality $\mathcal{F}_{\text{HCOM}}$ in Figure 1 and $\mathcal{F}_{OT}^{t,n}$ in Figure 2, further definitions can be found in the full version of this paper [DDGN14].

2.2 Linear Secret Sharing

In very short terms, a linear secret sharing scheme is a secret sharing scheme defined over a finite field $\mathbb{F}$, where the shares are computed as a linear function of the secret (consisting of one or more field elements) and some random field elements. A special case is Shamir's well known scheme. However, we need a more general model for our purposes. We follow the approach from [CDP12] and recall the definitions we need from their model.

Definition 1. *A linear secret sharing scheme $\mathcal{S}$ over the finite field $\mathbb{F}$ is defined by the following parameters: number of players n, secret length ℓ, randomness length e, privacy threshold t and reconstruction threshold r. Also, a $n \times (\ell + e)$ matrix M over $\mathbb{F}$ is given and $\mathcal{S}$ must have r-reconstruction and t-privacy as explained below. If $\ell > 1$, then $\mathcal{S}$ is called a* packed *linear secret sharing scheme.*

Let $d = \ell + e$ and let $\mathcal{P} = \{P_1, \ldots, P_n\}$ be the set of players, then the row number i of M, denoted by $\boldsymbol{m}_i$, is assigned to player P_i. If A is a player subset, then M_A denotes the matrix consisting of rows from M assigned to players in A. To share a secret $\boldsymbol{s} \in \mathbb{F}^\ell$, one first forms a column vector $\boldsymbol{f} \in \mathbb{F}^d$ where $\boldsymbol{s}$ appears in the first ℓ entries and with the last e entries chosen uniformly at random. The *share vector* of $\boldsymbol{s}$ in the scheme $\mathcal{S}$ is computed as $\boldsymbol{c} = M \cdot \boldsymbol{f}$ and its i-th component $\boldsymbol{c}[i]$ is the share given to the player P_i. We will use π_ℓ to denote the projection that outputs the first ℓ coordinates of a vector, *i.e.* $\pi_\ell(\boldsymbol{f}) = \boldsymbol{s}$.

Now, *t-privacy* means that for any player subset A of size at most t, the distribution of $M_A \cdot \boldsymbol{f}$ is independent of $\boldsymbol{s}$. It is easy to see that this is the case if and only if there exists, for each position j in $\boldsymbol{s}$, a *sweeping vector* $\boldsymbol{w}^{A,j}$. This is a column vector of d components such that $M_A \cdot \boldsymbol{w}^{A,j} = \boldsymbol{0}$ and $\pi_\ell(\boldsymbol{w}^{A,j})$ is a vector whose j-th entry is 1 while all other entries are 0.

Finally, *r-reconstruction* means that for any player subset B of size at least r, $\boldsymbol{s}$ is uniquely determined from $M_B \cdot \boldsymbol{f}$. It is easy to see that this is the case if and only if there exists, for each position j in $\boldsymbol{s}$, a *reconstruction vector* $\boldsymbol{r}_{B,j}$. This is a row vector of $|B|$ components such that for any $\boldsymbol{f} \in \mathbb{F}^d$, $\boldsymbol{r}_{B,j} \cdot M_B \cdot \boldsymbol{f} = \boldsymbol{f}[j]$, where $\boldsymbol{f}[j]$ is the j-th entry in $\boldsymbol{f}$.

A packed secret sharing scheme was constructed in Franklin and Yung [FY92]. However, to get our results, we will need a scheme that works over constant size fields, such an example can be found in [CDP12].

Multiplying Shares: for $\boldsymbol{v}, \boldsymbol{w} \in \mathbb{F}^k$, where $\boldsymbol{v} \otimes_i \boldsymbol{w} = (\boldsymbol{v}[i]\boldsymbol{w}[j])_{j \neq i}$, the vector $\boldsymbol{v} \otimes \boldsymbol{w} \in \mathbb{F}^{k^2}$ is defined by $\boldsymbol{v} \otimes \boldsymbol{w} = (\boldsymbol{v}[1]\boldsymbol{w}[1], \ldots, \boldsymbol{v}[k]\boldsymbol{w}[k], \boldsymbol{v} \otimes_1 \boldsymbol{w}, \ldots, \boldsymbol{v} \otimes_k \boldsymbol{w})$. If M is the matrix of the linear secret sharing scheme $\mathcal{S}$, we can define a new scheme $\widehat{\mathcal{S}}$ considering the matrix $\widehat{M}$, whose i-th row is the vector $\boldsymbol{m}_i \otimes \boldsymbol{m}_i$. Clearly $\widehat{M}$ has n rows and d^2 columns and for any $\boldsymbol{f}^1, \boldsymbol{f}^2 \in \mathbb{F}^d$ it holds that

$(M \cdot \boldsymbol{f}^1) * (M \cdot \boldsymbol{f}^2) = \widehat{M} \cdot (\boldsymbol{f}^1 \otimes \boldsymbol{f}^2)^\top$ where $*$ is just the Schur product (or componentwise product). Note that if t is the privacy threshold of $\mathcal{S}$, then the scheme $\widehat{\mathcal{S}}$ also has the t-privacy property. But in general it does not hold that the $\widehat{\mathcal{S}}$ has r-reconstruction. However, suppose that $\widehat{\mathcal{S}}$ has $(n-t)$-reconstruction, then $\mathcal{S}$ is said to have the t-*strong multiplication property*.

In particular, if $\mathcal{S}$ has the t-strong multiplication property, then for any player set A of size at least $n-t$ and for any index $j = 1, \ldots, \ell$ there exists a row vector $\widehat{\boldsymbol{r}}_{A,j}$ such that $\widehat{\boldsymbol{r}}_{A,j} \cdot \left[(M_A \cdot \boldsymbol{f}^1) * (M_A \cdot \boldsymbol{f}^2) \right] = \boldsymbol{s}^1[j]\boldsymbol{s}^2[j]$ for any $\boldsymbol{s}^1, \boldsymbol{s}^2 \in \mathbb{F}^\ell$.

3 Packed Verifiable Secret-Sharing

In a *Verifiable Secret-Sharing* scheme (VSS) a dealer distributes shares of a secret to the players in $\mathcal{P}$ in such a way that the honest players are guaranteed to get consistent shares of a well-defined secret or agree that the dealer cheated. In this section we present a packed verifiable secret sharing protocol that generalizes and combines the ideas of packed secret sharing from [FY92] and VSS based on polynomials in 2 variables from [BOGW88]. The protocol is not a full-blown VSS, as it aborts as soon as anyone complains, but this is all we need for our results. The proofs for all lemmas in this section can be found in the full version of this paper [DDGN14].

The protocol can be based on any linear secret-sharing scheme $\mathcal{S}$ over $\mathbb{F}$ as defined in Section 2. We assume an active adversary who corrupts t players and possibly the dealer, and we assume that at least r players are honest. The protocol will secret-share ℓ column vectors $\boldsymbol{s}^1, \ldots, \boldsymbol{s}^\ell \in \mathbb{F}^\ell$. In the following, F will be a $d \times d$ matrix with entries in $\mathbb{F}$ and for $1 \le i \le n$ we will define $\boldsymbol{h}^i = F \cdot \boldsymbol{m}_i^\top$ and $\boldsymbol{g}_i = \boldsymbol{m}_i \cdot F$. It is then clear that $\boldsymbol{m}_j \cdot \boldsymbol{h}^i = \boldsymbol{g}_j \cdot \boldsymbol{m}_i^\top$ for $1 \le i, j \le n$. We will use $\boldsymbol{f}^b$ to denote the b-th column of F. The protocol is shown in Figure 3.

Packed Verifiable Secret-Sharing Protocol π_{VSS}

1. Let $\boldsymbol{s}^1, \ldots, \boldsymbol{s}^\ell \in \mathbb{F}^\ell$ be the secrets to be shared. The dealer chooses a random $d \times d$ matrix F with entries in $\mathbb{F}$, subject to $\pi_\ell(\boldsymbol{f}^b) = \boldsymbol{s}^b$ for any $b = 1, \ldots, \ell$.
2. The dealer sends $\boldsymbol{h}^i$ and $\boldsymbol{g}_i$ to P_i.
3. Each player P_j sends $\boldsymbol{g}_j \cdot \boldsymbol{m}_i^\top$ to P_i, for $i = 1, \ldots, n$.
4. Each P_i checks, for $j = 1, \ldots, n$, that $\boldsymbol{m}_j \cdot \boldsymbol{h}^i$ equals the value received from P_j. He broadcasts *Accept* if all checks are OK, otherwise he broadcasts *Reject*.
5. If all players said *Accept*, then each P_j stores, for $b = 1, \ldots, \ell$, $\boldsymbol{g}_j[b]$ as his share of $\boldsymbol{s}^b$, otherwise the protocol aborts.

Fig. 3. The VSS protocol

For a column vector $\boldsymbol{v} \in \mathbb{F}^d$, we will say that $\boldsymbol{v}$ *shares* $\boldsymbol{s} \in \mathbb{F}^\ell$, if $\pi_\ell(\boldsymbol{v}) = \boldsymbol{s}$ and each honest player P_j holds $\boldsymbol{m}_j \cdot \boldsymbol{v}$. In other words, $\boldsymbol{c} = M \cdot \boldsymbol{v}$ forms a share vector of $\boldsymbol{s}$ in exactly the way we defined in the previous section. We now show some basic facts about π_{VSS}:

Lemma 1 (completeness). *If the dealer in π_{VSS} is honest, then all honest players accept and the column vector $\boldsymbol{f}^b$ shares $\boldsymbol{s}^b$ for any $b = 1, \ldots, \ell$.*

Lemma 2 (soundness). *If the dealer in π_{VSS} is corrupt, but no player rejects, then for $b = 1, \ldots, \ell$, there exists a column vector $\boldsymbol{v}^b$ and a bit string $\boldsymbol{s}^b$ such that $\boldsymbol{v}^b$ shares $\boldsymbol{s}^b$.[5]*

Finally, we show a strong privacy property guaranteeing that if we open any linear function of $\boldsymbol{s}^b$'s, then no further information on the $\boldsymbol{s}^b$'s is released. To be more precise about this, assume $T : \mathbb{F}^\ell \mapsto \mathbb{F}^{\ell'}$, where $\ell' \leq \ell$, is a surjective linear function. By $T(\boldsymbol{s}^1, \ldots, \boldsymbol{s}^\ell)$, we mean a tuple $(\boldsymbol{u}^1, \ldots, \boldsymbol{u}^{\ell'})$ of column vectors in $\mathbb{F}^\ell$ s.t. $\boldsymbol{u}^b[a] = T(\boldsymbol{s}^1[a], \ldots, \boldsymbol{s}^\ell[a])[b]$. Put differently, if we arrange the column vectors $\boldsymbol{s}^1, \ldots, \boldsymbol{s}^\ell$ in a $\ell \times \ell$ matrix, then what happens is that we apply T to each row, and let the $\boldsymbol{u}^b$'s be the columns in the resulting matrix. In a completely similar way, we define a tuple of ℓ' column vectors of length d by the formula $T(\boldsymbol{f}^1, \ldots, \boldsymbol{f}^\ell) = (\boldsymbol{w}^1, \ldots, \boldsymbol{w}^{\ell'})$. It is easy to see that if $\boldsymbol{f}^1, \ldots, \boldsymbol{f}^\ell$ share $\boldsymbol{s}^1, \ldots, \boldsymbol{s}^\ell$, then $\boldsymbol{w}^1, \ldots, \boldsymbol{w}^{\ell'}$ share $\boldsymbol{u}^1, \ldots, \boldsymbol{u}^{\ell'}$, since the players can apply T to the shares they received in the first place, to get shares of $\boldsymbol{u}^1, \ldots, \boldsymbol{u}^{\ell'}$. In the following we will abbreviate and use $T(F)$ to denote $T(\boldsymbol{f}^1, \ldots, \boldsymbol{f}^\ell)$.

Now, by *opening* $T(\boldsymbol{s}^1, \ldots, \boldsymbol{s}^\ell)$, we mean that the (honest) dealer makes $T(F)$ public, which allows anyone to compute $T(\boldsymbol{s}^1, \ldots, \boldsymbol{s}^\ell)$. We want to show that, in general, if $T(\boldsymbol{s}^1, \ldots, \boldsymbol{s}^\ell)$ is opened, then the adversary learns $T(\boldsymbol{s}^1, \ldots, \boldsymbol{s}^\ell)$ and no more information about $\boldsymbol{s}^1, \ldots, \boldsymbol{s}^\ell$. This is captured by Lemma 3. Suppose that $A = \{P_{i_1}, \ldots, P_{i_t}\}$ is a set of players corrupted by the adversary.

Lemma 3 (privacy). *Suppose the dealer in π_{VSS} is honest. Now, in case 1 suppose he executes π_{VSS} with input $\boldsymbol{s}^1, \ldots, \boldsymbol{s}^\ell$ and then opens $T(\boldsymbol{s}^1, \ldots, \boldsymbol{s}^\ell)$. In case 2, he executes π_{VSS} with input $\widetilde{\boldsymbol{s}}^1, \ldots, \widetilde{\boldsymbol{s}}^\ell$ and then opens $T(\widetilde{\boldsymbol{s}}^1, \ldots, \widetilde{\boldsymbol{s}}^\ell)$. If $T(\boldsymbol{s}^1, \ldots, \boldsymbol{s}^\ell) = T(\widetilde{\boldsymbol{s}}^1, \ldots, \widetilde{\boldsymbol{s}}^\ell)$, then the views of the adversary in the two cases are identically distributed.*

As the last step, we show an extra randomness property satisfied by the share vectors obtained by Protocol π_{VSS}. If C is a $a \times b$ matrix, define $\pi^\ell(C)$ as the $a \times \ell$ matrix given by the first ℓ columns of C and $\pi_\ell(C)$ as the $\ell \times b$ matrix given by the first ℓ rows of C. Note that, if V is a $d \times \ell$ matrix such that $\pi_\ell(V) = (\boldsymbol{s}^1, \ldots, \boldsymbol{s}^\ell)$, then the dealer might have chosen V as the first ℓ columns in his matrix F. We want to show that given the adversary's view, any V could have been chosen, as long as it is consistent with the adversary's shares of $\boldsymbol{s}^1, \ldots, \boldsymbol{s}^\ell$.

Lemma 4 (randomness of the share vectors). *Suppose that the dealer in π_{VSS} is honest and let $A = \{P_{i_1}, \ldots, P_{i_t}\}$ be a set of players corrupted by the adversary. If we define G_A as the matrix whose j-th row is $\boldsymbol{g}_{i_j}$, then all the $d \times \ell$ matrices V such that $\pi_\ell(V) = (\boldsymbol{s}^1, \ldots, \boldsymbol{s}^\ell)$ and $M_A \cdot V = \pi^\ell(G_A)$ are equally likely, even given the adversary's entire view.*

---

[5] Recall that this just means that $\pi_\ell(\boldsymbol{v}^b) = \boldsymbol{s}^b$ and secret sharing with $\boldsymbol{v}^b$ produces the shares held by the honest parties in the protocol, i.e., $(M\boldsymbol{v}^b)[j] = \boldsymbol{g}_j[b]$ for all honest P_j.

For the applications of π_{VSS} that we will show in Section 5, we will require some new specialized forms of π_{VSS}, which we describe in the following two sections.

3.1 Applying a Linear Map to All the Secrets

Let $\varphi : \mathbb{F}^\ell \to \mathbb{F}^\ell$ be a linear function. Suppose that the dealer executes two correlated instances of Protocol π_{VSS} in the following way: first the dealer executes π_{VSS} with input $s^1, \ldots, s^\ell$ choosing matrix F in step 1, later on, he executes π_{VSS} with input $\varphi(s^1), \ldots, \varphi(s^\ell)$ under the condition that the matrix chosen for the second instance, F_φ, satisfies $\pi_\ell(f^{\varphi,i}) = \varphi(\pi_\ell(f^i))$ for $i = 1, \ldots, d$. The dealer sends to P_i vectors h^i and g_i and also the vectors $h^{\varphi,i} = F_\varphi \cdot m_i^\top$, $g_{\varphi,i} = m_i \cdot F_\varphi$. The protocol is shown in figure 4.

Packed Verifiable Secret-Sharing Protocol for φ, π_{VSS}^φ

1. Let $s^1, \ldots, s^\ell \in \mathbb{F}^\ell$ be the secrets to be shared. The dealer chooses two random $d \times d$ matrices F, F_φ subject to $\pi_\ell(f^b) = s^b$ for any $b = 1, \ldots, \ell$ and $\pi_\ell(f^{\varphi,i}) = \varphi(\pi_\ell(f^i))$ for any $i = 1, \ldots, d$.
2. The dealer sends h^i, g_i, $h^{\varphi,i}$ and $g_{\varphi,i}$ to P_i.
3. Each player P_j sends $g_j \cdot m_i^\top$ and $g_{\varphi,j} \cdot m_i^\top$ to P_i, for $i = 1, \ldots, n$.
4. Each P_i checks, for $j = 1, \ldots, n$, that $m_j \cdot h^i$ and $m_j \cdot h^{\varphi,i}$ are equal to the values received from P_j and also that $\pi_\ell(\tilde{h}^i) = \varphi\left(\pi_\ell(h^i)\right)$. He broadcasts *Accept* if all checks are OK, otherwise he broadcasts *Reject*.
5. If all players said *Accept*, then each P_j stores, for $b = 1, \ldots, \ell$, $g_j[b]$ and $g_{\varphi,j}[b]$ as his share respectively of s^b and $\varphi(s^b)$, otherwise the protocol aborts.

Fig. 4. The VSS protocol for φ

The completeness of the π_{VSS}^φ protocol is trivial to prove. Moreover we will show in the following lemma 5 and 6, that also the properties of soundness and privacy are still valid for the π_{VSS}^φ protocol.

Lemma 5. *If the dealer in π_{VSS}^φ is corrupt, but no player rejects, then for any $b = 1, \ldots, \ell$ there exist column vectors v^b, $v^{\varphi,b}$ and s^b, $s^{\varphi,b}$ such that v^b shares s^b, $v^{\varphi,b}$ shares $s^{\varphi,b}$ and $\varphi\left(s^b\right) = s^{\varphi,b}$.*

Lemma 6. *Suppose the dealer in π_{VSS}^φ is honest. Now, in case 1 suppose the dealer executes π_{VSS}^φ with input $s^1, \ldots, s^\ell$ and in case 2, he executes π_{VSS}^φ with input $\tilde{s}^1, \ldots, \tilde{s}^\ell$. Let $A = \{P_{i_1}, \ldots, P_{i_t}\}$ be a set of players corrupted by the adversary, then the adversary's view in the two cases are identically distributed.*

Finally we show the randomness property satisfied by the pair of share vectors of s^i, $\varphi(s^i)$ obtained by the π_{VSS}^φ protocol.

Lemma 7. *Suppose that the dealer in π_{VSS}^φ is honest and let $A = \{P_{i_1}, \ldots, P_{i_t}\}$ be a set of players corrupted by the adversary. If we define G_A as the matrix* *whose j-th row is g_{i_j} and $G_{\varphi,A}$ as the matrix whose jth column is g_{φ,i_j}, then all the pairs of $d \times \ell$ matrices (V, V_φ) such that $\pi_\ell(V) = (s^1, \ldots, s^\ell)$, $\pi_\ell(V_\varphi) = (\varphi(s^1), \ldots, \varphi(s^\ell))$, $M_A \cdot V = \pi^\ell(G_A)$ and $M_A \cdot V_\varphi = \pi^\ell(G_{\varphi,A})$ are equally likely, even given the adversary's entire view.*

3.2   Sharing an All Zeros Vector

We are interested in modifying Protocol π_{VSS} in order to share several times just the vector 0^ℓ, *i.e.* the all zeros column vector in $\mathbb{F}^\ell$. Suppose that the $d \times d$ random matrix F chosen by the dealer has the first ℓ rows equal to zero. Let R be the $e \times d$ matrix formed by the last e rows of F, then

$$h^i = F \cdot m_i^\top = \left(0, \ldots, 0, R \cdot m_i^\top\right)^\top$$

$$g_i = m_i \cdot F = (m_i[\ell+1], \ldots, m_i[d]) \cdot R$$

Given the special form of the vectors h^i, the players can check not only that the shares are consistent, but also that they are consistent with 0^ℓ. Define $h^{0,i} = \left(R \cdot m_i^\top\right)^\top$, $m_{0,i} = (m_i[\ell+1], \ldots, m_i[d])$ and $M_{0,A}$ as the matrix whose rows are the vectors $m_{0,i}$ with $P_i \in A$. The protocol in this case is shown in Figure 5.

Packed Verifiable Secret-Sharing Protocol for 0^ℓ's π_{VSS}^0

1. The dealer chooses a random $e \times d$ matrix R with entries in $\mathbb{F}$,
2. The dealer sends $h^{0,i}$ and g_i to P_i.
3. Each player P_j sends $g_j \cdot m_i^\top$ to P_i, for $i = 1, \ldots, n$.
4. Each P_i checks that for $j = 1, \ldots, n$, $m_{0,j} \cdot h^{0,i}$ equals the value received from P_j. He broadcasts *Accept* if all checks are OK, otherwise he broadcasts *Reject*.
5. If all players said *Accept*, then each P_j stores, for $b = 1, \ldots, \ell$, $g_j[b]$ as his b-th share of 0^ℓ, otherwise the protocol aborts.

Fig. 5. The VSS protocol for 0^ℓ's

Again the completeness of Protocol π_{VSS}^0 is trivial. We will show the soundness property in Lemma 8, while privacy is not required in this special case.

Lemma 8. *If the dealer in π_{VSS}^0 is corrupt, but no player rejects, then there exist column vectors $v^1, \ldots, v^\ell$ each of which shares 0^ℓ.*

Finally we show that the randomness property that is satisfied by the share vectors obtained in Protocol π_{VSS} is also satisfied by the share vectors of 0^ℓ obtained by Protocol π_{VSS}^0.

Lemma 9. *Suppose that the dealer is honest and he executes Protocol π_{VSS}^0. Let $A = \{P_{i_1}, \ldots, P_{i_t}\}$ be a set of players corrupted by the adversary and define G_A as the matrix whose j-th row is g_{i_j}, then all the $e \times \ell$ matrices V such that $M_{0,A} \cdot V = \pi^\ell(G_A)$ are equally likely, even given the adversary's entire view.*

4   Low Overhead UC Commitments

In this section we introduce our construction of UC commitments with low overhead. A main ingredient will be the n-player VSS scheme from the previous section. We will use n as the security parameter. We will assume throughout that the underlying linear secret sharing scheme S is such that the parameters t and r are $\Theta(n)$, and furthermore that S has t-strong multiplication. We will call such an S a *commitment-friendly* linear secret sharing scheme.

The protocol first does a set-up phase where the sender executes the VSS scheme "in his head", where the secrets are random strings $r_1, \ldots, r_\ell$. The VSS is secure against t corrupted players. Next, he chooses n seeds $x_1, \ldots, x_n$ for a pseudorandom generator G, and $\mathcal{F}_{OT}^{t,n}$ is used to transfer a subset of t seeds to the verifier. Finally, the sender sends the view of each virtual VSS player to the receiver, encrypted with $G(x_1), \ldots, G(x_n)$ as "one-time pads". Note that the receiver can decrypt t of these views and check that they are consistent, and also he now knows t shares of each r_i.

To commit to $m \in \{0,1\}^\ell$, the sender picks the next unused secret r_η and sends $m + r_\eta$.

To open, the sender reveals m and the vector f^η (from the VSS) that shares r_η. The receiver can now compute all shares of r_η and check that they match those he already knows.

Intuitively, this is binding because the sender does not know which VSS players the receiver can watch. This means that the sender must make consistent views for most players, or be rejected immediately. But if most views are consistent, then the (partially encrypted) set of shares of r_η that was sent during set-up is almost completely consistent. Since the reconstruction threshold is smaller than n by a constant factor this means that the prover must change many shares to move to a different secret, and the receiver will notice this with high probability, again because the sender does not know which shares are already known to the receiver.

Hiding follows quite easily from security of the PRG G and privacy of the VSS scheme, since the receiver only gets t shares of any secret.

The Commit and Reveal phases of protocol π_{HCOM} are described in Figure 6 while the necessary steps for addition, multiplication and linear function evaluation are described separately in Section 5 for the sake of clarity.

The proof of the following theorem can be found in the full version of this paper [DDGN14].

Theorem 4. *Let $G : \{0,1\}^{\ell_{PRG}} \to \{0,1\}^{2(\ell+e)}$ be a pseudrandom generator and let π_{VSS} be a packed verifiable secret sharing scheme as described in Section 3 with parameters (M, r, t), based on a commitment-friendly secret sharing scheme. Then protocol π_{HCOM} UC-realizes $\mathcal{F}_{HCOM}$ in the $\mathcal{F}_{OT}^{t,n}(\ell_{PRG})$-hybrid model in the presence of static, active adversaries.*

Complexity. It is evident that in the set-up phase, or later, P_s could execute any number of instances of the VSS and send the resulting views of players encrypted with the seeds $\{x_i\}$, as long as we have a PRG with sufficient stretch. This way we can accommodate as many commitments as we want, while only using the OT-functionality once.[6] Therefore, the amortised cost of a commitment is essentially only what we pay after the OT has been done. We now consider what the cost will be per committed bit in communication and computation. Using the linear secret sharing scheme from [CDP12], we can get a commitment friendly secret sharing scheme over a constant size field, so this means that the communication overhead is constant.

As for computation, under plausible complexity assumptions, there exists a PRG where we pay only a constant number of elementary bit operations per output bit (see, e.g., [VZ12]), so the cost of computing the PRG adds only a constant factor overhead for both parties. As for the computation of P_r, let us consider the set-up phase first. Let C be the set of players watched by P_r, and let G_C, H_C be matrices where we collect the h^i and g_j's they have been assigned. Then what P_r wants to check is that $M_C H_C = G_C M_C^\top$. In [DZ13], a probabilistic method is described for checking such a relation that has complexity $O(n^2)$ field operations and fails with only negligible probability. This therefore also adds only a constant factor overhead because one VSS instance allows commitment to ℓ^2 bits which is $\Theta(n^2)$. Finally, in the reveal phase P_r computes $M f^\eta$ and verifies a few coordinates. If one can check $\Theta(n)$ such commitments simultaneously, the same trick from [DZ13] can be used, and we get an overall constant factor overhead for P_r. We note that checking many commitments in one go is exactly what we need for the application to non-interactive proofs we describe later.

For P_s, using the scheme from [CDP12], there is no way around doing standard matrix products which can be done in $O(n^{2+\sigma})$ complexity for $\sigma < 1$. This gives us overhead n^σ per committed bit.

Finally, if we use instead standard packed secret sharing based on polynomials, the field size must be linear in n, but on the other hand we can use FFT algorithms in our computations. This gives a poly-logarithmic overhead for both players in communication and computation.

5   Homomorphic Properties

In this section, we show how to implement the add, multiply and linear function commands in $\mathcal{F}_{\mathrm{HCOM}}$. As before, we assume a commitment-friendly linear secret sharing scheme $\mathcal{S}$.

We first need some notation: consider a single commitment as we defined it in the previous section and note that the data pertaining to that commitment consists of a vector f and the committed value m held by P_s, whereas P_r holds

---

[6] It is not hard to see that since a corrupt P_s looses as soon as P_r sees a single inconsistency, P_s cannot get any advantage from executing a VSS after other commitments have been done.

$m + \pi_\ell(f)$ as well as a subset of the coordinates of Mf. We will refer to the vector $m + \pi_\ell(f)$ as the *message field* of the commitment.

We will use $\text{com}_S(m, f)$ as a shorthand for all this data, where the subscript S refers to the fact that the matrix M of S defines the relation between the data of P_s and that of P_r. Whenever we write $\text{com}_S(m, f)$, this should also be understood as stating that the players in fact hold the corresponding data.

The expression $\text{com}_S(m, f) + \text{com}_S(m', f')$ means that both players add the corresponding vectors that they hold of the two commitments, and store the result. It is easy to see that we have

$$\text{com}_S(m, f) + \text{com}_S(m', f') = \text{com}_S(m + m', f + f')$$

Furthermore, $\text{com}_S(m, f) * \text{com}_S(m', f')$ means that the players compute the coordinate-wise product of corresponding vectors they hold and store the result. We have

$$\text{com}_S(m, f) * \text{com}_S(m', f') = \text{com}_{\hat{S}}(m * m', f \otimes f')$$

Note that $\hat{S}$ appears in the last term. Recall that the coordinates of $f \otimes f'$ are ordered such that indeed the vector $\pi_\ell(f) * \pi_\ell(f')$ appears in the first ℓ coordinates of $f \otimes f'$.

Now, in order to support the additional commands, we will augment the set-up phase of the protocol: in addition to π_{VSS}, P_s will execute π^0_{VSS} and π^φ_{VSS}. For π^0_{VSS} we use $\hat{S}$ as the underlying linear secret sharing scheme, where the other VSS schemes use S. Furthermore, we need an instance of π^φ_{VSS} for each linear function φ we want to support. As before, all the views of the virtual players are sent to P_r encrypted under the seeds x_i. P_r checks consistency of the views as well as the special conditions that honest players check in π^0_{VSS} and π^φ_{VSS}.

Note that if one instance of π_{VSS} has been executed, this allows us to extract data for ℓ commitments. Likewise, an execution of π^0_{VSS} allows us to extract ℓ commitments of form $\text{com}_{\hat{S}}(0^\ell, u)$ for a random u, where by default we set the message field to 0. Finally, having executed π^φ_{VSS}, we can extract ℓ pairs of form $\text{com}_S(r, f_r), \text{com}_S(\varphi(r), f'_r)$ where r is random such that $r = \pi_\ell(f_r)$ and $\varphi(r) = \pi_\ell(f'_r)$. Again, for these commitments we set the message field to 0. The protocols are shown in Figure 7.

Generalizations In the basic case we are committing to bit strings, and we note that we can trivially get negation of bits using the operations we already have: Given $\text{com}_S(m, f)$, P_s commits to 1^ℓ so we have $\text{com}_S(1^\ell, f')$, we output $\text{com}_S(m, f) + \text{com}_S(1^\ell, f')$ and P_s opens $\text{com}_S(1^\ell, f')$ to reveal 1^ℓ.

If we do the protocol over a larger field than $\mathbb{F}_2$, it makes sense to also consider multiplication of a commitment by a public constant. This is trivial to implement, both parties simply multiply their respective vectors by the constant.

Proof intuition. The protocol in Figure 7 can be proven secure by essentially the same techniques we used for the basic commitment protocol, but we need in addition the specific properties of π_{VSS}, π^0_{VSS} and π^φ_{VSS}. First of all, it is clear that in the case when the sender is corrupted and the receiver is honest, a simulator for this protocol can extract the messages (and share vectors) in the commitments by following the same procedure as the simulator for the basic commitment protocol. The specific properties of the VSS protocols π_{VSS}, π_{VSS}^0 and π_{VSS}^φ come into play when constructing a simulator for the case when the sender is honest and the receiver is corrupted. More details are presented in the full version of this paper [DDGN14].

6   Applications

Two-party One-sided Functionalities. In this section we consider applications of our implementation of $\mathcal{F}_{\text{HCOM}}$. We will implement a one-sided functionality where only one party P_s has input x and some verifier is to receive output y, where $y = C(x)$ for a Boolean circuit C.

The basic idea of this is straightforward: P_s commits to each bit in x and to each output from a gate in C that is produced when x is the input. Now we can use the commands of $\mathcal{F}_{\text{HCOM}}$ to verify for each gate that the committed output is the correct function of the inputs. Finally, P_s opens the commitment to the final output to reveal y.

However, we would like to exploit the fact that our commitments can contain ℓ-bit strings and support coordinate-wise operations on ℓ-bit strings in parallel. To this end, we can exploit the construction found in [DIK10] (mentioned in the introduction), that allows us to construct from C a new circuit C' computing the same function as C, but where C' can be computed using only operations in parallel on ℓ-bit blocks as well as $\log \ell$ different permutations of the bits in a block. The construction always works, but if C is well-formed, C' will be of size $O(\log(|C|)|C|)$.

With these observations, we can use $\mathcal{F}_{\text{HCOM}}$ operations to compute C' instead of C. The difference to the first simplistic idea is that now every position in a block is used for computation. Hence, the final protocol is non-interactive, assuming the very first step doing the OT has been done. This is because P_s, since he knows C, can predict which multiplications and permutation operations P_r will need to verify, so he can compute the required opening information for commitments and send them immediately. Moreover, if we use the linear secret sharing scheme from [CDP12] as the basis for commitments, then the size of the entire proof as well as of the verifier's computation will be of size $O(|C| \log |C|)$ for well formed circuits. If C is regular we will get complexity $O(|C|)$, since we can implement the rerouting between layers by evaluating the block functions directly. Thus we get the results claimed in Theorem 1.

Multiparty Computation. Due to space constraints the material on MPC based on our VSS (Theorem 2) is left for the full version of this paper [DDGN14].

Protocol π_{HCOM} in the $\mathcal{F}_{OT}^{t,n}(\ell_{PRG})$-hybrid model

Let $G : \{0,1\}^{\ell_{PRG}} \to \{0,1\}^{2(\ell+e)}$ be a pseudorandom generator and let π_{VSS} be a packed verifiable secret sharing scheme as described in Section 3 with parameters (M, r, t) based on a commitment-friendly linear secret sharing scheme. A sender P_s and a receiver P_r interact between themselves and with $\mathcal{F}_{OT}^{t,n}(\ell_{PRG})$ as follows:

Setup Phase: At the beginning of the protocol P_s and P_r perform the following steps and then wait for inputs.

1. For $i = 1, \ldots, n$, P_s uniformly samples a random string $x_i \in \{0,1\}^{\ell_{PRG}}$. P_s sends $(\mathsf{sender}, sid, ssid, x_1, \ldots, x_n)$ to $\mathcal{F}_{OT}^{t,n}(\ell_{PRG})$.
2. P_r uniformly samples a set of t indexes $c_1, \ldots, c_t \leftarrow [1, n]$ and sends $(\mathsf{receiver}, sid, ssid, c_1, \ldots, c_t)$ to $\mathcal{F}_{OT}^{t,n}$.
3. Upon receiving $(\mathsf{received}, sid, ssid)$ from $\mathcal{F}_{OT}^{t,n}$, P_s uniformly samples n random strings $r_i \leftarrow \{0,1\}^\ell$, $i = 1, \ldots, \ell$ and internally runs π_{VSS} using $r_1, \ldots, r_n$ as input, constructing n strings $((h^i)^\top, g_i)$, $i = 1, \ldots, n$ of length $2(\ell+e)$ from the vectors generated by π_{VSS}. P_s computes $((\widetilde{h}^i)^\top, \widetilde{g}_i) = ((h^i)^\top, g_i) + G(x_i)$ and sends $(sid, ssid, ((\widetilde{h}^1)^\top, \widetilde{g}_1), \ldots, ((\widetilde{h}^n)^\top, \widetilde{g}_n))$ to P_r.
4. Upon receiving $(\mathsf{received}, sid, ssid, x_{c_1}, \ldots, x_{c_t})$ from $\mathcal{F}_{OT}^{t,n}$ and $(sid, ssid, ((\widetilde{h}^1)^\top, \widetilde{g}_1), \ldots, ((\widetilde{h}^n)^\top, \widetilde{g}_n))$ from P_s, P_r computes $((h^{c_j})^\top, g_{c_j}) = ((\widetilde{h}^{c_j})^\top, \widetilde{g}_{c_j}) - G(x_{c_j}), 1 \leq j \leq t$ and uses the procedures of π_{VSS} to check that the shares $g_{c_1}, \ldots, g_{c_t}$ are valid, i.e. it checks that $m_j \cdot h^i = g_j \cdot m_i^\top$ for $i, j \in \{c_1, \ldots, c_t\}$. If all shares are valid P_r stores $(ssid, sid, ((h^{c_1})^\top, g_{c_1}), \ldots, ((h^{c_t})^\top, g_{c_t}))$, otherwise it halts.

Commit Phase:

1. Upon input $(\mathsf{commit}, sid, ssid, P_s, P_r, m)$, P_s chooses an unused[a] random string r_η, computes $\widetilde{m} = m + r_\eta$ and sends $(sid, ssid, \eta, \widetilde{m})$ to P_r.
2. P_r stores $(sid, ssid, \widetilde{m})$ and outputs $(\mathsf{receipt}, sid, ssid, P_s, P_r)$.

Reveal Phase:

1. Upon input $(\mathsf{reveal}, sid, ssid, P_s, P_r)$, to reveal a message m, P_s reveals the random string r_η by sending $(sid, ssid, m, f^\eta)$ to P_r.[b]
2. P_r receives $(sid, ssid, \eta, m, f^\eta)$, computes $M f^\eta = (\overline{g}_1[\eta], \ldots, \overline{g}_n[\eta])^\top$, checks that $g_j[\eta] = \overline{g}_j[\eta]$ for $j \in \{c_1, \ldots, c_t\}$ and that $m = \widetilde{m} - r_\eta$. If the shares pass this check, P_r outputs $(\mathsf{reveal}, sid, ssid, P_s, P_r, m)$. Otherwise, it rejects the commitment and halts.

---

[a] We say that a string r_η is unused if it has not been selected by P_s for use in any previous commitment.

[b] Recall that f^η denotes the η-th column of F, $\pi_\ell(f^\eta) = r_\eta$ and that $M f^\eta = (g_1[\eta], \ldots, g_n[\eta])^\top$, i.e. f^η determines the shares of r_η generated in the setup phase.

Fig. 6. Protocol π_{HCOM} in the $\mathcal{F}_{OT}^{t,n}(\ell_{PRG})$-hybrid model

---

Protocols for addition, multiplication and linear operations

Setup Phase: Is augmented by executions of π^0_{VSS} and π^φ_{VSS} as described in the text. Throughout, opening a commitment $\mathsf{com}_\mathcal{S}(m, f)$ means that P_s sends m, f and P_r verifies, as in π_{HCOM}.

Addition: Given commitments $\mathsf{com}_\mathcal{S}(m, f), \mathsf{com}_\mathcal{S}(m', f')$, output

$$\mathsf{com}_\mathcal{S}(m, f) + \mathsf{com}_\mathcal{S}(m', f') = \mathsf{com}_\mathcal{S}(m + m', f + f').$$

Multiplication: Given commitments $\mathsf{com}_\mathcal{S}(a, f_a), \mathsf{com}_\mathcal{S}(b, f_b)$, and $\mathsf{com}_\mathcal{S}(c, f_c)$ extract the next unused commitment from π^0_{VSS}, $\mathsf{com}_{\hat{\mathcal{S}}}(0^\ell, u)$. Form a default commitment $\mathsf{com}_\mathcal{S}(1^\ell, f_1)$, where $\pi_\ell(f_1) = 1^\ell$ and the other coordinates are 0. This can be done by only local computation. P_s opens the following commitment to reveal 0^ℓ:

$$\mathsf{com}_\mathcal{S}(a, f_a) * \mathsf{com}_\mathcal{S}(b, f_b) - \mathsf{com}_\mathcal{S}(c, f_c) * \mathsf{com}_\mathcal{S}(1^\ell, f_1) + \mathsf{com}_{\hat{\mathcal{S}}}(0^\ell, u) =$$
$$\mathsf{com}_{\hat{\mathcal{S}}}(a * b - c, f_a \otimes f_b - f_c \otimes f_1 + u)$$

Linear Function Given commitment $\mathsf{com}_\mathcal{S}(m, f)$, extract from π^φ_{VSS} the next unused pair $\mathsf{com}_\mathcal{S}(r, f_r), \mathsf{com}_\mathcal{S}(\varphi(r), f'_r)$. P_s opens $\mathsf{com}_\mathcal{S}(m, f) - \mathsf{com}_\mathcal{S}(r, f_r)$ to reveal $m - r$. Both parties compute $\varphi(m - r)$ and form a vector v such that $\pi_\ell(v) = \varphi(m - r)$ and the rest of the entries are 0. Output

$$\mathsf{com}_\mathcal{S}(\varphi(r), f'_r) + \mathsf{com}_\mathcal{S}(\varphi(m - r), v) = \mathsf{com}_\mathcal{S}(\varphi(m), f'_r + v)$$

Fig. 7. Protocol for homomorphic operations on commitments

String Commitment with Partial Individual Opening. Here we wish to implement a functionality $F^{N,r}_{com}$ that first allows P_s to commit to N bits and then to open up to r bits individually, where he can decide adaptively which bits to open. We do this in the correlated random bits model where a functionality is assumed that initially gives bit strings to P_s and P_r with some prescribed joint distribution, the implementation must be statistically secure with error probability 2^{-k}.

Note that our protocol can be seen as a protocol in this model if we let players start from the strings that are output by the PRG. In this case we get statistically secure commitments with error probability $2^{-\Theta(\ell)}$ (since ℓ is $\Theta(n)$). So we can choose ℓ to be $\Theta(k)$ and get the required error probability. Then one of our commitments can be realised while consuming $O(k) = O(\ell)$ correlated random bits.

Note that we can open a single bit in a commitment to a as follows: to open the j'th bit a_j the prover commits to e_j, a vector with 1 in position j and 0 elsewhere and to c which has a_j in position j and 0's elsewhere. Now the multiplication check is done on commitments to a, e_j and c, and P_s opens e_j and c. P_r does the obvious checks and extracts a_j. It is trivial to show that this is a secure way to reveal only a_j and we consume $O(\ell)$ correlated random bits since only a constant number of commitments are involved.

Now we can implement $F_{com}^{N,r}$ with $N = \ell^u$ and $r = \ell^{u-1}$ for some u, and the implementation is done by having P_s commit to the N bits in the normal way using r commitments, and when opening any single bit, we execute the above procedure. This consumes a total of $O(N + r\ell) = O(N)$ correlated random bits. Thus the consumption per bit committed to is $O(1)$. Furthermore, we have $r = N^{(u-1)/u} = N^{1-1/u}$, so we get the result of Theorem 3 by choosing a large enough u.

Acknowledgements. We thank Yuval Ishai for pointing out interesting applications of our results and Ignacio Cascudo for clarifying key facts about algebraic geometric secret sharing schemes.

References

App13.   Applebaum, B.: Pseudorandom generators with long stretch and low locality from random local one-way functions. SIAM Journal on Computing 42(5), 2008–2037 (2013)

BBDK00.   Beimel, A., Burmester, M., Desmedt, Y., Kushilevitz Computing, E.: functions of a shared secret. SIAM J. Discrete Math. 13(3), 324–345 (2000)

BCPV13.   Blazy, O., Chevalier, C., Pointcheval, D., Vergnaud, D.: Analysis and improvement of lindell's UC-secure commitment schemes. In: Jacobson, M., Locasto, M., Mohassel, P., Safavi-Naini, R. (eds.) ACNS 2013. LNCS, vol. 7954, pp. 534–551. Springer, Heidelberg (2013)

BCR86.   Brassard, G., Crepeau, C., Robert, J.-M.: Information theoretic reductions among disclosure problems. In: 27th Annual Symposium on Foundations of Computer Science, pp. 168–173 (1986)

BOGW88.   Ben-Or, M., Goldwasser, S., Wigderson, A.: Completeness theorems for non-cryptographic fault-tolerant distributed computation (extended abstract). In: STOC, pp. 1–10 (1988)

Can01.   Canetti, R.: Universally composable security: A new paradigm for cryptographic protocols. In: FOCS, pp. 136–145. IEEE Computer Society (2001)

CDM00.   Cramer, R., Damgård, I., Maurer, U.M.: General secure multi-party computation from any linear secret-sharing scheme. In: Preneel, B. (ed.) EUROCRYPT 2000. LNCS, vol. 1807, pp. 316–334. Springer, Heidelberg (2000)

CDP12.   Cramer, R., Damgård, I., Pastro, V.: On the amortized complexity of zero knowledge protocols for multiplicative relations. In: Smith, A. (ed.) ICITS 2012. LNCS, vol. 7412, pp. 62–79. Springer, Heidelberg (2012)

CF01.   Canetti, R., Fischlin, M.: Universally composable commitments. In: Kilian, J. (ed.) CRYPTO 2001. LNCS, vol. 2139, pp. 19–40. Springer, Heidelberg (2001)

CLOS02.   Canetti, R., Lindell, Y., Ostrovsky, R., Sahai, A.: Universally composable two-party and multi-party secure computation. In: STOC, pp. 494–503 (2002)

DDGN14.   Damgård, I., David, B., Giacomelli, I., Nielsen, J.B.: Compact VSS and efficient homomorphic UC commitments. Cryptology ePrint Archive: Report 2014/370 (2014), https://eprint.iacr.org/2014/370

DG03.   Damgård, I., Groth, J.: Non-interactive and reusable non-malleable commitment schemes. In: Larmore, L.L., Goemans, M.X. (eds.) STOC, pp. 426–437. ACM (2003)

DI06.   Damgård, I.B., Ishai, Y.: Scalable secure multiparty computation. In: Dwork, C. (ed.) CRYPTO 2006. LNCS, vol. 4117, pp. 501–520. Springer, Heidelberg (2006)

DIK+08.   Damgård, I., Ishai, Y., Krøigaard, M., Nielsen, J.B., Smith, A.: Scalable multiparty computation with nearly optimal work and resilience. In: Wagner, D. (ed.) CRYPTO 2008. LNCS, vol. 5157, pp. 241–261. Springer, Heidelberg (2008)

DIK10.   Damgård, I., Ishai, Y., Krøigaard, M.: Perfectly secure multiparty computation and the computational overhead of cryptography. In: Gilbert, H. (ed.) EUROCRYPT 2010. LNCS, vol. 6110, pp. 445–465. Springer, Heidelberg (2010)

DNO10.   Damgård, I., Nielsen, J.B., Orlandi, C.: On the necessary and sufficient assumptions for UC computation. In: Micciancio, D. (ed.) TCC 2010. LNCS, vol. 5978, pp. 109–127. Springer, Heidelberg (2010)

DZ13.   Damgård, I., Zakarias, S.: Constant-overhead secure computation of boolean circuits using preprocessing. In: Sahai, A. (ed.) TCC 2013. LNCS, vol. 7785, pp. 621–641. Springer, Heidelberg (2013)

FJN+13.   Frederiksen, T.K., Jakobsen, T.P., Nielsen, J.B., Nordholt, P.S., Orlandi, C.: Minilego: Efficient secure two-party computation from general assumptions. In: Johansson, T., Nguyen, P.Q. (eds.) EUROCRYPT 2013. LNCS, vol. 7881, pp. 537–556. Springer, Heidelberg (2013)

FY92.   Franklin, M.K., Yung, M.: Communication complexity of secure computation (extended abstract). In: STOC, pp. 699–710 (1992)

GIKW14.   Garay, J., Ishai, Y., Kumaresan, R., Wee, H.: On the complexity of uc commitments. In: EuroCrypt 2014 (to appear 2014)

IKOS08.   Ishai, Y., Kushilevitz, E., Ostrovsky, R., Sahai, A.: Cryptography with constant computational overhead. In: Dwork, C. (ed.) STOC, pp. 433–442. ACM (2008)

IKOS09.   Ishai, Y., Kushilevitz, E., Ostrovsky, R., Sahai, A.: Zero-knowledge proofs from secure multiparty computation. SIAM J. Comput. 39(3), 1121–1152 (2009)

IPS08.   Ishai, Y., Prabhakaran, M., Sahai, A.: Founding cryptography on oblivious transfer – efficiently. In: Wagner, D. (ed.) CRYPTO 2008. LNCS, vol. 5157, pp. 572–591. Springer, Heidelberg (2008)

Lin11.   Lindell, Y.: Highly-efficient universally-composable commitments based on the ddh assumption. In: Paterson, K.G. (ed.) EUROCRYPT 2011. LNCS, vol. 6632, pp. 446–466. Springer, Heidelberg (2011)

Nao91.   Naor, M.: Bit commitment using pseudorandomness. J. Cryptology 4(2), 151–158 (1991)

NP99.   Naor, M., Pinkas, B.: Oblivious transfer with adaptive queries. In: Wiener, M. (ed.) CRYPTO 1999. LNCS, vol. 1666, pp. 573–590. Springer, Heidelberg (1999)

PVW08.   Peikert, C., Vaikuntanathan, V., Waters, B.: A framework for efficient and composable oblivious transfer. In: Wagner, D. (ed.) CRYPTO 2008. LNCS, vol. 5157, pp. 554–571. Springer, Heidelberg (2008)

RTWW11.   Ranellucci, S., Tapp, A., Winkler, S., Wullschleger, J.: On the efficiency of bit commitment reductions. In: Lee, D.H., Wang, X. (eds.) ASIACRYPT 2011. LNCS, vol. 7073, pp. 520–537. Springer, Heidelberg (2011)

VZ12.   Vadhan, S., Zheng, C.J.: Characterizing pseudoentropy and simplifying pseudorandom generator constructions. In: Proceedings of the 44th symposium on Theory of Computing, pp. 817–836. ACM (2012)

Round-Optimal Password-Protected Secret Sharing and T-PAKE in the Password-Only Model

Stanislaw Jarecki[1], Aggelos Kiayias[2], and Hugo Krawczyk[3]

[1] University of California Irvine
stasio@ics.uci.edu
[2] National and Kapodistrian University of Athens
aggelos@kiayias.com
[3] IBM Research
hugo@ee.technion.ac.il

Abstract. In a *Password-Protected Secret Sharing (PPSS)* scheme with parameters (t, n) (formalized by Bagherzandi et al. [2]), a user Alice stores secret information among n servers so that she can later recover the information solely on the basis of her password. The security requirement is similar to a (t, n)-threshold secret sharing, i.e., Alice can recover her secret as long as she can communicate with $t+1$ honest servers but an attacker gaining access to t servers cannot learn any information about the secret. In particular, the system is secure against offline password attacks by an attacker controlling up to t servers. On the other hand, accounting for inevitable on-line attacks one allows the attacker an advantage proportional to the fraction of dictionary passwords tested in on-line interactions with the user and servers.

We present the first round-optimal PPSS scheme, requiring just one message from user to server and from server to user, and prove its security in the challenging password-only setting where users do not have access to an authenticated public key. The scheme uses an Oblivious PRF whose security we define using a UC-style ideal functionality for which we show concrete, truly practical realizations in the random oracle model as well as standard-model instantiations. As an important application we use this scheme to build the first single-round password-only Threshold-PAKE protocol in the CRS and ROM models for arbitrary (t, n) parameters with no PKI requirements for any party (clients or servers) and no inter-server communication. Our T-PAKE protocols are built by combining suitable key exchange protocols on top of our PPSS schemes. We prove T-PAKE security via a generic composition theorem showing the security of any such composed protocol.

1   Introduction

Remarkably, passwords have become a fundamental pillar of electronic security. That's quite a high task for these low-entropy easily-memorable easily-guessed short character strings. In spite of repeated evidence of their vulnerability to

P. Sarkar and T. Iwata (Eds.): ASIACRYPT 2014, PART II, LNCS 8874, pp. 233–253, 2014.

misuse and attack, passwords are still in widespread use and will probably remain as such for a long while. The portability of passwords makes them ubiquitous keys to access remote services, open computing devices, decrypt encrypted files, protect financial and medical information, etc. Replacing passwords with long keys requires storing these keys in devices that are not always available to the user and are themselves at risk of falling in adversarial hands, hence endangering these keys and the data they protect.

An increasingly common solution to the problem of data security and availability is to store the data itself, or at least the keys protecting its security, at a remote server, which in turn is accessed using a password. This requires full trust in this single server and the one password. In particular, compromising such a server (or just its password file) is sufficient to crack most passwords stored at it through an *off-line* dictionary attack. Indeed, loss of millions of passwords to such attacks are common news nowadays [31]. Unfortunately, off-line attacks are unavoidable in single-server scenarios. A natural approach to solving this problem is to distribute the above trust over a set of servers, for example by sharing information among these servers using a secret sharing scheme. However, how does the user access these servers? Using the same password in each of these servers makes the off-line password recovery attack even worse (as it can be performed against any of these servers) while memorizing a different password for each server is impractical.

PPSS. The above problem and a framework for solution is captured by the notion of **Password-Protected Secret Sharing** (PPSS), originated by the work of Ford and Kaliski [16] and Jablon [18] and formalized by Bagherzandi et al. [2]. In such a scheme, parametrized by (t, n), user Alice has some secret information sc that she wants to store and protect, and be able to later access on the basis of a single password pw. (Secret sc can represent any form of information, but it is best to think of it as a cryptographic key which protects some cryptographic capability.) The scheme has an *initialization phase* where Alice communicates with each one of a set of n servers $S_1, \ldots, S_n$ after which each server S_i stores some information ω_i associated with user Alice. When Alice needs to retrieve sc, she performs a *reconstruction protocol* by interacting with at least $t + 1$ servers where the only input from Alice is her password pw.

The main requirements from this protocol are, informally: (i) an attacker breaking into t servers cannot gain any information on sc other than by correctly guessing Alice's password and running an on-line attack with it (more on this below). It follows, in particular, that off-line attacks on the password are not possible as long as the attacker has not compromised more than t servers. In this case, the only avenue of attack against the secrecy of sc is for the attacker to select one value pw' from a given dictionary D of passwords (from which the user has selected a password at random) and check its validity by interacting with the user and servers using pw' as the password. If the overall number of interactions between the attacker and the user, and between the attacker and the servers, is q then we allow the attacker to break the semantic security of sc with advantage $q/|D|$. Moreover, we will require that "testing" a guessed password by impersonating the user to the servers will require interacting with $t + 1$ different servers. (ii) Soundness: Similarly, a compromise of up to t servers cannot allow an attacker to make the user reconstruct a wrong secret $\mathsf{sc'} \neq \mathsf{sc}$, except with probability proportional to the number of *on-line* interactions between the attacker and the user. This is a necessary exception as the attacker can isolate the user and simulate a run with the servers with a password $\mathsf{pw'}$ and secret $\mathsf{sc'}$ chosen by the adversary; what's required is that *only if* $\mathsf{pw'}$ happens to be the user's password will the attack succeed. Additionally, a desirable property is (iii) Robustness: Alice can correctly reconstruct sc as long as (a) no more than t servers are corrupted and (b) Alice communicates without disruptions with at least $t+1$ honest servers. Note that robustness can only be achieved if $2t+1 \leq n$ while the other properties do not impose such intrinsic limitation.

T-PAKE. While PPSS schemes have many uses such as for retrieving keys, credentials, data, and so on, the main PPSS application is for bootstrapping a *Threshold Password-Authenticated Key Exchange (T-PAKE)* [27]. In a (t, n) T-PAKE protocol, a user with a single password is to establish authenticated keys with any given subset of the n servers, such that security of the keys established with uncorrupted servers is guaranteed as long as there are no more than t corrupted servers. PPSS schemes make it possible to build T-PAKE protocols by combining the PPSS scheme with a regular key exchange protocol in a modular and generic way. This allows one to focus on the PPSS design which by virtue of being a much simpler primitive, e.g., avoiding the intricacies of the security of (password) authenticated key exchange protocols, is likely to result in simpler and stronger solutions, as is indeed demonstrated by our results below.

Prior/Concurrent Work and Our Contributions

For the general case of (t, n) parameters, Bagherzandi et al. [2] showed a PPSS scheme in the random oracle model (ROM) where the reconstruction protocol involves three messages between the user and a subset of $t+1$ servers (effectively 4 messages in the typical case that the user initiates the interaction). However, if any of these servers deviates from the correct execution of the protocol, the protocol needs to be re-run with a new subset of servers, which potentially increases the number of protocol rounds to $O(n)$. Another significant shortcoming of the PPSS solution from [2] is that it is secure only in the PKI model, namely, where the user can authenticate the public keys of the servers. Indeed, if the attacker can induce the user to run the protocol on incorrect public keys, the protocol of [2] becomes completely insecure. Thus, [2] leaves at least two open questions: Do PPSS protocols with *optimal single-round communication* exist (i.e., requiring a single message from user to server and single message from server to user), and can such protocols work in the *password-only model*, namely when the user does not have a guaranteed authentic public key.

We answer both questions in the affirmative by exhibiting a PPSS protocol for a general (t, n) setting with optimal single-round communication (in ROM) which works in the password-only model. Concurrently to our work, Camenisch et al. [7] have also presented a general (t, n) setting PPSS which works in the password-only model (and ROM). Their protocol sends 10 messages between the user and each server, its total communication complexity is $O(t^2)$, and the computational cost is more than 7 times the cost of our solution. Moreover, its robustness is fragile in the same way as that of [2], i.e. the user runs the reconstruction protocol with a chosen subset of $t+1$ players, and the protocol must be re-started if any server in this chosen group deviates. By contrast, the protocol we present has 2 messages and $O(n \log n)$ (worst-case) communication complexity, which reduces to $O(n)$ if the user caches $O(n)$ data between reconstruction protocol instances. Our protocol also has stronger robustness guarantee, namely, Alice recovers her shared secret sc in the single protocol instance as long as it has unobstructed communication with at least $t+1$ honest servers and if $2t+1 \le n$. While [7] formalize a UC functionality for PPSS (which they call "TPASS", for Threshold Password-Authenticated Secret Sharing) and prove their protocol to realize this functionality, we model the security of a PPSS scheme in the password-only setting with a game-based notion. We show that our game-based security notion is strong enough to imply the security of a natural T-PAKE construction built on top of a PPSS scheme. However, a PPSS satisfying a UC formalization might simplify the use of PPSS in the design of other cryptographic schemes, which leaves designing a UC secure PPSS with low message complexity and good robustness as an interesting open question.

Our PPSS construction is based on a novel version of so-called Oblivious Pseudorandom Function (OPRF) [17] and our contributions touch on three distinct elements, OPRF's, PPSS, and T-PAKE's, which we discuss next.

OPRF. The basic building block of our PPSS construction is a *Verifiable* Oblivious PRF (V-OPRF). Oblivious PRF (OPRF) was defined [17,20] as a protocol between two parties, a server and a user, where the first holds the key k for a PRF function f while the latter holds an argument x on which $f_k(\cdot)$ should be evaluated. At the end of the protocol the user learns $f_k(x)$ and is convinced that such value is properly evaluated while the server learns nothing. Formalizing the OPRF primitive in a way that can serve our application is not trivial. Indeed, the intuitive definition of OPRF [17,20] as the secure computation of a two-party functionality which on input pair (k, x) returns an output pair $(\perp, f_k(x))$, is limiting for at least three reasons: (1) It does not imply security when several OPRF instances are executed concurrently, as is the case in our PPSS construction; (2) It does not apply to our setting where the existence of authenticated channels cannot be assumed; and (3) It is not clear how to instantiate such functionality in the concurrent setting without on-line extractable zero-knowledge proofs of knowledge, which would add a significant overhead to any OPRF instantiation.

We overcome these issues via a novel formalization of the *verifiable* version of the OPRF primitive, V-OPRF, as an ideal functionality in the Universal Composability (UC) framework [10] for which we show several very efficient instantiations. Expressing V-OPRF in the UC framework is a delicate task, especially in the setting of interest to us where there are no authenticated channels. Our formalism enforces that the server who generates the PRF key k also produces a function descriptor π, which fixes a deterministic function f_π. (For honest servers, π is a commitment to k and the fixed function f_π is equal to the PRF f_k.) Then, in any (non-rejecting) execution of the V-OPRF protocol executed given the function descriptor π, the V-OPRF functionality verifies that the user's output is computed as $f_\pi(x)$. In other words, the V-OPRF functionality ensures *consistency* between V-OPRF instances executed under the same function descriptor π as well as *verifiability* that the output value is computed using the committed function f_π.

Our UC V-OPRF formalization bears interesting similarities to UC blind signatures [25,15,1]. In a nutshell, instead of on-line extraction of argument x from the (potentially malicious) client in every V-OPRF instance, a V-OPRF functionality issues a *ticket* for every instance executed under a given descriptor π. The user (or adversary) can then use these tickets to evaluate function f_π on inputs of their choice, but with the constraint that m tickets cannot be used to compute f_π values on more than m distinct inputs. Given this similarity, we observe that an efficient realization of V-OPRF can be achieved (in ROM) by hashing a deterministic blind signature-message pair.

We obtain three highly efficient variants of this design strategy, which provide three *single-round* V-OPRF instantiations in ROM, and we prove them UC-secure under "one-more" type of assumptions [3,21]. Specifically, we show such V-OPRF instantiations in ROM under a one-more Gap DH assumption on any group of prime order, a similar assumption on the group with a bilinear map, and a one-more RSA assumption. We also provide an efficient *standard model* V-OPRF construction for the Naor-Reingold PRF [30], based on the honest-but-curious OPRF protocol given by [17]. This protocol has four messages and is secure under Strong-RSA and the Decisional Composite Residuosity (DCR) assumptions. A single round standard-model (CRS) protocol is possible too but at a significant higher computational complexity.

We note that the UC formalization of the Verifiable Oblivious PRF functionality that is at the core of our security treatment is likely to have applications beyond this work. Indeed, OPRF's have been shown to be useful in a variety of scenarios, including Searchable Symmetric Encryption (SSE) schemes, e.g. [13,11], and secure two-party computation of Set Intersection [17,20,21].

PPSS. Our PPSS protocol is *password-only* in the Common Reference String (CRS) model, i.e. the user needs no other inputs than her password and a CRS string defining a non-malleable commitment scheme instance which can be part of the user's V-OPRF software. Our PPSS protocol is *single-round* in the hybrid model where parties can access the V-OPRF functionality. Given the V-OPRF instantiations discussed above, this implies three different instantiations of a single-round (i.e. two-message) PPSS schemes in ROM based on different one-more type of assumptions, and a four-message PPSS scheme in the CRS model.

Our PPSS construction follows the strategy of the early protocols of Ford and Kaliski [16] and Jablon [18] who treated the case of $t = n$: Secret-share the secret sc into shares $(s_1, \ldots, s_n)$, let each server S_i pick key k_i for a PRF f, and let $\mathbf{c} = (e_1, \ldots, e_n)$ where e_i is an encryption of s_i under $\rho_i = f_{k_i}(\mathsf{pw})$. Each server S_i stores $(k_i, \mathbf{c})$, and in the reconstruction protocol the user re-computes each ρ_i via an instance of a V-OPRF protocol with each server S_i on its input pw and S_i's input k_i. If the user also gets string $\mathbf{c}$ from the servers, the user can decrypt shares s_i using the ρ_i's and interpolate these shares to reconstruct sc. The first thing to note is that ciphertexts e_i must *not* be committing to the encryption key ρ_i. Otherwise, an adversary could test a password guess pw* in an interaction with a single V-OPRF instance (instead of requiring $t + 1$ interactions with $t + 1$ different servers as our security notion imposes on the attacker), by computing $\rho_i^* = f_{k_i}(\mathsf{pw}^*)$ and testing if decryption of e_i under ρ_i^* returns a plausible share value. We prevent such tests by sharing sc over a binary extension field $\mathbb{F} = GF(2^\ell)$, choosing a PRF f which maps onto ℓ-bit strings, and setting e_i to $s_i \oplus f_{k_i}(\mathsf{pw})$. Secondly, the above simple protocol can allow a malicious server S_i to find the user's password pw if S_i is not forced to use the same function f_{π_i} in each V-OPRF instance. Consider the OPRF protocol of [17] for the Naor-Reingold PRF $f_{k_i}(x) = g^v$ where $v = k_{i,0} \cdot \prod_{x_j=1} k_{i,j}$ for $k_i = (k_{i,0}, \ldots, k_{i,\ell})$ [30]. If in some PPSS instance, a misbehaving S_i uses key k_i' which differs from k_i on one index j, i.e. in one component $k_{i,j}$, S_i can conclude that the j-th bit of pw is 0 if the user recovers its secret correctly from such PPSS instance. Note that the adversary can learn whether user's secret is reconstructed correctly by observing any higher-level protocol which uses this secret, e.g. a T-PAKE protocol discussed below. We counter this attack by using the verifiability property of our V-OPRF functionality, which ensures that S_i computes the function committed in π_i, and by extending the user-related information stored by each server to $\omega = (\boldsymbol{\pi}, \mathbf{c}, C)$ where $\boldsymbol{\pi}$ is a vector of function descriptors $\pi_1, \ldots, \pi_n$ of each server, and C is a non-malleable commitment to the values $\boldsymbol{\pi}$, $\mathbf{c}$ and user's password pw. This commitment is the basis for ensuring that the on-line attacker playing the role of the servers can test at most one password guess per one reconstruction protocol instance. Note that the described solution requires $O(n)$ storage and bandwidth per server, but it is straightforward to reduce these to $O(\log n)$ using a Merkle tree commitment.

With an instantiation of V-OPRF in ROM we achieve a remarkably efficient reconstruction protocol without relying on PKI or secure channels. The user runs an optimal 2-message protocol with $t + 1$ (or more) servers, and in the case of our V-OPRF construction based on the one-more Gap DH assumption, the protocol involves just 2 exponentiations by the server and a total of $2t + 3$ multi-exponentiations for the user, employing the optimized ROM-based NIZK for discrete logarithm equality of [12], plus a few inexpensive operations. The (one-time) initialization stage is also very efficient, involving $2n + 1$ exponentiations for the user and 3 exponentiations per each server. Note that there is no inter-server communication in the protocol and that the user can communicate with each server independently, so it can be done in parallel and/or in any order without the servers being aware of each other. Moreover, the user can initiate the V-OPRF protocol with more than $t + 1$ servers, and it will reconstruct secret sc as long as $t + 1$ contacted servers reply with correct triple $\omega = (\boldsymbol{\pi}, \mathbf{c}, C)$ and complete the V-OPRF instances on function descriptors π_i in $\boldsymbol{\pi}$.

Our PPSS protocol in the password-only setting enjoys the following security hedging property: While avoiding the need to rely on the authenticity of servers' public keys held by the user is an important security property, when such public keys are available they can add significant security, because they render on-line attacks against a user ineffective and strengthen the security and the soundness properties of the PPSS scheme. Thus, to get the benefits of both worlds, both with and without the correctly functioning PKI, running a password-only PPSS protocol over PKI-authenticated links achieves the following: If the user has the correct servers' public keys, she gets the additional security benefits stated above, otherwise, if some or all of the public keys are either incorrect or missing, she still enjoys the security of the password-only setting.

T-PAKE. When composed with regular key exchange protocols, our PPSS scheme leads to the most efficient T-PAKE protocols to date even when compared to protocols that assume that the user carries a public key that it can use to authenticate the servers. Figure 1 summarizes the state of the art in T-PAKE protocols and how our protocols compare to this prior work. Interestingly, while there is a large body of work on single-server PAKE protocols (e.g. [4,23,24,5]) that has produced remarkable schemes, including one-round password-only protocols in the standard model, threshold PAKE has seen less progress, with most protocols showing disadvantages over a single-server PAKE. In particular, before our work, no single-round (t, n)-PAKE protocol was known, not even in the ROM and assuming PKI. Most protocols assume a public key carried by the client (making them non password-only) and all assume secure channels (or PKI) between servers. Even in the $n = t = 2$ case no one-round protocol was known, and all previously known protocols for this case require inter-server secure channels. Our work improves on these parameters achieving the best known properties in *all* the aspects reflected in the table.

In particular, *we achieve single-round password-only protocol in the CRS and ROM models for arbitrary (t, n) parameters with no PKI requirements for any party and no inter-server communication* (secure communication is only assumed when a user first registers with the servers). In addition, the protocol is computationally very efficient (and more so than any of the previous protocols, even for the (2,2) case). We also exhibit a password-only standard-model implementation of our scheme requiring two rounds of communication (4 messages in total) between client and servers. Our T-PAKE protocols are built by combining (existing) suitable key exchange protocols on top of our V-OPRF-based PPSS scheme. We prove T-PAKE security via a generic composition theorem showing the security of any such composed protocol.

Organization. In Section 2 we present the formalization of the V-OPRF functionality in the UC setting. In Section 3 we show an efficient realization of this functionality in the random oracle model (ROM) (further ROM and standard (CRS) model constructions are presented in the full version [19]). In Section 4 we define and formalize PPSS in the password-only model. In Section 5 we present an efficient PPSS realization which relies on the V-OPRF functionality. Finally, in Section 6 we consider T-PAKE schemes obtained by composing a

scheme	$(t+1, n)$	ROM/std	client	inter-server	msgs	total comm.	comp. C | S
BJKS [6]	$(2,2)$	ROM	PKI	PKI	7	$O(1)$	$O(1)$
KMTG [22]	$(2,2)$	Std/ROM	CRS	sec.chan.	≥ 5	$O(1)$	$O(1)$
CLN [9]	$(2,2)$	Std/ROM	CRS	PKI	8	$O(1)$	$O(1)$
DRG [14]	$t<n/3$	Std	CRS	sec.chan.	≥ 12	$O(n^3)$	$O(1)$ | $O(n^2)$
MSJ [27]	any	ROM	PKI	PKI	7	$O(n^2)$	$O(1)$ | $O(n)$
BJSL [2]	any	ROM	PKI	PKI	3	$O(t)$	$8t+17$ | 16
CLLN [7]	any	ROM	CRS	PKI	10	$O(t^2)$	$14t+24$ | $7t+28$
Our PPSS1	any	ROM	CRS	none	2	$O(t \log n)$	$2t+3$ | 2
Our PPSS2	any	Std	CRS	none	4	$O(\ell t \log n)$	$O(t\ell)$ | $O(l)$

Fig. 1. Comparison between PPSS/T-PAKE schemes. "PPSS1" and "PPSS2" refer to our PPSS scheme of Section 5 with V-OPRF instantiated, respectively, with protocol 2HashDH-NIZK of Section 3 (this instantiation is shown in Figure 5) and with protocol NR-V-OPRF (deferred to the full version [19]). The "total comm." column counts the number of transmitted group elements and other objects of length polynomial in the security parameter, like public-key signatures. Variable ℓ denotes the length of the password string (or its hash). The last column counts (multi)exponentiations in a prime-order group (except for our PPSS2 where exponentiations are modulo a Paillier modulus) performed by the client and each server in the reconstruction protocol. All costs in the last four rows refer to an optimistic scenario with no adversarial interference. With worst-case adversarial interference, for BJSL and CLLN all costs grow by the factor of $n-t$, while for our schemes costs grow by the factor of $\lfloor n/(t+1) \rfloor$.

PPSS scheme with a regular key-exchange protocol, and present a full specification of our most efficient instantiation of the PPSS and T-PAKE protocols. This version of the paper omits many details, constructions and proofs; please refer to the full version [19] for a complete presentation.

2   Functionality $\mathcal{F}_{\text{VOPRF}}$

The $\mathcal{F}_{\text{VOPRF}}$ functionality can be thought of as a collection of tables that are indexed by "labels" denoted by the function parameters π. Users may obtain values from these tables on inputs x of their choice without leaking any information about these inputs (and corresponding outputs) to the adversary. $\mathcal{F}_{\text{VOPRF}}$ generates these tables dynamically and fills them with random values on demand. Each table is associated by the functionality with a specific sender. In addition to the tables registered to honest senders, the adversary is allowed to register with $\mathcal{F}_{\text{VOPRF}}$ its own tables. Interacting with an adversary-registered table does not jeopardize the privacy of the user's input but naturally $\mathcal{F}_{\text{VOPRF}}$ will provide no pseudorandomness guarantee for the output derived from such tables. However, $\mathcal{F}_{\text{VOPRF}}$ will ensure that all adversarial tables are completely determined according to a deterministic function that is committed by the adversary at the time of the table's initialization in the form of a circuit M.

Functionality $\mathcal{F}_{\text{VOPRF}}$

Key generation:

On message (KeyGen, *sid*) from S, forward (KeyGen, *sid*, S) to adversary $\mathcal{A}^*$.
On message (Parameter, *sid*, S, π, M) from $\mathcal{A}^*$, ignore this call if param(S) is already defined. Otherwise, set param(S) = $\langle \pi \rangle$ and initialize tickets(π) = 0, and hist(π) to the empty string. If S is honest send (Parameter, *sid*, π) to party S, else parse M as a circuit with ℓ-bit output and insert (π, M) in CorParams.

V-OPRF evaluation:

On message (Eval, *sid*, S, x) from party U for sender S, record (U, x) and forward (Eval, *sid*, U, S) to $\mathcal{A}^*$.

On message (SenderComplete, *sid*, S) from $\mathcal{A}^*$ for some honest S output (SenderComplete, *sid*) to party S and set tickets(π) = tickets(π) + 1 for π s.t. $\langle \pi \rangle$ = param(S).

On message (UserComplete, *sid*, U, π, flag) from $\mathcal{A}^*$, recover (U, x) and:

 - If flag = $\top$ and $\langle \pi \rangle$ = param(S) for an honest S then: If tickets(π) $\leq$ 0 ignore the UserComplete request of $\mathcal{A}^*$. Otherwise: (1) if hist(π) includes a pair $\langle x, \rho' \rangle$, set $\rho = \rho'$, else sample ρ at random from $\{0, 1\}^\ell$ and enter $\langle x, \rho \rangle$ into hist(π); (2) Set tickets(π) = tickets(π) − 1 and output (Eval, π, ρ) to party U.
 - Else, if flag = $\bot$ then return (Eval, $\pi, \bot$) to U.
 - Else, if flag = $\top$ and π is such that $(\pi, M) \in$ CorParams for some circuit M, compute $\rho = M(x)$, enter $\langle x, \rho \rangle$ in hist(π), output (Eval, π, ρ) to party U.

Fig. 2. Verifiable Oblivious PRF functionality $\mathcal{F}_{\text{VOPRF}}$

A major consideration in our definition of $\mathcal{F}_{\text{VOPRF}}$ is to avoid the need for input extractability (from dishonest users) in the real-world realizations of the functionality. Such need is common in UC-defined functionalities but in our case it would disqualify the more efficient instantiations of $\mathcal{F}_{\text{VOPRF}}$ presented here. Thus, instead of resorting to input extraction requirements, we define a "ticket mechanism" that increases a ticket upon function evaluation at a sender and decreases it when this value is computed at the user (or the adversary). The functionality guarantees that tickets remain non-negative, namely, for any function parameter π registered with a honest sender S, the number of inputs on which users compute the function π is no more than the number of evaluations of the function at S.

Another important aspect of our $\mathcal{F}_{\text{VOPRF}}$ formalism is the way we handle the 1-1 relationship between a sender S and its function parameter π, where S is used to identify a sender and π describes this sender's committed function. The unique sender-function binding that is known to the functionality cannot be enforced in a real-world setting where users cannot validate such a binding as is the case when no authenticated channels (or other forms of authenticated information) are available to the user. Since these settings are common in our applications, we define $\mathcal{F}_{\text{VOPRF}}$ so that the user can provide a name of a sender

---

Parameters:
  Generator g of cyclic group of order m, hash functions $H_1(\cdot), H_2(\cdot), H_3(\cdot)$.
Key Generation:
  On (KEYGEN, sid), pick $k \in_R \mathbb{Z}_m$, set $y = g^k$, return $(\text{PARAMETER}, sid, y)$.
V-OPRF Evaluation:
  On message (EVAL, sid, S, x), pick $r \in_R \mathbb{Z}_m$ and send $a = H_1(x)^r$ to S.
  On message a from network entity U, check if $a \in \langle g \rangle$, compute $b = a^k$ and $\zeta = \text{NIZK}_{\text{EQ}}^{H_3}[g, y, a, b]$, and send $\langle y, b, \zeta \rangle$ to U.
  On message $\langle y, b, \zeta \rangle$ from party S, verify the NIZK ζ and $b \in \langle g \rangle$. If the tests pass return $(\text{EVAL}, y, H_2(y, x, b^{1/r}))$, else return $(\text{EVAL}, y, \perp)$.

Fig. 3. Protocol 2HashDH-NIZK

whose function it intends to compute but the result returned to the user applies a function π determined by the attacker, and is possibly different than the function associated to the requested S.

In spite of the above, note that if $\mathcal{F}_{\text{VOPRF}}$ is used in a context where the user knows a-priori a correspondence between S and π, the user can reject responses that are not consistent with it. We make essential use of this capability in our applications. Finally, note that $\mathcal{F}_{\text{VOPRF}}$ guarantees that value ρ obtained by the user is in the table π even though such table may not have been the user's original target. This provides $\mathcal{F}_{\text{VOPRF}}$ with a verifiability property which is verifier-dependent and may not be transferable to others; in particular, it is a weaker guarantee than the verifiability propery of verifiable random functions [29].

3   Efficient Realization of $\mathcal{F}_{\text{VOPRF}}$

We present a class of constructions for realizing $\mathcal{F}_{\text{VOPRF}}$ in the random oracle model. Our constructions share the following general structure: the receiver hashes and blinds her input and requests the sender's secret-key application on this blinded value. The receiver verifies the sender's response and then obtains the V-OPRF output by applying a second hash function. Due to the double hashing (which is essential in the security proof) we term the constructions with the "2Hash" prefix. For lack of space we present here only a single instance of this design methodology, protocol 2HashDH-NIZK in Figure 3, while our further ROM constructions, based on RSA or a group with a bilinear map, are presented in the full version [19]. Protocol 2HashDH-NIZK uses a non-interactive zero-knowledge proof $\text{NIZK}_{\text{EQ}}^{H_3}[g, y, a, b]$ of discrete logarithm equality $\text{DL}(g, y) = \text{DL}(a, b)$. This NIZK has to be straight-line simulatable and simulation sound, and it can be implemented with one multi-exponentiation for both the prover and the verifier using hash function H_3 modeled as a random oracle [12].

We will argue the security of the construction employing the following assumption: the (N, Q) One-more Gap DH assumption, states that for any PPT

$\mathcal{A}$ it holds that the following probability is negligible:

$$\mathsf{Prob}[\mathcal{A}^{(\cdot)^k, \mathsf{DDH}(\cdot,\cdot,\cdot,\cdot)}(g, g^k, g_1, \ldots, g_N) = \{(g_{j_s}, g_{j_s}^k) \mid s = 1, \ldots, Q+1\}]$$

where Q is the number of queries that $\mathcal{A}$ poses to the $(\cdot)^k$ oracle. The probability is taken over all choices of $g^k, g_1, \ldots, g_N$ which are assumed to be random elements of $\langle g \rangle$. We denote by $\epsilon_{\mathsf{omdh}, G}(N, Q)$ the maximum advantage of any PPT adversary against the assumption.

Theorem 1. *The 2HashDH-NIZK protocol over group G of order m UC-realizes $\mathcal{F}_{\mathsf{VOPRF}}$ per Fig. 2 in ROM assuming (i) the existence of PRF functions, (ii) the (N, Q) One-More Gap DH assumption on G where Q is the number of V-OPRF executions and $N = Q + q_1$ where q_1 is the number of $H_1(\cdot)$ queries.*

More precisely, for any adversary against 2HashDH-NIZK there is an ideal-world adversary (simulator) that produces a view that no environment can distinguish with advantage better than $q_S \cdot \epsilon_{\mathsf{omdh}, G}(N, Q) + q_3/m^2 + 2 \cdot q_U/m + N^2/m + \epsilon_{\mathsf{PRF}}(q_2)$ where q_S is the number of senders, q_U the number of users, q_2, q_3 are the number of queries to oracles H_2, H_3, and $\epsilon_{\mathsf{PRF}}(q_2)$ is the security of the PRF function against adversaries executing in comparable time and posing q_2 queries.

Proof. See full version [19]. □

Remark. We note that in the construction of Figure 3, the outgoing message that is constructed given an (EVAL, sid, S, x) command is independent of S. It is easy to see that security is preserved even if the user employs the same outgoing message for any sequence of consecutive (EVAL, sid, S_1, x), ... , (EVAL, sid, S_n, x) commands for any $n \geq 1$. We make essential use of this feature in the optimized protocol of Figure 5 where the user uses the same blinded input with all servers.

V-OPRF Constructions in the Standard Model. A 4-message realization of V-OPRF in the standard model based on the Strong RSA and DCR assumptions is presented in the full version [19].

4   Password-Protected Secret Sharing: Definitions

Our definition of PPSS adapts the PPSS notion of [2] to the CRS model, but also re-defines PPSS in terms of a *key derivation mechanism* rather than an encryption-style notion used in [2]. In other words, rather than used directly to semantically protect any message, a PPSS will generate and protect a random key. This change allows for better modularity, because the resulting key can be used not only for message encryption (and authentication), but also e.g. for an Authenticated Key Exchange. A Password-Protected Secret Sharing (PPSS) scheme in the CRS model is a protocol involving $n + 1$ parties, a user U, and n servers $S_1, \ldots, S_n$. A PPSS scheme is a tuple (ParGen, SKeyGen, Init, Rec), where ParGen and SKeyGen are randomized algorithms and Init and Rec are multi-paty protocols with the following syntax:

1) Algorithm ParGen generates string CRS for a given security parameter τ.
2) Each S_i runs SKeyGen(CRS) to generate private state σ_i and public param. π_i.

3) Protocol Init is executed by U and servers $S_1, \ldots, S_n$, where U runs algorithm U_{Init} on inputs a password $pw \in \{0,1\}^\tau$, global parameters CRS, and a vector of server's public parameters $\pi = (\pi_1, \ldots, \pi_n)$, while each S_i runs algorithm S_{Init} on input (CRS, σ_i, π_i). The outputs of Init is a τ-bit key K for U and a user-specific information ω_i for each server S_i.

4) Protocol Rec is executed by U and servers $S_1, \ldots, S_n$, where U runs algorithm U_{Rec} on (CRS, pw), and each S_i runs algorithm S_{Rec} on $(CRS, \sigma_i, \pi_i, \omega_i)$. Protocol Rec generates no output for the servers, while U outputs K' which is either a τ-bit string or a rejection symbol $\perp$.

The *correctness* requirement is that Rec returns the same key K which was generated in Init, i.e. that for any τ, any CRS output by $ParGen(1^\tau)$, any (σ_i, π_i) output by n instances of $SKeyGen(CRS)$, and any $pw \in \{0,1\}^\tau$, if $(K, \omega_1, \ldots, \omega_n)$ is the vector of outputs of Init executed on inputs (pw, CRS, π) for U and (CRS, σ_i, π_i) for each S_i, then U's local output in an instance of Rec executed on inputs (CRS, pw) for U and $(CRS, \sigma_i, \pi_i, \omega_i)$ for each S_i, is equal to K.

Server's User-Related State. We stress that the state $(\sigma_i, \pi_i, \omega_i)$ of each server S_i is stored *for each user separately*, and the PPSS security notion we define below assumes that each S_i stores a separate $(\sigma_i, \pi_i, \omega_i)$ tuple for each user account. Indeed, the security of the PPSS protocol we present in Section 5 would be decreased if S_i re-uses the same OPRF key, stored in σ_i in this PPSS protocol, across multiple user accounts. (Technically, the adversary would get additional oracle access to the same $S_{Rec}^\diamond$ oracle, see below, for each user account on which the server re-uses the same (σ_i, π_i) pair.) Consequently, if S_i wants to provide PPSS service to multiple users, it has to generate a separate (σ_i, π_i) pair for each user (these per-user keys can be derived internally by S_i using a PRF and a global PRF key applied to a user's identifier).

Security. We define security of a PPSS scheme in terms of adversary's advantage in distinguishing the key K output by U from a random string. We assume that the adversary sees CRS and the vector of server's public parameters π used in the initialization instance, as well as the private states $\sigma_B \triangleq \{\sigma_i\}_{i \in B}$ and $\omega_B \triangleq \{\omega_i\}_{i \in B}$ for some set B of corrupted servers, and that it has concurrent oracle access to instances of $U_{Rec}(CRS, pw)$ and $S_{Rec}(CRS, \sigma_i, \pi_i, \omega_i)$, for i in $\overline{B} \triangleq \{1, ..., n\} \setminus B$. We denote as $U_{Rec}^\diamond(CRS, pw, b, K^{(0)})$ an oracle which executes the interactive algorithm $U_{Rec}(CRS, pw)$, and when this algorithm terminates with a local output K, the $U_{Rec}^\diamond$ oracle (re-)sets K to $K^{(0)}$ if $b = 0$ and $K \neq \perp$, and then returns K to the caller. However, if $b = 1$ or $K = \perp$ then the caller receives the unmodified value K (to which we will refer as $K^{(1)}$) as it was output by the U_{Rec} instance. We denote as $S_{Rec}^\diamond(CRS, \sigma_{\overline{B}}, \pi, \omega_{\overline{B}})$ an oracle which on input $i \in \overline{B}$ executes the interactive algorithm $S_{Rec}(CRS, \sigma_i, \pi_i, \omega_i)$.

Intuitively, we should call an (t, n)-threshold PPSS scheme secure if for any password dictionary D, if pw is randomly chosen in D then the adversary's advantage in distinguishing the PPSS-protected key K from a random string (i.e., guessing b) is at most negligibly above $1/|D|$, the probability of guessing the password, times $q_u + \lfloor q_s/(t - t' + 1) \rfloor$, where q_u and q_s are the numbers, respectively, of the $\mathsf{U}_{\mathsf{Rec}}$ and $\mathsf{S}_{\mathsf{Rec}}$ protocol instances the adversary can interact with, and $t' \leq t$ is the number of corrupted servers. Factor $1/|D| \cdot \lfloor q_s/(t-t'+1) \rfloor$ corresponds to an inherent vulnerability due to on-line dictionary attacks: An adversary who learns the shares of $t' \leq t$ servers can test any password $\tilde{\mathsf{pw}}$ in D by running the user's protocol on $\tilde{\mathsf{pw}}$ interacting with $t - t' + 1$ uncorrupted servers. Factor $1/|D| \cdot q_u$ corresponds to an inherent vulnerability of password-authenticated protocols in the CRS model, because the adversary can run the initialization protocol Init on a password guess $\tilde{\mathsf{pw}}$ and then run the servers' protocol interacting with the user: If the user does not reject (by outputting $\perp$), the adversary can conclude that $\tilde{\mathsf{pw}} = \mathsf{pw}$.

To make the PPSS notion easier to use in applications it is important that the adversary sees the key-pseudorandomness challenge, either a real key or a random key, already after the initialization protocol Init, rather than only when this key is reconstructed in protocol Rec. (E.g. our T-PAKE constructions rely on this property.) To make sure that the PPSS-protected key remains pseudorandom in each key usage, whether after the initialization or after each reconstruction instance, we let the adversary see the key generated by Init as well as the key(s) output by every Rec instance. That is, at the end of Init and after each Rec instance the attacker is given $K^{(0)}$ if $b = 0$, but if $b = 1$ then the attacker is given the actual value of the key output by, respectively, $\mathsf{U}_{\mathsf{Init}}$ or $\mathsf{U}_{\mathsf{Rec}}$. Note that $K^{(0)}$ does not change across different reconstructions because it is fixed at the start of the experiment, while $K^{(1)}$ is determined by the actual outputs of $\mathsf{U}_{\mathsf{Init}}$ and $\mathsf{U}_{\mathsf{Rec}}$ instances. Importantly, note that this definition implies that in the real execution the reconstruction instances must output the same key that was created in the initialization or the attacker can trivially guess b. We further discuss this *soundness* property below.

Definition 1. *A PPSS scheme is (T, q_u, q_s, ϵ)-secure (for fixed threshold parameters (t, n) if for any $D \subseteq \{0,1\}^\tau$, any set $\mathsf{B} \subseteq \{1, ..., n\}$ of size $t' \leq t$, and any algorithm $\mathcal{A}$ with running time T, we have*

$$\mathbf{Adv}_{\mathcal{A}}^{\mathsf{ppss}} \leq \left(q_u + \left\lfloor \frac{q_s}{t - t' + 1} \right\rfloor \right) \cdot \frac{1}{|D|} + \epsilon \qquad (1)$$

where $\mathbf{Adv}_{\mathcal{A}}^{\mathsf{ppss}} = |p_{\mathcal{A}}^{(1)} - p_{\mathcal{A}}^{(0)}|$ and $p_{\mathcal{A}}^{(b)} = \Pr[b' = 1]$ in a game below, for $b \in \{0,1\}$:
(1) Choose pw at random in D, generate $\mathsf{CRS} \leftarrow \mathsf{ParGen}(1^\tau)$ and $(\sigma_i, \pi_i) \leftarrow$
$\mathsf{SKeyGen}(\mathsf{CRS})$ for $i \in \overline{\mathsf{B}}$. Give CRS and $\{\pi_i\}_{i \in \overline{\mathsf{B}}}$ to $\mathcal{A}$ and let $\mathcal{A}$ generate $\{\pi_i\}_{i \in \mathsf{B}}$.
(2) Run an instance of Init between U, which executes protocol $\mathsf{U}_{\mathsf{Init}}(\mathsf{CRS}, \mathsf{pw}, \pi_1,$
$..., \pi_n)$, and the servers, where each S_i for $i \in \overline{\mathsf{B}}$ executes protocol $\mathsf{S}_{\mathsf{Init}}(\mathsf{CRS}, \sigma_i,$
$\pi_i)$, while servers S_i for $i \in \mathsf{B}$ are controlled by adversary $\mathcal{A}$. The protocol proceeds on public channels, with $\mathcal{A}$ playing a man-in-the-middle on all communications. Denote U's output in this Init instance as $K^{(1)}$, and denote S_i's output, for $i \in \overline{\mathsf{B}}$, as ω_i. Choose $K^{(0)}$ at random in $\{0,1\}^\tau$. Give key $K^{(b)}$ to $\mathcal{A}$.
(3) Let $\mathcal{A}$ interact with q_u instances of $\mathsf{U}_{\mathsf{Rec}}^{\diamond}(\mathsf{CRS}, \mathsf{pw}, b, K^{(0)})$ and q_s instances of $\mathsf{S}_{\mathsf{Rec}}^{\diamond}(\mathsf{CRS}, \sigma_{\overline{\mathsf{B}}}, \pi, \omega_{\overline{\mathsf{B}}})$. Let b' be the final output of $\mathcal{A}$.

Secure Initialization. Note that in the above definition we assume that in the initalization protocol U runs the $\mathsf{U_{Init}}$ procedure on input a vector of public parameters $\boldsymbol{\pi} = (\pi_1, \ldots, \pi_n)$ where π_i for each $i \in \overline{\mathsf{B}}$ is the true output of SKeyGen executed by server S_i. In other words, we assume that the user runs the initialization procedure on correct (i.e. authentic) values π_i for the honest servers. This is equivalent to assuming that the user can authenticate, e.g. via the PKI, the servers with whom it wants to initialize the PPSS scheme. The requirement of authenticated channels between the user and the honest servers *during the initialiation protocol* is indeed neccessary, or otherwise the adversary would be able to pose as $t+1$ servers among $\mathsf{S}_1, \ldots, \mathsf{S}_n$ and recover U's secret sc from the initialization protocol. (A similar assumption on authenticity of servers' public keys in the initialization is also made in [8].)

Soundness. The above definition captures also a *soundness* property of a PPSS scheme, because it implies an upper-bound on the probability that an adversary causes any $\mathsf{U_{Rec}}$ instance to output $K' \notin \{K, \perp\}$ where K was an output of $\mathsf{U_{Init}}$. Assume algorithm $\mathcal{A}$ which outputs 0 if every key returned by $\mathsf{U^\diamond_{Rec}}$ oracle is either equal to $K^{(b)}$ which was output by $\mathsf{U_{Init}}$, or to $\perp$. Note that in the security experiment with $b = 0$, oracle $\mathsf{U^\diamond_{Rec}}$ always returns $K^{(0)}$ or $\perp$, so $p_\mathcal{A}^{(0)} = 0$. The security definition implies that $p_\mathcal{A}^{(1)} \leq \left(q_u + \left\lfloor \frac{q_s}{t-t'+1} \right\rfloor \right) \cdot \frac{1}{|D|} + \epsilon$, hence this is also an upper-bound on the probability that any $\mathsf{U_{Rec}}$ instance outputs K' which is neither $\perp$ nor K output in $\mathsf{U_{Init}}$.

Robustness. Another desirable property of a PPSS scheme is robustness, which we define as the requirement that the user reconstructs the key created in Init as long as it communicates without obstructions with at least $t+1$ non-corrupt servers and with at most t corrupt ones. This property is distinct from soundness in that it assumes that the adversary lets the user communicate with $t+1$ non-corrupt servers without interference. Note that this implies that the number t number of corrupt servers satisfies $t < n/2$, a restriction which is not imposed by either the security or the soundness properties.

5   A PPSS Protocol in the $\mathcal{F}_{\mathrm{VOPRF}}$-Hybrid World

We show a PPSS protocol based on any realization of the $\mathcal{F}_{\mathrm{VOPRF}}$ functionality. The protocol is shown in Figure 4 in the $\mathcal{F}_{\mathrm{VOPRF}}$-hybrid model (a specific instantiation based on the 2HashDH-NIZK V-OPRF of Fig. 3 is shown in Fig. 5). The protocol is secure in the CRS model and it assumes a pseudorandom generator and a computationally hiding, computationally binding, and non-malleable (with respect to decommitment) commitment scheme, which can be realized e.g. by a CCA-secure public key encryption, or by hashing the message together with a random nonce in ROM. To provide rationale for our design we first consider a subset of the protocol in Figure 4 and then show the necessity of some additional elements. In SKeyGen, each server S_i picks its public parameter π_i as the V-OPRF function descriptor, which in all our V-OPRF instantiations is a commitment to the private key of the underlying PRF (see Section 2). In protocol Init, on U's inputs a password pw and a vector of function descriptors $\boldsymbol{\pi} = (\pi_1, \ldots, \pi_n)$, which are authentically delivered to the user, user U picks a random key K, secret-shares it into shares $s_1, \ldots, s_n$, and then encrypts each s_i using one-time pad encryption under key $\rho_i = F_{\pi_i}(\mathsf{pw})$, computed in a V-OPRF instance with server S_i. The vector of function descriptors $\boldsymbol{\pi} = (\pi_1, \ldots, \pi_n)$ and the ciphertexts $\mathbf{c} = (c_1, \ldots, c_n)$, where $c_i = s_i \oplus F_{\pi_i}(\mathsf{pw})$, is public, and given to each server. At reconstruction the servers send these two vectors to U, who can recover $t + 1$ shares s_i, and interpolate them to recover K, after $t + 1$ V-OPRF instances in which U recomputes the values $\rho_i = F_{\pi_i}(\mathsf{pw})$ for $t + 1$ different i's.

This simplified protocol, however, is not secure. As we explain in the introduction, if the attacker learns whether the receiver recovers the shared key K correctly, the above protocol would enable a malicious server S_i to get information about user's password pw (including recovering it completely using binary search in an OPRF based on the Naor-Reingold PRF), by manipulating the function descriptor π_i in each OPRF instance executed by S_i in this reconstruction protocol. In fact, in the PPSS security model defined in section 4, the above simplified protocol allows a malicious server to recover $\boldsymbol{\pi}$ through an off-line dictionary search after a single instance of PPSS reconstruction. Note that our PPSS security model reveals the whole key K output in a PPSS reconstruction to the adversary, which models putting this key to an arbitrary usage by the higher-level protocol, e.g. by the T-PAKE scheme built from PPSS in Section 6. Now, if $\mathcal{A}$ sends to $\mathsf{U}_{\mathsf{Rec}}$ a vector of function descriptors π_i^* which correspond to PRF keys k_i^* which $\mathcal{A}$ creates, and if $\mathcal{A}$ learns the key K output by this $\mathsf{U}_{\mathsf{Rec}}$ instance, then $\mathcal{A}$ can stage an off-line dictionary attack running the user's reconstruction algorithm for every guess $\tilde{\mathsf{pw}}$ in the password dictionary D, and locally computing values $F_{\pi_i^*}(\tilde{\mathsf{pw}})$ using the PRF keys k_i^*. This is yet another reason why we need to extend the above protocol by adding a non-malleable commitment C that binds user's password pw to the reconstructed secret K. We accomplish this binding as follows: The CRS string will include an instance of a non-malleable commitment scheme COM. In the initialization procedure, the user secret-shares not the key K directly, but a random value s, and then it uses s as a PRG seed to derive the key K together with the commitment randomness r, and sets each state ω_i given to S_i to $(\boldsymbol{\pi}, \mathbf{c}, C)$ where $C = \mathsf{COM}((\mathsf{pw}, \boldsymbol{\pi}, \mathbf{c}); r)$. By the binding property of commitment COM, the adversary playing the role of the servers must commit to a password guess $\tilde{\mathsf{pw}}$ in value C it sends to the user, and the reconstruction procedure rejects unless the guess was right, i.e. unless $\tilde{\mathsf{pw}} = \mathsf{pw}$, disabling the off-line dictionary attack above. We need the non-malleability of the commitment scheme to forestall the possibility that the adversary modifies either the vector of function descriptors $\boldsymbol{\pi}$ or the ciphertexts $\mathbf{c}$, and hence in particular modifies the reconstructed key K, without guessing the password.

Communication Complexity, Robustness. In Figure 4 we show a PPSS scheme whose communication complexity is $O(n^2 \cdot \mathrm{poly}(\tau))$ where τ is a security parameter, because the protocol starts with each server S_i sending to U a tuple

---

Parameters: Security parameters τ and ℓ, binary extension field $\mathbb{F} = GF(2^\ell)$, session ID $sid = (\mathsf{S}_1, \ldots, \mathsf{S}_n)$, threshold parameters $t, n \in \mathbb{N}$.

ParGen(τ): Sets CRS as an instance of a non-malleable commitment COM.

SKeyGen(CRS): S_i sends (KEYGEN, sid) to $\mathcal{F}_{\text{VOPRF}}$ and sets π_i to π it receives in the response (PARAMETER, sid, π) from $\mathcal{F}_{\text{VOPRF}}$. The private state σ_i of S_i is the unique handle "S_i" has to the V-OPRF function F_{π_i} implemented by the ideal $\mathcal{F}_{\text{VOPRF}}$ functionality. (In all our V-OPRF instantiations σ_i is a PRF key and the function descriptor π_i is a commitment to it.)

$\mathsf{U}_{\text{Init}}(\text{CRS}, \text{pw}, \pi_1, \ldots, \pi_n) \rightleftharpoons \{\mathsf{S}_{\text{Init}}(\text{CRS}, \sigma_i, \pi_i)\}_{i=1}^n$:
Step 1. User U picks $s \leftarrow \mathbb{F}$ and generates $(s_1, \ldots, s_n)$ as a (t, n) Shamir's secret-sharing of s over field $\mathbb{F}$. (Indices $0, 1, \ldots, n$ used in Shamir's secret-sharing are encoded as some distinct field elements $\langle 0 \rangle_{\mathbb{F}}, \langle 1 \rangle_{\mathbb{F}}, \ldots, \langle n \rangle_{\mathbb{F}}$.) For $i = 1$ to n, U sends (EVAL, sid, S_i, pw) to $\mathcal{F}_{\text{VOPRF}}$.
Step 2. User U collects $\mathcal{F}_{\text{VOPRF}}$ responses $(\pi_1', \rho_1), \ldots, (\pi_n', \rho_n)$, and aborts if $\pi_i' \neq \pi_i$ for any i. If all parameters π_i' match those in the inputs, U computes $c_i \leftarrow s_i \oplus \rho_i$ for $i = 1$ to n, $\mathbf{c} \leftarrow (c_1, \ldots, c_n)$, $\boldsymbol{\pi} \leftarrow (\pi_1, \ldots, \pi_n)$, $[r\|K] \leftarrow G(s)$, $C \leftarrow \text{COM}((\text{pw}, \boldsymbol{\pi}, \mathbf{c}); r)$, sends $\omega = (\boldsymbol{\pi}, \mathbf{c}, C)$ to each S_i, and outputs K as a local output.

$\mathsf{U}_{\text{Rec}}(\text{CRS}, \text{pw}) \rightleftharpoons \{\mathsf{S}_{\text{Rec}}(\text{CRS}, \sigma_i, \pi_i, \omega_i)\}_{i \in S}$:
For each $i = 1, \ldots, n$, user U sends (EVAL, sid, S_i, pw) to $\mathcal{F}_{\text{VOPRF}}$ and initiates a run of the protocol Rec with S_i.
Each S_i responds by sending ω_i to U and (SENDERCOMPLETE, sid, S_i) to $\mathcal{F}_{\text{VOPRF}}$, and U collects $\mathcal{F}_{\text{VOPRF}}$ responses (π_i', ρ_i) and ω_i for each $i \in S$.
Let S be a subset of servers such that: (i) $|S| = t + 1$; (ii) there exists $\omega = (\boldsymbol{\pi}, \mathbf{c}, C)$ with $\boldsymbol{\pi} = (\pi_1, \ldots, \pi_n)$ and $\mathbf{c} = (c_1, \ldots, c_n)$ such that $\omega_i = \omega$ for all $\mathsf{S}_i \in S$; (iii) for all $\mathsf{S}_i \in S$, $\pi_i' = \pi_i$ and $\rho_i \neq \bot$. If no such subset exists output $\bot$ and halt.
Reconstruction: Set $u_i \leftarrow c_i \oplus \rho_i$ for all $i \in S$. Interpolate points $\{(\langle i \rangle_{\mathbb{F}}, u_i)\}_{i \in S}$ with a polynomial $U \in \mathbb{F}[x]$, and set $s \leftarrow U(\langle 0 \rangle_{\mathbb{F}})$. Compute $[r\|K] \leftarrow G(s)$. If $\text{COM}((\text{pw}, \boldsymbol{\pi}, \mathbf{c}); r) = C$ then output K, else output $\bot$.

Fig. 4. A PPSS scheme in the $\mathcal{F}_{\text{VOPRF}}$-hybrid-model

ω which contains n function descriptors π_i and n field elements c_i. The reason we do this is simplicity, plus we suspect that in most applications the number of servers n will be small enough that the $O(n^2)$ cost of this communication will not be significant in practice. However, for large n we can reduce the communication to $O(n \log n)$ using a Merkle Tree hash [28]. Each server S_i would then send only its own π_i, c_i values together with the co-path in the hash tree which allows U to agree on the set of $t + 1$ servers whose tree co-paths hash to the same root value. In practice U could also cash the ω vector as it does not change between Rec protocol instances, in which case the communication cost becomes $O(n)$. The communication cost can be decreased even further, to $O(t)$ group elements, at the cost of reducing robustness. The user could instigate V-OPRF instances with only $t + c$ servers instead of with all n, for any c between 1 and $n - t$. This would reduce bandwidth at the price of increasing the protocol costs in the case of an active attack: If just c among the $t + c$ servers U contacts are either corrupted or connected to U over corrupted links, the reconstruction attempt fails, and the user needs to instigate V-OPRF instances with the remaining servers.

Theorem 2. (PPSS Security) *Assuming commitment scheme* COM *is computationally hiding, computationally binding, and non-malleable (with respect to decommitment), and that G is a pseudorandom generator, the PPSS scheme in Figure 4 is (T, q_u, q_s, ϵ)-secure for $\epsilon = \epsilon_H + \epsilon_B + q_u \cdot \epsilon_{NM} + 4\epsilon_G$, where ϵ_H, ϵ_B, ϵ_{NM} and ϵ_G are the bounds implied by, respectively, computational hiding of* COM, *copmutational binding of* COM, *non-malleability of* COM *with respect to decommitment, and the pseudorandomness of G, on input sizes implied by the usage of* COM *and G in the PPSS scheme, for adversaries whose time is bounded by T plus the time taken by a single instance of* Init, q_u *instances of* U_{Rec}, *and q_s instances of* S_{Rec}.

Proof. See full version [19].

6 From PPSS to Single-Round T-PAKE

Composition of a PPSS scheme with a (regular) key-exchange protocol allows us to obtain very efficient one-round T-PAKE protocols with arbitrary threshold parameters and in the password-only CRS model, i.e. no PKI or secure channels are assumed. For lack of space we refer to the full version [19] for a general composition theorem proving the security of T-PAKE protocols built by this methodology. Here we only present examples of T-PAKE schemes obtained through this approach, and illustrate them with the most efficient T-PAKE instantiation, presented in Figure 5, resulting from our single-round PPSS of Section 5 implemented with the 2Hash-DH OPRF shown of Section 3.

T-PAKE via PPSS and Symmetric-Key KE. Let P be a (t, n)-PPSS protocol in the CRS model. To bootstrap a (t, n)-TPAKE protocol using P, each server S_i, $i = 1, \ldots, n$, generates its state pair (σ_i, π_i) and runs with client C the Init procedure of protocol P. As a result a user's secret, which we call K_C, is (t, n)-secret-shared among these servers under the protection of the PPSS scheme and the client's password pw. Next, client C uses key K_C to compute n keys $K_i = f_{K_C}(i), i = 1, \ldots, n$, where f is a pseudorandom function, and transmits each K_i (protected under the secure communication assumed at initialization) to the corresponding S_i who stores K_i in its client-specific $\zeta_i(C)$ state. Later, when a T-PAKE session at C is invoked, C runs the Rec procedure of protocol P with a sufficient number of servers to obtain K_C. C uses K_C to compute $K_1, \ldots, K_n$ and uses these keys as shared keys with the corresponding servers to exchange a session key. Any KE protocol that assumes pre-shared keys between pairs of parties can be used for this purpose. For example, C and S_i can compute their session key as $f_{K_i}(n_C, n_{S_i}, id_C, id_{S_i})$ where id_C, id_{S_i} stand for the identities of C and S_i respectively, and n_C, n_{S_i} are nonces exchanged between these parties that also serve as session identifiers. Note that when using a one-round PPSS scheme, the exchange of nonces can be piggybacked on top of the PPSS messages hence *preserving the single round complexity of the protocol* (with one additional message from C to S_i if key confirmation is desired). A full specification of this protocol based on the 2HashDH-NIZK V-OPRF is presented in Figure 5. One can also add forward secrecy to the protocol by using the shared key to authenticate a Diffie-Hellman exchange (also piggybacked on top of the two PPSS messages to preserve the single-round complexity).

T-PAKE via PPSS and public-key KE. The above scheme provides a full T-PAKE protocol with very little extra cost over the PPSS scheme. Its relative drawback is (as in any pre-shared key scheme) that the server needs to keep a per-client secret and also that it requires secrecy for the transmission of key K_i to S_i (otherwise, our PPSS scheme only needs authenticated channels during initialization). To avoid these secrecy requirements, key exchange protocols based on public keys of the parties can be accommodated on top of a PPSS as follows. At initialization, the client generates a pair of private and public keys, and obtains public keys for all its servers. C then generates a file (we call it a keystore in our formal treatment [19]) that includes its own key pair (with the private key encrypted under a key derived from K_C) and the servers' public keys. The keystore is stored at each server authenticated with a MAC computed by C using a key derived from K_C. In addition, each server stores C's public key. When a T-PAKE session is invoked at C, the client retrieves keystore from the servers and, after reconstructing K_C, uses this key to check the integrity of keystore and to decrypt its private key. With this information and the (authenticated) public keys of the servers contained in keystore, C is ready to perform the key exchange protocol. Similarly, the servers can use C's public key that they stored to bootstrap the public-key based key exchange. In particular, using a two-message KE protocol whose messages are independent of the parties private- and public-keys. (such as HMQV [26]), one obtains a single-round T-PAKE by piggy-backing the two KE messages on top of the two PPSS ones.

DH-Based Instantiation of PPSS and T-PAKE. For illustration and for the reader's convenience we describe in Figure 5 the specific instantiation of the PPSS and T-PAKE protocols based on the 2HashDH-NIZK V-OPRF, with the NIZK for DL equality implemented as in [12], and a symmetric-key KE scheme. We comment on some of our choices for this illustration. The initialization is presented for the case in which the client generates the servers' V-OPRF keys and computes all the values in the ω vector by itself. Another option, more in line with the formal description of the PPSS protocol from Figure 4, is for the servers to choose their own V-OPRF keys and engage in an V-OPRF computation with the client for generating the pads used to encrypt the shares s_i. One advantage of the latter option is that servers can save in the amount of secret memory and derive the V-OPRF keys for each user U using a single key $\hat{M}K$ and a PRF F, i.e., as $k_U = F_{MK}(U)$ (we are abusing the symbols U and S_i to denote the identities of these parties). This option is more useful with a PK-based KE, where servers do not need to store user-specific secrets (in contrast, the protocol from Figure 5 requires the server storing the session key with each user). User performance during reconstruction is improved by choosing a common value ρ

Parties: User U, Servers $S_1, \ldots, S_n$.

Public parameters and components: Security parameters τ and ℓ, threshold parameters $t, n \in \mathbb{N}, t \leq n$, field $\mathbb{F} = GF(2^\ell)$, cyclic group of prime order m with generator g; hash functions H_1, H_2, H_3, H_4, H_5 with ranges $\langle g \rangle, \{0,1\}^\ell, \{0,1\}^\tau, \mathbb{Z}_m, \mathbb{Z}_m$, respectively; pseudorandom generator G and $\boxed{\text{pseudorandom function family } f}$.

Initialization (secure channels between U and each server S_i are assumed only through initialization): User U performs the following steps:

1. Chooses $s \in_R \mathbb{F}$ and generates shares $(s_1, \ldots, s_n)$ as a (t, n) Shamir's secret-sharing of s over field $\mathbb{F}$.
2. For $i = 1, \ldots, n$, U chooses value $k_i \in_R \mathbb{Z}_m$ and sets $\pi_i = g^{k_i}$ and $c_i = s_i \oplus H_2(\pi_i, \mathsf{pw}, (H_1(\mathsf{pw}))^{k_i})$.
3. Sets $\mathbf{c} = (c_1, \ldots, c_n)$, $\boldsymbol{\pi} = (\pi_1, \ldots, \pi_n)$, $[r\|K] = G(s)$, $C = H_3(r, \mathsf{pw}, \boldsymbol{\pi}, \mathbf{c})$;
   $\boxed{\text{For } i = 1, \ldots, n, \text{ sets } K_i = f_K(S_i)}$.
4. For $i = 1, \ldots, n$, sends to server S_i the values $\omega_i = (\boldsymbol{\pi}, \mathbf{c}, C), k_i, \boxed{K_i}$.
5. U memorizes pw and erases all other information.

Each server S_i, $i = 1, \ldots, n$, stores $\omega_i, k_i, y_i = g^{k_i}, \boxed{K_i}$ in its U-specific storage ζ_i.

Reconstruction/Key Exchange

– User U initiates a key exchange session with servers $S_1, \ldots, S_n$ by sending to each S_i the value $a = (H_1(\mathsf{pw}))^\rho$ with $\rho \in_R \mathbb{Z}_m$ $\boxed{\text{and a nonce } \mu_i \in_R \{0,1\}^\tau}$.

– Upon receiving $(a, \boxed{\mu_i})$, server S_i checks that $a \in \langle g \rangle$ and if so, S_i retrieves k_i and $y_i = g^{k_i}$ from its U-specific storage $\zeta_i(U)$, picks $z \in_R \mathbb{Z}_m$, and computes $b_i = a^{k_i}$, $\gamma = H_4(g, y_i, a, b_i)$, $v_i = H_5(g, y_i, a, b_i, (g \cdot a^\gamma)^z)$, and $u_i = z + v_i \cdot k_i \bmod m$. S_i sends to U the values y_i, b_i, u_i, v_i as well as $\boxed{\text{a nonce } \mu_i' \in_R \{0,1\}^\tau \text{ and}}$ the value ω_i stored in $\zeta_i(U)$. $\boxed{S_i \text{ computes the session key with } U \text{ as } SK_i = f_{K_i}(\mu_i, \mu_i', U, S_i).}$

– Upon receiving values $b_i, u_i, v_i, \omega_i, \boxed{\mu_i'}$ from S_i, U proceeds as follows:

  – U chooses a subset of servers S for which the following conditions hold: (i) there is a value $\omega = (\boldsymbol{\pi}, \mathbf{c}, C)$ with $\boldsymbol{\pi} = (\pi_1, \ldots, \pi_n)$ and $\mathbf{c} = (c_1, \ldots, c_n)$ such that $\omega_i = \omega$ for all $S_i \in S$; (ii) $y_i = \pi_i$ for all $S_i \in S$; (iii) $b_i \in \langle g \rangle$ and the equality $v_i = H_5(g, y_i, a, b_i, (g \cdot a^\gamma)^{u_i} \cdot (y_i \cdot b_i^\gamma)^{-v_i})$ for $\gamma = H_4(g, y_i, a, b_i)$ holds for all $S_i \in S$; (iv) $|S| = t + 1$.
  – If no such subset exists U aborts. Else, set $s_i = c_i \oplus H_2(y_i, \mathsf{pw}, b_i^{1/\rho})$, for each $S_i \in S$, and reconstruct s from these s_i shares using polynomial interpolation.
  – Compute $[r\|K] = G(s)$. If $C \neq H_3(r, \mathsf{pw}, \boldsymbol{\pi}, \mathbf{c})$ then U aborts.
  – $\boxed{\text{For each } S_i \in S, \text{ set } K_i = f_K(S_i) \text{ and compute } SK_i = f_{K_i}(\mu_i, \mu_i', U, S_i).}$

Fig. 5. DH-based PPSS and T-PAKE Protocols (boxed text indicates key-exchange specific operations on top of PPSS)

for blinding the $H_1(\mathsf{pw})$ value sent to all servers. We stress that while we specifiy the actions of honest servers, corrupted ones can deviate from the protocol in any way they choose to. Finally, note that the protocol as presented does not include an explicit authentication mechanism. This can be easily added, for example, by server S_i adding the value $f_{K_i}(0, \mu_i, \mu_i')$ to its message and by U adding a third message with value $f_{K_i}(1, \mu_i', \mu_i)$ (in this case, the session key could be derived as $SK_i = f_{K_i}(2, \mu_i, \mu_i', \mathsf{U}, \mathsf{S}_i)$).

Acknowledgements. The second author was partly supported by ERC grant CODAMODA.

References

1. Abe, M., Ohkubo, M.: A framework for universally composable non-committing blind signatures. In: Matsui, M. (ed.) ASIACRYPT 2009. LNCS, vol. 5912, pp. 435–450. Springer, Heidelberg (2009)
2. Bagherzandi, A., Jarecki, S., Saxena, N., Lu, Y.: Password-protected secret sharing. In: ACM Conference on Computer and Communications Security (2011)
3. Bellare, M., Namprempre, C., Pointcheval, D., Semanko, M.: The one-more-rsa-inversion problems and the security of chaum's blind signature scheme. J. Cryptology 16(3), 185–215 (2003)
4. Bellare, M., Pointcheval, D., Rogaway, P.: Authenticated key exchange secure against dictionary attacks. In: Preneel, B. (ed.) EUROCRYPT 2000. LNCS, vol. 1807, p. 139. Springer, Heidelberg (2000)
5. Benhamouda, F., Blazy, O., Chevalier, C., Pointcheval, D., Vergnaud, D.: New techniques for sPHFs and efficient one-round PAKE protocols. In: Canetti, R., Garay, J.A. (eds.) CRYPTO 2013, Part I. LNCS, vol. 8042, pp. 449–475. Springer, Heidelberg (2013)
6. Brainard, J., Juels, A., Kaliski, B., Szydlo, M.: A new two-server approach for authentication with short secrets. In: 12th USENIX Security Symp. (2003)
7. Camenisch, J., Lehmann, A., Lysyanskaya, A., Neven, G.: Memento: How to reconstruct your secrets from a single password in a hostile environment. In: Garay, J.A., Gennaro, R. (eds.) CRYPTO 2014, Part II. LNCS, vol. 8617, pp. 256–275. Springer, Heidelberg (2014)
8. Camenisch, J., Lehmann, A., Lysyanskaya, A., Neven, G.: Memento: How to reconstruct your secrets from a single password in a hostile environment. In: Garay, J.A., Gennaro, R. (eds.) CRYPTO 2014, Part II. LNCS, vol. 8617, pp. 256–275. Springer, Heidelberg (2014)
9. Camenisch, J., Lysyanskaya, A., Neven, G.: Practical yet universally composable two-server password-authenticated secret sharing. In: ACM Conference on Computer and Communications Security, pp. 525–536 (2012)
10. Canetti, R.: Universally composable security: A new paradigm for cryptographic protocols. In: 42nd Annual Symposium on Foundations of Computer Science, pp. 136–145. IEEE Computer Society Press (2001)
11. Cash, D., Jarecki, S., Jutla, C., Krawczyk, H., Rosu, M., Steiner, M.: Highly-scalable searchable symmetric encryption with support for Boolean queries. Crypto 2013. Cryptology ePrint Archive, Report 2013/169 (March 2013)
12. Chow, S., Ma, C., Weng, J.: Zero-knowledge argument for simultaneous discrete logarithms. In: Thai, M.T., Sahni, S. (eds.) COCOON 2010. LNCS, vol. 6196, pp. 520–529. Springer, Heidelberg (2010)

13. Curtmola, R., Garay, J.A., Kamara, S., Ostrovsky, R.: Searchable symmetric encryption: improved definitions and efficient constructions. In: Juels, A., Wright, R.N., Vimercati, S. (eds.) ACM CCS 2006: 13th Conference on Computer and Communications Security, pp. 79–88. ACM Press (October/November 2006)
14. Di Raimondo, M., Gennaro, R.: Provably secure threshold password-authenticated key exchange. J. Comput. Syst. Sci. 72(6), 978–1001 (2006)
15. Fischlin, M.: Round-optimal composable blind signatures in the common reference string model. In: Dwork, C. (ed.) CRYPTO 2006. LNCS, vol. 4117, pp. 60–77. Springer, Heidelberg (2006)
16. Ford, W., Kaliski Jr., B.S.: Server-assisted generation of a strong secret from a password. In: WETICE, pp. 176–180 (2000)
17. Freedman, M.J., Ishai, Y., Pinkas, B., Reingold, O.: Keyword search and oblivious pseudorandom functions. In: Kilian, J. (ed.) TCC 2005. LNCS, vol. 3378, pp. 303–324. Springer, Heidelberg (2005)
18. Jablon, D.P.: Password authentication using multiple servers. In: Naccache, D. (ed.) CT-RSA 2001. LNCS, vol. 2020, p. 344. Springer, Heidelberg (2001)
19. Jarecki, S., Kiayias, A., Krawczyk, H.: Round-optimal password-protected secret sharing and t-pake in the password-only model. Cryptology ePrint Archive, Report 2014/650 (2014), http://eprint.iacr.org/
20. Jarecki, S., Liu, X.: Efficient oblivious pseudorandom function with applications to adaptive OT and secure computation of set intersection. In: Reingold, O. (ed.) TCC 2009. LNCS, vol. 5444, pp. 577–594. Springer, Heidelberg (2009)
21. Jarecki, S., Liu, X.: Fast secure computation of set intersection. In: Garay, J.A., De Prisco, R. (eds.) SCN 2010. LNCS, vol. 6280, pp. 418–435. Springer, Heidelberg (2010)
22. Katz, J., Mackenzie, P., Taban, G., Gligor, V.: Two-server password-only authenticated key exchange. In: Proc. Applied Cryptography and Network Security ACNS 2005 (2005)
23. Katz, J., Ostrovsky, R., Yung, M.: Efficient password-authenticated key exchange using human-memorable passwords. In: Pfitzmann, B. (ed.) EUROCRYPT 2001. LNCS, vol. 2045, p. 475. Springer, Heidelberg (2001)
24. Katz, J., Vaikuntanathan, V.: Round-optimal password-based authenticated key exchange. J. Cryptology 26(4), 714–743 (2013)
25. Kiayias, A., Zhou, H.-S.: Equivocal blind signatures and adaptive uc-security. In: Canetti, R. (ed.) TCC 2008. LNCS, vol. 4948, pp. 340–355. Springer, Heidelberg (2008)
26. Krawczyk, H.: HMQV: A high-performance secure diffie-hellman protocol. In: Shoup, V. (ed.) CRYPTO 2005. LNCS, vol. 3621, pp. 546–566. Springer, Heidelberg (2005)
27. MacKenzie, P.D., Shrimpton, T., Jakobsson, M.: Threshold password-authenticated key exchange. J. Cryptology 19(1), 27–66 (2006)
28. Merkle, R.C.: A digital signature based on a conventional encryption function. In: Pomerance, C. (ed.) CRYPTO 1987. LNCS, vol. 293, pp. 369–378. Springer, Heidelberg (1988)
29. Micali, S., Rabin, M.O., Vadhan, S.P.: Veriable random functions. In: 40th Annual Symposium on Foundations of Computer Science, pp. 120–130. IEEE Computer Society Press (October 1999)
30. Naor, M., Reingold, O.: Number-theoretic constructions of efficient pseudo-random functions. In: FOCS, pp. 458–467. IEEE Computer Society (1997)
31. New York Times. Russian Hackers Amass Over a Billion Internet Passwords, http://www.nytimes.com/2014/08/06/technology/ russian-gang-said-to-amass-more-than-a-billion-stolen-internet -credentials.html?_r=0 (August 5, 2005)

Secret-Sharing for NP

Ilan Komargodski*, Moni Naor*, and Eylon Yogev*

Weizmann Institute of Science
{ilan.komargodski,moni.naor,eylon.yogev}@weizmann.ac.il

Abstract. A computational secret-sharing scheme is a method that enables a dealer, that has a secret, to distribute this secret among a set of parties such that a "qualified" subset of parties can efficiently reconstruct the secret while any "unqualified" subset of parties cannot efficiently learn anything about the secret. The collection of "qualified" subsets is defined by a monotone Boolean function.

It has been a major open problem to understand which (monotone) functions can be realized by a computational secret-sharing scheme. Yao suggested a method for secret-sharing for any function that has a polynomial-size monotone circuit (a class which is strictly smaller than the class of monotone functions in P). Around 1990 Rudich raised the possibility of obtaining secret-sharing for all monotone functions in NP: In order to reconstruct the secret a set of parties must be "qualified" and provide a witness attesting to this fact.

Recently, Garg et al. [14] put forward the concept of witness encryption, where the goal is to encrypt a message relative to a statement $x \in L$ for a language $L \in$ NP such that anyone holding a witness to the statement can decrypt the message, however, if $x \notin L$, then it is computationally hard to decrypt. Garg et al. showed how to construct several cryptographic primitives from witness encryption and gave a candidate construction.

One can show that computational secret-sharing implies witness encryption for the same language. Our main result is the converse: we give a construction of a computational secret-sharing scheme for *any* monotone function in NP assuming witness encryption for NP and one-way functions. As a consequence we get a completeness theorem for secret-sharing: computational secret-sharing scheme for any *single* monotone NP-complete function implies a computational secret-sharing scheme for *every* monotone function in NP.

1 Introduction

A secret-sharing scheme is a method that enables a dealer, that has a secret piece of information, to distribute this secret among n parties such that a "qualified" subset of parties has enough information to reconstruct the secret while any "unqualified" subset of parties learns nothing about the secret. A monotone

---

* Research supported in part by a grant from the Israel Science Foundation, the I-CORE Program of the Planning and Budgeting Committee, BSF and IMOS. Moni Naor is the incumbent of the Judith Kleeman Professorial Chair.

collection of "qualified" subsets (i.e., subsets of parties that can reconstruct the secret) is known as an access structure, and is usually identified with its characteristic monotone function.[1] Besides being interesting in their own right, secret-sharing schemes are an important building block in many cryptographic protocols, especially those involving some notion of "qualified" sets (e.g., multi-party computation, threshold cryptography and Byzantine agreement). For more information we refer to the extensive survey of Beimel on secret-sharing schemes and their applications [4].

A significant goal in constructing secret-sharing schemes is to *minimize* the amount of information distributed to the parties. We say that a secret-sharing scheme is *efficient* if the size of all shares is polynomial in the number of parties and the size of the secret.

Secret-sharing schemes were introduced in the late 1970s by Blakley [8] and Shamir [32] for the *threshold access structure*, i.e., where the subsets that can reconstruct the secret are all the sets whose cardinality is at least a certain threshold. Their constructions were fairly efficient both in the size of the shares and in the computation required for sharing and reconstruction. Ito, Saito and Nishizeki [21] considered general access structures and showed that every monotone access structure has a (possibly *inefficient*) secret-sharing scheme that realizes it. In their scheme the size of the shares is proportional to the DNF (resp. CNF) formula size of the corresponding function. Benaloh and Leichter [7] proved that if an access structure can be described by a polynomial-size monotone *formula*, then it has an efficient secret-sharing scheme. The most general class for which secret-sharing is known was suggested by Karchmer and Wigderson [22] who showed that if the access structure can be described by a polynomial-size monotone *span program* (for instance, undirected connectivity in a graph), then it has an efficient secret-sharing scheme. Beimel and Ishai [5] proposed a secret-sharing scheme for an access structure which is conjectured to lie outside NC. On the other hand, there are no known lower bounds that show that there exists an access structure that requires only inefficient secret-sharing schemes.[2]

Computational Secret-Sharing. In the secret-sharing schemes considered above the security is guaranteed information theoretically, that is, even if the parties are computationally unbounded. These secret-sharing schemes are known as perfect secret-sharing schemes. A natural variant, known as computational secret-sharing schemes, is to allow only computationally limited dealers and parties, i.e., they are probabilistic algorithms that run in polynomial-time. More precisely, a

---

[1] It is most sensible to consider only *monotone* sets of "qualified" subsets of parties. A set M of subsets is called monotone if $A \in M$ and $A \subseteq A'$, then $A' \in M$. It is hard to imagine a meaningful method for sharing a secret to a set of "qualified" subsets that does not satisfy this property.

[2] Moreover, there are not even non-constructive lower bounds for secret-sharing schemes. The usual counting arguments (e.g., arguments that show that most functions require large circuits) do not work here since one needs to enumerate over the sharing and reconstruction algorithms whose complexity may be larger than the share size.

computational secret-sharing scheme is a secret-sharing scheme in which there exists an *efficient* dealer that generates the shares such that a "qualified" subset of parties can *efficiently* reconstruct the secret, however, an "unqualified" subset that pulls its shares together but has only limited (i.e., polynomial) computational power and attempts to reconstruct the secret should fail (with high probability). Krawczyk [25] presented a computational secret-sharing scheme for threshold access structures that is more efficient (in terms of the size of the shares) than the perfect secret-sharing schemes given by Blakley and Shamir [8,32]. In an unpublished work (mentioned in [4], see also Vinod et al. [33]), Yao showed an efficient computational secret-sharing scheme for access structures whose characteristic function can be computed by a polynomial-size monotone *circuit* (as opposed to the *perfect* secret-sharing of Benaloch and Leichter [7] for polynomial-size monotone *formulas*). Yao's construction assumes the existence of pseudorandom generators, which can be constructed from any one-way function [19]. There are access structures which are known to have an efficient *computational* secret-sharing schemes but are not known to have efficient *perfect* secret-sharing schemes, e.g., directed connectivity.[3] Yao's scheme does not include all monotone access structures with an efficient algorithm to determine eligibility. One notable example where no efficient secret-sharing is known is matching in a graph.[4] Thus, a major open problem is to answer the following question:

Which access structures have efficient computational secret-sharing schemes, and what cryptographic assumptions are required for that?

Secret-Sharing for NP. Around 1990 Steven Rudich raised the possibility of obtaining secret-sharing schemes for an even more general class of access structures than P: monotone functions in NP, also known as mNP.[5] An access structure that is defined by a function in mNP is called an mNP access structure. Intuitively, a secret-sharing scheme for an mNP access structure is defined (in the natural way) as following: for the "qualified" subsets there is a witness attesting to this fact and *given* the witness it should be possible to reconstruct the secret. On the other hand, for the "unqualified" subsets there is no witness, and so it should not be possible to reconstruct the secret. For example, consider the Hamiltonian access structure. In this access structure the parties correspond to edges of the complete undirected graph, and a set of parties X is said to be "qualified" if and

---

[3] In the access structure for directed connectivity, the parties correspond to edge slots in the complete *directed* graph and the "qualified" subsets are those edges that connect two distinguished nodes s and t.

[4] In the access structure for matching the parties correspond to edge slots in the complete graph and the "qualified" subsets are those edges that *contain* a perfect matching. Even though matching is in P, it is known that there is no monotone circuit that computes it [30].

[5] Rudich raised it in private communication with the second author around 1990 and was not written to the best of our knowledge; some of Rudich's results can be found in Beimel's survey [4] and in Naor's presentation [28].

only if the corresponding set of edges contains a Hamiltonian cycle and the set of parties knows a witness attesting to this fact.

Rudich observed that if NP $\neq$ coNP, then there is no *perfect* secret-sharing scheme for the Hamiltonian access structure in which the sharing of the secret can be done efficiently (i.e., in polynomial-time).[6] This (conditional) impossibility result motivates looking for *computational* secret-sharing schemes for the Hamiltonian access structure and other mNP access structures. Furthermore, Rudich showed that the construction of a computational secret-sharing schemes for the Hamiltonian access structure gives rise to a protocol for oblivious transfer. More precisely, Rudich showed that if one-way functions exist and there is a *computational* secret-sharing scheme for the Hamiltonian access structure (i.e., with efficient sharing and reconstruction), then efficient protocols for oblivious transfer exist.[7] In particular, constructing a computational secret-sharing scheme for the Hamiltonian access structure assuming one-way functions will resolve a major open problem in cryptography and prove that Minicrypt=Cryptomania, to use Impagliazzo's terminology [20].

In the decades since Rudich raised the possibility of access structures beyond P not much has happened. This changed with the work on witness encryption by Garg et al. [14], where the goal is to encrypt a message relative to a statement $x \in L$ for a language $L \in$ NP such that: Anyone holding a witness to the statement can decrypt the message, however, if $x \notin L$, then it is computationally hard to decrypt. Garg et al. showed how to construct several cryptographic primitives from witness encryption and gave a candidate construction.

A by-product of the proposed construction of Garg et al. was a construction of a computational secret-sharing scheme for a *specific* monotone NP-complete language. However, understanding whether one can use a secret-sharing scheme for any single (monotone) NP-complete language in order to achieve secret-sharing schemes for any language in mNP was an open problem. One of our main results is a positive answer to this question. Details follow.

Our Results. In this paper, we construct a secret-sharing scheme for *every* mNP access structure assuming witness encryption for NP and one-way functions. In addition, we give two variants of a formal definition for secret-sharing for mNP access structures (indistinguishability and semantic security) and prove their equivalence.

Theorem 1. *Assuming witness encryption for* NP *and one-way functions, there is an efficient computational secret-sharing scheme for every* mNP *access structure.*

We remark that if we relax the requirement of computational secret-sharing such that a "qualified" subset of parties can reconstruct the secret with very

---

[6] Moreover, it is possible to show that if NP $\not\subseteq$ coAM, then there is no *statistical* secret-sharing scheme for the Hamiltonian access structure in which the sharing of the secret can be done efficiently [28].

[7] The resulting reduction is *non*-black-box. Also, note that the results of Rudich apply for any other monotone NP-complete problem as well.

high probability (say, negligibly close to 1), then our scheme from Theorem 1 actually gives a secret-sharing scheme for every monotone functions in MA.

As a corollary, using the fact that a secret-sharing scheme for a language implies witness encryption for that language and using the completeness of witness encryption,[8] we obtain a completeness theorem for secret-sharing.

Corollary 1 (Completeness of Secret-Sharing). *Let L be a monotone language that is* NP-*complete (under Karp/Levin reductions) and assume that one-way functions exist. If there exists a computational secret-sharing scheme for the access structure defined by L, then there are computational secret-sharing schemes for every* mNP *access structure.*

1.1   On Witness Encryption and Its Relation to Obfuscation

Witness encryption was introduced by Garg et al. [14]. They gave a formal definition and showed how witness encryption can be combined with other cryptographic primitives to construct public-key encryption (with efficient key generation), identity-based encryption and attribute-based encryption. Lastly, Garg et al. presented a candidate construction of a witness encryption scheme which they assumed to be secure. In a more recent work, a new construction of a witness encryption scheme was proposed by Gentry, Lewko and Waters [16].

Shortly after the paper of Garg et al. [14] a candidate construction of indistinguishability obfuscation was proposed by Garg et al. [13]. An indistinguishability obfuscator is an algorithm that guarantees that if two circuits compute the same function, then their obfuscations are computationally indistinguishable. The notion of indistinguishability obfuscation was originally proposed in the seminal work of Barak et al. [2,3].

Recently, there have been two significant developments regarding indistinguishability obfuscation: first, candidate constructions for obfuscators for all polynomial-time programs were proposed [13,11,1,29,15] and second, intriguing applications of indistinguishability obfuscation when combined with other cryptographic primitives[9] have been demonstrated (see, e.g., [13,31,9]).

As shown by Garg et al. [13], indistinguishability obfuscation implies witness encryption for all NP, which, as we show in Theorem 1, implies secret-sharing for all mNP. In fact, using the completeness of witness encryption (see Footnote 8), even an indistinguishability obfuscator for 3CNF formulas (for which there is a simple candidate construction [10]) implies witness encryption for all NP. Understanding whether witness encryption is strictly weaker than indistinguishability obfuscation is an important open problem.

---

[8] Using standard Karp/Levin reductions between NP-complete languages, one can transform a witness encryption scheme for a single NP-complete language to a witness encryption scheme for any other language in NP.

[9] See [23] for a thorough discussion of the need in additional hardness assumptions on top of $i\mathcal{O}$.

1.2   Other Related Work

A different model of secret-sharing for mNP access structures was suggested by
Vinod et al. [33]. Specifically, they relaxed the requirements of secret-sharing
by introducing a semi-trusted third party T who is allowed to interact with the
dealer and the parties. They require that T does not learn anything about the
secret and the participating parties. In this model, they constructed an efficient
secret-sharing scheme for any mNP access structures (that is also efficient in
terms of the round complexity of the parties with T) assuming the existence of
efficient oblivious transfer protocols.

1.3   Main Idea

Let Com be a perfectly-binding commitment scheme. Let $M \in$ mNP be an access
structure on n parties $\mathcal{P} = \{p_1, \ldots, p_n\}$. Define M' to be the NP language that
consists of sets of n strings $c_1, \ldots, c_n$ as follows. $M'(c_1, \ldots, c_n) = 1$ if and only
if there exist $r_1, \ldots, r_n$ such that $M(x) = 1$, where $x = x_1 \ldots x_n$ is such that

$$\forall i \in [n]: \quad x_i = \begin{cases} 1 & \text{if } r_i \neq \perp \text{ and } \mathsf{Com}(i, r_i) = c_i, \\ 0 & \text{otherwise.} \end{cases}$$

For the language M' denote by $(\mathsf{Encrypt}_{M'}, \mathsf{Decrypt}_{M'})$ the witness encryption
scheme for M'. A secret-sharing scheme for the access structure M consists
of a setup phase in which the dealer distributes secret shares to the parties.
First, the dealer samples uniformly at random n openings $r_1, \ldots, r_n$. Then, the
dealer computes a witness encryption ct of the message S with respect to the
instance $(c_1 = \mathsf{Com}(1, r_1), \ldots, c_n = \mathsf{Com}(n, r_n))$ of the language M', namely ct $=$
$\mathsf{Encrypt}_{M'}((c_1, \ldots, c_n), S)$. Finally, the share of party p_i is set to be $\langle r_1, \mathsf{ct} \rangle$.

Clearly, if $\mathsf{Encrypt}_{M'}$ and Com are efficient, then the generation of the shares
is efficient. Moreover, the reconstruction procedure is the natural one: Given a
subset of parties $X \subseteq \mathcal{P}$ such that $M(X) = 1$ and a valid witness w, decrypt
ct using the shares of the parties X and w. By the completeness of the witness
encryption scheme, given a valid subset of parties X and a valid witness w the
decryption will output the secret S.

As for the *security* of this scheme, we want to show that it is impossible to
extract (or even learn anything about) the secret having a subset of parties X for
which $M(X) = 0$ (i.e., an "unqualified" subset of parties). Let X be such that
$M(X) = 0$ and let D be an algorithm that extracts the secret given the shares of
parties corresponding to X. Roughly speaking, we will use the ability to extract
the secret in order to solve the following task: we are given a list of n unopened
string commitments $c_1, \ldots, c_n$ and a promise that it either corresponds to the
values $A_0 = \{1, \ldots, n\}$ or it corresponds to the values $A_1 = \{n + 1, \ldots, 2n\}$ and
we need to decide which is the case. Succeeding in this task would break the
security guarantee of the commitment scheme.

We sample n openings $r_1, \ldots, r_n$ uniformly at random and create a new witness encryption ct$'$ such that ct$' = \mathsf{Encrypt}_{M'}((c_1', \ldots, c_n'), S)$ as above, where we

replace the commitments corresponding to parties not in X with commitments from the input as follows:

$$\forall i \in [n]: \quad c'_i = \begin{cases} \mathsf{Com}(i, r_i) & \text{if } \mathsf{p}_i \in X \\ c_i & \text{otherwise.} \end{cases}$$

For $i \in [n]$ we set the share of party p_i to be $\langle r_i, \mathsf{ct}' \rangle$. We run D with this new set of shares. If we are in the case where $c_1, \ldots, c_n$ corresponds to A_0, then D is unable to distinguish between ct and ct' and, hence, will be able to extract the secret. On the other hand, if $c_1, \ldots, c_n$ corresponds to A_1, then there is no valid witness to decrypt ct' (since the commitment scheme is perfectly-binding). Therefore, by the security of the witness encryption scheme, it is computationally hard to learn anything about the secret S from ct'. Hence, if D is able to extract the secret S, then we deduce that $c_1, \ldots, c_n$ correspond to A_0 and, otherwise we conclude that $c_1, \ldots, c_n$ correspond to A_1.

The above gives intuition for proving security in the non-uniform setting. To see this, we assume that there exists an X such that $M(X) = 0$ and the distinguisher D can extract the secret from the shares of X. Our security definition (see Section 3) is uniform and requires the distinguisher D to find such an X *and* extract the secret with noticeable probability. In the uniform case, we first run D to get X and must make sure that $M(X) = 0$. Otherwise, if $M(X) = 1$, in both cases (that $c_1, \ldots, c_n$ correspond to A_0 or to A_1) it is easy to extract the secret and thus we might be completely fooled. The problem is that M is a language in mNP and, in general, it could be hard to test whether $M(X) = 0$. We overcome this by sampling many subsets X and use D to estimate which one to use. For more information we refer to Section 4.1.

2   Preliminaries

We start with some general notation. We denote by $[n]$ the set of numbers $\{1, 2, \ldots, n\}$. Throughout the paper we use n as our security parameter. We denote by $\mathbf{U}_n$ the uniform distribution on n bits. For a distribution or random variable R we write $r \leftarrow R$ to denote the operation of sampling a random element r according to R. For a set S, we write $s \xleftarrow{R} S$ to denote the operation of sampling an s uniformly at random from the set S. We denote by $\mathsf{neg} : \mathbb{N} \to \mathbb{R}$ a function such that for every positive integer c there exists an integer N_c such that for all $n > N_c$, $\mathsf{neg}(n) < 1/n^c$.

2.1   Monotone NP

A function $f : 2^{[n]} \to \{0, 1\}$ is said to be **monotone** if for every $X \subseteq [n]$ such that $f(X) = 1$ it also holds that $\forall Y \subseteq [n]$ such that $X \subseteq Y$ it holds that $f(Y) = 1$.

A monotone Boolean circuits is a Boolean circuit with AND and OR gates (without negations). A **non-deterministic circuit** is a Boolean circuit whose inputs are divided into two parts: standard inputs and non-deterministic inputs.

A non-deterministic circuit accepts a standard input if and only if there is some setting of the non-deterministic input that causes the circuit to evaluate to 1. A monotone non-deterministic circuit is a non-deterministic circuit where the monotonicity requirement applies only to the standard inputs, that is, every path from a standard input wire to the output wire does not have a negation gate.

Definition 1 ([18]). *We say that a function L is in* mNP *if there exists a uniform family of polynomial-size monotone non-deterministic circuit that computes* L.

Lemma 1 ([18, Theorem 2.2]). mNP = NP ∩ mono, *where* mono *is the set of all monotone functions.*

2.2 Computational Indistinguishability

Definition 2. *Two sequences of random variables* $X = \{X_n\}_{n \in \mathbb{N}}$ *and* $Y = \{Y_n\}_{n \in \mathbb{N}}$ *are* **computationally indistinguishable** *if for every probabilistic polynomial-time algorithm A there exists an integer N such that for all* $n \geq N$,

$$|\Pr[A(X_n) = 1] - \Pr[A(Y_n) = 1]| \leq \mathsf{neg}(n).$$

where the probabilities are over X_n, Y_n *and the internal randomness of A.*

2.3 Secret-Sharing

A perfect (resp., computational) secret-sharing scheme involves a dealer who has a secret, a set of n parties, and a collection A of "qualified" subsets of parties called the access structure. A secret-sharing scheme for A is a method by which the dealer (resp., efficiently) distributes shares to the parties such that (1) any subset in A can (resp., efficiently) reconstruct the secret from its shares, and (2) any subset not in A cannot (resp., efficiently) reveal any partial information on the secret. For more information on secret-sharing schemes we refer to [4] and references therein.

Throughout this paper we deal with secret-sharing schemes for access structures over n parties $\mathcal{P} = \mathcal{P}_n = \{\mathsf{p}_1, \ldots, \mathsf{p}_n\}$.

Definition 3 (Access structure). *An* **access structure** M *on* $\mathcal{P}$ *is a monotone set of subsets of* $\mathcal{P}$. *That is, for all* $X \in M$ *it holds that* $X \subseteq \mathcal{P}$ *and for all* $X \in M$ *and* X' *such that* $X \subseteq X' \subseteq \mathcal{P}$ *it holds that* $X' \in M$.

We may think of M as a characteristic function $M : 2^{\mathcal{P}} \to \{0, 1\}$ that outputs 1 given as input $X \subseteq \mathcal{P}$ if and only if X is in the access structure.

Many different definitions for secret-sharing schemes appeared in the literature. Some of the definitions were not stated formally and in some cases rigorous security proofs were not given. Bellare and Rogaway [6] survey many of these different definitions and recast them in the tradition of provable-security cryptography. They also provide some proofs for well-known secret-sharing schemes that were previously unanalyzed. We refer to [6] for more information.

2.4 Witness Encryption

Definition 4 (Witness encryption [16]). *A witness encryption scheme for an* NP *language L (with a corresponding relation R) consists of the following two polynomial-time algorithms:*

> Encrypt$(1^\lambda, x, M)$*: Takes as input a security parameter* 1^λ*, an unbounded-length string x and an message M of polynomial length in* λ*, and outputs a ciphertext* ct.
> Decrypt(ct, w)*: Takes as input a ciphertext* ct *and an unbounded-length string w, and outputs a message M or the symbol* $\perp$*.*

These algorithms satisfy the following two conditions:

1. **Completeness (Correctness):** *For any security parameter* λ*, any* $M \in \{0,1\}^{\mathsf{poly}(\lambda)}$ *and any* $x \in L$ *such that* $R(x, w)$ *holds, we have that*

$$\Pr[\mathsf{Decrypt}(\mathsf{Encrypt}(1^\lambda, x, M), w) = M] = 1.$$

2. **Soundness (Security):** *For any probabilistic polynomial-time adversary A, there exists a negligible function* $\mathsf{neg}(\cdot)$*, such that for any* $x \notin L$ *and equal-length messages* M_1 *and* M_2 *we have that*

$$\left| \Pr[A(\mathsf{Encrypt}(1^\lambda, x, M_1)) = 1] - \Pr[A(\mathsf{Encrypt}(1^\lambda, x, M_2)) = 1] \right| \leq \mathsf{neg}(\lambda).$$

Remark. Our definition of Rudich secret-sharing (that is given in Section 3) is uniform. The most common definition of witness encryption in the literature is a non-uniform one (both in the instance and in the messages). To achieve our notion of security for Rudich secret-sharing it is enough to use a witness encryption scheme in which the messages are chosen uniformly.

2.5 Commitment Schemes

In our construction we need a non-interactive commitment scheme such that commitments of different strings has disjoint support. Since the dealer in the setup phase of a secret-sharing scheme is not controlled by an adversary (i.e., it is honest), we can relax the foregoing requirement and use non-interactive commitment schemes that work in the CRS (common random string) model, Moreover, since the domain of input strings is small (it is of size $2n$) issues of non-uniformity can be ignored. Thus, we use the following definition:

Definition 5 (Commitment scheme in the CRS model). *A polynomial-time computable function* Com$: \{0,1\}^\ell \times \{0,1\}^n \times \{0,1\}^m \to \{0,1\}^*$*, where* ℓ *is the length of the string to commit, n is the length of the randomness, m is the length of the CRS. We say that* Com *is a (non-interactive perfectly binding) commitment scheme in the CRS model if for any two inputs* $x_1, x_2 \in \{0,1\}^\ell$ *such that* $x_1 \neq x_2$ *it holds that:*

1. Computational Hiding: *Let* crs ← $\{0,1\}^m$ *be chosen uniformly at random. The random variables* Com($x_1, \mathbf{U}_n$, crs) *and* Com($x_2, \mathbf{U}_n$, crs) *are computationally indistinguishable (given* crs*).*

2. Perfect Binding: *With all but negligible probability over the CRS, the supports of the above random variables are disjoint.*

Commitment schemes that satisfy the above definition, in the CRS model, can be constructed based on any pseudorandom generator [26] (which can be based on any one-way functions [19]). For simplicity, throghout the paper we ignore the CRS and simply write Com($\cdot, \cdot$). We say that Com(x, r) is the commitment of the value x with the opening r.

3   The Definition of Rudich Secret-Sharing

In this section we formally define computational secret-sharing for access structures realizing monotone functions in NP, which we call *Rudich secret-sharing*. Even though secret-sharing for functions in NP were considered in the past [33,4,14], no formal definition was given.

Our definition consists of two requirements: completeness and security. The *completeness* requirement assures that a "qualified" subset of parties that wishes to reconstruct the secret and *knows* the witness will be successful. The *security* requirement guarantees that as long as the parties form an "unqualified" subset, they are unable to learn the secret.

Note that the security requirement stated above is possibly hard to check efficiently: For some access structures in mNP (e.g., monotone NP-complete problems) it might be computationally hard to verify that the parties form an "unqualified" subset. Next, in Definition 6 we give a *uniform* definition of secret-sharing for NP. In Section 3.1 we give an alternative definition and show their equivalence.

Definition 6 (Rudich secret-sharing). *Let* $M : 2^{\mathcal{P}} \to \{0,1\}$ *be an access structure corresponding to a language* $L \in$ mNP *and let* V_M *be a verifier for* L. *A secret-sharing scheme* $\mathcal{S}$ *for* M *consists of a setup procedure* SETUP *and a reconstruction procedure* RECON *that satisfy the following requirements:*

1. SETUP($1^n, S$) *gets as input a secret* S *and distributes a share for each party. For* $i \in [n]$ *denote by* $\Pi(S, i)$ *the random variable that corresponds to the share of party* p_i. *Furthermore, for* $X \subseteq \mathcal{P}$ *we denote by* $\Pi(S, X)$ *the random variable that corresponds to the set of shares of parties in* X.

2. *Completeness:*
   If RECON($1^n, \Pi(S, X), w$) *gets as input the shares of a "qualified" subset of parties and a valid witness, and outputs the shared secret. Namely, for* $X \subseteq \mathcal{P}$ *if* $M(X) = 1$, *then for any valid witness* w *such that* $V_M(X, w) = 1$, *it holds that:*

$$\Pr\left[\text{RECON}(1^n, \Pi(S, X), w) = S\right] = 1,$$

   where the probability is over the internal randomness of the scheme and of RECON.

3. *Indistinguishability of the Secret:*
   For every pair of probabilistic polynomial-time algorithms (Samp, D) *where* Samp(1^n) *defines a distribution over pairs of secrets* S_0, S_1, *a subset of parties* X *and auxiliary information* σ, *it holds that*

$$| \Pr[M(X) = 0 \wedge D(1^n, S_0, S_1, \Pi(S_0, X), \sigma) = 1] -$$
$$\Pr[M(X) = 0 \wedge D(1^n, S_0, S_1, \Pi(S_1, X), \sigma) = 1] | \leq \mathsf{neg}(n),$$

where the probability is over the internal randomness of the scheme, the internal randomness of D and the distribution $(S_0, S_1, X, \sigma) \leftarrow$ Samp(1^n). *That is, for every pair of probabilistic polynomial-time algorithms* (Samp, D) *such that* Samp *chooses two secrets* S_0, S_1 *and a subset of parties* $X \subseteq \mathcal{P}$, *if* $M(X) = 0$ *then D is unable to distinguish (with noticeable probability) between the shares of X generated by* SETUP(S_0) *and the shares of X generated by* SETUP(S_1).

Notation. For ease of notation, 1^n and σ are omitted when they are clear from the context.

3.1   An Alternative Definition: Semantic Security

The security requirement (i.e., the third requirement) of a Rudich secret-sharing scheme that is given in Definition 6 is phrased in the spirit of *computational indistinguishability.* A different approach is to define the security of a Rudich secret-sharing in the spirit of *semantic security.* As in many cases (e.g., encryption [17]), it turns out that the two definitions are equivalent.

Definition 7 (Rudich secret-sharing - semantic security version)
   Let $M : 2^{\mathcal{P}} \rightarrow \{0,1\}$ be an mNP *access structure with verifier V_M. A secret-sharing scheme $\mathcal{S}$ for M consists of a setup procedure* SETUP *and a reconstruction procedure* RECON *as in Definition 6 and has the following property instead of the* indistinguishability of the secret *property:*

3 *Unlearnability of the Secret:*
   For every pair of probabilistic polynomial-time algorithms (Samp, D) *where* Samp(1^n) *defines a distribution over a secret S, a subset of parties X and auxiliary information σ, and for every efficiently computable function $f : \{0,1\}^* \rightarrow \{0,1\}^*$ it holds that there exists a probabilistic polynomial-time algorithm D' (called a* simulator*) such that*

$$| \Pr[M(X) = 0 \wedge D(1^n, \Pi(S, X), \sigma) = f(S)] -$$
$$\Pr[M(X) = 0 \wedge D'(1^n, X, \sigma) = f(S)] | \leq \mathsf{neg}(n),$$

where the probability is over the internal randomness of the scheme, the internal randomness of D and D', and the distribution $(S, X, \sigma) \leftarrow$ Samp(1^n). *That is, for every pair of probabilistic polynomial-time algorithms* (Samp, D) *such that* Samp *chooses a secret S and a subset of parties $X \subseteq \mathcal{P}$, if $M(X) = 0$ then D is unable to learn anything about S that it could not learn without access to the secret shares of X.*

Theorem 2. *Definition 7 and Definition 6 are equivalent.*

We defer the proof of Theorem 2 to the full version of the paper [24].

3.2 Definition of Adaptive Security

Our definition of Rudich secret-sharing only guarantees security against static adversaries. That is, the adversary chooses a subset of parties before it sees any of the shares. In other words, the selection is done *independently* of the sharing process and hence, we may think of it as if the sharing process is done *after* Samp chooses X.

A stronger security guarantee would be to require that even an adversary that chooses its set of parties in an *adaptive* manner based on the shares it has seen so far is unable to learn the secret (or any partial information about it). Namely, the adversary chooses the parties one by one depending on the secret shares of the previously chosen parties.

The security proof of our scheme (which is given in Section 4) does not hold under this stronger requirement. It would be interesting to strengthen it to the adaptive case as well. One problem that immediately arises in an analysis of our scheme against adaptive adversaries is that of *selective decommitment* (cf. [12]), that is when an adversary sees a collection of commitments and can select a subset of them and receive their openings. The usual proofs of security of commitment schemes are not known to hold in this case.

4   Rudich Secret-Sharing from Witness Encryption

In this section we prove the main theorem of this paper. We show how to construct a Rudich secret-sharing scheme for any mNP access structure assuming witness encryption for NP and one-way functions.

Theorem 3. *[Theorem 1 Restated] Assuming witness encryption for NP and one-way functions, there is an efficient computational secret-sharing scheme for every mNP access structure.*

Let $\mathcal{P} = \{p_1, \ldots, p_n\}$ be a set of n parties and let $M : 2^{\mathcal{P}} \to \{0, 1\}$ be an mNP access structure. We view M either as a function or as a language. For a language L in NP let $(\mathsf{Encrypt}_L, \mathsf{Decrypt}_L)$ be a witness encryption scheme and let $\mathsf{Com} : [2n] \times \{0, 1\}^n \to \{0, 1\}^{q(n)}$ be a commitment scheme, where $q(\cdot)$ is a polynomial.

The Scheme. We define a language M' that is related to M as follows. The language M' consists of sets of n strings $\{c_i\}_{i \in [n]} \in \{0, 1\}^{q(n)}$ as follows. $M'(c_1, \ldots, c_n) = 1$ if and only if there exist $\{r_i\}_{i \in [n]}$ such that $M(x) = 1$, where $x \in \{0, 1\}^n$ is such that

$$\forall i \in [n]: \quad x_i = \begin{cases} 1 & \text{if } r_i \neq \perp \text{ and } \mathsf{Com}(i, r_i) = \mathsf{c}_i, \\ 0 & \text{otherwise.} \end{cases}$$

For every $i \in [n]$, the *share* of party p_i is composed of 2 components: (1) $r_i \in \{0,1\}^n$ - an opening of a commitment to the value i, and (2) a witness encryption ct. The witness encryption encrypts the secret S with respect to the commitments of all parties $\{\mathsf{c}_i = \mathsf{Com}(i, r_i)\}_{i \in [n]}$. To reconstruct the secret given a subset of parties X, we simply decrypt ct given the corresponding openings of X and the witness w that indeed $M(X) = 1$. The secret-sharing scheme is formally described in Figure 1.

The Rudich Secret-Sharing Scheme $\mathcal{S}$ for M

The SETUP Procedure:

Input: A secret S.

Let M' be the language as described above, and let $(\mathsf{Encrypt}_{M'}, \mathsf{Decrypt}_{M'})$ be a witness encryption for M' (see Definition 4).

1. For $i \in [n]$:
   (a) Sample uniformly at random an opening $r_i \in \{0,1\}^n$.
   (b) Compute the commitment $\mathsf{c}_i = \mathsf{Com}(i, r_i)$.
2. Compute $\mathsf{ct} \leftarrow \mathsf{Encrypt}_{M'}((\mathsf{c}_1, \ldots, \mathsf{c}_n), S)$.
3. Set the share of party p_i to be $\Pi(S, i) = \langle r_i, \mathsf{ct} \rangle$.

The RECON Procedure:

Input: A non-empty subset of parties $X \subseteq \mathcal{P}$ together with their shares and a witness w of X for M.

1. Let ct be the witness encryption in the shares of X.
2. For any $i \in [n]$ let $r_i' = \begin{cases} r_i & \text{if } \mathsf{p}_i \in X \\ \perp & \text{otherwise.} \end{cases}$
3. Output $\mathsf{Decrypt}_{M'}(\mathsf{ct}, (r_1', \ldots, r_n', w))$.

Fig. 1. Rudich secret-sharing scheme for NP

Observe that if the witness encryption scheme and Com are both efficient, then the scheme is efficient (i.e., **SETUP** and **RECON** are probabilistic polynomial-time algorithms). **SETUP** generates n commitments and a witness encryption of polynomial size. **RECON** only decrypts this witness encryption.

Completeness. The next lemma states that the scheme is complete. That is, whenever the scheme is given a qualified $X \subseteq \mathcal{P}$ and a valid witness w of X, it is possible to successfully reconstruct the secret.

Lemma 2. *Let $M \in \mathsf{NP}$ be an mNP access structure. Let $\mathcal{S} = \mathcal{S}_M$ be the scheme from Figure 1 instantiated with M. For every subset of parties $X \subseteq \mathcal{P}$ such that $M(X) = 1$ and any valid witness w it holds that*

$$\Pr\left[\mathsf{RECON}(\Pi(S,X),w) = S\right] = 1.$$

Proof. Recall the definition of the algorithm RECON from Figure 1: RECON gets as input the shares of a subset of parties $X = \{\mathsf{p}_{i_1}, \ldots, \mathsf{p}_{i_k}\}$ for $k, i_1, \ldots, i_k \in [n]$ and a valid witness w. Recall that the shares of the parties in X consist of k openings for the corresponding commitments and a witness encryption ct. RECON decrypts ct given the openings of parties in X and the witness w.

By the completeness of the witness encryption scheme, the output of the decryption procedure on ct, given a valid X and a valid witness, is S (with probability 1). $\square$

Indistinguishability of the Secret. We show that our scheme is secure. More precisely, we show that given an "unqualified" set of parties $X \subseteq \mathcal{P}$ as input (i.e., $M(X) = 0$), with overwhelming probability, any probabilistic polynomial-time algorithm cannot distinguish the shared secret from another.

To this end, we assume towards a contradiction that such an algorithm exists and use it to efficiently solve the following task: given two lists of n commitments and a promise that one of them corresponds to the values $\{1, \ldots, n\}$ and the other corresponds to the values $\{n+1, \ldots, 2n\}$, identify which one corresponds to the values $\{1, \ldots, n\}$. The following lemma shows that solving this task efficiently can be used to break the hiding property of the commitment scheme.

Lemma 3. *Let $\mathsf{Com} \colon [2n] \times \{0,1\}^n \to \{0,1\}^{q(n)}$ be a commitment scheme where $q(\cdot)$ is a polynomial. If there exist $\varepsilon = \varepsilon(n) > 0$ and a probabilistic polynomial-time algorithm D for which*

$$|\Pr[D(\mathsf{Com}(1, \mathbf{U}_n), \ldots, \mathsf{Com}(n, \mathbf{U}_n)) = 1] -$$
$$\Pr[D(\mathsf{Com}(n, \mathbf{U}_n), \ldots, \mathsf{Com}(2n, \mathbf{U}_n)) = 1]| \geq \varepsilon,$$

then there exist a probabilistic polynomial-time algorithm D' and $x, y \in [2n]$ such that

$$|\Pr[D'(\mathsf{Com}(x, \mathbf{U}_n)) = 1] - \Pr[D'(\mathsf{Com}(y, \mathbf{U}_n)) = 1]| \geq \varepsilon/n.$$

The proof of the lemma follows from a standard hybrid argument. See details in the full version of the paper [24].

At this point we are ready to prove the security of our scheme. That is, we show that the ability to break the security of our scheme translates to the ability to break the commitment scheme (using Lemma 3).

Lemma 4. *Let* $\mathcal{P} = \{p_1, \ldots, p_n\}$ *be a set of n parties. Let $M : 2^{\mathcal{P}} \to \{0,1\}$ be an* mNP *access structure. If there exist a non-negligible $\varepsilon = \varepsilon(n)$ and a pair of probabilistic polynomial-time algorithms* (Samp, D) *such that for* $(S_0, S_1, X) \leftarrow$ Samp(1^n) *it holds that*

$$\Pr[M(X) = 0 \wedge D(S_0, S_1, \Pi(S_0, X)) = 1]$$
$$- \Pr[M(X) = 0 \wedge D(S_0, S_1, \Pi(S_1, X)) = 1] \geq \varepsilon,$$

then there exists a probabilistic algorithm D' that runs in polynomial-time in n/ε such that for sufficiently large n

$$|\Pr[D'(\text{Com}(1, \mathbf{U}_n), \ldots, \text{Com}(n, \mathbf{U}_n)) = 1] -$$
$$\Pr[D'(\text{Com}(n+1, \mathbf{U}_n), \ldots, \text{Com}(2n, \mathbf{U}_n)) = 1]| \geq \varepsilon/10 - \text{neg}(n).$$

The proof of Lemma 4 appears in Section 4.1.

Using Lemma 4 we can prove Theorem 3, the main theorem of this section. The *completeness* requirement (Item 2 in Definition 6) follows directly from Lemma 2. The *indistinguishability of the secret* requirement (Item 3 in Definition 6) follows by combining Lemmas 3 and 4 together with the hiding property of the commitment scheme. Section 4.1 is devoted to the proof of Lemma 4.

4.1   Main Proof of Security

Let M be an mNP access structure, (Samp, D) be a pair of algorithms and $\varepsilon > 0$ be a function of n, as in the Lemma 4. We are given a list of (unopened) string commitments $c_1, \ldots, c_n \in \{\text{Com}(z_i, r)\}_{r \in \{0,1\}^n}$, where for $Z = \{z_1, \ldots, z_n\}$ either $Z = \{1, \ldots, n\} \triangleq A_0$ or $Z = \{n+1, \ldots, 2n\} \triangleq A_1$. Our goal is to construct an algorithm D' that distinguishes between the two cases (using Samp and D) with non-negligible probability (that is related to ε). Recall that Samp chooses two secrets S_0, S_1 and $X \subseteq \mathcal{P}$ and then D gets as input the secret shares of parties in X for one of the secrets. By assumption, for $(S_0, S_1, X) \leftarrow$ Samp(1^n) we have that

$$|\Pr[M(X) = 0 \wedge D(S_0, S_1, \Pi(S_0, X)) = 1] -$$
$$\Pr[M(X) = 0 \wedge D(S_0, S_1, \Pi(S_1, X)) = 1]| \geq \varepsilon. \tag{1}$$

Roughly speaking, the algorithm D' that we define creates a new set of shares using $c_1, \ldots, c_n$ such that: If $c_1, \ldots, c_n$ are commitments to $Z = A_0$ then D is able to recover the secret; otherwise, (if $Z = A_1$) it is computationally hard to recover the secret. Thus, D' can distinguish between the two cases by running D on the new set of shares and acting according to its output.

We begin by describing a useful subroutine we call $\mathsf{D_{ver}}$. The inputs to $\mathsf{D_{ver}}$ are n string commitments $c_1, \ldots, c_n$, two secrets S_0, S_1 and a subset of $k \in [n]$ parties X. Assume for ease of notations that $X = \{p_1, \ldots, p_k\}$. $\mathsf{D_{ver}}$ first chooses b uniformly at random from the set $\{0,1\}$ and samples uniformly at random n openings $r_1, \ldots, r_n$ from the distribution $\mathbf{U}_n$. Then, $\mathsf{D_{ver}}$ computes the witness encryption ct'_b of the message S_b with respect to the instance

$\mathsf{Com}(1, r_1), \ldots, \mathsf{Com}(k, r_k), \mathsf{c}_{k+1}, \ldots, \mathsf{c}_n$ of M' (see Figure 1) and sets for every $i \in [n]$ the share of party p_i to be $\Pi'(S_b, i) = \langle r_i, \mathsf{ct}'_b \rangle$. Finally, $\mathsf{D}_{\mathsf{ver}}$ emulates the execution of D on the set of shares of X ($\Pi'(S_b, X)$). If the output of D equals to b, then $\mathsf{D}_{\mathsf{ver}}$ outputs 1 (meaning the input commitments correspond to $Z = A_0$); otherwise, $\mathsf{D}_{\mathsf{ver}}$ outputs 0 (meaning the input commitments correspond to $Z = A_1$).

The naïve implementation of D' is to run Samp to generate S_0, S_1 and X, run $\mathsf{D}_{\mathsf{ver}}$ with the given string commitments, S_0, S_1 and X, and output accordingly. This, however, does not work. To see this, recall that the assumption (eq. (1)) only guarantees that D is able to distinguish between the two secrets when $M(X) = 0$. However, it is possible that with high probability (yet smaller than $1 - 1/\mathsf{poly}(n)$) over Samp it holds that $M(X) = 1$, in which we do not have any guarantee on D. Hence, simply running Samp and $\mathsf{D}_{\mathsf{ver}}$ might fool us in outputting the wrong answer.

The first step to solve this is to observe that, by the assumption in eq. (1), Samp generates an X such that $M(X) = 0$ with (non-negligible) probability at least ε. By this observation, notice that by running Samp for $\Theta(n/\varepsilon)$ iterations we are assured that with very high probability (specifically, $1 - \mathsf{neg}(n)$) there exists an iteration in which $M(X) = 0$. All we are left to do is to recognize in which iteration $M(X) = 0$ and only in that iteration we run $\mathsf{D}_{\mathsf{ver}}$ and output accordingly.

However, in general it might be computationally difficult to test for a given X whether $M(X) = 0$ or not. To overcome this, we observe that we need something much simpler than testing if $M(X) = 0$ or not. All we actually need is a procedure that we call B that checks if $\mathsf{D}_{\mathsf{ver}}$ is a good distinguisher (between commitments to A_0 and commitments to A_1) for a given X. On the one hand, by the assumption, we are assured that this is indeed the case if $M(X) = 0$. On the other hand, if $M(X) = 1$ and $\mathsf{D}_{\mathsf{ver}}$ is biased, then simply running $\mathsf{D}_{\mathsf{ver}}$ and outputting accordingly is enough.

Thus, our goal is to estimate the bias of $\mathsf{D}_{\mathsf{ver}}$. The latter is implemented efficiently by running $\mathsf{D}_{\mathsf{ver}}$ independently $\Theta(n/\varepsilon)$ times on both inputs (i.e., with $Z = A_0$ and with $Z = A_1$) and counting the number of "correct" answers.

Recapping, our construction of D' is as follows: D' runs for $\Theta(n/\varepsilon)$ iterations such that in each iteration it runs $\mathsf{Samp}(1^n)$ and gets two secrets S_0, S_1 and a subset of parties X. Then, it estimates the *bias* of $\mathsf{D}_{\mathsf{ver}}$ for that specific X (independently of the input). If the bias is large enough, D' evaluates $\mathsf{D}_{\mathsf{ver}}$ with the input of D', the two secrets S_0, S_1 and the subset of parties X and outputs its output. The formal description of D' is given in Figure 2.

Analysis of D'. We defer the detailed analysis of D' to the full version of the paper [24].

The algorithm D'

Input: A sequence of commitments $c_1, \ldots, c_n$ where $\forall i \in [n]$: $c_i \in \{\mathsf{Com}(z_i, r)\}_{r \in \{0,1\}^n}$ and for $Z = \{z_1, \ldots, z_n\}$ either $Z = \{1, \ldots, n\} \triangleq A_0$ or $Z = \{n+1, \ldots, 2n\} \triangleq A_1$.

1. Do the following for $T = n/\varepsilon$ times:
   (a) $S_0, S_1, X \leftarrow \mathsf{Samp}(1^n)$.
   (b) Run $\mathsf{bias} \leftarrow \mathsf{B}(S_0, S_1, X)$.
   (c) If $\mathsf{bias} = 1$:
       i. Run $\mathsf{resD} \leftarrow \mathsf{D}_{\mathsf{ver}}(c_1, \ldots, c_n, S_0, S_1, X)$.
       ii. Output resD (and HALT).
2. Output 0.

The sub-procedure B

Input: Two secrets S_0, S_1 and a subset of parties $X \subseteq \mathcal{P}$.

1. Set $q_0, q_1 \leftarrow 0$. Run $T_{\mathsf{B}} = 4n/\varepsilon$ times:
   (a) $q_0 \leftarrow q_0 + \mathsf{D}_{\mathsf{ver}}(\mathsf{Com}(1, \mathbf{U}_n), \ldots, \mathsf{Com}(n, \mathbf{U}_n), S_0, S_1, X)$.
   (b) $q_1 \leftarrow q_1 + \mathsf{D}_{\mathsf{ver}}(\mathsf{Com}(n+1, \mathbf{U}_n), \ldots, \mathsf{Com}(2n, \mathbf{U}_n), S_0, S_1, X)$.
2. If $|q_0 - q_1| > n$, output 1.
3. Output 0.

The sub-procedure $\mathsf{D}_{\mathsf{ver}}$

Input: A sequence of commitments $c_1, \ldots, c_n$, two secrets S_0, S_1 and a subset of parties $X \subseteq \mathcal{P}$.

1. Choose $b \in \{0, 1\}$ uniformly at random.
2. For $i \in [n]$: Sample $r_i \xleftarrow{R} \mathbf{U}_n$ and let $c'_i = \begin{cases} \mathsf{Com}(i, r_i) & \text{if } \mathsf{p}_i \in X \\ c_i & \text{otherwise.} \end{cases}$
3. Compute $\mathsf{ct}'_b \leftarrow \mathsf{Encrypt}_{M'}((c'_1, \ldots, c'_n), S_b)$.
4. For $i \in [n]$ let the new share of party p_i be $\Pi'(S_b, i) = \langle r_i, \mathsf{ct}'_b \rangle$.
5. Return 1 if $D(S_0, S_1, \Pi'(S_b, X)) = b$ and 0 otherwise.

Fig. 2. The description of the algorithm D'

5 Conclusions and Open Problems

We have shown a construction of a secret-sharing scheme for any mNP access structure. In fact, our construction yields the first candidate computational secret-sharing scheme for *all* monotone functions in P (recall that not every monotone function in P can be computed by a polynomial-size monotone circuit, see e.g., Razborov's lower bound for matching [30]). Our construction only requires witness encryption scheme for NP.

We conclude with several open problems:

- Is there a secret-sharing scheme for mNP that relies only on standard hardness assumptions, or at least falsifiable ones [27]?
- Is there a way to use secret-sharing for monotone P to achieve secret-sharing for monotone NP (in a black-box manner)?
- Construct a Rudich secret-sharing scheme for every access structure in mNP that is secure against *adaptive* adversaries (see Section 3.2 for a discussion). Under a stronger assumption, i.e., extractable witness encryption (in which if an algorithm is able to decrypt a ciphertext, then it is possible to extract a witness), Zvika Brakerski observed that our construction is secure against adaptive adversaries as well.
- Show a completeness theorem (similarly to Corollary 1) for secret-sharing schemes that are also secure against *adaptive* adversaries, as defined in Section 3.2.

Acknowledgements. We are grateful to Amit Sahai for suggesting to base our construction on witness encryption. We thank Zvika Brakerski for many helpful discussions and insightful ideas. The second author thanks Steven Rudich for sharing with him his ideas on secret sharing beyond P. We thank the anonymous referees for many helpful remarks.

References

1. Barak, B., Garg, S., Kalai, Y.T., Paneth, O., Sahai, A.: Protecting obfuscation against algebraic attacks. In: Nguyen, P.Q., Oswald, E. (eds.) EUROCRYPT 2014. LNCS, vol. 8441, pp. 221–238. Springer, Heidelberg (2014)
2. Barak, B., Goldreich, O., Impagliazzo, R., Rudich, S., Sahai, A., Vadhan, S.P., Yang, K.: On the (im)possibility of obfuscating programs. In: Kilian, J. (ed.) CRYPTO 2001. LNCS, vol. 2139, pp. 1–18. Springer, Heidelberg (2001)
3. Barak, B., Goldreich, O., Impagliazzo, R., Rudich, S., Sahai, A., Vadhan, S.P., Yang, K.: On the (Im)possibility of obfuscating programs (Preliminary version appeared in CRYPTO 2001). In: Kilian, J. (ed.) CRYPTO 2001. LNCS, vol. 2139, p. 1. Springer, Heidelberg (2001)
4. Beimel, A.: Secret-sharing schemes: A survey. In: Chee, Y.M., Guo, Z., Ling, S., Shao, F., Tang, Y., Wang, H., Xing, C. (eds.) IWCC 2011. LNCS, vol. 6639, pp. 11–46. Springer, Heidelberg (2011)
5. Beimel, A., Ishai, Y.: On the power of nonlinear secret-sharing. SIAM Journal on Discrete Mathematics 19(1), 258–280 (2005)
6. Bellare, M., Rogaway, P.: Robust computational secret sharing and a unified account of classical secret-sharing goals. In: ACM Conference on Computer and Communications Security, pp. 172–184. ACM (2007)
7. Benaloh, J.C., Leichter, J.: Generalized secret sharing and monotone functions. In: Goldwasser, S. (ed.) CRYPTO 1988. LNCS, vol. 403, pp. 27–35. Springer, Heidelberg (1990)

8. Blakley, G.R.: Safeguarding cryptographic keys. In: Proceedings of the AFIPS National Computer Conference, vol. 22, pp. 313–317 (1979)
9. Boneh, D., Zhandry, M.: Multiparty key exchange, efficient traitor tracing, and more from indistinguishability obfuscation. In: Garay, J.A., Gennaro, R. (eds.) CRYPTO 2014, Part I. LNCS, vol. 8616, pp. 480–499. Springer, Heidelberg (2014)
10. Brakerski, Z., Rothblum, G.N.: Black-box obfuscation for d-CNFs. In: ITCS, pp. 235–250. ACM (2014)
11. Brakerski, Z., Rothblum, G.N.: Virtual black-box obfuscation for all circuits via generic graded encoding. In: Lindell, Y. (ed.) TCC 2014. LNCS, vol. 8349, pp. 1–25. Springer, Heidelberg (2014)
12. Dwork, C., Naor, M., Reingold, O., Stockmeyer, L.J.: Magic functions. Journal of the ACM 50(6), 852–921 (2003)
13. Garg, S., Gentry, C., Halevi, S., Raykova, M., Sahai, A., Waters, B.: Candidate indistinguishability obfuscation and functional encryption for all circuits. In: FOCS, pp. 40–49 (2013)
14. Garg, S., Gentry, C., Sahai, A., Waters, B.: Witness encryption and its applications. In: STOC, pp. 467–476. ACM (2013)
15. Gentry, C., Lewko, A.B., Sahai, A., Waters, B.: Indistinguishability obfuscation from the multilinear subgroup elimination assumption. IACR Cryptology ePrint Archive 2014, 309 (2014)
16. Gentry, C., Lewko, A.B., Waters, B.: Witness encryption from instance independent assumptions. In: Garay, J.A., Gennaro, R. (eds.) CRYPTO 2014, Part I. LNCS, vol. 8616, pp. 426–443. Springer, Heidelberg (2014)
17. Goldwasser, S., Micali, S.: Probabilistic encryption. Journal of Computer and System Sciences 28(2), 270–299 (1984)
18. Grigni, M., Sipser, M.: Monotone complexity. In: Proceedings of LMS Workshop on Boolean Function Complexity, vol. 169, pp. 57–75. Cambridge University Press (1992)
19. Håstad, J., Impagliazzo, R., Levin, L.A., Luby, M.: A pseudorandom generator from any one-way function. SIAM J. Comput. 28(4), 1364–1396 (1999)
20. Impagliazzo, R.: A personal view of average-case complexity. In: Structure in Complexity Theory Conference, pp. 134–147. IEEE Computer Society (1995)
21. Ito, M., Saito, A., Nishizeki, T.: Multiple assignment scheme for sharing secret. Journal of Cryptology 6(1), 15–20 (1993)
22. Karchmer, M., Wigderson, A.: On span programs. In: Structure in Complexity Theory Conference, pp. 102–111. IEEE Computer Society (1993)
23. Komargodski, I., Moran, T., Naor, M., Pass, R., Rosen, A., Yogev, E.: One-way functions and (im)perfect obfuscation. IACR Cryptology ePrint Archive 2014, 347 (2014), to appear in FOCS 2014
24. Komargodski, I., Naor, M., Yogev, E.: Secret-sharing for NP. IACR Cryptology ePrint Archive 2014, 213 (2014)
25. Krawczyk, H.: Secret sharing made short. In: Stinson, D.R. (ed.) CRYPTO 1993. LNCS, vol. 773, pp. 136–146. Springer, Heidelberg (1994)
26. Naor, M.: Bit commitment using pseudorandomness. Journal of Cryptology 4(2), 151–158 (1991)
27. Naor, M.: On cryptographic assumptions and challenges. In: Boneh, D. (ed.) CRYPTO 2003. LNCS, vol. 2729, pp. 96–109. Springer, Heidelberg (2003)
28. Naor, M.: Secret sharing for access structures beyond P (2006), slides: http://www.wisdom.weizmann.ac.il/~naor/PAPERS/minicrypt.html

29. Pass, R., Seth, K., Telang, S.: Indistinguishability obfuscation from semantically-secure multilinear encodings. In: Garay, J.A., Gennaro, R. (eds.) CRYPTO 2014, Part I. LNCS, vol. 8616, pp. 500–517. Springer, Heidelberg (2014)

30. Razborov, A.A.: Lower bounds for the monotone complexity of some Boolean functions. Dokl. Ak. Nauk. SSSR 281, 798–801 (1985), english translation in: Soviet Math. Dokl. Vol. 31, pp. 354-357 (1985)

31. Sahai, A., Waters, B.: How to use indistinguishability obfuscation: deniable encryption, and more. In: STOC, pp. 475–484. ACM (2014)

32. Shamir, A.: How to share a secret. Communications of the ACM 22(11), 612–613 (1979)

33. Vinod, V., Narayanan, A., Srinathan, K., Pandu Rangan, C., Kim, K.: On the power of computational secret sharing. In: Johansson, T., Maitra, S. (eds.) INDOCRYPT 2003. LNCS, vol. 2904, pp. 162–176. Springer, Heidelberg (2003)

Tweaks and Keys for Block Ciphers: The TWEAKEY Framework

Jérémy Jean, Ivica Nikolić, and Thomas Peyrin

Division of Mathematical Sciences, School of Physical and Mathematical Science, Nanyang Technological University, Singapore
{JJean, INikolic, Thomas.Peyrin}@ntu.edu.sg

Abstract. We propose the TWEAKEY framework with goal to unify the design of tweakable block ciphers and of block ciphers resistant to related-key attacks. Our framework is simple, extends the key-alternating construction, and allows to build a primitive with arbitrary tweak and key sizes, given the public round permutation (for instance, the AES round). Increasing the sizes renders the security analysis very difficult and thus we identify a subclass of TWEAKEY, that we name STK, which solves the size issue by the use of finite field multiplications on low hamming weight constants. Overall, this construction allows a significant increase of security of well-known authenticated encryptions mode like ΘCB3 from birthday-bound security to full security, where a regular block cipher was used as a black box to build a tweakable block cipher. Our work can also be seen as advances on the topic of secure key schedule design.

Keywords: tweak, block cipher, key schedule, authenticated encryption.

1 Introduction

Block ciphers are among the most scrutinized cryptographic primitives, used in many constructions as basic secure bricks that ensure data encryption and/or authenticity. In the last few decades, a lot of research has been conducted on this topic, and it is believed that building a secure and efficient block cipher is now a well-understood problem. In particular, designs that allowed to prove their security against classical differential or linear attacks have been a very important step forward, and have been incorporated in the current main worldwide standard AES-128 [34]. This topic is mature and the community has recently been focusing on other directions, such as the possibility to build ciphers dedicated to very constrained environments [9, 12, 23].

The security of the block ciphers, both Feistel and Substitution-Permutation networks, has been well studied when the key is fixed and secret, however, when the attacker is allowed to ask for encryption or decryption with different (and related) keys the situation becomes more complicated. In the past, many published ciphers have been broken in this so-called *related-key model* [4, 5] and it has even been demonstrated that the Advanced Encryption Standard (AES) has

P. Sarkar and T. Iwata (Eds.): ASIACRYPT 2014, PART II, LNCS 8874, pp. 274–288, 2014.

flaws in this model [6,7]. It is known how to design a cryptographically good permutation composed of several iterated rounds, but when it comes to keying this permutation with subkeys generated by the key schedule, it is hard to ensure that the overall construction remains secure. Most key schedule constructions are ad-hoc, in the sense that the designers came up with a key schedule that is quite different from the internal permutation of the cipher, in a hope that no meaningful structure is created by the interaction of the two components. This is the case of PRESENT [9] or AES [34] ciphers, where the key schedule is purposely made different from the round function. Some key schedules can be very weak but fast and lightweight (like in LED [23], where many rounds are required to ensure security against related-key attacks), while some can be very strong but slow (like in the internal cipher of the WHIRLPOOL hash function [3]). In order to partially ease this task of deriving a good schedule, some automatic tools analyzing the resistance of the ciphers against simple related-key differential attacks have been developed [8, 21, 32].

The Hasty Pudding cipher [37], proposed to the AES competition organized by the NIST, permitted the user to insert an additional input to the classical key and plaintext pair, called *spice* by the designers of this cipher. This extra input T, later renamed as *tweak*, was supposed to be completely public and to randomize the instance of the block cipher: to different values of T correspond different and independent families of permutations E_K. This feature was formalized in 2002 by Liskov et al. [30,31], who showed that tweakable block ciphers are valuable building blocks if retweaking (changing the tweak value) is less costly than changing its secret key. Tweakable block ciphers (see MERCY [13], for example) found many different utilizations in cryptography, such as disk encryption where each block is ciphered with the same key, but the block index is used as tweak value.

Simple constructions of a tweakable block cipher $E_K(T, P)$ based on a block cipher $E_K(P)$, like XORing the tweak into the key input and/or message input, are not satisfactory. For example, only XORing the tweak into the key input would result in an undesirable property that $E_K(T, P) = E_{K \oplus X}(T \oplus X, P)$. Liskov et al. propose instead to use universal hash families for that purpose. The XE and XEX constructions [36] (and the follow-up standard XTS [19]) are based on finite field multiplications in $GF(2^n)$, and present the particularity of being efficient if sequential tweaks are used. Nonetheless, even with such feature, these scheme might not be really efficient as the cipher execution is not negligible compared to a finite field multiplication in $GF(2^n)$ (for example when AES is the internal block cipher and the scheme implementation uses AES-NI instructions). More importantly, *these methods ensure only security up to the birthday-bound (relative to the block cipher size)*. This can be a problem as the main block cipher standards only have 64- or 128-bit block size. Minematsu [33] partially overcomes this limitation by proving beyond birthday-bound security for his design, but at the expense of a very reduced efficiency. The same observation applies to more recent beyond-birthday constructions such as [29, 38]. Overall, *none of the state-of-the-art block-cipher-based schemes provide both efficiency and beyond birthday-bound security.*

Ad-hoc constructions would be a solution, with the obvious drawback that security proofs regarding the construction would be very hard to obtain. So far, this direction has seen a surprisingly low number of proposals. The NIST SHA-3 competition for hash functions triggered a few, like SKEIN [20] (with its ad-hoc internal tweakable block cipher Threefish) and BLAKE2 [2]. It is interesting to note that both are Addition-Rotation-XOR (ARX) functions and thus offer less possibility of proofs with regard to classical differential-linear attacks. As of today, it remains an open problem to design an ad-hoc AES-like tweakable block cipher, which in fact would be very valuable for authenticated encryption as AES-NI instruction sets guarantee extremely fast software implementations. Such a primitive would enable very efficient authenticated encryption with beyond birthday-bound security and proof regarding the mode of operation.

Liskov et al. proposed to separate the roles of the secret key (which provides uncertainty to the adversary) from that of the tweak (which provides independent variability) – interestingly, almost all tweakable block cipher proposals (except Threefish) follow this rule. This might be seen as counter intuitive as it is required the tweak input to be somehow more efficient than the key input, but at the same time the security requirement on the tweak seem somehow stronger than on the key, since the attacker can fully control the former (even though tweak-recovery attacks are irrelevant). We argue in this article that, in practice, when one designs a block cipher these two inputs should be considered almost the same, as incorporating a tweak and a secret key shares in fact a lot of common ground, especially for the large family of key-alternating ciphers.

Our Contributions. In this article, we bring together key schedule design and tweak input handling for block ciphers in a common framework that we call TWEAKEY (Section 2). The idea is to provide a simple framework to design a tweakable block cipher with any key and any tweak sizes. Our construction is very simple and can be seen as a natural extension of key-alternating ciphers: a subtweakey (i.e. a value obtained from the key and the tweak inputs) is incorporated into the internal state at every round of the iterative cipher. One advantage of such a framework is that one can obtain a tweakable single-key block cipher or a double-key length block cipher with the very same primitive.

Not all instances of TWEAKEY are secure and, in particular, the case where the key and tweak material is treated exactly the same way does not lead to a secure cipher. However, handling the key and the tweak material the same way would be attractive in terms of performance, implementation, but more importantly it would greatly simplify the security analysis, which is currently the main difficulty designers have to face when constructing an ad-hoc tweakable block cipher. Indeed, the main challenge is to evaluate the appropriate number of rounds required to make the cipher secure – when the tweak size t and key size k are too large this problem becomes infeasible. We propose a solution in Section 3 and we give a subclass of TWEAKEY for AES-like ciphers, named STK (for Superposition TWEAKEY), where the key and the tweak materials are treated almost the same way – the small difference between the linear key and the tweak schedules is sufficient to remove the aforementioned weakness. Due to the structure of STK, the security analysis is rendered much easier, and the number of rounds can be kept small. The STK construction leads to promising performances: in [24], a complete 128-bit tweak 128-bit key 128-bit block cipher proposal Deoxys-BC based on the AES round function is proposed as an instance of the STK construction. It is faster and more lightweight than other tentatives to build a tweakable block cipher from AES-128. When used in ⊖CB3 [28] authenticated encryption, Deoxys-BC runs at about 1.3 c/B on the latest Intel processors. This has to be compared to OCB3, which runs at 0.7-0.88 c/B when instantiated with AES-128, but only ensures birthday-bound security. Alternatively, Deoxys-BC could be a replacement for AES-256, which has related-key issues as shown in [7]. The STK construction offers a very lightweight tweakey schedule (only composed of a substitution of bits), that even allows the key to be hardwired in hardware implementations. Similarly, one can mention Joltik-BC: a lightweight instance of the STK construction as a 64-bit tweak 64/128-bit key 64-bit tweakable block cipher.

In [26], the problem of tweaking AES-128 without altering the key schedule is handled. The authors introduce Kiasu-BC as part of the TWEAKEY framework as a way to securely introduce a 64-bit tweak in the 10-round AES-128 block cipher.

2   The TWEAKEY Framework

In this section, we introduce the TWEAKEY construction framework that allows to add a tweak of (almost) any length to a key-alternating block cipher and/or to extend the key space of the block cipher to (almost) any size. In some sense, one can view the TWEAKEY framework as a simple generalization of key-alternating ciphers, offering more flexibility with regards to tweak and/or key sizes. Similarly to key-alternating ciphers, we emphasize that not all TWEAKEY instances are secure. We give in later sections natural instances of TWEAKEY that lead to secure ciphers.

2.1   Key-Alternating Ciphers

A symmetric primitive like a block cipher E is usually built upon a smaller building block f that is iterated a certain number of times – we refer to such a function f as a *round function*. Usually f is cryptographically weak, but its iterations bring security to E. The number r of iterations heavily depends on the targeted security of E, the structure of f, its differential properties, its algebraic degree, etc. In general, the function f takes two inputs: the first is the state, while the second is a round-dependent parameter called *round key* or *subkey*. The round keys are obtained by the expansion of a master secret K with an expansion (key schedule) algorithm: $K \rightarrow (K_0, \ldots, K_r)$. Formally, for a non-negative $i < r$, we write $f(s_i, K_i) = s_{i+1}$ the function that transforms the state s_i in one round into the state s_{i+1}, with the use of the round key K_i. Initially,

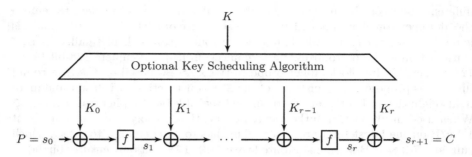

Fig. 1. Key-alternating cipher: the function f is applied r times, surrounded by subkey mixing operations

the state s_0 is set to the plaintext value P, and state s_r at the output of the r-th round is the ciphertext C .

As a subclass of iterated block ciphers, we consider further the particular case of *key-alternating block ciphers*, which specify how the round keys are used (see Figure 1). The concept has been initially introduced by Daemen in [14, 16] and has later been reused in many block cipher designs, e.g. [9, 15, 23]. Specifically, we say that E is a key-alternating cipher when the general form $f(s_i, K_i) = s_{i+1}$ for $i < r$ becomes $f(s_i \oplus K_i) = s_{i+1}$, where the current state s_i and the incoming round key K_i are XORed prior to the application of the round function f. Moreover, a final round key K_r is added after the r applications of f to produce the ciphertext. The soundness of such a construction has been theoretically studied recently in [1, 10].

2.2 Tweakable Block Ciphers

The concept of tweakable block ciphers goes back to the Hasty Pudding cipher [37], and has later been formalized by Liskov, Rivest and Wagner in [30, 31], where they suggest to move the randomization of symmetric primitives brought by the high-level operations of the modes directly at the block-cipher level. The signature of standard block ciphers can be described as $E : \{0,1\}^k \times \{0,1\}^n \to \{0,1\}^n$ where an n-bit plaintext P is transformed into an n-bit ciphertext $C = E(K, M)$ using a k-bit key K. On top on these inputs, tweakable block ciphers introduce an additional t-bit parameter T called tweak (see Figure 2). The signature for a tweakable block cipher therefore becomes $E : \{0,1\}^k \times \{0,1\}^t \times \{0,1\}^n \to \{0,1\}^n$, the ciphertext $C = E(K, T, P)$ where the tweak T does not need to be secret and thus can be placed in the public domain. Similarly to a regular block cipher where $E(K, \cdot)$ is a permutation for all $K \in \{0,1\}^k$, a tweakable block cipher preserves this behavior as $E(K, T, \cdot)$ is a permutation for all $(K, T) \in \{0,1\}^k \times \{0,1\}^t$.

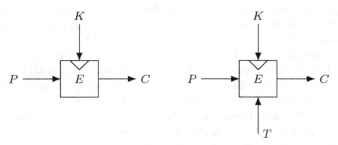

(a) Regular Block Cipher.   (b) Tweakable Block Cipher.

Fig. 2. Types of ciphers

Usually, the security notion expected from a tweakable block cipher is to be indistinguishable from a tweakable random permutation (a family of independent random permutations parameterized by T). It is important to note that the security model considers that the attacker has full control over both the message and the tweak inputs.

Adversarial Model. Besides the classical single-key attack model, a typical model for block ciphers is the related-key model, where the adversary can ask for encryption/decryption of plaintext/ciphertext with a key related to the original one. In this article, we only consider the relation between the keys and tweaks to be the classical XOR difference, and refer to [17] for more details on this so-called *key access scheme*. Similarly to the related-key model, the related-tweak model denotes a situation where the adversary can ask for encryption/decryption of plaintext/ciphertext with a tweak related to the original one, while the key remains the original one. Continuing further, we can also combine these two models and consider the related-key related-tweak adversarial model. Moreover, instead of related-key or related-tweak model, one can consider open-key and/or open-tweak models, where the adversary has full control over the key/tweak. This model is reasonable to consider as in practice an active adversary might have a full control over the tweak. For the key, this model might be interesting when the block cipher is used in a hash function setting, where message blocks are usually inserted in the key input of the inner block cipher of the compression function. Since in this article we do not always separate key and tweak input, we sometimes denote *related-tweakey* or *open-tweakey* to refer to related-key related-tweak or open-key open-tweak model, respectively.

2.3 The TWEAKEY Construction

In theory, for a tweakable block cipher the distinction between the tweak input and the key input is clear: the former is public and can be fully controlled by the attacker, while the later is secret. This might indicate that in practice the tweak input must be handled more carefully then the key input, since the attacker is given more power[1]. However, from the point of view of applications, what is intrinsically required for a tweakable block cipher is that computing consecutive cipher calls with different random tweak values should be very efficient, while not necessarily required for the key input. This tends to indicate that, in the contrary, the tweak input should not use more computations than the key input.

This contradiction regarding the proportion of computations between the tweak and key inputs should make tweakable block cipher designers handle both inputs almost equivalently (we note that this is the case for example in Threefish [20]). Moving in this direction, we introduce the TWEAKEY framework, that tries to bridge the gap between key and tweak inputs by providing a unified vision. This framework can be seen as a direct extension of the key-alternating cipher construction. As of today, building a tweakable block cipher with a key-alternating approach has never been considered, but we note that Goldenberg et al. [22] studied how to insert a tweak input inside a Luby-Rackoff cipher from a theoretical point of view.

The term *tweakey* refers to an input that can be both tweak or key material, *without distinction*. Using our framework, the obvious advantage is that one can leverage the work already done on key schedule design in order to build proper tweak schedule, or tweakey schedule more generally.

The TWEAKEY construction is a framework to build a n-bit tweakable block cipher with t-bit tweak and k-bit key. It consists of two states: the n-bit internal state s and the $(t + k)$-bit tweakey state tk, and we denote respectively as s_i and tk_i their values throughout the rounds. The state s_0 is initialized with the plaintext P (or ciphertext C for decryption), and tk_0 is initialized with the tweak and key material. Then, the cipher is composed of r successive rounds each composed of three steps:

- a subtweakey extraction function g from the tweakey state, and incorporation of this subtweakey to the internal state (for ease of description, we consider that the subtweakey incorporation is done with a simple XOR, but this can be trivially extended to other operations),
- an internal state update permutation f,
- a tweakey state update function h.

This can be summarized as: $s_{i+1} = f(s_i \oplus g(tk_i))$ followed by $tk_{i+1} = h(tk_i)$. At the end, the last subtweakey is incorporated to the last internal state and $s_r \oplus g(tk_r)$ represents the ciphertext C (or plaintext P for decryption). The subtweakeys are usually of size n bits, but they might be smaller. The framework is depicted in Figure 3.

---

[1] One may argue that key recovery attacks are not to be considered for the tweak input, which makes the tweak and the key inputs fundamentally different. However, from a designer perspective, it seems easier to protect against key-recovery attacks, than against a known-key distinguisher. For example, for most ciphers, more rounds can be attacked in the open-key model than in the related-key model.

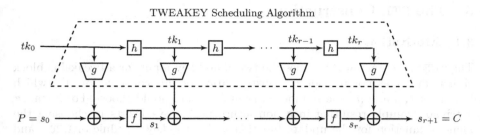

Fig. 3. The TWEAKEY framework

Increasing the amount of tweak or key material obviously renders the task of the designer much more complex in terms of security analysis. To separate these situations, we denote TK-p the class of tweakable block ciphers when one handles $p \times n$ of tweakey material. For example, a simple single-key cipher would fit in TK-1, while an n-bit key, n-bit tweak block cipher (or for a double-key cipher with no tweak input) would fit in TK-2. By extension, a public permutation would fit in TK-0. The tweakey material can be any amount of key and/or tweak. A tweak-only cipher can be an interesting primitive as well, for example when building a compression function (the members of the MD-SHA hash function family would actually fit in our framework, the subtweakey having smaller size than n).

We emphasize that TWEAKEY is only a framework and, as such, will not guarantee a secure cipher. It is up to the designer to ensure picking a proper TWEAKEY instance. The functions f, g and h must be chosen along with the number of rounds r such that no known attack can apply on the resulting primitives. More precisely, this must be true for any choice of the tweak/key size tradeoff inside the tweakey input. A natural way to achieve this while keeping the same f, g and h would be to set the number of rounds as the maximal number of required rounds over all the possible tweak/key size tradeoffs. By known attacks, we refer in particular to classical differential/linear attacks, even in related-tweakey or open-tweakey model, or meet-in-the-middle techniques. Moreover, the key schedule is often used to break inherent symmetries from the internal state update function and to break round similarities (for example in the case of AES), hence this has to be taken in account as well.

Cipher Instances Separation. Since the tweak and key material are not made distinct in our framework, one might argue that since the tweakable block cipher is always the same whatever is the amount of key or tweak inputs, there are some obvious relations between these different versions. If the designer would prefer to avoid these properties, this can be easily and securely done for example by encoding the various cipher versions on a few bits of the tweakey state (with two distinct key/tweak sizes versions, one tweak bit would then have to be booked for that matter). Nevertheless, in the rest of this article, we do not consider related-cipher attacks [39].

3   The STK Construction

3.1   Motivation

The TWEAKEY framework unifies the tweak and key input for a tweakable block cipher, but does not provide real instantiation of this construction, i.e. which functions f, g and h (and number of rounds r) one should choose. For instance, a trivial example resulting in a non-secure primitive consists in choosing the identity function for the update function h (i.e. the key schedule of LED), and defining g as the XOR of all n-bit tweakey words. In such a case, regardless of the choice of the function f, the construction would not be secure as cancellations of the tweakey words would lead to outputs of g consisting of zero bits.

One of the main causes for the low number of ad-hoc tweakable block ciphers is the fact that adding a tweak input makes the security analysis much harder. Building a block cipher secure in the related-key model is already not an easy task, and by incorporating an additional tweak or a double key, the task becomes even more difficult. In the case of AES, there exists tools [8, 21] to analyze the best differential characteristics in the related-key model, but they mainly work for TK-1. As soon as we switch to bigger keys or add tweak inputs, like TK-2 or TK-3, the searches might become infeasible, unless very good characteristics exist to speed up the search with branching cuts, which would mean that the cipher is insecure.

One research direction that we follow in this article consists in finding a construction within the TWEAKEY framework that simplifies this analysis. A potential and natural solution would be that all $p = (t + k)/n$ n-bit words of tweakey are handled the same way (i.e. the function h is symmetric with regards to the p n-bit words of the tweakey state of TK-p), and that g simply XORs all these n-bit tweakey state words to the internal state. The security analysis is simplified as any analysis independently performed on one of the n-bit words of tweakey will hold for the other words as well (and thus the tools working for TK-1 could now do the analysis even for TK-p with $p > 1$). The problem is to understand what happens when all words are considered together as their interaction might cause potential weaknesses (e.g. if we insert differences in all the tweakey words). For example, assume we would like to build an AES-like cipher with double key: this would fit in TK-2 as $k = 2n$ and $t = 0$. If the two n-bit tweakey words were treated equivalently, we could use the differential characteristic search tools to assess the security of the primitive with regards to classical differential attacks, and then use this information to pick an appropriate number of rounds. However, there is an obvious weakness if we strictly follow this strategy: starting with the two tweakey words equal would lead to zero being XORed to the internal state every round, since their value would always cancel each other in the XOR. Using constants to separate the two words would work, but only if the h function is strongly non-linear, which is something we would like to avoid for efficiency reasons. In fact, we would like to push even further the efficiency incentive and only consider nibble-wise substitutions for the h function.

In the remaining of the section, we propose a simple solution to overcome this issue for AES-like ciphers. The basic idea is to minimize possible differences cancellations between tweakey words by using small field multiplications. Following this mechanism still allows to apply the existing differential characteristic search tools, while avoiding the trivial characteristic in the tweakey scheduling algorithm.

3.2   The STK (Superposition TWEAKEY) Construction

The STK construction is a subclass of the TWEAKEY framework for AES-like ciphers defined over a finite field $GF(2^c)$. Recall that $p = (t + k)/n$ denotes the number of n-bit words in the tweakey state composed of t-bit tweak and k-bit key. Assuming that the AES-like S-Box operates on c bits (thus we have n/c nibbles in a n-bit word), the STK construction further specifies the f, g and h functions as follows (also see Figure 4):

- the function g simply XORs all the p n-bit words of the tweakey state to the internal state (AddRoundTweakey, denoted ART), and then XORs a round-dependent constant C_i,
- the function h first applies the same nibble position substitution function h' to each of the p n-bit words of the tweakey state, and then multiply each c-bit cell of the j-th n-bit word by a nonzero coefficient α_j in the finite field $GF(2^c)$ (with $\alpha_i \neq \alpha_j$ for all $1 \leq i \neq j \leq p$)
- the function f is an AES-like round.

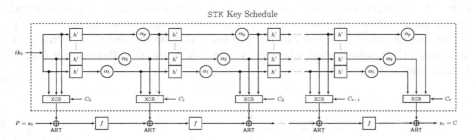

Fig. 4. The STK construction: example with TK-p

3.3   Rationale behind the STK Construction

Most automated differential analysis tools for AES-like ciphers (e.g., [8, 18, 21]) use truncated differential representation to make feasible the search for differential characteristics. In the truncated difference representation [27], the exact value of a difference in a nibble is not specified. Rather, only the presence (active nibble) or absence (inactive nibble) of a difference is kept track of. In the STK construction, the different subtweakey words will have precisely the same truncated representation of the difference if the input tweakey words have the same difference. The reason behind this is that they all apply the same functions g and h, which are completely independent of the tweakey word considered. This feature already significantly simplifies the analysis for the designer, since a simple TK-1 differential analysis (already known to be possible with the current tools) will ensure the security for all situations in which only a single tweakey word contains a difference. Having all the tweakey words treated almost equivalently is therefore very helpful for the designer.

The issue, however, is to understand what happens when differences are placed in several tweakey words at the same place (in the same nibbles). In particular, the difficulty lies in the cancellations that might happen in the nibbles at the output of g (recall that g will XOR all the subtweakey word to the state). These cancellations are the reason why having exactly the same update function for all tweakey words leads to a design that is not secure. The trick we use is to apply a nibble-wise multiplication with a distinct coefficient α_j for all tweakey words. This prevents the large number of cancellation of differences in a particular nibble position at the output of g. To explain this, first observe that as we apply the very same nibble position substitution function h' to each of the p tweakey words, the relative position of the nibble between the tweakey words is always the same (i.e. two nibbles at the same position inside their tweakey word will always keep that property). Thus, we can divide the tweakey nibbles into n/c fully independent subgroups (according to the nibble position in the n-bit tweakey words), and to each of these subgroups will correspond one and only one nibble at the output of g at every round. More precisely, in each subgroup, we have p input nibbles $\mathbf{x} = [x_1, \ldots, x_p]$ (one in each tweakey word) and $r+1$ output nibbles $\mathbf{y} = [y_0, \ldots, y_r]$ (since we have to generate $r+1$ sub-tweakeys). Our STK construction ensures that whenever a non-null difference is inserted in the input nibbles of the subgroup, there will always be at least $r+1-p$ active output nibbles. These output nibbles $\mathbf{y}$ can be expressed in terms of $\mathbf{x}$ by using a right-matrix multiplication $\mathbf{y} = \mathbf{x} \times \mathbf{V}$ with the following $p \times (r+1)$ Vandermonde matrix:

$$\mathbf{V} = \left(\alpha_i^j\right)_{i,j} = \begin{pmatrix} \alpha_1^0 & \alpha_1^1 & \cdots & \alpha_1^r \\ \alpha_2^0 & \alpha_2^1 & \cdots & \alpha_2^r \\ \vdots & \vdots & \ddots & \vdots \\ \alpha_p^0 & \alpha_p^1 & \cdots & \alpha_p^r \end{pmatrix},$$

In order to minimize the number of nonzero elements in $\mathbf{y}$ for $\mathbf{x} \neq 0$, we need to ensure that all the columns in $\mathbf{V}$ are linearly independent. This is true as long as the α_i coefficients, $1 \leq i \leq p$ are pairwise distinct. Using for example the specific distinct coefficients $\alpha_j = j \in GF(2^c)$, $1 \leq j < 2^c$, in TK-p, $1 \leq p \leq c-1$, then at most p elements of $\mathbf{y}$ can be zero for $\mathbf{x} \neq 0$, which is the property that we targeted.

To summarize, when we deal with differences in several tweakey words (which is supposedly very hard to analyze due to the important number of nibbles), the study of the STK construction is again the same as for a classical TK-1 analysis, except that at most p active output nibbles can be erased in each subgroup. This extra constraint in the search is rather easy to include in the existing analysis tools [8, 21] and this is precisely why we believe the STK construction to be interesting. It has been created with this criteria in mind, so as to ease a systematic cryptographic analysis by existing tools, rather than only relying on ad-hoc constructions, which are *de facto* more difficult to evaluate.

As a side note, the constants C_i in the STK construction prevent obvious issues regarding symmetries in the internal state for an AES-like cipher, as the RCON constants do for the original AES key scheduling algorithm. The choice of these constants are left at the discretion of the designers, but one could recommend for instance to use the AES RCON constants, based of the exponentiation of 2 in $GF(2^c)$, or the exponentiation of any other primitive element in that field.

The nibble positions permutation h' is also left at the discretion of the designers, but it must be carefully chosen so as to provide the best resistance against classical differential/linear attacks. This will permit the designers to safely choose an appropriate number of rounds r. This number will of course strongly depend on the amount of tweakey material, since more tweakey material makes it harder for the designer to create a secure tweakable block cipher. Our analysis tools indicate that using identity function instead of h' would lead to designs that require a great number of rounds. Therefore, we recommend h' to be a nibble positions permutation so as to prevent the existence of very good differential characteristics, but yet remaining a very efficient function to compute.

3.4 Performances

The performance of the STK construction is very high due to the simple transformations used in the schedules – all of them are linear and lightweight. The cost of the nibble position permutation h' is very low, however, the choice of the coefficients α_j might have a significant impact on the performances. For optimal efficiency, one should typically use $\alpha_1 = 1$ and $\alpha_2 = 2$ in the case of TK-2. For larger instances, TK-p with $p > 2$, one could use powers of 2 as coefficients α_j in order to maintain high efficiency in the computations of the coefficients multiplications. In most of the applications, the tweak is changed more frequently than the key. For instance, in a number of authenticated encryption schemes, the key is the same across different calls to the tweakable cipher, while the tweak is different in each call. Thus, it is reasonable to make the tweak schedule more efficient than the key schedule. Therefore, the tweak schedule should use the most efficiently implementable coefficients α_j ($\alpha_1 = 1$ would be the first choice). However, for some particular use-cases, it can be better to assign coefficient $\alpha_1 = 1$ to a key input. Indeed, for hardware implementations, it might be very valuable in certain scenarios to hard-wire the key in order to greatly reduce the area required (this is a feature of several lightweight ciphers). Yet, this would be possible only if the key input is not modified during the execution of the entire cipher and this is ensured only if $\alpha_1 = 1$ is assigned to this key input. The efficiency of the STK constructions can best be measured in term of key/tweak agility, i.e. how well the construction behaves when the key and/or the tweak are frequently changed. Due to the very low number of transformations, and all being completely linear, this construction has obviously one of the simplest possible schedules.

4   Conclusion

We have introduced the TWEAKEY framework, which helps designers to build a secure tweakable block cipher by bringing together key schedule design and tweak input. Inside this framework, we have identified a new type of construction, named STK, that is simple and generic and which provides efficient schemes, as shown by the two STK instances Deoxys-BC and Joltik-BC. We have also shown how to directly tweak the AES-128 block cipher, with the very simple and extremely efficient Kiasu-BC tweakable block cipher. The three candidates Kiasu [26], Joltik [25] and Deoxys [24] to the CAESAR authenticated encryption competition by the same authors are based on three instances of either the TWEAKEY or the STK constructions and are claimed secured against classical class of cryptanalytic attacks, as differential and meet-in-the-middle attacks.

We believe this work opens many questions and future works. First, it would be interesting to prove the soundness of our framework and the STK construction. Namely, can we generalize the recent proofs done on key-alternating ciphers? Secondly, we believe that several nibble positions permutation h' might be of particular interest for the STK construction. The search space is quite large, thus a smart method in order to prune bad candidates is necessary, as well as very optimized search tools. This problem is actually even more complex, since the best permutation for TK-i might not necessarily be the best for TK-j with $i \neq j$. Then, a very valuable advance would be to find a way to tweak directly the AES-128 (keeping the original key schedule) with a 128-bit tweak, since the best achievable option to date only handles up to 64-bit tweak (Kiasu-BC). Our searches led us to the conclusion that this seems quite hard to achieve. Finally, the problem of designing a simple, secure and efficient key schedule for AES-like ciphers remains an open problem. Is it possible to find an efficient key schedule that could lead to simple human-readable proofs on the minimal number of active S-Boxes in a differential characteristic in the related-tweakey model?

References

1. Shamir, A.: How to share a secret. Communications of the ACM 22(11), 612–613 (1979)
2. Aumasson, J.-P., Neves, S., Wilcox-O'Hearn, Z., Winnerlein, C.: BLAKE2: Simpler, smaller, fast as MD5. In: Jacobson, M., Locasto, M., Mohassel, P., Safavi-Naini, R. (eds.) ACNS 2013. LNCS, vol. 7954, pp. 119–135. Springer, Heidelberg (2013)
3. Barreto, P.S.L.M., Rijmen, V.: The WHIRLPOOL Hashing Function. Submitted to NESSIE (September 2000)
4. Biham, E.: New Types of Cryptanalytic Attacks Using Related Keys. In: Helleseth, T. (ed.) EUROCRYPT 1993. LNCS, vol. 765, pp. 398–409. Springer, Heidelberg (1994)

5. Biham, E., Dunkelman, O., Keller, N.: A Unified Approach to Related-Key Attacks. In: Nyberg, K. (ed.) FSE 2008. LNCS, vol. 5086, pp. 73–96. Springer, Heidelberg (2008)
6. Biryukov, A., Khovratovich, D.: Related-Key Cryptanalysis of the Full AES-192 and AES-256. In: Matsui, M. (ed.) ASIACRYPT 2009. LNCS, vol. 5912, pp. 1–18. Springer, Heidelberg (2009)
7. Biryukov, A., Khovratovich, D., Nikolić, I.: Distinguisher and Related-Key Attack on the Full AES-256. In: Halevi, S. (ed.) CRYPTO 2009. LNCS, vol. 5677, pp. 231–249. Springer, Heidelberg (2009)
8. Biryukov, A., Nikolić, I.: Automatic Search for Related-Key Differential Characteristics in Byte-Oriented Block Ciphers: Application to AES, Camellia, Khazad and Others. In: Gilbert, H. (ed.) EUROCRYPT 2010. LNCS, vol. 6110, pp. 322–344. Springer, Heidelberg (2010)
9. Bogdanov, A.A., Knudsen, L.R., Leander, G., Paar, C., Poschmann, A., Robshaw, M., Seurin, Y., Vikkelsoe, C.: PRESENT: An Ultra-Lightweight Block Cipher. In: Paillier, P., Verbauwhede, I. (eds.) CHES 2007. LNCS, vol. 4727, pp. 450–466. Springer, Heidelberg (2007)
10. Bogdanov, A., Knudsen, L.R., Leander, G., Standaert, F.-X., Steinberger, J., Tischhauser, E.: Key-Alternating Ciphers in a Provable Setting: Encryption Using a Small Number of Public Permutations. In: Pointcheval, D., Johansson, T. (eds.) EUROCRYPT 2012. LNCS, vol. 7237, pp. 45–62. Springer, Heidelberg (2012)
11. Canetti, R., Garay, J.A. (eds.): CRYPTO 2013, Part I. LNCS, vol. 8042. Springer, Heidelberg (2013)
12. De Cannière, C., Dunkelman, O., Knežević, M.: KATAN and KTANTAN — A Family of Small and Efficient Hardware-Oriented Block Ciphers. In: Clavier, C., Gaj, K. (eds.) CHES 2009. LNCS, vol. 5747, pp. 272–288. Springer, Heidelberg (2009)
13. Crowley, P.: Mercy: A Fast Large Block Cipher for Disk Sector Encryption. In: Schneier, B. (ed.) FSE 2000. LNCS, vol. 1978, pp. 49–63. Springer, Heidelberg (2001)
14. Daemen, J., Govaerts, R., Vandewalle, J.: Correlation Matrices, vol. 35, pp. 275–285
15. Daemen, J., Knudsen, L.R., Rijmen, V.: The Block Cipher SQUARE. In: Biham, E. (ed.) FSE 1997. LNCS, vol. 1267, pp. 149–165. Springer, Heidelberg (1997)
16. Daemen, J., Rijmen, V.: The Design of Rijndael: AES - The Advanced Encryption Standard. Springer (2002)
17. Daemen, J., Rijmen, V.: On the related-key attacks against AES. Proceedings of the Romanian Academy, Series A 13(4), 395–400 (2012)
18. Derbez, P., Fouque, P.-A., Jean, J.: Improved Key Recovery Attacks on Reduced-Round AES in the Single-Key Setting. In: Johansson, T., Nguyen, P.Q. (eds.) EUROCRYPT 2013. LNCS, vol. 7881, pp. 371–387. Springer, Heidelberg (2013)
19. Dworkin, M.J.: SP 800-38E. Recommendation for Block Cipher Modes of Operation: The XTS-AES Mode for Confidentiality on Storage Devices (2010)
20. Ferguson, N., Lucks, S., Schneier, B., Whiting, D., Bellare, M., Kohno, T., Callas, J., Walker, J.: The SKEIN Hask Function Family (2009)
21. Fouque, P.-A., Jean, J., Peyrin, T.: Structural evaluation of AES and chosen-key distinguisher of 9-round AES-128. In: Canetti, R., Garay, J.A. (eds.) CRYPTO 2013, Part I. LNCS, vol. 8042, pp. 183–203. Springer, Heidelberg (2013)
22. Goldenberg, D., Hohenberger, S., Liskov, M., Schwartz, E.C., Seyalioglu, H.: On Tweaking Luby-Rackoff Blockciphers. In: Kurosawa, K. (ed.) ASIACRYPT 2007. LNCS, vol. 4833, pp. 342–356. Springer, Heidelberg (2007)

23. Guo, J., Peyrin, T., Poschmann, A., Robshaw, M.: The LED Block Cipher. In: Preneel, B., Takagi, T. (eds.) CHES 2011. LNCS, vol. 6917, pp. 326–341. Springer, Heidelberg (2011)

24. Jean, J., Nikolić, I., Peyrin, T.: Deoxys v1.1, Submission to the CAESAR competition (2014), http://www1.spms.ntu.edu.sg/~syllab/Deoxys

25. Jean, J., Nikolić, I., Peyrin, T.: Joltik v1.1, Submission to the CAESAR competition (2014), http://www1.spms.ntu.edu.sg/~syllab/Joltik

26. Jean, J., Nikolić, I., Peyrin, T.: Kiasu v1.1, Submission to the CAESAR competition (2014), http://www1.spms.ntu.edu.sg/~syllab/Kiasu

27. Knudsen, L.R.: Truncated and Higher Order Differentials, vol. 35, pp. 196–211

28. Krovetz, T., Rogaway, P.: The Software Performance of Authenticated-Encryption Modes. In: Joux, A. (ed.) FSE 2011. LNCS, vol. 6733, pp. 306–327. Springer, Heidelberg (2011)

29. Landecker, W., Shrimpton, T., Terashima, R.S.: Tweakable Blockciphers with Beyond Birthday-Bound Security. In: Safavi-Naini, R., Canetti, R. (eds.) CRYPTO 2012. LNCS, vol. 7417, pp. 14–30. Springer, Heidelberg (2012)

30. Liskov, M., Rivest, R.L., Wagner, D.: Tweakable Block Ciphers. In: Yung, M. (ed.) CRYPTO 2002. LNCS, vol. 2442, pp. 31–46. Springer, Heidelberg (2002)

31. Liskov, M., Rivest, R.L., Wagner, D.: Tweakable Block Ciphers. Journal of Cryptology 24(3), 588–613 (2011)

32. Matsui, M.: On Correlation between the Order of S-Boxes and the Strength of DES. In: De Santis, A. (ed.) EUROCRYPT 1994. LNCS, vol. 950, pp. 366–375. Springer, Heidelberg (1995)

33. Minematsu, K.: Beyond-Birthday-Bound Security Based on Tweakable Block Cipher. In: Dunkelman, O. (ed.) FSE 2009. LNCS, vol. 5665, pp. 308–326. Springer, Heidelberg (2009)

34. National Institute of Standards and Technology (NIST): Advanced Encryption Standard (AES). FIPS PUB 197, U.S. Department of Commerce (November 2001)

35. Preneel, B. (ed.): FSE 1994. LNCS, vol. 1008. Springer, Heidelberg (1995)

36. Rogaway, P.: Efficient Instantiations of Tweakable Blockciphers and Refinements to Modes OCB and PMAC. In: Lee, P.J. (ed.) ASIACRYPT 2004. LNCS, vol. 3329, pp. 16–31. Springer, Heidelberg (2004)

37. Schroeppel, R.: The Hasty Pudding Cipher (1998)

38. Shrimpton, T., Terashima, R.S.: A Modular Framework for Building Variable-Input-Length Tweakable Ciphers. In: Sako, K., Sarkar, P. (eds.) ASIACRYPT 2013, Part I. LNCS, vol. 8269, pp. 405–423. Springer, Heidelberg (2013)

39. Wu, H.: Related-Cipher Attacks. In: Deng, R.H., Qing, S., Bao, F., Zhou, J. (eds.) ICICS 2002. LNCS, vol. 2513, pp. 447–455. Springer, Heidelberg (2002)

Memory-Demanding Password Scrambling

Christian Forler, Stefan Lucks, and Jakob Wenzel

Bauhaus-Universität Weimar, Germany
{Christian.Forler,Stefan.Lucks,Jakob.Wenzel}@uni-weimar.de

Abstract. Most of the common password scramblers hinder password-guessing attacks by "key stretching", e.g., by iterating a cryptographic hash function many times. With the increasing availability of cheap and massively parallel off-the-shelf hardware, iterating a hash function becomes less and less useful. To defend against attacks based on such hardware, one can exploit their limitations regarding to the amount of fast memory for each single core. The first password scrambler taking this into account was `scrypt`. In this paper we mount a cache-timing attack on `scrypt` by exploiting its password-dependent memory-access pattern. Furthermore, we show that it is possible to apply an efficient password filter for `scrypt` based on a malicious garbage collector. As a remedy, we present a novel password scrambler called CATENA which provides both a password-independent memory-access pattern and resistance against garbage-collector attacks. Furthermore, CATENA instantiated with the here introduced (G, λ)-DBH operation satisfies a certain time-memory tradeoff called λ-memory-hardness, i.e., using only $1/b$ the amount of memory, the time necessary to compute the password hash is increased by a factor of b^λ. Finally, we introduce a more efficient instantiation of CATENA based on a bit-reversal graph.

Keywords: password hashing, memory-hard, cache-timing attack.

1 Introduction

Passwords[1] are user-memorizable secrets for user authentication and cryptographic key derivation. Typical (user-chosen) passwords suffer from low entropy and can be attacked by trying out all possible candidates in order of likelihood. In some cases, the security of interactive password-based authentication and key derivation can be enhanced by dedicated cryptographic protocols defeating "offline" password guessing where an adversary is in possession of the password hashes (see [3] for an early example). Otherwise, the best protection are cryptographic password scramblers, performing *key stretching*. This means to hash the password with an intentionally slow one-way hash function to delay the adversary, without inconveniencing the user.

Traditional password scramblers, e.g., `md5crypt` [10] or `sha512crypt` [7], iterate a cryptographic primitive (a block cipher or a hash function) many times.

---

[1] We do not distinguish "passwords" from "passphrases" or "PINs".

P. Sarkar and T. Iwata (Eds.): ASIACRYPT 2014, PART II, LNCS 8874, pp. 289–305, 2014.

An adversary who has b computing units (*cores*) can easily try out b different passwords in parallel. With recent technological trends, such as the availability of graphical processing units (GPUs) with hundreds of cores [13], the question of how to slow down such adversaries becomes pressing. Fast memory is expensive. Thus, each core of a GPU (or any other cheap and massively parallel machine) possesses a very limited amount of fast memory ("cache"). Therefore, a defense against massively-parallel attacks on cheap hardware is to consume plenty of memory to cause a large amount of cache misses, up to the limit available to the user. Modern password scramblers allow to adjust the required memory by a logarithmic security parameter g, called *garlic* (memory-cost factor). A required property for such memory-consuming algorithms is *memory-hardness*, i.e., assume that a password scrambler uses $S = 2^g$ units of memory and an adversary has less than S units of memory for each core. Then, the attack must slow down greatly. The first password scrambler that took this into account was scrypt [16], which inherited these features from its underlying function called ROMix. However, a memory-consuming password scrambler might suffer from a new problem. If the memory-access pattern depends on the password, this pattern may leak during a *cache-timing attack*.

Contribution. In this paper we present two side-channel attacks against ROMix (1) a cache-timing attack exploiting its password-dependent memory-access pattern. This attack requires a spy process that runs on the defender's machine, without access to the internal memory of ROMix and (2) we show that ROMix is vulnerable to garbage-collector attacks. Both attacks should be considered severe since they might put the usage of memory-consuming password scramblers at high risk.

As a remedy, we introduce CATENA, a memory-consuming password-scrambling framework which is resistant against the mentioned side-channel attacks. Furthermore, we present two instantiations whose workflow can be represented as directed acyclic graphs with bounded indegree. One is based on an adapted variant of the bit-reversal graph and the other one is based on an adapted variant of the double-butterfly graph (which is on the other hand constructed by putting two fast Fourier transformations (FFT) back-to-back).

Outline. Section 2 introduces two notions of memory-hardness. Section 3 describes our side-channel attacks against scrypt. In Section 4, we introduce our novel password-scrambling framework CATENA. Section 5 describes two instantiations, namely CATENA-BRG and CATENA-DBG, and Section 6 discusses their security. Section 7 concludes our work. Finally, in Appendix A we state the main difference between this version and our original submission.

2   Memory-Hardness

To describe memory requirements, we adopt and slightly change the notion from [16].

Definition 1 (Memory-Hard Function [16])
Let g denote the memory-cost factor. For all $\alpha > 0$, a memory-hard function f can be computed on a Random Access Machine using $S(g)$ space and $T(g)$ operations, where $S(g) \in \Omega(T(g)^{1-\alpha})$.

Thus, for $S \cdot T = G^2$ with $G = 2^g$, using b cores, we have $(1/b \cdot S) \cdot (b \cdot T) = G^2$. A formal generalization of this notion is given in the following.

Definition 2 (λ-Memory-Hard Function)
Let g denote the memory-cost factor. For a λ-memory-hard function f, which is computed on a Random Access Machine using $S(g)$ space and $T(g)$ operations with $G = 2^g$, it holds that $T(g) = \Omega(G^{\lambda+1}/S(g)^\lambda)$.

Thus, we have $(1/b \cdot S^\lambda) \cdot (b \cdot T) = G^{\lambda+1}$.

Remark 1. Note that for a λ-memory-hard function f, the relation $S(g) \cdot T(g)$ is always in $\Omega(G^{\lambda+1})$, i.e., it holds that if S decreases, T has to increase, and vice versa.

λ-Memory-Hard vs. Sequential Memory-Hard. In [16], Percival introduced the notion of sequential memory-hardness (SMH), which is satisfied by his introduced password scrambler scrypt. An algorithm is sequential memory-hard if an adversary has no computational advantage from the use of multiple CPUs. This means that one does not gain any advantage from parallelism, i.e., running such an algorithm on b cores, one needs b times the memory required for one core. On the other hand, a λ-memory-hard function satisfies a certain time-memory tradeoff. Thus, if only $\mathcal{O}(1/b)$ times the memory is available, one needs $\mathcal{O}(b^\lambda)$ times the computational effort, independent of the number of cores.

3   Side-Channel Attacks on scrypt

Algorithm 1 describes the scrypt password scrambler and its core operation ROMix. For pre- and post-processing, scrypt invokes the one-way function PBKDF2 [9] to support inputs and outputs of arbitrary length. ROMix uses a hash function H with n output bits, where n is the size of a cache line (at current machines usually 64 bytes). To support hash functions with smaller output sizes, [16] proposes to instantiate H by a function called BlockMix, which we will not elaborate on. For our security analysis of ROMix, we model H as a random oracle.

ROMix takes two inputs: an initial state x that depends on both password and salt, and the array size G that defines the required storage. One can interpret $\log_2(G)$ as the garlic of scrypt. In the *expand phase* (Lines 20–23), ROMix initializes an array v. More detailed, the array cells $v_0, \ldots, v_{G-1}$ are set to $x, H(x), \ldots, H(\ldots(H(x)))$, respectively. In the *mix phase* (Lines 24–27), ROMix updates x depending on v_j. The sequential memory-hardness comes from the way how the index j is computed, depending on the current value of x, i.e., $j \leftarrow x \bmod G$. After G updates, the final value of x is returned and undergoes the post-processing.

Algorithm 1. The `scrypt` algorithm and its core operation `ROMix` [16]

scrypt	ROMix
Input: pwd {Password}	**Input:** x {Initial State}
$\quad s$ {Salt}	$\quad G$ {Cost Parameter}
$\quad G$ {Cost Parameter}	**Output:** x {Hash Value}
Output: x {Password Hash}	20: **for** $i = 0, \ldots, G - 1$ **do**
10: $x \leftarrow$ PBKDF2$(pwd, s, 1, 1)$	21: $\quad v_i \leftarrow x$
11: $x \leftarrow$ ROMix(x, G)	22: $\quad x \leftarrow H(x)$
12: $x \leftarrow$ PBKDF2$(pwd, x, 1, 1)$	23: **end for**
13: **return** x	24: **for** $i = 0, \ldots, G - 1$ **do**
	25: $\quad j \leftarrow x \bmod G$
	26: $\quad x \leftarrow H(x \oplus v_j)$
	27: **end for**
	28: **return** x

Algorithm 2. `ROMixMC`, performing `ROMix` with G/K storage

Input: x {Initial State},	7: **for** $i = 0, \ldots, G - 1$ **do**
$\quad G$ {1st Cost Parameter},	8: $\quad j \leftarrow x \bmod G$
$\quad K$ {2nd Cost Parameter}	9: $\quad \ell \leftarrow K \cdot \lfloor j/K \rfloor$
Output: x {Hash Value}	10: $\quad y \leftarrow v_\ell$
1: **for** $i = 0, \ldots, G - 1$ **do**	11: $\quad$ **for** $m = \ell + 1, \ldots, j$ **do**
2: $\quad$ **if** $i \bmod K = 0$ **then**	12: $\quad\quad y \leftarrow H(y)$ { Invariant: $y \leftarrow v_m$ }
3: $\quad\quad v_i \leftarrow x$	13: $\quad$ **end for**
4: $\quad$ **end if**	14: $\quad x \leftarrow H(x \oplus y)$
5: $\quad x \leftarrow H(x)$	15: **end for**
6: **end for**	16: **return** x

A minor issue of `scrypt` is its use of the password pwd as one of the inputs for post-processing. Thus, it has to stay in storage during the entire password-scrambling process. This is risky if there is any chance that the memory can be compromised while `scrypt` is running. Compromising the memory should not happen, anyway, but this issue could easily be fixed without any known bad effect on the security of `scrypt`, e.g., one could replace Line 12 of Algorithm 1 by $x \leftarrow$ PBKDF2$(x, s, 1, 1)$.

Below, we will attack `scrypt` from the hardware side. The general idea of side-channel attacks against cryptographic algorithms is not new [11], neither is the usage of a spy process for cache-timing attacks [15]. But, to the best of our knowledge, we are the first to apply this approach to `scrypt`, or to password-hashing in general.

3.1   Brief Analysis of `ROMix`

In the following we introduce a way to run `ROMix` with less than G units of storage. Suppose we only have $S < G$ units of storage for the values in v. For convenience, we assume G is a multiple of S and set $K \leftarrow G/S$. The memory-constrained algorithm `ROMixMC` (see Algorithm 2) generates the same result as

ROMix with less than G storage places and is $\Theta(K)$ times slower than ROMix. From the array v, we will only store the values $v_0, v_K, v_{2K}, \ldots, v_{(S-1)K}$ – using all the S available memory units. At Line 9, the variable ℓ is assigned to the highest multiple of K less or equal to j. By verifying the invariant at Line 12, one can easily see that ROMixMC computes the same hash value as the original ROMix, except that v_j is computed on-the-fly, beginning with v_ℓ. These computations call H $(K-1)/2$ times on average. Thus, the mix phase of ROMixMC is about $\Theta(K)$ times slower than the mix phase of ROMix, which dominates the workload for ROMixMC.

Next, we briefly discuss why ROMixMC is sequentially memory-hard (for the full proof see [16]). The intuition is as follows: the indices j are determined by the output of the random oracle H and thus, uniformly distributed random values over $\{0, \ldots, G-1\}$. With no way to anticipate the next j, the best approach is to minimize the size of the "gaps", i.e., the number of consecutively unknown v_j. This is indeed what ROMixMC does, by storing one v_i every K-th step.

3.2   Cache-Timing Attacks

The Spy Process. The idea to compute a "random" index j and then ask for the value v_j, which is so useful for sequential memory-hardness, may be exploited to mount a cache-timing attack against scrypt. Consider a spy process that runs on the same machine as scrypt. This spy process cannot read the internal memory of scrypt, but shares its cache memory with ROMix:

1. The spy process interrupts ROMix, just before entering the mix phase (Line 24 of Algorithm 1) and overwrites the (entire) cache with arbitrary values w_i to flush out all ROMix' values v_j.
2. The spy process allows ROMix to perform a few more iterations of the mix loop (Line 24–27).
3. The spy process interrupts ROMix again. Now it reads the w_i, measuring precisely how long each read operation takes. If the corresponding v_j has been used by ROMix in the second step, a "cache-miss" occurs, wich makes reading w_i slow. Else, w_i is likely to be still cached, and reading it is likely to be fast.

So, the spy process can tell an adversary the indices j for which v_j has been read during the first few iterations of the mix loop (Lines 24–27 of Algorithm 1). Given this information, we can attack scrypt.

First Cache-Timing Attack. Let x be the output of the operation PBKDF2$(pwd, s, 1, 1)$, where pwd denotes the current password candidate and s the salt. Then, we can sieve the password candidates as follows:

1. Run the expand phase of ROMix, without storing the values v_i, i.e., skip Line 21 of Algorithm 1.
2. Compute the index $j \leftarrow x \bmod G$.
3. If j is one of the indices were read by ROMix, then store pwd in a list. Otherwise, conclude that pwd is a wrong password.

This sieve can run in parallel on any number of cores, where each core tests another password candidate *pwd*. Note that each core needs only a small and constant amount of memory, i.e., the data structure to decide if j is one of the indices being read with v_j, which can be shared among all cores. Thus, we can use exactly the kind of hardware, that scrypt was designed to hinder.

Let r denote the number of iterations the loop in Lines 24–27 of ROMix performed, before the second interrupt from the spy process. Thus, we have a list of r indices j used by ROMix. The probability for a false password to survive is r/G.

We can further improve the attack. Assuming $r \ll G$, we may have space to store the r values v_j that were actually used by ROMix on each core. This allows us to simulate the first r iterations of the loop in Lines 24–27. We can discard a password candidate immediately if the simulation tries to read any v_j which is not on our short list. The probability for a false password candidate to survive is now down to $(r/G)^r$.

Second Cache-Timing Attack. It may be more realistic to assume the second interrupt to be late, perhaps after the *mix phase* of ROMix. So, the loop in Lines 24–27 of ROMix was run $r = G$ times and, on average, each v_i has been read once. Actually, some values have been read several times, and we expect about $(1/e)G \approx 0.37\,G$ array elements v_i not to have been read at all. At a first look, we can eliminate about 37 % of the false password candidates – a small gain for such hard work.

In the following we introduce a way to push the attack further, inspired by Algorithm 2, the memory-constrained ROMixMC. Our second cache-timing attack on scrypt only needs the smallest possible amount of memory: $S = 1, K = G/S = G$, and thus, we only have to store v_0. Like ROMixMC, we will compute the values v_j on-the-fly when needed. Unlike ROMixMC, we will stop execution whenever one of our values j is such that v_j has not been read by ROMix (according to the information from our spy process). Thus, if only the first j has not been read, we immediately stop the execution without any on-the-fly computation; if the first j has been read, but not the second, we need one on-the-fly computation of v_j, and so forth. Since a fraction (i.e., $1/e$) of all values v_i was not read, we will need about $1/(1 - 1/e) \approx 1.58$ on-the-fly computations of some v_j, each at the average price of $(G-1)/2$ calls of H. Additionally, each iteration needs one call to H for computing $x \leftarrow H(x \oplus v_j)$. Including the work for the expand phase, with G calls to H, the expected number of calls to reject a wrong password is about

$$G + 1.58 \cdot \left(1 + \frac{G-1}{2}\right) \approx 1.79\,G.$$

As it turns out, rejecting a wrong password with constant memory is faster than computing ordinary ROMix with all the required storage, which actually makes $2G$ calls to H, without computing any v_i on-the-fly. We stress that the ability to abort the computation, thanks to the information gathered by the spy process, is crucial.

3.3 The Garbage-Collector Attack

Memory-demanding password scramblers such as ROMix defend against a massively-parallel password-cracking approach, since the required memory is proportional to the number of passwords scrambled in parallel.

But, memory-demanding password scrambling may also open the gates for new attack opportunities for the adversary. If we allocate a huge block of memory for password scrambling, holding $v_0, v_1, \ldots, v_{G-1}$, this memory becomes "garbage" after the password scrambler has terminated, and will be collected for reuse, eventually. One usually assumes that the adversary learns the hash of the secret. The *garbage-collector attack* assumes that the adversary additionally learns the memory content, i.e., the values v_i, after termination of the password scrambler.

For ROMix, the value $v_0 = H(x)$ is a plain hash of the original secret x. Hence, a malicious garbage collector can completely bypass ROMix and search directly for x with $H(x) = v_0$, implying that each password candidate can be checked in time and memory complexity of $\mathcal{O}(1)$. Furthermore, if the adversary fails to learn v_0, but any of the other values $v_i = H(v_{i-1})$, the computational effort grows to $\mathcal{O}(i)$, but the memory complexity is still $\mathcal{O}(1)$.

Thus, ROMix does not provide much defense against garbage-collector attacks. A possible countermeasure would be to overwrite $v_0, \ldots, v_{G-1}$ after running ROMix. But, this step might be removed by a compiler due to optimization, since it is algorithmically ineffective.

3.4 Discussion

Currently, the attacks above are of theoretical nature. The garbage-collector attack requires the adversary to be able to read the memory occupied by ROMix, after its usage. Whereas the cache-timing attack requires to (1) run a spy process on the machine ROMix is running, (2) interrupt ROMix twice at the right points of time, and (3) precisely measure the timings of memory reads. Moreover, other processes running on the same machine can add a huge amount of noise to the cache timings. It is not clear if a "real" server can ever be attacked that way.

However, in an idealized "laboratory" setting, the applicability of cache-timing attacks against ROMix has been demonstrated [2]. The idealized conditions included execution rights on the system.

Remark 2. Even without knowing the password hash at all,

1. the adversary can find out when the password has been changed,
2. and the adversary can mount a password-guessing attack,

just from knowing the memory-access pattern.

Note that severe security issues can be caused by the second point. Consider any offline attack on the password. When passwords are hashed using an old-style password hash function, e.g., md5crypt [10], the adversary needs to first read the file containing the password hash. Without the password hash, mounting an offline attack is not possible. Even plaintext passwords are safe from offline adversaries which are not capable of reading the file containing the plaintext passwords. But, using the seemingly strong password hash function scrypt may enable offline password cracking, even when the adversary fails to ever learn the password hash.

4   Catena – A Memory-Hard Password Scrambler

In this section we introduce our password scrambler CATENA. First, we specify CATENA and explain its properties regarding to password hashing, i.e., client-independent update and server relief. Thereupon, we present two instantiations of CATENA, called CATENA-BRG and CATENA-DBG. Both instances are designed to provide a high resilience against cache-timing attacks, and the latter naturally defends against garbage-collector attacks, whereas the former provides this kind of resistance only for $\lambda \geq 2$.

4.1   Specification

A formal definition is shown in Algorithm 3, where the function F_λ (see Line 3) is a placeholder for a certain instantiation. The password-dependent input of H is appended to a prefix c, which denotes the iteration counter (garlic factor). Note that a secure password scrambler must satisfy preimage security. CATENA inherits the preimage security from the underlying hash function H. Next, we discuss the tweak and two further novel features of CATENA.

Tweak. The parameter t is an additional multi-byte value which is given by:

$$t \leftarrow \lambda \,||\, |s| \,||\, H(AD),$$

The first byte λ defines together with the value g (see above) the security parameters for CATENA. The 32-bit value $|s|$ denotes the total length of the salt in bits. Finally, the n-bit value $H(AD)$ is the hash of the associated data AD, which can contain additional information like hostname, user-ID, name of the company, or the IP of the host, with the goal to customize the password hashes. Note that the order of the values does not matter as long as tey are fixed for a certain application.

The tweak is processed together with the secret password and the salt (see Line 1 of Algorithm 3). Thus, t can be seen as a weaker version of a salt increasing the additional computational effort for an adversary when using different values. Furthermore, it allows to differentiate between different applications of CATENA, and can depend on all possible input data. Note that one can easily provide unique tweak values (per user) when including the user-ID in the associated data.

Algorithm 3. Catena

Input: λ {Depth}, pwd {Password}, t {Tweak} s {Salt}, g {Garlic}, F_λ {Instance}
Output: x {Hash of the Password}
  1: $x \leftarrow H(t \parallel pwd \parallel s)$
  2: **for** $c = 1, \ldots, g$ **do**
  3:     $x \leftarrow F_\lambda(c, x)$
  4:     $x \leftarrow H(c \parallel x)$
  5: **end for**
  6: **return** x

4.2 Properties

Client-Independent Update. Its sequential structure enables CATENA to provide client-independent updates. Let $h \leftarrow \text{CATENA}_\lambda(pwd, t, s, g, F_\lambda)$ be the hash of a specific password pwd, where t, s, g, and F_λ denote tweak, the salt, the garlic, and the instantiation, respectively. After increasing the security parameter from g to $g' = g + 1$, we can update the hash value h without user interaction by computing:

$$h' = H(g' \parallel F_\lambda(g', h)).$$

It is easy to see that the equation $h' = \text{CATENA}_\lambda(pwd, t, s, g', F_\lambda)$ holds.

Server Relief. In the last iteration of the **for**-loop in Algorithm 3, the client has to omit the last invocation of the hash function H (see Line 4). The current output of CATENA is then transmitted to the server. Next, the server computes the password hash by applying the hash function H. Thus, the vast majority of the effort (memory usage and computational time) for computing the password hash is handed over to the client, freeing the server. This enables someone to deploy CATENA even under restricted environments or when using constrained devices – or when a single server has to handle a huge amount of authentication requests.

5 Instantiations

In this section we introduce two concrete instantiations of CATENA: CATENA-BRG and CATENA-DBG.

5.1 Catena-BRG

For CATENA-BRG, F_λ is implemented by the (G, λ)-Bit-Reversal Hashing $((G, \lambda)$-BRH) algorithm, which is based on the bit-reversal permutation.

Definition 3 (Bit-Reversal Permutation τ). *Fix a number $k \in \mathbb{G}$ and represent $i \in \mathbb{Z}_{2^k}$ as a binary k-bit number, $(i_0, i_1, \ldots, i_{k-1})$. The bit-reversal permutation $\tau : \mathbb{Z}_{2^k} \to \mathbb{Z}_{2^k}$ is defined by*

$$\tau(i_0, i_1, \ldots, i_{k-1}) = (i_{k-1}, \ldots, i_1, i_0).$$

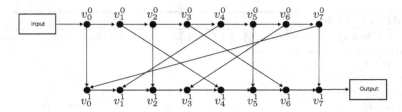

Fig. 1. An $(8,1)$-BRG

Algorithm 4. (G, λ)-Bit-Reversal Hashing $((G, \lambda)$-BRH$)$

Input: g {Garlic}, x {Value to Hash}, λ {Depth}, H {Hash Function}
Output: x {Password Hash}
 1: $v_0 \leftarrow H(x)$
 2: **for** $i = 1, \ldots, 2^g - 1$ **do**
 3:     $v_i \leftarrow H(v_{i-1})$
 4: **end for**
 5: **for** $k = 1, \ldots, \lambda$ **do**
 6:     $r_0 \leftarrow H(v_0 \parallel v_{2^g-1})$
 7:     **for** $i = 1, \ldots, 2^g - 1$ **do**
 8:         $r_i \leftarrow H(r_{i-1} \parallel v_{\tau(i)})$
 9:     **end for**
10:     $v \leftarrow r$
11: **end for**
12: **return** r_{2^g-1}

The bit-reversal permutation τ defines the (G, λ)-Bit-Reversal Graph.

Definition 4 ((G, λ)-Bit-Reversal Graph). *Fix a natural number g, let $\mathcal{V}$ denote the set of vertices, and $\mathcal{E}$ the set of edges within this graph. Then, a (G, λ)-bit-reversal graph $\Pi_g^\lambda(\mathcal{V}, \mathcal{E})$ consists of $(\lambda + 1) \cdot 2^g$ vertices*

$$\{v_0^0, \ldots, v_{2^g-1}^0\} \cup \{v_0^1, \ldots, v_{2^g-1}^1\} \cup \cdots \cup \{v_0^{\lambda-1}, \ldots, v_{2^g-1}^{\lambda-1}\} \cup \{v_0^\lambda, \ldots, v_{2^g-1}^\lambda\},$$

and $(2\lambda + 1) \cdot 2^g - 1$ edges as follows:

- *$(\lambda+1) \cdot (2^g - 1)$ edges $v_{i-1}^j \rightarrow v_i^j$ for $i \in \{1, \ldots, 2^g - 1\}$ and $j \in \{0, 1, \ldots, \lambda\}$.*
- *$\lambda \cdot 2^g$ edges $v_i^j \rightarrow v_{\tau(i)}^{j+1}$ for $i \in \{0, \ldots, 2^g - 1\}$ and $j \in \{0, 1, \ldots, \lambda - 1\}$.*
- *λ additional edges $v_{2^g-1}^j \rightarrow v_0^{j+1}$ where $j \in \{0, \ldots, \lambda - 1\}$.*

For example, Figure 1 illustrates an (8,1)-BRG. Note that this graph is almost identical – except for one additional edge $e = (v_7^0, v_0^1)$ – to the bit-reversal graph presented by Lengauer and Tarjan in [12].

Bit-Reversal Hashing. The (G, λ)-Bit-Reversal Hashing function is defined in Algorithm 4. It requires $\mathcal{O}(2^g)$ invocations of a given hash function H for a

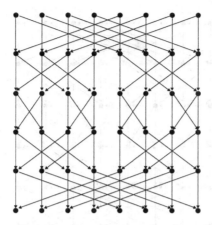

Fig. 2. A Cooley-Tukey FFT graph with eight input and output vertices

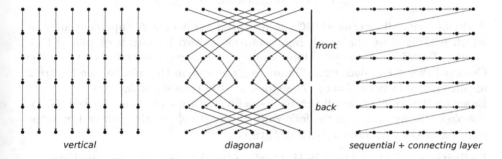

Fig. 3. Types of edges as we use them in our definitions

fixed value of x. The three inputs (g, x, λ) of (G, λ)-BRH represent the garlic $g = \log_2(G)$, the value to process, and the depth, respectively. Thus, g specifies the required units of memory. Moreover, incrementing g by one doubles the time and memory effort for computing the password hash.

5.2 Catena-DBG

Note that a (G, λ)-Double-Butterfly Graph is based on a stack of λ G-superconcentrators. The following definition of a G-superconcentrator is a slightly adapted version of that introduced in [12].

Definition 5 (G-Superconcentrator). *A directed acyclic graph $\Pi(\mathcal{V}, \mathcal{E})$ with a set of vertices $\mathcal{V}$ and a set of edges $\mathcal{E}$, a bounded indegree, G inputs, and G outputs is called a G-superconcentrator if for every k such that $1 \leq k \leq G$ and for every pair of subsets $V_1 \subset \mathcal{V}$ of k inputs and $V_2 \subset \mathcal{V}$ of k outputs, there are k vertex-disjoint paths connecting the vertices in V_1 to the vertices in V_2.*

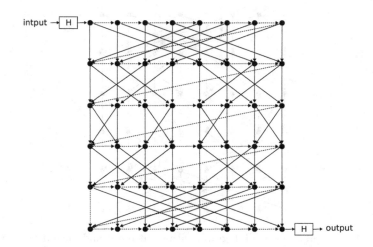

Fig. 4. An $(8, 1)$-double-butterfly graph

A double-butterfly graph (DBG) is a special form of a G-superconcentrator which is defined by the graph representation of two back-to-back placed Fast Fourier Transformations [5]. More detailed, it is a representation of twice the Cooley-Tukey FFT algorithm [6] omitting one row in the middle (see Figure 2 for an example where $G = 8$). Therefore, a DBG consists of $2g$ rows.

Based on the double-butterfly graph, we define the sequential and stacked (G, λ)-double-butterfly graph. In the following, we denote $v_{i,j}^k$ as the j-th vertex in the i-th row of the k-th double-butterfly graph.

Definition 6 ((G, λ)-**Double-Butterfly Graph**). *Fix a natural number $g \geq 1$ and let $G = 2^g$. Then, the (G, λ)-Double-Butterfly Graph $\Pi(\mathcal{V}, \mathcal{E})$ consists of $2^g \cdot (\lambda \cdot (2g - 1) + 1)$ vertices*

- $\{v_{0,0}^k, \ldots, v_{0,2^g-1}^k\} \cup \ldots \cup \{v_{2g-2,0}^k, \ldots, v_{2g-2,2^g-1}^k\}$ *for $1 \leq k \leq \lambda$ and*
- $\{v_{2g-1,0}^\lambda, \ldots, v_{2g-1,2^g-1}^\lambda\}$,

and $\lambda \cdot (2g - 1) \cdot (3 \cdot 2^g) + 2^g - 1$ edges

- *vertical:* $2^g \cdot (\lambda \cdot (2g - 1))$ *edges*

$$(v_{i,j}^k, v_{i+1,j}^k) \text{ for } 0 \leq i \leq 2g - 2, 0 \leq j \leq 2^g - 1, \text{ and } 1 \leq k \leq \lambda,$$

- *diagonal:* $2^g \cdot \lambda \cdot g + 2^g \cdot \lambda \cdot (g - 1)$ *edges*

$$(v_{i,j}^k, v_{i+1,j\oplus 2^{g-1-i}}^k) \text{ for } 0 \leq i \leq g - 1, 0 \leq j \leq 2^g - 1, \text{ and } 1 \leq k \leq \lambda.$$
$$(v_{i,j}^k, v_{i+1,j\oplus 2^{i-(g-1)}}^k) \text{ for } g \leq i \leq 2g - 2, 0 \leq j \leq 2^g - 1, \text{ and } 1 \leq k \leq \lambda.$$

- *sequential:* $(2^g - 1) \cdot (\lambda \cdot (2g - 1) + 1)$ *edges*

$$(v_{i,j}^k, v_{i,j+1}^k) \quad \text{for} \quad 1 \leq i \leq 2g - 1, 0 \leq j \leq 2g - 2, 1 \leq k \leq \lambda, \text{ and}$$
$$(v_{2g-1,j}^\lambda, v_{2g-1,j+1}^\lambda) \quad \text{for} \quad 0 \leq j \leq 2g - 2$$

Algorithm 5. (G, λ)-Double-Butterfly Hashing

Input: g {Garlic}, x {Value to hash}, λ {Depth}, H {Hash Function}
Output: x {Password Hash}
1:   $v_0 \leftarrow H(x)$
2:   **for** $i = 1, \ldots, 2^g - 1$ **do**
3:     $v_i \leftarrow H(v_{i-1})$
4:   **end for**
5:   **for** $k = 1, \ldots, \lambda$ **do**
6:     **for** $i = 1, \ldots, 2g - 1$ **do**
7:       $r_0 \leftarrow H(v_{2^g-1} \oplus v_0 \;\|\; v_{\sigma(g,i-1,0)})$
8:       **for** $j = 1, \ldots, 2^g - 1$ **do**
9:         $r_i \leftarrow H(r_{i-1} \oplus v_i \;\|\; v_{\sigma(g,i-1,j)})$
10:       **end for**
11:       $v \leftarrow r$
12:     **end for**
13: **end for**
14: **return** v_{2^g-1}

– *connecting layer:* $\lambda \cdot (2g - 1)$ *edges*

$$(v_{i,2^g-1}^k, v_{i+1,0}^k) \quad for \quad 1 \le k \le \lambda, \quad 0 \le i \le 2g - 2.$$

Figure 3 illustrates the individual types of edges we used in our Definition above. Moreover, an example for $G = 8$ and $\lambda = 1$ can be seen in Figure 4.

Double-Butterfly Hashing. The (G, λ)-double-butterfly hashing operation is defined in Algorithm 5. The structure is based on a (G, λ)-double-butterfly graph. Note that the function σ (see Lines 7 and 9) is given by

$$\sigma(g, i, j) = \begin{cases} j \oplus 2^{g-1-i} & \text{if } 0 \le i \le g - 1, \\ j \oplus 2^{i-(g-1)} & \text{otherwise.} \end{cases}$$

Thus, σ determines the indices of the vertices of the diagonal edges.

   Since the security of CATENA in terms of password hashing is based on a time-memory tradeoff, it is desired to implement it in an efficient way, making it possible to increase the required memory. We recommend to use BLAKE2b [1] as the underlying hash function, implying a block size of 1024 bits with 512 bits of output. Thus, it can process two input blocks within one compression function call. This is suitable for CATENA-BRG since a bit-reversal graph satisfies a fixed indegree of at most 2. When considering CATENA-DBG, we cannot simply concatenate the inputs to H while keeping the same performance per hash function call, i.e., three inputs to H require two compression function calls, which is a strong slow-down in comparison to (G, λ)-BRG. Therefore, we compute $H(X, Y, Z) = H(X \oplus Y \;\|\; Z)$ instead of $H(X, Y, Z) = H(X \;\|\; Y \;\|\; Z)$ obtaining the same performance as CATENA-BRG per hash function call. Obviously, this doubles the probability of input collision. Nevertheless, for a 512-bit hash function the advantage for an adversary is still negligible.

Based on the approach above, the number of hash function calls to compute Row r_i from Row r_{i-1} is the same for CATENA-BRG and CATENA-DBG. Moreover, for both instantiations it holds that the number of hash function calls is equal to the number of compression function calls. More detailed, the (G, λ)-BRG requires $2^g - 1 + \lambda \cdot 2^g$ calls to H and the (G, λ)-DBG requires $2^g - 1 + \lambda \cdot (2g - 1) \cdot 2^g$ calls to H. It is easy to see, that the performance of CATENA-DBG in comparison to CATENA-BRG is decreased by a logarithmic factor.

6   Security

In this section we discuss the security of CATENA-BRG and CATENA-DBG against side-channel attacks. Furthermore, we discuss the memory-hardness of both instantiations.

6.1   Resistance against Side-Channel Attacks

Straightforward implementations of either CATENA-BRG or CATENA-DBG provide neither a password-dependent memory-access pattern nor password-dependent branches. Therefore, both instantiations are resistant against cache-timing attacks.

Considering a malicious garbage collector, each of Algorithms 4 and 5 exposes the arrays v and r. Both arrays are overwitten multiple times. Therefore, CATENA-DBG is resistant against garbage-collector attacks. *Thus, any variant of* CATENA *with some fixed* $\lambda \geq 2$ *is at least as resistant to garbage-collector attacks as the same variant with* $\lambda - 1$ *in the absence of a malicious garbage collector.*

6.2   Memory-Hardness

In 1970, Hewitt and Paterson introduced a method for analyzing time-memory tradeoffs (TMTOs) on directed acyclic graphs [14], called *pebble game*. While their method has been known for decades, it was recently used in a cryptographic context, e.g., [8]. In general, a pebble game is a common model to derive and analyze TMTOs as shown in [17,18,19,20,21].

The pebble-game model is restricted to DAGs with bounded in-degree and can be seen as a single-player game. Let $\Pi(\mathcal{V}, \mathcal{E})$ be a DAG and let $G = |\mathcal{V}|$ be the number of vertices within $\Pi(\mathcal{V}, \mathcal{E})$. In the setup phase of the game, the player gets S pebbles (tokens) with $S \leq G$. A pebble can be placed (*pebble*) or be removed (*unpebble*) from a vertex $v \in \mathcal{V}$ under certain requirements:

1. A pebble may be removed from a vertex v at any time.
2. A pebble can be placed on a vertex v if all predecessors of the vertex v are marked.
3. If all immediate predecessors of an unpebbled vertex v are marked, a pebble may be moved from a predecessor of v to v.

A *move* is the application of either the second or the third action stated above. The goal of the game is to pebble Π, i.e., to mark all vertices of the graph Π at least once. The total amount of moves represent the computational costs.

Catena-BRG. In [12], Lengauer and Tarjan have already proven the lower bound of pebble movements for a $(G,1)$-bit-reversal graph.

Theorem 1 (Lower Bound for a $(G,1)$-BRG [12]). *If $S \geq 2$, then, pebbling the bit-reversal graph $\Pi_g(\mathcal{V},\mathcal{E})$ consisting of $G = 2^g$ input nodes with S pebbles takes time*

$$T > \frac{G^2}{16S}.$$

Biryukov and Khovratovich have shown in [4] that stacking more than one bit-reversal graph only adds some linear factor to the quadratic time-memory tradeoff. Hence, a (G,λ)-BRG with $\lambda > 1$ does not achieve the properties of a λ-memory-hard function.

Catena-DBG. Likewise, the authors of [12] analyzed the time-memory tradeoff for a stack of λ G-superconcentrators. Since the double-butterfly is a special form of a G-superconcentrators their bound also holds for (G,λ)-DBG.

Theorem 2 (Lower Bound for a (G,λ)-Superconcentrator [12]). *Pebbling a (G,λ)-superconcentrator using $S \leq G/20$ black and white pebbles requires T placements such that*

$$T \geq G\left(\frac{\lambda G}{64S}\right)^{\lambda}.$$

Discussion. For scenarios where a quadratic time-memory tradeoff is sufficient, we recommend the efficient CATENA-BRG with either $\lambda = 1$ or – if garbage-collector attacks pose a relevant threat – with $\lambda = 2$. Note that the benefit of greater values for λ is very limited since the costs for pebbling the bit-reversal graph remain quadratic. For scenarios that require a higher time-memory trade-off, we highly recommend the λ-memory-hard CATENA-DBG with $\lambda = 2$ or $\lambda = 3$, which is sufficient for most practical applications.

We have to point out that the computational effort for (G,λ)-DBH with reasonable values for G, e.g., $G \in [2^{17}, 2^{21}]$, may stress the patience of many users since the number of vertices and edges grows logarithmic with G. Thus, it remains an open research problem to find a (G,λ)-superconcentrator – or any other λ-memory-hard function – that can be computed more efficiently than a (G,λ)-DBH.

7   Conclusion

In this paper we introduced a new class of side-channel attacks, called garbage-collector attack, which bases on a malicious garbage collector. We showed that the common password scrambler `scrypt` is vulnerable to this kind of attacks. Furthermore, we presented a (theoretical) cache-timing attack on `scrypt` that exploits its password-dependent memory-access pattern. Both attacks enable an adversary to construct a *memoryless* password filter that enables massively-parallel password-guessing attacks. Moreover, we show that our attacks work even without knowledge of the password hash. All regular implementations, i.e., implementations that are not hardened against side-channel attacks, of password scramblers with a password-dependent memory access pattern appear to be vulnerable to these attacks.

As a remedy, we introduced a novel password-scrambling framework CATENA. We presented two instantiations with a password-independent memory-access pattern: CATENA-BRG and CATENA-DBG. The former is more efficient and 1-memory-hard, whereas the latter is less efficient but offers a higher level of security, i.e., λ-memory-hardness. Finally, we want to emphasize that CATENA-BRG and CATENA-DBG inherit their security from well-analyzed structures, namely bit-reversal and double-butterfly graphs.

Acknowledgement. Thanks to B. Cox, E. List, C. Percival, A. Peslyak, S. Thomas, S. Touset, and the reviewers of the ASIACRYPT'14 for their helpful hints, comments, and fruitful discussions. Finally, we thank A. Biryukov and D. Khovratovich for pointing out a flaw in our proof of the time-memory tradeoff for the (G, λ)-BRH operation by providing a tradeoff cryptanalysis [4].

References

1. Aumasson, J.-P., Neves, S., Wilcox-O'Hearn, Z., Winnerlein, C.: BLAKE2: Simpler, Smaller, Fast as MD5. In: Jacobson, M., Locasto, M., Mohassel, P., Safavi-Naini, R. (eds.) ACNS 2013. LNCS, vol. 7954, pp. 119–135. Springer, Heidelberg (2013)
2. Barsuhn, A.: Cache-Timing Attack on scrypt. Bauhaus-Universität Weimar, Bachelor Dissertation (December 2013)
3. Bellovin, S.M., Merrit, M.: Encrypted Key Exchange: Password-Based Protocols Secure Against Dictionary Attacks. In: Proceedings of the IEEE Symposium on Research in Security and Privacy, Oakland (1992)
4. Biryukov, A., Khovratovich, D.: Tradeoff cryptanalysis of Catena. PHC mailing list: discussions@password-hashing.net
5. Bradley, W.F.: Superconcentration on a Pair of Butterflies. CoRR abs/1401.7263 (2014)
6. Cooley, J.W., Tukey, J.W.: An algorithm for the machine calculation of complex Fourier series. Math. Comput. 19, 297–301 (1965)
7. Drepper, U.: Unix crypt using SHA-256 and SHA-512, http://www.akkadia.org/drepper/SHA-crypt.txt (accessed May 16, 2013)

8. Dziembowski, S., Kazana, T., Wichs, D.: Key-evolution schemes resilient to space-bounded leakage. In: Rogaway, P. (ed.) CRYPTO 2011. LNCS, vol. 6841, pp. 335–353. Springer, Heidelberg (2011)
9. Kaliski, B.: RFC 2898 - PKCS #5: Password-Based Cryptography Specification Version 2.0. Technical report, IETF (2000)
10. Kamp, P.-H.: The history of md5crypt, http://phk.freebsd.dk/sagas/md5crypt.html (accessed May 16, 2013)
11. Kocher, P.C.: Timing Attacks on Implementations of Diffie-Hellman, RSA, DSS, and Other Systems. In: Koblitz, N. (ed.) CRYPTO 1996. LNCS, vol. 1109, pp. 104–113. Springer, Heidelberg (1996)
12. Lengauer, T., Tarjan, R.E.: Asymptotically tight bounds on time-space trade-offs in a pebble game. J. ACM 29(4), 1087–1130 (1982)
13. Nvidia. Nvidia GeForce GTX 680 - Technology Overview (2012)
14. Paterson, M.S., Hewitt, C.E.: Comparative schematology. In: Dennis, J.B. (ed.) Record of the Project MAC conference on concurrent systems and parallel computation, Chapter Computation schemata, pp. 119–127. ACM, New York (1970)
15. Percival, C.: Cache Missing for Fun and Profit. BDSCan (2004)
16. Percival, C.: Stronger Key Derivation via Sequential Memory-Hard Functions. Presented at BSDCan (May 2009)
17. Savage, J., Swamy, S.: Space-time trade-offs on the FFT algorithm. IEEE Transactions on Information Theory 24(5), 563–568 (1978)
18. Savage, J.E., Swamy, S.: Space-Time Tradeoffs for Oblivious Interger Multiplications. In: Maurer, H.A. (ed.) ICALP 1979. LNCS, vol. 71, pp. 498–504. Springer, Heidelberg (1979)
19. Sethi, R.: Complete Register Allocation Problems. SIAM J. Comput. 4(3), 226–248 (1975)
20. Swamy, S., Savage, J.E.: Space-Time Tradeoffs for Linear Recursion. In: POPL, pp. 135–142 (1979)
21. Tompa, M.: Time-Space Tradeoffs for Computing Functions, Using Connectivity Properties of their Circuits. In: STOC, pp. 196–204 (1978)

A   Changelog

Based on the cryptanalysis provided by Biryukov and Khovratovich in [4], we decided to provide a slightly changed version in comparison to our submitted version. The major changes are (1) removing the flawed proof for λ-memory-hardness of a (G, λ)-BRG and (2) providing a new instance called CATENA-DBG based on a (G, λ)-double-butterfly graph (variant of a stack of λ Double-Butterfly Graphs (DBG)).

Side-Channel Analysis of Multiplications in GF(2^{128})

Application to AES-GCM

Sonia Belaïd[1], Pierre-Alain Fouque[2], and Benoît Gérard[3]

[1] École normale supérieure, Paris, France
and Thales Communications & Security, Gennevilliers, France
[2] Université de Rennes 1, Rennes, France
and Institut Universitaire de France, France
[3] DGA–MI and IRISA, Rennes, France
Sonia.Belaid@ens.fr,
{fouque,benoit.gerard}@irisa.fr

Abstract. In this paper, we study the side-channel security of the field multiplication in GF(2^n). We particularly focus on GF(2^{128}) multiplication which is the one used in the authentication part of AES-GCM but the proposed attack also applies to other binary extensions. In a hardware implementation using a 128-bit multiplier, the full 128-bit secret is manipulated at once. In this context, classical DPA attacks based on the divide and conquer strategy cannot be applied. In this work, the algebraic structure of the multiplication is leveraged to recover bits of information about the secret multiplicand without having to perform any key-guess. To do so, the leakage corresponding to the writing of the multiplication output into a register is considered. It is assumed to follow a Hamming weight/distance leakage model. Under these particular, yet easily met, assumption we exhibit a nice connection between the key recovery problem and some classical coding and Learning Parities with Noise problems with certain instance parameters. In our case, the noise is very high, but the length of the secret is rather short. In this work we investigate different solving techniques corresponding to different attacker models and eventually refine the attack when considering particular implementations of the multiplication.

Keywords: Field Multiplication, Authenticated Encryption, AES-GCM, Side-Channel.

1 Introduction

The multiplication in GF(2^{128}) is used in several cryptographic algorithms to diffuse a secret parameter. Two widely deployed examples are the authentication encryption mode AES-GCM and the mode of operation OCB. While it is important to guarantee the security of such algorithms against *black-box* attacks, e.g. using the knowledge of the inputs and outputs, it becomes mandatory to thwart *side-channel attacks* for an industrial use.

P. Sarkar and T. Iwata (Eds.): ASIACRYPT 2014, PART II, LNCS 8874, pp. 306–325, 2014.
© International Association for Cryptologic Research 2014

The main motivation of this work is to show that such multiplication, although manipulating huge part of the secret at once, can be attacked by a side-channel adversary. Hence, in a major part of this work we consider that the multiplication is an atomic operation (that is performed using a 128-bit multiplier) what is the worst case for an attacker. In an additional part we show that, as one may expect, considering designs having intermediate results indeed provides more leakages and thus lead to more powerful attacks.

As already mentioned, we focus on the application to AES-GCM. Proposed by McGrew and Viega in [25] and standardized by NIST since 2007, this authenticated encryption algorithm aims to provide both confidentiality and integrity. It combines an encryption based on the widely used AES algorithm in counter mode and an authentication based on the GHASH function involving multiplications in GF(2^{128}). This latter one mixes ciphertexts, potential additional data and a secret parameter derived from the encryption key to produce a tag. The security of the algorithm has been analyzed by many authors but despite significant progress in these attacks [28,16,30] there is currently no real attack on this mode. The most efficient attacks are the ones described by Ferguson when the tag is very short (32 bits) [14] and by Joux when the nonces are reused [18].

In the particular case of AES-GCM, attacking the multiplier will provide to the attacker the knowledge of the authentication key H. Due to the malleability of the counter mode, the knowledge of the authentication key induces a huge security breach and thus protecting the multiplier is of real importance in contexts where a side-channel attacker is considered. Notice that if the multiplication is protected, then the simple additional countermeasure that consists in masking the tag register is enough to thwart the proposed attack.

Related Work. Some of the algorithms that we consider here come from the coding theory and we think it is a nice view to cast many side-channel attacks. Indeed, a secret value H for instance is encoded as the different leakage values obtained by the adversary. Usually, these leakages allow to recover H, but here for 128 bits, the Hamming weight does not give enough information. Moreover, we only get noisy versions of the leakage values and these values form the codeword with errors. The errors are independent from each other and the noise level is rather high as in many stream cipher cryptanalysis. Given these values, the goal of the adversary is to recover the original message H and the adversary faces a classical decoding problem.

As for the AES-GCM, Jaffe describes in [17] a very efficient Differential Power Analysis attack on its encryption counter mode. Basically, the main idea is to use a DPA attack on the two first rounds of the AES block cipher. Then, as most of the plaintext is the same between two evaluations, it is possible to recover the secret key by guessing parts of the first and second round subkeys. This attack is particularly efficient since it also allows to recover the counter if it is hidden. However, the implementations of AES are now well protected using masking and many papers proposed such countermeasures [8,29,9,15], so that we can assume that it is not possible to recover the secret key on the encryption part.

Our Contributions. In this paper, we consider a particular leakage model where only the storage of values leaks information to the adversary. We assume that each time a value is written in the large register, a noisy version of the Hamming distance or the Hamming weight of this value can be known by the adversary. For instance, in the context of the AES-GCM authentication, the first time the register is written, we can learn the Hamming weight of the multiplication result between the authentication key H and some known value M. Our key point is that the least significant bit of the Hamming weight can be expressed as a linear function of the bits of H. If we are able to find 128 such equations, then it is easy to recover H. However, in side-channel attacks, we only access to noisy versions of the Hamming weight and then, the problem becomes more difficult. Classically, this problem has been known as the Learning Parities with Noise (LPN) [7] and it is famous to have many applications in cryptography. We then consider many attacker models, according to whether the inputs M are known, can be chosen and repeated. If we consider only the tag generation algorithm, additional authentication data can be input to the encryption and these values are first authenticated. We think that this model is powerful and allows us to consider many attacks on different implementations. For instance, since we only consider the writing in the accumulator of the polynomial evaluation, we do not take into account the way the multiplication is implemented and our attack also works even though the multiplication is protected against side-channel attacks.

In the first part of this paper, we consider inputs that are non controlled by the adversary. This is the case for instance in AES-GCM with the authentication of encrypted messages. We show through practical experiments that the proposed attacks may even be successful for reasonable levels of noise if averaging traces is allowed. Then, we consider methods to choose the input values for instance for the additional data or for the tag verification algorithm (e.g., with ciphertexts whose tag is incorrect), so that to improve the basic attacks. In the final part, we discuss three examples. The first one is the mode of operation OCB in which the same multiplication is used and lead to the same attacks. The specificity in this algorithm comes from the uncontrolled messages which are actually advantageously structured for our needs. The second one is a multiplication used in a context of re-keying, which is an alternative of masking. A new secret key is computed for each encryption through a multiplication which is performed differently. The previous attacks do not work in this model. The last example consider classical implementations of the field multiplication in $\mathsf{GF}(2^{128})$ to show how the inner steps can improve the complexities of the aforementioned attacks.

2   Backgrounds, Leakage and Attacker Models

2.1   AES-GCM Description

AES-GCM is an authenticated encryption algorithm which aims to provide both confidentiality and integrity. It combines an encryption based on the widely used AES algorithm in counter mode and an authentication with the Galois mode.

The so-called *hash key* H used for the authentication is derived from the *encryption key* K as $H = \mathsf{AES}_K(0^{128})$. The *ciphertext* of the encryption is denoted as $C_1, \ldots, C_n$ where the blocks C_i have length 128 bits except C_n which is of size u ($u \leqslant 128$). Similarly, the additional *authenticated data* is composed of 128-bit blocks $A_1, \ldots, A_m$ where the last one has size ν ($\nu \leqslant 128$). Eventually, we denote by $(X_i)_{0 \leqslant i \leqslant m+n+1}$ the intermediate results of function GHASH with X_{m+n+1} being the output exclusively ored with an encryption of the initial counter to form the tag. Figure 1 illustrates the procedure with the previously defined notations. For the sake of simplicity we will use a single letter M for both kinds of outputs

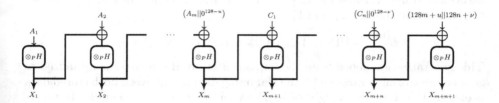

Fig. 1. AES-GCM authentication

(A_i or C_i). Then, the definition of the GHASH function can be simply described by the following recursion

$$X_{i+1} = (X_i \oplus M_{i+1}) \otimes_P H, \tag{1}$$

where $\otimes_P$ is the Galois Field multiplication described below.

Galois Field Multiplication. The multiplication $\otimes_P H$ is performed in the field GF(2^{128}) between 128-bit data. For AES-GCM, the Galois' extension is defined by the primitive polynomial $P(Y) = Y^{128} + Y^7 + Y^2 + Y + 1$. We denote by α a root of this polynomial P. An element Q in GF(2^{128}) can be represented by a vector of coefficient $(q_0, q_1, \ldots, q_{127})$ where $Q = \sum_{0 \leqslant i < 128} q_i \alpha^i$. In the following we denote by Q either the element of GF(2^{128}) or the corresponding vector. To avoid ambiguities we will differentiate field multiplication ($\otimes_P$) from matrix/vector multiplication ($\cdot$). Since attacks we present in the paper heavily rely on the linearity of multiplication in GF(2^{128}), we will use a matrix representation of the multiplication. Namely, let Q and R be two elements of GF(2^{128}), the result of the multiplication $Q \otimes_P R$ can be seen as a matrix/vector product $Q_P \cdot R$ where Q_P is a matrix obtained by concatenating columns representing coefficients of $Q \otimes_P \alpha^i$:

$$Q_P = \begin{pmatrix} q_0 & q_{127} & \cdots & q_1 \oplus q_{127} \oplus q_{126} \\ q_1 & q_0 \oplus q_{127} & \cdots & q_2 \oplus q_{123} \oplus q_1 \oplus q_{127} \oplus q_{122} \\ \vdots & \vdots & \ddots & \vdots \\ q_{127} & q_{126} & \cdots & q_0 \oplus q_{127} \oplus q_{126} \oplus q_{121} \end{pmatrix}.$$

2.2 Attacker Context

Leakage Model. A usual assumption (see for instance [24]) when there is no information on the implementation is to consider that all the variables V_i written in the registers of a cryptographic computation leak the sum of their *Hamming weight* (HW) and a *independent noise* ε_σ which follows a Gaussian distribution with a null mean and standard deviation σ (denoted by $\mathcal{N}(0, \sigma)$):

$$L_i^{(\mathsf{HW})} = \mathsf{HW}(V_i) + \varepsilon_\sigma, \quad \varepsilon_\sigma \sim \mathcal{N}(0, \sigma).$$

A common generalization of this leakage model when the attacker is given the successive stored variables is to consider the *Hamming distance* (HD) between two consecutive data V_{i-1} and V_i:

$$L_i^{(\mathsf{HD})} = \mathsf{HD}(V_i, V_{i-1}) + \varepsilon_\sigma = \mathsf{HW}(V_i \oplus V_{i-1}) + \varepsilon_\sigma.$$

This generalization depends on the implementation. If a register is initialized to zero before storing a variable V_i, the Hamming distance between both stored data is exactly the Hamming weight of V_i. However, in the case of a sum for instance, we can reasonably assume that the new computed variable overwrites the stored one (intermediate result), leaking the Hamming distance between them. In the following and in order to cover most embedded devices, we consider both models.

Attacker Model. Now we defined the models for information leakage, we discuss the attacker capabilities. From the axiom "Only Computation Leaks" of Micali and Reyzin [27], we only give the attacker the leakage of the manipulated data. Furthermore, in the most part of this paper, we restrict the leaking data to the multiplication's output to cover all the implementations. We now discuss the three characteristics that define the attacker model.

Known/Chosen Inputs. For the known operands of the Galois field multiplication, we consider the two classical attacker models namely the known message model (e.g., ciphertexts) and the chosen message model (e.g., additional data to authenticate). These two models will be respectively considered in Section 3 and Section 4.

Limited/Unlimited Queries. The attacker may face limitation in the number of queries. Such limitation may be due to time constraints but we may also consider an attacker querying for forged tag verifications in which case an error-counter may limit the number of invalid tag verifications.

Enabled/Disabled Averaging. Eventually, the attacker may be able to average traces obtained for the same computation. This is the case in the chosen messages setting but it may also be the case in the known messages setting when for instance the first blocks to authenticate have a specific format. If such feature is available, then the attacker may execute λ times each computation and average the corresponding traces. Since the leakage model considers an additive Gaussian noise, this decreases the standard deviation of the noise from σ to $\sigma/\sqrt{\lambda}$.

Attack Paths. We present hereafter the key idea of this paper and some preliminary results.

Main Observation. The cornerstone of the attacks presented in Section 3 and Section 4 is the fact that the less significant bit (further referred to as LSB) of the Hamming weight of a variable (equivalently distance between two variables) is a linear function of its bits. While a side-channel attacker generally uses a divide-and-conquer strategy to recover small parts of the key by making guesses, it is not possible anymore as the size of chunks gets large. This prevents attackers from targeting whole 128-bit variables. Nevertheless, in the particular case where the intermediate variable is the output of a linear function involving a public input and the key, then it means that the LSB of the Hamming weight is a linear function of this input and the key. If we denote by $\mathsf{lsb}_0\left(\mathsf{HW}(M \otimes_P H)\right)$ (or also b_0) the bit 0 of the Hamming weight of the product $M \otimes_P H$, we get

$$\mathsf{lsb}_0\left(\mathsf{HW}(M \otimes_P H)\right) = \bigoplus_{0 \leqslant i \leqslant 127} \left(\bigoplus_{0 \leqslant j \leqslant 127} (M \otimes_P \alpha^i)_j \right) h_i. \qquad (2)$$

This is precisely what is exploited in the attacks we present. Obviously this work can also be applied to any algorithm in which such multiplication appears and is not restricted to AES-GCM.

First Block. Observing Equation 1, we see that we only know the input of the multiplication with H for the first block of data X_1 since the input of further blocks will depend on H. Moreover, since X_0 is zero, we are in a context where Hamming distance and Hamming weight leakages are equivalent and we thus only refer to the Hamming weight in the following. While the linearity of its parity bit is a very good thing from an attacker point of view, the drawback is that this value is highly influenced by the measurement noise. Assume that we use the following decision procedure to guess the bit from a leakage value,

$$\widetilde{b_0} \stackrel{\text{def}}{=} \mathsf{lsb}_0\left(\lceil \mathsf{HW}(M \otimes_P H) + \varepsilon_\sigma \rfloor\right),$$

where $\lceil \cdot \rfloor$ is the rounding operator. Then, we obtain a noisy bit of information that we can model as follows

$$\widetilde{b_0} = \mathsf{lsb}_0\left(\mathsf{HW}(M \otimes_P H)\right) \oplus b_{\mathcal{N}}. \qquad (3)$$

where the error-bit $b_{\mathcal{N}}$ is the potential error due to the Gaussian noise. This error-bit follows a Bernoulli distribution with a parameter p such that the probability of no error is

$$1 - p = \sum_{i=-\infty}^{\infty} \int_{2i-0.5}^{2i+0.5} \phi_{0,\sigma}(t)dt \qquad (4)$$

with $\phi_{0,\sigma}(x) = e^{-\frac{x^2}{2\sigma^2}}/(\sigma\sqrt{2\pi})$, $\forall x \in \mathbb{R}$ the probability density function of the Gaussian law with null mean and standard deviation σ. In Table 1 we provide a few values of this Bernoulli parameter for several standard deviations. Note that we generally evaluate the complexity of an attack according to the Signal-to-Noise Ratio which is the ratio between the signal variance and the noise variance. For 8-bit implementations[1], we consider this SNR around 0.2 [23,4] which is a typical value both for hardware [19] and software implementations [11]. It corresponds to a signal variance of 8 (with the chosen leakage model) and a noise variance of 10 (standard deviation around 3). While we do not have reference measurements for 128-bit implementations, we can assume that the noise standard deviation is close, that is around 3.

Other Blocks. The generation of traces is expensive for an attacker and in some models it may also be limited. Therefore the number of traces is generally the main criteria when evaluating the complexity of an attack. In the context of the AES-GCM, the authentication is performed through a chained sequence of multiplications. This is quite frustrating for an attacker to only consider the first block when so much information is available. We will discuss in Section 3 and Section 4 how to exploit some of the following multiplications to obtain more bits of information from a single trace.

Other Leakage Bits. As mentioned above, we only exploit the LSB of the leakage for the attacks because it directly depends on a linear combination of the key bits. However, it is also strongly impacted by the noise which involves multiple errors in the system. In this paragraph, we discuss the complexities of considering further bits of leakage. We first focus on the impact of noise on each of them. In this purpose, Table 1 gives the values of the parameters of the Bernoulli law followed by each one of them. They are computed using (4) with i varying from $\lfloor -7\sigma \rfloor$ to $\lceil 7\sigma \rceil$ to capture at least $\left(1f00 - 2.56 \cdot 10^{-10}\right)\%$ of the values. The results are directly related to the number of errors in the system which decrease

Table 1. Bernoulli parameter p for different levels of noise with all $\varepsilon \ll 10^{-9}$

std dev σ	Bernoulli parameter p						
	1st bit	2nd bit	3rd bit	4th bit	5th bit	6th bit	7th bit
0.5	$3.1\ 10^{-1}$	$2.7\ 10^{-3}$	ε	ε	ε	ε	ε
1	$0.5 - 4.6\ 10^{-3}$	$1.3\ 10^{-1}$	$4.7\ 10^{-4}$	ε	ε	ε	ε
2	$0.5 - 1.7\ 10^{-9}$	$3.8\ 10^{-1}$	$8.0\ 10^{-2}$	$1.8\ 10^{-4}$	ε	ε	ε
3	$0.5 - \varepsilon$	$4.3\ 10^{-1}$	$2.3\ 10^{-1}$	$1.2\ 10^{-2}$	$2.0\ 10^{-7}$	ε	ε
4	$0.5 - \varepsilon$	$4.5\ 10^{-1}$	$3.2\ 10^{-1}$	$6.1\ 10^{-2}$	$1.1\ 10^{-4}$	ε	ε
5	$0.5 - \varepsilon$	$4.6\ 10^{-1}$	$3.7\ 10^{-1}$	$1.3\ 10^{-1}$	$1.9\ 10^{-3}$	ε	ε

---

[1] Notice that our attacks also work on 8-bit implementations where they are more efficient since the attacker can capture intermediate leakage on 8-bit values.

together with the increase of the bits indices. However, the resulting systems are made of equations of higher degrees (exponential with the index):

$$b_i = \bigoplus_{0 \leqslant j_1 < \cdots < j_{2^i} \leqslant 127} \left(\prod_{1 \leqslant \ell \leqslant 2^i} \bigoplus_{0 \leqslant k \leqslant 127} (M \otimes_P \alpha^k)_{j_\ell} \, h_k \right), \quad \forall \, 0 \leqslant i \leqslant 7.$$

and thus are more complicated to solve. In particular, the methods capturing the errors removal like LPN and linear decoding unfortunately do not apply on non-linear systems[2]. We thus have to consider first a solver on the error-free system of equations and then complete its complexity with the errors removal. To the best of our knowledge, one of the most efficient solver is the algorithm F5 [13] provided by Faugere and based on the Gröbner bases. While the solving complexity of the (error-free) quadratic system may be reasonable, it gets computationally impractical when considering the most significant bits[3].

3 Known Inputs

As described in the previous section, for each observed first multiplication, an attacker obtains a noisy Hamming weight value of the output. The LSB of the Hamming weight being linearly dependent on the key (see Equation 2), the attacker can gather many measurements to form a linear system having the authentication key H as solution. In this section we discuss different techniques to solve this noisy linear system. First we propose a simple procedure that allows the attacker to recover the key. Then, we investigate enhancements and other techniques that help decreasing the attack complexity in presence of higher noise.

3.1 Naive Attack

From Equation 2, the (noisy) linear system formed from t messages $(M^{(\ell)})_{0 \leqslant \ell \leqslant t}$ can be written as follows:

$$\mathcal{S} = \begin{cases} \bigoplus_{0 \leqslant i \leqslant 127} \left(\bigoplus_{0 \leqslant j \leqslant 127} (M^{(0)} \otimes_P \alpha^i)_j \right) h_i = \widetilde{b_0}^{(0)} \\ \qquad \cdots \\ \bigoplus_{0 \leqslant i \leqslant 127} \left(\bigoplus_{0 \leqslant j \leqslant 127} (M^{(t-1)} \otimes_P \alpha^i)_j \right) h_i = \widetilde{b_0}^{(t-1)}. \end{cases} \tag{5}$$

The values $\widetilde{b_0}^{(\ell)}$ are obtained as in Equation 3 that is we simply round the leakage observations to the closest integer and extract the least significant bits. Once the system $\mathcal{S}$ is correctly defined in GF(2), we can efficiently and directly solve it (e.g., calling mzd_solve_left of the library m4ri [1]) if these two conditions are fulfilled:

---

[2] A linearised system would be involve too many variables to be efficiently solved.
[3] The complexities related to each bit can be computed with the package [6].

i) $\mathcal{S}$ contains at least as many linearly independent equations as the number of unknown variables (128),

ii) there is no error in the bits $\widetilde{b_0}^{(\ell)}$ (i.e., $\widetilde{b_0}^{(\ell)} = b_0^{(\ell)}$).

First, the probability of obtaining a full-rank matrix from k k-bit messages, is $\prod_{0 \leqslant i \leqslant k-1} \left(1 - 2^{i-k}\right)$. In our context (that is $k = 128$), this probability is close to 0.3. To obtain a full-ranked matrix with high probability (say 0.9) the number of additional messages m must satisfy $1 - \prod_{1 \leqslant j \leqslant 2^m} \left(1 - \prod_{0 \leqslant i \leqslant k-1} \left(1 - 2^{i-k}\right)\right) \geqslant$ 0.9. For $k = 128$, a single additional message ($m = 1$) makes the equation holds. Note however that the full rank condition does not need to be fulfilled to recover the key. If it is not the attacker should first recover a solution of the system and the kernel of the linear application then test all the solutions to eventually recover the key.

Second, we consider the negative impact of the measurement noise. The latter introduces errors in the system which thus cannot be solved with classical techniques. A simple (naive) solution is to consider that one of the $\widetilde{b_0}^{(i)}$'s is erroneous and to solve k times the system with the k possible vectors $\widetilde{b_0} \oplus \alpha^i$. If the key is not found we can incrementally test all other numbers of errors until the correct key is found. Notice that the inversion is only done once: solving the system with a different vector $\widetilde{b_0}$ only requires a matrix/vector multiplication. If e errors are made among the k messages, then the correct key will be found in at most $C_k^{(e)}$ matrix/vector products:

$$C_k^{(e)} = \sum_{i=0}^{e} \binom{k}{i}. \tag{6}$$

When the number of errors grows, it quickly becomes computationally hard. For instance, for $e = 6$ and $k = 128$, $C_k^{(e)} \approx 2^{32}$. In the next section we investigate techniques to decrease the number of errors in $\mathcal{S}$.

3.2   Improved Attack

In this section, we propose two improvements for the attack. The first one consists in an optimal decision to guess the Hamming weight LSB. This criterium can also be used to advantageously select 128 traces among many more to limit the errors. The second improvement is to show that an attacker can actually use the leakage obtained from the two first multiplications and not only from the first one.

Reducing the Noise

An Optimal Decision Rule. We propose here to use the LLR statistics (for Log Likelihood Ratio) to derive a bit value $\widehat{b_0}$ *in average* closer to b_0 than $\widetilde{b_0}$ from a leaked Hamming weight. This statistics is extensively used in classical cryptanalysis (an application to linear cryptanalysis can be found in [2]) since the

Neyman-Pearson lemma states that for a binary hypothesis test the optimal[4] decision rule is to compare the LLR to a threshold. The LLR of a leakage ℓ is given by:

$$\mathsf{LLR}(\ell) = \log(\mathbb{P}[b_0 = 0|\ell]) - \log(\mathbb{P}[b_0 = 1|\ell]).$$

The bit b_0 is equally likely equal to 0 or 1 since we have an *a priori* uniform distribution for the secret. Thus, using Bayes relation we obtain that

$$\mathsf{LLR}(\ell) = \log\left(\mathbb{P}[\ell|b_0 = 0]\right) - \log\left(\mathbb{P}[\ell|b_0 = 1]\right)$$

$$\text{with} \quad \mathbb{P}[\ell|b_0 = i] = \sum_{w=0}^{128} \mathbb{P}[\ell|b_0 = i, \mathsf{HW}(\ell) = w]\mathbb{P}[\mathsf{HW}(\ell) = w].$$

If the result of $\mathsf{LLR}(\ell)$ is positive it means that the parity bit is likely to be equal to 0. Otherwise, we should assume it is equal to 1. We thus define:

$$\widehat{b_0} \stackrel{\text{def}}{=} \begin{cases} 0 & \text{if } \mathsf{LLR}(\ell) \geqslant 0, \\ 1 & \text{otherwise.} \end{cases}$$

Such technique will decrease the error rate since it boils down to select the most probable value for b_0. Unfortunately it turns out that it has a small impact on the number of errors made, but its combination with the following technique will be useful as illustrated in Figure 2.

Selecting Traces. Nevertheless, when more than k traces are available, it would be of interest to only select the k most reliable ones to decrease the number of errors in the system. Basically, we would like to take into account the confidence we have in a given bit. For instance, assuming a 0 parity bit from a leakage 64.01 seems more reliable than for a leakage equal to 64.49. Interestingly, the higher the absolute value of the LLR is for a given trace, the more confident we are in the choice. Therefore, an attacker should select the n samples with the highest LLR values to form the system. The point is that those k samples may not be linearly independent. Two solutions can then be used:

i) one may only consider a subset of these k samples, solve the system and brute force the remaining bits,

ii) or one may choose k linearly independent samples from the highest LLR values.

Finding the set of k linearly independent samples maximizing this sum is a combinatorial optimization problem which may be quite hard, thus we use a simple "first come/first selected" algorithm that provides a set of k samples. The algorithm iteratively looks for the sample with the highest absolute LLR value that increases the system rank. Figure 2 represents the averaged experimental values (on 10.000 samples) of the Bernoulli parameter p for 500 messages in different scenarios. The black curve represents the use of function *round* to

---

[4] For more precisions about this lemma and the meaning of optimal refer to [10].

fix $\widetilde{b_0}$. The green one represents the LLR without selection of the best traces while the blue one integers this selection among the first linearly independent traces. Eventually, the red curve models the optimal with the 128 best traces (having the highest LLR values but not necessarily linearly independent). As mentioned, we can observe that the use of the LLR does not significantly improve the attack (black and green curves very close). However, the chosen selection of the best LLR allows the attack to resist higher levels of noise: 0.4 (resp. 0.5) instead of 0.3 (resp. 0.4) to achieve the same Bernoulli parameter. Notice that we cannot distinguish the red curve from the blue one. This proximity means that while the "first come/first selected" approach is not optimal it is not worth working on a refined algorithm since the improvement will be bounded by the distance between both curves.

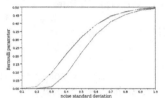
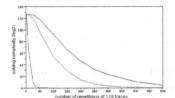

Fig. 2. Bernoulli parameter with rounding (black), LLR (green), traces selection (blue) and best LLR traces (red)

Fig. 3. Solving complexities for several repetitions numbers with $\sigma = 1$ (blue), $\sigma = 3$ (red) and $\sigma = 4$ (black)

Averaging Traces. In the context where the attacker can monitor few multiplications with the same input, we can also consider another commonly used method which consists in averaging the traces. As claimed in Section 2 and experimentally confirmed in Section 3.3, repeating the traces m times allows to divide the noise standard deviation by $\sqrt{m}$. Figure 3 gives the complexity of removing the errors (averaged from 10,000 tests computed from Eq. 6) according to the number of repetitions of 128 traces for several levels of noise. Note that the full complexity of the attack also includes the system solving (a single inversion in k^3 and $C_k^{(e)} - 1$ matrix/vector products in k^2). Considering it, we can claim that with less than 2^{16} traces (i.e., 500 repetitions), the attacker can practicaly recover the key for σ^2 up to 10 ($\sigma \approx 3$).

Saving Traces with Further Blocks. Up to now we only considered the first multiplication since not knowing H implies not knowing the input of the second multiplication (indeed $X_2 = (M_1 \otimes_P H \oplus M_2) \otimes_P H$). Nonetheless, re-writing this equality as

$$X_2 = M_1 \otimes_P H^2 \oplus M_2 \otimes_P H,$$

we observe that X_2 also is a linear function of H since squaring is linear over GF(2). Denoting by S the matrix corresponding to the squaring operation, then

$$X_2 = (M_1 \cdot S \oplus M_2) \otimes_P H.$$

Thus a linear relation can also be obtained from the second multiplication substituting $M_1 \cdot S \oplus M_2$ to M in Equation 2. And this is also true in the Hamming distance model since $X_1 \oplus X_2 = (M_1 \cdot S \oplus M_1 \oplus M_2) \otimes_P H$. This observation is of great importance since it significantly improves the complexity of the attacks with a number of required traces divided by a factor two.

3.3   Experimental Results

We illustrate here the necessity of averaging and we confirm the corresponding decreasing of the noise by giving the results obtained on real experiments.

Settings. We implemented the GHASH function on the Virtex-5 FPGA of a SASEBO board and acquired traces from an EM probe. We obtained 10^5 traces that we separated in two $5 \cdot 10^4$ trace sets (Set 1 and Set 2). We then built templates using Set 1 and a projection obtained using the same technique as in [12]. Afterwards we performed the first part of the attack (that is guessing parity bits of Hamming weights with the LLR technique) using this template. We then attacked both sets of traces.

Results. In Table 2 we provide the results we obtained. For a given number of averaging (*av.*) we report (i) the noise standard deviation σ, (ii) the simulated error rate obtained from this standard deviation (10^7 simulations) and (iii) the error rates obtained when applying the template to Set 1 and Set 2.

Table 2. Noise levels and error rates obtained from EM traces

av.	σ	error rates			av.	σ	error rates		
		simul	Set 1	Set 2			simul	Set 1	Set 2
1	1.958	n/c	n/c	n/c	6	0.770	0.466	0.454	0.467
2	1.287	n/c	n/c	n/c	8	0.637	0.414	0.407	0.457
3	1.063	0.5 - 2.26 10^{-3}	n/c	n/c	10	0.579	0.378	0.370	0.422
4	0.882	0.486	0.483	0.495	12	0.520	0.333	0.338	0.404

First, we see that doubling the number of averaging roughly leads to a reduction of noise standard deviation by a factor $\sqrt{2}$ as we would expect. Second, the attack performs better on the first set since it is the one that have been used to build templates. For Set 1, the error rates actually correspond to theoretical approximations based on the noise standard deviation. We also see that the error rates obtained for Set 2, while obviously deviating from expected values, are significantly decreasing with the number of averaging. Indeed, when averaging is possible, the obtained features show that an attack can be mount easily. As one can see, Table 2 does not contain data for error rates corresponding to less than 4 averagings. This is due to the fact that the deviation from 0.5 is too small to be estimated using 50,000 traces. We did not managed to get more traces since experiments with higher levels of averaging confirm our predictions.

3.4   Solving the System with More Errors and Advanced Algorithms

There are many algorithms to recover the authentication key from noisy Hamming weight LSBs. In the case where more than n multiplications are observed, the attacker will obtain an overdefined linear system. In other words, the attacker will get redundant linear relations involving bits of the key H. Guessed LSBs extracted from leaking multiplication can thus be seen as forming a noisy codeword that encodes the authentication key H using the code defined by the linear relations of the form of Equation 2. Recovering the key is then equivalent to decoding the noisy codeword.

Learning Parities with Noise Algorithms. The Learning Parities with Noise (LPN) problem is the problem of recovering $x \in GF(2)^k$ given many pairs $(a, (a, s) \oplus e)$ with $a \in GF(2)^k$ samples randomly chosen, (a, s) denotes the scalar product and e is a Bernoulli distribution of parameter p. The most efficient algorithms to solve the LPN problem are based on the Blum-Kalai-Wasserman algorithm [7]. This algorithm tries to perform Gaussian elimination in a smart way but cancelling many bits with one single xor. The idea is to use many samples and xoring those that have many bits in common. However, this algorithm is exponential in the number of samples, time and memory of order $2^{O(k/\log k)}$ where k is the size of the secret values. This algorithm has been improved by Fouque and Levieil in [21] but it allows to reduce the constant in the exponent. In practice, it requires a huge number of samples but since here the size is relatively short $k = 128$, we could use such algorithms. However, since the noise involves a Bernoulli parameter p getting closer to $1/2$, it expects 2^{40} bytes of memory and 2^{34} queries when the standard deviation σ equals 0.5, while it grows to 2^{241} bytes of memory and 2^{334} queries when σ equals 2. Lyubashevsky gave in [22] a variant of BKW with running time $2^{O(k/\log\log k)}$ for $k^{1+\varepsilon}$ samples. A further modification proposed more recently by Kirchner in [20] achieved better runtimes for small values of p. This algorithm runs in time $O(2^{\sqrt{k}})$ with $O(k)$ samples when $p = O(1/\sqrt{k})$.

Linear Decoding. Since inputs are not controlled by the attacker, the corresponding linear code is random. Decoding over random linear codes is known to be a hard problem (NP problem). The currently best algorithm that solves this problem is the one presented by Becker *et al.* in [3] which has complexity $O(2^{0.0494n})$ (where n is the code length). Nevertheless, using such algorithm only makes sense if the noise is low enough to ensure that the actual key-codeword is the closest to the noisy word obtained by the attacker. Indeed if the noise is too high, then the channel capacity will decrease below the code rate and thus the closest codeword to the obtained noisy one may not be the one the attacker looks for. Using the binary symmetric channel model[5] we obtain that for a standard deviation σ of 0.5 the code length should be at least $\frac{128}{0.107} \approx 1200$ which would yield a complexity $2^{59.28}$ using [3]. Obviously the attacker has better using less than 1200 relations and test more than a single key candidate. To do so she will need a list-decoding algorithm. For cryptographic parameters (that is a key that can be very badly ranked), the only known solution is to see the linear code as a punctured Reed-Muller code and to use a Fast Walsh transform to obtain probabilities for each possible codeword. Since this technique has complexity $O(k2^k)$ with k the code dimension, it is not straightforwardly applicable here. We discuss in Section 4 how we can take profit of controlling inputs to use such decoding algorithm.

3.5 Complexities Evaluation

In this section, we built a system of equations from a new trick, that is the use of a single leakage bit. Then, we discussed methods to solve it involving step by step decoding (Eq. 6) and LLR statistics and the existing tools: LPN and linear decoding. We now propose a comparison of the methods complexities through Table 3 both in terms of number of samples C_s and of computation time C_t. As for the LLR method combined with the step by step error removal (Equation 6) the time complexity C_t includes not only the errors removal $C_k^{(e)}$ but also the the linear system solving (a single inversion in k^3 and $C_k^{(e)} - 1$ matrix/vector products in k^2). As for the number of samples C_s, it is divided by two in all methods thanks to the smart use of the second GCM block X_2. We remark that for low levels of noise (at least until $\sigma = 0.4$), linear decoding is the best method to choose both in terms of number of samples and time complexity. Afterwards, it depends on the number of available samples. Concretely, the more traces we have the less time we need.

Table 3. Complexities of recovering the key with LLR and Eq. 6, LPN and linear decoding according to the level of noise

σ Method	0.1 C_s/C_t	0.2 C_s/C_t	0.3 C_s/C_t	0.4 C_s/C_t	0.5 C_s/C_t
LLR and Eq. 6	$2^8/2^{21}$	$2^8/2^{21}$	$2^8/2^{22}$	$2^8/2^{65}$	$2^8/2^{107}$
LPN (LF Algo)	$2^7/2^{21}$	$2^7/2^{23}$	$2^{26}/2^{28}$	$2^{32}/2^{34}$	$2^{48}/2^{50}$
Linear decoding	$2^6/2^6$	$2^6/2^7$	$2^7/2^{11}$	$2^8/2^{25}$	$2^9/2^{62}$

---

[5] That is using the aforementioned Bernoulli parameter as error probability.

4   Chosen Inputs

Let us now consider techniques that may be used to recover the key in the model where the attacker is able to control multiplication inputs. A first idea is that in such context averaging should be considered as obviously enabled[6] and thus measurement noise could be decreased by repeating inputs. Two other ideas are:

i) structuring the messages to make the system easier to solve,
ii) choosing messages to be able to exploit more than two multiplications.

The following is dedicated to the discussion of these two ideas.

4.1   Structured Messages

In Section 3 we saw that recovering the key could be seen as a decoding problem. The difficulty arose from the fact that the linear code corresponding to our attack is random and have a high dimension (128). Assuming the attacker is now able to control inputs of the multiplication, she may choose the underlying code.

Choice of the Code. The question is now *which code should we use?* As a cryptanalyst the requirements for a linear code may be different from the one found in coding theory.

List Decoding. First, an attacker aims at recovering the key. She has computing power and can enumerate many key candidates before finding the correct one. Such a feature means that a list decoding algorithm should be available for the chosen code. Moreover, the list size is not of the same order of magnitude that can be found in coding theory. Ideally, we would like to obtain a list of all key candidates ordered by probabilities of being the correct one. Obviously such a list cannot be created since its size would be 2^{128}. Nevertheless, using the key enumeration algorithm of [31], an attacker can enumerate keys from ordered lists of key chunks. If the linear code underlying the attack is a concatenated code then such algorithm can be used. Indeed, the corresponding matrix of the system would be a block diagonal matrix. Each block corresponds to a smaller linear code that may be fully decoded, that is the attacker obtains a list of all possible keys with the corresponding probabilities.

Soft Information. Second, since the noise may be high, we would like to take profit of the whole available information and not only consider obtained bits $\tilde{b}_0$ or $\hat{b}_0$. We illustrated in Section 3.2 the gain obtained when considering LLR statistics to take into account the relations reliabilities. Here, we would like a code which decoding algorithm can exploit such soft information.

---

[6] Except maybe in pathological cases.

Taking into account the aforementioned constraints, we opted for a concatenating code of smaller random linear codes. The latters can efficiently be decoded using a Fast Walsh Transform (FWT) as mentioned in Section 2. We thus aim at obtaining a matrix corresponding to the system $\mathcal{S}$ of the form

$$\begin{pmatrix} \boxed{\mathcal{S}_0} & & \\ & \boxed{\mathcal{S}_1} & \\ & & \ddots \end{pmatrix} \cdot \begin{pmatrix} H \end{pmatrix} = \begin{pmatrix} \widehat{b_0} \\ \vdots \\ \widehat{b_t} \end{pmatrix}. \tag{7}$$

Generating Structured Inputs. To generate the inputs that yield a matrix similar to the one in (7), the attacker has to consider the application

$$\varphi : M \mapsto \left(\bigoplus_{0 \leqslant j \leqslant 127} \mathsf{lsb}_j \left(M \otimes_P \alpha^0 \right), \ldots, \bigoplus_{0 \leqslant j \leqslant 127} \mathsf{lsb}_j \left(M \otimes_P \alpha^{127} \right) \right)$$

that maps an input M to the corresponding vector of coefficients for the system $\mathcal{S}$. To generate the bloc $\mathcal{S}_c$, she chooses inputs in the kernel of $\varphi_{|\mathcal{I}_c}$ where indices in $\mathcal{I}_c$ correspond to columns outside block $\mathcal{S}_c$. A basis of these kernels are efficiently computed using Gauss eliminations.

Simulations. To illustrate the method results, we give two graphs. The left one presents the averaged rank of the correct key among all the 2^{128} possible ones from the key chunks probabilities according to the noise standard deviations for 256 samples (blue) and 1024 samples (red). The right one is a security graph [32] which draws the evolution of the bounds of the correct key rank according to the number of samples for $\sigma = 0.5$.

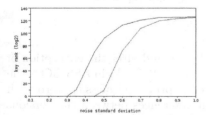

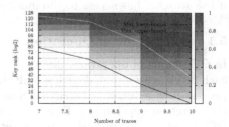

Fig. 4. Key rank for 256 (blue) and 1024 (red) samples

Fig. 5. Security graph for $\sigma = 0.5$

4.2 Saving Traces

A second way, for the attacker, to take profit of the control she has on inputs is to leverage on Ferguson observation [14]. During the specification process of AES-GCM, Ferguson observed that it was possible to obtain a tag that is linearly dependent on the authentication key H in the particular case where the polynomial corresponding to the tag only has non-zero coefficients in positions where the exponent of H is a power of two. This observation relies on the linearity of the squaring operation as mentioned in Section 3.

We saw that this observation allows to exploit the two first multiplications but if the attacker has the control on the inputs she can choose them to do more. Again, this trick can be used either in case of Hamming weight or Hamming distance. The only limitation is that the number of blocks to authenticate grows exponentially in the number of exploitable multiplications. The trade-off will depend on the available time for getting traces and on a potential limitation in the number of queries. To illustrate this we show how an attacker can exploit 3 multiplications in a single trace. From Eq. 1, we obtain the expression of the four first X_i's when M_2 is set to 0:

$$X_1 = M_1 \otimes_P H, \qquad\qquad X_3 = M_1 \otimes_P H^3 \oplus M_3 \otimes_P H,$$
$$X_2 = M_1 \otimes_P H^2, \qquad\qquad X_4 = M_1 \otimes_P H^4 \oplus M_3 \otimes_P H^2 \oplus M_4 \otimes_P H.$$

We see that relations obtained from X_1, X_2 and X_4 only involve power-of-two of H which means that the relation is a linear function of H. For instance $X_4 = (M_1 \cdot S \cdot S \oplus M_3 \cdot S \oplus M_4) \otimes_P H$, and because she knows S and can choose M_4, the attacker can obtain the input of its choice for the fourth multiplication.

5 Other Applications

In this section, we discuss different applications of the presented attacks. We first consider the OCB mode of operation on which the proposed attacks allow to recover the masks. Then, we look at another multiplication on which our attacks unpractical. Finally, we give some hints on the complexity of our attacks if the attacker has access to the inner parts of the multiplication.

5.1 OCB Mode of Operation

In OCB mode of operation, the masks added before and after the encryption of each block are computed with a multiplication in $GF(2^{128})$. As in AES-GCM, the process uses a secret constant computed by encrypting the message zero with the secret encryption key. Despite some small differences between the versions OCB1 and OCB2, in both cases, a secret value $(E_K(0^{128}))$ is multiplied by a power of two. Thus, the scenario is easier than for the uncontrolled setting since the messages are very sparse. Plus, their shape is close to the one considered in the controlled setting. As a consequence, the secret constant can be recovered in OCB1 and OCB2 at least as well as (but generally easier than) in AES-GCM and allows to recover the masks which are supposed to protect the encryption.

5.2   Re-keying

In [26], Medwed et al. propose to multiply known uniformly distributed 128-bit messages r with a 128-bit secret master key k to generate session keys in the context of re-keying. Each resulting session key is then used for a single encryption of a plaintext block. Doing so, only the generation of the session keys is required to resist Differential Power Analysis attacks. Therefore, the authors only mask this operation but the resulting session key can still leak its (noisy) Hamming weight. The context is thus the same than in AES-GCM but the multiplication is different, the variables being defined on $\mathsf{GF}(2^8)[y]/y^{16}+1$. Re-using the matrix/vector modelization, we can represent the message r being multiplied with the key k by a (16×16) matrix R_p as follows:

$$R_p = \begin{pmatrix} r_0 & r_{m-1} & \cdots & r_1 \\ r_1 & r_0 & \cdots & r_2 \\ \vdots & \vdots & \ddots & \vdots \\ r_{m-1} & r_{m-2} & \cdots & r_0 \end{pmatrix}$$

with the r_i in $\mathsf{GF}(2^8)$. From this matrix, we can easily write the equation involving the LSB b_0 of the Hamming weight of the result:

$$\mathsf{lsb}_0\left(\mathsf{HW}\left[\left(\bigoplus_{0 \leqslant i \leqslant m-1} r_i\right) \cdot \left(\bigoplus_{0 \leqslant j \leqslant m-1} k_j\right)\right]\right) = b_0$$

with $\cdot$ the multiplication in $\mathsf{GF}(2^8)$. As we can see, only the sum of all the key bytes can be recovered if the attack is successful. However, no individual key bit can be determined. If we extended the attack to more leakage bits, we could (at most) successively recover all the bits of the Hamming weight of the key. It is worth noting that the non-applicability of the attack directly comes from the multiplication's polynomial. Any polynomial with an even number of monomials makes the attack fail when considering only the LSB of the Hamming weight.

5.3   Specific Implementations

Previously, we considered a secure multiplication on $\mathsf{GF}(2^{128})$ for which we just had access to the result. Doing so, we covered all the multiplier implementations like [5] including the protected ones (e.g., with masking). We now show that making assumptions on the multiplier implementation improves the efficiency of our attack. As explained in [25,33], a usual method to implement a multiplier is to split one of the two operands in smaller blocks (a (128×128)-bit multiplier does not generally fit the area requirements) and perform intermediate multiplications which are progressively accumulated. In the full version of this paper, we focus on two such multipliers with one or the other operand being split. When the secret key is split, the attacker can follow a divide-and-conqueer strategy to recover each block. Nevertheless, when the message is split, the attacker cannot practically enumerate all the possible secret keys. In this case the scenario is still easier than the generic one since we focus on sparse messages of at most n bits.

References

1. Rix, A.W., Beerends, J.G., Hollier, M.P., Hekstra, A.P.: Perceptual evaluation of speech quality (PESQ) – a new method for speech quality assessment of telephone networks and codecs. In: Proceedings ICASSP, pp. 749–752 (2001)
2. Baignères, T., Junod, P., Vaudenay, S.: How Far Can We Go Beyond Linear Cryptanalysis? In: Lee, P.J. (ed.) ASIACRYPT 2004. LNCS, vol. 3329, pp. 432–450. Springer, Heidelberg (2004)
3. Becker, A., Joux, A., May, A., Meurer, A.: Decoding Random Binary Linear Codes in $2^{n/20}$: How $1 + 1 = 0$ Improves Information Set Decoding. In: Pointcheval, D., Johansson, T. (eds.) EUROCRYPT 2012. LNCS, vol. 7237, pp. 520–536. Springer, Heidelberg (2012)
4. Belaïd, S., Grosso, V., Standaert, F.-X.: Masking and leakage-resilient primitives: One, the other(s) or both? Cryptology ePrint Archive, Report 2014/053 (2014), http://eprint.iacr.org/2014/053
5. Bernstein, D.J.: Bernstein. Faster binary-field multiplication and faster binary-field macs. In: SAC. LNCS, Springer, Heidelberg (2014)
6. Bettale, L.: Magma Package: Hybrid Approach for Solving Multivariate Polynomial Systems over Finite Fields, http://www-polsys.lip6.fr/~bettale/hybrid/
7. Blum, A., Kalai, A., Wasserman, H.: Noise-tolerant learning, the parity problem, and the statistical query model. J. ACM 50(4), 506–519 (2003)
8. Chari, S., Jutla, C.S., Rao, J.R., Rohatgi, P.: Towards Sound Approaches to Counteract Power-Analysis Attacks. In: Wiener, M. (ed.) CRYPTO 1999. LNCS, vol. 1666, pp. 398–412. Springer, Heidelberg (1999)
9. Coron, J.-S.: Higher Order Masking of Look-Up Tables. In: Nguyen, P.Q., Oswald, E. (eds.) EUROCRYPT 2014. LNCS, vol. 8441, pp. 441–458. Springer, Heidelberg (2014)
10. Cover, T.M., Thomas, J.A.: Information theory. Wiley series in communications. Wiley (1991)
11. Durvaux, F., Renauld, M., Standaert, F.-X., van Oldeneel tot Oldenzeel, L., Veyrat-Charvillon, N.: Efficient Removal of Random Delays from Embedded Software Implementations Using Hidden Markov Models. In: Mangard, S. (ed.) CARDIS 2012. LNCS, vol. 7771, pp. 123–140. Springer, Heidelberg (2013)
12. Durvaux, F., Standaert, F.-X., Veyrat-Charvillon, N., Mairy, J.-B., Deville, Y.: Efficient selection of time samples for higher-order DPA with projection pursuits. IACR Cryptology ePrint Archive, 2014:412 (2014)
13. Faugère, J.-C.: A new efficient algorithm for computing Gröbner bases without reduction to zero F5. In: International Symposium on Symbolic and Algebraic Computation Symposium - ISSAC (2002)
14. Ferguson, N.: Authentication weaknesses in GCM (2005), http://csrc.nist.gov/groups/ST/toolkit/BCM/
15. Grosso, V., Prouff, E., Standaert, F.-X.: Efficient Masked S-Boxes Processing – A Step Forward –. In: Pointcheval, D., Vergnaud, D. (eds.) AFRICACRYPT. LNCS, vol. 8469, pp. 251–266. Springer, Heidelberg (2014)
16. Handschuh, H., Preneel, B.: Key-Recovery Attacks on Universal Hash Function Based MAC Algorithms. In: Wagner, D. (ed.) CRYPTO 2008. LNCS, vol. 5157, pp. 144–161. Springer, Heidelberg (2008)
17. Jaffe, J.: A first-order DPA attack against AES in counter mode with unknown initial counter. In: Paillier, P., Verbauwhede, I. (eds.) CHES 2007. LNCS, vol. 4727, pp. 1–13. Springer, Heidelberg (2007)

18. Joux, A.: Authentication Failures in NIST version of GCM (2006), http://csrc.nist.gov/CryptoToolkit/modes/
19. Katashita, T., Satoh, A., Kikuchi, K., Nakagawa, H., Aoyagi, M.: Evaluation of DPA Characteristics of SASEBO for Board Level Simulation (2010)
20. Kirchner, P.: Improved generalized birthday attack. Cryptology ePrint Archive, Report 2011/377 (2011), http://eprint.iacr.org/2011/377
21. Levieil, É., Fouque, P.-A.: An Improved LPN Algorithm. In: De Prisco, R., Yung, M. (eds.) SCN 2006. LNCS, vol. 4116, pp. 348–359. Springer, Heidelberg (2006)
22. Lyubashevsky, V.: The Parity Problem in the Presence of Noise, Decoding Random Linear Codes, and the Subset Sum Problem. In: Chekuri, C., Jansen, K., Rolim, J.D.P., Trevisan, L. (eds.) APPROX 2005 and RANDOM 2005. LNCS, vol. 3624, pp. 378–389. Springer, Heidelberg (2005)
23. Mangard, S.: Hardware Countermeasures against DPA – A Statistical Analysis of Their Effectiveness. In: Okamoto, T. (ed.) CT-RSA 2004. LNCS, vol. 2964, pp. 222–235. Springer, Heidelberg (2004)
24. Mangard, S., Oswald, E., Popp, T.: Power analysis attacks - revealing the secrets of smart cards. Springer (2007)
25. McGrew, D.A., Viega, J.: The Galois/Counter Mode of Operation, GCM (Day 2005)
26. Medwed, M., Standaert, F.-X., Großschädl, J., Regazzoni, F.: Fresh Re-keying: Security against Side-Channel and Fault Attacks for Low-Cost Devices. In: Bernstein, D.J., Lange, T. (eds.) AFRICACRYPT 2010. LNCS, vol. 6055, pp. 279–296. Springer, Heidelberg (2010)
27. Micali, S., Reyzin, L.: Physically Observable Cryptography. In: Naor, M. (ed.) TCC 2004. LNCS, vol. 2951, pp. 278–296. Springer, Heidelberg (2004)
28. Procter, G., Cid, C.: On Weak Keys and Forgery Attacks Against Polynomial-Based MAC Schemes. In: Moriai, S. (ed.) FSE 2013. LNCS, vol. 8424, pp. 287–304. Springer, Heidelberg (2014)
29. Rivain, M., Prouff, E.: Provably Secure Higher-Order Masking of AES. In: Mangard, S., Standaert, F.-X. (eds.) CHES 2010. LNCS, vol. 6225, pp. 413–427. Springer, Heidelberg (2010)
30. Saarinen, M.-J.O.: Cycling Attacks on GCM, GHASH and Other Polynomial MACs and Hashes. In: Canteaut, A. (ed.) FSE 2012. LNCS, vol. 7549, pp. 216–225. Springer, Heidelberg (2012)
31. Veyrat-Charvillon, N., Gérard, B., Renauld, M., Standaert, F.-X.: An Optimal Key Enumeration Algorithm and Its Application to Side-Channel Attacks. In: Knudsen, L.R., Wu, H. (eds.) SAC 2012. LNCS, vol. 7707, pp. 390–406. Springer, Heidelberg (2013)
32. Veyrat-Charvillon, N., Gérard, B., Standaert, F.-X.: Security Evaluations beyond Computing Power. In: Johansson, T., Nguyen, P.Q. (eds.) EUROCRYPT 2013. LNCS, vol. 7881, pp. 126–141. Springer, Heidelberg (2013)
33. Yang, B., Mishra, S., Karri, R.: A high speed architecture for galois/counter mode of operation (GCM). Cryptology ePrint Archive, Report 2005/146 (2005), http://eprint.iacr.org/2005/146

Higher-Order Threshold Implementations

Begül Bilgin[1,2], Benedikt Gierlichs[1], Svetla Nikova[1],
Ventzislav Nikov[3], and Vincent Rijmen[1]

[1] KU Leuven, ESAT-COSIC and iMinds, Belgium
{name.surname}@esat.kuleuven.be
[2] University of Twente, EEMCS-SCS, The Netherlands
[3] NXP Semiconductors, Belgium
venci.nikov@gmail.com

Abstract. Higher-order differential power analysis attacks are a serious threat for cryptographic hardware implementations. In particular, glitches in the circuit make it hard to protect the implementation with masking. The existing higher-order masking countermeasures that guarantee security in the presence of glitches use multi-party computation techniques and require a lot of resources in terms of circuit area and randomness. The Threshold Implementation method is also based on multi-party computation but it is more area and randomness efficient. Moreover, it typically requires less clock-cycles since all parties can operate simultaneously. However, so far it is only provable secure against 1st-order DPA. We address this gap and extend the Threshold Implementation technique to higher orders. We define generic constructions and prove their security. To illustrate the approach, we provide 1st, 2nd and 3rd-order DPA-resistant implementations of the block cipher KATAN-32. Our analysis of 300 million power traces measured from an FPGA implementation supports the security proofs.

1   Introduction

Differential power analysis (DPA) attacks as introduced by Kocher et al. [19] exploit unintentional information leakage of a device's internal processing through its power consumption. Over the years, many types of countermeasures against DPA attacks have been proposed. One family of countermeasures is called masking and consists in computing the algorithm on a randomized representation of the data. For this purpose, the data is split in several shares that are processed sequentially or in parallel. A DPA attack that exploits the information leakage of several shares jointly, be it by combining the leakage from several points in time or by analyzing higher-order statistical moments of the leakage at one point in time, is a higher-order DPA (HO-DPA) attack [6,24].

It is preferable to protect the implementation of a cryptographic algorithm with a higher-order masking countermeasure, where $d > 1$ random masks are used to generate $d + 1$ shares of a variable, since 2nd-order DPA attacks can be relatively inexpensive to mount. It is well known that the number of measurements required for a HO-DPA attack to succeed scales exponentially in the noise standard deviation, the exponent being $d + 1$ [6,33].

P. Sarkar and T. Iwata (Eds.): ASIACRYPT 2014, PART II, LNCS 8874, pp. 326–343, 2014.
© International Association for Cryptologic Research 2014

In a secure masking, all $d + 1$ shares are necessary to re-construct the variable. Such a secure masking is called d^{th}-order masking and leads to a dth-order secure implementation in software. An implementation of the same secure masking in CMOS-like glitchy hardware, on the other hand, will typically be insecure [22,23,27].

Related Work. Several masking schemes that are secure against HO-DPA have been proposed so far, e.g. [8,9,14,17,18,34,35]. However the scheme in [17] is shown to be insecure in [10] and [9] discovers and proposes a fix to a leak in [35]. Only one scheme claims to be secure against HO-DPA even in the presence of glitches [34], based on separation of the operations in the time domain. Nevertheless, implementing this scheme within the defined models is a challenging task. Moradi and Mischke [26] provided practical evidence that a simple separation of the operations in the time domain alone is not sufficient when the shares of a sensitive variable are processed in consecutive clock cycles.

Threshold Implementation (TI) is a masking scheme based on secret sharing and multi-party computation [29,30,31]. It provides provable security against 1st-order DPA even in the presence of glitches. The only requirement is that the shares leak independently, but this requirement holds for all masking schemes.

So far, several S-boxes and symmetric-key algorithms have been implemented with this method and the security claim has been confirmed in practical experiments [2,28,32]. However, it has also been confirmed that a TI is vulnerable to HO-DPA [2,25].

Contribution. So far, the theory of TI and its practical security is limited to counteract 1st-order DPA. In this work, we define *Higher-Order Threshold Implementation (HO-TI)* to thwart HO-DPA. We define generic constructions and use results from CHES 2010, EUROCRYPT 2010 and 2014 to prove their security. We provide a relation between 1st-order DPA secure implementations of 4×4 S-boxes in the alternating group with 5-shares provided in [4] and 2nd-order DPA secure implementations of these S-boxes. To illustrate the HO-TI approach in a comprehensible example, we provide 1st, 2nd and 3rd-order DPA-resistant implementations of the block cipher KATAN-32. Our analysis of 300 million power traces measured from an FPGA implementation supports the security proofs.

2   Theory of HO-TI

We use lower case characters to refer to elements of a finite field and upper case characters to describe vectors and vector functions. Stochastic variables are described by the superscript $\$$. The probability that $x^{\$}$ takes the value x is $Pr(x^{\$} = x)$. In order to implement a function $f(x) = y$ from $\mathcal{F}^n$ to $\mathcal{F}^m$ with TI, we first split each variable x into s shares x_i where $i \in \{1, 2, \ldots, s\}$ by means of Boolean masking, such that the XOR sum of these shares is equal to the variable itself ($x = \sum_i x_i$). For all values x with $Pr(x^\$ = x) > 0$, let $Sh(x)$ denote the set of valid share vectors X for x:

$$Sh(x) = \{X \in \mathcal{F}^{ns} \mid x_1 \oplus x_2 \oplus \cdots \oplus x_s = x\}.$$

We use the terms sharing or masking interchangeably for a valid share vector X and use the term s-sharing of x to emphasize the number of shares. $Pr(X^\$ = X \mid x^\$ = x)$ denotes the probability that $X^\$ = X$ when the unmasked value equals x, taken over all auxiliary inputs of the masking.

In a TI, f is implemented as a vector of functions F that takes X as input. Each function in this vector is called a component function and represented by f_i where $i \in \{1, \ldots, s\}$. From now on, we use the term sharing of the function to describe F and s-sharing of f to emphasize the number of component functions of F. F must satisfy the following property for a correct implementation.

Property 1 (Correctness). $\forall y \in \mathcal{F}^m$, $\forall X \in Sh(x)$ and $\forall Y \in Sh(y)$; $F(X) = Y \iff f(x) = y$.

We call each share of X an input share and each share of Y an output share.

2.1 HO-TI of an Arbitrary Function

Like for other masking schemes, the masking of the input of a shared function F must be uniform. We call a masking X of a variable x uniform if and only if $Pr(X^\$ = X \mid x^\$ = x)$ is equal to the same constant p for each X and $\sum_X Pr(X^\$ = X \mid x^\$ = x) = Pr(x^\$ = x)$. If the input is a uniform masking, then the shared function F must satisfy the following property in order to achieve security against d^{th}-order DPA.

Property 2 (d^{th}-order non-completeness). Any combination of up to d component functions f_i of F must be independent of at least one input share.

One can see that this d^{th}-order non-completeness property is equivalent to the non-completeness property defined in [31] for 1st-order DPA resistance when $d = 1$. We define a TI that satisfies Property 1 and Property 2 as d^{th}-order TI.

In 2010, two different works at EUROCRYPT and CHES [13,35] show a correspondence between the HO-DPA attack model and the so-called "probing model" where the d probing model considers an adversary that is allowed to observe the value of up to d intermediate wires of the circuit during the computation. Moreover, at EUROCRYPT 2014, this probing model is used by [12] to prove security against HO-DPA, and a relation between the probing model and the noisy leakage model is provided in [8]. We make use of the following results.

Lemma 1. *The attack order in a higher-order DPA corresponds to the number of wires that are probed in the circuit (per unmasked bit).*

This lemma implies that if a circuit is perfectly secure against d probes, then combining d power consumption points as in a dth-order DPA will reveal no information. Since TI operates on the component functions in parallel and does not separate these operations in the time domain, this is equivalent to security against DPA exploiting the dth-order statistical moment. However, it should be noted that the models considered in the mentioned papers do not take glitches into account. Thanks to the TI separation of the component functions we are able to use their models and results, and prove stronger security in the presence of glitches.

Theorem 1. *If the input masking X of the shared function F is a uniform masking and F is a dth-order TI then the dth statistical moment of the power consumption of a circuit implementing F is independent of the unmasked input value x even if the inputs are delayed or glitches occur in the circuit.*

Proof. By Lemma 1, it is sufficient to prove that an adversary who can probe d wires does not get any information about x. By construction, if F satisfies Property 1 and Property 2, an adversary who probes d or less wires will get information from all but at least one input share, which is independent of the input.   □

2.2   On the Number of Shares

The storage of the state of a symmetric key algorithm and hence the storage of the sharing of the state is typically the most expensive part in terms of area in a hardware implementation. In all masking schemes, the number of shares required increases with the order of DPA to protect against. Considering that DPA is a powerful attack especially against constrained devices, defining a higher-order masking that has a small area footprint, therefore with the minimum number of shares, becomes important.

An affine function $f(x) = y$ can be implemented with $s \geq d + 1$ component functions to thwart d^{th}-order DPA. One possible way to generate F is to define the first component function to be $f_1(x_1) = y_1 = f(x_1)$ and the rest of the component functions to be $f_i(x_i) = y_i$ where f_i is equal to f without constant terms and $2 \leq i \leq s$. To give an example $f(x) = 1 + x$ can be implemented with the following component functions:

$$f_1(x_1) = 1 + x_1 \text{ and } f_i(x_i) = x_i, \text{ where } i \in \{2, \dots, s\}.$$

However, the minimum number of shares required increases together with the nonlinearity. When the whole cryptographic algorithm is considered, one way to construct a TI is to adapt the number of shares to be minimum for each component of the algorithm, and to decrease or increase the number of shares as required. This approach is partially applied in [2]. Even though this method may lead to a relatively small circuit, it raises the problem to generate fresh randomness to be able to increase the number of shares. Another method is to keep the number of shares constant as much as possible as in [29,32] to avoid using fresh randomness. We adopt the second idea and try to keep the number of input shares that are used by a sharing of a nonlinear operation as small as possible. It is also possible to have different numbers of input and output shares to a nonlinear operation. This idea was already mentioned for 1st-order TI in [3]. Unlike that particular case where the number of input shares s_{in} is greater than the number of output shares s_{out}, we require $s_{out} \geq s_{in}$ to avoid using fresh randomness for increasing the number of shares back to s_{in}. A way to decrease the number of shares without using extra randomness will be discussed in the following subsection.

Theorem 2. *There always exist a dth-order TI of a function of degree t that requires $s_{in} \geq t \times d + 1$ input and $s_{out} \geq \binom{s_{in}}{t}$ output shares.*

Proof. Consider, without loss of generality, the product $x^1 x^2 x^3 \ldots x^t$ of first t variables where $\mathcal{F}^n \ni x = (x^1, x^2, \ldots x^n)$ and $x^j \in \mathcal{F}$. We represent the sharing of each variable x^j as x_i^j where $i \in \{1, \ldots, s_{in}\}$. Then,

$$x^1 x^2 x^3 \ldots x^t = (x_1^1 + x_2^1 + \cdots + x_{s_{in}}^1) \ldots (x_1^t + x_2^t + \cdots + x_{s_{in}}^t)$$
$$= (x_1^1 x_1^2 \ldots x_1^t) + (x_1^1 x_1^2 \ldots x_2^t) + \ldots + (x_{s_{in}}^1 x_{s_{in}}^2 \ldots x_{s_{in}}^t).$$

To satisfy the correctness each term in the above sum should exist in (or belong to) at least one component function. This can be done in the following way. Let each component function use only t different shares such that any t combination of s_{in} shares is used by only one component function. Hence any combination of up to d component functions carries information from at most $t \times d$ shares. To achieve the non-completeness property, $s_{in} > t \times d$ which implies the equation $s_{in} \geq t \times d + 1$ for the number of input shares. With the given sharing, there exist $\binom{s_{in}}{t}$ different ways of choosing t combinations of s_{in} shares and placing them in component functions. Hence, this sharing needs $s_{out} \geq \binom{s_{in}}{t}$ component functions. The proof can be extended to all degree t terms. □

Theorem 2 shows that the number of input shares of a function depends linearly on the order of security for a TI. Moreover, the required number of input and output shares given in Theorem 2 corresponds to the number of shares for $d = 1$ in [31].

We point out that a TI using the number of shares defined in the previous theorem is not the only possible construction. Moreover, the theorem does not imply that the number of output shares or the total number of input and output shares $(s_{in} + s_{out})$ are minimized. As an example, consider $y = f(a, b, b) = 1 + a + bc$ where $y, a, b, c \in \mathcal{F}$. For a 2nd-order TI of f, by Theorem 2, one requires $s_{in} = 5$ input shares which implies $s_{out} = 10$ output shares. One of the many alternatives for constructing the component functions for that scenario is

$$
\begin{aligned}
y_1 &= 1 + a_2 + b_2 c_2 + b_1 c_2 + b_2 c_1 & y_2 &= a_3 + b_3 c_3 + b_1 c_3 + b_3 c_1 \\
y_3 &= a_4 + b_4 c_4 + b_1 c_4 + b_4 c_1 & y_4 &= a_1 + b_1 c_1 + b_1 c_5 + b_5 c_1 \\
y_5 &= b_2 c_3 + b_3 c_2 & y_6 &= b_2 c_4 + b_4 c_2 \qquad (1) \\
y_7 &= a_5 + b_5 c_5 + b_2 c_5 + b_5 c_2 & y_8 &= b_3 c_4 + b_4 c_3 \\
y_9 &= b_3 c_5 + b_5 c_3 & y_{10} &= b_4 c_5 + b_5 c_4 \, .
\end{aligned}
$$

If we do not fix $s_{in} = 5$, we can also construct a 2nd-order TI with $s_{in} = 6$ input and $s_{out} = 7$ output shares as described in Appendix A.2. It is still an open question to find a lower bound for $s_{in} + s_{out}$.

The component functions provided in Equation (1) for a 2nd-order TI of a degree two function are constructed in a systematic way following the proof of Theorem 2. Namely, they are constructed with $s_{in} = 2 \times 2 + 1 = 5$ input shares and each component function uses one of the $\binom{5}{2} = 10$ possible combinations of $t = 2$ shares exactly. When this construction is reduced to achieve 1st-order DPA security, one gets the equation given in [29] which is repeated in Appendix A.1 for completeness.

Component functions for functions of higher degrees and/or other security levels can be derived with the same construction. We provide an example of a 3rd-order TI of f in Appendix A.3.

2.3   Decreasing the Number of Shares

With the construction described in the previous section, we see that the number of output shares becomes greater than the number of input shares when $d > 1$. To avoid further increase in shares and hence in area, we need to decrease the number of shares. This decrease can be done by combining different shares with an affine function as described in the following theorem.

Theorem 3. *Given $s_{in} \geq d + r$ input shares where $r \geq 1$ that are not necessarily uniform masking but secure against $(d + r - 1)^{st}$-order DPA, any sharing G that combines any r of the input shares linearly in one component function and keeps the rest of the input shares unchanged, is secure against d^{th}-order DPA.*

Proof. We represent the variable a with $s_{in} \geq d + r$ shares for a given d, that are not necessarily a uniform masking. Assume that this initial masking of a is secure against $(d + r - 1)^{st}$-order DPA. That implies that combining any $d + r - 1$ shares does not reveal the unmasked value a. Consider $s_{in} - 1$ component functions: the first component function combines the first two input shares linearly, without loss of generality, the other component functions each take one share as input and output it unchanged, i.e. $g_1 = a_1 + a_2$ and $g_{i-1} = a_i$ for $3 \leq i \leq s_{in}$. This construction satisfies both Property 1 and Property 2 for $(d + r - 2)^{nd}$-order security and one needs $s_{in} - 1 \geq d + r - 1$ shares to reveal the unmasked variable. Moreover, the component function g_1 only uses a balanced gate. Namely, a 2×1 XOR gate whose output changes with probability 1 for any input bit change, independent of the input value. Hence, even though the input is not uniform, this sharing of g will not leak information. A mere $r - 1$ repetition of this procedure gives a sharing with $d + 1$ shares that satisfies Property 1 and Property 2 and that is hence d^{th}-order DPA secure. Moreover, since there are only balanced gates involved, one can combine this repetitive construction in one step.   □

Remark 1. To satisfy Property 2, the nonlinear operation generating the sharing for a mentioned in the proof of Theorem 3 and the operation to decrease the number of shares should be separated by registers.

Given Theorem 3, one can decrease the number of shares from s_{out} to s_{in} as follows. Let the nonlinear function we want to share be $f(x) = y$ with d^{th}-order TI sharing $F(X) = Y$ such that X is an s_{in}-sharing and Y is an s_{out}-sharing. Consider another sharing $G(Y) = Z$ of a function g as defined in Theorem 3 where Z is again an s_{in}-sharing, and G is a dth-order TI. It is not necessarily required that the input sharing of G is uniform. As an example, for Equation (1) which represents a 2nd-order TI of a quadratic function, one possible way to decrease the shares such that X and Z are represented with the same number of shares is given below.

$$z_i = y_i, \text{ where } i < 5 \text{ and } z_5 = y_5 + y_6 + y_7 + y_8 + y_9 + y_{10}. \tag{2}$$

With this TI, it is important to make sure that Remark 1 is applied by using registers after the nonlinear operation F.

2.4   On Uniformity

We have proved that a function f can be implemented in a way that is secure against d^{th}-order DPA if Property 1 and Property 2 are satisfied and the masking of the input is uniform (Theorem 1). Hence, we need to make sure that the input to a shared function K of a nonlinear function k which follows $H = G \circ F$ is also a uniform masking unless it is equal to the exceptional linear case defined in Theorem 3. This is equivalent to saying that H should be a uniform sharing of the function h as defined by the following property.

Property 3 (Uniform sharing of functions). The sharing H of h is uniform if and only if $\forall z \in \mathcal{F}^m, \forall (z_1, z_2, \ldots, z_{s_{out_z}}) \in Sh(z), \forall x \in \mathcal{F}^n$ with $h(x) = z$ and $s_{out_z} \geq d + 1$:

$$|\{(x_1, x_2, \ldots, x_{s_{in}}) \in Sh(x) | H(x_1, x_2, \ldots, x_{s_{in}}) = (z_1, z_2, \ldots, z_{s_{out_z}})\}| = \frac{\mathcal{F}^{n(s_{in}-1)}}{\mathcal{F}^{m(s_{out_z}-1)}}.$$

We call a dth-order TI that is a uniform sharing, a uniform d^{th}-order TI. Unfortunately, we do not know a straight forward way to generate the component functions with s_{min} input and s_{out} output shares provided in Theorem 2 so that this property holds (unlike the other two properties) for any Boolean function. Hence, a sharing should be explicitly checked to satisfy Property 3. In this paper, we recall a uniform sharing of an AND and an XOR gate that is secure against 1st-order DPA in Equation (6) which is equal to the formula derived in [31]. Moreover, we provide uniform sharings that are secure against 2nd and 3rd-order DPA by the sharings of $H = G \circ F$ generated from Equation (1) and Equation (8) together with Equation (2). Note that Equations (1) and (8) alone are not uniform. We found these sharings with a guided computer search. In the following section, we will also provide a way to construct uniform 2nd-order TI of 4×4 S-boxes in the alternating group with 5 input and 10 output shares.

2.5   Constructing 2nd-Order TI of Some 4×4 S-Boxes

The majority of the S-boxes used in lightweight implementations are 4×4 S-boxes, therefore it is important to provide hardware implementations of these S-boxes secure against HO-DPA.

In [4], it is shown that all 4×4 S-boxes that are in the alternating group (S-boxes that can be represented as an even number of transpositions, e.g. PRESENT [5], KLEIN [15] and NOEKEON S-boxes and half of the Optimal S-boxes [21]) can be decomposed into quadratic S-boxes. Moreover, these S-boxes (represented as $s(x)$ or one of their affine equivalents $s'(x) = a(s(b(x)))$ s.t. a and b are affine permutations) have a uniform 1st-order TI with 5 input and output shares with *direct sharing*. To be more precise, if the sharing in Equation (3) (given for $f(a, b, c) = 1 + a + bc$) is applied to each term of the vectorial Boolean functions, the resulting TI is 1st-order DPA-resistant and uniform.

$$
\begin{aligned}
y_1 &= 1 + a_2 + b_2c_2 + b_2c_3 + b_3c_2 + b_2c_4 + b_4c_2 \\
y_2 &= a_3 + b_3c_3 + b_3c_4 + b_4c_3 + b_3c_5 + b_5c_3 \\
y_3 &= a_4 + b_4c_4 + b_4c_5 + b_5c_4 + b_4c_1 + b_1c_4 \\
y_4 &= a_5 + b_5c_5 + b_5c_1 + b_1c_5 + b_5c_2 + b_2c_5 \\
y_5 &= a_1 + b_1c_1 + b_1c_2 + b_2c_1 + b_1c_3 + b_3c_1 .
\end{aligned}
\tag{3}
$$

Generating the component functions as in Equation (4) for any of the mentioned S-boxes would lead to 2nd-order TI with $s_{in} = 5$ and $s_{out} = 10$.

$$
\begin{aligned}
y_1 &= 1 + a_2 + b_2c_2 + b_2c_3 + b_3c_2 & y_2 &= b_2c_4 + b_4c_2 \\
y_3 &= a_3 + b_3c_3 + b_3c_4 + b_4c_3 & y_4 &= b_3c_5 + b_5c_3 \\
y_5 &= a_4 + b_4c_4 + b_4c_5 + b_5c_4 & y_6 &= b_4c_1 + b_1c_4 \\
y_7 &= a_5 + b_5c_5 + b_5c_1 + b_1c_5 & y_8 &= b_5c_2 + b_2c_5 \\
y_9 &= a_1 + b_1c_1 + b_1c_2 + b_2c_1 & y_{10} &= b_1c_3 + b_3c_1 .
\end{aligned}
\tag{4}
$$

If the sharing G of $g(y) = z$ described in Section 2.3 is generated as $g_i = y_{2i-1} + y_{2i}$ for $i \leq 5$, the overall sharing $H(X) = G(F(X))$ of the S-box (or one of its affine equivalent) is uniform since the sharing H is equivalent to the sharing given in Equation (3). Hence, we can construct uniform 2nd-order TI of all 4×4 S-boxes in the alternating group.

3   Implementation

We recall the block cipher KATAN and propose HO-TIs of it. We provide the area requirements of these implementations in the Faraday Standard Cell Library FSA0A_C_Generic_Core which is based on UMC $0.18\mu m$ GenericII Logic Process with 1.8V voltage. We verify the functionality of the implementations with ModelSim and synthesize using Synopsys Design Vision D-201-.03-SP4 without any optimization.

3.1   KATAN

KATAN [11] is a family of block ciphers that is designed to be efficient in hardware. The family has three variants with 32, 48 or 64-bit state size. All these variants use an 80-bit key, hence have the same security level. A plaintext block, of the same size as the state, is loaded into the state to start an encryption. After 254 rounds, the content of the state is taken as the ciphertext. The round operation is very similar for all variants and has only a few AND and XOR gates.

Our main consideration is to show how to instantiate a higher-order TI of a simple algorithm and to analyze its side channel leakage. For this reason, we implement the smallest variant of KATAN with 32-bit state size and focus on encryption. A description of one round of KATAN-32 is provided in Appendix B.

3.2   TI of KATAN

We describe a general TI that has s_{in} input shares, the same number of shares in the state and s_{out} output shares for nonlinear operations. An example of a 2nd-order TI of one round KATAN-32 where $s_{in} = 5$ and $s_{out} = 10$ is depicted in Figure 1 (z coordinate refers to s_{in} different shares of the state). In all these versions, we use the same unshared key schedule for simplicity.

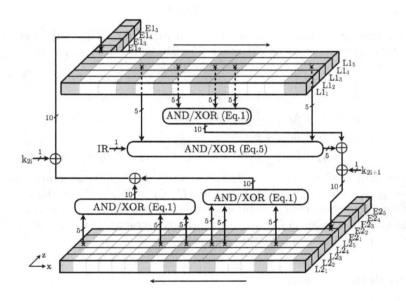

Fig. 1. Description of 2nd-order TI of one round of KATAN-32

We assume that the plaintext has a uniform masking with s_{in} shares that is provided as input. Each share will be split into two chunks of 13 and 19 bits that will be written to the registers $L1_j$ and $L2_j$ respectively where $j \leq s_{in}$. Since we already know how to implement an AND and an XOR gate with a uniform TI, we split the operations of the round update accordingly. For all the AND/XOR blocks except the one that receives IR, we use the TI in Equation (1) (resp. Eqn. (6) and Eqn.(8) for 1st and 3rd-order TI) which takes s_{in} input shares. For the AND/XOR block that receives IR we use the sharing

$$y_i = a_i + IR \times b_i \text{ where } i \leq s_{in} \tag{5}$$

because we do not share the round counter (and hence IR).

The XOR of two AND/XOR blocks is applied over s_{out} shares or over the first s_{in} shares if the output of the AND/XOR block that receives IR is involved. Similarly, the key is introduced only in the first of the s_{out} shares. This s_{out}-sharing is written to the first bit of the $L1$ and $L2$ registers respectively which have s_{out} shares only for the first bits. One can think of it as having an extension of $s_{out} - s_{in}$ shares for those bits in addition to the s_{in} shares of the state. In the next clock cycle, the s_{out} shares in the first bits of the $L1$ and $L2$ registers are reduced to s_{in} shares as described in Section 2.3 and written as the second bits. This implementation does not increase the number of clock cycles compared to the unprotected KATAN-32 implementation. In Table 1, we show the area requirements of these implementations in NAND gate equivalents. The gate counts for the round function include the decrease of the number of shares by means of Equation (2). The key register is included in the gate count of the key schedule together with the LFSR update.

Table 1. Synthesis results for plain and TI of KATAN-32

	State Array	Round Function	Key Schedule	Control	Other	Total
Plain	170	54	444	64	270	1002
1st-order TI	510	135	444	64	567	1720
2nd-order TI	900	341	444	64	807	2556
3rd-order TI	1330	760	444	64	941	3539

4 Analysis

We implement our 2nd-order TI of Katan-32 on a SASEBO-G board [1] using Xilinx ISE version 10.1 to evaluate its leakage characteristics in practice. The board features two Xilinx Virtex-II Pro FPGA devices: we implement the 2nd-order TI of Katan-32 in the crypto FPGA (xc2vp7) while the control FPGA (xc2vp30) handles I/O with the measurement PC and other equipment including the random number generation. We use the "keep hierarchy" constraint when we generate the bitstream for the crypto FPGA to prevent the tools from optimizing over module boundaries. This is to prevent the tools from merging component functions and to reduce the chance for crosstalk. The key is hard-coded in the Katan-32 implementation. The PRNG on the control FPGA is implemented as AES-128 in CTR mode. To start an encryption, we share the plaintext in 5 shares using random numbers from the PRNG and send the shares to the Katan-32 implementation. When the PRNG is turned off, it outputs zeros.

We measure the power consumption of the crypto FPGA during the first 12 rounds of Katan-32 encryption as the voltage drop over a 1Ω resistor in the FPGA core GND line. The output of the passive probe is sampled with a Tektronix DPO 7254C digital oscilloscope at 1GS/s sampling rate and 1mV/div amplitude resolution. We provide the FPGA with a stable 3 MHz clock signal and use synchronized clocks to obtain high-quality measurements.

The main goal of our evaluation is not to demonstrate that the implementation resists state-of-the-art attacks that exploit the 1st or 2nd statistical moment of the leakage distributions, but beyond that to demonstrate that there is no evidence of leakage in these moments of the leakage distributions, exploitable by state-of-the-art attacks or not. Obviously achieving this goal is much more demanding than resistance to known attacks, but it directly corresponds to our claims regarding provable security. We narrow the evaluation to univariate attacks because our implementation processes all component functions in parallel.

We use *leakage detection* to evaluate our implementation. Contrary to the classical approach of testing whether a given attack is successful, this approach decouples the detection of leakage from its exploitation. For our purpose we use the *non-specific* t-test based fixed versus random leakage detection methodology of [7,16], see Appendix C for a brief introduction.

For all tests we obtain two sets of measurements. For the first set, we fix the plaintext to some chosen value. We denote this set $\mathcal{S}_0$. For the second set, the plaintexts are uniformly distributed and random. We denote this set $\mathcal{S}_{random}$. We obtain the measurements for both sets interleaved and in a random order, i.e. before each measurement we flip a coin, to avoid any deterministic or time-dependent external and internal influences on the test result.

We compute Welch's (two-tailed) t-test

$$t = \frac{\mu(\mathcal{S}_0) - \mu(\mathcal{S}_1)}{\sqrt{\frac{\sigma^2(\mathcal{S}_0)}{|\mathcal{S}_0|} + \frac{\sigma^2(\mathcal{S}_1)}{|\mathcal{S}_1|}}}$$

(where $\mu()$ is the sample mean, $\sigma^2()$ is the sample variance and $|\cdot|$ denotes the sample size) to determine if the samples in both sets were drawn from populations with the same mean (or from the same population). The null hypothesis is that the samples in both sets were drawn from populations with the same mean. In our context, this means that the TI is effective. The alternative hypothesis is that the samples in both sets were drawn from populations with different means. In our context, this means that the TI is not effective.

At each point in time, the test statistic t together with the degrees of freedom ν, computed with the Welch-Satterthwaite equation

$$\nu = \frac{(\sigma^2(\mathcal{S}_0)/|\mathcal{S}_0| + \sigma^2(\mathcal{S}_1)/|\mathcal{S}_1|)^2}{(\sigma^2(\mathcal{S}_0)/|\mathcal{S}_0|)^2/(|\mathcal{S}_0| - 1) + (\sigma^2(\mathcal{S}_1)/|\mathcal{S}_1|)^2/(|\mathcal{S}_1| - 1)},$$

allow to compute a p value to determine if there is sufficient evidence to reject the null hypothesis at a particular significance level $(1-\alpha)$. The p value expresses the probability of observing the measured (or a greater) difference by chance if the null hypothesis was true. In other words, small p values give evidence to reject the null hypothesis.

While this evaluation methodology relieves us from choosing certain parameters such as targeted intermediate value, power model and distinguisher, it does not resolve all such issues. As in any evaluation, the tests are limited to the number of measurements at hand and one has to choose a threshold to decide if an observed difference is statistically significant or not. Nevertheless, as we demonstrate below this type of evaluation is very data-efficient, i.e. a small number of measurements is required to provide evidence of leakage, and a decision threshold can be motivated with some basic experiments.

To calibrate our threshold value we apply the test methodology to two groups of 10 000 measurements each for which we know that the null hypothesis is true. For the first group of measurements we switch off the PRNG and use the same fixed plaintext for both sets, i.e. all measurements in both sets are samples from the same population and the only cause of variance is noise. We compute the t statistic, record its greatest absolute value and repeat the experiment 100 times on a random split of the measurements in this group. The highest absolute t value we observed was 4.7944. For the second group we switch on the PRNG and use random plaintexts for both sets, i.e. the measurements in both sets are samples from distributions with the same mean and high variance. We repeat the analysis and the highest absolute t value we observed was 4.8608. Based on these results and the recommendation in [7] we select the significance threshold ± 4.5. For large sample sizes, observing a single t value greater/smaller than ± 4.5 roughly corresponds to a 99.999% probability of the null hypothesis being false.

To confirm that our setup works correctly and to get some reference values we first evaluate the implementation with the PRNG switched off. Figure 2 shows the t values of fixed versus random tests with two different fixed plaintexts (left and right) and for the $1^{st}, 2^{nd}$ and 3rd statistical moment of the distributions (for the higher-order moments we pre-process the traces to expose the desired standardized moment before we apply the t-test, e.g. for the 2nd moment we center and then square the traces). Horizontal lines mark the ± 4.5 thresholds.

The plots clearly show that there is sufficient evidence of leakage in all cases, as there are multiple and systematic crossings of the thresholds. Comparing the plots on the left hand side with the plots on the right hand side, we see that the "shape" of the t curve depends on the fixed plaintext value. This is no longer true when we switch on the PRNG, because all shares of the input are random. We used 1 000 measurements (500 for fixed and 500 for random plaintext) to generate these plots, but less than 100 measurements are required to see evidence of leakage in the 1st statistical moment.

Now we switch on the PRNG and repeat the evaluation with a randomly chosen fixed plaintext using 300 million measurements (150M for fixed, 150M for random, done in a temperature controlled environment). Figure 3 (top left and

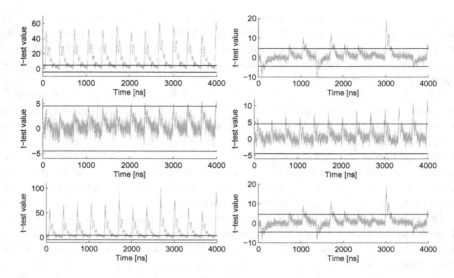

Fig. 2. Fixed versus random t-test evaluation results with PRNG switched off; left: for fixed plaintext 0x00000000, right: for a randomly chosen fixed plaintext; from top to bottom: $1^{st}, 2^{nd}$ and 3rd-order statistical moment; 1 000 measurements

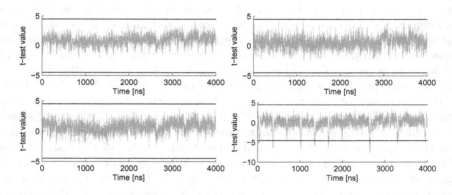

Fig. 3. Fixed versus random t-test evaluation results with PRNG switched on for a randomly chosen fixed plaintext; from top left, top right, to bottom right: $1^{st}, 2^{nd}, 3^{rd}$ and 5th statistical moment; 300 million measurements

right) shows plots of the t values for the 1st and 2nd moment. As expected there is not sufficient evidence of leakage. But as mentioned earlier, one may always wonder if the number of measurements at hand is sufficient. For completeness, we also provide evaluation results of the 3rd and 5th moment. The 3rd moment is the smallest moment for which our implementation does not provide provable security in the combinational logic (Property 2) and the 5th moment is the smallest moment for which our implementation does not provide provable security in the memory elements (the state is shared in at least 5 shares). Therefore we may be able to detect leakage in these moments. Figure 3 (bottom left and right) shows plots of the t values.

While there is not sufficient evidence of leakage also in the 3rd moment, we can see multiple and systematic crossings of the threshold in the 5th moment. This result suggests that we use enough measurements, and that we should be able to detect leakage in the lower-order moments, if there was any. Together, the results support our claim regarding provable 2nd-order DPA resistance.

One may wonder why we do not detect leakage in the 3rd moment. Several explanations are possible but their careful investigation is beyond the scope of this paper.

5   Conclusion

Research on HO-DPA attacks shows that these attacks are realistic threats, and advances in the field can only increase the attack potential. It is therefore desirable to have masking schemes that can be implemented securely at any order. In hardware implementations, glitches make this a challenging task. TI is a masking technique that provides provable security even in the presence of glitches, but the method is limited to 1st-order DPA resistance. We address this gap and extend the technique to higher orders. We define generic constructions, prove their security and provide exemplary 1st, 2nd and 3rd-order DPA-resistant implementations of the block cipher KATAN-32. Our analysis of 300 million power traces from a 2nd-order DPA-resistant implementation in an FPGA with a leakage detection test does not show significant evidence of leakage and supports the security proofs. We also show that this method can be straightaway applied to generate 2nd-order TI of 4×4 S-boxes in the alternating group.

Acknowledgements. This work has been supported in part by the Research Council of KU Leuven (OT/13/071), by the FWO (G.0550.12), by the Hercules foundation (AKUL/11/19) and by GOA (tense). B. Bilgin was partially supported by the FWO project G0B4213N and Benedikt Gierlichs is a Postdoctoral Fellow of the Research Foundation - Flanders (FWO).

References

1. AIST. Side-channel Attack Standard Evaluation BOard,
   http://staff.aist.go.jp/akashi.satoh/SASEBO/en/
2. Bilgin, B., Gierlichs, B., Nikova, S., Nikov, V., Rijmen, V.: A more efficient AES threshold implementation. In: Pointcheval, D., Vergnaud, D. (eds.) AFRICACRYPT. LNCS, vol. 8469, pp. 267–284. Springer, Heidelberg (2014)
3. Bilgin, B., Nikova, S., Nikov, V., Rijmen, V., Stütz, G.: Threshold implementations of all 3 ×3 and 4 ×4 S-boxes. Cryptology ePrint Archive,
   http://eprint.iacr.org/
4. Bilgin, B., Nikova, S., Nikov, V., Rijmen, V., Stütz, G.: Threshold implementations of all 3 ×3 and 4 ×4 S-boxes. In: Prouff, E., Schaumont, P. (eds.) CHES 2012. LNCS, vol. 7428, pp. 76–91. Springer, Heidelberg (2012)

5. Bogdanov, A.A., Knudsen, L.R., Leander, G., Paar, C., Poschmann, A., Robshaw, M., Seurin, Y., Vikkelsoe, C.: PRESENT: An ultra-lightweight block cipher. In: Paillier, P., Verbauwhede, I. (eds.) CHES 2007. LNCS, vol. 4727, pp. 450–466. Springer, Heidelberg (2007)
6. Chari, S., Jutla, C.S., Rao, J.R., Rohatgi, P.: Towards sound approaches to counteract power-analysis attacks. In: Wiener, M. (ed.) CRYPTO 1999. LNCS, vol. 1666, pp. 398–412. Springer, Heidelberg (1999)
7. Cooper, J., De Mulder, E., Goodwill, G., Jaffe, J., Kenworthy, G., Rohatgi, P.: Test vector leakage assessment (TVLA) methodology in practice. In: International Cryptographic Module Conference (2013), http://icmc-2013.org/wp/wp-content/uploads/2013/09/goodwillkenworthtestvector.pdf
8. Coron, J.-S.: Higher order masking of look-up tables. In: Nguyen, P.Q., Oswald, E. (eds.) EUROCRYPT 2014. LNCS, vol. 8441, pp. 441–458. Springer, Heidelberg (2014)
9. Coron, J.-S., Prouff, E., Rivain, M., Roche, T.: Higher-order side channel security and mask refreshing. In: FSE 2013 (2013) (to appear)
10. Coron, J.-S., Prouff, E., Roche, T.: On the use of shamir's secret sharing against side-channel analysis. In: Mangard, S. (ed.) CARDIS 2012. LNCS, vol. 7771, pp. 77–90. Springer, Heidelberg (2013)
11. De Cannière, C., Dunkelman, O., Knežević, M.: KATAN and KTANTAN — A family of small and efficient hardware-oriented block ciphers. In: Clavier, C., Gaj, K. (eds.) CHES 2009. LNCS, vol. 5747, pp. 272–288. Springer, Heidelberg (2009)
12. Duc, A., Dziembowski, S., Faust, S.: Unifying leakage models: From probing attacks to noisy leakage. In: Nguyen, P.Q., Oswald, E. (eds.) EUROCRYPT 2014. LNCS, vol. 8441, pp. 423–440. Springer, Heidelberg (2014)
13. Faust, S., Rabin, T., Reyzin, L., Tromer, E., Vaikuntanathan, V.: Protecting circuits from leakage: the computationally-bounded and noisy cases. In: Gilbert, H. (ed.) EUROCRYPT 2010. LNCS, vol. 6110, pp. 135–156. Springer, Heidelberg (2010)
14. Genelle, L., Prouff, E., Quisquater, M.: Thwarting higher-order side channel analysis with additive and multiplicative maskings. In: Preneel, B., Takagi, T. (eds.) CHES 2011. LNCS, vol. 6917, pp. 240–255. Springer, Heidelberg (2011)
15. Gong, Z., Nikova, S., Law, Y.W.: KLEIN: A new family of lightweight block ciphers. In: Juels, A., Paar, C. (eds.) RFIDSec 2011. LNCS, vol. 7055, pp. 1–18. Springer, Heidelberg (2012)
16. Goodwill, G., Jun, B., Jaffe, J., Rohatgi, P.: A testing methodology for side channel resistance validation. NIST non-invasive attack testing workshop (2011), http://csrc.nist.gov/news_events/non-invasive-attack-testing-workshop/08_Goodwill.pdf
17. Goubin, L., Martinelli, A.: Protecting AES with Shamir's Secret Sharing Scheme. In: Preneel, B., Takagi, T. (eds.) CHES 2011. LNCS, vol. 6917, pp. 79–94. Springer, Heidelberg (2011)
18. Kim, H., Hong, S., Lim, J.: A fast and provably secure higher-order masking of AES S-box. In: Preneel, B., Takagi, T. (eds.) CHES 2011. LNCS, vol. 6917, pp. 95–107. Springer, Heidelberg (2011)
19. Kocher, P.C., Jaffe, J., Jun, B.: Differential power analysis. In: Wiener, M. (ed.) CRYPTO 1999. LNCS, vol. 1666, pp. 388–397. Springer, Heidelberg (1999)

20. Mather, L., Oswald, E., Bandenburg, J., Wójcik, M.: Does my device leak information? An *a priori* statistical power analysis of leakage detection tests. In: Sako, K., Sarkar, P. (eds.) ASIACRYPT 2013, Part I. LNCS, vol. 8269, pp. 486–505. Springer, Heidelberg (2013)

21. Leander, G., Poschmann, A.: On the classification of 4 bit S-boxes. In: Carlet, C., Sunar, B. (eds.) WAIFI 2007. LNCS, vol. 4547, pp. 159–176. Springer, Heidelberg (2007)

22. Mangard, S., Popp, T., Gammel, B.M.: Side-channel leakage of masked CMOS gates. In: Menezes, A. (ed.) CT-RSA 2005. LNCS, vol. 3376, pp. 351–365. Springer, Heidelberg (2005)

23. Mangard, S., Pramstaller, N., Oswald, E.: Successfully attacking masked AES hardware implementations. In: Rao, J.R., Sunar, B. (eds.) CHES 2005. LNCS, vol. 3659, pp. 157–171. Springer, Heidelberg (2005)

24. Messerges, T.S.: Using second-order power analysis to attack DPA resistant software. In: Paar, C., Koç, Ç.K. (eds.) CHES 2000. LNCS, vol. 1965, pp. 238–251. Springer, Heidelberg (2000)

25. Moradi, A.: Statistical tools flavor side-channel collision attacks. In: Pointcheval, D., Johansson, T. (eds.) EUROCRYPT 2012. LNCS, vol. 7237, pp. 428–445. Springer, Heidelberg (2012)

26. Moradi, A., Mischke, O.: On the simplicity of converting leakages from multivariate to univariate. In: Bertoni, G., Coron, J.-S. (eds.) CHES 2013. LNCS, vol. 8086, pp. 1–20. Springer, Heidelberg (2013)

27. Moradi, A., Mischke, O., Eisenbarth, T.: Correlation-enhanced power analysis collision attack. In: Mangard, S., Standaert, F.-X. (eds.) CHES 2010. LNCS, vol. 6225, pp. 125–139. Springer, Heidelberg (2010)

28. Moradi, A., Poschmann, A., Ling, S., Paar, C., Wang, H.: Pushing the limits: A very compact and a threshold implementation of AES. In: Paterson, K.G. (ed.) EUROCRYPT 2011. LNCS, vol. 6632, pp. 69–88. Springer, Heidelberg (2011)

29. Nikova, S., Rechberger, C., Rijmen, V.: Threshold implementations against side-channel attacks and glitches. In: Ning, P., Qing, S., Li, N. (eds.) ICICS 2006. LNCS, vol. 4307, pp. 529–545. Springer, Heidelberg (2006)

30. Nikova, S., Rijmen, V., Schläffer, M.: Secure hardware implementation of non-linear functions in the presence of glitches. In: Lee, P.J., Cheon, J.H. (eds.) ICISC 2008. LNCS, vol. 5461, pp. 218–234. Springer, Heidelberg (2009)

31. Nikova, S., Rijmen, V., Schläffer, M.: Secure hardware implementation of nonlinear functions in the presence of glitches. J. Cryptology 24(2), 292–321 (2011)

32. Poschmann, A., Moradi, A., Khoo, K., Lim, C.-W., Wang, H., Ling, S.: Side-channel resistant crypto for less than 2300 GE. J. Cryptology 24(2), 322–345 (2011)

33. Prouff, E., Rivain, M.: Masking against side-channel attacks: A formal security proof. In: Johansson, T., Nguyen, P.Q. (eds.) EUROCRYPT 2013. LNCS, vol. 7881, pp. 142–159. Springer, Heidelberg (2013)

34. Prouff, E., Roche, T.: Higher-order glitches free implementation of the AES using secure multi-party computation protocols. In: Preneel, B., Takagi, T. (eds.) CHES 2011. LNCS, vol. 6917, pp. 63–78. Springer, Heidelberg (2011)

35. Rivain, M., Prouff, E.: Provably secure higher-order masking of AES. In: Mangard, S., Standaert, F.-X. (eds.) CHES 2010. LNCS, vol. 6225, pp. 413–427. Springer, Heidelberg (2010)

36. Standaert, F.-X., Malkin, T., Yung, M.: A unified framework for the analysis of side-channel key recovery attacks. In: Joux, A. (ed.) EUROCRYPT 2009. LNCS, vol. 5479, pp. 443–461. Springer, Heidelberg (2009)

A   Component Functions of the Sharing F of $f(a, b, c) = 1 + a + bc$

A.1 1st-Order TI and $s_{in} = 3$

$$
\begin{aligned}
y_1 &= 1 + a_2 + b_2c_2 + b_1c_2 + b_2c_1 \\
y_2 &= a_3 + b_3c_3 + b_2c_3 + b_3c_2 \\
y_3 &= a_1 + b_1c_1 + b_1c_3 + b_3c_1
\end{aligned}
\tag{6}
$$

A.2 2nd-Order TI and $s_{in} = 6$

$$
\begin{aligned}
y_1 &= 1 + a_2 + b_2c_2 + b_1c_2 + b_2c_1 + b_1c_3 + b_3c_1 + b_2c_3 + b_3c_2 \\
y_2 &= a_3 + b_3c_3 + b_3c_4 + b_4c_3 + b_3c_5 + b_5c_3 \\
y_3 &= a_4 + b_4c_4 + b_2c_4 + b_4c_2 + b_2c_6 + b_6c_2 \\
y_4 &= a_5 + b_5c_5 + b_1c_4 + b_4c_1 + b_1c_5 + b_5c_1 \\
y_5 &= b_2c_5 + b_5c_2 + b_4c_5 + b_5c_4 \\
y_6 &= a_6 + b_6c_6 + b_3c_6 + b_6c_3 + b_4c_6 + b_6c_4 \\
y_7 &= a_1 + b_1c_1 + b_1c_6 + b_6c_1 + b_5c_6 + b_6c_5
\end{aligned}
\tag{7}
$$

A.3 3rd-Order TI and $s_{in} = 7$

$$
\begin{aligned}
y_1 &= 1 + a_2 + b_2c_2 + b_1c_2 + b_2c_1 & y_2 &= a_3 + b_3c_3 + b_1c_3 + b_3c_1 \\
y_3 &= a_4 + b_4c_4 + b_1c_4 + b_4c_1 & y_4 &= a_5 + b_5c_5 + b_1c_5 + b_5c_1 \\
y_5 &= a_6 + b_6c_6 + b_1c_6 + b_6c_1 & y_6 &= a_1 + b_1c_1 + b_1c_7 + b_7c_1 \\
y_7 &= b_2c_3 + b_3c_2 & y_8 &= b_2c_4 + b_4c_2 \\
y_9 &= b_2c_5 + b_5c_2 & y_{10} &= b_2c_6 + b_6c_2 \\
y_{11} &= a_7 + b_7c_7 + b_2c_7 + b_7c_2 & y_{12} &= b_3c_4 + b_4c_3 \\
y_{13} &= b_3c_5 + b_5c_3 & y_{14} &= b_3c_6 + b_6c_3 \\
y_{15} &= b_3c_7 + b_7c_3 & y_{16} &= b_4c_5 + b_5c_4 \\
y_{17} &= b_4c_6 + b_6c_4 & y_{18} &= b_4c_7 + b_7c_4 \\
y_{19} &= b_5c_6 + b_6c_5 & y_{20} &= b_5c_7 + b_7c_5 . \\
y_{21} &= b_6c_7 + b_7c_6
\end{aligned}
\tag{8}
$$

B   Unmasked KATAN-32

The description of one round of KATAN-32 is provided in Fig. 4 where each block represents one bit. 32-bit plaintext is divided into two chunks of 13 and 19 bits and written to the registers $L1$ and $L2$ respectively. In every round, several bits are used to update the first bits of the registers together with a one bit shift to the right for $L1$ and to the left for $L2$. The bit depicted by IR is the last bit of a round counter that decides irregularly if the fourth bit of $L1$ is used for the round update or not. k_{2i} and k_{2i+1} are the $2i^{\text{th}}$ and $(2i+1)^{\text{st}}$ bits of the 80-bit key for rounds $i \leq 40$. For the rest of the rounds they are generated from the original key by an LFSR.

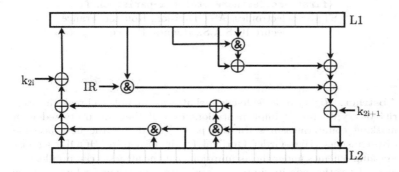

Fig. 4. Description of one round of KATAN-32

C T-test Based Fixed versus Random Leakage Detection

The t-test based fixed versus random leakage detection methodology has two main ingredients: first, chosen inputs allow to generate two sets of measurements for which intermediate values in the implementation have a certain difference. Without making an assumption about how the implementation leaks, a safe choice is to keep the intermediate values fixed for one set of measurements, while they take random values for the second set. The test is *specific*, if particular intermediate values or transitions in the implementation are targeted (e.g. S-box input, S-box output, Hamming distance in a round register, etc.). This type of testing requires knowledge of the device key and carefully chosen inputs. On the other hand, the test is *non-specific* if *all* intermediate values and transitions are targeted at the same time. This type of testing only requires to keep all inputs to the implementation fixed for one set of measurements, and to choose them randomly for the second set. Obviously, the non-specific test is extremely powerful. The second ingredient is a simple, robust and efficiently computable statistical test to determine if the two sets of measurements are significantly different. Contrary to the information-theoretic metric of Standaert et al. [36] and the mutual-information-based leakage detection tests explored in [20] the t-test based approach evaluates a specific statistical moment of the measured distributions.

Masks Will Fall Off
Higher-Order Optimal Distinguishers

Nicolas Bruneau[1,2], Sylvain Guilley[1,3], Annelie Heuser[1,*], and Olivier Rioul[1]

[1] Télécom ParisTech, Institut Mines-Télécom, CNRS LTCI
Department Comelec, Paris, France
{firstname.lastname}@telecom-paristech.fr
[2] STMicroelectronics, AST Division, Rousset, France
[3] Secure-IC S.A.S., Rennes, France

Abstract. Higher-order side-channel attacks are able to break the security of cryptographic implementations even if they are protected with masking countermeasures. In this paper, we derive the best possible distinguishers (High-Order Optimal Distinguishers or HOOD) against masking schemes under the assumption that the attacker can profile. Our exact derivation admits simple approximate expressions for high and low noise and shows to which extent the optimal distinguishers reduce to known attacks in the case where no profiling is possible. From these results, we can explain theoretically the empirical outcome of recent works on second-order distinguishers. In addition, we extend our analysis to any order and to the application to masked tables precomputation. Our results give some insight on which distinguishers have to be considered in the security analysis of cryptographic devices.

Keywords: Side-channel analysis, higher-order masking, masking tables, higher-order optimal distinguisher (HOOD), template attack.

1 Introduction

In order to secure embedded devices against side-channel attacks, masking schemes have been introduced. Recent works have shown provable protections with a security parameter d, such that each sensitive variable is secured with d random masks [4]. The computation is carried out in such a way that the knowledge of any tuple of d intermediate variables does not disclose any information on any sensitive variable. Accordingly, all distinguishers using up to d leakages will fail to recover the correct key. A successful attack would be a $(d+1)$th-order CPA, which uses combination functions to transform the measured leakage and the prediction on each share into a single value in order to compute Pearson correlation coefficients [4, 10, 13, 14, 18, 19, 23].

---

[*] Annelie Heuser is a Google European Fellow in the field of Privacy and is partially founded by this fellowship.

P. Sarkar and T. Iwata (Eds.): ASIACRYPT 2014, PART II, LNCS 8874, pp. 344–365, 2014.

Using combination functions to fit to a known tool like CPA looks more like an engineering recipe than the optimal solution. Yet, it is shown by Prouff et al. [16] that, in the case of second-order attacks, using the normalized product combination function combined with an optimal prediction function is the most efficient solution among all known combination functions. Even more, Standaert et al. [20] showed that the information loss induced from the combination functions vanishes for high noise.

In [14] Oswald and Mangard introduced several template-based attacks on masking schemes. Among them is the so-called template-based DPA attack, which extends the traditional template attack from Chari et al. [5] to first-order masking schemes. Besides, their approach classifies measurements according to the key, and not according to sensitive variables.

A slightly different scenario has been analyzed by Tunstall et al. in [22], where the authors study the security of masking tables for software implementations as defined in [1]. The authors suggested a two-stage CPA: first, for each individual trace, extract the mask during the precomputation and, second, use this knowledge about the mask to reveal the secret key using a vertical attack.

In this article we tackle the questions *"what is the best possible distinguisher in case of profiling?"* and *"how far are they from known practical distinguishers?"*.

In particular, we derive optimal higher-order distinguishers against higher-order masking schemes in case profiling is possible. Here, optimality means maximizing the success rate. Starting from second-order optimal distinguisher we derive approximations for high and low noise and recover known attacks. In particular, we show to what extent the optimal second-order distinguisher can be translated into a second-order CPA attack using combination functions. Given these results for second-order we extend our analysis to $(d + 1)$th-order distinguisher against dth-order masking schemes.

Additionally, we investigate the scenario of masking tables as in [22]. We derive the optimal attack against masking tables and again derive approximations for the scenario of high and low noise which results in new attacks and compare it to the two-stage CPA.

2   Preliminaries

2.1   Masking Countermeasure and Notations

Even though many different masking schemes have been investigated so far, which clearly differ in their strength, the principle of attacking is equivalent. A masking scheme is characterized by the number of random masks that are used per sensitive variable. In the following we consider a dth-order masked implementation where we assume that the masks are uniformly distributed over a space $\mathcal{M}$. Calligraphic letters (e.g., $\mathcal{X}$) denote sets, capital letters (e.g., X) denote random variables taking values in these sets, and the corresponding lowercase letters (e.g., x) denote their realizations. Let k^* denote the secret cryptographic key, k any possible key hypothesis from the keyspace $\mathcal{K}$, and T be the input or ciphertext of the cryptographic algorithm. The mapping $f : (\mathcal{T}, \mathcal{K}, \mathcal{M}) \rightarrow \mathbb{F}_2^n$ maps the input or ciphertext $t \in \mathcal{T}$, a key hypothesis $k \in \mathcal{K}$ and the mask $m \in \mathcal{M}$ to an internally processed variable in some space $\mathbb{F}_2^n$ that is assumed to relate to the measured leakage X, where n is the number of bits. Generally it is assumed that f is known to the attacker. The measured leakage X can then be written as

$$X = \varphi(f(T, k^*, M)) + N, \tag{1}$$

where N denotes an independent—not necessarily Gaussian—additive noise with zero mean and where φ a device-specific deterministic function. In this paper we start by assuming that φ is known to the attacker due to profiling to consider the most powerful attack. We then show to which extend and scenarios this assumption can be relaxed while still achieving the same efficiency of the attack.

Specifically, in a dth-order masking scheme the implementation is protected with d masks with corresponding leakages

$$X^{(\omega)} = \varphi^{(\omega)}(f^{(\omega)}(T^{(\omega)}, k^*, M^{(\omega)})) + N^{(\omega)}, \tag{2}$$

with $\omega \in \{0, \ldots, d\}$ and $M^{(\omega)} \in \mathcal{M}^{(\omega)}$ where the $\mathcal{M}^{(\omega)}$ does not need to be equal in general. Accordingly, a dth-order masking scheme can be broken using $(d + 1)$th-order distinguishers by targeting $d + 1$ shares. For simplification we denote $Y(T^{(\omega)}, k, M^{(\omega)}) = \varphi^{(\omega)}(f^{(\omega)}(T^{(\omega)}, k, M^{(\omega)}))$.

Example 1 (First-order software masking). For example a first-order masking scheme ($d = 1$) might leak with

$$X^{(0)} = \mathsf{HW}[M] + N^{(0)}, \tag{3}$$

$$X^{(1)} = \mathsf{HW}[\mathsf{Sbox}[T \oplus k^*] \oplus M] + N^{(1)}, \tag{4}$$

with $\mathsf{Sbox} : \mathbb{F}_2^8 \rightarrow \mathbb{F}_2^8$ being the AES Substitution box and $T^{(1)} = T$ uniformly distributed over $\mathbb{F}_2^8$ (and $T^{(0)}$ is non-existent). Thus, $\varphi^{(0)}(\cdot) = \varphi^{(1)}(\cdot) = \mathsf{HW}[\cdot]$ (the Hamming weight function), $M^{(0)} = M^{(1)} = M$, $f^{(0)}(T, k, M) = M$ and $f^{(1)}(T, k, M) = \mathsf{Sbox}[T \oplus k] \oplus M$.

Example 2 (Tables pre-computation). Again when assuming a Hamming weight leakage model, a masking scheme using Sbox recomputation [11] might leak with

$$X^{(\omega)} = \mathsf{HW}[\omega \oplus M] + N^{(\omega)}, \qquad \forall \omega \in \{0, 1, \ldots, 2^n - 1\} \cong \mathbb{F}_2^n \tag{5}$$

$$X^{(2^n)} = \mathsf{HW}[T \oplus k^* \oplus M] + N^{(2^n)}. \tag{6}$$

A detailed description will be given in Sect. 5.

Definition 1 (Perfect masking (dth-order) [2]). *Let us denote the random variables $F^{(\omega)}(t, k) = f^{(\omega)}(t, k, M^{(\omega)})$ for $\omega \in \{0, \ldots, d\}$ and a fixed pair (t, k). A masking scheme is* perfect *at dth-order if the joint distribution of maximum d of $F^{(0)}(t, k), \ldots, F^{(d)}(t, k)$ is identically distributed of any pair $(t, k) \in \mathcal{T} \times \mathcal{K}$.*

Note that dth-order security implies $1\text{st}, 2\text{nd}, \ldots, (d-1)$th-order security.

Proposition 1. *If a masking scheme is perfect, then whatever function ψ, $\sum_{m^{(\omega)} \in \mathcal{M}^{(\omega)}} \psi(f^{(\omega)}(t, k, m^{(\omega)}))$ is constant for any pair (t, k) for any $0 \leq \omega \leq d$.*

Proof. Let $\tilde{t}, \tilde{k}$ be any value in $\mathcal{T}$ and $\mathcal{K}$, respectively. As the masking scheme is perfect up to dth-order (which implies 1st-order) the distribution of $f^{(\omega)}(t, k, M^{(\omega)})$ is equivalent to $f^{(\omega)}(\tilde{t}, \tilde{k}, M^{(\omega)})$, hence $\psi(f^{(\omega)}(t, k, M^{(\omega)}))$ and $\psi(f^{(\omega)}(\tilde{t}, \tilde{k}, M^{(\omega)}))$ have the same distribution. In particular, the sum of realizations is identical.   □

In our setup we assume that the attacker is able to measure q i.i.d. measurements. All values indexed by $i \in \{1, \ldots, q\}$ are in bold face (e.g. $\mathbf{a} = (a_1, \ldots, a_q) \in \mathcal{A}^q$ for $a_i \in \mathcal{A}$). Values indexed by the intermediate variable index (ω) $(a^{(\omega)} \in \mathcal{A})$ are denoted by $a^{(\star)} = (a^{(0)}, \ldots, a^{(d)}) \in \mathcal{A}^{d+1}$. Moreover, $a_i^{(\omega)} \in \mathcal{A}$ and $\mathbf{a}^{(\star)} \in \mathcal{A}^{q \times (d+1)}$.

Note that contrary to $\mathbf{a}$, the vectors along (ω) can be linked, e.g., $\bigoplus_{\omega=0}^{d} M^{(\omega)} = 0$ in Example 1 or $\forall \omega \in \{0, \ldots, 2^n - 1\} \cup \{2^n\}$, $M^{(\omega)} = M^{(0)}$ in Example 2. Thus the set of admissible masks, denoted by $\mathcal{M}^{(\star)}$, is a subset of the Cartesian product over all $\mathcal{M}^{(\omega)}$. Additionally, regarding the noise, we have that $\forall (i, \omega) \neq (i', \omega')$, $N_i^{(\omega)}$ is independent of $N_{i'}^{(\omega')}$.

We write $\mathbb{P}(m) = \mathbb{P}(M = m)$ for discrete probability distributions, p for densities, and when the random variable X is conditioned by the event $Y = y$, we use the notation $p_k(X | Y = y)$ to recall that y depends on a (fixed) key guess k. As the model is known to the attacker, we also have: $p_k(X | Y = y) = p_k(X | T = t, M = m)$ when $y = y(t, k, m)$. Indeed, owing to Eq. (2), Y is a *sufficient statistic* for k [8]. We then use $p_k(x | t, m)$ to denote $p(X = x | Y = \varphi(f(t, k, m)))$. We denote the scalar product between $\mathbf{x}$ and $\mathbf{y}$ by $\langle \mathbf{x} | \mathbf{y} \rangle = \sum_{i=1}^{q} x_i y_i$, the Euclidean norm by $\|\mathbf{x}\|_2 = \sqrt{\langle \mathbf{x} | \mathbf{x} \rangle}$ and the componentwise product by $\mathbf{x} \cdot \mathbf{y} = (x_1 y_1, \ldots, x_q y_q)$. Given a function $g(k)$, we use the notation $\arg \max_k \ g(k)$ to denote the value of k that maximizes $g(k)$. Finally, $\mathbb{1}_E : E \to \{0, 1\}$ denotes the indicator function of the set E.

2.2 Combination Functions for Higher-Order CPA Attacks

In order to conduct a second-order CPA attack, two kinds of *combination functions*, i.e., $c_X : \mathcal{X}^{d+1} \to \mathbb{R}$ and $c_Y : \mathcal{T}^{d+1} \to \mathbb{R}$, are required. However, this seems to be more inspired from an engineering perspective –an "act from necessity"– than a sound mathematical tool to maximize the success. That the use of combination functions comes with information loss was already pointed out by Mangard and Oswald and Mangard [14] and Standaert et al. [20]. The history and selection of combination functions is indeed epic, where the literature mostly concentrated on second-order CPA ($d = 1$).

The most prominent function to combine the leakages is the *product* combining function $c_X^{\mathrm{prod}}(X^{(0)}, X^{(1)}) = X^{(0)} \cdot X^{(1)}$ introduced by Chari et al. in [4] and the *absolute difference* $c_X^{\mathrm{diff}}(X^{(0)}, X^{(1)}) = |X^{(0)} - X^{(1)}|$ by Messerges in [13]. Oswald and Mangard [14] proposed (for e.g. the setting given in Example 1) an even more exotic combination function and corresponding prediction function[1]:

---

[1] Note that these are the corrected formulas given in [16].

$$c_X^{\sin}(X^{(0)}, X^{(1)}) = \sin\left((X^{(0)} - X^{(1)})^2\right) \tag{7}$$

$$c_Y^{\sin}(T) = -89.95 \, \sin(\mathsf{HW}[Y])^3 - 7.82 \, \sin(\mathsf{HW}[Y])^2 + 67.66 \, \sin(\mathsf{HW}[Y]), \tag{8}$$

where $Y = Y(T, k, 0)$. Contrary to what was suggested in previous papers, Prouff et al. [16] showed that all these combination functions should be accompanied by $c_Y^{\mathrm{opt}}(T^{(0)}, T^{(1)}) = \mathbb{E}\{c_X^*(Y^{(0)}, Y^{(1)})|T^{(0)}, T^{(1)}\}$ to maximize the absolute value of correlation, where the expectation is taken over the mask M and c_X^* denotes the same combination function as c_X but defined as a map $\mathcal{Y}^{d+1} \to \mathbb{R}$. Moreover, the *normalized product function*, i.e., $c_X^{\mathrm{n\text{-}prod}}(X^{(0)}, X^{(1)}) = (X^{(0)} - \mathbb{E}\{X^{(0)}\}) \cdot (X^{(1)} - \mathbb{E}\{X^{(1)}\})$ is shown to be the most efficient of all known combination functions when considering a Hamming weight leakage model.

3 Optimal Distinguisher for Second-Order Attacks

3.1 Motivation

As highlighted in Subsect. 2.2 the introduction of combination functions for second-order CPA is more a necessary evil than an optimized procedure to maximize success. In [20] the authors empirically showed that a combination function always goes hand in hand with information loss. However, the authors depicted that for large noise the second-order CPA with the normalized product function $c_X^{\mathrm{n\text{-}prod}}(X^{(0)}, X^{(1)})$ becomes (nearly) equivalent to the maximum likelihood distinguisher applied to the joint distribution.

This observation might not be obvious in theory since correlation is only an appropriate statistical tool when the underlying noise is Gaussian. Unfortunately, when multiplying two Gaussian distributions, as it is done for $c_X^{\mathrm{n\text{-}prod}}(X^{(0)}, X^{(1)})$, does clearly *not* result in a Gaussian distribution.

Thus, our aim is to precisely state the higher-order optimal distinguisher (HOOD) expression for second-order when the attacker has full information about underlying the leakage and determine when this knowledge can be lessened to relate the expression to second-order CPA. This will help to understand known empirical results [9,14,16,20]. In particular, we investigate low and high noise scenarios to see which combination function from the pool described in Subsect. 2.2 would be a reasonable choice.

3.2 Explicit Derivations

In [14] the authors state various template attacks against first-order masking schemes. The most efficient is a straightforward extension of the classical template attack [5] over all pairs (t, k). Our approach goes in a similar direction: we utilize the joint distribution of both leakages $X^{(0)}$ and $X^{(1)}$ without using a combination function, which gives us the optimal second-order distinguisher maximizing the success rate.

Theorem 2 (Second-order HOOD). *If the model (i.e., $\varphi^{(\omega)}$) is known to the attacker for all ω, then the* second-order HOOD *is*

$$\mathcal{D}^2_{opt}(\mathbf{x}^{(\star)}, \mathbf{t}^{(\star)}) = \arg\max_{k \in \mathcal{K}} p_k(\mathbf{x}^{(\star)}|\mathbf{t}^{(\star)}) \tag{9}$$

$$= \arg\max_{k \in \mathcal{K}} \prod_{i=1}^{q} \sum_{m^{(\star)} \in \mathcal{M}^{(\star)}} \mathbb{P}(m^{(\star)}) \prod_{\omega=0}^{1} p_k(x_i^{(\omega)}|t_i^{(\omega)}, m^{(\omega)}). \tag{10}$$

Note that as the attacker knows the model he is able to compute the required probability distributions and densities.

Proof. Let us denote the key guess of any second-order distinguisher by $\hat{k} = \arg\max_{k \in \mathcal{K}} \mathcal{D}^2(\mathbf{x}^{(\star)}, \mathbf{t}^{(\star)})$. Then, using a frequentist approach we start from the success probability $\mathbb{P}_S$ over all possible secret keys k

$$\mathbb{P}_S = \frac{1}{|\mathcal{K}|} \sum_{k \in \mathcal{K}} \mathbb{P}(\hat{k} = k) = \frac{1}{|\mathcal{K}|} \sum_{k \in \mathcal{K}} \sum_{\mathbf{t}^{(\star)}} \mathbb{P}(\mathbf{t}^{(\star)}) \, \mathbb{P}(\hat{k} = k | \mathbf{T}^{(\star)} = \mathbf{t}^{(\star)}) \tag{11}$$

$$= \frac{1}{|\mathcal{K}|} \sum_{k \in \mathcal{K}} \sum_{\mathbf{t}^{(\star)} \in \mathcal{T}^{q \times (d+1)}} \mathbb{P}(\mathbf{t}^{(\star)}) \int_{\mathcal{X}^{q \times (d+1)}} p_k(\mathbf{x}^{(\star)}|\mathbf{t}^{(\star)}) \mathbb{1}_{k=\hat{k}} \, d\mathbf{x}^{(\star)} \tag{12}$$

$$= \frac{1}{|\mathcal{K}|} \sum_{k \in \mathcal{K}} \sum_{\mathbf{t}^{(\star)} \in \mathcal{T}^{q \times (d+1)}} \mathbb{P}(\mathbf{t}^{(\star)}) \int_{\mathcal{X}^{q \times (d+1)}} p_{\hat{k}}(\mathbf{x}^{(\star)}|\mathbf{t}^{(\star)}) \, d\mathbf{x}^{(\star)}. \tag{13}$$

In Eq. (11), we have to compute $\mathbb{P}(\hat{k} = k | \mathbf{t}^{(\star)})$ where $\hat{k} = \hat{k}(\mathbf{x}^{(\star)}, \mathbf{t}^{(\star)}) = \arg\max_{k \in \mathcal{K}} \mathcal{D}^2(\mathbf{x}^{(\star)}, \mathbf{t}^{(\star)})$ is a function of $\mathbf{x}^{(\star)}$ and $\mathbf{t}^{(\star)}$. This is therefore a probability on the random variable $\mathbf{X}^{(\star)}$ knowing $\mathbf{T}^{(\star)} = \mathbf{t}^{(\star)}$, which follows the density $p_k(\mathbf{x}^{(\star)}|\mathbf{t}^{(\star)})$. Like for every probability taken on a random variable with density, the required probability is the integral of density over the events. So $\mathbb{P}(\hat{k}(\mathbf{x}^{(\star)}, \mathbf{t}^{(\star)}) = k | \mathbf{t}^{(\star)}) = \int p_k(\mathbf{x}^{(\star)}|\mathbf{t}^{(\star)}) \, d\mathbf{x}^{(\star)}$, where the integral is taken over all $\mathbf{x}^{(\star)}$ such that $\hat{k}(\mathbf{x}^{(\star)}, \mathbf{t}^{(\star)}) = k$; this is the indicator function inside the integral in Eq. (12).

Now, $\mathbb{P}(\mathbf{t}^{(\star)})$ is independent of the key. Thus, for each given sequence $\mathbf{x}^{(\star)}, \mathbf{t}^{(\star)}$ maximizing the success rate amounts to choose $k = \hat{k}$ such that $p_k(\mathbf{x}^{(\star)}|\mathbf{t}^{(\star)})$ is maximized. Moreover,

$$p_k(\mathbf{x}^{(\star)}|\mathbf{t}^{(\star)}) = \prod_{i=1}^{q} p_k(x_i^{(\star)}|t_i^{(\star)}) \tag{14}$$

$$= \prod_{i=1}^{q} \sum_{m^{(\star)} \in \mathcal{M}^{(\star)}} \mathbb{P}(m^{(\star)}) \, p_k(x_i^{(\star)}|t_i^{(\star)}, m^{(\star)}) \tag{15}$$

$$= \prod_{i=1}^{q} \sum_{m^{(\star)} \in \mathcal{M}^{(\star)}} \mathbb{P}(m^{(\star)}) \prod_{\omega=0}^{1} p_k(x_i^{(\omega)}|t_i^{(\omega)}, m^{(\omega)}). \tag{16}$$

We used from (15) to (16) that $N_i^{(\omega)}$ is i.i.d. across the values of $i = \{1, \ldots, q\}$ and independent for $\omega = \{0, 1\}$. Accordingly, $\arg\max_{k \in \mathcal{K}}$ of Eq. (16) forms the optimal distinguisher $\mathcal{D}^2_{opt}(\mathbf{x}^{(\star)}, \mathbf{t}^{(\star)})$. $\qquad\square$

Remark 1. To simplify our notation, we assume in the following that the masks at each order are drawn from the same space $\mathcal{M}$, with uniform probability $\mathbb{P}(M = m) = 1/|\mathcal{M}|$ and only one text byte is manipulated with the masks, as in software implementations (cf. Ex. 1). That is, $\forall i\ t_i^{(0)} = t_i^{(1)} = t_i$ and, moreover, there is only one mask $m^{(0)} = m^{(1)} = m$. Accordingly, Eq. (10) simplifies to

$$\mathcal{D}^2_{opt}(\mathbf{x}^{(\star)}, \mathbf{t}) = \arg\max_{k \in \mathcal{K}} \prod_{i=1}^{q} \sum_{m \in \mathcal{M}} p_k(x_i^{(0)}|t_i, m) \cdot p(x_i^{(1)}|t_i, m). \tag{17}$$

However, all our results hereafter can be easily extended to the scenario without simplifications.

As it is most often assumed that the noise distribution at the manipulation of each share is Gaussian (e.g., [14, 16]), we further deduce Eq. (17) for Gaussian noise.

Proposition 3 (Second-order HOOD for Gaussian noise). *Assuming that* $N^{(\omega)} \sim \mathcal{N}(O, \sigma^{(\omega)2})$ *then the second-order optimal distinguisher becomes*

$$\mathcal{D}^{2,G}_{opt}(\mathbf{x}^{(0)}, \mathbf{x}^{(1)}, \mathbf{t}) = \arg\max_{k \in \mathcal{K}} \prod_{i=1}^{q} \sum_{m \in \mathcal{M}} \exp\left\{ -\frac{1}{2} \left(\frac{-2x_i^{(0)}y^{(0)}(t_i, k, m) + y^{(0)}(t_i, k, m)^2}{\sigma^{(0)2}} \right.\right.$$
$$\left.\left. + \frac{-2x_i^{(1)}y^{(1)}(t_i, k, m) + y^{(1)}(t_i, k, m)^2}{\sigma^{(1)2}} \right) \right\}. \tag{18}$$

Proof. In this case $p_k(x_i^{(\omega)}|t_i, m)$ is the 1D Gaussian density with mean $y^{(\omega)}$ (t_i, k, m) and standard deviation $\sigma^{(\omega)}$. Removing all constants gives us the required formula. □

As a next step we give approximations for high noise and low noise.

Corollary 4 (Second-order HOOD for high Gaussian noise). *When considering that* $\mathbb{E}\{y(T, m, k)\} = \mathbb{E}\{\varphi(f(T, m, k))\}$ *is independent of the choice of* $k \in \mathcal{K}$ *(owing to Proposition 1)[2], which is given in case of high noise since a large number of measurements q is considered, then the distinguishing rule simplifies to*

$$\mathcal{D}^{2,G,\sigma\uparrow}_{opt}(\mathbf{x}^{(\star)}, \mathbf{t}) = \arg\max_{k \in \mathcal{K}} \prod_{i=1}^{q} \sum_{m \in \mathcal{M}} \exp\left\{ \frac{x_i^{(0)}y^{(0)}(t_i, k, m)}{\sigma^{(0)2}} + \frac{x_i^{(1)}y^{(1)}(t_i, k, m)}{\sigma^{(1)2}} \right\}. \tag{19}$$

Proof. In Eq. (18) we can now remove the terms $y^{(\omega)}(t_i, k, m)^2$ for $\omega = \{0, 1\}$ because, $\mathbb{E}\{y(T, m, k)\} = \mathbb{E}\{\varphi(f(T, m, k))\}$ is independent of $k \in \mathcal{K}$. This gives Eq. (19). □

---

[2] This assumption has been also made in [21].

Remark 2. We see that the exact optimal distinguishing expression (Eq. (18)) operates in the *direct scale*, such that the function φ (including its scaling factor) and thus the exact relationship between X and Y has to be known. Whereas the distinguisher for high noise (Eq. (19)) operates in the *proportional scale*[3], thus the relationship between X and Y has only to be known up to an irrelevant affine law. That is to say, the attacker shall know that $X^{(\omega)} = a^{(\omega)}\varphi^{(\omega)}(f^{(\omega)}(T^{(\omega)}, k, m)) + b^{(\omega)} + N^{(\omega)}$ with unknown $a^{(\omega)}, b^{(\omega)} \in \mathbb{R}$. For more information on direct and proportional scales we refer the reader to [24].

Remark 2 already gives a hint about a possible relationship between the second-order HOOD and second-order CPA for high noise, which we will discuss in the next subsection.

Proposition 5 (Second-order HOOD for low Gaussian noise). *Assuming that both shares have the same low noise standard deviation $\sigma = \sigma^{(0)} = \sigma^{(1)}$ then the optimal distinguisher reduces at first order to*

$$\mathcal{D}_{opt}^{2,G,\sigma\downarrow}(\mathbf{x}^{(\star)}, \mathbf{t}) = \arg\min_{k \in \mathcal{K}} \sum_{i=1}^{q} \max_{m \in \mathbb{F}_2^n} (x_i^{(0)} - y^{(0)}(t_i, k, m))^2 + (x_i^{(1)} - y^{(1)}(t_i, k, m))^2.$$

$$(20)$$

Proof. Starting from Eq. (18) and using $y_i^{(\omega)} = y^{(\omega)}(t_i, k, m)$ we have

$$\hat{k} = \arg\max_{k \in \mathcal{K}} \prod_{i=1}^{q} \sum_{m \in \mathbb{F}_2^n} \exp\left\{ -\frac{1}{\sigma^{(0)2}}(x_i^{(0)} - y_i^{(0)})^2 - \frac{1}{\sigma^{(1)2}}(x_i^{(1)} - y_i^{(1)})^2 \right\}. \quad (21)$$

Now as $\sigma = \sigma^{(0)} = \sigma^{(1)}$ and as the sum over exponential reduces at first order to the minimum we have the first order approximation for $\sigma \to 0$

$$= \arg\max_{k \in \mathcal{K}} \prod_{i=1}^{q} \min_{m \in \mathbb{F}_2^n} \exp\left\{ -(x_i^{(0)} - y_i^{(0)}(t_i, k, m))^2 - (x_i^{(1)} - y_i^{(1)}(t_i, k, m))^2 \right\}.$$

Applying the logarithm that is strictly monotonous increasing yields

$$= \arg\max_{k \in \mathcal{K}} \sum_{i=1}^{q} \min_{m \in \mathbb{F}_2^n} \left(-(x_i^{(0)} - y_i^{(0)}(t_i, k, m))^2 - (x_i^{(1)} - y_i^{(1)}(t_i, k, m))^2 \right) \quad (22)$$

$$= \arg\min_{k \in \mathcal{K}} \sum_{i=1}^{q} \max_{m \in \mathbb{F}_2^n} \left((x_i^{(0)} - y_i^{(0)}(t_i, k, m))^2 + (x_i^{(1)} - y_i^{(1)}(t_i, k, m))^2 \right). \qquad \square$$

Interestingly, one can see directly from Eq. (20) that the optimal distinguisher for low noise cannot be rewritten as correlation with any combination functions and moreover that it operates in the direct scale. Even more, the nature of distinguisher seems not very intuitive.

---

[3] But not *anti-proportional* scale, or in other words, the "sign" has to be known.

3.3   Comparison with Second-Order CPA

Proposition 6 (Relationship between second-order HOOD for high noise and second-order CPA). *The second-order HOOD for high noise can be approximated as*

$$\mathcal{D}_{opt}^{2,G,\sigma\uparrow} \approx \arg\max_k \langle \mathbf{x}^{(0)} \cdot \mathbf{x}^{(1)} | \sum_{m \in \mathcal{M}} y^{(0)}(\mathbf{t}, k, m) y^{(1)}(\mathbf{t}, k, m) \rangle, \qquad (23)$$

which is all the more equivalent as the noise gets larger. Accordingly, as the noise is larger, the closer the optimal distinguishing rule to second-order CPA with

$$c_X^{n\text{-}prod}(X^{(0)}, X^{(1)}) = (X^{(0)} - \mathbb{E}\{X^{(0)}\}) \cdot (X^{(1)} - \mathbb{E}\{X^{(1)}\}) \ and \qquad (24)$$

$$c_Y^{opt}(Y^{(\star)}) = \mathbb{E}\{c_X^*(Y^{(0)}(T, k, M), Y^{(1)}(T, k, M)) | T\}. \qquad (25)$$

Proof. We use the first order Taylor expansion $\exp\{\varepsilon\} = 1 + \varepsilon + O(\varepsilon^2)$. Note that this approximation is all the better as ε is close to zero and thus as the argument of $\exp\{\cdot\}$ his high. Starting from Eq. (19) and using $y_i^{(\omega)} = y^{(\omega)}(t_i, k, m)$, we have

$$\mathcal{D}^{2,G,\sigma\uparrow}(\mathbf{x}^{(0)}, \mathbf{x}^{(1)}, \mathbf{t}) = \arg\max_k \prod_{i=1}^{q} \sum_{m \in \mathcal{M}} \exp\left\{\frac{1}{\sigma^{(0)^2}} x_i^{(0)} y_i^{(0)}\right\} \exp\left\{\frac{1}{\sigma^{(1)^2}} x_i^{(1)} y_i^{(1)}\right\}$$

$$\approx \arg\max_k \prod_{i=1}^{q} \sum_{m \in \mathcal{M}} \left(1 + \frac{1}{\sigma^{(0)^2}} x_i^{(0)} y_i^{(0)}\right) \left(1 + \frac{1}{\sigma^{(1)^2}} x_i^{(1)} y_i^{(1)}\right) \qquad (26)$$

$$= \arg\max_k \prod_{i=1}^{q} \sum_{m \in \mathcal{M}} \left(1 + \frac{1}{\sigma^{(0)^2} \sigma^{(1)^2}} x_i^{(0)} y_i^{(0)} x_i^{(1)} y_i^{(1)} + \frac{1}{\sigma^{(0)^2}} x_i^{(0)} y_i^{(0)} + \frac{1}{\sigma^{(1)^2}} x_i^{(1)} y_i^{(1)}\right).$$
$$(27)$$

In Eq. (27), owing to the perfect masking definition, the terms $\sum_{m \in \mathcal{M}} x_i^{(0)} y_i^{(0)}$ and $\sum_{m \in \mathcal{M}} x_i^{(1)} y_i^{(1)}$ are constant (const$^{(0)}$ and const$^{(1)}$). Additionally, as the logarithm function is increasing, we consider the logarithm of the product, and we use the approximation $\ln\{1 + \varepsilon\} = \varepsilon + \mathcal{O}(\varepsilon^2)$, (reciprocal of the previous Taylor's expansion of the exponential function), which is again all the better as ε is close to zero and thus for high noise. Accordingly,

$$\mathcal{D}^{2,G,\sigma\uparrow}(\mathbf{x}^{(0)}, \mathbf{x}^{(1)}, \mathbf{t}) \approx \arg\max_k \ln \prod_{i=1}^{q} \sum_{m \in \mathcal{M}} \left(1 + \frac{1}{\sigma^{(0)^2} \sigma^{(1)^2}} x_i^{(0)} y^{(0)}(t_i, k, m) \cdot\right.$$

$$\left. x_i^{(1)} y^{(1)}(t_i, k, m) + \frac{1}{\sigma^{(0)^2}} x_i^{(0)} y^{(0)}(t_i, k, m) + \frac{1}{\sigma^{(1)^2}} x_i^{(1)} y^{(1)}(t_i, k, m)\right) \qquad (28)$$

$$= \arg\max_k \sum_{i=1}^{q} \ln\left\{1 + \sum_{m \in \mathcal{M}} \frac{1}{\sigma^{(0)^2} \sigma^{(1)^2}} x_i^{(0)} y_i^{(0)} x_i^{(1)} y_i^{(1)} + \text{const}^{(0)} + \text{const}^{(1)}\right\}$$

$$\approx \arg\max_k \sum_{i=1}^{q} x_i^{(0)} x_i^{(1)} \sum_{m \in \mathcal{M}} y^{(0)}(t_i, k, m) y^{(1)}(t_i, k, m) + \text{const}^{(0)} + \text{const}^{(1)}$$

$$= \arg\max_k \langle \mathbf{x}^{(0)} \cdot \mathbf{x}^{(1)} | \sum_{m \in \mathcal{M}} y^{(0)}(\mathbf{t}, k, m) y^{(1)}(\mathbf{t}, k, m) \rangle. \qquad (29)$$

Note that we can remove the $\text{const}^{(0)}, \text{const}^{(1)}$ as they do not depend on the key guess. For large number of measurements (resulting from large noise) the $\arg\max_{k \in \mathcal{K}}$ of the correlation coefficient can be simplified as

$$\arg\max_{k \in \mathcal{K}} \frac{\langle \mathbf{x} - \overline{\mathbf{x}} | \mathbf{y}(k) \rangle}{\|\mathbf{x} - \overline{\mathbf{x}}\|_2 \cdot \|\mathbf{y}(k) - \overline{\mathbf{y}(k)}\|_2} \approx \arg\max_{k \in \mathcal{K}} \langle \mathbf{x} - \overline{\mathbf{x}} | \mathbf{y}(k) \rangle. \qquad (30)$$

Accordingly, if $\mathbf{x}^{(0)}$ and $\mathbf{x}^{(1)}$ are centered, then in Eq. (29) $c_X^{n\text{-}prod} = \mathbf{x}^{(0)} \cdot \mathbf{x}^{(1)}$, and $c_Y^{opt} = \sum_{m \in \mathcal{M}} y^{(0)}(\mathbf{t}, k, m) y^{(1)}(\mathbf{t}, k, m)$ is the *optimal prediction function*.

$\square$

As correlation is a measure in the proportional scale, we can relax our assumptions made about the knowledge of the attacker. More precisely, he does not need to know $y^{(0)}$ and $y^{(1)}$ exactly but any linear transformation $l^{(\omega)}(y^{(\omega)}) = ay^{(\omega)} + b$, as it is most often assumed in the literature [16, 20]. Yet, in Prop. 6 we do not recover the absolute value of the correlation, thus, for second-order CPA the "sign" must be known and taking the absolute value does not result in an equivalence for high noise, which is also empirically validated in our experiments in Subsect. 3.4.

Remark 3. Prouff et al. illustrated in [16] that for large noise the improved (i.e., centered) product combining function has the best efficiency among the known combination functions, which is inline with our findings in Prop. 6. Moreover, we can claim that the improved product combining function is the most efficient among *all* combining functions for high noise as it becomes equivalent to the optimal second-order distinguisher. Moreover, our study is not restricted to a particular HW or HD leakage model scenario as in the previous studies.

Remark 4. The determination of optimal combination functions is a vivid research topic. As already mentioned, the optimality of the centered product amongst all combination functions has been conjectured by Prouff et al. in [16]. Afterwards, mathematical arguments for optimality were given by Carlet et al. [3], and independently by Ding et al. in [7].

Remark 5. As underlined in [20], the function to be maximized in Eq. (23) is a straightforward generalization of Pearson's correlation coefficient to the case of three random variables: $X^{(0)}$, $X^{(1)}$, and $\mathbb{E}\{y^{(0)}(T, k, M) y^{(1)}(T, k, M) | T\}$, where the expectation is taken over M.

3.4 Experimental Validation

For our experimental validation we used simulations of a first-order masking scheme where each share is leaking in the Hamming weight model to be able to directly compare our results to previous publications conducting the same setting [16, 20] (see Example 1). We simulated the noise arising from a Gaussian distributions $N \sim \mathcal{N}(0, \sigma^2)$ for $\sigma = \sigma^{(0)} = \sigma^{(1)} \in \{0.5, 4\}$. To be reliable we conducted 500 independent experiments with uniformly distributed k^* to compute the empirical success rate. Moreover, when plotting the empirical success

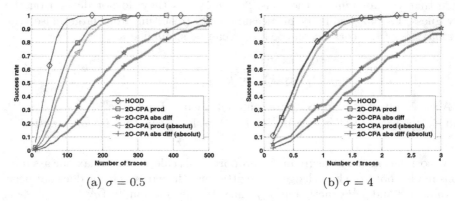

(a) $\sigma = 0.5$ (b) $\sigma = 4$

Fig. 1. Success rate for second-order attacks

rate, we additionally highlight the standard deviation of the success rate by error bars. If the error bars do not overlap, we can unambiguously conclude that one distinguisher is better than the other [12].

For our simulations we calculated the second-order HOOD and second-order CPA with the normalized product and the absolute difference combination function as described in Subsect. 2.2. For a low value of σ, $\mathcal{D}_{opt}^{2,G}$ (HOOD) clearly outperforms 2nd-order CPA (2O-CPA) independent of the combination functions (see Fig. 1a), which is inline with our theoretical analysis and the empirical analysis in [20]. For high values of σ, the second-order HOOD and second-order CPA with the normalized product combining function become equivalently efficient (see Fig. 1b), which coincides with Prop. 6. Note that as said before, taking the absolute value of the correlation is not equivalent to HOOD, which is confirmed in Fig. 1b.

4   Higher-Order Optimal Distinguisher (HOOD) for Any Order

The claim in [16] that the normalized product combining function $c_X^{n\text{-}prod}$ in combination with c_Y^{opt} is optimal[4] was only done for $d = 1$. We now extent our investigation to $(d + 1)$th-order distinguishers in order to analyze if the assumption can straightforwardly be generalized.

Theorem 7 (General HOOD). *When* $\varphi^{(\omega)} : \mathbb{F}_2^n \to \mathbb{R}$ *is known for all* ω, $N_i^{(\omega)}$ *i.i.d. across values of* $i = \{1, \ldots, q\}$ *and independent across the values of* $\omega = \{0, \ldots, d\}$, *then the* general higher-order optimal distinguisher *is*

$$\mathcal{D}_{opt}^d(\mathbf{x}^{(\star)}, \mathbf{t}^{(\star)}) = \arg\max_{k \in \mathcal{K}} \prod_{i=1}^{q} \sum_{m^{(\star)} \in \mathcal{M}^{(\star)}} \mathbb{P}(m^{(\star)}) \prod_{\omega=0}^{d} p_k(x_i^{(\omega)} | t_i^{(\omega)}, m^{(\omega)}). \quad (31)$$

---

[4] Note again, that the authors used the absolute correlation coefficient of the correct key as a measure of optimality; not the success rate.

Proof. The proof is a straightforward extension of proof of Theorem 2. □

Proposition 8 (HOOD for Gaussian noise). *Under the same assumptions as in Theorem 7 and additionally assuming Gaussian noise, i.e., $N_i^{(\omega)} \sim \mathcal{N}(0, \sigma_\omega^2)$, Eq. (31) becomes*

$$
\mathcal{D}_{opt}^{d,G}(\mathbf{x}^{(\star)}, \mathbf{t}) = \arg\max_{k \in \mathcal{K}} \sum_{i=1}^{q} \log\left\{ \sum_{m^{(\star)} \in \mathcal{M}^{(\star)}} \exp\left\{ \sum_{\omega=0}^{d} \frac{1}{\sigma^{(\omega)2}} \left(x_i^{(\omega)} y_i^{(\omega)} - \frac{1}{2} y_i^{(\omega)2} \right) \right\} \right\}.
$$

(32)

Proof. As $p_k(x_i^{(\omega)} | t_i, m) = p_{k,N^{(\omega)}}(x_i^{(\omega)} - y^{(\omega)}(t_i, k, m))$ we have

$$
\arg\max_{k \in \mathcal{K}} \prod_{i=1}^{q} \sum_{m^{(\star)} \in \mathcal{M}^{(\star)}} \mathbb{P}(m^{(\star)}) \prod_{\omega=0}^{d} p_k(x_i^{(\omega)} | t_i, m^{(\omega)})
$$

(33)

$$
= \arg\max_{k \in \mathcal{K}} \prod_{i=1}^{q} \sum_{m^{(\star)} \in \mathcal{M}^{(\star)}} \mathbb{P}(m^{(\star)}) \prod_{\omega=0}^{d} \frac{1}{\sqrt{2\pi}\sigma^{(\omega)}} \exp\left\{ -\frac{1}{2\sigma^{(\omega)2}} (x_i^{(\omega)} - y^{(\omega)}(t_i, k, m))^2 \right\}.
$$

Now, removing all key-independent constants yields

$$
\arg\max_{k \in \mathcal{K}} \prod_{i=1}^{q} \sum_{m \in \mathcal{M}} \prod_{\omega=0}^{d} \exp\left\{ -\frac{1}{2\sigma^{(\omega)2}} (x_i^{(\omega)} - y^{(\omega)}(t_i, k, m))^2 \right\}
$$

Now, as the product of $\exp\{\cdot\}$ is the $\exp\{\cdot\}$ of the sum and expanding the square and removing the key-independent factor $x_i^{(\omega)2}$ gives the required equation. □

Proposition 9 (HOOD for high Gaussian noise). *For high Gaussian noise (low SNR) we can further approximate the HOOD to*

$$
\mathcal{D}_{opt}^{d,G,\sigma\uparrow}(\mathbf{x}^{(\star)}, \mathbf{t}) = \arg\max_{k} \prod_{i=1}^{q} \sum_{m \in \mathcal{M}} \exp\left\{ \sum_{\omega=0}^{d} \frac{1}{\sigma^{(\omega)2}} x_i^{(\omega)} y^{(\omega)} \right\},
$$

(34)

and as $\sigma^{(\omega)}$ becomes large Eq. (34) becomes closer to $(d+1)$th-order CPA with

$$
c_X^{n\text{-}prod}(X^{(\star)}) = \prod_{\omega=0}^{d} (X^{(\omega)} - \mathbb{E}\{X^{(\omega)}\}) \quad and \quad c_Y^{opt}(Y^{(\star)}) = \mathbb{E}\{c_X^*(Y^{(\star)}(M, k))\}.
$$

Proof. As in the case of $d = 1$, we use the first-order Taylor expansion $\exp\{\varepsilon\} = 1 + \varepsilon + O(\varepsilon^2)$. Starting from Eq. (32), we have

$$
\mathcal{D}^{d,G,\sigma\uparrow}(\mathbf{x}^{(0)}, \mathbf{x}^{(1)}, \mathbf{t}) = \arg\max_{k} \prod_{i=1}^{m} \sum_{m \in \mathcal{M}} \prod_{\omega=0}^{d} \exp\left\{ \frac{1}{2\sigma^{(\omega)2}} x_i^{(\omega)} y_i^{(\omega)} \right\}
$$

(35)

$$
\approx \arg\max_{k} \prod_{i=1}^{m} \sum_{m \in \mathcal{M}} \prod_{\omega=0}^{d} \left(1 + \frac{1}{2\sigma^{(\omega)2}} x_i^{(\omega)} y_i^{(\omega)} \right).
$$

(36)

Now, in Eq. (36) when factorizing the product over ω all the terms not depending on all shares $1, \ldots, d$ simultaneously, i.e., $\prod_{\omega=0}^{d} x_i^{(\omega)} y_i^{(\omega)}$, do not depend on the key due to the perfect masking definition. Moreover, following the same argumentation as for Prop. 6, we recover that if $\forall \omega \ \mathbf{x}^{(\omega)}$ are centered, then $c_X^{n\text{-}prod} = \prod_{\omega=0}^{d} \mathbf{x}^{(\omega)}$, and $c_Y^{opt} = \sum_{m \in \mathcal{M}} \prod_{\omega=0}^{d} y^{(\omega)}(\mathbf{t}, k, m)$ is the *optimal prediction function* for higher-order CPA. $\qquad \square$

Proposition 9 shows that the normalized production combination function combined with the optimal prediction function is therefore not only optimal for dth-order CPA in case of $d = 1$ but for any value of d.

Proposition 10 (HOOD for low Gaussian noise). *For low noise variance $\sigma = \sigma^{(0)} = \cdots = \sigma^{(d)}$ the optimal distinguisher (Eq. (32)) is simplified to*

$$\mathcal{D}_{opt}^{d,G,\sigma\downarrow}(\mathbf{x}^{(\star)}, \mathbf{t}) = \arg\min_{k \in \mathcal{K}} \sum_{i=1}^{q} \max_{m \in \mathcal{M}} \sum_{\omega=0}^{d} (x_i^{(\omega)} - y_i^{(\omega)})^2 \qquad (37)$$

$$= \arg\min_{k \in \mathcal{K}} \sum_{i=1}^{q} \max_{m \in \mathcal{M}} \|x_i^{(\star)} - y_i^{(\star)}\|_2^2. \qquad (38)$$

Proof. The proof is a straightforward extension of the proof for Prop. 5. $\qquad \square$

5   HOOD for Precomputation Masking Tables

5.1   Classical Attacks

We now consider the attack of a masking scheme using Sbox recomputation as described in [11]. Appendix A provides a description of the underlying algorithm.

It is noteworthy that the traditional approach to reduce the multiplicity of leakage samples by a combination $c_X : \mathcal{X}^d \to \mathbb{R}$ *would fail* in the setup of masking tables. Indeed, the combination functions are usually considered symmetric into its arguments, meaning that any swap of the inputs does not affect the combination. This (tacit) hypothesis has been made, for instance, for

- the absolute difference $c_X^{diff}(X^{(\star)}) = (|X^{(0)} - X^{(1)}| = |X^{(1)} - X^{(0)}|)$, and
- the centered product $c_X^{n\text{-}prod}(X^{(\star)}) = ((X^{(0)} - \mathbb{E}\{X^{(0)}\})(X^{(1)} - \mathbb{E}\{X^{(1)}\}) = (X^{(1)} - \mathbb{E}\{X^{(1)}\})(X^{(0)} - \mathbb{E}\{X^{(0)}\}))$.

We assume here that the attacker applies the combination function on the leakages occurring during the Sbox recomputation (see Alg. 1), i.e., the attacker gains 2^n leakages

$$X^{(0)} = \varphi^{(0)}(M) + N^{(0)} \qquad (39)$$

$$X^{(1)} = \varphi^{(1)}(M \oplus 1) + N^{(1)} \qquad (40)$$

$$\vdots$$

$$X^{(2^n - 1)} = \varphi^{(2^n - 1)}(M \oplus (2^n - 1)) + N^{(2^n - 1)}, \qquad (41)$$

and would apply e.g., $c_X^{diff}(X^{(\star)})$ or $c_X^{n\text{-}prod}(X^{(\star)})$. Additionally, he measures the leakage $X^{(2^n)} = \varphi^{(2^n)}(T \oplus k \oplus M) + N^{(2^n)}$ and finally combines it with the previous combined leakages as $\bar{c}_X(X^{(2^n)}, c_X(X^{(0)}, \ldots, X^{(2^n-1)}))$.

Following the methodology in [16] and assuming an equal leakage function on each share[5], i.e., $\varphi = \varphi^{(0)} = \cdots = \varphi^{(2^n)}$, the optimal function to combine the predictions would then be

$$c_Y^{opt} = \mathbb{E}\{\bar{c}_X^*(c_X^*(\varphi(M), \varphi(M \oplus 1), \ldots, \varphi(M \oplus (2^n - 1))), \varphi(t \oplus k \oplus M))\} \quad (42)$$

$$= \frac{1}{2^n} \sum_{m \in \mathbb{F}_2^n} \bar{c}_X^*(c_X^*(\varphi(m), \varphi(m \oplus 1), \ldots, \varphi(m \oplus (2^n - 1))), \varphi(t \oplus k \oplus m))$$

$$= \frac{1}{2^n} \sum_{m' \in \mathbb{F}_2^n} \bar{c}_X^*(c_X^*(\varphi(m' \oplus k), \ldots, \varphi(m' \oplus k \oplus (2^n - 1))), \varphi(t \oplus m')) \quad (43)$$

$$= \frac{1}{2^n} \sum_{m' \in \mathbb{F}_2^n} \bar{c}_X^*(c_X^*(\varphi(M'), \varphi(m' \oplus 1), \ldots, \varphi(M' \oplus (2^n - 1))), \varphi(t \oplus m')). \quad (44)$$

In Eq. (43), we change m for $m' = m \oplus k$ and in Eq. (44), the input terms at position ζ are replaced with those at position $\zeta \oplus k$ (because of the symmetry property of c). Accordingly, c_Y^{opt} does not depend on the key k and is even constant as the same operation can be done on $t \oplus k$, therefore higher-order CPA fails.

Of course, the Sbox precomputation masking scheme can be attacked by various attacks (e.g., the classic means, collision attacks, second-order attacks) that concentrate on specific stages of Alg. 1. However, a better attack would consist in using altogether all the leakages from the Sbox recomputation with one (or more) of the samples used during the computation proper (starting from line 8, when the key is involved). One example of such strategy has been exposed in [22], which we label as 2-stage CPA attack.

Definition 2 (2-stage CPA attack [22])

$$2{\times}\text{CPA}^{mt}(\mathbf{x}, \mathbf{t}) = \arg\max_{k \in \mathcal{K}} \rho(\mathbf{x}^{(2^n)}, y^{(2^n)}(\mathbf{t}, k, \hat{\mathbf{m}})), \quad (45)$$

where $\forall i \ \hat{m}_i$ is the mask that maximizes the correlation between $x_i^{(\omega)}$ and $y_i^{(\omega)} = \omega \oplus m_i$ for $\omega \in [0, 2^n[$. This attack is a synergy between a horizontal and a vertical attack. For each trace (separately $\forall i$), the first attack in Eq. (45) consists in recovering the mask during the precomputation (lines 2 to 5 in Appendix A). Second, a regular CPA using a model in which both the plaintext t and the mask m are assumed as public knowledge is launched. Even if the mask $\hat{m}$ is not recovered correctly for each trace (since 2^n leakage samples during the precomputation can be seen as small), it can be expected that the value of the

---

[5] This assumption is reasonable for software implementation, which is the adequate scenario for masking tables.

mask is recovered by the first horizontal attack probabilistically well enough for it to be biased, i.e., better guessed than random. This gives a rough idea of the proof of soundness for this attack.

Nonetheless, this attack is probably not the most efficient, as it uses separately the information available from the Sbox precomputation and from the leakage of the AES algorithm proper. The next subsection investigates the optimal attack and gives approximation for high and low noise.

5.2   HOOD for Precomputation Masking Tables

When using masking tables (Alg. 1) the attacker first has all leaking samples during the precomputation, i.e., $y_i^{(\omega)} = \varphi(\omega \oplus m)$ that are independent of i for $0 \le \omega \le (2^n - 1)$, and, second, the leakage arising from the combination of the mask m, plaintext t_i and the key, i.e., $y_i^{(2^n)} = \varphi(t_i \oplus k \oplus m)$. Thus, all terms for $\omega \ne 2^n$ do not depend on the key and the higher-order optimal distinguisher from Eq. (32) can be further deduced.

Theorem 11 (HOOD for masking tables). *When $\varphi : \mathbb{F}_2^n \to \mathbb{R}$ is known, $N_i^{(\omega)} \sim \mathcal{N}(0, \sigma_\omega^2)$ and i.i.d. across values of $i = \{1, \ldots, q\}$ and independent across the values of $\omega = \{0, \ldots, 2^n\}$, then the higher-order optimal distinguisher against masking tables takes the form*

$$\mathcal{D}_{opt}^{mt,G}(\mathbf{x}^{(\star)}, \mathbf{t}) =$$

$$\arg\max_{k \in \mathcal{K}} \sum_{i=1}^{q} \log \left\{ \sum_{m \in \mathbb{F}_2^n} \exp \left\{ \sum_{\omega \in \mathbb{F}_2^n} \frac{1}{\sigma^{(\omega)2}} \left(x_i^{(\omega)} \varphi(\omega \oplus m) - \frac{1}{2}\varphi^2(\omega \oplus m) \right) \right. \right.$$
$$\left. \left. + \frac{1}{\sigma^{(2^n)2}} \left(x_i^{(2^n)} \varphi(t_i \oplus m \oplus k) - \frac{1}{2}\varphi^2(t_i \oplus m \oplus k) \right) \right\} \right\}. \tag{46}$$

Proof. Straightforward computation from Eq. (32) yields

$$\arg\max_{k \in \mathcal{K}} \prod_{i=1}^{q} \sum_{m \in \mathbb{F}_2^n} \prod_{\omega \in \mathbb{F}_2^n} \exp \left\{ \frac{1}{\sigma^{(\omega)2}} \left(x_i^{(\omega)} y_i^{(\omega)} - \frac{1}{2} y_i^{(\omega)2} \right) \right\} \tag{47}$$

$$= \arg\max_{k \in \mathcal{K}} \sum_{i=1}^{q} \log \left\{ \sum_{m \in \mathbb{F}_2^n} \exp \left\{ \sum_{\omega \in \mathbb{F}_2^n} \frac{1}{\sigma^{(\omega)2}} \left(x_i^{(\omega)} y_i^{(\omega)} - \frac{1}{2} y_i^{(\omega)2} \right) \right\} \right\} \tag{48}$$

Now plugging the respective leakages as described in Subsect. 5.2 gives

$$= \arg\max_{k \in \mathcal{K}} \sum_{i=1}^{q} \log \left\{ \sum_{m \in \mathbb{F}_2^n} \exp \left\{ \sum_{\omega \in \mathbb{F}_2^n} \frac{1}{\sigma^{(\omega)2}} \left(x_i^{(\omega)} \varphi(\omega \oplus m) - \frac{1}{2}\varphi^2(\omega \oplus m) \right) \right. \right.$$
$$\left. \left. + \frac{1}{\sigma^{(2^n)2}} \left(x_i^{(2^n)} \varphi(t_i \oplus m \oplus k) - \frac{1}{2}\varphi^2(t_i \oplus m \oplus k) \right) \right\} \right\}. \qquad \square \tag{49}$$

Proposition 12 (HOOD for masking tables for low SNR). *For large Gaussian noise (or low SNR) the distinguisher becomes*

$$\mathcal{D}_{opt}^{mt,G,\sigma\uparrow}(\mathbf{x}^{(\star)}, \mathbf{t}) =$$

$$\arg\max_{k\in\mathcal{K}} \sum_{\omega\in\mathbb{F}_2^n} \frac{1}{\sigma^{(\omega)2}} \sum_{i=1}^{q} \begin{pmatrix} x_i^{(\omega)} x_i^{(2^n)} \sum_m \varphi(\omega\oplus m)\varphi(t_i\oplus k\oplus m) \\ -\frac{1}{2}x_i^{(2^n)} \sum_m \varphi(t_i\oplus k\oplus m)\varphi(\omega\oplus m)^2 \\ -\frac{1}{2}x_i^{(\omega)} \sum_m \varphi(\omega\oplus m)\varphi(t_i\oplus k\oplus m)^2 \\ +\frac{1}{4}\sum_m \varphi(\omega\oplus m)^2\varphi(t_i\oplus k\oplus m)^2 \end{pmatrix}. \tag{50}$$

Proof. Due to the lack of space we neglect the term $\arg\max_{k\in\mathcal{K}}$ in front of each line. Starting from Eq. (32) we use again the first-order Taylor expansion $\exp\{\varepsilon\} = 1 + \varepsilon + O(\varepsilon^2)$. So,

$$\prod_{i=1}^{q} \sum_{m\in\mathbb{F}_2^n} \prod_{\omega=0}^{2^n} \left(1 + \frac{1}{\sigma^{(\omega)2}}\left(x_i^{(\omega)}y_i^{(\omega)} - \frac{1}{2}y_i^{(\omega)2}\right) + \frac{1}{2\sigma^{(\omega)4}}\left(x_i^{(\omega)}y_i^{(\omega)} - \frac{1}{2}y_i^{(\omega)2}\right)^2\right).$$

Furthermore, an expansion at second-order gives

$$\prod_{i=1}^{q} \sum_{m\in\mathbb{F}_2^n} \left(1 + \sum_{\omega=0}^{2^n} \frac{1}{\sigma^{(\omega)2}}\left(x_i^{(\omega)}y_i^{(\omega)} - \frac{1}{2}y_i^{(\omega)2}\right) + \frac{1}{2\sigma^{(\omega)4}}\left(x_i^{(\omega)}y_i^{(\omega)} - \frac{1}{2}y_i^{(\omega)2}\right)^2\right.$$

$$\left. + \sum_{\omega\neq\omega'}^{2^n} \frac{1}{\sigma^{(\omega)2}\sigma^{(\omega')2}}\left(x_i^{(\omega)}y_i^{(\omega)} - \frac{1}{2}y_i^{(\omega)2}\right)\left(x_i^{(\omega')}y_i^{(\omega')} - \frac{1}{2}y_i^{(\omega')2}\right)\right). \tag{51}$$

From the perfect masking condition (see Prop. 1), the first-order term

$$\sum_{m\in\mathbb{F}_2^n} \sum_{\omega=0}^{2^n} \frac{1}{\sigma^{(\omega)2}}\left(x_i^{(\omega)}y_i^{(\omega)} - \frac{1}{2}y_i^{(\omega)2}\right) = \sum_{\omega=0}^{2^n} \frac{1}{\sigma^{(\omega)2}}\left(x_i^{(\omega)}\sum_{m\in\mathbb{F}_2^n} y_i^{(\omega)} - \frac{1}{2}\sum_{m\in\mathbb{F}_2^n} y_i^{(\omega)2}\right)$$

is constant as well as

$$\sum_{m\in\mathbb{F}_2^n} \sum_{\omega=0}^{2^n} \frac{1}{2\sigma^{(\omega)4}}\left(x_i^{(\omega)}y_i^{(\omega)} - \frac{1}{2}y_i^{(\omega)2}\right)^2 \tag{52}$$

$$= \sum_{\omega=0}^{2^n} \frac{1}{2\sigma^{(\omega)4}}\left(x_i^{(\omega)2}\sum_{m\in\mathbb{F}_2^n} y_i^{(\omega)2} + \frac{1}{4}\sum_{m\in\mathbb{F}_2^n} y_i^{(\omega)4} - x_i^{(\omega)}\sum_{m\in\mathbb{F}_2^n} y_i^{(\omega)3}\right). \tag{53}$$

The other terms in ω, ω' can be written as

$$2\sum_{\omega<\omega'}^{2^n} \frac{1}{\sigma^{(\omega)2}\sigma^{(\omega')2}}\left(x_i^{(\omega)}x_i^{(\omega')}\sum_{m\in\mathbb{F}_2^n} y_i^{(\omega)}y_i^{(\omega')} - \frac{1}{2}x_i^{(\omega')}\sum_{m\in\mathbb{F}_2^n} y_i^{(\omega')}y_i^{(\omega)2}\right.$$

$$\left. -\frac{1}{2}x_i^{(\omega)}\sum_{m\in\mathbb{F}_2^n} y_i^{(\omega)}y_i^{(\omega')2} + \frac{1}{4}\sum_{m\in\mathbb{F}_2^n} y_i^{(\omega)2}y_i^{(\omega')2}\right). \tag{54}$$

Moreover, all terms involving only combinations of $\omega < d = 2^n$ do not depend on the key, thus we can further simplify to the required equation

$$\sum_{\omega \in \mathbb{F}_2^n} \frac{1}{\sigma^{(\omega)2}} \left(\sum_{i=1}^{q} x_i^{(\omega)} x_i^{(2^n)} \sum_{m \in \mathbb{F}_2^n} y^{(\omega)} y^{(2^n)} - \frac{1}{2} x_i^{(2^n)} \sum_{m \in \mathbb{F}_2^n} y^{(2^n)} y^{(\omega)2} \right. \tag{55}$$

$$\left. - \frac{1}{2} x_i^{(\omega)} \sum_{m \in \mathbb{F}_2^n} y^{(\omega)} y^{(2^n)2} + \frac{1}{4} \sum_{m \in \mathbb{F}_2^n} y^{(\omega)2} y^{(2^n)2} \right). \qquad \square$$

Proposition 13 (Relationship between HOOD and CPA for masking tables). *When all noise variances are equal, i.e., $\sigma = \sigma^{(\omega)} \; \forall \omega$, Eq. (50) further simplifies to*

$$\mathcal{D}_{opt}^{mt,G,\sigma\uparrow}(\mathbf{x}^{(\star)}, \mathbf{t}) = \arg\max_{k \in \mathcal{K}} \sum_{\omega \in \mathbb{F}_2^n} \sum_{i=1}^{q} \left(x_i^{(\omega)} x_i^{(2^n)} \sum_{m \in \mathbb{F}_2^n} \varphi(\omega \oplus m) \varphi(t_i \oplus k \oplus m) \right.$$

$$\left. - \frac{1}{2} x_i^{(\omega)} \sum_{m \in \mathbb{F}_2^n} \varphi(\omega \oplus m) \varphi^2(t_i \oplus k \oplus m) \right), \tag{56}$$

which becomes close to a combination of higher-order CPAs, i.e.,

$$\mathcal{D}_{C\text{-}CPA}^{mt,\sigma\uparrow}(\mathbf{x}^{(\star)}, \mathbf{t}) = \arg\max_{k \in \mathcal{K}} \sum_{\omega \in \mathbb{F}_2^n} \rho(c_X^{n\text{-}prod}(\mathbf{x}^{(\omega)}, \mathbf{x}^{(2^n)}), c_Y^{opt}(\mathbf{y}^{(\omega)}, \mathbf{y}^{(2^n)})) \tag{57}$$

$$- \frac{1}{2} \rho(\mathbf{x}^{(\omega)}, c_Y^{opt}(\mathbf{y}^{(\omega)}, \mathbf{y}^{(2^n)2})).$$

Proof. If all the variances are equal we have

$$\sum_{\omega \in \mathbb{F}_2^n} \frac{\varphi^2(\omega \oplus m)}{\sigma^{(\omega)}} = \frac{1}{\sigma} \sum_{\omega \in \mathbb{F}_2^n} \varphi^2(\omega \oplus m) = \frac{1}{\sigma} \sum_{\omega \in \mathbb{F}_2^n} \varphi^2(\omega). \tag{58}$$

So, regarding the second term in Eq. (50) we have

$$\sum_{\omega \in \mathbb{F}_2^n} \frac{1}{\sigma^{(\omega)2}} \sum_{i=1}^{q} x_i^{(2^n)} \sum_{m \in \mathbb{F}_2^n} \varphi(t_i \oplus k \oplus m) \varphi(\omega \oplus m)^2 \tag{59}$$

$$= \sum_{i=1}^{q} x_i^{(2^n)} \sum_{m \in \mathbb{F}_2^n} \varphi(t_i \oplus k \oplus m) \sum_{\omega \in \mathbb{F}_2^n} \frac{1}{\sigma^{(\omega)2}} \varphi(\omega \oplus m)^2 \tag{60}$$

$$= \sum_{i=1}^{q} x_i^{(2^n)} \sum_{m \in \mathbb{F}_2^n} \varphi(t_i \oplus k \oplus m) \sum_{\omega \in \mathbb{F}_2^n} \frac{1}{\sigma^2} \varphi(\omega)^2 \tag{61}$$

$$= \sum_{i=1}^{q} x_i^{(2^n)} \sum_{m \in \mathbb{F}_2^n} \varphi(t_i \oplus m) \sum_{\omega \in \mathbb{F}_2^n} \frac{1}{\sigma^2} \varphi(\omega)^2, \tag{62}$$

which clearly does not depend on the key k. The same goes for the fourth term, which proofs the first part. Now, rewriting Eq. (56) gives

$$\arg\max_{k\in\mathcal{K}} \sum_{\omega\in\mathbb{F}_2^n} \langle \mathbf{x}^{(\omega)}\mathbf{x}^{(2^n)} \mid \sum_{m\in\mathbb{F}_2^n} \varphi(\omega\oplus m)\varphi(\mathbf{t}\oplus k\oplus m)\rangle$$
$$-\langle \frac{1}{2}\mathbf{x}^{(\omega)} \mid \sum_{m\in\mathbb{F}_2^n} \varphi(\omega\oplus m)\varphi^2(\mathbf{t}\oplus k\oplus m)\rangle, \tag{63}$$

and using the same argumentation as in the proof of Prop. 9 gives the required formula from the second part. □

Interestingly, instead of using one CPA to recover the mask and one to recover the secret key (see Def. 2) we recover that the best methodology is to attack each share $\omega < 2^n$ with $\omega = 2^n$ (minus a regulation term) and then use a combination of all attacks. Note again that we can make the same relaxations about the leakage model as done in Subsect. 3.3.

Remark 6. For low noise, we can straightforwardly use Prop. 10, which is validated in our empirical results.

5.3   Experimental Validation

To empirically validate our theoretical results we use simulations of a first order masking scheme with precomputation tables. We target the xor operation in the precomputation phase and the AddRoundKey of the algorithm (see line 3 and line 8 of Alg. 1 in Appendix A).

Thus, we have the same leakages as depicted in Examples 2, where for computationally reasons for all distinguishers we only target four bits ($n = 4$).

Remark 7. Targeting the AddRoundKey phase has some advantages. First, it allows to perform the evaluation on only four bits without the loss of generality of using a four bits Sbox. Second, in the Sbox precomputation algorithm of Coron [6] the output masks are different for each entry of the Sbox and could therefore not be combined with the mask of the precomputation table. However, as in our analysis the attacker can still take advantage of the 2^n leakages of the masked inputs of the Sbox combined with the leakage of the AddRoundKey operation.

Similarly to the previous experiments, T is uniformly distributed over $\mathbb{F}_2^4$ and the noise is arising from a Gaussian distribution $N \sim \mathcal{N}(0, \sigma^2)$ for $\sigma = \sigma^{(0)} = \ldots = \sigma^{(16)} \in \{0.5, 5\}$. Again to compute the success rate we conducted 500 independent experiments with uniformly distributed k^* and shaded the success rate with error bars.

Figure 2 shows the success rates. For low noise ($\sigma = 0.5$) the optimal distinguisher (HOOD) and its approximation for low noise (HOOD-low) perform similar and better than the $2nd$-order CPA (2O-CPA) with normalized product

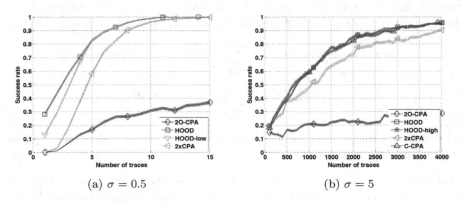

(a) $\sigma = 0.5$ (b) $\sigma = 5$

Fig. 2. Success rate for masking tables

combination function and the 2-stage CPA in Eq. 45 (2xCPA). Naturally, all distinguishers outperform 2nd-order CPA as it only utilizes two leakages $X^{(0)}$ and $X^{(256)}$. For higher noise ($\sigma = 5$) the HOOD and its approximation for high noise (HOOD-high) perform better than the 2-stage CPA (2xCPA) and 2nd-order CPA. Moreover it can be noticed that the distinguisher based on combinations of CPA (Eq. (57)) (C-CPA) and the optimal ones are equally efficient. Accordingly, we have empirically validated that our new distinguisher approximated from the HOOD is valid for high noise and more efficient than the two-stage CPA. In particular, it requires around 1000 traces less to reach $\hat{\mathbb{P}}_S = 90\%$ for $\sigma = 5$.

6 Conclusions and Perspectives

We have found the optimal distinguishers for higher-order masking, and especially, analyzed the application of second-order distinguisher and distinguisher against masking tables. This gives the first theoretical proof that for a high noise non-profiled second-order CPA becomes as efficient as the optimal distinguisher in terms of success rate. In particular, we explain that the normalized product combining function with the optimal prediction function [16] is sound and the optimal one among all (known and unknown) combination functions. We furthermore extended this result to $(d + 1)$th-order distinguisher, which has not been analyzed before. For low noise, the optimal distinguisher does not reduce to any kind of correlation. In the application of masking tables we provide a new distinguisher based on correlation whose again is as efficient as the optimal distinguisher in case of high noise. Naturally, this new distinguisher outperforms all known (non-profiled) distinguisher for this application. Given all these results we theoretically and empirically show that for high noise the security analysis with non-profiled distinguisher is sufficient as it coincides with the optimal distinguisher.

These results raise various new perspectives. First of all, our methodology of starting from the optimal distinguisher and deriving approximated distinguisher could be applied to other scenarios. One application, for example, could be the scenario used in [17]. Moreover, future work should deal with the exact analysis of the impact of noise on the masking efficiency in a theoretical manner. This comes along with an analysis of the impact of the number of shares, in particular, with an investigation of the arguments done in [15,23] about exponential attack complexity.

Acknowledgement. The authors would like to thank the anonymous reviewers for their valuable comments and suggestions to improve the quality of the paper.

References

1. Akkar, M.-L., Giraud, C.: An Implementation of DES and AES, Secure against Some Attacks. In: Koç, Ç.K., Naccache, D., Paar, C. (eds.) CHES 2001. LNCS, vol. 2162, pp. 309–318. Springer, Heidelberg (2001)

2. Blömer, J., Guajardo, J., Krummel, V.: Provably Secure Masking of AES. In: Handschuh, H., Hasan, M.A. (eds.) SAC 2004. LNCS, vol. 3357, pp. 69–83. Springer, Heidelberg (2004)

3. Carlet, C., Freibert, F., Guilley, S., Kiermaier, M., Kim, J.-L., Solé, P.: Higher-order CIS codes. IEEE Transactions on Information Theory 60(9), 5283–5295 (2014)

4. Chari, S., Jutla, C.S., Rao, J.R., Rohatgi, P.: Towards Sound Approaches to Counteract Power-Analysis Attacks. In: Wiener, M. (ed.) CRYPTO 1999. LNCS, vol. 1666, pp. 398–540. Springer, Heidelberg (1999)

5. Chari, S., Rao, J.R., Rohatgi, P.: Template Attacks. In: Kaliski Jr., B.S., Koç, Ç.K., Paar, C. (eds.) CHES 2002. LNCS, vol. 2523, pp. 13–28. Springer, Heidelberg (2003)

6. Coron, J.-S.: Higher order masking of look-up tables. In: Nguyen, P.Q., Oswald, E. (eds.) EUROCRYPT 2014. LNCS, vol. 8441, pp. 441–458. Springer, Heidelberg (2014)

7. Ding, A.A., Zhang, L., Fei, Y., Luo, P.: A Statistical Model for Higher Order DPA on Masked Devices. Cryptology ePrint Archive, Report 2014/433 (June 2014), http://eprint.iacr.org/2014/433/ (to appear at CHES 2014)

8. Fisher, R.A.: On the mathematical foundations of theoretical statistics. Philosophical Transactions of the Royal Society of London, A 222, 309–368 (1922)

9. Gierlichs, B., Batina, L., Preneel, B., Verbauwhede, I.: Revisiting Higher-Order DPA Attacks: In: Pieprzyk, J. (ed.) CT-RSA 2010. LNCS, vol. 5985, pp. 221–234. Springer, Heidelberg (2010)

10. Joye, M., Paillier, P., Schoenmakers, B.: On Second-Order Differential Power Analysis. In: Rao, J.R., Sunar, B. (eds.) CHES 2005. LNCS, vol. 3659, pp. 293–308. Springer, Heidelberg (2005)

11. Kocher, P.C., Jaffe, J., Jun, B.: Differential Power Analysis. In: Wiener, M. (ed.) CRYPTO 1999. LNCS, vol. 1666, pp. 388–397. Springer, Heidelberg (1999)

12. Maghrebi, H., Rioul, O., Guilley, S., Danger, J.-L.: Comparison between Side-Channel Analysis Distinguishers. In: Chim, T.W., Yuen, T.H. (eds.) ICICS 2012. LNCS, vol. 7618, pp. 331–340. Springer, Heidelberg (2012)

13. Messerges, T.S.: Using Second-Order Power Analysis to Attack DPA Resistant Software. In: Paar, C., Koç, Ç.K. (eds.) CHES 2000. LNCS, vol. 1965, pp. 238–251. Springer, Heidelberg (2000)

14. Oswald, E., Mangard, S.: Template Attacks on Masking—Resistance Is Futile. In: Abe, M. (ed.) CT-RSA 2007. LNCS, vol. 4377, pp. 243–256. Springer, Heidelberg (2006)

15. Prouff, E., Rivain, M.: Masking against Side-Channel Attacks: A Formal Security Proof. In: Johansson, T., Nguyen, P.Q. (eds.) EUROCRYPT 2013. LNCS, vol. 7881, pp. 142–159. Springer, Heidelberg (2013)

16. Prouff, E., Rivain, M., Bevan, R.: Statistical Analysis of Second Order Differential Power Analysis. IEEE Trans. Computers 58(6), 799–811 (2009)

17. Roche, T., Lomné, V.: Collision-Correlation Attack against Some 1st-Order Boolean Masking Schemes in the Context of Secure Devices. In: Prouff, E. (ed.) COSADE 2013. LNCS, vol. 7864, pp. 114–136. Springer, Heidelberg (2013)

18. Schramm, K., Paar, C.: Higher Order Masking of the AES. In: Pointcheval, D. (ed.) CT-RSA 2006. LNCS, vol. 3860, pp. 208–225. Springer, Heidelberg (2006)

19. Standaert, F.X., Peeters, E., Quisquater, J.-J.: On the masking countermeasure and higher-order power analysis attacks. In: International Conference on Information Technology: Coding and Computing, ITCC 2005, vol. 1, pp. 562–567 (2005)

20. Standaert, F.-X., Veyrat-Charvillon, N., Oswald, E., Gierlichs, B., Medwed, M., Kasper, M., Mangard, S.: The World is Not Enough: Another Look on Second-Order DPA. In: Abe, M. (ed.) ASIACRYPT 2010. LNCS, vol. 6477, pp. 112–129. Springer, Heidelberg (2010)

21. Thillard, A., Prouff, E., Roche, T.: Success through Confidence: Evaluating the Effectiveness of a Side-Channel Attack. In: Bertoni, G., Coron, J.-S. (eds.) CHES 2013. LNCS, vol. 8086, pp. 21–36. Springer, Heidelberg (2013)

22. Tunstall, M., Whitnall, C., Oswald, E.: Masking Tables – An Underestimated Security Risk. In: Moriai, S. (ed.) FSE 2013. LNCS, vol. 8424, pp. 425–444. Springer, Heidelberg (2014)

23. Waddle, J., Wagner, D.: Towards Efficient Second-Order Power Analysis. In: Joye, M., Quisquater, J.-J. (eds.) CHES 2004. LNCS, vol. 3156, pp. 1–15. Springer, Heidelberg (2004)

24. Whitnall, C., Oswald, E., Standaert, F.-X.: The Myth of Generic DPA... and the Magic of Learning. In: Benaloh, J. (ed.) CT-RSA 2014. LNCS, vol. 8366, pp. 183–205. Springer, Heidelberg (2014)

A   Algorithm of Masking Tables

---

input : t, one byte of plaintext, and k, one byte of key
output: The application of AddRoundKey and SubBytes on t, i.e.,
$\qquad S(t \oplus k)$

1 $m \leftarrow_{\mathcal{R}} \mathbb{F}_2^n$, $m' \leftarrow_{\mathcal{R}} \mathbb{F}_2^n$ // Draw of random input and output masks ;
2 **for** $\omega \in \{0, 1, \dots, 2^n - 1\}$ **do** // Sbox masking
3 $\quad$ $z \leftarrow \omega \oplus m$ // Masked input ;
4 $\quad$ $z' \leftarrow S[\omega] \oplus m'$ // Masked output ;
5 $\quad$ $S'[z] \leftarrow z'$ // Creating the masked Sbox entry ;
6 **end**
7 $t \leftarrow t \oplus m$ // Plaintext masking ;
8 $t \leftarrow t \oplus k$ // Masked AddRoundKey ;
9 $t \leftarrow S'[t]$ // Masked SubBytes ;
10 $t \leftarrow t \oplus m'$ // Demasking ;
11 **return** t

Algorithm 1. Beginning of a block cipher masked by Sbox precomputation

We have indicated the words length of all data as n, typically, $n = 8$ bit for AES. Two random masks m and m' are drawn initially from $\mathbb{F}_2^n$ and all the data manipulated by the algorithm will be exclusive-ored with one of the two masks.

Masking the plaintext is straightforward (see line 7). Key addition can be done safely as a second step, as the plaintext is already masked (see line 8). Passing through the Sbox is less obvious, as this operation is non-linear. Therefore, the Sbox is recomputed masked, as shown on lines 2 to 5: a new table S', that has also size $2^n \times n$ bits, is required for this purpose. In the Sbox precomputation step (lines 2 to 5), the key byte k is not manipulated. The leakage only concerns the mask.

Black-Box Separations for One-More (Static) CDH and Its Generalization[*]

Jiang Zhang[1], Zhenfeng Zhang[1,**], Yu Chen[2], Yanfei Guo[1], and Zongyang Zhang[3]

[1] Trusted Computing and Information Assurance Laboratory,
State Key Laboratory of Computer Science,
Institute of Software, Chinese Academy of Sciences, P.R. China
[2] State Key Laboratory of Information Security,
Institute of Information Engineering, Chinese Academy of Sciences, P.R. China
[3] National Institute of Advanced Industrial Science and Technology (AIST), Japan
{jiangzhang09,cycosmic,zongyang.zhang}@gmail.com,
{zfzhang,guoyanfei}@tca.iscas.ac.cn

Abstract. As one-more problems are widely used in both proving and analyzing the security of various cryptographic schemes, it is of fundamental importance to investigate the hardness of the one-more problems themselves. Bresson *et al.* (CT-RSA '08) first showed that it is difficult to rely the hardness of some one-more problems on the hardness of their "regular" ones. Pass (STOC '11) then gave a stronger black-box separation showing that the hardness of some one-more problems cannot be based on standard assumptions using black-box reductions. However, since previous works only deal with one-more problems whose solution can be efficiently checked, the relation between the hardness of the one-more (static) CDH problem over non-bilinear groups and other hard problems is still unclear. In this work, we give the first impossibility results showing that black-box reductions cannot be used to base the hardness of the one-more (static) CDH problem (over groups where the DDH problem is still hard) on any standard hardness assumption. Furthermore, we also extend the impossibility results to a class of generalized "one-more" problems, which not only subsume/strengthen many existing separations for traditional one-more problems, but also give new separations for many other interesting "one-more" problems.

1 Introduction

The first one-more problem, n-RSA, was introduced by Bellare *et al.* [4] for proving the security of the Chaum's RSA-based blind signature scheme [17]. Formally, the n-RSA problem asks an algorithm to invert the RSA-function at $n + 1$ random points

---

[*] Jiang Zhang, Zhenfeng Zhang and Yanfei Guo are sponsored by the National Basic Research Program of China under Grant No. 2013CB338003, and the National Natural Science Foundation of China (NSFC) under Grant No. 61170278, 91118006. Yu Chen is sponsored by NSFC under Grant No. 61303257, and the Strategic Priority Research Program of CAS under Grant No. XDA06010701. Zongyang Zhang is an International Research Fellow of JSPS and his work is in part supported by NSFC under grant No. 61303201.

[**] Corresponding author.

P. Sarkar and T. Iwata (Eds.): ASIACRYPT 2014, PART II, LNCS 8874, pp. 366–385, 2014.

with at most n calls to an RSA-inversion oracle. In particular, it is the regular RSA problem when $n = 0$. Similar to the n-RSA problem, Bellare *et al.* [4] also suggested that a class of one-more inversion problems can be formulated for any family of *one-way* functions, which basically asks an algorithm to invert a one-way function at some random points with a bounded number of queries (*i.e.*, less than the number of given points) to an inversion oracle. The hardness assumption on this class of problems aims to capture the intuition that an algorithm cannot gain advantage from the inversion oracle other than making "trivial" use of it. Instantiated with the discrete logarithm (DL) function, the one-more DL problem, n-DL, was given in [5]. Later, Boldyreva [8] constructed a secure blind signature scheme based on the hardness of the one-more static CDH problem (or chosen-target CDH problem [8]). Roughly, the one-more static CDH problem (n-sDH for short) defined in [8] is to solve $n + 1$ static Diffie-Hellman (sDH) instances [12] with at most n queries to an sDH solution oracle.[1]

The one-more inversion problems not only make it possible to find security proofs for many classical cryptographic constructions [7,6,2,3,13,19], but are also used to illustrate the impossibilities of proving the security of some other cryptographic schemes such as [36,27,42,25], even though the original intention of introducing them is to "prove security". Due to plenty of fruitful results, many cryptographic researchers also put effort into studying the hardness of one-more inversion problems, "*to see how they relate to other problems and to what extent we can believe in them as assumptions*" [5]. In CRYPTO '08, Garg *et al.* [27] raised it as a major open question "*to understand relationship between the DL problem and the n-DL problem*". Earlier in the same year, Bresson *et al.* [10] and Brown [11] presented the first evidence that one-more inversion problems seem to be weaker than their "regular" ones. Specifically, they showed that the hardness of some $(n + 1)$-P problem (*e.g.*, $(n + 1)$-RSA) cannot be based on the hardness of "its own" n-P problem (*e.g.*, n-RSA) using some "restricted" black-box reductions. Later, Pass [37] showed that black-box reductions cannot be used to base the hardness of a special kind of one-more inversion problems (what was called one-more problems based on homomorphic certified permutations [37]) on any standard assumption. However, all the above impossibility results explicitly require the underlying problem P to be efficiently verifiable, and thus cannot apply to the n-sDH problem over groups where the DDH problem is hard.[2]

1.1 Our Results

In this paper, we present the first impossibility results showing that black-box reductions cannot be used to base the hardness of the n-sDH problem (over groups where the DDH problem is hard) on any standard hardness assumption $\mathcal{C}$. In particular, the assumption $\mathcal{C}$ itself can be n'-sDH problem for smaller n'. Technically, we construct a meta-reduction (*i.e.*, "reduction against the reduction" [9,24,28,21]) which directly

---

[1] The notation "n-CDH" is used in [10] instead of "n-sDH". We use "n-sDH" because the n-CDH problem can be defined directly based on the CDH problem (instead of the sDH problem), and our separation results apply to such n-CDH problems as well.

[2] We note that the computational complexity between the n-sDH problem and the DL problem over specific groups has also been studied in the literature [31,34,29].

breaks the assumption $\mathcal{C}$ by interacting with any black-box reduction from $\mathcal{C}$ to the n-sDH problem. Due to the nice feature of meta-reductions [28,37], our results also apply to black-box reductions that may make non-black-box use of the assumption $\mathcal{C}$. Then, we extend our proof techniques to obtain separation results for a class of more generalized "one-more" problems, which not only subsume/strengthen many existing separations for the traditional one-more problems (*e.g.*, n-RSA, n-DL, and the unforgeability of blind signatures), but also give new separations for many other interesting "one-more" problems.

Throughout the paper, the security of a cryptographic problem P is defined via a game between a challenger $\mathcal{C}(P)$ and an adversary $\mathcal{A}$. In particular, the challenger $\mathcal{C}(P)$ provides the adversary $\mathcal{A}$ with a stateful (and possibly unbounded) oracle Orcl. The actual behavior of the oracle Orcl is determined by the description of P. We say that a problem P is *non-interactive* if its oracle Orcl $= \perp$. Sometimes, we will slightly abuse the notation, and use $\mathcal{C}$ to denote both the problem and its associated challenger. A hard cryptographic problem $\mathcal{C}$ is said to be t-*round*, if the number of the messages exchanged between the Orcl and the adversary $\mathcal{A}$ is at most t (which might be a priori bounded polynomial in the security parameter).

Impossibility for One-More Static CDH Problems. By a nice observation that Cash *et al.*'s *trapdoor test* (for the twin Diffie-Hellman problems [16]) allows some form of verification for the CDH problem, we give the first black-box separations for the n-sDH problem over general groups by carefully injecting the trapdoor test technique into our meta-reduction. The difficulty of this approach lies in the fact that the trapdoor test does not really allow us to publicly and efficiently check the validity of any single CDH tuple (since it can only check whether or not two carefully prepared tuples are both CDH tuples by using some private coins. Especially, if one of the two tuples is not a CDH tuple, it cannot determine which one is not). We overcome this difficulty by designing an unbounded adversary $\mathcal{A}$ with "delay verifications" and a meta-reduction $\mathcal{M}$ with "dynamic decisions" on whether or not to use, and how to use the trapdoor test in simulating $\mathcal{A}$ to the reduction $\mathcal{R}$. Formally, we have the following theorem.

Theorem 1. *There is no black-box reduction $\mathcal{R}$ for basing the hardness of the n-sDH problem on any $t(k)$-round hard problem $\mathcal{C}$ (or else $\mathcal{C}$ could be solved efficiently), where k is the security parameter and $n = 2 \cdot \omega(k + t + 1)$.*

Since we consider very general black-box reductions, the requirement on $n = 2 \cdot \omega(k+t+1)$ seems a bit loose. However, if one would like to consider a class of restricted black-box reductions—single-instance reductions [26,25], a tighter separation result for $n \geq 2(t + 1)$ can be achieved.

Black-Box Separations for Generalized "One-More" Problems. A natural extension of the traditional one-more problem is defined by "relaxing" the requirement on the oracle. Formally, we consider a class of generalized "one-more" problems, where each problem is associated with two non-interactive subproblems P_1 and P_2. Here, we do not require $P_1 = P_2$. For any integer $n \geq 0$, we denote n-(P_1, P_2) as the problem which asks an algorithm to solve $n + 1$ random P_1 instances with at most n calls to a

P_2 oracle (*e.g.*, $P_1 = CDH$, $P_2 = DL$). Obviously, the traditional one-more problem is a special case of our generalization with $P_1 = P_2$. In particular, we briefly denote it as n-P_1 if $P_1 = P_2$, which coincides with traditional notations (*e.g.*, n-DL). Now, we consider a class of n-(P_1, P_2) problems that there exists an efficient reduction T from P_1 to P_2 (*e.g.*, from CDH to DL) with the following two properties:

- T solves one P_1 instance by using at most γ (non-adaptive) queries to a P_2 oracle, where γ is a constant;
- T always correctly solves its input P_1 instance after obtaining γ correct responses from the P_2 oracle, and outputs "$\perp$" if one of the γ responses is incorrect with overwhelming probability (we remark that this condition implicity require that T can somehow verify the correctness of all the γ responses as a whole, but it is not required to determine which one of the responses is incorrect, *e.g.*, we have $\gamma = 2$ for the n-sDH problem).

Then, similar separation results also hold for such class of n-(P_1, P_2) problems if, in addition, P_1 has unique solution [23,37] and P_2 is randomly self-reducible [1]. Note that here we still do not explicitly require P_2 to be efficiently verifiable as for the sDH problem. Formally, we have the following theorem.

Theorem 2. *If there exists an efficient reduction* T *from* P_1 *to* P_2 *with the above two properties,* P_1 *has unique solution and* P_2 *is randomly self-reducible, then there is no black-box reduction* $\mathcal{R}$ *for basing the hardness of the* n-(P_1, P_2) *problem on any* $t(k)$-*round hard problem* $\mathcal{C}$ *(or else* $\mathcal{C}$ *could be solved efficiently), where* k *is the security parameter and* $n = \gamma \cdot \omega(k + t + 1)$.

Like the discussion after Theorem 1, if only single-instance reductions are considered, we can get a tighter separation result for $n \geq \gamma \cdot (t+1)$. Note that for the traditional n-DL, n-RSA and n-sDH over gap Diffie-Hellman groups [35] where $P_1 = P_2$, there is a natural reduction T with $\gamma = 1$. The above theorem indeed subsumes/strengthens existing separations for those problems in the literature [10,37]. Since our generalized "one-more" problem also captures the "one-more unforgeability" of blind signatures and many other interesting "one-more" problems (*e.g.*, n-(CDH, DL)), our results actually give a broad separation for some of those problems. For instance, one can directly define the one-more CDH problem, n-CDH, based on the CDH problem instead of the sDH problem, and our impossibility results apply to the n-CDH problem.

1.2   The Idea behind Our Impossibility Results

To better illustrate our techniques, we start from a simple *vanilla* reduction $\mathcal{R}$ (depicted in Fig.1) from the traditional one-more problem n_1-sDH to n_2-sDH (*i.e.*, $P_1 = P_2 = $ sDH) for integers $n_2 > n_1 \geq 0$, which only runs a single instance of the n_2-sDH adversary $\mathcal{A}$ without rewinding [10,26]. Concretely, upon receiving the challenge n_1-sDH instance y from $\mathcal{C}$, the reduction $\mathcal{R}$ invokes a *single instance* of $\mathcal{A}$ by simulating an n_2-sDH challenger $\mathcal{C}'$, and tries to find the solution x of y by interacting with $\mathcal{A}$.

Intuitively, if the adversary $\mathcal{A}$ can somehow see the n_1-sDH instance input y of $\mathcal{R}$, it can directly solve them by using its own n_2 sDH queries to $\mathcal{R}$. This intuition is actually

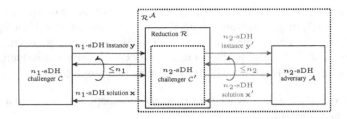

Fig. 1. A single-instance reduction $\mathcal{R}$ from n_1-sDH to n_2-sDH, where $n_2 > n_1 \geq 0$

the basic idea of [10], which constructed a meta-reduction $\mathcal{M}$ that runs $\mathcal{R}$ with its own n_1-sDH instance, and simulates an n_2-sDH adversary $\mathcal{A}$ to $\mathcal{R}$. However, this approach has two technical barriers. First, the sDH queries that $\mathcal{A}$ is allowed to make might not be in the same group or have the same public parameters as the input $\mathbf{y}$ of $\mathcal{R}$. Second, $\mathcal{R}$ can cheat $\mathcal{M}$ by returning random group elements if DDH is hard in the considered group. To get around these two barriers, the authors [10] put on additional restrictions on $\mathcal{R}$ (*e.g.*, algebraic or parameter-invariant [10]), and considered the n-sDH problem over gap Diffie-Hellman groups [35] where the DDH problem is easy.

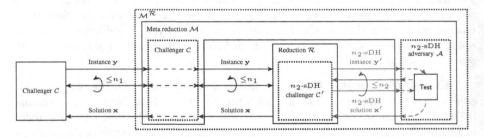

Fig. 2. Our meta-reduction $\mathcal{M}$ against $\mathcal{R}$ from n_1-round hard problem $\mathcal{C}$ to n_2-sDH problem, where $n_2 > n_1 \geq 0$

We remove the restrictions on $\mathcal{R}$ by using rewinding technique, which allows our meta-reduction $\mathcal{M}$ (depicted in Fig.2) to directly solve $\mathbf{y}'$ (*i.e.*, to make $n_2 + 1$ sDH queries to $\mathcal{R}$ for all the instances in $\mathbf{y}'$), and outputs whatever $\mathcal{R}$ returns as the solution to its own challenge instance $\mathbf{y}$. In this case, $\mathcal{M}$ actually does not care about what $\mathbf{y}$ is. Without loss of generality, we simply denote $\mathbf{y}$ as an instance of any n_1-round hard problem $\mathcal{C}$. The requirement $n_2 \geq n_1 + 1$ is still needed to ensure that there is at least one query that $\mathcal{R}$ answers without having interactions with its own challenger $\mathcal{C}$. In other words, we have to guarantee that there is at least one chance that $\mathcal{M}$ can safely rewind $\mathcal{R}$ without affecting the external interactions between $\mathcal{R}$ and $\mathcal{C}$.

To deal with the n-sDH problem over general groups where DDH is hard, we use a good observation on the remarkable **trapdoor test** algorithm (denoted by Test hereafter) introduced by Cash *et al.* [16] for twin Diffie-Hellman problems. Informally, given an element $y \in \mathbb{G}$, the algorithm outputs another uniformly distributed element

$z \in \mathbb{G}$ together with some private coins r. Then, for any elements $h, f_1, f_2 \in \mathbb{G}$, the Test algorithm can use r to determine whether or not both (y, h, f_1) and (z, h, f_2) are CDH tuples with overwhelming probability. Briefly, the Test algorithm cannot *publicly* check the validity of any single CDH tuple, but it can determine whether or not two carefully prepared tuples are both CDH tuples (by using the private coins r). Especially, if one of them is not a CDH tuple, the algorithm cannot determine which one is not. This "inability" of the Test algorithm poses an obstacle when we try to use it in our meta-reduction $\mathcal{M}$ to prevent the reduction $\mathcal{R}$ from cheating, since $\mathcal{R}$ might also notice this. To overcome this obstacle (*i.e.*, to hide the use of the Test algorithm from $\mathcal{R}$), we first present an unbound adversary $\mathcal{A}$ (against the n_2-sDH problem) with "delay verifications" such that it delays the verification of the odd-numbered response to the point immediately after obtaining the next even-numbered response, and checks the validity of the responses "two by two". Then, we construct a meta-reduction $\mathcal{M}$ that carefully tracks all the "private coins"used by the Test algorithm, and (statistically) hides the two sDH queries needed by the Test algorithm into its own sDH queries to the reduction $\mathcal{R}$. This requires the meta-reduction $\mathcal{M}$ to make "dynamic decisions" on whether or not to use, and how to use the Test algorithm in preparing each sDH query to $\mathcal{R}$.

To finally establish the separation results for general black-box reductions (*i.e.*, without any additional restrictions on $\mathcal{R}$), we have to deal with two technical issues. First, $\mathcal{R}$ might rewind the (unbounded) adversary $\mathcal{A}$ to obtain extra advantage. This is circumvented by designing a "magical" adversary $\mathcal{A}$ such that it performs "deterministically" [37,25]. Second, $\mathcal{R}$ might invoke many instances of $\mathcal{A}$, a naive rewinding of $\mathcal{R}$ will result in an exponential running-time due to "nested rewindings" [22,20]. We deal with this problem by making use of recursive rewinding techniques [41,37], which allow our meta-reduction $\mathcal{M}$ to cleverly find a "safe rewinding chance" and cancel a rewinding when it has to do too much work [38,20,15].

In all, we finally separate the n-sDH problem from any other (priori bounded) polynomial round hard problems. The impossibility results for generalized "one-more" problems can be analogously obtained if there is a reduction T for the underlying problem which can play a similar role as the Test algorithm for the n-sDH problem.

1.3   Related Work, Comparison and Discussion

Relation to Bresson et al. [10]. In CT-RSA '08, Bresson *et al.* [10] studied the relations between the traditional one-more inversion problems and their "regular" ones, and showed that the hardness of the traditional n-P problem cannot be based on the $(n-1)$-P problem using some "restricted" (*e.g.*, algebraic or parameter-invariant [10]) black-box reductions. Concretely, they showed that a class of restricted black-box reductions cannot be used to base the hardness of n-DL, n-RSA, and n-sDH over gap Diffie-Hellman groups [35], on their corresponding one-more problems with less oracle queries, *e.g.*, $(n-1)$-DL, $(n-1)$-RSA, $(n-1)$-sDH over gap Diffie-Hellman groups. As discussed in Section 1.2, the restrictions on $\mathcal{R}$ in their results seem unavoidable since their meta-reductions heavily rely on the "direct" connections between the challenge n_1-P instance input of the reduction $\mathcal{R}$ and the n_2-P instance output by $\mathcal{R}$, where $n_2 \geq n_1 + 1$. This is also the reason why separations of the one-more problems from other hard problems cannot be derived. For comparison, our meta-reduction makes use of the "rewinding" technique and the "inner" connections between the instances in the n-P problem and its associated oracle queries, which allows us to separate the n-P problem from any other (priori bounded) polynomial round hard problems. In particular, we rule out the existence of general black-box reductions (*i.e.*, without imposing any other restrictions on $\mathcal{R}$) for sufficiently large n.

Relation to Pass [37]. In STOC '11, Pass [37] presented a broad separation result showing that the security of constant-round, public-coin, (generalized) computational special-sound arguments for unique witness relations cannot be based on any standard assumption. Pass's results apply to many well-known cryptographic problems such as the traditional one-more inversion problems and the security of the two-move *unique* blind signatures (*i.e.*, each message has a unique signature for a fixed verification key). In particular, Pass showed that the hardness of n-DL and n-RSA cannot be based on any t-round standard assumption using black-box reductions if $n = \omega(k + 2t + 1)$, where k is the security parameter.[3] The use of recursive rewinding in this work is inspired by Pass [37], which makes our meta-reduction have an analogous structure to that in [37] and a similar requirement on $n = \gamma \cdot \omega(k + t + 1)$.

Since Pass's proof [37] crucially relies on the fact that the underlying problem is publicly and efficiently verifiable, their results cannot apply to the n-sDH problem over general groups. Especially, since the Test algorithm [16] does not really allow us to publicly and efficiently check the validity of any single CDH tuple (as discussed in Section 1.2), one cannot trivially use Pass's separation results and the Test algorithm to obtain our impossibility results in a "black-box fashion". Actually, our results are achieved by carefully combining many known techniques in the literature (*e.g.*, [38,20,15,37,25]) and new techniques such as *"delay verifications"* and *"dynamic decisions"* in our meta-reduction. We also extend our proof techniques to a class of generalized "one-more" problems, which allows us to obtain separation results for many interesting "one-more" problems such as the security of a class of two-move blind signatures.

Other Related Work. Fischlin and Schröder [26] showed that a class of "restricted" black-box reductions cannot be used to prove the security of three-move blind signatures based on any hard *non-interactive* problem. Katz *et al.* [33] showed that there is no black-box construction of blind signatures from one-way permutations. Both results overlap with ours in the context of two-move blind signatures, and we strengthen the separation result (in this context) by proving that the security of a class of two-move blind signatures (including non-black-box constructions) cannot be based on any polynomial round hard problem using black-box reductions.

2   Preliminaries

Let $|x|$ denote the length of a string x, and $|S|$ denote the size of a set S. Denote $x\|y$ as the bit concatenation of two strings $x, y \in \{0, 1\}^*$. We use the notation $\leftarrow$ to indicate the output of some algorithm, and the notation $\leftarrow_r$ to denote randomly choosing elements

---

[3] The factor "2" before t is because a knowledge extractor, whose behavior might dependent on the distribution of its its input transcripts, is used [37]. Please refer to [37] for details.

from some distribution (or the uniform distribution over some finite set). For example, if $\mathcal{A}$ is a probabilistic algorithm, $z \leftarrow \mathcal{A}(x, y, \ldots; r)$ means that the output of algorithm $\mathcal{A}$ with inputs $x, y, \ldots$, and randomness r is z. When r is unspecified, we mean running $\mathcal{A}$ with uniformly random coins. We say that $\mathcal{A}$ is a PPT algorithm if it runs in probabilistic polynomial-time.

The natural security parameter throughout the paper is k, and all other quantities are implicit functions of k. We use standard notation O and ω to classify the growth of functions. We say that a function $f(k)$ is negligible if for any constant $c > 0$, there exists an N such that $f(k) < 1/k^c$ for all $k > N$.

2.1 Cryptographic Problems

In this subsection, we recall several definitions of cryptographic problems.

Definition 1 (Cryptographic Problem). *A cryptographic problem* $P = ($PGen, IGen, Orcl, Vrfy$)$ *consists of four algorithms:*

- *The parameter generator* PGen *takes as input the security parameter* 1^k, *outputs a public parameter param, which specifies the instance space* $\mathcal{Y}$ *and the solution space* $\mathcal{X}$, *in brief, param* $\leftarrow$ PGen(1^k).
- *The instance generator* IGen *takes as input the public parameter param, outputs an instance* $y \in \mathcal{Y}$, *i.e.,* $y \leftarrow$ IGen$(param)$.
- *The stateful oracle algorithm* Orcl$(param, \cdot)$ *takes as input a query* $q \in \{0, 1\}^*$, *returns a response* r *for* q *or a special symbol* $\perp$ *if* q *is an invalid query.*
- *The deterministic verification algorithm* Vrfy *takes as inputs the public parameter, an instance* $y \in \mathcal{Y}$ *and a candidate solution* $x \in \mathcal{X}$, *returns* 1 *if and only if* x *is a correct solution of* y, *else returns* 0.

Throughout this paper, we implicitly assume it is easy to check whether an element y (resp., x) is in $\mathcal{Y}$ (resp., $\mathcal{X}$). We say that a cryptographic problem $P = ($PGen, IGen, Orcl, Vrfy$)$ is efficiently verifiable if Vrfy is running in polynomial-time. When Orcl $= \perp$, we say that P is a *non-interactive* problem, and denote $P = ($PGen, IGen, Vrfy$)$ in brief. Usually, the two algorithms PGen and IGen are also required to be PPT algorithms, but we do not explicitly need the requirements in this paper.

Definition 2 (Hard Cryptographic Problem). *Let* k *be the security parameter. A cryptographic problem* $P = ($PGen, IGen, Orcl, Vrfy$)$ *is said to be hard with respect to a threshold function* $\mu(k)$, *if for all PPT algorithm* $\mathcal{A}$, *the advantage of* $\mathcal{A}$ *in the security game with the challenger* $\mathcal{C}$ *(who provides inputs to* $\mathcal{A}$ *and answers* $\mathcal{A}$'s *oracle queries) is negligible in* k:

$$\mathrm{Adv}_{P,\mathcal{A}}(1^k) = \Pr[param \leftarrow \mathrm{PGen}(1^k), y \leftarrow \mathrm{IGen}(param);$$
$$x \leftarrow \mathcal{A}^{\mathrm{Orcl}(param, \cdot)}(y) : \mathrm{Vrfy}(param, y, x) = 1] - \mu(k).$$

Usually, $\mu(k) = 0$ is used for computational problems (*e.g.*, DL, CDH, n-DL), and $\mu(k) = 1/2$ is used for decisional problems (*e.g.*, DDH, DBDH).

As in [37], we also put on restrictions on the number of interactions between $\mathcal{A}$ and the oracle in the game, and a hard cryptographic problem is said to be t-*round* if the number of the messages exchanged (via oracle queries) between $\mathcal{C}$ and $\mathcal{A}$ is at most t.

2.2 Black-Box Reductions

A black-box reduction $\mathcal{R}$ from a cryptographic problem P_1 to another cryptographic problem P_2 is a PPT oracle algorithm such that $\mathcal{R}^{\mathcal{A}}$ solves P_1 whenever $\mathcal{A}$ solves P_2 with non-negligible probability. In addition to normally communicating with $\mathcal{A}$, $\mathcal{R}$ also has many powers such as restarting or "rewinding" $\mathcal{A}$. Black-box reductions often take advantage of these features. For example, $\mathcal{R}$ could make use of "rewind" to get out of a "bad condition" by first rewinding $\mathcal{A}$ to a previous state and then trying some different choices [14,20].

3 Black-Box Separations of One-More Static CDH Problems

In this section, we present the first separation results for one-more static CDH problems over general groups where the DDH problem may still be hard.

Let k be a security parameter, $\mathbb{G}$ be a group of prime order $q \geq 2^{2k}$, and g be a generator of $\mathbb{G}$. For any two group elements $A = g^a, B = g^b$, we denote $CDH(A, B) = g^{ab} = A^b = B^a$. Recall that the n-sDH problem is asking an algorithm to solve $n + 1$ static Diffie-Hellman (sDH) instances [12] with at most n queries to an oracle that solves sDH problems. Formally, given parameters $param = (k, \mathbb{G}, q, g, h)$ and $n + 1$ group elements $\mathbf{y} = (y_1, \ldots, y_{n+1})$, the algorithm is asked to output $\mathbf{x} = (x_1, \ldots, x_{n+1})$ such that $x_i = CDH(y_i, h)$ for all $i = \{1, \ldots, n + 1\}$, with at most n queries to an oracle $sDH(\cdot, h)$.

Now, we recall the trapdoor test algorithm (with compatible notations) in the following lemma, please refer to [16] for details.

Lemma 1 (Trapdoor Test [16]). *Let $\mathbb{G}$ be a cyclic group of prime order q, generated by $g \in \mathbb{G}$. Let $y \in \mathbb{G}$ be an element of $\mathbb{G}$, and $r, r' \in \mathbb{Z}_q$ are uniformly distributed over $\mathbb{Z}_q$. Define $z = g^{r'}/y^r$. Then, for any elements $h, f_1, f_2 \in \mathbb{G}$, we have: 1) z is uniformly distributed over $\mathbb{G}$; 2) y and z are independent, then the probability that the truth value of*

$$f_1^r f_2 = h^{r'} \tag{1}$$

does not agree with the truth value of

$$f_1 = CDH(y, h) \wedge f_2 = CDH(z, h) \tag{2}$$

is at most $1/q$; moreover, if (2) holds, then (1) certainly holds.

Note that the probability in the above lemma is over the random choices of r and r', and is independent of the choices of h, f_1 and f_2. This fact is very important for our separation results. For simplicity, we denote the PPT algorithm in the above lemma as Test, and assume that it works in two phases. In the initial phase, it takes the parameter $param = (k, \mathbb{G}, q, g, h)$, a group element $y \in \mathbb{G}$, and randomness $r, r' \in \mathbb{Z}_q$ as inputs, returns a group element $z = g^{r'}/y^r \in \mathbb{G}$, i.e., $z \leftarrow$ Test$(\mathbf{init}, param, y; r, r')$. In the finish phase, it takes another two elements (f_1, f_2) as inputs, returns a bit $e \in \{0, 1\}$ that indicates whether the condition $f_1^r f_2 = h^{r'}$ holds, in brief, $e \leftarrow$ Test$(\mathbf{finish}, param, y, z, f_1, f_2; r, r')$. Besides, we say that the algorithm Test *fails* if it returns 1, but at least one of the two tuples (y, h, f_1) and (z, h, f_2) is not a CDH tuple. By Lemma 1, the probability that the Test algorithm *fails* is at most $1/q$ (which is negligible in the security parameter k), where the probability is over random choices of $r, r' \leftarrow_r \mathbb{Z}_q$.

3.1 An Unbounded Adversary

In this subsection, we present an unbounded adversary $\mathcal{A}$ (depicted in Fig. 3) that solves the n-sDH problem for $n \geq 2$. Informally, the adversary $\mathcal{A}$ only makes random queries to its oracle, and delays the verification of the odd-numbered response to the point immediately after obtaining the next even-numbered response from its oracle (*i.e.*, it verifies the responses from its oracle two by two). Besides, $\mathcal{A}$ always makes even number of queries to its oracle, and omits the last query if n is odd. As in [37,25], a random function G is used by $\mathcal{A}$ to generate its inner random coins with its own view as input.

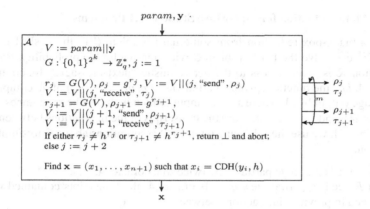

Fig. 3. The adversary $\mathcal{A}$ uses a random function G to generate all its internal randomness

Description of $\mathcal{A}$. Given the public parameter $param = (k, \mathbb{G}, q, g, h)$ with security parameter k, and an n-sDH instance $\mathbf{y} = (y_1, \ldots, y_{n+1})$, $\mathcal{A}$ is asked to compute the solution of $\mathbf{y}$, with at most n queries to its sDH$(\cdot, h)$ oracle. Let $V = param\|\mathbf{y}$, *i.e.*, the view of $\mathcal{A}$. Then, the adversary $\mathcal{A}$ randomly chooses a function G from all functions $\{0,1\}^{2^k} \to \mathbb{Z}_q^*$, and let $m = \lfloor \frac{n}{2} \rfloor$.

For $j = \{1, 3, \ldots, 2m-1\}$, $\mathcal{A}$ computes $r_j \leftarrow G(V)$, and $\rho_j = g^{r_j}$. Then, it updates the view $V := V\|(j, \text{"send"}, \rho_j)$ and sends an external sDH query with ρ_j. After obtaining the solution τ_j of ρ_j, $\mathcal{A}$ updates the view $V := V\|(j, \text{"receive"}, \tau_j)$. Then, it makes another sDH query ρ_{j+1} in the same way as ρ_j by using randomness $r_{j+1} \leftarrow G(V)$, and obtains τ_{j+1} from its sDH oracle. If $\tau_j \neq h^{r_j}$ or $\tau_{j+1} \neq h^{r_{j+1}}$, $\mathcal{A}$ returns $\bot$ and aborts; else let $j := j + 2$.

After completing $2m$ sDH queries without abort, $\mathcal{A}$ computes $x_i = \text{CDH}(y_i, h)$ for all $i \in \{1, \ldots, n+1\}$ by brute-force search (which is not necessarily done in polynomial time), and outputs $\mathbf{x} = (x_1, \ldots, x_{n+1})$ as the solution to its own challenge $\mathbf{y} = (y_1, \ldots, y_{n+1})$.

The use of the random function G brings us two benefits. First, the distribution of each sDH query of $\mathcal{A}$ is uniformly random over $\mathbb{G}$, which allows our meta-reduction $\mathcal{M}$ to (statistically) hide its real "intention". Second, it is hard for the reduction $\mathcal{R}$ to obtain (significant) advantage by rewinding $\mathcal{A}$, since $\mathcal{A}$ always generates the same sDH query at the same view, and a random sDH query at a freshly different view.

Besides, the way that the adversary $\mathcal{A}$ verifies the responses from its oracle two by two is very crucial for our separation results, which allows our meta-reduction $\mathcal{M}$ to "delay" the verification of the odd-numbered sDH response from the reduction $\mathcal{R}$ (actually, it cannot efficiently do the verification if DDH is hard), and to embed two sDH queries needed by the Test algorithm into two consecutive sDH queries to $\mathcal{R}$.

3.2   The Meta-reduction for One-More Static CDH Problems

Let $\mathcal{R}$ be a black-box reduction from some hard problem $\mathcal{C}$ to the n-sDH problem, namely, $\mathcal{R}^{\mathcal{A}}$ can solve the hard problem $\mathcal{C}$ with non-negligible probability. Basically, the reduction $\mathcal{R}$ is given access to the deterministic "next-messages" function of the adversary $\mathcal{A}$, i.e., the function, given a partial view $(x, m_1, \ldots, m_j)$ of $\mathcal{A}$, computes the next message output by $\mathcal{A}$, where x is the input (which includes the randomness that $\mathcal{R}$ chooses for $\mathcal{A}$), and $(m_1, \ldots, m_j)$ are the transcripts of the interactions between $\mathcal{R}$ and $\mathcal{A}(x)$. As in [37], we use the following (standard) assumptions about $\mathcal{R}$ to simplify our presentation:

- $\mathcal{R}$ never feeds the same partial view twice to its oracle $\mathcal{A}$;
- When $\mathcal{R}$ feeds a partial view q to its oracle $\mathcal{A}$, the transcripts contained in q are generated in previous interactions between $\mathcal{R}$ and $\mathcal{A}$.

Both of the two assumptions are without loss of generality, since the oracle $\mathcal{A}$ is a "deterministic function" and we can easily modify $\mathcal{R}$ to satisfy the two conditions. Besides, in order to better illustrate how our meta-reduction works, and how the Test algorithm is injected, we denote instance $\mathcal{A}_i$ as a copy of $\mathcal{A}$ on a specified input x (i.e., $\mathcal{A}(x)$). In particular, a unique positive integer i is used for each different input x (recall that x includes the randomness that $\mathcal{R}$ chooses for $\mathcal{A}$).

In the concurrent zero-knowledge protocols [41,40,38], the term "**slot**" usually denotes the point where a rewinding is possible. We adapt this notion to our case, which is slightly different from that in [37]. Intuitively, a **slot** in our context always opens with an odd-numbered query and ends with the response of the immediately followed even-numbered query. Formally, let V_R be the view of $\mathcal{R}$, which includes all the messages sent and received by $\mathcal{R}$ in the interactions with both the adversary $\mathcal{A}$ and the challenger $\mathcal{C}$. A partial view of V_R is a prefix of V_R. For some integer $j \geq 1$, we say that a **slot** s_i of $\mathcal{A}_i$ opens at a partial view $V_o^{s_i}$ if $\mathcal{A}_i$ sends the $(2j - 1)$-th query q_1 to $\mathcal{R}$ immediately after $V_o^{s_i}$ (note that q_1 must be a "fresh" query of $\mathcal{A}_i$ by our simplification assumption), and we say that a **slot** s_i of $\mathcal{A}_i$ closes at a partial view $V_c^{s_i}$ if $\mathcal{A}_i$ receives a response r_2 to the $2j$-th query q_2 from $\mathcal{R}$ at the end of $V_c^{s_i}$. In particular, we denote the partial view $V_o^{s_i}$ as the **opening** of s_i, and the partial view $V_c^{s_i}$ as the **closing** of s_i. Since a **slot** may open without being closed (i.e., $\mathcal{R}$ may never respond to some query q_i sent by $\mathcal{A}_i$), we uniquely identify a **slot** s_i by its **opening**. We also denote a particular closed **slot** s_i as a pair of its **opening** and **closing** for convenience, i.e., $s_i = (V_o^{s_i}, V_c^{s_i})$.

Definition 3 (Good Slot). *Let* $M = M(k)$ *be the maximum number of the messages sent and received by the reduction* $\mathcal{R}$ *on input the security parameter* k. *For any positive integer* d, *we say that a closed **slot*** $s_i = (V_o^{s_i}, V_c^{s_i})$ *(of* $\mathcal{A}_i$) *is* d-***good*** *if the following two conditions hold:*

- *Between the time that* s_i *opens and the time that* s_i *closes,* $\mathcal{R}$ *does not interact with the challenger* $\mathcal{C}$;
- *Between the time that* s_i *opens and the time that* s_i *closes, the number of **slots** that open is at most* $\frac{M}{n^d}$.

Informally, the definition of a **good** slot brings us three benefits. First, since the reduction $\mathcal{R}$ does not interact with the challenger $\mathcal{C}$ during the **slot**, rewinding the reduction $\mathcal{R}$ to the opening of a **good** slot will not affect the interactions between $\mathcal{R}$ and $\mathcal{C}$. Second, each **slot** always contains two consecutive (fresh) queries, which allows our meta-reduction $\mathcal{M}$ to embed two sDH queries in a **slot**. Third, each **slot** always opens with an odd-numbered query, $\mathcal{M}$ can delay the verification of the response to the first query, and simultaneously check the validity of the two responses from $\mathcal{R}$ (by using the Test algorithm).

We make use of recursive rewinding techniques in [41,38,20,15] to rule out general black-box reductions (*i.e.*, without any additional restrictions on the reductions), which was recently introduced by Pass [37] to give impossibility results for a class of witness-hiding protocols. Basically, we provide the meta-reduction $\mathcal{M}$ with many rewinding chances (*i.e.*, **slots**), and let $\mathcal{M}$ be always "on the lookout" for **good slots** to rewind $\mathcal{R}$ such that the rewinding will not "blow up" the running time of $\mathcal{M}$ too much (*e.g.*, running in an exponential time). Formally, we have the following theorem.

Theorem 3 (Black-Box Separations for One-More Static CDH Problems). *For any integers* $t(k)$ *and* $n = 2 \cdot \omega(k + t + 1)$, *there is no black-box reduction* $\mathcal{R}$ *for basing the hardness of the* n-sDH *problem on any* t-*round hard problem* $\mathcal{C}$ *(or else* $\mathcal{C}$ *could be solved efficiently), where* k *is the security parameter.*[4]

Proof. We now proceed to give the description of our meta-reduction $\mathcal{M}$, which recursively calls a procedure SOLVE to rewind the reduction $\mathcal{R}$. In the simulation of the adversary $\mathcal{A}$ to $\mathcal{R}$, the meta-reduction $\mathcal{M}$ has to maintain three technical tables $\mathcal{L}_1$, $\mathcal{L}_2$ and $\mathcal{L}_3$. Informally, the first table $\mathcal{L}_1$ is used to record all the n-sDH instances (that $\mathcal{M}$ has to solve) and the corresponding solutions (that have been found by $\mathcal{M}$). The other two tables are used to successfully inject the Test algorithm into the simulation, where table $\mathcal{L}_2$ records the randomness used by the Test algorithm (for each target sDH instance) and table $\mathcal{L}_3$ records the randomness used to re-randomize each sDH query made by the Test algorithm. For functionality, table $\mathcal{L}_1$ is used in a "call by reference" method, namely, the changes of $\mathcal{L}_1$ at the $(d + 1)$-th recursive level will be reflected at the d-th recursive level. But this is not required for table $\mathcal{L}_2$ and table $\mathcal{L}_3$. Formally,

---

[4] Basically, the constant '2' can be safely absorbed by the asymptotic function '$\omega(\cdot)$'. We leave it here mainly because it is introduced by our "delay verification" proof technique, which is different from previous results, *e.g.*, [37]. We also hope this can give a clear relation between the results in Theorem 3 and its generalized results in Theorem 4.

- Table $\mathcal{L}_1$ consists of four tuples $(i, param_i \| \mathbf{y}_i, b_i, \mathbf{x}_i)$ and indicates that 1) $\mathcal{R}$ invokes the i-th instance of $\mathcal{A}$ with parameter $param_i$ and $n + 1$ sDH instances $\mathbf{y}_i = (y_{i,1}, \cdots, y_{i,n+1})$ as inputs; 2) $\mathcal{M}$ has found the first b_i solutions of $\mathbf{y}_i$, and stored them in $\mathbf{x}_i$ (*i.e.*, $|\mathbf{x}_i| = b_i$). Thus, $\mathcal{L}_1 = \perp$ when $\mathcal{R}$ is invoked (by $\mathcal{M}$), and $(b_i = 0, \mathbf{x}_i = \perp)$ when a new instance of $\mathcal{A}$ is invoked (by $\mathcal{R}$).
- Table $\mathcal{L}_2$ consists of six tuples $(i, t_i, y_{i,t_i}, r_{i,t_i}, r'_{i,t_i}, z_{i,t_i})$ and indicates that the i-th instance of $\mathcal{A}$ prepares an "aided" element z_{i,t_i} with randomness r_{i,t_i} and r'_{i,t_i}, and aims to find the solution of the t_i-th sDH instance y_{i,t_i}, *i.e.*, $z_{i,t_i} \leftarrow$ Test($\mathbf{init}, param_i, y_{i,t_i}; r_{i,t_i}, r'_{i,t_i}$).
- Table $\mathcal{L}_3$ consists of five tuples $(i, t_i, \delta, u_i, \rho_i)$ that indicates the actual query made by $\mathcal{A}_i$, where $\delta \in \{0, 1\}$ and $u_i \in \mathbb{Z}_q^*$. If $\delta = 0$, it means that $\mathcal{A}_i$ sends an odd-numbered sDH query with $\rho_i = y_{i,t_i}^{u_i}$. Else if $\delta = 1$, it means that $\mathcal{A}_i$ sends an even-numbered sDH query with $\rho_i = z_{i,t_i}^{u_i}$.

Description of $\mathcal{M}$. Given a security parameter k, a description of $\mathcal{C}$ instance, let $V_R = k \| \mathcal{C}$, $\mathcal{M}$ runs $\mathcal{R}$ with $\mathcal{C}$, and executes SOLVE($1^k, 0, V_R, \perp, \perp, \perp$) to simulate the unbounded adversary $\mathcal{A}$. Whenever $\mathcal{R}$ outputs the solution of $\mathcal{C}$, $\mathcal{M}$ outputs it and halts. Let $c = \lceil \log_k M \rceil$, for each level $0 \le d \le c$, procedure SOLVE works as follows:

SOLVE($1^k, d, V, \mathcal{L}_1, \mathcal{L}_2, \mathcal{L}_3$):
On input a security parameter k, the current recursive level d, the partial view V of $\mathcal{R}$ and three tables $\mathcal{L}_1, \mathcal{L}_2$ and $\mathcal{L}_3$, let $v = V$ and repeat the following steps:

1. If $d = 0$ and $\mathcal{R}$ makes external interactions with $\mathcal{C}$, simply relay the messages between $\mathcal{R}$ and $\mathcal{C}$, and update the view v.
2. If $d > 0$ and $\mathcal{R}$ attempts to make external interactions with $\mathcal{C}$ or the number of **slots** opened after V in v exceeds $\frac{M}{k^d}$, cancel the current recursive level and return. (Note that this case happens if and only if the probability that V becomes the **opening** of a d-**good** slot is non-negligible. Thus, the algorithm can simply cancel the current recursive rewinding at level d and return to the $(d-1)$-th level whenever the slot starting from V cannot be d-**good** anymore.)
3. If $\mathcal{R}$ feeds $\mathcal{A}$ with a partial view which only contains the input message, denote $(param, \mathbf{y})$ as the corresponding n-sDH instance.[5] Choose an unused smallest positive integer i for this instance $\mathcal{A}_i$ of $\mathcal{A}$, and add $(i, param \| \mathbf{y}, 0, \perp)$ to $\mathcal{L}_1$. Finally, update the view v.
4. If $\mathcal{R}$ feeds $\mathcal{A}_i$ with a partial view which contains a response τ_i to a previous sDH instance query ρ_i from $\mathcal{A}_i$, update the view v and proceed as follows:
   - If ρ_i is the $(2j-1)$-th query of $\mathcal{A}_i$ for some $j \ge 1$, continue;
   - If ρ_i is the $2j$-th query of $\mathcal{A}_i$ for some $j \ge 1$, let (ρ'_i, τ'_i) and (ρ_i, τ_i) be the last two consecutive pairs of query and response of $\mathcal{A}_i$, retrieve $(i, t_i, 0, u'_i, \rho'_i)$ and $(i, t_i, 1, u_i, \rho_i)$ from $\mathcal{L}_3$, and $(i, t_i, y_{i,t_i}, r_{i,t_i}, r'_{i,t_i}, z_{i,t_i})$ from $\mathcal{L}_2$, such that $\rho'_i = y_{i,t_i}^{u'_i}$ and $\rho_i = z_{i,t_i}^{u_i}$, and compute $f_1 = (\tau'_i)^{1/u'_i}, f_2 = (\tau_i)^{1/u_i}$. Then, if Test($\mathbf{finish}, param_i, y_{i,t_i}, z_{i,t_i}, f_1, f_2; r_{i,t_i}, r'_{i,t_i}$) $= 0$, abort the simulation of $\mathcal{A}_i$. Otherwise, retrieve $(i, param_i \| \mathbf{y}_i, b_i, \mathbf{x}_i)$ from

---

[5] Recall that the input message contains the n-sDH instance and the randomness that $\mathcal{R}$ chooses for $\mathcal{A}$. Besides, $\mathcal{R}$ never fed the same input to $\mathcal{A}$ before by our simplification assumptions.

$\mathcal{L}_1$. If $t_i = b_i + 1$, update the tuple $(i, param_i \| \mathbf{y}_i, b_i, \mathbf{x}_i)$ in $\mathcal{L}_1$ by letting $b_i = t_i$ and $\mathbf{x}_i = \mathbf{x}_i \| f_1$ (or else, we must have $t_i \leq b_i$, which means the solution of y_{i,t_i} has been found previously). Finally, let $s_i = (V_o^{s_i}, v)$ be the **slot** with **closing** v, and distinguish the following cases:

- If s_i opened before V (*i.e.*, s_i did not open at the current recursive level), continue;
- Else if $V_o^{s_i} = V$ (*i.e.*, s_i opened at V) and $d > 0$, end the current recursive level and return.[6]
- Else if $V_o^{s_i} \neq V$ (*i.e.*, s_i opened after V) and s_i is a $(d+1)$-***good*** **slot**, repeat the procedure SOLVE($1^k, d+1, V_o^{s_i}, \mathcal{L}_1, \mathcal{L}_2, \mathcal{L}_3$) until $b_i = n+1$.
- Otherwise, continue;

5. If $\mathcal{R}$ is expecting a message from $\mathcal{A}_i$, retrieve $(i, b_i, param_i \| \mathbf{y}_i, \mathbf{x}_i)$ from $\mathcal{L}_1$ (recall that $param_i = (k, \mathbb{G}, p, g, h)$), and proceed as follows:

   - If $\mathcal{A}_i$ has completed $2m = 2 \cdot \lfloor \frac{n}{2} \rfloor$ sDH queries (*i.e.*, $\mathcal{A}_i$ has to send the solution of $\mathbf{y}_i$ to $\mathcal{R}$), send $\mathbf{x}_i$ to $\mathcal{R}$ if $b_i = n + 1$ (*i.e.*, the solution $\mathbf{x}_i$ of $\mathbf{y}_i$ has been found), else output "**fail**" and halt.
   - Else if it is the $(2j - 1)$-th query for some $j \geq 1$, let $t_i = b_i + 1$ if $b_i < n + 1$, else $t_i = n + 1$. If there is no tuple $(i, t_i, y_{i,t_i}, *, *, *)$ in $\mathcal{L}_2$, choose $r_{i,t_i}, r'_{i,t_i} \leftarrow_r \mathbb{Z}_q$, compute $z_{i,t_i} \leftarrow \text{Test}(\textbf{init}, param_i, y_{i,t_i}; r_{i,t_i}, r'_{i,t_i})$, and add the tuple $(i, t_i, y_{i,t_i}, r_{i,t_i}, r'_{i,t_i}, z_{i,t_i})$ to $\mathcal{L}_2$. Then, choose $u_i \leftarrow_r \mathbb{Z}_q^*$, send $\rho_i = y_{i,t_i}^{u_i}$ to $\mathcal{R}$, and add the tuple $(i, t_i, 0, u_i, \rho_i)$ to $\mathcal{L}_3$.
   - Else if it is the $2j$-th query for some $j \geq 1$, let ρ_i be the $(2j - 1)$-th query of $\mathcal{A}_i$, retrieve $(i, t_i, 0, u_i, \rho_i)$ and $(i, t_i, y_{i,t_i}, *, *, z_{i,t_i})$ from $\mathcal{L}_3$ and $\mathcal{L}_2$, respectively, such that $\rho_i = y_{i,t_i}^{u_i}$. Then, choose $u'_i \leftarrow_r \mathbb{Z}_q^*$, send $\rho'_i = z_{i,t_i}^{u'_i}$ to $\mathcal{R}$, and add the tuple $(i, t_i, 1, u'_i, \rho'_i)$ to $\mathcal{L}_3$.

   Finally, update the view v accordingly.
6. If $\mathcal{R}$ returns the solution of $\mathcal{C}$, output the solution and halt.

Remark 1. By our simplification assumptions, $\mathcal{R}$ never feeds the same partial view twice to $\mathcal{A}$. This allows $\mathcal{M}$ to simply prepare each sDH query (on behalf of $\mathcal{A}$) by using freshly chosen randomness, since our unbounded adversary $\mathcal{A}$ (in Section 3.1) uses a random function G to deterministically generate its "inner" randomness by using the interaction transcripts with $\mathcal{R}$ as input.

To show that our meta-reduction $\mathcal{M}$ can efficiently solve problem $\mathcal{C}$ (with non-negligible probability), we only have to show that 1) $\mathcal{M}$ perfectly simulates the unbounded adversary $\mathcal{A}$ except with negligible probability; 2) $\mathcal{M}$ runs in expected polynomial time. We prove the two claims in the following two lemmas.

Lemma 2. *$\mathcal{M}$ perfectly simulates the unbounded adversary $\mathcal{A}$ except with negligible probability.*

---

[6] This means that the rewinding at the partial view V (at level d) is successful, and the algorithm returns to the $(d - 1)$-th level to check whether it has found all the solutions of $\mathbf{y}_i$ for instance $\mathcal{A}_i$ (*i.e.*, to check whether $b_i = n + 1$).

Proof. Since $\mathcal{M}$ always randomizes its sDH queries to $\mathcal{R}$ (on behalf of $\mathcal{A}$) by using uniformly chosen randomness from $\mathbb{Z}_q^*$ (*i.e.*, u_i or u_i' in step 5), the distribution of these queries is essentially the same to that of $\mathcal{A}$ (which always makes random sDH queries to $\mathcal{R}$). We finish the proof of this lemma by proving the following two cases: 1) If $\mathcal{R}$ cheats in the interactions (*i.e.*, by returning a false answer to an sDH query), $\mathcal{M}$ will reject it in the same way as $\mathcal{A}$ except with negligible probability; 2) If $\mathcal{R}$ does not cheat, $\mathcal{M}$ can find the correct solution $\mathbf{x}_i$ of the n-sDH input $\mathbf{y}_i$ of instance $\mathcal{A}_i$ from table $\mathcal{L}_1$ (*i.e.*, it will not output "**fail**" in step 5).

For the first case, since both the meta-reduction $\mathcal{M}$ and the unbounded adversary $\mathcal{A}$ do not immediately check the validity of any odd-numbered response, the simulation of $\mathcal{A}$ after receiving an odd-numbered response from $\mathcal{R}$ (*i.e.*, the first case in step 4) is essentially the same as that of the real adversary $\mathcal{A}$. After receiving an even-numbered query, the unbounded adversary $\mathcal{A}$ will check the validity of the previous two consecutive responses, and will always return $\perp$ and abort if one of the previous two responses is invalid. As for the meta-reduction $\mathcal{M}$, it always embeds the first query of the Test algorithm in an odd-numbered query, and the second query of the Test algorithm in the immediately followed even-numbered query, and then checks the two consecutive responses by using the Test algorithm. Obviously, if the Test algorithm does not *fail*, $\mathcal{M}$ can perfectly detect whether $\mathcal{R}$ cheats or not, which is essentially the same as $\mathcal{A}$. Now, we show that the probability that the Test algorithm *fails* at least one time is negligible. Recall that the total number of the messages sent and received by $\mathcal{R}$ is bounded by $M(k)$, for any partial view v of $\mathcal{R}$, there are at most $M(k)$ messages in v. Thus, $\mathcal{M}$ has to run the Test algorithm at most $(n+1)M(k)$ times (since $\mathcal{R}$ can invoke at most $M(k)$ instances of $\mathcal{A}$, and for each instance $\mathcal{A}_i$, $\mathcal{M}$ has to run the Test algorithm at most $(n+1)$ times). Since $\mathcal{M}$ always independently and randomly chooses the randomness for each time running of the Test algorithm (in step 5), the probability that the Test algorithm will *fail* at least one time is at most $\frac{(n+1)M(k)}{q}$ by Lemma 1, which is negligible in k.

For the second case, let v be the partial view of $\mathcal{R}$, at which $\mathcal{R}$ expects the simulated instance $\mathcal{A}_i$ to provide the solution of its input n-sDH $\mathbf{y}_i$. Since the unbounded adversary $\mathcal{A}$ always sequentially makes $2m = 2 \cdot \lfloor \frac{n}{2} \rfloor$ sDH queries, there must exist at least m **slots** of $\mathcal{A}_i$ in the partial view v. Recall that the total number of recursive levels is bounded by $c = \lceil \log_k M \rceil$ (which is a constant if $\mathcal{R}$ runs in polynomial time), there must exist some recursive level d such that there are at least $\frac{m}{c+1}$ **slots** of $\mathcal{A}_i$ in the partial view v (by the pigeon hole principle). Since $n = 2 \cdot \omega(k + t + 1)$, we have $\frac{m}{c+1} \geq k + t + 1$ for sufficiently large k. Hereafter, we always assume that there is at least $k + t + 1$ **slots** of $\mathcal{A}_i$ at level d. By the definition of SOLVE, the total number of **slots** opened at level d is at most $\frac{M}{k^d}$ (this obviously holds at level $d = 0$). Thus, there are at least $t+1$ **slots** of $\mathcal{A}_i$ that contain at most $\frac{M}{k^{d+1}}$ **slots** (or else there are at least $k+1$ **slots** of $\mathcal{A}_i$ that contain more than $\frac{M}{k^{d+1}}$ **slots**, which makes the total number of **slots** opened at level d exceeds $\frac{M}{k^d}$, and the recursive rewinding at level d will be canceled). Since $\mathcal{C}$ is t-round, there is at least one **slot** of $\mathcal{A}_i$ (in those $t + 1$ **slots**) during which $\mathcal{R}$ has no interactions with the challenger $\mathcal{C}$. Obviously, such a **slot** is $(d + 1)$-***good***, thus will be rewound. In all, we have proven that for each complete instance $\mathcal{A}_i$, there must exist at least one ***good*** **slot** that would be rewound. Since $\mathcal{M}$ will end the recursive calls to SOLVE at the opening of a $(d+1)$-**good** **slot** in step 4 until $b_i = n+1$, $\mathcal{M}$ can always find the solution $\mathbf{x}_i$ of the input n-sDH instance $\mathbf{y}_i$ of $\mathcal{A}_i$ from table $\mathcal{L}_1$ before it has to send the solution of $\mathbf{y}_i$ to $\mathcal{R}$ (*i.e.*, $\mathcal{M}$ will not output "**fail**" in step 5). Since $\mathcal{R}$ does not cheat in this case, $\mathbf{x}_i$ must be the correct (unique) solution of $\mathbf{y}_i$. This completes the proof of Lemma 2.

Lemma 3. $\mathcal{M}$ *runs in expected polynomial time.*

Proof. We estimate the maximum running time of $\mathcal{M}$ that never outputs "**fail**" and halts in the simulation. Since the total number of the messages sent and received by $\mathcal{R}$ is bounded by $M(k)$, there are at most $M(k)$ **good** slots that might be rewound at each recursive level. Let v be a partial view (at level d) immediately after which a **slot** s opens. Then, let δ be the probability that s becomes $(d+1)$-**good**, where the probability is over all the randomness used in the interactions between $\mathcal{M}$ and $\mathcal{R}$ after v. Now, assume that we arrived at a partial view v' such that $s = (v, v')$ is $(d+1)$-**good**, then if $\mathcal{M}$ rewinds $\mathcal{R}$ to v (i.e., rewinding $\mathcal{R}$ to the opening of s at recursive level $d + 1$), the probability that the slot becomes $(d + 1)$-**good** is essentially close to δ. This is because the behavior of the simulated adversary (by $\mathcal{M}$) after rewinding $\mathcal{R}$ to v (i.e., at level $d + 1$) is almost identically distributed to that starting from v at level d. Since δ is non-negligible (or else it is unlikely to arrive at the partial view v'), $\mathcal{M}$ can expect to obtain a $(d + 1)$-**good** **slot** with probability negligibly close to 1 by rewinding $\mathcal{R}$ polynomial times at v. Let $p(k)$ be such a polynomial. Then, $\mathcal{M}$ can expect to find the solution of some $y_{i,j}$ for $\mathcal{A}_i$ with probability negligibly close to 1 by rewinding $\mathcal{R}$ at most $p(k)$ times. Thus, for each of those **good** slots, the expected number of rewindings is bounded by $(n + 1)p(k)$ (recall that $\mathcal{M}$ has to find the solutions of $(n + 1)$ sDH instances for each $\mathcal{A}_i$), and the total number of rewindings at each recursive level is expected at most $(n + 1)p(k)M(k)$. By an induction computation, we have the expected number of the messages sent and received by $\mathcal{M}$ is bounded by $((n + 1)p(k)M(k))^{c+1}$, which is a polynomial in k if $M(k)$ is a polynomial in k. This completes the proof of Lemma 3.

4   More Black-Box Separations for One-More Problems

Let $P_1 = (\mathsf{PGen}, \mathsf{IGen}_1, \mathsf{Vrfy}_1)$ and $P_2 = (\mathsf{PGen}, \mathsf{IGen}_2, \mathsf{Vrfy}_2)$ be two non-interactive cryptographic problems with the same parameter generator PGen. Now, we give the definition of the generalized "one-more" problems.

Definition 4 (Generalized "One-More" Problems). *For any integer $n \geq 0$, the generalized "one-more" problem n-$(P_1, P_2) = (\mathsf{PGen}, \mathsf{IGen}, \mathsf{Orcl}, \mathsf{Vrfy})$ associated with two subproblems P_1 and P_2 is defined as follows:*

- *The parameter generator $\mathsf{PGen}(1^k)$ algorithm returns the public parameter param for P_1 and P_2.*
- *The instance generator $\mathsf{IGen}(param)$ independently runs the $\mathsf{IGen}_1(param)$ algorithm $n + 1$ times to generate $(n + 1)$ random P_1 instances $\{y_i\}_{i\in\{1,...,n+1\}}$, and returns an instance $\mathbf{y} = (y_1, \ldots, y_{n+1})$.*

- *The stateful oracle algorithm* Orcl *takes as inputs a public parameter* $param$, *and a* P_2 *instance* y, *returns the solution* x *of* y *or a special symbol* $\perp$ *if* y *is an invalid query. (If* P_2 *is not a unique solution problem, the oracle can return one of the candidate solutions in a predefined rule, e.g., the first one in lexicographic order.)*
- *The verification algorithm* Vrfy *takes the public parameter* $param$, *an instance* $\mathbf{y} = (y_1, \ldots, y_{n+1})$, *and a candidate solution* $\mathbf{x} = (x_1, \ldots, x_{n+1})$ *as inputs, returns 1 if and only if* $\mathsf{Vrfy}_1(param, y_i, x_i) = 1$ *for all* $i \in \{1, \ldots, n+1\}$.

In particular, if $P_1 = P_2$, *we briefly denote* n-(P_1, P_2) *as* n-P_1. *Besides, the* n-(P_1, P_2) *problem is said to be hard if for any PPT adversary* $\mathcal{A}$, *the advantage of* $\mathcal{A}$ *in solving all the* $(n + 1)$ *random* P_1 *instances in* $\mathbf{y}$ *with at most* n P_2 *queries to the oracle* Orcl *is negligible.*

This class of problems not only subsumes the traditional one-more problems (where $P_1 = P_2$, *e.g.*, n-RSA, n-DL), the unforgeability of two-move blind signatures (where it is likely $P_1 \neq P_2$) and so forth, but also encompasses many other interesting problems that have not been (well) studied in the literature. For example, to solve $n + 1$ BDH instances by using n DL queries, *i.e.*, n-$(\mathrm{BDH}, \mathrm{DL})$ in our notation [18].

Our separation results in Theorem 3 can be generalized to the n-(P_1, P_2) problem if there is a PPT reduction T which can be used to solve one P_1 instance by only using γ (non-adaptive) queries to a P_2 oracle, where γ can be any constant (*e.g.*, $\gamma = 2$ for the n-sDH problem). In particular, we assume that reduction T works in two phases: Given the public parameter $param$, a P_1 instance y, and a randomness r, it enters into the **query** phase and outputs a vector of P_2 instances $\mathbf{z} = (z_1, \ldots, z_\gamma)$. In brief, $\mathbf{z} \leftarrow \mathsf{T}(\mathbf{query}, param, y; r)$. After being fed back with the solution vector $\mathbf{f} = (f_1, \ldots, f_\gamma)$ of $\mathbf{z}$ where f_i is a candidate solution of z_i, T enters into the **finish** phase, and returns the solution x of y or a special symbol $\perp$. In brief, $x/\perp \leftarrow \mathsf{T}(\mathbf{finish}, param, y, \mathbf{z}, \mathbf{f}; r)$. Informally, we say that a reduction T is a "promise reduction" if it always tries to return a correct answer, otherwise returns $\perp$ to indicate that "some of its inputs are invalid".

Definition 5 (Promise Reduction). *Let* Ω_t *be the randomness space of* T, *we say that a reduction* T *from* P_1 *to* P_2 *is a promise reduction if it satisfies the following two properties:*

Efficient Computability: *There is a PPT algorithm that computes* T.

Correctness-Preserving: *Fixing the parameter* $(param, y)$, *for any* $\mathbf{z} \leftarrow \mathsf{T}(\mathbf{query}, param, y; r)$, *any candidate solutions* $\mathbf{f}$ *of* $\mathbf{z}$, *and* $x \leftarrow \mathsf{T}(\mathbf{finish}, param, y, \mathbf{z}, \mathbf{f}; r)$, *we have*

- *If* $\mathsf{Vrfy}_2(param, z_i, f_i) = 1$ *holds for all* $i \in \{1, \ldots, \gamma\}$, *we have that* $x \neq \perp$ *and* $\mathsf{Vrfy}_1(param, y, x) = 1$ *hold;*
- *If there exists* $i \in \{1, \ldots, \gamma\}$ *such that* $\mathsf{Vrfy}_2(param, z_i, f_i) = 0$, *then we have* $x = \perp$ *with overwhelming probability;*

where the probabilities are over the random choice of $r \leftarrow_r \Omega_t$.

Remark 2. The first requirement on the correctness-preserving property of T can be relaxed to "hold with non-negligible probability" if P_1 is efficiently verifiable (*i.e.*, Vrfy_1 is a PPT algorithm). Since we can repeat the reduction T polynomial times (at a cost of slightly increasing the running time of $\mathcal{M}$) to get a correct solution with probability negligibly close to 1. The strong requirement is used here for simplicity.

Theorem 4 (Black-Box Separation for Generalized "One-More" Problems). *For integer $n > 0$, let n-(P_1, P_2) be defined as in Definition 4. If P_1 has unique solution, P_2 is randomly self-reducible and there is a promise reduction T from P_1 to P_2 with at most γ queries. Then, there is no black-box reduction $\mathcal{R}$ for basing the hardness of the n-(P_1, P_2) problem on any $t(k)$-round hard problem $\mathcal{C}$ (or else $\mathcal{C}$ could be solved efficiently), where k is the security parameter and $n = \gamma \cdot \omega(k + t + 1)$.*

The proof is very similar to the proof of Theorem 3, we defer it to the full version.

Remark 3. The requirement on $n = \gamma \cdot \omega(k+t+1)$ is needed to successfully apply "recursive rewinding" [41,38,20,15] to rule out general black-box reductions. However, if one would like to consider restricted black-box reductions—single-instance reductions [26,25], a tighter separation result for $n \geq \gamma \cdot (t + 1)$ can be achieved.

Since our generalized "one-more" problems abstract many interesting problems, Theorem 4 actually gives a very broad impossibility result for a large class of problems. For the three traditional one-more inversion problems we have the following corollary.

Corollary 1. *There is no black-box reduction $\mathcal{R}$ for basing the hardness of n-DL, n-RSA, or n-sDH over gap Diffie-Hellman groups on any $t(k)$-round hard problem $\mathcal{C}$ (or else $\mathcal{C}$ could be solved efficiently), where k is the security parameter and $n = \omega(k + t + 1)$.*

Actually, our separation results naturally apply to the n-CDH problem, which can be directly defined based on CDH problems as in Definition 4 (*i.e.*, $P_1 = P_2 = \text{CDH}$).

Corollary 2. *There is no black-box reduction $\mathcal{R}$ for basing the hardness of the n-CDH problem over general groups on any $t(k)$-round hard problem $\mathcal{C}$ (or else $\mathcal{C}$ could be solved efficiently), where k is the security parameter and $n = 2 \cdot \omega(k + t + 1)$.*

Since the one-more unforgeability of blind signatures can be treated as a standard generalized "one-more" problems, our separation results actually apply to a class of two-move blind signatures [17,8,30] that are statistically blinding [32,39] and allow statistical signature-derivation check [26]. We defer the details to the full version.

Corollary 3. *If a two-move (unique) blind signature BS is statistically blinding and allows statistical signature-derivation check, then there is no black-box reduction $\mathcal{R}$ for basing the one-more unforgeability of BS on any polynomially round hard problem $\mathcal{C}$ (or else $\mathcal{C}$ could be solved efficiently).*

Besides, our results also apply to many other interesting problems that may not have been (well) studied in the literature. For example, if P_2 is the DL problem, P_1 can be any other DL-based problems such as CDH, DDH and BDH.

Finally, we clarify that our impossibility result only rules out black-box reductions from other (standard) hard problems to this class of problems, it does not mean the problems in the class are easy to solve, or there are no non-black-box reductions basing the hardness of these problems on other (standard) hard problems. In fact, some of them might be very useful in proving or analyzing cryptographic constructions.

Acknowledgments. We thank Yi Deng, Phong Q. Nguyen, and the anonymous reviewers of CRYPTO 2014 and ASIACRYPT 2014 for their helpful comments and suggestions on earlier versions of our paper.

References

1. Abadi, M., Feigenbaum, J., Kilian, J.: On hiding information from an oracle. Journal of Computer and System Sciences 39(1), 21–50 (1989)
2. Bellare, M., Namprempre, C., Neven, G.: Security proofs for identity-based identification and signature schemes. In: Cachin, C., Camenisch, J.L. (eds.) EUROCRYPT 2004. LNCS, vol. 3027, pp. 268–286. Springer, Heidelberg (2004)
3. Bellare, M., Namprempre, C., Neven, G.: Security proofs for identity-based identification and signature schemes. Journal of Cryptology 22(1), 1–61 (2009)
4. Bellare, M., Namprempre, C., Pointcheval, D., Semanko, M.: The power of RSA inversion oracles and the security of Chaum's RSA-based blind signature scheme. In: Syverson, P.F. (ed.) FC 2001. LNCS, vol. 2339, pp. 309–328. Springer, Heidelberg (2002)
5. Bellare, M., Namprempre, C., Pointcheval, D., Semanko, M.: The one-more-RSA-inversion problems and the security of Chaum's blind signature scheme. Journal of Cryptology 16(3), 185–215 (2003)
6. Bellare, M., Neven, G.: Transitive signatures: new schemes and proofs. IEEE Transactions on Information Theory 51(6), 2133–2151 (2005)
7. Bellare, M., Palacio, A.: GQ and Schnorr identification schemes: Proofs of security against impersonation under active and concurrent attacks. In: Yung, M. (ed.) CRYPTO 2002. LNCS, vol. 2442, pp. 162–177. Springer, Heidelberg (2002)
8. Boldyreva, A.: Threshold signatures, multisignatures and blind signatures based on the Gap-Diffie-Hellman-group signature scheme. In: Desmedt, Y.G. (ed.) PKC 2003. LNCS, vol. 2567, pp. 31–46. Springer, Heidelberg (2002)
9. Boneh, D., Venkatesan, R.: Breaking RSA may not be equivalent to factoring. In: Nyberg, K. (ed.) EUROCRYPT 1998. LNCS, vol. 1403, pp. 59–71. Springer, Heidelberg (1998)
10. Bresson, E., Monnerat, J., Vergnaud, D.: Separation results on the "one-more" computational problems. In: Malkin, T. (ed.) CT-RSA 2008. LNCS, vol. 4964, pp. 71–87. Springer, Heidelberg (2008)
11. Brown, D.R.L.: Irreducibility to the one-more evaluation problems: More may be less. Cryptology ePrint Archive, Report 2007/435 (2007)
12. Brown, D.R.L., Gallant, R.P.: The static Diffie-Hellman problem. Cryptology ePrint Archive, Report 2004/306 (2004)
13. Canard, S., Gouget, A., Traoré, J.: Improvement of efficiency in (unconditional) anonymous transferable e-cash. In: Tsudik, G. (ed.) FC 2008. LNCS, vol. 5143, pp. 202–214. Springer, Heidelberg (2008)
14. Canetti, R., Gennaro, R., Jarecki, S., Krawczyk, H., Rabin, T.: Adaptive security for threshold cryptosystems. In: Wiener, M. (ed.) CRYPTO 1999. LNCS, vol. 1666, pp. 98–115. Springer, Heidelberg (1999)
15. Canetti, R., Lin, H., Pass, R.: Adaptive hardness and composable security in the plain model from standard assumptions. In: FOCS, pp. 541–550 (2010)
16. Cash, D., Kiltz, E., Shoup, V.: The twin Diffie-Hellman problem and applications. In: Smart, N.P. (ed.) EUROCRYPT 2008. LNCS, vol. 4965, pp. 127–145. Springer, Heidelberg (2008)
17. Chaum, D.: Blind signatures for untraceable payments. In: CRYPTO, pp. 199–203 (1982)
18. Chen, Y., Huang, Q., Zhang, Z.: Sakai-ohgishi-kasahara identity-based non-interactive key exchange scheme, revisited. In: Susilo, W., Mu, Y. (eds.) ACISP 2014. LNCS, vol. 8544, pp. 274–289. Springer, Heidelberg (2014)
19. De Cristofaro, E., Tsudik, G.: Practical private set intersection protocols with linear complexity. In: Sion, R. (ed.) FC 2010. LNCS, vol. 6052, pp. 143–159. Springer, Heidelberg (2010)
20. Deng, Y., Goyal, V., Sahai, A.: Resolving the simultaneous resettability conjecture and a new non-black-box simulation strategy. In: FOCS, pp. 251–260 (2009)
21. Dodis, Y., Haitner, I., Tentes, A.: On the instantiability of hash-and-sign RSA signatures. In: Cramer, R. (ed.) TCC 2012. LNCS, vol. 7194, pp. 112–132. Springer, Heidelberg (2012)

22. Dwork, C., Naor, M., Sahai, A.: Concurrent zero-knowledge. Journal of the ACM 51(6), 851–898 (2004)
23. Fiore, D., Schröder, D.: Uniqueness is a different story: Impossibility of verifiable random functions from trapdoor permutations. In: Cramer, R. (ed.) TCC 2012. LNCS, vol. 7194, pp. 636–653. Springer, Heidelberg (2012)
24. Fischlin, M.: Black-box reductions and separations in cryptography. In: Mitrokotsa, A., Vaudenay, S. (eds.) AFRICACRYPT 2012. LNCS, vol. 7374, pp. 413–422. Springer, Heidelberg (2012)
25. Fischlin, M., Fleischhacker, N.: Limitations of the meta-reduction technique: The case of Schnorr signatures. In: Johansson, T., Nguyen, P.Q. (eds.) EUROCRYPT 2013. LNCS, vol. 7881, pp. 444–460. Springer, Heidelberg (2013)
26. Fischlin, M., Schröder, D.: On the impossibility of three-move blind signature schemes. In: Gilbert, H. (ed.) EUROCRYPT 2010. LNCS, vol. 6110, pp. 197–215. Springer, Heidelberg (2010)
27. Garg, S., Bhaskar, R., Lokam, S.V.: Improved bounds on security reductions for discrete log based signatures. In: Wagner, D. (ed.) CRYPTO 2008. LNCS, vol. 5157, pp. 93–107. Springer, Heidelberg (2008)
28. Gentry, C., Wichs, D.: Separating succinct non-interactive arguments from all falsifiable assumptions. In: STOC, pp. 99–108 (2011)
29. Granger, R.: On the static Diffie-Hellman problem on elliptic curves over extension fields. In: Abe, M. (ed.) ASIACRYPT 2010. LNCS, vol. 6477, pp. 283–302. Springer, Heidelberg (2010)
30. Herranz, J., Laguillaumie, F.: Blind ring signatures secure under the chosen-target-CDH assumption. In: Katsikas, S.K., López, J., Backes, M., Gritzalis, S., Preneel, B. (eds.) ISC 2006. LNCS, vol. 4176, pp. 117–130. Springer, Heidelberg (2006)
31. Joux, A., Lercier, R., Naccache, D., Thomé, E.: Oracle-assisted static Diffie-Hellman is easier than discrete logarithms. In: Parker, M.G. (ed.) Cryptography and Coding 2009. LNCS, vol. 5921, pp. 351–367. Springer, Heidelberg (2009)
32. Juels, A., Luby, M., Ostrovsky, R.: Security of blind digital signatures. In: Kaliski Jr., B.S. (ed.) CRYPTO 1997. LNCS, vol. 1294, pp. 150–164. Springer, Heidelberg (1997)
33. Katz, J., Schröder, D., Yerukhimovich, A.: Impossibility of blind signatures from one-way permutations. In: Ishai, Y. (ed.) TCC 2011. LNCS, vol. 6597, pp. 615–629. Springer, Heidelberg (2011)
34. Koblitz, N., Menezes, A.: Another look at non-standard discrete log and Diffie-Hellman problems. Cryptology ePrint Archive, Report 2007/442 (2007)
35. Okamoto, T., Pointcheval, D.: The Gap-problems: A new class of problems for the security of cryptographic schemes. In: Kim, K.-C. (ed.) PKC 2001. LNCS, vol. 1992, pp. 104–118. Springer, Heidelberg (2001)
36. Paillier, P., Vergnaud, D.: Discrete-log-based signatures may not be equivalent to discrete log. In: Roy, B. (ed.) ASIACRYPT 2005. LNCS, vol. 3788, pp. 1–20. Springer, Heidelberg (2005)
37. Pass, R.: Limits of provable security from standard assumptions. In: STOC, pp. 109–118 (2011)
38. Pass, R., Venkitasubramaniam, M.: On constant-round concurrent zero-knowledge. In: Canetti, R. (ed.) TCC 2008. LNCS, vol. 4948, pp. 553–570. Springer, Heidelberg (2008)
39. Pointcheval, D., Stern, J.: Security arguments for digital signatures and blind signatures. Journal of Cryptology 13(3), 361–396 (2000)
40. Prabhakaran, M., Rosen, A., Sahai, A.: Concurrent zero knowledge with logarithmic round-complexity. In: FOCS, pp. 366–375 (2002)
41. Richardson, R., Kilian, J.: On the concurrent composition of zero-knowledge proofs. In: Stern, J. (ed.) EUROCRYPT 1999. LNCS, vol. 1592, pp. 415–431. Springer, Heidelberg (1999)
42. Seurin, Y.: On the exact security of Schnorr-type signatures in the random oracle model. In: Pointcheval, D., Johansson, T. (eds.) EUROCRYPT 2012. LNCS, vol. 7237, pp. 554–571. Springer, Heidelberg (2012)

Black-Box Separations for Differentially Private Protocols[*]

Dakshita Khurana, Hemanta K. Maji, and Amit Sahai

Department of Computer Science, UCLA and Center for Encrypted Functionalities,
Los Angeles, USA
{dakshita,hmaji,sahai}@cs.ucla.edu

Abstract. We study the maximal achievable accuracy of distributed differentially private protocols for a large natural class of boolean functions, in the computational setting.

In the information theoretic model, McGregor et al. [FOCS 2010] and Goyal et al. [CRYPTO 2013] demonstrate several functionalities whose differentially private computation results in much lower accuracies in the distributed setting, as compared to the client-server setting.

We explore lower bounds on the computational assumptions under which this accuracy gap can possibly be reduced for two-party boolean output functions. In the distributed setting, it is possible to achieve optimal accuracy, i.e. the maximal achievable accuracy in the client-server setting, for any function, if a semi-honest secure protocol for oblivious transfer exists. However, we show the following strong impossibility results:

- For *any* general boolean function and fixed level of privacy, the maximal achievable accuracy of any (fully) black-box construction based on existence of key-agreement protocols is at least a constant smaller than optimal achievable accuracy. Since key-agreement protocols imply the existence of one-way functions, this separation also extends to one-way functions.
- Our results are tight for the AND and XOR functions. For AND, there exists an accuracy threshold such that any accuracy up to the threshold can be information theoretically achieved; while no (fully) black-box construction based on existence of key-agreement can achieve accuracy beyond this threshold. An analogous statement is also true for XOR (albeit with a different accuracy threshold).

Our results build on recent developments in black-box separation techniques for functions with private input [1,16,27,28]; and translate information theoretic impossibilities into black-box separation results.

Keywords: Differentially Private Protocols, Computational Complexity, Random Oracle, Key-agreement Protocols, Black-box Separation.

[*] Research supported in part from a DARPA/ONR PROCEED award, NSF grants 1228984, 1136174, 1118096, and 1065276, a Xerox Faculty Research Award, a Google Faculty Research Award, an equipment grant from Intel, and an Okawa Foundation Research Grant. This material is based upon work supported by the Defense Advanced Research Projects Agency through the U.S. Office of Naval Research under Contract N00014-11-1-0389. The views expressed are those of the author and do not reflect the official policy or position of the Department of Defense, the National Science Foundation, or the U.S. Government.

P. Sarkar and T. Iwata (Eds.): ASIACRYPT 2014, PART II, LNCS 8874, pp. 386–405, 2014.

1 Introduction

Differential privacy [7] provides strong input privacy guarantees to individuals participating in a statistical query database. Consider the quintessential example of trying to publish some statistic computed on a database holding confidential data hosted by a trusted server [31]. For example, consider a query that checks if there is an empirical correlation between smoking and lung cancer instances from the medical records of patients stored at a hospital. The server wants to provide *privacy* guarantees to each record holder as well as help the client compute the statistic *accurately*. Even in this setting, where privacy concerns lie at the server's end only, it is clear that privacy and accuracy are antagonistic to each other. The tradeoff between accuracy and privacy is non-trivial and well understood only for some classes of functions (for e.g. [30,15]). For any level of privacy, we refer to the maximal achievable accuracy in the client-server setting for a particular functionality, as the *optimal accuracy*.

In the distributed setting, where multiple mutually distrusting servers host parts of the database, privacy concerns are further aggravated. Continuing the previous example, consider the case of two hospitals interested in finding whether a correlation exists between smoking and lung cancer occurrences by considering their combined patient records. In such a setting, we want the servers to engage in a protocol, at the end of which the privacy of each record of both the servers is guaranteed without a significant loss in accuracy. Note that the privacy requirements must be met for both servers, *even given their view of the protocol transcript, not just the computed output*; thus, possibly, necessitating an additional loss in accuracy.

At a basic level, we wish to study privacy-accuracy tradeoffs that arise in the distributed setting. Following [15], in order to obtain results for a wide class of functions, we focus on the computation of functions with Boolean output, with accuracy defined (very simply) as the probability that the answer is correct. The intuition that privacy in the distributed setting is more demanding is, in fact, known to be true in the information theoretic setting: For any fixed level of privacy, it was shown that for all boolean functions that the maximal achievable accuracy in the distributed setting is significantly lower than the optimal accuracy achievable in the client-server setting [15], as long as the boolean function depends on both server's inputs. But in the computational setting, this gap vanishes if a (semi-honest[1]) protocol for oblivious-transfer exists. The two servers would then be able to use secure multi-party computation [14] to simulate the client-server differentially private computation, thereby achieving optimal accuracy on the union of their databases. Although this computational assumption suffices, it is not at all clear whether this assumption is *necessary* as well.

Indeed, this is a fascinating question because even for very simple functions, like XOR, that require no computational assumptions to securely compute in

---

[1] In this work, as in previous works on distributed differential privacy, we restrict ourselves to the semi-honest setting where all parties follow the specified protocol, but remember everything they have seen when trying to break privacy.

the semi-honest setting, the question of differentially private computation is non-trivial. Could there be any simple functions that can be computed differentially privately with weaker assumptions? For the general class of boolean output functions, our paper considers the following problem:

"What are the computational assumptions under which there exist distributed differentially private protocols for boolean f with close to optimal accuracy?"

Goyal et al. [15] showed that for any boolean function such that both parties' inputs influence the outcome, achieving close to optimal accuracy would imply the existence of one-way functions. Could one-way functions also be *sufficient* to achieve optimal accuracy for certain simple functions?

Our results give evidence that the answer is *no*. Indeed, we provide evidence that achieving optimal accuracy for *any* boolean function that depends on both parties' inputs is not possible based on one-way functions. We go further and provide similar evidence that this goal is not possible even based on the existence of key-agreement protocols (which also implies one-way functions; and, thus, is a stronger computational assumption). More precisely, we show a (fully black-box) separation [35] of the computational assumptions necessary to bridge the accuracy gap from the existence of key-agreement protocols. A black-box separation between two cryptographic primitives has been widely acknowledged as strong evidence that they are distinct [23]. Indeed, we note that a black-box separation is particularly meaningful in the context of protocols with guarantees only against *semi-honest* adversaries, like the differentially private protocols we consider in this work. (Recall that an impossibility result like ours is strongest when it applies to the *weakest* security setting possible – this is why we focus on just semi-honest security.) This is because the most common non-black-box techniques used in cryptography typically apply only to the setting of malicious adversaries: for example, cryptographic proof systems like zero-knowledge proofs are sometimes applied in a non-black-box manner in order for a party to prove that it behaves honestly. However, in the semi-honest security context, such proofs are never needed since even adversarial parties must follow the protocol as specified. We crucially employ recently developed separation techniques for protocols with private inputs from key-agreement protocols [27,28].

Our work is reminiscent of, but also quite different from, the work of Haitner et al. [16], who proved that the information theoretic impossibility of accurate distributed differentially private evaluation of the inner-product functionality [30] could be extended to a black-box separation result from one-way functions. Our results are different both qualitatively and technically: Qualitatively, our results differ in that they apply to the wide class of all boolean functions where the output of the function is sensitive to both parties' inputs. Furthermore, we show separations from key-agreement protocols as well. Moreover, our separation results for extremely simple binary functions like AND and XOR show that differentially private distributed computation even of very simple functions may also require powerful computational assumptions.

At a technical level, a crucial ingredient of our proofs is the recently developed toolset of [27,28] which deal with *private inputs of parties* even in presence of the "idealized key-agreement oracle," while Haitner et al. [16] *adapt* the analysis of McGregor et al. [30] to a setting where the input is part of the local random tape of parties, i.e. parties have no private inputs.

1.1 Our Contribution

Before we elaborate upon our results, we briefly summarize what is known so far about accuracy gaps in boolean distributed differentially private computation.

Suppose Alice and Bob have inputs x and y, respectively; and they are interested in computing $f(x, y)$ in a differentially private manner in the distributed setting. An ε-differentially private protocol for some functionality f ensures that the probability of Alice's views conditioned on y and y' are $\lambda := \exp(\varepsilon)$ multiplicatively-close to each other, where y and y' represented as bit-strings differ only in one coordinate (i.e. they are *adjacent* inputs). Let x and y be the private inputs of parties Alice and Bob respectively. A protocol between them is α-accurate if for any x and y, the output of the protocol agrees with $f(x, y)$, with probability at least α.

For boolean functions, the optimal accuracy (in the client-server model) is $\alpha_\varepsilon^* := \frac{\lambda}{(\lambda+1)}$, where $\lambda = \exp(\varepsilon)$.[2] Goyal et al. [15] showed that, in the information theoretic setting, $f = \mathsf{AND}$ can only be computed ε-differentially privately up to accuracy $\alpha_\varepsilon^{(\mathsf{AND})} := \frac{\lambda(\lambda^2+\lambda+2)}{(\lambda+1)^3}$. Similarly, when $f = \mathsf{XOR}$ the maximal achievable accuracy is $\alpha_\varepsilon^{(\mathsf{XOR})} := \frac{(\lambda^2+1)}{(\lambda+1)^2}$. Note that $\alpha_\varepsilon^{(\mathsf{XOR})} < \alpha_\varepsilon^{(\mathsf{AND})} < \alpha_\varepsilon^*$, for any finite $\varepsilon > 0$. By observing that any boolean function f which is sensitive to both parties' inputs either contains an embedded XOR or AND[3] [3], the maximal achievable accuracy is bounded by:

$$\alpha_\varepsilon^{(f)} := \begin{cases} \alpha_\varepsilon^{(\mathsf{XOR})}, & \text{if } f \text{ contains an embedded } \mathsf{XOR} \\ \alpha_\varepsilon^{(\mathsf{AND})}, & \text{otherwise.} \end{cases} \qquad (1)$$

Note that in the computational setting, if semi-honest secure protocol for oblivious-transfer exists then we can achieve accuracy $\alpha = \alpha_\varepsilon^*$ for any boolean f. We explore the necessary computational assumptions for which this gap in accuracy in the distributed and client-server setting vanishes. Although Goyal et al. [15] showed that achieving close to optimal accuracy implies one-way functions, we show that it is highly unlikely that such constructions can solely be based on one-way functions. In fact, we show a (fully) black-box separation from a weaker variant of differential privacy, namely *computational differential privacy* (see Section 2).

---

[2] In the client-server setting, any boolean function f can be computed ε-differentially privately by evaluating a suitably noisy version of f.

[3] We say that f contains an embedded XOR if there exists $x_0, x_1, y_0, y_1, z_0, z_1$ such that $f(x_a, y_b) = z_{\mathsf{XOR}(a,b)}$ for all $a, b \in \{0, 1\}$. Similarly, we define en embedded AND. Note that embedded OR is identical to embedded AND (by interchanging z_0 and z_1).

Informal Theorem 1. *For any boolean f and privacy threshold $\varepsilon > 0$, there exists a constant $c > 0$ such that any ε-differentially private α-accurate evaluation of f (in the distributed setting) which uses key-agreement protocols in fully black-box manner cannot have accuracy $\alpha > (\alpha_\varepsilon^* - c)$, where $\alpha_\varepsilon^* = \frac{\lambda}{(\lambda+1)}$ and $\lambda = \exp(\varepsilon)$.*

Further, our result is tight for $f \in \{\mathsf{AND}, \mathsf{XOR}\}$ and, in fact, a stronger lower bound is exhibited. We show that for $f \in \{\mathsf{AND}, \mathsf{XOR}\}$: 1) In the information theoretic setting, it is possible to ε-differentially privately α-accurately evaluate f in the distributed setting [15], if $\alpha \leqslant \alpha_\varepsilon^{(f)}$, and 2) In the computational setting, it is impossible to construct (by using key-agreement protocols in black-box manner) an ε-differentially private α-accurate evaluation of f, for $\alpha \geqslant \alpha_\varepsilon^{(f)} + 1/\mathsf{poly}(\kappa)$ (where, κ is the statistical security parameter). In fact, this gives a (fully) black-box separation of a weaker notion of differential privacy, namely *computational differential privacy* (see Section 2). Note that it suffices to just consider $f \in \{\mathsf{AND}, \mathsf{XOR}\}$ because the maximal achievable accuracy for a general boolean function is bounded in terms of $\alpha_\varepsilon^{(\mathsf{AND})}$ and $\alpha_\varepsilon^{(\mathsf{XOR})}$. As a primer, we begin with the separation result from existence of one-way functions.

Separation from One-Way Functions. Random oracles serve as an idealization of one-way functions because they cannot be inverted at non-negligible fraction of their image by any algorithm whose query complexity is polynomial in query-length of the random oracle [23,12].

Suppose there exists a purported ε-differentially private α-accurate protocol for $f \in \{\mathsf{AND}, \mathsf{XOR}\}$ in the random oracle world, where parties have unbounded computational power and their query complexity is at most n. We show that if $\alpha \geqslant \alpha_\varepsilon^{(f)} + \sigma$ then one of the parties could perform additional $\mathsf{poly}(n/\sigma\varepsilon)$ queries to the random oracle and break the ε-differential privacy of the protocol. The existence of this strategy relies on the recent progress of "Eavesdropper strategies in the random oracle setting" for protocols with private inputs [27]. For more details, refer to Imported Theorem 1.

This impossibility result easily translates into a fully black-box separation as defined in [35]. This translation of impossibility in the random-oracle model into a black-box separation uses techniques introduced in [23,13,1,5,16,27].

Informal Theorem 2 (Separation from One-way Functions). *For $f \in \{\mathsf{AND}, \mathsf{XOR}\}$, $\varepsilon > 0$ and $\alpha \geqslant \alpha_\varepsilon^{(f)} + 1/\mathsf{poly}(\kappa)$, where κ is the security parameter, there cannot exist an ε-differentially private α-accurate protocol for f in the distributed setting which uses one-way functions in fully black-box manner.*

Note that this separation also extends to primitives which can be constructed from one-way functions in black-box manner, like pseudorandom generators [21,18,19] and digital signatures/universal one-way hash functions [33,36,26]. Moreover, it is also applicable to other computational primitives like ideal-ciphers [4,20] (which are indifferentiable [29] from random oracles) and one-way permutations (which themselves cannot be based on one-way function [37,25]).

Separation from Public-Key Encryption. To show a similar separation result from key-agreement protocols, it suffices to show a separation from public-key encryption; because public-key encryption is equivalent to two-round key agreement which in turn directly implies (any round) key-agreement protocols. Before we proceed further, we introduce the idealization of public-key encryption as an oracle [13].

Our public-key encryption oracle is a triplet of correlated oracles $\mathsf{PKE} \equiv (\mathsf{Gen}, \mathsf{Enc}, \mathsf{Dec})$. The key-generation oracle Gen is a length tripling random oracle which maps $sk \in \{0,1\}^n$ to $pk \in \{0,1\}^{3n}$, i.e. $\mathsf{Gen}(sk) = pk$. The encryption oracle, is a collection of 2^{3n} independent length-tripling oracles which maps a message m, using a public-key $pk \in \{0,1\}^{3n}$, to a cipher text c, i.e. $\mathsf{Enc}(m; pk) = c$. The decryption oracle Dec decrypts a cipher text $c \in \{0,1\}^{3n}$ using a secret key $sk \in \{0,1\}^n$. It maps it to (the lexicographically first) m such that $\mathsf{Gen}(sk) = pk$ and $\mathsf{Enc}(m; pk) = c$; otherwise outputs $\perp$, i.e. $\mathsf{Dec}(c, sk) \in \{m, \perp\}$.

This oracle is too powerful and yields a semi-honest secure protocol for oblivious-transfer (see discussion in [13]). Thus, it cannot be used to show the intended separation result. An additional Test oracle is provided, which allows testing of whether pk lies in the range of the Gen oracle, and whether c lies in the range of the Enc oracle with public key pk. Intuitively, the Test oracle can be thought of as part of Gen and Enc oracles themselves. Such oracles with *image-testability* are referred to as *image-testable random oracles* (ITRO) [28].

To tackle the decryption oracle, we follow the technique introduced by [28]. Suppose there exists a purported ε-differentially private α-accurate protocol for f in the PKE-oracle world. Then there exists an $(\varepsilon + \gamma)$-differentially private $(\alpha - \gamma)$-accurate protocol for f in the "PKE minus decryption oracle" world, i.e. in the $(\mathsf{Gen}, \mathsf{Enc})$ oracle world (with implicitly included Test oracles), with query complexity $\mathsf{poly}(n/\gamma\varepsilon)$ and identical round complexity. The slight loss in parameter γ can be made arbitrarily small $1/\mathsf{poly}(n)$.

Finally, similar to the separation from one-way functions, we show that if $(\alpha - \gamma) \geqslant \alpha_{\varepsilon+\gamma}^{(f)} + (\sigma/2)$ then one of the parties can perform $\mathsf{poly}(n/\sigma\gamma\varepsilon)$ queries and violate the $(\varepsilon + \gamma)$-differential privacy of this protocol. This part of the result crucially relies on the recently proven result of [28] which shows that image-testable random oracles *mimics* several properties of random-oracles and the "eavesdropper strategies" in the random oracle model extend to (collections of) image-testable random oracles as well. Hence, we have the following result.

Informal Theorem 3 (Separation from Key-Agreement). *For $f \in$ {AND, XOR}, $\varepsilon > 0$ and $\alpha \geqslant \alpha_\varepsilon^{(f)} + 1/\mathsf{poly}(\kappa)$, where κ is the security parameter, there cannot exist an ε-differentially private α-accurate protocol for f in the distributed setting which uses key-agreement protocols in fully black-box manner.*

We emphasize that our negative results not only hold for ε-differential privacy, but also hold for a weaker (ε, δ)-indistinguishability based computational differential privacy (see Section 2 for definition). For a precise statement refer to our main theorem, Theorem 1.

1.2 Related Work

Differential Privacy. Differential privacy [8,7,10,11,6] has been popular as a strong privacy guarantee to participants of statistical databases. In settings where the database could possibly be split among various parties, Dwork et al. [9] obtained distributed differential privacy via SFE and secure noise generation. Subsequently, [2] studied trade-offs between distributed privacy and SFE. A computational relaxation of differential privacy was defined by Mironov et al. [31], that would help improve the range of achievable accuracies while still maintaining this relaxed notion of privacy.

A gap in the maximal achievable accuracy of differentially private protocols, between the client-server and distributed settings, was first observed by McGregor et al. [30] for specific large functions such as the inner product and hamming distance. Recently, Goyal et al. [15] showed the existence of a constant information theoretic gap between the accuracies of boolean output functions, in the client-server and distributed settings. They also showed that any hope of bridging this gap necessitates the assumption that one-way functions exist.

Black-Box Separations. Impagliazzo and Luby [22] showed that most non-trivial cryptographic primitives imply existence of one-way functions. Subsequently, it turned out that several primitives like pseudorandom generators [21,18] and digital signatures/universal one-way hash functions [33,36] can indeed be constructed from one-way functions; thus, establishing equivalence of these primitives to existence of one-way functions. It is highly unlikely, on the other hand, that primitives like key-agreement [23] protocols and semi-honest secure oblivious-transfer protocol [13] can be securely constructed from one-way functions using black-box construction. A black-box separation result between two cryptographic primitives is widely acknowledged as an evidence that they should be treated as separate computational assumptions.

Reingold et al. [35] formally defined (several variants of) black-box separations. And Gertner et al. [12] provided a technique to translate information theoretic impossibility results in random oracle model into unconditional black-box separation results.

Recently, there has been significant progress in black-box separation techniques where parties have private inputs due to [27,28]. They show that if semi-honest secure function evaluation of any two-party deterministic function exists by using one-way functions or key-agreement protocols in black-box manner then there exists a semi-honest secure protocol for that function in the information theoretic plain model itself. Haitner et al. [16] show that the information theoretic impossibility of evaluating the inner-product functionality both differentially privately and accurately [30], in the client-server model, can be translated into a black-box separation result from one-way functions.

1.3 Technical Outline

Our black-box separation results are a consequence of amalgamation of the following techniques: 1) Information theoretic lower bounds for ε-differentially private α-accurate protocols for $f \in \{\mathsf{AND}, \mathsf{XOR}\}$ in the distributed setting [15], and 2) Recent progress in black-box separation techniques as introduced in [1,16,27,28]. Our separation from key-agreement protocols especially relies on the recent results of [28]. We essentially show that based on computational assumptions like "existence of one-way functions" and "existence of (any round) key-agreement protocol" it is highly unlikely to construct ε-differentially private α-accurate protocols for $f \in \{\mathsf{AND}, \mathsf{XOR}\}$, if $\alpha \geqslant \alpha_\varepsilon^{(f)}$.

Henceforth, we shall assume that $f \in \{\mathsf{AND}, \mathsf{XOR}\}$ and understand the computational assumptions necessary to realize ε-differentially private α-accurate protocols for f, where $\alpha > \alpha_\varepsilon^{(f)}$.

Information Theoretic Result. Before we begin, we sketch an intuitive summary of the proof technique of Goyal et al. [15]. They leveraged the Markov-chain property of distribution of next-message function in the information theoretic setting, i.e. the next message sent by a party is *solely* a (deterministic) function of its current view. Suppose the public transcript generated thus far is m. Then, using this Markov-chain property of protocols in the information theoretic setting and the fact that they begin with independent views, one can obtain the following *protocol compatibility constraint*: $\Pr[m|x, y] \cdot \Pr[m|x', y'] = \Pr[m|x, y'] \cdot \Pr[m|x', y]$, for private inputs $x, x' \in \mathcal{X}$ and $y, y' \in \mathcal{Y}$. By considering every complete transcript m, the protocol compatibility constraint implies a set of constraints. For every privacy parameter $\varepsilon \geqslant 0$, they show that there exists an ε-differentially private α-accurate protocol for f, if $\alpha \in [0, \alpha_\varepsilon^{(f)}]$.

Separation from One-Way Functions. Although the result presented in this section is subsumed by our main theorem, we feel that an independent presentation of this result adds clarity to the overall proof.

Suppose we have a (purportedly) ε-differentially private α-accurate protocol for f in the random oracle model, where each party performs at most n private queries to the random oracle. A random oracle randomly maps κ-length bit-strings to κ-length bit-strings, where κ is the statistical security parameter. Assume that $\alpha \geqslant \alpha_\varepsilon^{(f)} + \sigma$, where $\sigma = 1/\mathsf{poly}(\kappa)$. To show a black-box separation result from one-way functions, we need to show that if α is significantly larger than $\alpha_\varepsilon^{(f)}$, then differential privacy must be violated by one of the parties.

But, the Markov-chain property (upon which the information theoretic characterization crucially relies) is not a priori guaranteed in the random oracle model. So, a logical starting point is to consider an algorithm which perform additional queries to the random oracle to kill correlations between parties and ensures this property (with high probability), cf. [23,1,5,16,27]. For any $\rho > 0$, there exists a (deterministic) algorithm Eve_ρ which performs additional $\mathsf{poly}(n/\rho)$ queries to the random oracle based on the public transcript; and appends the sequence of query-answer pairs to the current transcript. This Eve_ρ ensures that when she stops, the joint view of Alice and Bob is ρ-close to a product distribution with $(1 - \rho)$ probability. Being agnostic to the private inputs used by the parties, Eve_ρ can ensure this Markov-chain property only when, for any complete transcript m, the probabilities $\Pr[y|m]$ and $\Pr[x|m]$, for every $x \in \{0,1\}$ and $y \in \{0,1\}$, is at least a constant [27].

Note that the ε-differential privacy constraint implies that $\Pr[x|m]$ and $\Pr[x'|m]$ are $\lambda = \exp(\varepsilon)$ (multiplicative) approximations of each other for all adjacent x and x'. If f is a function such that both parties' inputs influence the output, then it has an embedded AND or XOR minor in its truth table. Let $\mathcal{X}$ and $\mathcal{Y}$ be the respective input sets of Alice and Bob such that f restricted to $\mathcal{X} \times \mathcal{Y}$ is an AND or XOR minor. Given such a minor, our negative result shall exhibit violation of the differential privacy guarantee. So, for all our negative results we have $|\mathcal{X}| = |\mathcal{Y}| = 2$. Consequently, $\Pr[x|m]$ is a constant for every $x \in \mathcal{X}$; otherwise the complete transcript m is a witness to violation of ε-differential privacy. Analogously, the same holds for every $y \in \mathcal{Y}$.

For Alice inputs in $\mathcal{X}$ and Bob inputs in $\mathcal{Y}$, for any $\rho > 0$, there exists Eve_ρ with query complexity $\mathsf{poly}(n/\rho)$ such that, with probability $(1-\rho)$ over the generated public transcript, the joint view of Alice-Bob is ρ-close to a product distribution. Now, consider the *augmented protocol* where the original ε-differentially private α-accurate protocol is augmented with Eve_ρ, who adds her sequence of query-answer pairs to the public transcript. In this augmented protocol, we show that ε-differentially private α-accurate protocol implies $\alpha \leqslant \alpha_{\varepsilon,\rho}^{(f)}$, which can be made arbitrarily close to $\alpha_\varepsilon^{(f)}$ by choosing suitably small value of ρ. Intuitively, this result relies on the fact that the polytope of feasible solutions to the constraints in the information theoretic setting cannot change significantly if each of them has bounded slope and is weakened slightly (see full version for details). When $\alpha = \alpha_\varepsilon^{(f)} + \sigma$, where $\sigma = 1/\mathsf{poly}(n)$, by choosing suitably small $\rho = \mathsf{poly}(\sigma\varepsilon)$, one of the parties can violate the ε-differential privacy guarantee by performing $\mathsf{poly}(n/\rho)$ additional queries to the random oracle.

This technique is applied in a significantly sophisticated manner to show the separation from key-agreement protocols.

Separation from Key-Agreement. We show a separation from public-key encryption, which is equivalent to a 2-round key-agreement protocol. Separation from 2-round key-agreement implies separation from (any round) key-agreement protocols. This separation relies on the recent results pertaining to the "ideal public-key encryption oracle" (PKE-oracle, introduced by [13]) as shown in [28].

Our result depends on two technical results proven in [28]. First, they show that, against semi-honest adversaries, queries to the decryption-oracle of PKE-oracle are (nearly) useless; and, finally, the PKE-oracle minus the decryption-oracle (closely) mimics properties of (collection of) random oracles.

The first part shows that if there exists an ε-differentially private α-accurate protocol for f in the PKE-oracle world, then there exists another (closely related) $(\varepsilon + \gamma)$-differentially private $(\alpha - \gamma)$-differentially private protocol for f in the "PKE-oracle minus the decryption-oracle" world with query complexity $\mathsf{poly}(n/\gamma)$. Here, the parameter γ can be made arbitrarily small $1/\mathsf{poly}(n)$.

Finally, we use the property that "PKE-oracle minus decryption-oracle" is *similar* to the random oracle world [28]. We use the fact that, relative to this oracle, there exists an Eve_ρ which can make the joint distribution of Alice-Bob joint views ρ-close to product with high probability. Since $(\alpha - \gamma) > \alpha_{\varepsilon+\gamma,\rho}^{(f)}$, one of the parties can violate the $(\varepsilon + \gamma)$-differential privacy of the protocol.

Overall, if δ is at least $\alpha_\varepsilon^{(\text{AND})} + \sigma$, where $\sigma = 1/\text{poly}(n)$, we can choose $\gamma, \rho = \text{poly}(\sigma\varepsilon)$ to show that the ε-differential privacy is violated by performing only $\text{poly}(n/\sigma\varepsilon)$ queries to the PKE-oracle. In fact, our final theorem rules out a stronger form of differentially private protocols, namely, (ε, δ)-computational differential privacy (see Section 2 for definitions). Intuitively, $\delta = 0$ corresponds to the previously discussed notion of ε-differential privacy. Our final theorem is:

Theorem 1. *For any boolean function f whose output is sensitive to both parties' inputs, $\varepsilon > 0$ and $\lambda = e^\varepsilon$, define $\alpha_\varepsilon^{(f)}$ as follows:*

$$\alpha_\varepsilon^{(f)} := \begin{cases} \alpha_\varepsilon^{(\text{XOR})} = \frac{\lambda^2+1}{(\lambda+1)^2}, & \text{if } f \text{ contains an embedded XOR} \\ \alpha_\varepsilon^{(\text{AND})} = \frac{\lambda(\lambda^2+\lambda+2)}{(\lambda+1)^3}, & \text{otherwise.} \end{cases}$$

Then for any $\alpha \geqslant \alpha_\varepsilon^{(f)} + \sigma$, where $\sigma = 1/\text{poly}(\kappa)$ and κ is the statistical security parameter, there exists a $\hat{\delta} = \text{poly}(\sigma\varepsilon)$ such that any (ε, δ)-computational differentially private α-accurate protocol for f in the distributed setting constructed in a fully black-box manner from key-agreement protocols must have $\delta \geqslant \hat{\delta}$. Further, when $f \in \{\text{AND}, \text{XOR}\}$ and $\varepsilon > 0$, there exists an ε-differentially private α-accurate protocol for f, if $\alpha \leqslant \alpha_\varepsilon^{(f)}$.

The negative result rules out fully-BB constructions of ε indistinguishable computationally differentially private (ε-IND-CDP) α-accurate protocols with $\alpha > \alpha_\varepsilon^{(f)}$, based on existence of key agreement. The second part of the theorem (the positive result) is with respect to the stronger notion of ε-differential privacy.

An overview of the separation from one-way functions is provided in Section 3. An overview of the proof of Theorem 1 is presented in Section 4. Compplete proofs are deferred to the full version.

2 Preliminaries

We introduce important definitions in this section, with details in the full version.

Differential Privacy. The following definitions of differential privacy are provided for the distributed setting:

Definition 1 ((ε, δ)-Differential Privacy). *A two-party protocol Π is (ε, δ)-differentially private, referred to as (ε, δ)-DP, if for any subset S of Alice-views, for all Alice inputs x and for any pair of adjacent[4] Bob inputs y, y', we have:*

$$\Pr[S|x, y] \leqslant \exp(\varepsilon) \cdot \Pr[S|x, y'] + \delta$$

The same condition also holds for adjacent Alice inputs x, x' and all Bob's inputs y with respect to Bob private views.

---

[4] Two inputs are adjacent if they differ only in one coordinate.

Definition 2 ((ε, δ)-(IND)-Computational Differential Privacy). *A two-party protocol Π is (ε, δ)-computational differentially private, referred to as (ε, δ)-IND-CDP, if for any efficient adversary $\mathcal{A}$, for all Alice inputs x and any pair of adjacent Bob inputs y, y', we have:*

$$\Pr[\mathcal{A}(V_A, 1^\kappa) = 1 | x, y] \leqslant \exp(\varepsilon) \cdot \Pr[\mathcal{A}(V_A, 1^\kappa) = 1 | x, y'] + \delta$$

The same condition also holds for adjacent Alice inputs x, x' and all Bob's inputs y, with respect to Bob private views.

We refer $(\varepsilon, \mathsf{negl}(\kappa))$-IND-CDP as ε-IND-CDP, defined first in [31]. We note that this indistinguishability based definition is weaker than the simulation based one (SIM-CDP privacy [31]). Our separations hold even for this weaker differential privacy definition. In the above definition, the protocol Π, ε and δ are parameterized by the security parameter κ as well, but is not explicitly mentioned for ease of presentation. Without loss of generality, we assume that ε is not an increasing function (of κ); and in all our analysis we shall have δ as a decreasing function.

Accuracy. Following [15] we measure the accuracy of two-party protocols in evaluating a boolean function as follows:

Definition 3 (α-Accuracy). *A two party protocol Π evaluates a function f α-accurately, if, for every private input x and y of Alice and Bob respectively, the output of the protocol is identical to $f(x, y)$ with probability at least α.*

Information theoretic bounds on the maximal achievable accuracy for ε-DP protocols computing the AND and XOR functions, are known in the Plain Model [15]. Define $\lambda = \exp(\varepsilon)$, then $\alpha_\varepsilon^{(\mathsf{AND})} = \frac{\lambda(\lambda^2 + \lambda + 2)}{(\lambda+1)^3}$ is the maximal achievable accuracy of any protocol for the AND function, and $\alpha_\varepsilon^{(\mathsf{XOR})} = \frac{\lambda^2 + 1}{(\lambda+1)^2}$, is the maximal achievable accuracy of any protocol for the XOR function.

Black-Box Separations. We use the definition of fully black-box construction as introduced by Reingold et al. [35]. To show a separation of (ε, δ)-IND-CDP α-accurate protocol from key-agreement protocols, we need to show existence of an oracle relative to which key-agreement protocol exists but there exists an adversary which violates the (purported) (ε, δ)-IND-CDP guarantee.

3   Separation from One-Way Functions

Our main result shows a separation from key-agreement protocols. Despite the fact that the separation from one-way functions will be subsumed by our separation from key-agreement protocols, we present this result separately because it is conceptually simpler and captures several of the crucial ideas required to show such black-box separation results.

For $\varepsilon > 0$ differential privacy parameter, suppose $\alpha \in [\alpha_\varepsilon^{(f)} + 1/\mathsf{poly}(\kappa), \alpha_\varepsilon^*]$. We shall show that, for such choices of α, we cannot construct ε-IND-CDP α-accurate protocols for boolean f, in the information theoretic random oracle world. It suffices to show this result for $f \in \{\mathsf{AND}, \mathsf{XOR}\}$. This is done by showing an impossibility result in the random oracle model against information theoretic adversaries but with polynomially bounded query complexity. However, we shall show existence of an adversary who can break the ε-IND-CDP.

3.1   Notations and Definitions

We introduce some notations for our separation result. For security parameter κ, let $\mathbb{O}_\kappa$ denote the set of all functions from $\{0,1\}^\kappa \to \{0,1\}^\kappa$.

We will consider *private-input randomized two party protocols* Π, such that Alice and Bob have access to a common random oracle $O \xleftarrow{\$} \mathbb{O}_\kappa$. As in the plain model, parties send messages to each other in alternate rounds, starting with Alice in the first round. However, they have (private) access to a common random oracle.

For odd i, at the beginning of the ith round, Alice queries the random oracle multiple times based on her current view (private input x, local randomness r_A, private query-answer pairs and the transcript $m^{(i-1)}$ so far). She appends the new set of query-answer pairs $P_{A,i}$ to her partial sequence of query-answers. The complete set of private query-answers at this point is denoted by $P_A^{(i)}$. She then computes her next-message m_i as a function of her current view, $(x, r_A, m^{(i-1)}, P_A^{(i)})$. The ith round ends when she sends message m_i. Her view at the end of round i is $V_A^{(i)} \equiv (x, r_A, m^{(i)}, P_A^{(i)})$. Similarly, Bob queries the oracle followed by computing and sending his message in even rounds as a function of his view. His view at the end of round i is (analogously) defined to be $V_B^{(i)} \equiv (y, r_B, m^{(i)}, P_B^{(i)})$. At the end of n rounds, both parties locally obtain outputs as an efficiently computable deterministic function out of their view, $z_A = \mathsf{out}(V_A^{(n)})$ and $z_B = \mathsf{out}(V_B^{(n)})$. We note at this point, that we our analysis will only be over functions with boolean output, such that $z_A, z_B \in \{0,1\}$. Our underlying sample space in the random oracle world is the joint distribution over Alice-Bob views when $r_A, r_B \sim \mathbf{U}$ and $O \xleftarrow{\$} \mathbb{O}_\kappa$.

Two-Party Protocols in the Random Oracle World. Before we present our separation result, we need to introduce the notion of *public-query* strategy and *augmentation of a protocol* with a public-query strategy.

Definition 4 (Public Query strategy). *A public query strategy is a deterministic algorithm, which, after every round of the protocol, queries the oracle multiple times based on the transcript generated thus far. It then adds this sequence of query-answers to the transcript being generated.*

Definition 5 (Augmented Protocol). *Given a protocol Π, the augmented protocol $\Pi^+ := (\Pi, Eve)$ denotes Π augmented with a public query strategy "Eve"* *which generates public query-answer sequences after every message in Π and appends them to the protocol transcript after the messages in Π.*

Now, we define the views of parties (Alice, Bob and Eve) in an augmented protocol $\Pi^+:=(\Pi,\text{Eve})$. The protocol Π proceeds with parties sending messages in alternate rounds and Eve appending query-answer pairs after the message of the underlying protocol Π is sent.

Formally, consider an odd i. Alice is supposed to generate the message m_i in round i. Round i *begins* with Alice querying the random oracle based on her view $(x, r_A, m^{(i-1)}, P_A^{(i-1)}, P_E^{(i-1)})$, where $P_E^{(i-1)}$ is the sequence of query-answer pairs added by Eve thus far. Alice performs additional queries $P_{A,i}$ and sends the next message m_i. Thereafter, the public query strategy Eve performs additional queries to the random oracle and adds the corresponding sequence of query-answer pairs $P_{E,i}$ to the transcript. This marks the end of round i. At this point, the views of parties Alice, Bob and Eve are: $V_A^{(i)} \equiv (x, r_A, m^{(i)}, P_A^{(i)}, P_E^{(i)})$, $V_B^{(i)} \equiv (y, r_B, m^{(i)}, P_B^{(i)}, P_E^{(i)})$ and $V_E^{(i)} \equiv (m^{(i)}, P_E^{(i)})$, respectively.

(ε, δ)-IND-CDP in the Random Oracle Model

Definition 6 ((ℓ, n) Two-party Protocol). *An (ℓ, n) two-party protocol is a two-party protocol of round complexity at most n such that both parties have query complexity at most ℓ.*

Definition 7 ((ε, δ)-IND-CDP (ℓ, n) Protocol)
A two-party protocol Π is (ε, δ)-IND-CDP if for any computationally unbounded adversary (but polynomial-query complexity) $\mathcal{A}$ and any pair of adjacent Bob inputs y, y', we have:

$$\Pr[\mathcal{A}^O(V_A, 1^\kappa) = 1 | y] \leqslant \exp(\varepsilon) \cdot \Pr[\mathcal{A}^O(V_A, 1^\kappa) = 1 | y'] + \delta$$

The same condition also holds for adjacent Alice inputs x, x' with respect to Bob private views.

We emphasize that the adversary $\mathcal{A}$ gets access to an oracle O with respect to which the view V_A is generated. Accuracy is defined identically as in the plain model.

Remark: We briefly motivate the reasons behind choosing $\mathcal{A}$ as computationally unbounded adversary with polynomially bounded query complexity. Consider a world where "random oracle plus PSPACE" oracle is provided. A computationally bounded adversary in that oracle world shall correspond to an unbounded computational power adversary with polynomially bounded query complexity in the random oracle world. Therefore, we define ε-IND-CDP with respect to such adversaries because we shall exhibit such an adversary to show the separation from one-way functions. Note that we allow the adversary $\mathcal{A}$ to perform additional queries to the random oracle, because, in the computational setting, a computationally bounded adversary can perform additional queries to the one-way function itself.

We shall use the following definition on "closeness to product distribution."

Definition 8 (Close to Product Distribution). *A joint distribution* $(\mathbf{X}, \mathbf{Y})$ *is ρ-close to product distribution if* $\boldsymbol{\Delta}\left((\mathbf{X}, \mathbf{Y}), \mathbf{X} \times \mathbf{Y}\right) \leqslant \rho$. *Here,* $\mathbf{X}$ *and* $\mathbf{Y}$ *are the respective marginal distributions.*

3.2 Imported Results

The crux of the information theoretic bounds derived by [15] was the leveraging of an important Markov-chain property of the distribution of the next-message function of parties in the information theoretic setting. More specifically, the next message sent by a party is *solely* a deterministic function of its current view. Then, using the Markov chain property of protocols in the information theoretic plain model, it is easy to conclude that if the views of both parties were independent before protocol execution, they remain independent conditioned on the public transcript $m^{(n)}$. For any private inputs $x, x' \in \mathcal{X}$ and $y, y' \in \mathcal{Y}$, the following *protocol compatibility constraint* can be obtained directly:

$$\Pr[m^{(n)}|x, y] \cdot \Pr[m^{(n)}|x', y'] = \Pr[m^{(n)}|x', y] \cdot \Pr[m^{(n)}|x, y']$$

We begin with the observation that this constraint is not guaranteed a-priori in the information theoretic random oracle world. Intuitively, the views of both parties may be correlated via the common random oracle and not just the transcript. However, there are algorithms which query the random oracle polynomially many times to obtain independent views [23,1,5,16,27]. The state of the art (where parties have private inputs) is due to [27], from where we import the following theorem.

Imported Theorem 1 (Independence of Views in RO World [27]). *Given any two-party (ℓ, n) protocol Π (where parties have private inputs), there exists a public query strategy Eve_ρ which performs at most* $\mathsf{poly}(n\ell/\rho)$ *queries such that in the augmented protocol $\Pi^+ := (\Pi, Eve_\rho)$, with probability $(1 - \rho)$ over $V_E \sim \mathbf{V}_E$, we have: For all $(x, y) \in \mathcal{X} \times \mathcal{Y}$, if $\Pr[x, y|V_E] > \rho$, then $(\mathbf{V}_A, \mathbf{V}_B|V_E, x, y)$ is ρ-close to product distribution, i.e.*

$$\boldsymbol{\Delta}\left((\mathbf{V}_A, \mathbf{V}_B|V_E, x, y), (\mathbf{V}_A|V_E, x) \times (\mathbf{V}_B|V_E, y)\right) \leqslant \rho$$

3.3 Impossibility in the RO World

Instead of a key agreement enabling oracle, if we just have a random oracle, it suffices to show the following lemma:

Lemma 1 (Key Lemma for RO-Separation). *Suppose $f \in \{\mathsf{AND}, \mathsf{XOR}\}$. Consider any $\varepsilon > 0$, $\alpha \in [\alpha_\varepsilon^{(f)} + \sigma, \alpha_\varepsilon^*]$ and (positive) decreasing δ. If there exists an (ε, δ)-IND-CDP α-accurate protocol for f in the information theoretic random oracle world, then there exists a public query strategy Eve_ρ with query complexity $\mathsf{poly}(n\ell/\rho)$, where $\rho = \sigma^2 \varepsilon / \exp(2\varepsilon)$, such that in the augmented protocol $\Pi^+ := (\Pi, Eve_\rho)$, (at least) one of the following is true:*

1. *There exists* $(\hat{y}, \hat{y}', \hat{x})$ *so that: With probability* $\tilde{\delta} = \mathsf{poly}(\sigma)$ *over* $V_E \sim \mathbf{V}_E$ *we have:*

$$\Pr[V_E | \hat{x}, \hat{y}] > \exp(\varepsilon) \cdot \Pr[V_E | \hat{x}, \hat{y}'] + \delta' \Pr[V_E] \ ,$$

   where $\delta' = \mathsf{poly}(\sigma)$.

2. *There exists* $(\hat{x}, \hat{x}', \hat{y})$ *so that: With probability* $\tilde{\delta} = \mathsf{poly}(\sigma)$ *over* $V_E \sim \mathbf{V}_E$ *we have:*

$$\Pr[V_E | \hat{y}, \hat{x}] > \exp(\varepsilon) \cdot \Pr[V_E | \hat{y}, \hat{x}'] + \delta' \Pr[V_E] \ ,$$

   where $\delta' = \mathsf{poly}(\sigma)$.

Proof Overview: Let p_{V_E} denote the probability of obtaining public transcript V_E over the sample space. Let $p_{V_E | x, y}$ denote the probability of obtaining public transcript V_E from Π, when inputs of Alice and Bob are $x \in \mathcal{X}$, $y \in \mathcal{Y}$.

We first observe that if some input occurs with very low probability, then ε-IND-CDP can be trivially broken. Therefore, we can directly invoke Imported Theorem 1 such that Eve_ρ generates a close-to product distribution on the views of both parties with high probability. This gives an approximate protocol compatibility constraint on most transcripts.

Next, we observe that if the views of parties are nearly independent, then with high probability, for any inputs $x, x' \in \mathcal{X}$ and $y, y' \in \mathcal{Y}$ the distributions $p_{V_E | x, y} \cdot p_{V_E | x', y'}$ and $p_{V_E | x, y'} \cdot p_{V_E | x', y}$ must be close. We obtain the following equation (refer full version for proof),

$$p_{V_E | x, y} \cdot p_{V_E | x', y'} = p_{V_E | x, y'} \cdot p_{V_E | x', y} \pm 96 \rho p_{V_E}^2$$

Next, using the differential privacy constraint we mimic the proof of Goyal et al. [15] to obtain that for some transcript V_E, for some adjacent (x, y, y'), there are $\hat{\delta} = \mathsf{poly}(\sigma)$ transcripts such that for $\delta' = \mathsf{poly}(\sigma)$:

$$p_{V_E | x, y} > \lambda p_{V_E | x, y'} + \delta' p_{V_E}$$

Using averaging arguments, it is possible to show the existence of a tuple $(\hat{y}, \hat{y}', \hat{x})$ or $(\hat{x}, \hat{x}', \hat{y})$ satisfying the conditions of the lemma.   □

4   Separation from Key-Agreement Protocols

For $\varepsilon > 0$ differential privacy parameter, suppose $\alpha \in [\alpha_\varepsilon^{(f)} + 1/\mathsf{poly}(\kappa), \alpha_\varepsilon^*]$. In this section, we shall show that, for such choices of α, there exists an oracle relative to which public-key encryption exists but ε-IND-CDP α-accurate protocols for boolean f do not exist. It suffices to show this result for $f \in \{\mathsf{AND}, \mathsf{XOR}\}$. This is done by showing an impossibility result in the key agreement world against information theoretic adversaries but with polynomially bounded query complexity. However, we shall show existence of an adversary who can break the ε-IND-CDP.

Note that public-key encryption is equivalent to 2-round key-agreement protocols and hence this separation translates into a separation of non-trivial (ε, δ)-differentially private protocols for AND or XOR from (any round) key-agreement protocols.

4.1   Notations and Definitions

We give some notation and definitions. These definitions were introduced in [28].

Oracle Classes

Image-Testable Random Oracle Class. This is the set $\mathbb{O}_\kappa$ consisting of all possible pairs of correlated oracles $O \equiv (R, T)$ of the form:

- $R : \{0,1\}^\kappa \to \{0,1\}^{3\kappa}$ which is a *length-tripling (injective) random oracle*.
- $T : \{0,1\}^{3\kappa} \to \{0,1\}$ which is a *test oracle* defined by: $T(\beta) = 1$ if there exists $\alpha \in \{0,1\}^\kappa$ such that $R(\alpha) = \beta$; otherwise $T(\beta) = 0$.

The length of a query uniquely determines whether it is a query to the R oracle (called R-query) or to the T oracle (called T-query).

Keyed Version of Image-Testable Random Oracle Class. Given a set $\mathbb{K}$ of keys, consider oracle $O^{(\mathbb{K})}$ such that for every $k \in \mathbb{K}$, $O^{(k)} \in \mathbb{O}_k$ (the class of image-testable random oracles). A query is parsed as $\langle k, q \rangle$, the answer to which is $O^{(k)}(q)$. Let $\mathbb{O}_k^{(\mathbb{K})}$ denote the set of all possible oracles $O^{(\mathbb{K})}$. Then, $\mathbb{O}_k^{(\mathbb{K})}$ is the keyed version of the class of image-testable random oracles.

Public Key Encryption Oracle Class. We define a class of "PKE-enabling" oracles, from [28]. With access to this oracle, a semantically secure PKE scheme can be readily constructed, yet we shall show that it does not give (ε, δ)-IND-CDP protocols with any better than information theoretic accuracy. This oracle, called $\mathbb{PKE}_\kappa$ is a collection of oracles (Gen, Enc, Test$_1$, Test$_2$, Dec) defined as follows:

- Gen: A length-tripling injective random oracle $\{0,1\}^\kappa \to \{0,1\}^{3\kappa}$ that takes as input a secret key sk and returns the corresponding public key pk, i.e., Gen(sk) = pk. A public key pk is *valid* only if it is in the range of Gen.
- Enc: A collection of keyed length-tripling injective random oracles, with keys in $\{0,1\}^{3\kappa}$. For each $pk \in \{0,1\}^{3\kappa}$, the oracle implements a random injective function $\{0,1\}^\kappa \to \{0,1\}^{3\kappa}$. When queried on any (possibly invalid) random public key pk, the oracle provides the corresponding ciphertext $c \in \{0,1\}^{3\kappa}$.
- Test$_1$: This is a function that tests if a public key pk is *valid*, that is, it returns 1 if and only if pk is in the range of Gen
- Test$_2$: This is a function that tests if a public key and ciphertext pair is *valid*, i.e., it returns 1 if and only if c is in the range of the Enc oracle keyed by pk.
- Dec: This is the decryption oracle, $\{0,1\}^\kappa \times \{0,1\}^{3\kappa} \to \{0,1\}^\kappa \cup \{\perp\}$, which takes as input a secret key, ciphertext pair and returns the unique m, such that Enc(Gen(sk), m) = c. If such an m does not exist, it returns $\perp$.

We note that Enc produces ciphertexts for public key pk irrespective of whether there exists sk satisfying Gen (sk) = pk. This is crucial because we want the key set $\mathbb{K}$ to be defined independent of the Gen oracle.

We also note that $\mathbb{PKE}_\kappa$ without Dec is exactly the same as the image-testable random oracle $\mathbb{O}_k^{(\mathbb{K})}$, with $\mathbb{K} = \{0,1\}^{3\kappa} \cup \{\perp\}$. This fact will be used very crucially in the sections that follow, where we compile out the Decryption oracle and work with the resulting image-testable random oracle $\mathbb{O}_k^{(\mathbb{K})}$.

Our Setting. We will consider *private-input randomized two party protocols* Π, such that Alice and Bob have access to a common oracle PKE_κ. As in the plain model, parties send messages to each other in alternate rounds, starting with Alice in the first round. However, they have (private) access to a the common PKE_κ oracle consisting of $(\mathsf{Gen}, \mathsf{Enc}, \mathsf{Test}_1, \mathsf{Test}_2, \mathsf{Dec})$.

The views with respect to the $\mathbb{PKE}_\kappa$ oracle remain the same as views in the random oracle world. Our underlying sample space in the random oracle world is the joint distribution over Alice-Bob views when $r_A, r_B \sim \mathbf{U}$ and $\mathsf{PKE}_\kappa \sim \mathbb{PKE}_\kappa$.

We use the definition of (ε, δ)-IND-CDP protocols in the oracle world and accuracy of protocols as introduced in previous section.

4.2   Compiling Out the Decryption Oracle

Using the query techniques of [28], for any arbitrarily small γ, it is possible to construct an $(\varepsilon + \gamma, \delta + \gamma)$ differentially private protocol with accuracy $\alpha - \gamma$, that uses only the family of image testable random oracles $\mathbb{O}_k^{(\mathbb{K})}$ oracle from an (ε, δ) differentially private protocol that uses the $\mathbb{PKE}_k$ oracle.

Imported Theorem 2 (Decryption Queries are Useless [28]). *Suppose Π is an (ℓ, n) (ε, δ)-differentially private α-accurate protocol for f in the $\mathbb{PKE}_\kappa$ oracle world. For every $\gamma > 0$, there exists a protocol Π' in the* $(\mathsf{Gen}, \mathsf{Enc}, \mathsf{Test}_1, \mathsf{Test}_2)$ *oracle world which is an $(\varepsilon + \gamma, \delta + \gamma)$ differentially private $(\alpha - \gamma)$-accurate* $(\mathsf{poly}(n\ell/\gamma), n)$ *protocol for f.*

4.3   Impossibility in ITRO World

Recall that the PKE-oracle without the decryption oracle is in fact a collection of keyed image-testable random oracles, where the key-set is $\mathbb{K} = \{0,1\}^{3\kappa} \cup \{\bot\}$. We import the following result of eavesdropper strategy:

Imported Theorem 3 (Independence of Views in ITRO World [28]). *For any key-set $\mathbb{K}$ and any (ℓ, n) protocol Π (where parties have private inputs), there exists a public query strategy Eve_ρ which performs at most $\mathsf{poly}(\ell/\rho)$ queries such that in the augmented protocol $\Pi^+ := (\Pi, \mathsf{Eve}_\rho)$, the following holds over the views of Eve_ρ, when $V_E \sim \mathbf{V}_E$, with probability at least $(1 - \rho)$: For all $(x, y) \in \mathcal{X} \times \mathcal{Y}$, if $\Pr[x, y | V_E] > \rho$ then $(\mathbf{V}_A, \mathbf{V}_B | V_E, x, y)$ is ρ-close to product distribution, i.e.*

$$\Delta\left((\mathbf{V}_A, \mathbf{V}_B | V_E, x, y), (\mathbf{V}_A | V_E, x) \times (\mathbf{V}_B | V_E, y)\right) \leqslant \rho$$

This gives us exactly the same independence characterization as Section 3.3, and we can obtain the following Lemma for the ITRO world (analogously to the random oracle world).

Lemma 2 (Key Lemma for ITRO-Separation). *Suppose $f \in \{\mathsf{AND}, \mathsf{XOR}\}$. Consider any $\varepsilon > 0$, $\alpha \in [\alpha_\varepsilon^{(f)} + \sigma, \alpha_\varepsilon^*]$ and (positive) decreasing δ. For any key-set $\mathbb{K}$, if there exists an (ε, δ)-IND-CDP α-accurate protocol for f in the image-testable random oracle world with respect to key-set $\mathbb{K}$, then there exists a public* *query strategy Eve$_\rho$ with query complexity* $\mathsf{poly}(n\ell/\rho)$, *where* $\rho = \sigma^2\varepsilon/\exp(2\varepsilon)$, *such that in the augmented protocol* $\Pi^+:=(\Pi, Eve_\rho)$, *(at least) one of the following is true:*

1. *There exists* $(\hat{y}, \hat{y}', \hat{x})$ *so that: With probability* $\tilde{\delta} = \mathsf{poly}(\sigma)$ *over* $V_E \sim \mathbf{V}_E$ *we have:*
$$\Pr[V_E|\hat{x}, \hat{y}] > \exp(\varepsilon) \cdot \Pr[V_E|\hat{x}, \hat{y}'] + \delta'\Pr[V_E]\ ,$$
where $\delta' = \mathsf{poly}(\sigma)$.
2. *There exists* $(\hat{x}, \hat{x}', \hat{y})$ *so that: With probability* $\tilde{\delta} = \mathsf{poly}(\sigma)$ *over* $V_E \sim \mathbf{V}_E$ *we have:*
$$\Pr[V_E|\hat{y}, \hat{x}] > \exp(\varepsilon) \cdot \Pr[V_E|\hat{y}, \hat{x}'] + \delta'\Pr[V_E]\ ,$$
where $\delta' = \mathsf{poly}(\sigma)$.

4.4   Impossibility in Key Agreement World

To prove Theorem 1, it suffices to show the following Lemma:

Lemma 3 (Key Lemma for KA-Separation). *Suppose* $f \in \{\mathsf{AND}, \mathsf{XOR}\}$. *Consider any* $\varepsilon > 0$, $\alpha \in [\alpha_\varepsilon^{(f)} + \sigma, \alpha_\varepsilon^*]$ *and (positive) decreasing* δ. *If there exists an* (ε, δ)-IND-CDP α-accurate protocol for f in the $\mathbb{PKE}_\kappa$ world, then for $\gamma = \sigma^3$, the corresponding protocol Π' as defined in Imported Theorem 2 is an $(\varepsilon + \gamma, \delta + \gamma)$-IND-CDP $(\alpha - \gamma)$-accurate $(\mathsf{poly}(n\ell/\gamma), n)$ protocol in $\mathbb{O}_k^{(\mathbb{K})}$, where $\mathbb{K} = \{0, 1\}^{3\kappa} \cup \{\bot\}$. Then, there exists a public query strategy Eve$_\rho$ with query complexity $\mathsf{poly}(n\ell/\gamma\rho)$, where $\rho = \sigma^2\varepsilon/\exp(2\varepsilon)$, such that in the augmented protocol $\Pi'^+:=(\Pi', Eve_\rho)$, $\delta + \gamma > \gamma^{5/6}$.*

Proof Overview: Note that we can use Imported Theorem 2 to compile any given two-party (ε, δ)-IND-CDP (ℓ, n) protocol Π in the key agreement world with accuracy $\alpha > \alpha_\varepsilon^{(\mathsf{AND})} + \sigma$ for the AND function (resp. $\alpha > \alpha_\varepsilon^{(\mathsf{XOR})} + \sigma$ for the XOR function), to an $(\varepsilon + \gamma, \delta + \gamma)$-IND-CDP (ℓ, n) protocol Π' with accuracy $(\alpha - \gamma)$ in the ITRO world (which closely mimics the RO world).

In fact, while moving from the key agreement to the ITRO world, there is a γ-loss in protocol accuracy and a corresponding (say γ') increase in maximal achievable accuracy. These parameters can be carefully tied to σ such that setting $\gamma + \gamma' = \sigma^6$, helps obtain $\delta + \gamma > \gamma^{5/6}$, thereby giving $\delta = poly(\sigma)$ transcripts violating the differential privacy constraint.

In fact, we can show (refer full version) that if $\sigma = 1/\mathsf{poly}(\kappa)$, it is possible to construct an adversary that breaks $(\varepsilon+\gamma)$-IND-CDP of the $(\varepsilon+\gamma, \delta+\gamma)$-IND-CDP (ℓ, n) protocol Π' in the ITRO world, with accuracy $(\alpha - \gamma)$ according to Definition 2. This gives a contradiction and completes the proof.   □

References

1. Barak, B., Mahmoody, M.: Merkle puzzles are optimal - an $O(n^2)$-query attack on any key exchange from a random oracle. In: Halevi (ed.) [17], pp. 374–390

2. Beimel, A., Nissim, K., Omri, E.: Distributed private data analysis: Simultaneously solving how and what. In: Wagner, D. (ed.) CRYPTO 2008. LNCS, vol. 5157, pp. 451–468. Springer, Heidelberg (2008)
3. Chor, B., Kushilevitz, E.: A zero-one law for boolean privacy (extended abstract). In: Johnson (ed.) [24], pp. 62–72.
4. Coron, J.-S., Patarin, J., Seurin, Y.: The random oracle model and the ideal cipher model are equivalent. In: Wagner, D. (ed.) CRYPTO 2008. LNCS, vol. 5157, pp. 1–20. Springer, Heidelberg (2008)
5. Dachman-Soled, D., Lindell, Y., Mahmoody, M., Malkin, T.: On the black-box complexity of optimally-fair coin tossing. In: Ishai, Y. (ed.) TCC 2011. LNCS, vol. 6597, pp. 450–467. Springer, Heidelberg (2011)
6. Dinur, I., Nissim, K.: Revealing information while preserving privacy. In: Neven, F., Beeri, C., Milo, T. (eds.) PODS, pp. 202–210. ACM (2003)
7. Dwork, C.: Differential privacy. In: Bugliesi, M., Preneel, B., Sassone, V., Wegener, I. (eds.) ICALP 2006. LNCS, vol. 4052, pp. 1–12. Springer, Heidelberg (2006)
8. Dwork, C.: A firm foundation for private data analysis. Commun. ACM 54(1), 86–95 (2011)
9. Dwork, C., Kenthapadi, K., McSherry, F., Mironov, I., Naor, M.: Our data, ourselves: Privacy via distributed noise generation. In: Vaudenay, S. (ed.) EUROCRYPT 2006. LNCS, vol. 4004, pp. 486–503. Springer, Heidelberg (2006)
10. Dwork, C., McSherry, F., Nissim, K., Smith, A.: Calibrating noise to sensitivity in private data analysis. In: Halevi, S., Rabin, T. (eds.) TCC 2006. LNCS, vol. 3876, pp. 265–284. Springer, Heidelberg (2006)
11. Dwork, C., Nissim, K.: Privacy-preserving datamining on vertically partitioned databases. In: Franklin, M. (ed.) CRYPTO 2004. LNCS, vol. 3152, pp. 528–544. Springer, Heidelberg (2004)
12. Gennaro, R., Gertner, Y., Katz, J., Trevisan, L.: Bounds on the efficiency of generic cryptographic constructions. SIAM J. Comput. 35(1), 217–246 (2005)
13. Gertner, Y., Kannan, S., Malkin, T., Reingold, O., Viswanathan, M.: The relationship between public key encryption and oblivious transfer. In: FOCS, pp. 325–335. IEEE Computer Society (2000)
14. Goldreich, O., Micali, S., Wigderson, A.: How to play any mental game or a completeness theorem for protocols with honest majority. In: Aho, A.V. (ed.) STOC, pp. 218–229. ACM (1987)
15. Goyal, V., Mironov, I., Pandey, O., Sahai, A.: Accuracy-privacy tradeoffs for two-party differentially private protocols. In: Canetti, R., Garay, J.A. (eds.) CRYPTO 2013, Part I. LNCS, vol. 8042, pp. 298–315. Springer, Heidelberg (2013)
16. Haitner, I., Omri, E., Zarosim, H.: Limits on the usefulness of random oracles. In: Sahai, A. (ed.) TCC 2013. LNCS, vol. 7785, pp. 437–456. Springer, Heidelberg (2013)
17. Halevi, S. (ed.): CRYPTO 2009. LNCS, vol. 5677. Springer, Heidelberg (2009)
18. Håstad, J.: Pseudo-random generators under uniform assumptions. In: Ortiz (ed.) [34], pp. 395–404.
19. Håstad, J., Impagliazzo, R., Levin, L.A., Luby, M.: A pseudorandom generator from any one-way function. SIAM J. Comput. 28(4), 1364–1396 (1999)
20. Holenstein, T., Künzler, R., Tessaro, S.: Equivalence of the random oracle model and the ideal cipher model, revisited. In: STOC (2011)
21. Impagliazzo, R., Levin, L.A., Luby, M.: Pseudo-random generation from one-way functions (extended abstracts). In: Johnson (ed.) [24], pp. 12–24
22. Impagliazzo, R., Luby, M.: One-way functions are essential for complexity based cryptography (extended abstract). In: FOCS, pp. 230–235. IEEE (1989)

23. Impagliazzo, R., Rudich, S.: Limits on the provable consequences of one-way permutations. In: Johnson (ed.) [24], pp. 44–61
24. Johnson, D.S. (ed.): Proceedings of the Twenty-First Annual ACM Symposium on Theory of Computing, Seattle, Washington, USA, May 15-17. ACM (1989)
25. Kahn, J., Saks, M.E., Smyth, C.D.: A dual version of Reimer's inequality and a proof of Rudich's conjecture. In: IEEE Conference on Computational Complexity, pp. 98–103 (2000)
26. Katz, J., Koo, C.-Y.: On constructing universal one-way hash functions from arbitrary one-way functions. IACR Cryptology ePrint Archive, 2005:328 (2005)
27. Mahmoody, M., Maji, H.K., Prabhakaran, M.: Limits of random oracles in secure computation. In: ITCS (2014)
28. Mahmoody, M., Maji, H.K., Prabhakaran, M.: On the power of public-key encryption in secure computation. In: Lindell, Y. (ed.) TCC 2014. LNCS, vol. 8349, pp. 240–264. Springer, Heidelberg (2014)
29. Maurer, U.M., Renner, R., Holenstein, C.: Indifferentiability, impossibility results on reductions, and applications to the random oracle methodology. In: Naor (ed.) [32], pp. 21–39
30. McGregor, A., Mironov, I., Pitassi, T., Reingold, O., Talwar, K., Vadhan, S.P.: The limits of two-party differential privacy. In: FOCS, pp. 81–90. IEEE Computer Society (2010)
31. Mironov, I., Pandey, O., Reingold, O., Vadhan, S.P.: Computational differential privacy. In: Halevi (ed.) [17], pp. 126–142
32. Naor, M. (ed.): TCC 2004. LNCS, vol. 2951. Springer, Heidelberg (2004)
33. Naor, M., Yung, M.: Universal one-way hash functions and their cryptographic applications. In: Johnson (ed.) [24], pp. 33–43
34. Ortiz, H. (ed.): Proceedings of the 22nd Annual ACM Symposium on Theory of Computing, Baltimore, Maryland, USA, May 13-17. ACM (1990)
35. Reingold, O., Trevisan, L., Vadhan, S.P.: Notions of reducibility between cryptographic primitives. In: Naor (ed.) [32], pp. 1–20
36. Rompel, J.: One-way functions are necessary and sufficient for secure signatures. In: Ortiz (ed.) [34], pp. 387–394
37. Rudich, S.: Limits on the Provable Consequences of One-way Functions. PhD thesis, University of California at Berkeley (1988)

Composable Security of Delegated Quantum Computation

Vedran Dunjko[1,2,*], Joseph F. Fitzsimons[3,4], Christopher Portmann[5,6], and Renato Renner[5]

[1] School of Informatics, University of Edinburgh, Edinburgh EH8 9AB, U.K.
[2] Division of Molecular Biology, Ruđer Bošković Institute,
Bijenička Cesta 54, P.P. 180, 10002 Zagreb, Croatia
[3] Singapore University of Technology and Design,
20 Dover Drive, 138682, Singapore
[4] Centre for Quantum Technologies, National University of Singapore,
Block S15, 3 Science Drive 2, 117543, Singapore
`joe.fitzsimons@nus.edu.sg`
[5] Institute for Theoretical Physics, ETH Zurich,
8093 Zurich, Switzerland
`{chportma,renner}@phys.ethz.ch`
[6] Group of Applied Physics, University of Geneva,
1211 Geneva, Switzerland

Abstract. Delegating difficult computations to remote large computation facilities, with appropriate security guarantees, is a possible solution for the ever-growing needs of personal computing power. For delegated computation protocols to be usable in a larger context — or simply to securely run two protocols in parallel — the security definitions need to be composable. Here, we define composable security for delegated quantum computation. We distinguish between protocols which provide only *blindness* — the computation is hidden from the server — and those that are also *verifiable* — the client can check that it has received the correct result. We show that the composable security definition capturing both these notions can be reduced to a combination of several distinct "trace-distance-type" criteria — which are, individually, non-composable security definitions.

Additionally, we study the security of some known delegated quantum computation protocols, including Broadbent, Fitzsimons and Kashefi's Universal Blind Quantum Computation protocol. Even though these protocols were originally proposed with insufficient security criteria, they turn out to still be secure given the stronger composable definitions.

* Now at: Institute for Theoretical Physics, University of Innsbruck, Technikerstraße 25, A-6020 Innsbruck, Austria. `vedran.dunjko@uibk.ac.at`

P. Sarkar and T. Iwata (Eds.): ASIACRYPT 2014, PART II, LNCS 8874, pp. 406–425, 2014.

1   Introduction

1.1   Background

It is unknown in what form quantum computers will be built. One possibility is that large quantum servers may take a role similar to that occupied by massive superclusters today. They would be available as important components in large information processing clouds, remotely accessed by clients using their home-based simple devices. The issue of the security and the privacy of the computation is paramount in such a setting.

Childs [16] proposed the first such delegated quantum computation (DQC) protocol, which hides the computation from the server, i.e., the computation is blind. This was followed by Arrighi and Salvail [2], who introduced a notion of verifiability — checking that the server does what is expected — but only for a restricted class of public functions. In recent years, this problem has gained a lot of interest, with many papers proposing new protocols, e.g., [1,11,15,19–21, 28,31–36,41], and even small-scale experimental realizations [7,8].

However, with the exception of recent work by Broadbent, Gutoski and Stebila [12], none of the previous DQC papers consider the *composability* of the protocol. They prove security by showing that the states held by the client and server fulfill some local condition: the server's state must not contain any information about the input and the client's final state must either be the correct outcome or an error flag. Even though this means that the server cannot — from the information leaked during a single execution of the protocol in an isolated environment — learn the computation or produce a wrong output without being detected, it does not guarantee any kind of security in any realistic setting. In particular, if a server treats two requests simultaneously or if the delegated computation is used as part of a larger protocol (such as the quantum coins of Mosca and Stebila [37]), these works on DQC cannot be used to infer security. A *composable security* framework must be used for a protocol to be secure in an arbitrary environment. In the following, we use the expression *local* to denote the non-composable security conditions previously used for DQC. This term is chosen, because these criteria consider the state of a (local) subsystem, instead of the global system as seen by a distinguisher in composable security.[1]

In fact, exactly these local properties have been proven to be insufficient to define secure communication. There exist protocols which are shown to both encrypt and authenticate messages by fulfilling local criteria equivalent to the ones used in DQC — the scheme is secure if the eavesdropper obtains no information about the message from the ciphertext and authentic if the receiver either gets the original message or an error flag. But if the eavesdropper learns whether the message was transmitted faithfully or not, she learns some information about this message [9,27,30]. Since any secure communication protocol can be seen as delegated computation for the identity operation — Eve is required to apply the

---

[1] Standard terms for various forms of non-composable security, e.g., *stand-alone* or *sequential*, have precise definitions which do not apply to these security criteria.

identity operation to the message, but may cheat and try to learn or modify it — there is a strict gap between security of DQC and previously used local criteria.[2]

Composable frameworks have the further advantage that they require the interaction between different entities to be modeled explicitly, and often make hidden assumptions apparent. For example, it came as a surprise when Barrett et al. [6] showed that device independent quantum key distribution (DIQKD) is insecure if untrusted devices (with internal memory) are used more than once. It is however immediate when one models the security of DIQKD in a composable framework, that existing security proofs make the assumption that devices are used only once. Another example, the security definitions of zero-knowledge protocols [22] and coin expansion [26] make the assumption that the dishonest party executes his protocol without interaction with the environment.[3] By explicitly modeling this restriction,[4] these proofs can be lifted to a composable framework. This has been used by, e.g., Unruh [44], who explicitly limits the number of parallel executions of a protocol to achieve security in the bounded storage model.

Correctly defining the security of a cryptographic task is fundamental for a protocol and proof to have any usefulness or even meaning. In this paper we solve this problem for DQC, which has been open since the first version of Childs's work [16] was made available in 2001.

1.2   Scope and Security of DQC

A common feature of all DQC protocols is that the client, while not being capable of full-blown quantum computation, has access to limited quantum-enriched technology, which she needs to interact with the server. One of the key points upon which the different DQC protocols vary, is the complexity and the technical feasibility of the aforementioned quantum-enriched technology. In particular, in the proposal of Childs [16], the client has quantum memory, and the capacity to perform local Pauli operations. The protocol of Arrighi and Salvail [2] requires the client to have the ability to generate relatively involved superpositions of multi-qubit states, and perform a family of multi-qubit measurements.

---

[2] An alternative example of this gap is as follows. The task is to compute a witness for a positive instance of an NP problem, and we do so with the following protocol: the server simply picks a witness at random and sends it to the client. Although the protocol does not achieve completeness, it appears to be sound: the protocol obviously does not leak any information about the input, since no information is sent from the client to the server. The client can also verify that the solution received is correct, and never accepts a wrong answer. But if the server ever learns whether the witness was accepted — e.g., it is composed with another protocol which makes this information public — he learns something about the input. If there are only two choices for the input with distinct witnesses, he learns exactly which one was used.

[3] The security definitions for these two problems are instances of what is generally known as *stand-alone security* [23].

[4] This can be done by introducing a resource — e.g., a trusted third party — that runs whatever circuits Alice and Bob give it in an isolated system, then returns the transcript of the protocol to both players.

Aharonov, Ben-Or and Eban [1], for the purposes of studying quantum prover interactive proof systems, considered a DQC protocol in which the client has a constant-sized quantum computer. The blind DQC protocol proposed by Broadbent, Fitzsimons and Kashefi [11] has arguably the lowest requirements on the client. In particular, she does not need any quantum memory,[5] and is only required to prepare single qubits in separable states randomly chosen from a small finite set analogous to the BB84 states.[6] Alternatively, Morimae and Fujii [32,35] propose a DQC protocol in which the client only needs to measure the qubits she receives from the server to perform the computation.

A second important distinction between these protocols is in the types of problems the protocol empowers the client to solve. Most protocols, e.g., [1,11, 16,20,32,35], allow a client to perform universal quantum computation, whereas in [2] the client is restricted to the evaluation of random-verifiable[7] functions.

Finally, an important characteristic of these protocols is the flavor of security guaranteed to the client. Here, one is predominantly interested in two distinct features: privacy of computation (generally referred to as blindness) and verifiability of computation. Blindness characterizes the degree to which the computational input and output, and the computation itself, remain hidden from the server. This is the main security concern of, e.g., [11, 16, 35]. Verifiability ensures that the client has means of confirming that the final output of the computation is correct. In addition to blindness, some form of verifiability is given by, e.g., [1, 2, 20, 32]. These works do however not concern themselves with the cryptographic soundness of their security notions. In particular, none of them consider the issue of composability of DQC. A notable exception is the recent work of Broadbent, Gutoski and Stebila [12], who, independently from our work, prove that a variant of the DQC protocol of Aharonov, Ben-Or and Eban [1] provides composable security.[8]

1.3 Composable Security

The first frameworks for defining composable security were proposed independently by Canetti [13,14] and by Backes, Pfitzmann and Waidner [3,4,39], who dubbed them *Universally Composable (UC) security* and *Reactive Simulatability*, respectively. These security notions have been extended to the quantum setting by Ben-Or and Mayers [10] and Unruh [42,43].

---

[5] This holds in the case of classical input and output. If quantum inputs and/or outputs are considered, then the client has to be able to apply a quantum one-time pad to the input state, and also decrypt a quantum one-time pad of the output state.

[6] The states needed by the protocol of [11] are $\{(|0\rangle + e^{ik\pi/4}|1\rangle)/\sqrt{2}\}_k$ for $k \in \{0, \dots, 7\}$.

[7] Roughly speaking, a function f is random-verifiable if pairs of instances and solutions $(x, f(x))$ can be generated efficiently, where x is sampled according to the uniform distribution from the function's domain.

[8] The work of Broadbent et al. [12] is on one-time programs. Their result on the composability of DQC is obtained by modifying their main one-time program protocol and security proof so that it corresponds to a variant of the DQC protocol from [1].

More recently, Maurer and Renner proposed a new composable framework, Abstract Cryptography (AC) [29]. Unlike its predecessors that use a bottom-up approach to defining models of computation, algorithms, complexity, efficiency, and then security of cryptographic schemes, the AC approach is top-down and axiomatic, where lower abstraction levels inherit the definitions and theorems (e.g., a composition theorem) from the higher level, but the definition or concretization of low levels is not required for proving theorems at the higher levels. In particular, it is not hard-coded in the security notions of AC whether the underlying computation model is classical or quantum, and this framework can be used equally for both.

Even though these frameworks differ considerably in their approach, they all share the common notion that composable security is defined by the distance between the real world setting and an ideal setting in which the cryptographic task is accomplished in some perfect way. We use AC in this work, because it simplifies the security definitions by removing many notions which are not necessary at that level of abstraction. But the same results could have been proven using another framework, e.g., a quantum version of UC security [43].

1.4 Results

In this paper, we define a composable framework for analyzing the security of delegated quantum computing, using the aforementioned AC framework [29]. We model DQC in a generic way, which is independent of the computing requirements or universality of the protocol, and encompasses to the best of our knowledge all previous work on DQC. We then define composable blindness and composable verifiability in this framework. The security definitions are thus applicable to any DQC protocol fitting in our model.

We study the relations between local security criteria used in previous works [1,2,11,16,20,32,35] and composable security of DQC. We show that by strengthening the existing notion of local-verifiability, we can close the gap between these local criteria and composable security of DQC. To do this we introduce the notion of *independent* local-verifiability. Intuitively, this captures the idea that the acceptance probability of the client should not depend on the input or computation performed, but rather only on the activities of the (dishonest) server. Our main theorem is as follows.

Theorem 1. *If a DQC protocol implementing a unitary[9] transformation provides ε_{bl}-local-blindness and ε_{ind}-independent ε_{ver}-local-verifiability for all inputs $\psi_{A_C A_Q}$, where A_C is classical and A_Q is quantum, then it is δN^2-secure, where $\delta = 4\sqrt{2\varepsilon_{ver}} + 2\varepsilon_{bl} + 2\varepsilon_{ind}$ and $N = \dim \mathcal{H}_{A_Q}$.*

Note that by choosing the parameters such that δ is exponentially small in the size of the quantum input ($\log N$) negates the factor N^2 blow-up in the overall error (see also Remark 13).

---

[9] Any quantum operation can be written as a unitary on a larger system, effectively allowing this theorem to apply to all quantum operations, see Remark 12.

Proving that a DQC protocol is secure then reduces to proving that these local criteria are satisfied.[10] For instance, the protocols of Morimae [32] and Fitzsimons and Kashefi [20] are shown to satisfy definitions of local-correctness, local-blindness and local-verifiability, equivalent to the ones considered here. To prove that these protocols are secure, it only remains to show that they also satisfy the stronger notion of *independent* local-verifiability introduced in this work, which we sketch in Sect. 6.1.

Finally, we analyze the security of two protocols — Broadbent, Fitzsimons and Kashefi [11] and Morimae and Fujii [35] — that do not provide any form of verifiability, so the generic reduction cannot be used. Instead we directly prove that both these protocols satisfy the definition of composable blindness, without verifiability (Theorems 14 and 15).

Interestingly — and somewhat unexpectedly — even though the local security definitions used in previous works are insufficient to guarantee composable security, the previously proposed protocols studied in this work are all still secure given the stronger security notions.

1.5   Structure of This Paper

In Sect. 2 we introduce two-party protocols and distance measures that we use in this work.[11] In Sect. 3 we explain delegated quantum computation, and model composable security for such protocols. In Sect. 4 we show that composable verifiability (which encompasses blindness) is equivalent to the distance between the real protocol and some ideal map that simultaneously provides both local-blindness and local-verifiability. This map is however still more elaborate than local criteria used in previous works. In Sect. 5 we break this map down into individual notions of local-blindness and independent local-verifiability, and prove that these are sufficient to achieve security. Finally, in Sect. 6 we look at the security of some existing protocols. We first discuss how our results can be applied to protocols that already provide local-verifiability. Then we prove that the DQC protocols of Broadbent, Kashefi and Fitzsimons [11] and Morimae and Fujii [35] are composably blind.

---

[10] This is similar in nature to the result on the composable security of quantum key distribution (QKD) [40], which shows that a QKD protocol that satisfies definitions of *robustness*, *correctness* and *secrecy* is secure in a composable sense. These individual notions are all expressed with trace-distance-type criteria, e.g., a QKD protocol is ε-secret if $(1 - p_{\text{abort}})\|\rho_{KE} - \tau_K \otimes \rho_E\|_{\text{tr}} \leq \varepsilon$, where p_{abort} is the probability of aborting, ρ_{KE} the joint state of the final key and the eavesdropper's system and τ_K is the fully mixed state. To prove that a QKD protocol is secure, it is thus sufficient to prove that it satisfies these individual notions.

[11] These are an instantiation of the abstract systems defined in AC. We refer to the full version of this work [18, Sections 2] for an introduction to the AC framework, that is essential to understand the details of the current paper.

2   Quantum Systems

2.1   Two-Party Protocols

A two-party protocol can in be modeled by a sequence of CPTP maps $\pi_A = \{\mathcal{E}_i : \mathcal{L}(\mathcal{H}_{AC}) \to \mathcal{L}(\mathcal{H}_{AC})\}_i$ and $\pi_B = \{\mathcal{F}_i : \mathcal{L}(\mathcal{H}_{CB}) \to \mathcal{L}(\mathcal{H}_{CB})\}_i$, where A and B are Alice and Bob's registers, and C represents a communication channel. Initially Alice and Bob place their inputs in their registers, and the channel C is in some fixed state $|0\rangle$. The players then apply successively their maps to their respective registers and the channel. For example, in the first round Alice applies $\mathcal{E}_1$ to the joint system AC, and sends C to Bob, who applies $\mathcal{F}_1$ to CB, and returns C to Alice. Then she applies $\mathcal{E}_2$, etc.

Such a sequence of maps, $\{\mathcal{E}_i : \mathcal{L}(\mathcal{H}_{AC}) \to \mathcal{L}(\mathcal{H}_{AC})\}_i$, has been called a *quantum strategy* by Gutoski and Watrous [24, 25] and a *quantum N-comb* by Chiribella, D'Ariano and Perinotti [17]. In particular, these authors derived independently a concise representation of combs/strategies in terms of the Choi-Jamiołkowski isomorphism. They also define the appropriate distance measure between two combs/strategies, corresponding to the optimal distinguishing advantage, which we sketch in the next section.

2.2   Distance Measures

The trace distance between two states ρ and σ is given by $D(\rho, \sigma) = \frac{1}{2}\|\rho - \sigma\|_{\mathrm{tr}}$, where $\|\cdot\|_{\mathrm{tr}}$ denotes the trace norm and is defined as $\|A\|_{\mathrm{tr}} := \mathrm{tr}\sqrt{A^\dagger A}$. If $D(\rho, \sigma) \leq \varepsilon$, we say that the two states are ε-close and often write $\rho \approx_\varepsilon \sigma$. This corresponds to the distinguishing advantage between two resources $\mathcal{R}$ and $\mathcal{S}$, which take no input and produce ρ and σ, respectively, as output: the probability of a distinguisher guessing correctly whether he holds $\mathcal{R}$ or $\mathcal{S}$ is exactly $\frac{1}{2} + \frac{1}{2}D(\rho, \sigma)$.

Another common metric which corresponds to the distinguishing advantage between resources of a certain type is the diamond norm. If the resources $\mathcal{R}$ and $\mathcal{S}$ take an input $\rho \in \mathcal{S}(\mathcal{H}_A)$ and produce an output $\sigma \in \mathcal{S}(\mathcal{H}_B)$, the distinguishing advantage between these resources is the diamond distance between the correspond maps $\mathcal{E}, \mathcal{F} : \mathcal{L}(\mathcal{H}_A) \to \mathcal{L}(\mathcal{H}_B)$. A distinguisher can generate a state ρ_{AR}, input the A part to the resource, and try to distinguish between the resulting states $\mathcal{E}(\rho_{AR})$ and $\mathcal{F}(\rho_{AR})$. We have $d(\mathcal{R}, \mathcal{S}) = \diamond(\mathcal{E}, \mathcal{F}) = \frac{1}{2}\|\mathcal{E} - \mathcal{F}\|_\diamond$, where

$$\|\varPhi\|_\diamond := \max\{\|(\varPhi \otimes \mathrm{id}_R)(\rho)\|_{\mathrm{tr}} : \rho \in \mathcal{S}(\mathcal{H}_{AR})\}$$

is the diamond norm. Note that the maximum of the diamond norm can always be achieved for a system R with $\dim \mathcal{H}_R = \dim \mathcal{H}_A$. Here too, we sometimes write $\mathcal{E} \approx_\varepsilon \mathcal{F}$ if two maps are ε-close.

If the resources considered are halves of two player protocols, say π_i or π_j, the above reasoning can be generalized for obtaining the distinguishing advantage. The distinguisher can first generate an initial state $\rho \in \mathcal{S}(\mathcal{H}_{AR})$—which for convenience we define as a map on no input $\rho := \mathcal{D}_0()$ — and input the A part of the state into the resource. It receives some output ρ_{CR} from the resource, can apply some arbitrary map $\mathcal{D}_1 : \mathcal{L}(\mathcal{H}_{CR}) \to \mathcal{L}(\mathcal{H}_{CR})$ to the state, and input the C part of the new state in the resource. Let it repeat this procedure with different maps $\mathcal{D}_i$ until the end of the protocol, after which it holds one of two states: φ_{AR} if it had access to π_i and ψ_{AR} if it had access to π_j. The trace distance $D(\varphi_{AR}, \psi_{AR})$ defines the advantage the distinguisher has of correctly guessing whether it was interacting with π_i or π_j, and by maximizing this over all possible initial inputs $\rho_{AR} = \mathcal{D}_0()$, and all subsequent maps $\{\mathcal{D}_i : \mathcal{L}(\mathcal{H}_{CR}) \to \mathcal{L}(\mathcal{H}_{CR})\}_i$, the distinguishing advantage between these resources becomes

$$d(\pi_i, \pi_j) = \max_{\{\mathcal{D}_i\}_i} D(\varphi_{AR}, \psi_{AR}). \tag{1}$$

This has been studied by both Gutoski [24] and Chiribella et al. [17], and we refer to their work for more details.

3 Delegated Quantum Computation

In the (two-party) delegated quantum computation (DQC) model, Alice asks a server, Bob, to execute some quantum computation for her. Intuitively, Alice plays the role of a client, and Bob the part of a computationally more powerful server. Alice has several security concerns. She wants the protocol to be blind, that is, she wants the server to execute the quantum computation without learning anything about the input other than what is unavoidable, e.g., an upper bound on its size, and possibly whether the output is classical or quantum. She may also want to know if the result sent to her by Bob is correct, which we refer to as verifiability.

In Sect. 3.1 we model the ideal resource that a DQC protocol constructs and the structure of a generic DQC protocol. And in Sect. 3.2 we give the corresponding security definitions. This section uses the AC cryptography nomenclature, which is explained in detail in the full version [18].

3.1 DQC Model

Ideal Resource. To model the security (and correctness) of a delegated quantum computation protocol, we need to model the ideal delegated computation resource S that we wish to build. We start with an ideal resource that provides blindness, and denote it S^{blind}.

The task Alice wants to be executed is provided as an input to the resource S^{blind} at the A-interface. It could be modeled as having two parts, some quantum state ψ_{A_1} and a classical description Φ_{A_2} of some quantum operation that she wants to apply to ψ, i.e., she wishes to compute $\Phi(\psi)$. This can alternatively be seen as applying a universal computation $\mathcal{U}$ to the input $\psi_{A_1} \otimes |\Phi\rangle\langle\Phi|_{A_2}$. We adopt this view in the remainder of this paper, and model the resource as performing some fixed computation $\mathcal{U}$ on an input ψ_A that may be part quantum and part classical.[12]

Any DQC protocol must reveal to the server an upper bound on the size of the computation it is required to execute. Other information might also be made intentionally available, such as whether the output of the computation is classical or quantum. Although one could imagine a generic DQC model in which these "permitted leaks" are entangled with the rest of the input, we restrict our considerations to classical information, i.e., a subsystem of the input ψ_A is classical[12] and contains a string $\ell^{\psi_A} \in \{0,1\}^*$ that is copied and provided to the server Bob at the start of the protocol, so that he may set up the required resources and programs for the computation. Alternatively, this string can be taken to be some fixed publicly available information, not modeled explicitly. We do so in the following sections to simplify the notation, but prefer make it explicit in this section so as not to hide the fact that some information about the input is always given to the server.

The ideal resource S^{blind} thus takes this input ψ_A at its A-interface, and, if Bob does not activate his filtered functionalities — which can be modeled by a bit b, set to 0 by default, and which a simulator σ_B can flip to 1 to signify that it is activating the cheating interface — S^{blind} outputs $\mathcal{U}(\psi_A)$. This ensures both correctness and universality (in the case where $\mathcal{U}$ is a universal computation). Alternatively, S^{blind} can be restricted to work for inputs corresponding to a certain class of computational problems, if we desire a construction only designed for such a class.

If the cheating B-interface is activated, the ideal resource outputs a copy of the string ℓ^{ψ_A} at this interface. Bob also has another filtered functionality, one which allows him to tamper with the final output. The most general operation he could perform is to give S^{blind} a quantum state ψ_B — which could be entangled with Alice's input ψ_A — along with the description of some map $\mathcal{E} : \mathcal{L}(\mathcal{H}_{AB}) \to \mathcal{L}(\mathcal{H}_A)$, and ask it to output $\mathcal{E}(\psi_{AB})$ at Alice's interface. Since S^{blind} only captures blindness, but says nothing about Bob's ability to manipulate the final output, we define it to perform this operation and output any $\mathcal{E}(\psi_{AB})$ at Bob's request. This is depicted in Fig. 1 with the filtered functionalities in gray.

Definition 2. *The ideal DQC resource* S^{blind} *which provides both correctness and blindness takes an input* ψ_A *at Alice's interface, but no honest input at Bob's interface. Bob's filtered interface has a control bit* b, *set by default to 0, which he can flip to activate the other filtered functionalities. The resource* S^{blind} *then outputs the permitted leak* ℓ^{ψ_A} *at Bob's interface, and accepts two further*

---

[12] Alternatively, the input can be modeled as entirely quantum, and both Alice and the ideal resource first measure the part of the input that should be classical, before executing π_A and the universal computation $\mathcal{U}$, respectively. This corresponds to plugging an extra measurement converter into the A-interfaces of both the real and ideal systems (that converts the quantum input into a classical-quantum input), which can only decrease the distance between the real and ideal systems, i.e., increase the security.

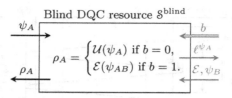

Fig. 1. An ideal DQC resources. The client Alice has access to the left interface, and the server Bob to the right interface. The double-lined input flips a bit set by default to 0. The functionalities provided at Bob's interface are grayed to signify that they are accessible only to a cheating server. If Bob is honest, this interface is obstructed by a filter, which we denote by $\perp_B$ in the following. $\mathcal{S}^{\text{blind}}$ provides blindness — it only leaks the permitted information at Bob's interface — but allows Bob to choose Alice's output.

inputs, a state ψ_B and map description $|\mathcal{E}\rangle\langle\mathcal{E}|$. If $b = 0$, it outputs the correct result $\mathcal{U}(\psi_A)$ at Alice's interface; otherwise it outputs Bob's choice, $\mathcal{E}(\psi_{AB})$.

A DQC protocol is verifiable if it provides Alice with a mechanism to detect a cheating Bob and output an error flag `err` instead of some incorrect computation. This is modeled by weakening Bob's filtered functionality: an ideal DQC resource with verifiability, $\mathcal{S}^{\text{blind}}_{\text{verif}}$, only allows Bob to input one classical bit c, which specifies whether the output should be $\mathcal{U}(\psi_A)$ or some error state $|\text{err}\rangle$, which by construction is orthogonal to the space of valid outputs. The ideal resource thus never outputs a wrong computation. This is illustrated in Fig. 2.

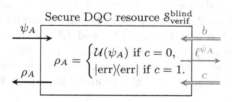

Fig. 2. Another ideal DQC resources. $\mathcal{S}^{\text{blind}}_{\text{verif}}$ provides both blindness and verifiability — in addition to leaking only the permitted information, it never outputs an erroneous computation result.

Definition 3. *The ideal DQC resource $\mathcal{S}^{\text{blind}}_{\text{verif}}$ which provides correctness, blindness and verifiability takes an input ψ_A at Alice's interface, and two filtered control bits b and c (set by default to 0). If $b = 0$, it simply outputs $\mathcal{U}(\psi_A)$ at Alice's interface. If $b = 1$, it outputs the permitted leak ℓ^{ψ_A} at Bob's interface, then reads the bit c, and conditioned on its value, it either outputs $\mathcal{U}(\psi_A)$ or $|\text{err}\rangle$ at Alice's interface.*

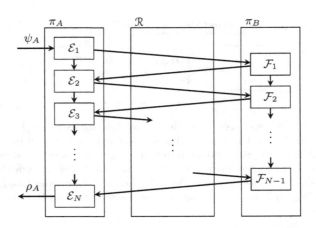

Fig. 3. A generic run of a DQC protocol. Alice has access to the left interface and Bob to the right interface. The entire system builds one CPTP operation which maps ψ_A to ρ_A.

Concrete Setting. In the concrete (or real) setting, the only resource that Alice and Bob need is a (two-way) communication channel $\mathcal{R}$. Alice's protocol π_A receives ψ_A as an input on its outside interface. It then communicates through $\mathcal{R}$ with Bob's protocol π_B, and produces some final output ρ_A. For the sake of generality we assume that the operations performed by π_A and π_B, and the communication between them, are all quantum. Of course, a protocol is only useful if Alice has very few quantum operations to perform, and most of the communication is classical. However, to model security, it is more convenient to consider the most general case possible, so that it applies to all possible protocols.

As described in Sect. 2.1, their protocols can be modeled by a sequence of CPTP maps $\{\mathcal{E}_i : \mathcal{L}(\mathcal{H}_{AC}) \to \mathcal{L}(\mathcal{H}_{AC})\}_{i=1}^{N}$ and $\{\mathcal{F}_i : \mathcal{L}(\mathcal{H}_{CB}) \to \mathcal{L}(\mathcal{H}_{CB})\}_{i=1}^{N-1}$. We illustrate a run of such a protocol in Fig. 3. The entire system consisting of the protocol (π_A, π_B) and the channel $\mathcal{R}$ is a map which transforms ψ_A into ρ_A. If both players played honestly and the protocol is correct, this should result in $\rho_A = \mathcal{U}(\psi_A)$.

In the following, when we refer to a DQC protocol, we simply mean any protocol satisfying the model of Fig. 3. Whether the protocol actually performs delegated quantum computation depends on whether it satisfies the correctness condition, which we define in Sect. 3.2.

3.2   Security of DQC

Applying the AC security definition (see the full version, [18, Definition 2.1]) to the DQC model from the previous section, we get that a protocol π constructs a blind quantum computation resource $\mathcal{S}^{\mathrm{blind}}$ from a communication channel $\mathcal{R}$ within ε if there exists a simulator σ_B such that

$$\pi_A \mathcal{R} \pi_B \approx_\varepsilon \mathcal{S}^{\mathrm{blind}} \perp_B \quad \text{and} \quad \pi_A \mathcal{R} \approx_\varepsilon \mathcal{S}^{\mathrm{blind}} \sigma_B, \tag{2}$$

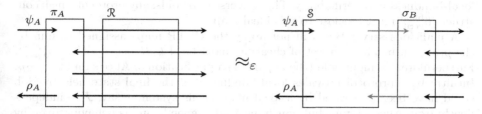

Fig. 4. An illustration of the second terms of (2) and (3). If a distinguisher cannot guess with advantage greater than ε whether it is interacting with the real construct on the left or the ideal construct on the right, the two are ε-close and the protocol ε-secure against a cheating Bob.

where $\perp_B$ is a filter which obstructs Bob's cheating interface.[13] The fist condition in (2) captures the correctness of the protocol, and we say that a protocol provides ε-*correctness* if this condition is fulfilled. The second condition, which we illustrate in Fig. 4, measures the security. If it is fulfilled, we have ε-*blindness*. If $\varepsilon = 0$ we say that we have *perfect blindness*.

Likewise in the case of verifiability, the ideal resource $\mathcal{S}_{\mathrm{verif}}^{\mathrm{blind}}$ is constructed by π from $\mathcal{R}$ if there exists a simulator σ_B such that,

$$\pi_A \mathcal{R} \pi_B \approx_\varepsilon \mathcal{S}_{\mathrm{verif}}^{\mathrm{blind}} \perp_B \quad \text{and} \quad \pi_A \mathcal{R} \approx_\varepsilon \mathcal{S}_{\mathrm{verif}}^{\mathrm{blind}} \sigma_B. \tag{3}$$

The first condition from (3) is identical to the first condition of (2), and captures ε-correctness. The second condition in (3) (also illustrated by Fig. 4) guarantees both blindness and verifiability, and if it is satisfied we say that the we have ε-*blind-verifiability*.

Note that the exact metrics used to distinguish between the resources from (2) and (3) are defined in Sect. 2.2. $\pi_A \mathcal{R} \pi_B$ and $\mathcal{S} \perp_B$ — as can be seen from their depictions in Figs. 3 and 2 (with a filter blocking the cheating interface of the latter) — are resources which implement a single map, so the diamond distance corresponds to the distinguishing advantage. $\pi_A \mathcal{R}$ and $\mathcal{S} \sigma_B$ are half of two-party protocols, so the distinguishing metric corresponds to the distance between quantum strategies/combs introduced by Gutoski and Watrous [24, 25] and Chiribella et al. [17], and described in Sect. 2.2.

4   Blind and Verifiable DQC

Finding a simulator to prove the security of a protocol can be challenging. In this section we reduce the task of proving that a DQC protocol constructs the ideal resource $\mathcal{S}_{\mathrm{verif}}^{\mathrm{blind}}$ to proving that the map implemented by the protocol is close to some ideal map that intuitively provides some form of

---

[13] These equations are to be interpreted graphically: π_A is plugged into the left interface of $\mathcal{R}$, and π_B is plugged into the right interface, see the illustrations in Figs. 3 and 4 or the full version [18] for further explanations.

local-blindness-and-verifiability. The converse also holds: any protocol which constructs $S_{\text{verif}}^{\text{blind}}$ must be close to this ideal map.

A malicious server Bob will not apply the CPTP maps assigned to him by the protocol, but his own set of cheating maps $\{\mathcal{F}_i : \mathcal{L}(\mathcal{H}_{CB}) \to \mathcal{L}(\mathcal{H}_{CB})\}_{i=1}^{N-1}$. Furthermore, he might hold (the B part of) a purification of Alice's input, ψ_{ABR}. Intuitively, a protocol provides local-blindness[14] if the final state held by Bob could have been generated by a local map on his system — say, $\mathcal{F}$ — independently from Alice's input, but which naturally depends on his behavior given by the maps $\{\mathcal{F}_i\}_i$. It provides local-verifiability[14] if the final state held by Alice is either the correct outcome or some error flag. Combining the two gives an ideal map of the from $\mathcal{U} \otimes \mathcal{F}^{\text{ok}} + \mathcal{E}^{\text{err}} \otimes \mathcal{F}^{\text{err}}$, where $\mathcal{F}^{\text{ok}}$ and $\mathcal{F}^{\text{err}}$ break $\mathcal{F}$ down in two maps which result in the correct outcome and an error flag, respectively.

Definition 4 (local-blind-verifiability). *We say that a DQC protocol provides ε-local-blind-verifiability, if, for all adversarial behaviors $\{\mathcal{F}_i\}_i$, there exist two completely positive, trace non-increasing maps $\mathcal{F}_B^{ok}$ and $\mathcal{F}_B^{err}$, such that*

$$\mathcal{P}_{AB} \approx_\varepsilon \mathcal{U}_A \otimes \mathcal{F}_B^{ok} + \mathcal{E}_A^{err} \otimes \mathcal{F}_B^{err}, \qquad (4)$$

where $\mathcal{P}_{AB} : \mathcal{L}(\mathcal{H}_{AB}) \to \mathcal{L}(\mathcal{H}_{AB})$ is the map corresponding to a protocol run with Alice behaving honestly and Bob using his cheating operations $\{\mathcal{F}_i\}_i$, and $\mathcal{E}_A^{err}$ discards the A system and produces an error flag $|err\rangle\langle err|$ orthogonal to all possible valid outputs. We say that the protocol provides ε-local-blind-verifiability for a set of initial states $\mathcal{B}$, if (4) holds when applied to these states, i.e., for all $\psi_{ABR} \in \mathcal{B}$,

$$\mathcal{P}_{AB}(\psi_{ABR}) \approx_\varepsilon (\mathcal{U}_A \otimes \mathcal{F}_B^{ok} + \mathcal{E}_A^{err} \otimes \mathcal{F}_B^{err})(\psi_{ABR}).$$

Remark 5. For simplicity, this definition assumes the allowed leaks (e.g., input size, computation size) to be fixed, and applies to all protocols $\mathcal{P}_{AB}$ tailored for inputs with an identical leak (e.g., identical size). These leaks could be explicitly modeled by allowing the maps $\mathcal{F}_B^{ok}$ and $\mathcal{F}_B^{err}$ to depend on them.

We now state the main theorem of this section, namely that it is both necessary and sufficient for a DQC protocol to satisfy Definition 4 to be blind-verifiable, i.e., to satisfy the second condition of Equation (3). A proof is is given in the full version [18]. In order to construct $S_{\text{verif}}^{\text{blind}}$, a DQC protocol also needs to be ε-correct, that is, satisfy the first condition from Equation (3). We show in Appendix A that this is fulfilled, if, when Bob behaves honestly, Equation (4) is satisfied for $\mathcal{F}_B^{ok} = \text{id}_B$ and $\mathcal{F}_B^{err} = 0$.

Theorem 6. *Any DQC protocol which provides ε-local-blind-verifiability is 2ε-blind-verifiable. And any DQC protocol which is ε-blind-verifiable provides ε-local-blind-verifiability.*

---

[14] We provide formal definitions of local-blindness and local-verifiability in Sect. 5.

5   Reduction to Local Criteria

Although the notion of local-blind-verifiability defined in the previous section captures the security of DQC in a single equation, it is still more elaborate than existing definitions found in the literature, that treat blindness and verifiability separately.

In this section we provide separate definitions for these local notions, and strengthen local-verifiability by requiring that the server Bob be able to infer on his own whether the client Alice will reject his response — learning whether Alice did reject will then not provide him with any information that he could not obtain on his own. We then show that in the case where Bob does not hold a state entangled with the input (e.g., when the input is entirely classical), these notions are sufficient to obtain local-blind-verifiability with a similar error parameter. In the case where Bob's system is entangled to Alice's input, we show that the same holds, albeit with an error increased by a factor $(\dim \mathcal{H}_{A_Q})^2$, where A_Q is the subsystem of Alice's input which is quantum.

This can be used to show that the protocol of Fitzsimons and Kashefi [20] and Morimae [32], which have already been analyzed using (insufficient) local criteria, are secure. We provide a proof sketch of the missing steps for both these protocols in the full version of this paper [18].

Local-blindness can be seen as a simplification of local-blind-verifiability, in which we ignore Alice's outcome and only check that Bob's system could have been generated locally, i.e., is independent from Alice's input (and output).

Definition 7 (Local-blindness). *A DQC protocol provides ε-local-blindness, if, for all adversarial behaviors $\{\mathcal{F}_i\}_i$, there exists a CPTP map $\mathcal{F} : \mathcal{L}(\mathcal{H}_B) \to \mathcal{L}(\mathcal{H}_B)$ such that*

$$\mathrm{tr}_A \circ \mathcal{P}_{AB} \approx_\varepsilon \mathcal{F} \circ \mathrm{tr}_A, \tag{5}$$

where $\circ$ is the composition of maps, tr_A the operator that trace out the A-system, and $\mathcal{P}_{AB} : \mathcal{L}(\mathcal{H}_{AB}) \to \mathcal{L}(\mathcal{H}_{AB})$ is the map corresponding to a protocol run with Alice behaving honestly and Bob using his cheating operations $\{\mathcal{F}_i\}_i$. We say that the protocol provides ε-local-blindness for a set of initial states $\mathcal{B}$, if (5) holds when applied to these states, i.e., for all $\psi_{ABR} \in \mathcal{B}$,

$$\mathrm{tr}_A \circ \mathcal{P}_{AB}(\psi_{ABR}) \approx_\varepsilon \mathcal{F} \circ \mathrm{tr}_A(\psi_{ABR}).$$

Likewise, local-verifiability can also be seen as a simplification of local-blind-verifiability, in which we ignore Bob's system and only check that Alice holds either the correct outcome or an error flag $|\mathrm{err}\rangle$, which by construction is orthogonal to any possible valid output. In the following we define local-verifiability only for the case where Bob's system is not entangled to Alice's input, since otherwise the correct outcome depends on Bob's actions, and cannot be modeled by describing Alice's system alone.[15]

---

[15] The resulting definition is equivalent to that of [20] and non-composable authentication definitions [5], which bound the probability of projecting the outcome on the space of invalid results.

Definition 8 (Local-verifiability). *A DQC protocol provides ε-local-verifiability, if, for all adversarial behaviors $\{\mathcal{F}_i\}_i$ and all initial states $\psi_{AR_1} \otimes \psi_{R_2B}$, there exists a $0 \leq p^\psi \leq 1$ such that*

$$\rho^\psi_{AR_1} \approx_\varepsilon p^\psi (\mathcal{U} \otimes \mathrm{id}_{R_1})(\psi_{AR_1}) + (1 - p^\psi)|err\rangle\langle err| \otimes \psi_{R_1}, \tag{6}$$

where $\rho^\psi_{AR_1}$ is the final state of Alice and the first part of the reference system. We say that the protocol provides ε-local-verifiability for a set $\mathcal{B}$ of initial states in product form, if (6) holds for all $\psi_{AR_1} \otimes \psi_{R_2B} \in \mathcal{B}$.

As mentioned in Sect. 1, local-blindness and local-verifiability together do not provide the security guarantees one expects from DQC. This seems to be because the verification procedure can depend on the input (as in the example from Footnote 2), and thus if Bob learns the result of this measurement, he learns something about the input. This motivates us to define a stronger notion, in which Bob can reconstruct on his own whether the output will be accepted — the outcome of Alice's verification procedure must thus be independent of her input. To do this, we introduce a new qubit in a system $\bar{B}$, which contains a copy of the information whether Alice accepts or rejects, i.e., for a final state

$$\rho^\psi_{ARB} = \phi^{ok}_{ARB} + |err\rangle\langle err| \otimes \phi^{err}_{RB}, \tag{7}$$

we define

$$\rho^\psi_{ARB\bar{B}} := \phi^{ok}_{ARB} \otimes |ok\rangle\langle ok| + |err\rangle\langle err| \otimes \phi^{err}_{RB} \otimes |err\rangle\langle err|. \tag{8}$$

Note that (8) can be generated from (7) by introducing a system $\bar{B}$ in the state $|ok\rangle$ and changing its value to $|err\rangle$ conditioned on A being in the state $|err\rangle$. Let $\mathcal{Q}_{A\bar{B}} : \mathcal{L}(\mathcal{H}_A) \to \mathcal{L}(\mathcal{H}_{A\bar{B}})$ be such an operation, i.e., $\rho^\psi_{ARB\bar{B}} = \mathcal{Q}_{A\bar{B}}(\rho^\psi_{ARB})$. Equation (7) can then be recovered from (8) by tracing out the system $\bar{B}$.

The notion of verifiability is strengthened by additionally requiring that leaking this system $\bar{B}$ to the adversary does not provide him with more information about the input, i.e., Bob could (using alternative maps) generate the system $\bar{B}$ on his own.

Definition 9. *A DQC protocol provides $\bar{\varepsilon}$-independent ε-local-verifiability, if, in addition to providing ε-local-verifiability, for all adversarial behaviors $\{\mathcal{F}_i : \mathcal{L}(\mathcal{H}_{CB}) \to \mathcal{L}(\mathcal{H}_{CB})\}_i$ there exist alternative maps $\{\mathcal{F}'_i : \mathcal{L}(\mathcal{H}_{CB\bar{B}}) \to \mathcal{L}(\mathcal{H}_{CB\bar{B}})\}_i$ (for an initially empty system $\bar{B}$), such that*

$$\mathrm{tr}_A \circ \mathcal{Q}_{A\bar{B}} \circ \mathcal{P}_{AB} \approx_{\bar{\varepsilon}} \mathrm{tr}_A \circ \mathcal{P}'_{AB\bar{B}}, \tag{9}$$

where $\circ$ is the composition of maps, $\mathcal{P}_{AB} : \mathcal{L}(\mathcal{H}_{AB}) \to \mathcal{L}(\mathcal{H}_{AB})$ and $\mathcal{P}'_{AB\bar{B}} : \mathcal{L}(\mathcal{H}_{AB}) \to \mathcal{L}(\mathcal{H}_{AB\bar{B}})$ are the maps corresponding to runs of the protocol with Alice being honest and Bob using maps $\{\mathcal{F}_i\}_i$ and $\{\mathcal{F}'_i\}_i$ respectively, and $\mathcal{Q}_{A\bar{B}} : \mathcal{L}(\mathcal{H}_A) \to \mathcal{L}(\mathcal{H}_{A\bar{B}})$ is a map which generates from A a system $\bar{B}$ holding a copy of the information whether Alice accepts or rejects. We say that a protocol *provides $\bar{\varepsilon}$-independent ε-local-verifiability for a set of initial states $\mathcal{B}$, if the same conditions hold for all states in $\mathcal{B}$, i.e., if we have ε-local-verifiability for $\mathcal{B}$, and if for all $\psi_{ABR} \in \mathcal{B}$,*

$$\text{tr}_A \circ \mathcal{Q}_{A\bar{B}} \circ \mathcal{P}_{AB}(\psi_{ABR}) \approx_{\bar{\varepsilon}} \text{tr}_A \circ \mathcal{P}'_{AB\bar{B}}(\psi_{ABR}).$$

Remark 10. By the triangle inequality, if a protocol provides both ε-local-blindness and $\bar{\varepsilon}$-independent ε'-local-verifiability, then there exists a map $\mathcal{F}'$: $\mathcal{L}(\mathcal{H}_B) \to \mathcal{L}(\mathcal{H}_{B\bar{B}})$ such that

$$\text{tr}_A \circ \mathcal{Q}_{A\bar{B}} \circ \mathcal{P}_{AB} \approx_{\varepsilon+\bar{\varepsilon}} \mathcal{F}' \circ \text{tr}_A . \tag{10}$$

We are now ready to state the main theorem, namely that the above local definitions are sufficient to achieve composable security.

Theorem 11 (Theorem 1 restated). *If a DQC protocol implementing a unitary transformation provides ε_{bl}-local-blindness and ε_{ind}-independent ε_{ver}-local-verifiability for all inputs $\psi_{A_C A_Q}$, where A_C is classical and A_Q is quantum, then it is δN^2-blind-verifiable, where $\delta = 4\sqrt{2\varepsilon_{ver}} + 2\varepsilon_{bl} + 2\varepsilon_{ind}$ and $N = \dim \mathcal{H}_{A_Q}$. If additionally it provides ε_{cor}-local-correctness,[16] it constructs S_{verif}^{blind} from a communication channel within $\varepsilon = \max\{\delta N^2, \varepsilon_{cor}\}$.*

Independent local-verifiability makes a statement about Alice's system at the end of the protocol — it is either in the correct state or contains an error flag. Local-blindness makes a statement about Bob's system at the end of the protocol — it contains no information about the input. To prove Theorem 11, we need to combine these two definitions to make a statement about the joint system of Alice and Bob at the end of the protocol, equivalent to local-blind-verifiability (Definition 4). The result then follows from Theorem 6.

The main idea of the proof is to show that, in the case of an input in product form between Alice and Bob, Uhlmann's theorem can be used to extend the statement about Alice's system being close to ideal to a joint AB system. We then show that if the input is entangled between Alice and Bob, the error can increase at most by a multiplicative factor of N^2. A complete proof is given in the full version [18].

Remark 12. This theorem only hold for protocols that construct a DQC resource for which the implemented operation $\mathcal{U}$ is unitary. Since any quantum operation can be written as a unitary on a larger system [38], this effectively allows the theorems to apply to any CPTP operation $\mathcal{E}$ as long as the necessary qubits for the unitary implementation are appended to the in- and outputs. For example, instead of defining universal computation as a unitary, most papers — e.g., [11, 20, 32, 35] — describe how to perform any (arbitrary) unitary operation U_x on any arbitrary input ρ_{in}. By appending the description x of the unitary U_x to the input and output, this is equivalent to applying the unitary transformation $\mathcal{U} := \sum_x U_x \otimes |x\rangle\langle x|$ to the input $\rho_{\text{in}} \otimes |x\rangle\langle x|$.

---

[16] See Definition 16 in Appendix A.

Remark 13. If the input is entirely classical (e.g., the client wants to factor a number), the failure ε is polynomial in the error parameters of the different local criteria, and the reduction is tight. If the input is quantum, the failure is multiplied by the dimension squared of the quantum (sub)system, and the errors of the local criteria need to be exponentially small in the size of the quantum input to compensate.

6   Existing Protocols

6.1   Applying the Security Reduction

The definitions of local-blindness and local-verifiability used in this work are equivalent to those used to prove local-security for most protocols in the literature, e.g., by Fitzsimons and Kashefi [20] and Morimae [32]. To prove that such protocols are secure, it remains to show that they satisfy the stronger definition of *independent* local-verifiability introduced in this work. We sketch in this section that this is the case for [20] and [32], and refer to the full version [18] for a longer discussion.

Both these works achieve local-verifiability by introducing randomly positioned *trap qubits* in the protocol: these are states which are independent of Alice's input, and for which she knows the outcome of the operation that the server, Bob, should perform. If the server does not trigger any of the traps, then with high probability he is running the correct program [20,32].

This technique used to achieve local-verifiability also provides independent local-verifiability, because the position of the traps and whether they get trigged are independent of the input. Thus, Bob could run the protocol on his own — without knowing Alice's input and choosing himself the position of the trap qubits — and would end up holding exactly the same bit as Alice that decides if the output is accepted or rejected.

6.2   Blindness

We present in this section the security results for two different DQC protocols proposed in the literature: we show that they construct the ideal blind quantum computation resource S^{blind} defined in Definition 2. The protocols and proofs appear in the full version [18], we only give a brief overview here. Note that since these protocols do not provide verifiability, we cannot use the generic results from Sect. 5 to prove that they are blind.

In the DQC protocol of Broadbent, Fitzsimons and Kashefi [11], Alice hides the computation by encrypting all the communication with a one-time pad. The main idea of the security proof is for the simulator to replace the encrypted states sent to the distinguisher by halves of EPR pairs. It then forwards the other halves to the ideal DQC resource, which gate teleports the real inputs. The distinguisher is then oblivious to whether it is interacting with the real protocol or the ideal resource and simulator.

Theorem 14. *The DQC protocol of Broadbent, Fitzsimons and Kashefi [11] provides perfect blindness.*

Morimae and Fujii [34] proposed a DQC protocol with one-way communication from Bob to Alice, in which Alice simply measures each qubit she receives, one at a time. We show that the general class of protocols with one-way communication is perfectly blind.

Theorem 15. *Any DQC protocol π with one-way communication from Bob to Alice provides perfect blindness.*

A Correctness

Intuitively, a protocol is correct if, when Bob behaves honestly, Alice ends up with the correct output. This must also hold with respect to a purification of the input.

Definition 16. *A DQC protocol provides ε-local-correctness, if, when both parties behave honestly, for all initial states ψ_{AR}, the map implemented by the protocol on Alice's input, $\mathcal{P}_A : \mathcal{L}(\mathcal{H}_A) \to \mathcal{L}(\mathcal{H}_A)$ is*

$$\mathcal{P}_A \approx_\varepsilon \mathcal{U}. \tag{11}$$

It is straightforward, that this is equivalent to the composable notion defined in Equations (2) and (3) in Sect. 3.2.

Lemma 17. *A DQC protocol which provides ε-local-correctness is also ε-correct.*

A proof is given in the full version [18].

Acknowledgments

This material is based on research supported in part by the Singapore National Research Foundation under NRF Award No. NRF-NRFF2013-01. VD acknowledges the support of the EPSRC Doctoral Prize Fellowship. Initial part of this work was performed while VD was at Heriot-Watt University, Edinburgh, supported by EPSRC (grant EP/E059600/1). CP and RR are supported by the Swiss National Science Foundation (via grant No. 200020-135048 and the National Centre of Competence in Research 'Quantum Science and Technology') and the European Research Council – ERC (grant No. 258932).

References

1. Aharonov, D., Ben-Or, M., Eban, E.: Interactive proofs for quantum computations. In: Proceedings of Innovations in Computer Science, ICS 2010, pp. 453–469 (2010)
2. Arrighi, P., Salvail, L.: Blind quantum computation. International Journal of Quantum Information 4(05), 883–898 (2006)
3. Backes, M., Pfitzmann, B., Waidner, M.: A general composition theorem for secure reactive systems. In: Naor, M. (ed.) TCC 2004. LNCS, vol. 2951, pp. 336–354. Springer, Heidelberg (2004)

4. Backes, M., Pfitzmann, B., Waidner, M.: The reactive simulatability (RSIM) framework for asynchronous systems. Information and Computation 205(12), 1685–1720 (2007), Extended version of [39]
5. Barnum, H., Crépeau, C., Gottesman, D., Smith, A., Tapp, A.: Authentication of quantum messages. In: Proceedings of the 43rd Symposium on Foundations of Computer Science, FOCS 2002, pp. 449–458. IEEE (2002)
6. Barrett, J., Colbeck, R., Kent, A.: Memory attacks on device-independent quantum cryptography. Physical Review Letters 110, 010503 (2013)
7. Barz, S., Fitzsimons, J.F., Kashefi, E., Walther, P.: Experimental verification of quantum computation. Nature Physics (2013)
8. Barz, S., Kashefi, E., Broadbent, A., Fitzsimons, J.F., Zeilinger, A., Walther, P.: Demonstration of blind quantum computing. Science 335(6066), 303–308 (2012)
9. Bellare, M., Namprempre, C.: Authenticated encryption: Relations among notions and analysis of the generic composition paradigm. In: Okamoto, T. (ed.) ASIACRYPT 2000. LNCS, vol. 1976, pp. 531–545. Springer, Heidelberg (2000)
10. Ben-Or, M., Mayers, D.: General security definition and composability for quantum & classical protocols (2004), http://www.arxiv.org/abs/quant-ph/0409062 (eprint)
11. Broadbent, A., Fitzsimons, J., Kashefi, E.: Universal blind quantum computation. In: Proceedings of the 50th Symposium on Foundations of Computer Science, FOCS 2009, pp. 517–526. IEEE Computer Society (2009)
12. Broadbent, A., Gutoski, G., Stebila, D.: Quantum one-time programs. In: Canetti, R., Garay, J.A. (eds.) CRYPTO 2013, Part II. LNCS, vol. 8043, pp. 344–360. Springer, Heidelberg (2013)
13. Canetti, R.: Universally composable security: A new paradigm for cryptographic protocols. In: Proceedings of the 42nd Symposium on Foundations of Computer Science, FOCS 2001, pp. 136–145. IEEE (2001)
14. Canetti, R.: Universally composable security: A new paradigm for cryptographic protocols. Cryptology ePrint Archive, Report 2000/067 (2013), http://eprint.iacr.org/2000/067, updated version of [13]
15. Chien, C.H., Meter, R.V., Kuo, S.Y.: Fault-tolerant operations for universal blind quantum computation (2013), http://www.arxiv.org/abs/1306.3664 (eprint)
16. Childs, A.M.: Secure assisted quantum computation. Quantum Information & Computation 5(6), 456–466 (2005)
17. Chiribella, G., D'Ariano, G.M., Perinotti, P.: Theoretical framework for quantum networks. Physical Review A 80, 022339 (2009)
18. Dunjko, V., Fitzsimons, J., Portmann, C., Renner, R.: Composable security of delegated quantum computation (2014), http://www.arxiv.org/abs/1301.3662 (eprint)
19. Dunjko, V., Kashefi, E., Leverrier, A.: Universal blind quantum computing with weak coherent pulses. Physical Review Letters 108, 200502 (2012)
20. Fitzsimons, J., Kashefi, E.: Unconditionally verifiable blind computation (2012), http://www.arxiv.org/abs/1203.5217 (eprint)
21. Giovannetti, V., Maccone, L., Morimae, T., Rudolph, T.G.: Efficient universal blind computation. Physical Review Letters 111, 230501 (2013)
22. Goldreich, O.: Foundations of Cryptography: Volume 1, Basic Tools. Cambridge University Press, New York (2001)
23. Goldreich, O.: Foundations of Cryptography: Volume 2, Basic Applications. Basic Applications, vol. 2. Cambridge University Press, New York (2004)
24. Gutoski, G.: On a measure of distance for quantum strategies. Journal of Mathematical Physics 53(3), 032202 (2012)

25. Gutoski, G., Watrous, J.: Toward a general theory of quantum games. In: Proceedings of the 39th Symposium on Theory of Computing, STOC 2007, pp. 565–574. ACM (2007)
26. Hofheinz, D., Müller-Quade, J., Unruh, D.: On the (im-)possibility of extending coin toss. In: Vaudenay, S. (ed.) EUROCRYPT 2006. LNCS, vol. 4004, pp. 504–521. Springer, Heidelberg (2006)
27. Krawczyk, H.: The order of encryption and authentication for protecting communications (or: How secure is SSL?). In: Kilian, J. (ed.) CRYPTO 2001. LNCS, vol. 2139, pp. 310–331. Springer, Heidelberg (2001)
28. Mantri, A., Pérez-Delgado, C.A., Fitzsimons, J.F.: Optimal blind quantum computation. Physical Review Letters 111, 230502 (2013)
29. Maurer, U., Renner, R.: Abstract cryptography. In: Proceedings of Innovations in Computer Science, ICS 2010, pp. 1–21. Tsinghua University Press (2011)
30. Maurer, U., Tackmann, B.: On the soundness of authenticate-then-encrypt: Formalizing the malleability of symmetric encryption. In: Proceedings of the 17th ACM Conference on Computer and Communication Security, pp. 505–515. ACM (2010)
31. Morimae, T.: Continuous-variable blind quantum computation. Physical Review Letters 109, 230502 (2012)
32. Morimae, T.: Verification for measurement-only blind quantum computing. Physical Review A 89, 060302 (2014)
33. Morimae, T., Dunjko, V., Kashefi, E.: Ground state blind quantum computation on AKLT state (2010), http://www.arxiv.org/abs/1009.3486 (eprint)
34. Morimae, T., Fujii, K.: Blind topological measurement-based quantum computation. Nature Communications 3, 1036 (2012)
35. Morimae, T., Fujii, K.: Blind quantum computation protocol in which alice only makes measurements. Physical Review A 87, 050301 (2013)
36. Morimae, T., Koshiba, T.: Composable security of measuring-Alice blind quantum computation (2013), http://www.arxiv.org/abs/1306.2113 (eprint)
37. Mosca, M., Stebila, D.: Quantum coins. In: Error-Correcting Codes, Finite Geometries and Cryptography. Contemporary Mathematics, vol. 523, pp. 35–47. American Mathematical Society (2010)
38. Nielsen, M.A., Chuang, I.L.: Quantum Computation and Quantum Information. Cambridge University Press (2000)
39. Pfitzmann, B., Waidner, M.: A model for asynchronous reactive systems and its application to secure message transmission. In: IEEE Symposium on Security and Privacy, pp. 184–200. IEEE (2001)
40. Portmann, C., Renner, R.: Cryptographic security of quantum key distribution (2014), http://www.arxiv.org/abs/1409.3525(eprint)
41. Sueki, T., Koshiba, T., Morimae, T.: Ancilla-driven universal blind quantum computation. Physical Review A 87, 060301 (2013)
42. Unruh, D.: Simulatable security for quantum protocols (2004), ŕlhttp://www.arxiv.org/abs/quant-ph/0409125 (eprint)
43. Unruh, D.: Universally composable quantum multi-party computation. In: Gilbert, H. (ed.) EUROCRYPT 2010. LNCS, vol. 6110, pp. 486–505. Springer, Heidelberg (2010)
44. Unruh, D.: Concurrent composition in the bounded quantum storage model. In: Paterson, K.G. (ed.) EUROCRYPT 2011. LNCS, vol. 6632, pp. 467–486. Springer, Heidelberg (2011)

All-But-Many Encryption

A New Framework for Fully-Equipped UC Commitments

Eiichiro Fujisaki

NTT Secure Platform Laboratories, Tokyo, Japan
fujisaki.eiichiro@lab.ntt.co.jp

Abstract. We present a general framework for constructing non-interactive universally composable (UC) commitment schemes that are secure against adaptive adversaries in the non-erasure model under a reusable common reference string. Previously, such "fully-equipped" UC commitment schemes have been known only in [5, 6], with *strict* expansion factor $O(\kappa)$; meaning that to commit λ bits, communication strictly requires $O(\lambda\kappa)$ bits, where κ denotes the security parameter. Efficient construction of a fully-equipped UC commitment scheme is a long-standing open problem. We introduce new abstraction, called *all-but-many encryption* (ABME), and prove that it captures a fully-equipped UC commitment scheme. We propose the first fully-equipped UC commitment scheme with *optimal expansion factor* $\Omega(1)$ from our ABME scheme related to the DCR assumption. We also provide an all-but-many lossy trapdoor function (ABM-LTF) [18] from our DCR-based ABME scheme, with a better lossy rate than [18].

1 Introduction

1.1 Motivating Application: Fully-Equipped UC Commitments

Universal composability (UC) framework [4] guarantees that if a protocol is proven secure in the UC framework, it remains secure even if it is run concurrently with arbitrary (even insecure) protocols. This composable property gives a designer a fundamental benefit, compared to the classic definitions, which only guarantee that a protocol is secure if it is run in the standalone setting. UC commitments are an essential ingredient to construct high level UC-secure protocols, which imply UC zero-knowledge protocols [5,10] and UC oblivious transfer [6], thereby meaning that any UC-secure two-party and multi-party computations can be realized in the presence of UC commitments. Since UC commitments cannot be realized without an additional set-up assumption [5], the common reference string (CRS) model is widely used. A commitment scheme consists of a two-phase protocol between two parties, a committer and a receiver. In the commitment phase, a committer gives a receiver the digital equivalent of a *sealed envelope* containing value x, and, in the opening phase, the committer reveals x in a way that the receiver can verify it. From the original concept, it is required that a committer cannot change the value inside the envelope (*binding property*), whereas the receiver can learn nothing about x (*hiding property*) unless

P. Sarkar and T. Iwata (Eds.): ASIACRYPT 2014, PART II, LNCS 8874, pp. 426–447, 2014.

the committer helps the receiver opens the envelope. Informally, a UC commitment scheme maintains the above binding and hiding properties under *any concurrent composition with arbitrary protocols*. To achieve this, a UC commitment scheme requires *equivocability* and *extractability* at the same time. Informally, equivocability of UC commitments in the CRS model can be interpreted as follows: An algorithm (called the simulator) that takes the secret behind the CRS string can generate an *equivocal* commitment that can be opened to any value. On the other hand, extractability can be interpreted as the ability of the simulator extracting the contents of a commitment generated by any adversarial algorithm, even after the adversary saw many equivocal commitments generated by the simulator.

Several factors as shown below feature UC commitments:

Non-Interactivity. If an execution of a commitment scheme is completed, simply by sending each one message from the committer to the receiver both in the commitment and opening phases, then it is called *non-interactive*; otherwise, interactive. From a practical viewpoint, non-interactivity is definitely favorable – non-interactive protocols are much easier to implement and more resilient to real threats such as denial of service attacks. Even from a theoretical viewpoint, non-interactive protocols generally make security proofs simpler.

CRS Re-usability. The CRS model assumes that CRS strings are generated in a trusted way and given to every party. For practical use, it is very important that a global single CRS string can be fixed beforehand and it can be *re-usable* in an unbounded number of executions of cryptographic protocols. Otherwise, a new CRS string must be set up in a trusted way every time when a new execution of a protocol is invoked.

Adaptive Security. If an adversary decides to corrupt parities only before a protocol starts, it is called a static adversary. On the other hand, if an adversary can decide to corrupt parties at any point in the executions of protocols, it is called an *adaptive* adversary. The attacks of adaptive adversaries are more realistic in the real world. So, adaptive UC security is more desirable.

Non-Erasure Model. When a party is corrupted, its complete inner state is revealed, including the randomness being used. Some protocols are only proven UC-secure under the assumption that the parties can securely erase their inner states at any point of an execution. However, reliable erasure is a difficult task on a real system. So, it is desirable that a *non-erasure* protocol is proven secure.

1.2 Previous Works

Canetti and Fischlin [5] presented the first UC secure commitment schemes. One of their proposals is "fully-equipped" – *non-interactive and adaptively secure in the non-erasure model under a global re-usable common reference string*. By construction, however, the proposal strictly requires, to commit to λ-bit secret, $O(\lambda\kappa)$ bits in communication and $O(\lambda)$ modular exponentiations in computation. Canetti et al. [6] also proposed another fully-equipped UC commitment scheme only from (enhanced) trapdoor permutations. It requires general non-interactive zero knowledge proofs and is simply inefficient.

So far, these two have been the only known fully-equipped UC commitment schemes. The known subsequent constructions of UC commitments [3,8,10,13,20, 22] have improved efficiency, but *sacrifice* at least one or a few requirements[1]. Efficient construction of a fully-equipped UC commitment scheme is a long-standing open problem.

1.3   Our Contribution

The UC framework is complicated with many subtleties. Therefore, it is desirable to translate the essence of basic UC secure protocols into simple cryptographic primitives. We introduce special tag-based public key encryption (Tag-PKE) that we call *all-but-many encryption* (ABME), and prove that it implies "fully-equipped" UC commitments. We propose a compact ABME scheme related to the DCR assumption and thereby the first fully-equipped UC commitment scheme with optimal expansion factor $\Omega(1)$. To commit λ bit, it requires $\Omega(\kappa)$ bits and a constant number of modular exponentiations. We also present an all-but-many lossy trapdoor function (ABM-LTF) [18] from our DCR-based ABME scheme, with a better lossy rate than [18].

In the full version [14], we present an ABME scheme from the DDH assumption with overhead $\Omega(\kappa/\log \kappa)$, which is slightly better than the prior work (with $\Omega(\kappa)$). We also present a fully-equipped UC commitment scheme from a *weak* ABME scheme under the general assumption (where (enhanced) trapdoor permutations exist), which is far more efficient than the prior scheme [6] under the same assumption.

Our Approach: All-But-Many Encryption. In an ABME scheme, a secret-key holding user (i.e., the simulator in the UC framework) can generate a *fake* ciphertext, which can be opened to any message with consistent randomness. On the other hand, it must be infeasible for a secret-key non-holding user (i.e., the adversary in the UC framework) (1) to distinguish a fake ciphertext from a *real* (honestly generated) ciphertext, even after the message and randomness are revealed, and (2) to produce a fake ciphertext (on a fresh tag) even given many fake ciphertexts.

To realize such a scheme, we divide its functionality into two primitives, called *probabilistic pseudo random functions* (PPRF) and *extractable sigma protocols* (extΣ). The former is a kind of a probabilistic version of a pseudo random function (family) in the public parameter model. The latter is special sigma (i.e., canonical 3-round public-coin HVZK) protocols [7] with some extractability. The concept of extractable sigma protocols is not completely new. A weaker notion, called *weak* extractable sigma protocols, appears in [15] to construct a few (interactive) simulation sound trapdoor commitment (SSTC) schemes. See

---

[1] Only [22] and [13] satisfy all but one requirement. [22] does not satisfy CRS reusability, whereas [13] does not support the non-erasure model.

also [16, 17, 21] for SSTC. This paper requires a stronger notion and its realization, which employed in a different framework. If two primitives are successfully combined, an ABME scheme can be constructed. We discuss more in the
following.

Probabilistic Pseudo-Random Function (PPRF). A PPRF $= (\mathsf{Gen}^{\mathsf{spl}}, \mathsf{Spl})$ is a
probabilistic version of a pseudo random function family in the public param-
eter model. $\mathsf{Gen}_{\mathsf{spl}}(1^\kappa)$ generates a pair of public-key/seed (pk, w), and A PPT
algorithm Spl takes (pk, w, t) and outputs (or *samples*) $u \leftarrow \mathsf{Spl}(pk, w, t)$. Let
$L_{pk}(t) = \{u | \exists (w, v) : u = \mathsf{Spl}(pk, w, t, v)\}$. Informally, a PPRF requires that (a)
u looks pseudo-random on any t (*pseudo randomness*) and (b) it is infeasible
for any adversary to find u^* in *some super set*, $\widehat{L}_{pk}(t^*)$, of $L_{pk}(t^*)$ on any fresh
t^*, even after it has access to oracle $\mathsf{Spl}(pk, w, \cdot)$ (*unforgeability on* $\widehat{L}_{pk}$), where
$\widehat{L}_{pk} := \{(t, u) | u \in \widehat{L}_{pk}(t)\}$. The super set $\widehat{L}_{pk}$ will be clear later.

Extractable Sigma Protocols. An extractable sigma protocol is a special sigma
protocol *associated with a language-generation algorithm and a decryption algorithm*. Recall the sigma protocols [7]. A sigma protocol Σ on NP language L is
a canonical 3-round public coin interactive proof system such that the prover
can convince the verifier that he knows the witness w behind common input
$x \in L$, where the prover first sends commitment a; the verifier sends back challenge (public-coin) e; the prover responds with z; and the verifier finally accepts
or rejects the conversation (a, e, z) on x. A sigma protocol is associated with a
simulation algorithm $\mathsf{sim}\Sigma$ that takes x (regardless of whether $x \in L$ or not)
and challenge e, and produces an accepting conversation $(a, e, z) \leftarrow \mathsf{sim}\Sigma(x, e)$
without witness w. It is guaranteed that, if $x \in L$, the distributions (a, e, z)
produced by $\mathsf{sim}\Sigma(x, e)$ on random e is statistically indistinguishable from the
transcript generated between two honest parties, called *honest-verifier statistically zero knowledge* (HVSZK). If $x \notin L$, for every a, there is unique e if there
is an accepting conversation (a, e, z), which is called *special soundness*.

An extractable sigma protocol $\mathsf{ext}\Sigma = (\mathsf{Gen}^{\mathsf{ext}}, \Sigma, \mathsf{Dec})$ uses two more algorithms: The language-generation algorithm $\mathsf{Gen}^{\mathsf{ext}}$ outputs a pair of public/secret
keys, (pk, sk), where pk determines two disjoint sets L_{pk} and L_{pk}^{ext}. Here sigma
protocol Σ works on L_{pk} and the decryption algorithm Dec works on L_{pk}^{ext}, meaning that $\mathsf{Dec}(sk, x, a)$ outputs challenge e if $x \in L_{pk}^{\mathsf{ext}}$ and if an accepting conversation (a, e, z) exists on x. Due to special soundness, e is uniquely determined if
$x \notin L_{pk}$. Therefore, the decryption algorithm is well defined.

Combining them. Suppose $\mathsf{ext}\Sigma$ and PPRF are so well combined that, for
$(L_{pk}, L_{pk}^{\mathsf{ext}})$ generated by $\mathsf{Gen}^{\mathsf{ext}}$, L_{pk} is the language derived from PPRF and
PPRF is unforgeable on $\widehat{L}_{pk}$ $(:= U'_{pk} \backslash L_{pk}^{\mathsf{ext}})$, where U'_{pk} denotes the entire set
with respects to pk. We can then transform the extractable sigma protocol into
an ABME scheme in the similar way that a sigma protocol is converted to an
instance-dependent commitment scheme [2, 19]. To encrypt message e on tag t,
a sender picks random u, runs $\mathsf{sim}\Sigma$ on instance (t, u) with challenge e, to get
$(a, e, z) \leftarrow \mathsf{sim}\Sigma(pk, (t, u), e)$, and finally outputs (t, u, a). Due to unforgeability
of PPRF, it holds that $(t, u) \in U'_{pk} \backslash \widehat{L}_{pk}$ with an overwhelming probability. Then,
e is uniquely determined given $((t, u), a)$, as long as an accepting conversation

(a, e, z) exists on (t, u). By our precondition, we can decrypt (t, u, a) using sk, as $e = \mathsf{Dec}(sk, (t, u), a)$ because $(t, u) \in L_{pk}^{\mathsf{ext}}$. On the other hand, a fake ciphertext on tag t is produced using (w, v) as follows: one sets $u := \mathsf{Spl}(pk, w, t; v)$, with random v, where $(t, u) \in L_{pk}$, and computes a, as similarly as an honest prover computes the first message on common input (t, u) with witness (w, v). To open a to e, he produces the third message z in the sigma protocol. It is obvious by construction that he can open a to any e because $(t, u) \in L_{pk}$.

Realizing Extractable Sigma Protocols. Although sigma protocols (with HVSZK) exist on *many* NP languages, it is not known how to extract the challenge as discussed above. The following is our key observation to realize the functionality. Sigma protocols are often implemented on Abelian groups associated with homomorphic maps, in which the first message of such sigma protocols implies a system of linear equations with e and z. Hence, there is a matrix derived from the linear systems. Due to completeness and special soundness, there is an invertible (sub) square matrix if and only if $x \notin L_{pk}$ (provided that the linear system is defined in a finite field). Therefore, if one knows the contents of the matrix, one can solve the linear systems when $x \notin L_{pk}$ and obtain e if its length is logarithmic. Suppose for instance that L_{pk} is the DDH language – it does not form a PPRF, but a good toy case to explain how to extract the challenge. Let $(g_1, g_2, h_1, h_2) \notin L_{pk}$, meaning that $x_1 \neq x_2$ where $x_1 := \log_{g_1}(h_1)$ and $x_2 := \log_{g_2}(h_2)$. The first message (A_1, A_2) of a canonical sigma protocol on L_{pk} implies linear equations

$$\begin{pmatrix} a_1 \\ a_2 \end{pmatrix} = \begin{pmatrix} 1 & x_1 \\ \alpha & \alpha x_2 \end{pmatrix} \begin{pmatrix} z \\ e \end{pmatrix} \tag{1}$$

where $A_1 = g_1^{a_1}$, $A_2 = g_2^{a_2}$, and $g_2 = g_1^{\alpha}$. The above matrix is invertible if and only if $(g_1, g_2, h_1, h_2) \notin L_{pk}$. We note that e is expressed as a linear combination of a_1 and a_2, i.e., $\beta_1 a_1 + \beta_2 a_2$, where the coefficients are determined by the matrix. Therefore, if the decryption algorithm takes (α, x_1, x_2) and the length of e is logarithmic, it can find out e by checking whether $g_1^e = A_1^{\beta_1} A_2^{\beta_2}$ or not. In the case when a partial information on the values of the matrix is given, the decryption algorithm can still find logarithmic-length e if the matrix is made so that e can be expressed as a *linear* combination of *unknown values* – the unknown values do not appear with a quadratic form or a more degree of forms in the equations.

In some case, we might be able to invert a homomorphic map, such as $f(a) = g^a$, using trapdoor f^{-1}. Then, the decryption algorithm can obtain (a_1, a_2) as well as the entire values of the matrix and hence extract the entire (polynomial-length) e. This happens in our DCR based implementation. In the case, the equivalent condition that the matrix is invertible is, not $x \notin L_{pk}$, but $x \notin \widehat{L}_{pk}$ for some superset $\widehat{L}_{pk}$, since the corresponding linear system is defined not on a finite field, but on a finite ring, such as $\mathbb{Z}_{n^d}$. This means that we require unforgeability on $\widehat{L}_{pk}$, so as to make an adversary output $x = (t, u)$ in $L_{pk}^{\mathsf{ext}} = U'_{pk} \backslash \widehat{L}_{pk}$.

1.4   Other Related Works

Simulation-based selective opening CCA (SIM-SO-CCA) secure PKE [12] is related to ABME, but both are incomparable. Indeed, the SIM-SO-CCA secure PKE scheme proposed in [12] does not satisfy the notion of ABME. On the other hand, ABME does not satisfy SIM-SO-CCA PKE, because it does not support CCA security. Although the scheme in [12] could be tailored to a fully-equipped UC commitment scheme, it cannot overcome the barrier of expansion factor $O(\kappa)$, because it strictly costs $O(\lambda\kappa)$ bits to encrypt λ bit.

Hofheinz has presented the notion of all-but-many lossy trapdoor function (ABM-LTF) [18], mainly to construct indistinguishable-based selective opening CCA (IND-SO-CCA) secure PKE schemes. ABM-LTF is lossy trapdoor function (LTF) [25] with (unbounded) *many* lossy tags. The relation between ABM-LTF and ABME is a generalized analogue of LTF and lossy encryption [1, 24] with unbounded many loss tags. However, unlike the other primitives, ABME always enjoys an *efficient* "opening" algorithm that can open a ciphertext on a "lossy" tag to any message with consistent randomness. Hofheinz has proposed two instantiations. One is related to the DCR assumption and the other is based on pairing groups of a composite order. In the DCR-based ABM-LTF, lossy tags are an analogue of Waters signatures defined in DJ PKE. Such tags are carefully embedded in a matrix so that it can be non-invertible if tags are lossy; otherwise invertible. We were inspired by the lossy tag idea and have generalized it as PPRF. In the latest e-print version [18], Hofheinz has proven that his DCR-based ABM-LTF can be converted to a SIM-SO-CCA PKE scheme. To realize this, an opening algorithm for ABM-LTF is needed, and he converted his DCR-based ABM-LTF into one with an opening algorithm, by sacrificing efficiency. We note that ABM-LTF with an opening algorithm meets the notion of ABME. We will show in Sect. 8 how Hofheinz's DCR-based ABM-LTF is converted to an ABME scheme. Its expansion factor is $\Omega(1)$. However, compared to our DCR-based ABME scheme in Sect. 7, Hofheinz's ABM-LTF based ABME scheme is rather inefficient for practical use. Indeed, its expansion rate of ciphertext length per message length is ≥ 31. In addition, you must use a modulus of $\geq n^6$. On the other hand, our DCR-based ABME scheme has a small expansion rate of $(5 + 1/d)$ and you can use modulus of n^{d+1} for any $d \geq 1$. We compare them in Sect. 8. We remark that Hofheinz has not shown that his DCR-based ABM-LTF can be converted to a UC commitment scheme.

2   Preliminaries

We write PPT and DPT algorithms to denote probabilistic polynomial-time and deterministic poly-time algorithms, respectively. For random variables, X_κ and Y_κ, ranging over $\{0, 1\}^\kappa$, the (statistical) distance between X_κ and Y_κ is defined as $\mathsf{Dist}(X_\kappa, Y_\kappa) \triangleq \frac{1}{2} \cdot |\Pr_{s \in \{0,1\}^\kappa}[X = s] - \Pr_{s \in \{0,1\}^\kappa}[Y = s]|$. We say that two probability ensembles, $X = \{X_\kappa\}_{\kappa \in \mathbb{N}}$ and $Y = \{Y_\kappa\}_{\kappa \in \mathbb{N}}$, are statistically indistinguishable (in κ), denoted $X \overset{s}{\approx} Y$, if $\mathsf{Dist}(X_\kappa, Y_\kappa) = \mathsf{negl}(\kappa)$. We say that

X and Y are computationally indistinguishable (in κ), denoted $X \overset{c}{\approx} Y$, if for every PPT D (with one-bit output), $\{D(1^\kappa, X_\kappa)\}_{\kappa \in \mathbb{N}} \overset{s}{\approx} \{D(1^\kappa, Y_\kappa)\}_{\kappa \in \mathbb{N}}$.

3 Building Blocks: Definitions

We now formally define probabilistic pseudo random functions and extractable sigma protocols.

3.1 Probabilistic Pseudo Random Function (PPRF)

$\mathsf{PPRF} = (\mathsf{Gen}^{\mathsf{spl}}, \mathsf{Spl})$ consists of the following two algorithms:

- $\mathsf{Gen}^{\mathsf{spl}}$, the key generation algorithm, is a PPT algorithm that takes 1^κ as input, creates pk and picks up $w \leftarrow \mathsf{KSP}^{\mathsf{spl}}_{pk}$ to outputs (pk, w), where pk uniquely determines $\mathsf{KSP}^{\mathsf{spl}}_{pk}$.
- Spl, the sampling algorithm, is a PPT algorithm that takes (pk, w) and $t \in \{0, 1\}^\kappa$, picks up inner random coins $v \leftarrow \mathsf{COIN}^{\mathsf{spl}}$, and outputs u.

Here we require that pk determines set U_{pk}. Let us define $U'_{pk} = \{0, 1\}^\kappa \times U_{pk}$, $L_{pk}(t) = \{u \in U_{pk} \,|\, \exists w, \exists v : u = \mathsf{Spl}(pk, w, t; v)\}$, and $L_{pk} = \{(t, u) \,|\, t \in \{0, 1\}^\kappa \text{ and } u \in L_{pk}(t)\}$. We are only interested in the case that L_{pk} is relatively small in U'_{pk}, in order to avoid sampling from U'_{pk} by chance. We require that PPRFs satisfy the following security requirements:

Efficiently Samplable and Explainable Domain: For every pk given by $\mathsf{Gen}^{\mathsf{spl}}$, set U is efficiently samplable and explainable [12], that is, there is an efficient sampling algorithm on U that takes pk and random coins R and outputs u uniformly from U_{pk}. In addition, for every $u \in U_{pk}$, there is an efficient explaining algorithm that takes pk and u and outputs random coins R behind u, where R is uniformly distributed subject to $sample(U_{pk}; R) = u$.

Pseudo Randomness: Any adversary A, given pk generated by $\mathsf{Gen}^{\mathsf{spl}}(1^\kappa)$, cannot distinguish whether it has had access to $\mathsf{Spl}(pk, w, \cdot)$ or $U(\cdot)$. Here U is the following oracle: If $\mathsf{Spl}(pk, w, \cdot)$ is a deterministic algorithm, $U : \{0, 1\}^\kappa \to U_{pk}$ is a random oracle. (Namely, it returns the same (random) value on the same input.) If $\mathsf{Spl}(pk, w, \cdot)$ is probabilistic, then $U(\cdot)$ picks up a fresh randomness $u \overset{U}{\leftarrow} U_{pk}$ for each query t. We say that PPRF is *pseudo random* if, for all non-uniform PPT A, $\mathsf{Adv}^{\mathsf{prf}}_{\mathsf{PPRF}, A}(\kappa) = \left| \Pr[\mathsf{Expt}^{\mathsf{prf}}_{\mathsf{PPRF}, A}(\kappa) = 1] - \Pr[\mathsf{Expt}^{\mathsf{prf}}_{U, A}(\kappa) = 1] \right|$ is negligible in κ, where $\mathsf{Expt}^{\mathsf{prf}}_{\mathsf{PPRF}, A}(\kappa)$ and $\mathsf{Expt}^{\mathsf{prf}}_{U, A}(\kappa)$ are defined in Fig. 1.

Unforgeability (on $\widehat{L}_{pk}$): Let $\widehat{L}_{pk}(t)$ be some super set of $L_{pk}(t)$. Let $\widehat{L}_{pk} = \{(t, u) \,|\, t \in \{0, 1\}^\kappa \text{ and } u \in \widehat{L}_{pk}(t)\}$. We define the game of unforgeability on $\widehat{L}_{pk}$ as follows: An adversary A takes pk generated by $\mathsf{Gen}^{\mathsf{spl}}(1^\kappa)$ and may have access to $\mathsf{Spl}(pk, w, \cdot)$. The aim of the adversary is to output $(t^*, u^*) \in \widehat{L}_{pk}$ such that t^* has not been queried. We say that PPRF is *unforgeable* on $\widehat{L}_{pk}$ if, for all

$$\begin{array}{|l|}\hline \mathsf{Expt}^{\mathsf{prf}}_{\mathsf{PPRF},A}(\kappa): \\ \quad (pk, w) \leftarrow \mathsf{Gen}^{\mathsf{spl}}(1^\kappa) \\ \quad b \leftarrow A^{\mathsf{Spl}(pk,w,\cdot)}(pk) \\ \quad \text{return } b. \\\hline\end{array} \qquad \begin{array}{|l|}\hline \mathsf{Expt}^{\mathsf{prf}}_{U,A}(\kappa): \\ \quad (pk, w) \leftarrow \mathsf{Gen}^{\mathsf{spl}}(1^\kappa) \\ \quad b \leftarrow A^{U(\cdot)}(pk) \\ \quad \text{return } b. \\\hline\end{array}$$

Fig. 1. The experiments, $\mathsf{Expt}^{\mathsf{prf}}_{\mathsf{PPRF},A}(\kappa)$ and $\mathsf{Expt}^{\mathsf{prf}}_{U,A}(\kappa)$

$$\begin{array}{|l|}\hline \mathsf{Expt}^{\mathsf{euf}\text{-}\widehat{L}}_{\mathsf{PPRF},A}(\kappa): \\ \quad (pk, w) \leftarrow \mathsf{Gen}^{\mathsf{spl}}(1^\kappa) \\ \quad (t^*, u^*) \leftarrow A^{\mathsf{Spl}(pk,w,\cdot)}(pk) \\ \quad \text{If } t^* \text{ has not been queried} \\ \qquad \text{and } u^* \in \widehat{L}_{pk}(t^*), \\ \quad \text{return } 1; \text{ otherwise } 0. \\\hline\end{array} \qquad \begin{array}{|l|}\hline \mathsf{Expt}^{\mathsf{seuf}\text{-}\widehat{L}}_{\mathsf{PPRF},A}(\kappa): \\ \quad (pk, w) \leftarrow \mathsf{Gen}^{\mathsf{spl}}(1^\kappa) \\ \quad (t^*, u^*) \leftarrow A^{\mathsf{Spl}(pk,w,\cdot)}(pk) \\ \quad (t^*, u^*) \notin \mathcal{QA} \\ \qquad \text{and } u^* \in \widehat{L}_{pk}(t^*), \\ \quad \text{return } 1; \text{ otherwise } 0. \\\hline\end{array}$$

Fig. 2. The experiments of unforgeability (in the left) and strong unforgeability (in the right)

non-uniform PPT A, $\mathsf{Adv}^{\mathsf{euf}\text{-}\widehat{L}}_{\mathsf{PPRF},A}(\kappa) = \Pr[\mathsf{Expt}^{\mathsf{euf}\text{-}\widehat{L}}_{\mathsf{PPRF},A}(\kappa) = 1]$ (where $\mathsf{Expt}^{\mathsf{euf}\text{-}\widehat{L}}_{\mathsf{PPRF},A}$ is defined in Fig. 2) is negligible in κ.

In some application, we require a stronger requirement, where in the same experiment above, it is difficult for the adversary to output (t^*, u^*) in $\widehat{L}_{pk}$, which did not appear in the query/answer list $\mathcal{QA}$. We say that PPRF is *strongly unforgeable* on $\widehat{L}_{pk}$ if, for all non-uniform PPT A, $\mathsf{Adv}^{\mathsf{seuf}\text{-}\widehat{L}}_{\mathsf{PPRF},A}(\kappa) = \Pr[\mathsf{Expt}^{\mathsf{seuf}\text{-}\widehat{L}}_{\mathsf{PPRF},A}(\kappa) = 1]$ (where $\mathsf{Expt}^{\mathsf{seuf}\text{-}\widehat{L}}_{\mathsf{PPRF},A}$ is defined in Fig. 2) is negligible in κ.

We remark that (strong) unforgeability implies (1) that $\widehat{L}_{pk}$ should be small enough in U'_{pk} to avoid sampling from $\widehat{L}_{pk}$ by chance, and (2) that, if Spl is a DPT algorithm and $\widehat{L}_{pk} = L_{pk}$, it is implied by pseudo randomness.

3.2 Extractable Sigma Protocol

An extractable sigma protocol, $\mathsf{ext}\Sigma = (\mathsf{Gen}^{\mathsf{ext}}, \mathsf{com}\Sigma, \mathsf{ch}\Sigma, \mathsf{ans}\Sigma, \mathsf{sim}\Sigma, \mathsf{Vrfy}, \mathsf{Dec})$ is a sigma protocol, associated with two algorithms, $\mathsf{Gen}^{\mathsf{ext}}$ and Dec, with the following properties.

- $\mathsf{Gen}^{\mathsf{ext}}$ is an PPT algorithm that takes 1^κ and outputs (pk, sk), such that pk defines the entire set U'_{pk}, and two sub disjoint sets, L_{pk} and L^{ext}_{pk}, i.e., $L_{pk} \cup L^{\mathsf{ext}}_{pk} \subset U'_{pk}$ and $L_{pk} \cap L^{\mathsf{ext}}_{pk} = \emptyset$. We also require that L_{pk} determines binary efficiently recognizable set R_{pk} such that $L_{pk} = \{x | \exists w : (x, w) \in R_{pk}\}$.
- $\mathsf{com}\Sigma$ is a PPT algorithm that takes pk and $(x, w) \in R_{pk}$, picks up inner coins r_a, and outputs a.
- $\mathsf{ch}\Sigma(pk)$ is a publicly-samplable set determined by pk.

- ansΣ is a DPT algorithm that takes (pk, x, r_a, e), where $e \in \mathsf{ch}\Sigma(pk)$, and outputs z.
- Vrfy is a DPT algorithm that accepts or rejects (pk, x, a, e, z).
- simΣ is a PPT algorithm that takes (pk, x, e) and outputs $(a, e, z) = \mathsf{sim}\Sigma$ $(pk, x, e; r_z)$, where $r_z \leftarrow \mathsf{COIN}^{\mathsf{sim}}$. *We additionally require that* $r_z = z$. Namely, $(a, e, r_z) = \mathsf{sim}\Sigma(pk, x, e; r_z)$.
- Dec is a DPT algorithm that takes (sk, x, a) and outputs e or $\perp$.

We require that extΣ satisfies the following properties:

Completeness: For every $(pk, sk) \in \mathsf{Gen}^{\mathsf{ext}}(1^\kappa)$, every $(x, w) \in R_{pk}$, every r_a (in an appropriate specified domain) and every $e \in \mathsf{ch}\Sigma(pk)$, it always holds that $\mathsf{Vrfy}(x, \mathsf{com}\Sigma(x, w; r_a), e, \mathsf{ans}\Sigma(x, w, r_a, e)) = 1$.

Special Soundness: For every $(pk, sk) \in \mathsf{Gen}^{\mathsf{ext}}(1^\kappa)$, every $x \in U'_{pk} \backslash L_{pk}$ and every a, there is *unique* $e \in \mathsf{ch}\Sigma(pk)$ if there is an accepting conversation for a on x. We say that a pair of two different accepting conversations for the same a on x, i.e., (a, e, z) and (a, e', z'), with $e \neq e'$, is a *collision* on x.

Enhanced Honest-Verifier Statistical Zero-Knowledgeness (eHVSZK): For every $(pk, sk) \in \mathsf{Gen}^{\mathsf{ext}}(1^\kappa)$, every $(x, w) \in R_{pk}$, and every $e \in \mathsf{ch}\Sigma(pk)$, the following ensembles are statistically indistinguishable in κ:

$$\{(\mathsf{com}\Sigma(pk, x, w; r_a), e, \mathsf{ans}\Sigma(pk, x, w, r_a, e))\}_{(pk, sk) \in \mathsf{Gen}^{\mathsf{ext}}(1^\kappa),\ (x, w) \in R_{pk},\ e \in \mathsf{ch}\Sigma(pk),\ \kappa \in \mathbb{N}.}$$

$$\overset{s}{\approx} \{\mathsf{sim}\Sigma(pk, x, e; r_z)\}_{(pk, sk) \in \mathsf{Gen}^{\mathsf{ext}}(1^\kappa), (x, w) \in R_{pk}, e \in \mathsf{ch}\Sigma(pk), \kappa \in \mathbb{N}}$$

Here the probability of the left-hand side is taken over random variable r_z and the right-hand side is taken over random variable r_a. We remark that since $(a, e, r_z) = \mathsf{sim}\Sigma(pk, x, e; r_z)$, we have $\mathsf{Vrfy}(pk, x, a, e, z) = 1$ if and only if $(a, e, z) = \mathsf{sim}\Sigma(pk, x, e; z)$. *Therefore, one can instead use* simΣ *to verify* (a, e, z) *on* x.

Extractability: For every $(pk, sk) \in \mathsf{Gen}^{\mathsf{ext}}(1^\kappa)$, every $x \in L_{pk}^{\mathsf{ext}}$, and every a such that there is an accepting conversation for a on x, Dec always outputs $e = \mathsf{Dec}(sk, x, a)$ such that (a, e, z) is an accepting conversation on x. We note that, when $x \notin L_{pk}$, e is unique given a, due to the special soundness property. Therefore, the extractability is well defined because $L_{pk} \cap L_{pk}^{\mathsf{ext}} = \emptyset$.

4   ABM Encryption

All-but-many encryption scheme ABM.Enc $=$ (ABM.gen, ABM.spl, ABM.enc, ABM.dec, ABM.col) consists of the following algorithms:

- ABM.gen is a PPT algorithm that takes 1^κ and outputs $(pk, (sk, w))$, where pk defines a set U_{pk}. We let $U'_{pk} = \{0, 1\}^\kappa \times U_{pk}$. pk also determines two disjoint sets, L_{pk}^{td} and L_{pk}^{ext}, such that $L_{pk}^{\mathsf{td}} \cup L_{pk}^{\mathsf{ext}} \subset U'_{pk}$.

- ABM.spl is a PPT algorithm that takes (pk, w, t), where $t \in \{0, 1\}^\kappa$, picks up inner random coins $v \leftarrow \mathsf{COIN}^{\mathsf{spl}}$, and computes $u \in U_{pk}$. We write $L_{pk}^{\mathsf{td}}(t)$ to denote the image of ABM.spl on t under pk, i.e.,

$$L_{pk}^{\mathsf{td}}(t) := \{u \in U_{pk} \mid \exists w, \exists v : u = \mathsf{ABM.spl}(pk, w, t; v)\}.$$

We require $L_{pk}^{\mathsf{td}} = \{(t, u) \mid t \in \{0, 1\}^\kappa \text{ and } u \in L_{pk}^{\mathsf{td}}(t)\}$. We set $\widehat{L}_{pk}^{\mathsf{td}} := U_{pk}' \backslash L_{pk}^{\mathsf{ext}}$. Since $L_{pk}^{\mathsf{td}} \cap L_{pk}^{\mathsf{ext}} = \emptyset$, we have $L_{pk}^{\mathsf{td}} \subseteq \widehat{L}_{pk}^{\mathsf{td}} \subset U_{pk}'$.
- ABM.enc is a PPT algorithm that takes pk, $(t, u) \in U_{pk}'$, and message $x \in \mathsf{MSP}$, picks up inner random coins $r \leftarrow \mathsf{COIN}^{\mathsf{enc}}$, and computes $c = \mathsf{ABM.enc}^{(t,u)}(pk, x; r)$, where MSP denotes the message space uniquely determined by pk, whereas $\mathsf{COIN}^{\mathsf{enc}}$ denotes the inner coin space uniquely determined by pk and x^2.
- ABM.dec is a DPT algorithm that takes sk, (t, u), and ciphertext c, and outputs $x = \mathsf{ABM.dec}^{(t,u)}(sk, c)$.
- $\mathsf{ABM.col} = (\mathsf{ABM.col}_1, \mathsf{ABM.col}_2)$ is a pair of PPT and DPT algorithms, respectively, such that
  - $\mathsf{ABM.col}_1$ takes $(pk, (t, u), w, v)$ and outputs $(c, \xi) \leftarrow \mathsf{ABM.col}_1^{(t,u)}(pk, w, v)$, where $v \in \mathsf{COIN}^{\mathsf{spl}}$.
  - $\mathsf{ABM.col}_2$ takes $((t, u), \xi, x)$, with $x \in \mathsf{MSP}$, and outputs $r \in \mathsf{COIN}^{\mathsf{enc}}$.

We require that all-but-many encryption schemes satisfy the following properties:

1. **Adaptive All-but-many property:** $(\mathsf{ABM.gen}, \mathsf{ABM.spl})$ is a probabilistic pseudo random function (PPRF) as defined in Sect. 3.1 with unforgeability on $\widehat{L}_{pk}^{\mathsf{td}} (= U_{pk}' \backslash L_{pk}^{\mathsf{ext}})$.
2. **Dual mode property:**
   - **(Decryption mode)** For every $\kappa \in \mathbb{N}$, every $(pk, (sk, w)) \in \mathsf{ABM.gen}(1^\kappa)$, every $(t, u) \in L_{pk}^{\mathsf{ext}}$, and every $x \in \mathsf{MSP}$, it always holds that

   $$\mathsf{ABM.dec}^{(t,u)}(sk, \mathsf{ABM.enc}^{(t,u)}(pk, x)) = x.$$

   - **(Trapdoor mode)** Define the following random variables: $\mathsf{dist}^{\mathsf{enc}}(t, pk, sk, w, x)$ denotes random variable (u, c, r) defined as follows: $v \leftarrow \mathsf{COIN}^{\mathsf{spl}}; u = \mathsf{ABM.spl}(pk, w, t; v); r \leftarrow \mathsf{COIN}^{\mathsf{enc}}; c = \mathsf{ABM.enc}^{(t,u)}(pk, x; r)$. $\mathsf{dist}^{\mathsf{col}}(t, pk, sk, w, x)$ denotes random variable (u, c, r) defined as follows: $v \leftarrow \mathsf{COIN}^{\mathsf{spl}}; u = \mathsf{ABM.spl}(pk, w, t; v); (c, \xi) = \mathsf{ABM.col}_1^{(t,u)}(pk, w, v); r = \mathsf{ABM.col}_2^{(t,u)}(\xi, x)$. Then, the following ensembles are statistically indistinguishable in κ:

---

[2] We allow the inner coin space to depend on messages to be encrypted, in order to be consistent with our weak ABM encryption scheme from general assumption appeared in the full version [14]

$$\left\{ \mathsf{dist}^{\mathsf{enc}}(t, pk, sk, w, x) \right\}_{(pk,(sk,w))\in\mathsf{ABM.gen}(1^\kappa), t\in\{0,1\}^\kappa, x\in\mathsf{MSP}, \kappa\in\mathbb{N}}$$

$$\overset{s}{\approx} \left\{ \mathsf{dist}^{\mathsf{col}}(t, pk, sk, w, x) \right\}_{(pk,(sk,w))\in\mathsf{ABM.gen}(1^\kappa), t\in\{0,1\}^\kappa, x\in\mathsf{MSP}, \kappa\in\mathbb{N}}$$

We say that a ciphertext c on (t, u) under pk is *valid* if there exist $x \in \mathsf{MSP}$ and $r \in \mathsf{COIN}^{\mathsf{enc}}$ such that $c = \mathsf{ABM.enc}^{(t,u)}(pk, x; r)$. We say that a valid ciphertext c on (t, u) under pk is *real* if $(t, u) \in L_{pk}^{\mathsf{ext}}$, otherwise *fake*. We remark that as long as c is a real ciphertext, regardless of how it is generated, there is only one consistent x in MSP and it is equivalent to $\mathsf{ABM.dec}^{(t,u)}(sk, c)$.

5 ABME from extΣ on Language Derived from PPRF

Let $\mathsf{PPRF} = (\mathsf{Gen}^{\mathsf{spl}}, \mathsf{Spl})$ be a PPRF and let $\mathsf{ext}\Sigma = (\mathsf{Gen}^{\mathsf{ext}}, \Sigma, \mathsf{Dec})$ be an extractable sigma protocol.

Assume the following conditions hold.

- The first output of $\mathsf{Gen}^{\mathsf{ext}}(1^\kappa)$ is distributed identically to the first output of $\mathsf{Gen}^{\mathsf{spl}}(1^\kappa)$.
- For every L_{pk} generated by $\mathsf{Gen}^{\mathsf{ext}}$, L_{pk} is the language derived from PPRF; namely, $L_{pk} = \{(t, u) \mid \exists (w, v) : t \in \{0,1\}^\kappa, u = \mathsf{Spl}(pk, w, t; v)\}$.
- For $(L_{pk}, L_{pk}^{\mathsf{ext}}, U_{pk}')$ generated by $\mathsf{Gen}^{\mathsf{ext}}$, PPRF is unforgeable on $\widehat{L}_{pk}$, where $\widehat{L}_{pk} := U_{pk}' \backslash L_{pk}^{\mathsf{ext}}$.

Then, we can construct an ABME scheme as described in Fig. 3.

- $\mathsf{ABM.gen}(1^\kappa)$ runs $\mathsf{Gen}^{\mathsf{ext}}(1^\kappa)$ to output (pk, sk). It chooses $w \leftarrow \mathsf{KSP}_{pk}^{\mathsf{spl}}$ and finally outputs $(pk, (sk, w))$. We note that by a precondition the distribution of pk from $\mathsf{Gen}^{\mathsf{ext}}(1^\kappa)$ is identical to that of $\mathsf{Gen}^{\mathsf{spl}}(1^\kappa)$.
- $\mathsf{ABM.spl}(pk, w, t; v)$ outputs $u := \mathsf{Spl}(pk, w, t; v)$ where $v \overset{\mathsf{U}}{\leftarrow} \mathsf{COIN}^{\mathsf{spl}}$.
- $\mathsf{ABM.enc}^{(t,u)}(pk, m; r)$ runs $(a, m, r) \leftarrow \mathsf{sim}\Sigma(pk, (t, u), m; r)$ to return the first output a, where $r \overset{\mathsf{U}}{\leftarrow} \mathsf{COIN}_{pk}^{\mathsf{enc}} (:= \mathsf{COIN}_{pk}^{\mathsf{sim}})$.
- $\mathsf{ABM.dec}^{(t,u)}(sk, c)$ outputs $m = \mathsf{Dec}(sk, (t, u), c)$.
- $\mathsf{ABM.col}_1^{(t,u)}(pk, w, v; r_a)$ outputs (c, ξ) such that $c := \mathsf{com}\Sigma(pk, (t, u), (w, v); r_a)$, and $\xi := (pk, t, u, w, v, r_a)$.
- $\mathsf{ABM.col}_2^{(t,u)}(\xi, m)$ outputs $r := \mathsf{ans}\Sigma(pk, (t, u), w, v, r_a, m)$, where $\xi = (pk, t, u, w, v, r_a)$.

Fig. 3. ABME from extΣ on language derived from PPRF

By construction, the adaptive all-but-many property holds in the resulting scheme. The dual mode property also holds because: (a) If $(t, u) \in L_{pk}^{\mathsf{ext}}$, the first output of $\mathsf{sim}\Sigma(pk, (t, u), m)$ is perfectly binding to challenge m due to special soundness (because $L_{pk}^{\mathsf{ext}} \subset U_{pk}' \backslash L_{pk}^{\mathsf{td}}$, with $L_{pk}^{\mathsf{td}} := L_{pk}$), and m can be extracted given $(pk, (t, u), a)$ using sk due to extractability. (b) If $(t, u) \in L_{pk}^{td}$, ABM.col runs the real sigma protocol with witness (w, v). Therefore, it can produce a fake commitment that can be opened in any way, while it is statistically indistinguishable from that of the simulation algorithm $\text{sim}\Sigma$ (that is run by ABM.enc), due to enhanced HVSZK. Therefore, the resulting scheme is ABME.

6 Fully-Equipped UC Commitment from ABME

We show that ABME implies fully-equipped UC commitment.

We work in the standard universal composability (UC) framework of Canetti [4]. We concentrate on the same model in [5] where the network is asynchronous, the communication is public but ideally authenticated, and the adversary is adaptive in corrupting parties and is active in its control over corrupted parties. Any number of parties can be corrupted and parties cannot erase any of their inner state. We consider UC commitment schemes that can be used repeatedly under a single common reference string. The multi-commitment ideal functionality $\mathcal{F}_{\text{MCOM}}$ from [6] is the ideal functionality of such commitments, which is given in Figure 4.

A fully-equipped UC commitment scheme is constructed as follows: A trusted party chooses and puts pk of ABME in the common reference string. In the commit phase, committer P_i takes tag $t = (\text{sid}, \text{ssid}, P_i, P_j)$ and message x committed to. It then picks up random u from U_{pk} and compute an ABM encryption $c = \text{ABM.enc}^{(t,u)}(pk, x; r)$ to send (t, u, c) to receiver P_j, which outputs $(\text{receipt}, \text{sid}, \text{ssid}, P_i, P_j)$. In the reveal phase, P_i sends (x, r) to P_j and P_j accepts if and only if $c = \text{ABM.enc}^{(t,u)}(pk, x; r)$. If P_j accepts, he outputs x, otherwise do nothing. The formal description is given in the full version [14].

Theorem 1. *The proposed commitment scheme from ABME UC-securely realizes the $\mathcal{F}_{\text{MCOM}}$ functionality in the $\mathcal{F}_{\text{CRS}}$-hybrid model in the presence of adaptive adversaries in the non-erasure model.*

Proof (Sketch). The formal proof is given in the full version [14]. We here sketch the essence. We consider the man-in-the-middle attack, where we show that the view of environment $\mathcal{Z}$ in the real world (in the CRS model) can be simulated in the ideal world. Let P_i, P_j be honest players and let $P_{i'}$ be a corrupted player controlled by adversary $\mathcal{A}$. In the man-in-the-middle attack, $P_{i'}$ (i.e., $\mathcal{A}$) is simultaneously participating in the left and right interactions. In the left interactions, $\mathcal{A}$ interacts with P_i, as playing the role of the receiver. In the right interactions, $\mathcal{A}$ interacts with P_j, as playing the role of the committer.

The following sketch corresponds to security proof in the (static) man-in-the-middle attack. It is not difficult to handle the adaptive case if this case has been proven secure.

In the Ideal World, $\mathcal{A}$ actually interacts with simulator $\mathcal{S}$ in both interactions, where $\mathcal{S}$ pretends to be P_i and P_j respectively. In the left interactions, environment $\mathcal{Z}$ sends $(\text{commit}, \text{sid}, \text{ssid}, P_i, P_{i'}, x)$ to the ideal commitment functionality $\mathcal{F}_{\text{MCOM}}$ (via honest P_i). After receiving $(\text{receipt}, \text{sid}, \text{ssid}, P_i, P_{i'})$ from

$\mathcal{F}_{\text{MCOM}}$, $\mathcal{S}$ starts the commitment protocol as the committer without given message x. It sends to $\mathcal{A}$ (u, c) on $t = (\text{sid}, \text{ssid}, P_i, P_{i'})$ as computed in Table 1. In the decommitment phase when $\mathcal{Z}$ sends $(\text{open}, \text{sid}, \text{ssid})$ to $\mathcal{F}_{\text{MCOM}}$ (via honest P_i), $\mathcal{S}$ receives x from $\mathcal{F}_{\text{MCOM}}$ and then computes $r = \text{ABM.col}_2^{(t,u)}(\xi, x)$ to send (t, x, r) to $\mathcal{A}$. In the right interactions, $\mathcal{S}$ receives (t', u', c') from $\mathcal{A}$ where $t' = (\text{sid}', \text{ssid}', P_{i'}, P_j)$. It then extracts $\tilde{x} = \text{ABM.dec}^{(t', u')}(sk, c')$ to send to $\mathcal{F}_{\text{MCOM}}$. $\mathcal{F}_{\text{MCOM}}$ then sends $(\text{receipt}, \text{sid}, \text{ssid}, P_{i'}, P_j)$ to environment $\mathcal{Z}$ (via honest P_j). In the decommitment phase when $\mathcal{A}$ opens (t', u', c') correctly with (x', r'), $\mathcal{S}$ sends $(\text{open}, \text{sid}, \text{ssid})$ to $\mathcal{F}_{\text{MCOM}}$; otherwise, do nothing. Upon receiving $(\text{open}, \text{sid}, \text{ssid})$, if the same $(\text{sid}, \text{ssid}, ..)$ was previously recorded, $\mathcal{F}_{\text{MCOM}}$ sends **stored** $\tilde{x}$ to environment $\mathcal{Z}$ (via honest P_j); otherwise, do nothing. We note that in the ideal world, honest parties convey inputs from $\mathcal{Z}$ to the ideal functionalities and vice versa. The view of $\mathcal{Z}$ consists of the view of $\mathcal{A}$ plus the value sent by $\mathcal{F}_{\text{MCOM}}$.

In Hybrid$^{\mathcal{F}_{\text{crs}}}$ (**the real world in the CRS model**), $\mathcal{A}$ interacts with real (committer) P_i in the left interactions, and real (receiver) P_j in the righ interactions. In the right interactions, at the end of the decommitment phase, P_j sends x' to $\mathcal{Z}$ if $\mathcal{A}$ has opened (t', u', c') correctly with (x', r'). The view of $\mathcal{Z}$ consists of the view of $\mathcal{A}$ plus the value sent by P_j.

The goal is to prove that the two views of $\mathcal{Z}$ above are computationally indistinguishable. As usual, we consider a sequence of hybrid games on which the probability spaces are identical, but we change the rules of games step by step. See Table 1 for summary.

Hybrid Game 1 is identical to the ideal world except that in the left interactions, at the beginning of the commitment phase, $\mathcal{S}$ (as P_i) is given message x on tag $t = (\text{sid}, \text{ssid}, P_i, P_{i'})$ by $\mathcal{F}_{\text{MCOM}}$. $\mathcal{S}$ computes $u \leftarrow \text{ABM.spl}(pk, w, t)$, and $c = \text{ABM.enc}^{(t,u)}(pk, x; r)$, picking up random r, to send (t, u, c) to adversary $\mathcal{A}$. In the decommitment phase, $\mathcal{S}$ sends (t, x, r) to $\mathcal{A}$.

Hybrid Game 2 is identical to Hybrid Game 1 except that in the right interactions, after receiving (t', u', c'), $\mathcal{S}_2$ sends ϵ to $\mathcal{F}_{\text{MCOM}}$. In the decommitment phase when $\mathcal{A}$ opens (t', u', c') correctly with (x, r), $\mathcal{S}$ sends $(\text{open}, \text{sid}, \text{ssid}, x')$ to $\mathcal{F}_{\text{MCOM}}$. $\mathcal{F}_{\text{MCOM}}$ sends x' to environment $\mathcal{Z}$ (via ideal $\tilde{P}_j$), instead of sending ϵ.

Hybrid Game 3 is identical to Hybrid Game 2 except that in the left interactions, $\mathcal{S}$ instead picks up random $u \leftarrow U_{pk}$ and computes $c = \text{ABM.enc}^{(t,u)}(pk, x; r)$, to send (t, u, c) to $\mathcal{A}$.

[Ideal $\Rightarrow$ Hybrid^1] The two views of $\mathcal{Z}$ between the ideal world and Hybrid^1 are statistically close, due to the trapdoor mode property.

[Hybrid$^1 \Rightarrow$ Hybrid^2] We note that the distance of the two views of $\mathcal{Z}$ between Hybrid^1 and Hybrid^2 is bounded by the following event. Let BD$_I$ denote the event in Hybrid Game I ($I \in \{1, 2\}$) that $\mathcal{S}$ receives a fake ciphertext (t', u', c') from $\mathcal{A}$, i.e., $(t', u') \in L_{pk}^{\text{td}}$, in the right intersections. If this event does not occur, the view of $\mathcal{Z}$ in both games are identical, which means $\neg \text{BD}_1 = \neg \text{BD}_2$. Hence, the distance of the views of $\mathcal{Z}$ in the two games is bounded by $\Pr[\text{BD}]$, where

$BD := BD_1 = BD_2$. We then evaluate $\Pr[BD]$ in Hybrid Game 2. (We note that we might not generally evaluate the probability in Hybrid Game 1, because S must decrypt (t', u', c'), which seems that it needs sk, but knowing sk implies some information on w.) We want to suppress $\Pr[BD]$ by using the assumption that (ABM.gen, ABM.spl) is unforgeable on $\widehat{L}_{pk}^{td}$. In Hybrid Game 2, we can construct an adversary B that breaks unforgeability of (ABM.gen, ABM.spl) on $\widehat{L}_{pk}^{td}$ as follows. In the left and right interactions, B simulates the role of S and interacts with A. B uses ABM.spl$(pk, w, \cdot)$ as oracle to play the role of S in the left interaction. After A halts, B outputs (t', u') at random from the communication with A in the right interactions. We note that, since the communication channel is fully authenticated, it holds that $t' \neq t$ for all t, t', because $t = (\star, \star, P_i, P_{i'})$ and $t' = (\star, \star, P_{i'}, P_j)$. If $(t', u') \in \widehat{L}_{pk}^{td}$, B succeeds in breaking unforgeability on $\widehat{L}_{pk}^{td}$, which is upper-bounded by some negligible function. Since event BD occurs at most with the success probability of B. Hence, its probability is negligible, too.

[Hybrid^2 $\Rightarrow$ Hybrid^3] It is obvious by construction that the distance of the two views of Z between Hybrid^2 and Hybrid^3 is bounded by the advantage of pseudo-randomness of (ABM.gen, ABM.spl).

[Hybrid^3 $\Rightarrow$ Hybrid$^{\mathcal{F}_{MCOM}}$] By construction, the two views of Z between Hybrid^3 and Hybrid$^{\mathcal{F}_{MCOM}}$ are identical.

Therefore, the two views of Z between the ideal world and Hybrid$^{\mathcal{F}_{MCOM}}$ are computationally close. ∎

Table 1. The man-in-the-midle attack in the hybrid games

Games	$P_i(S) \xrightarrow{(t,u,c)}$	Corr. $P_{i'}(A) \xrightarrow{(t',u',c')}$	$P_j(S) \xrightarrow{(t',\tilde{x})}$	$\mathcal{F}_{MCOM}$
Ideal	$u = $ ABM.spl$(pk, w, t; v)$ $(c, \xi) = $ ABM.col$_1^{(t,u)}(pk, w, v)$ open: $x, r = $ ABM.col$_2^{(t,u)}(\xi, x)$	(t', u', c') open: (x', r')	$\tilde{x} = $ ABM.dec$^{(t',u')}(sk, c')$	$\tilde{x}$
Hybrid^1	$u \leftarrow $ ABM.spl(pk, w, t) $c = $ ABM.enc$^{(t,u)}(pk, x, r)$ open: x, r	(t', u', c') open: (x', r')	$\tilde{x} = $ ABM.dec$^{(t',u')}(sk, c')$	$\tilde{x}$
Hybrid^2	$u \leftarrow $ ABM.spl(pk, w, t) $c = $ ABM.enc$^{(t,u)}(pk, x, r)$ open: x, r	(t', u', c') open: (x', r')	$\tilde{x} = \epsilon$	x'
Hybrid^3	$u \leftarrow U_{pk}$ $c = $ ABM.enc$^{(t,u)}(pk, x, r)$ open: x, r	(t', u', c') open: (x', r')	$\tilde{x} = \epsilon$	x'
	$P_i \xrightarrow{(t,u,c)}$	Corr. $P_{i'}(A) \xrightarrow{(t',u',c')}$	P_j	P_j
Hybrid$^{\mathcal{F}_{crs}}$	$u \leftarrow U_{pk}$ $c = $ ABM.enc$^{(t,u)}(pk, x, r)$ open: x, r	(t', u', c') open: (x', r')		x'

Here $t = (\mathsf{sid}, \mathsf{ssid}, P_i, P_{i'})$ and $t' = (\mathsf{sid}', \mathsf{ssid}', P_{i'}, P_j)$. The view of Z consists of the view of A plus the contents in the rightest column.

7   Compact ABME from Damgård-Jurik PKE

Damgård-Jurik PKE. Let $\Pi = (\mathbf{K}, \mathbf{E}, \mathbf{D})$ be a tuple of algorithms of Damgård-Jurik (DJ) PKE [9]. A public key of DJ PKE is $pk_{\mathsf{dj}} = (n, d)$ and the corresponding secret-key is $sk_{\mathsf{dj}} = (p, q)$ where $n = pq$ is a composite number of distinct odd primes, p and q, and $1 \le d < p, q$ is a positive integer (when $d = 1$ it is Paillier PKE [23]). We often write $\Pi^{(d)}$ to clarify parameter d. We let $g := (1 + n)$. To encrypt message $x \in \mathbb{Z}_{n^d}$, one computes $\mathbf{E}_{pk_{\mathsf{dj}}}(x; R) = g^x R^{n^d} \pmod{n^{d+1}}$ where $R \leftarrow \mathbb{Z}_n^{\times}$ [3]. For simplicity, we write $\mathbf{E}(x)$ instead of $\mathbf{E}_{pk_{\mathsf{dj}}}(x)$, if it is clear. DJ PKE is enhanced additively homomorphic, meaning that, for every $x_1, x_2 \in \mathbb{Z}_{n^d}$ and every $R_1, R_2 \in \mathbb{Z}_n^{\times}$, one can efficiently compute R such that $\mathbf{E}(x_1 + x_2; R) = \mathbf{E}(x_1; R_1) \cdot \mathbf{E}(x_2; R_2)$. Actually it can be done by computing $R = g^{\gamma} R_1 R_2 \pmod{n}$, where γ is an integer such that $x_1 + x_2 = \gamma n^d + ((x_1 + x_2) \bmod n^d)$. It is known that $\mathbb{Z}_{n^{d+1}}^{\times}$ is isomorphic to $\mathbb{Z}_{n^d} \times \mathbb{Z}_n^{\times}$ (the product of a cyclic group of order n^d and a group of order $\phi(n)$), and, for any $d < p, q$, element $g = (1 + n)$ has order n^d in $\mathbb{Z}_{n^{d+1}}^{\times}$ [9]. Therefore, $\mathbb{Z}_{n^{d+1}}^{\times}$ is the image of $\mathbf{E}(\cdot; \cdot)$. We note that it is known that $\mathbb{Z}_{n^{d+1}}^{\times}$ is *efficiently samplable and explainable* [10,12]. It is also known that DJ PKE is IND-CPA if the DCR assumption holds true [9].

Construction Idea. $(\mathsf{ABM.gen}, \mathsf{ABM.spl})$ below forms Waters-like signature scheme based on DJ PKE, where there is no verification algorithm and the signatures look pseudo random assuming that DJ PKE is IND-CPA. We then construct an extractable sigma protocol on the language derived from $(\mathsf{ABM.gen}, \mathsf{ABM.spl})$, as discussed in Sect. 5. Here, the decryption algorithm works only when the matrix below in (3) is invertible, which is equivalent to that $(t, (u_r, u_t)) \in L_{pk}^{\mathsf{ext}}$, where $L_{pk}^{\mathsf{ext}} =$

$$\{(t, (u_r, u_t)) \mid \mathbf{D}(u_t) \not\equiv x_1 x_2 + y(t)\mathbf{D}(u_r) \bmod p \ \wedge \ \mathbf{D}(u_t) \not\equiv x_1 x_2 + y(t)\mathbf{D}(u_r) \bmod q\}.$$

Therefore, we require that $(\mathsf{ABM.gen}, \mathsf{ABM.spl})$ should be unforgeable on $\widehat{L}_{pk}^{\mathsf{td}} (= U_{pk}' \backslash L_{pk}^{\mathsf{ext}})$. To prove this, we additionally require two assumptions on DJ PKE, called *the non-multiplication assumption* and *the non-trivial divisor assumption*, described in Appendix C. The first one is an analogue of the DH assumption in an additively homomorphic encryption. If we consider unforgeability on L_{pk}^{td}, this assumption suffices, but we require unforgeability on $\widehat{L}_{pk}^{\mathsf{td}}$. Then we need the latter assumption, too. These two assumptions are originally introduced in [18] to obtain a DCR-based ABM-LTF.

---

[3] In the original scheme, R is chosen from $\mathbb{Z}_{n^{d+1}}^{\times}$. However, since $\mathbb{Z}_n^{\times}$ is isomorphic to the cyclic group of order n^d in $\mathbb{Z}_{n^{d+1}}^{\times}$ by mapping $R \in \mathbb{Z}_n^{\times}$ to $R^{n^d} \in \mathbb{Z}_{n^{d+1}}^{\times}$, we can instead choose R from $\mathbb{Z}_n^{\times}$.

7.1 ABME from Damgård-Jurik with Optimal Expansion Factor $\Omega(1)$

- ABM.gen(1^κ): It gets $(pk_{dj}, sk_{dj}) \leftarrow \mathbf{K}(1^\kappa)$ (the key generation algorithm for DJ PKE), where $pk_{dj} = (n, d)$ and $sk_{dj} = (p, q)$. It then picks up $x_1, x_2 \xleftarrow{\cup} \mathbb{Z}_{n^d}$, $R_1, R_2 \xleftarrow{\cup} \mathbb{Z}_{n^{d+1}}^\times$, and computes $g_1 = \mathbf{E}(x_1; R_1)$ and $g_2 = \mathbf{E}(x_2; R_2)$. It then picks up $\tilde{h} \leftarrow \mathbf{E}(1)$ and computes $\boldsymbol{h} = (h_0, \ldots, h_\kappa)$ such that $h_j := \tilde{h}^{y_j}$ where $y_j \xleftarrow{\cup} \mathbb{Z}_{n^{d+1}}$ for $j = 0, 1, \ldots, \kappa$. Let $H(t) = h_0 \prod_{i=1}^\kappa h_i^{t_i} \pmod{n^{d+1}}$ and let $y(t) = y_0 + \sum_{i=1}^\kappa y_i t_i \pmod{n^d}$, where $(t_0, \ldots, t_\kappa)$ represents the bit string of t. We note that $H(t) = \tilde{h}^{y(t)}$. It outputs $(pk, (sk, w))$ where $pk := (n, d, g_1, g_2, \boldsymbol{h})$, $sk := (p, q)$ and $w := x_2$, where we define $U'_{pk} := \{0,1\}^\kappa \times (\mathbb{Z}_{n^{d+1}}^\times)^2$ that contains the disjoint sets of L_{pk}^{td} and L_{pk}^{ext} as described below.

- ABM.spl($pk, x_2, t; (r, R_r, R_t)$): It chooses $r \leftarrow \mathbb{Z}_{n^d}$ and outputs $u := (u_r, u_t)$ such that $u_r := \mathbf{E}(r; R_r)$ and $u_t := g_1^{x_2} \mathbf{E}(0; R_t) \cdot H(t)^r$ where $R_r, R_t \leftarrow \mathbb{Z}_{n^{d+1}}^\times$. We let $L_{pk}^{td} = \{(t, (u_r, u_t)) \mid \exists (x_2, (r, R_r, R_t)) : u_r = \mathbf{E}(r,; R_r) \text{ and } u_t = g_1^{x_2} \mathbf{E}(0; R_t) H(t)^r\}$. We then define $L_{pk}^{ext} = \{(t, (u_r, u_t)) \mid \mathbf{D}(u_t) \not\equiv x_1 x_2 + y(t)\mathbf{D}(u_r) \bmod p \wedge \mathbf{D}(u_t) \not\equiv x_1 x_2 + y(t)\mathbf{D}(u_r) \bmod q\}$. Since $(t, (u_r, u_t)) \in L_{pk}^{td}$ holds if and only if $\mathbf{D}(u_t) \equiv x_1 x_2 + y(t)\mathbf{D}(u_r) \pmod{n^d}$, it implies that $\mathbf{D}(u_t) \equiv x_1 x_2 + y(t)\mathbf{D}(u_r) \pmod{n}$. Hence, $L_{pk}^{td} \cap L_{pk}^{ext} = \emptyset$.

- ABM.enc$^{(t, (u_r, u_t))}(pk, m; (z, s, R_A, R_a, R_b))$: To encrypt message $m \in \mathbb{Z}_{n^d}$, it chooses $z, s \xleftarrow{\cup} \mathbb{Z}_{n^d}$ and computes $A := g_1^z H(t)^s u_t^m R_A^{n^d} \pmod{n^{d+1}}$, $a := \mathbf{E}(z; R_a) \cdot g_2^m \pmod{n^{d+1}}$ and $b := \mathbf{E}(s; R_b) \cdot u_r^m \pmod{n^{d+1}}$, where $R_A, R_a, R_b \xleftarrow{\cup} \mathbb{Z}_{n^{d+1}}^\times$. It outputs $c := (A, a, b)$ as the ciphertext of m on $(t, (u_r, u_t))$.

- ABM.dec$^{(t, (u_r, u_t))}(sk, c)$: To decrypt $c = (A, a, b)$, it outputs

$$m := \frac{x_1 \mathbf{D}(a) + y(t)\mathbf{D}(b) - \mathbf{D}(A)}{x_1 x_2 - (\mathbf{D}(u_t) - y(t)\mathbf{D}(u_r))} \bmod n^d. \tag{2}$$

- ABM.col$_1^{(t, (u_r, u_t))}(pk, x_2, (r, R_r, R_t))$: It picks up $\omega, \eta \xleftarrow{\cup} \mathbb{Z}_{n^d}$, $R'_A, R'_a, R'_b \xleftarrow{\cup} \mathbb{Z}_{n^{d+1}}^\times$. It then computes $A := g_1^\omega \cdot H(t)^\eta \cdot {R'_A}^{n^d} \pmod{n^{d+1}}$, $a := g^\omega {R'_a}^{n^d} \pmod{n^{d+1}}$, and $b := g^\eta {R'_b}^{n^d} \pmod{n^{d+1}}$. It outputs $c := (A, a, b)$ and $\xi := (x_2, (r, R_r, R_t), (u_r, u_t), \omega, \eta, R'_A, R'_a, R'_b)$.

- ABM.col$_2(\xi, m)$: To open c to m, it computes $z = \omega - m x_2 \bmod n^d$, $s = \eta - mr \bmod n^d$, $\alpha = \lfloor (\omega - m x_2 - z)/n^d \rfloor$, and $\beta = \lfloor (\eta - mr - s)/n^d \rfloor$. It then sets $R_A := R'_A \cdot R_t^{-m} \cdot g_1^\alpha \cdot H(t)^\beta \pmod{n^{d+1}}$, $R_a := R'_a \cdot R_2^{-m} \cdot g^\alpha \pmod{n^{d+1}}$, and $R_b := R'_b \cdot R_r^{-m} \cdot g^\beta \pmod{n^{d+1}}$. It outputs (z, s, R_A, R_a, R_b), where $A = g_1^z H(t)^s u_t^m R_A^{n^d} \pmod{n^{d+1}}$, $a = \mathbf{E}(z; R_a) \cdot g_2^m \pmod{n^{d+1}}$, and $b = \mathbf{E}(s; R_b) \cdot u_r^m \pmod{n^{d+1}}$.

We note that ABM.col runs a canonical sigma protocol on L_{pk}^{td} to prove that the prover knows $(x_2, (r, R_r, R_t))$ such that $u_r = \mathbf{E}_{pk}(r; R_r)$ and $u_t = g_1^{x_2} \mathbf{E}_{pk}(0; R_t) H(t)^r$. Hence, the trapdoor mode works correctly when $(t, (u_r, u_t))$

$\in L_{pk}^{td}$. On the contrary, ABM.enc runs a simulation algorithm of the sigma protocol with message (challenge) x. Notice that (A, a, b) implies the following linear system on $\mathbb{Z}_{n^d}$,

$$\begin{pmatrix} \mathbf{D}(A) \\ \mathbf{D}(a) \\ \mathbf{D}(b) \end{pmatrix} = \begin{pmatrix} x_1 & y(t) & \mathbf{D}(u_t) \\ 1 & 0 & x_2 \\ 0 & 1 & \mathbf{D}(u_r) \end{pmatrix} \begin{pmatrix} z \\ s \\ m \end{pmatrix} \tag{3}$$

The matrix is invertible if

$$\mathbf{D}(u_t) \neq (x_1 x_2 + y(t)\mathbf{D}(u_r)) \pmod{p} \text{ and } \mathbf{D}(u_t) \neq (x_1 x_2 + y(t)\mathbf{D}(u_r)) \pmod{q},$$

which means that $(t, (u_r, u_t)) \in L_{pk}^{ext}$. Hence, the decryption mode works correctly.

Lemma 1 (Implicit in [18]). (ABM.gen, ABM.spl) *is PPRF with unforgeability on* $\widehat{L}_{pk}^{td}(= U_{pk}' \backslash L_{pk}^{ext})$, *under the assumptions, 3, 4, and 5.*

The proof is given in the full version [14]. By this lemma, we have:

Theorem 2. *The scheme constructed as above is an ABME scheme if the DCR assumption (Assumption 3), the non-tirvial divisor assmuption (Assumption 4), and the non-multiplication assumption (Assumption 5) hold true.*

This scheme has a ciphertext consisting of only 5 group elements (including (u_r, u_t)) and optimal expansion factor $\Omega(1)$. This scheme requires a public-key consisting of $\kappa + 3$ group elements along with some structure parameters.

8   ABM-LTF Based ABME and Vice Versa

Hofheinz [18] has presented the notion of all-but-many lossy trapdoor function (ABM-LTF). We provide the definition in Appendix B. We remark that ABM-LTF requires that, in our words, (ABM.gen, ABM.spl) be *strongly* unforgeable, whereas ABME only requires it be unforgeable. However, as shown in [18], unforgeable PPRF can be converted into strongly unforgeable PPRF via a chameleon commitment scheme. Therefore, this difference is not important. We note that we can regard Hofheinz's DCR-based ABM-LTF (with only unforgeability) as a special case of our DCR-based ABME scheme by fixing a part of the coin space as $(R_A, R_a, R_b) = (1, 1, 1)$. Although the involved matrix of his original scheme is slightly different from ours, the difference is not essential. In the end, we can regard Hofheinz's DCR-based ABM-LTF as

$$\mathsf{ABM.eval}^{(t,(u_r,u_t))}(pk, (m, z, s)) := \mathsf{ABM.enc}^{(t,(u_r,u_t))}(pk, m; (z, s, 1, 1, 1)),$$

where (m, z, s) denotes a message. This ABM-LTF has $((d-1)\log n)$-lossyness. In the latest e-print version [18], Hofheinz has shown that his DCR-based ABM-LTF can be converted to SIM-SO-CCA PKE. To construct it, Hofheinz implicitly considered the following PKE scheme such that

$$\mathsf{ABM.enc}^{(t,(u_r,u_t))}(pk, M; (m, z, s)) := (\mathsf{ABM.eval}^{(t,(u_r,u_t))}(pk, (m, z, s)),$$
$$M \oplus H(m, z, s)),$$

where H is a suitable 2-universal hash function from $(\mathbb{Z}_{n^d})^3$ to $\{0,1\}^\kappa$ ($\kappa <$ n). According to the analysis in Sect. 7.2 in [18], if $d \geq 5$, it can open an ciphertext arbitrarily using Barvinok's alogorithm, when $(t, (u_r, u_t)) \in L^{\mathsf{loss}}$. Then it turns out ABME in our words. For practical use, it is rather inefficient, because its expansion rate of ciphertext length per message length is ≥ 31, and the modulus of $\geq n^6$ is required. The opening algorithm is also costly. Table 2 shows comparison.

Table 2. Comparison among ABMEs

ABME	expansion factor	ciphertext-length	message-length	pk-length
ABME from [18]	$\geq 31^*$	$(5(d+1)+1)\log n$	$\log n$	$(\kappa+3)d\log n$
Sect. 7.1 ($d \geq 1$)	$5+1/d$	$5(d+1)\log n$	$d\log n$	$(\kappa+3)d\log n$

$* : d \geq 5$ is needed.

On the contrary, our DCR-based ABME (strengthened with strong unforgeability) can be converted to ABM-LTF. Remember that $(A, a, b) =$ $\mathsf{ABM.enc}^{(t,(u_r,u_t))}$ $(pk, m; (z, s, R_A, R_a, R_b))$. It is obvious that we can extract not only message m but (z, s) by inverting the corresponding matrix, but we point out that we can further retrieve (R_A, R_a, R_b), too. This mean that our DCR based ABME turns out ABM-LTF. Indeed, after extracting (m, z, s) from (A, a, b), we have $(R_A)^{n^d}, (R_a)^{n^d}, (R_b)^{n^d}$ in $\mathbb{Z}_{n^{d+1}}^\times$. We remark that R_A, R_a, R_b lie not in $\mathbb{Z}_{n^{d+1}}^\times$ but in $(\mathbb{Z}/n\mathbb{Z})^\times$. So, letting $\alpha = r^{n^d} \bmod n^{d+1}$ where $r \in (\mathbb{Z}/n\mathbb{Z})^\times$, $r = \alpha^{(n^d)^{-1}} \bmod n$ is efficiently solved by $\phi(n)$. Thus, our DCR based ABME turns out ABM-LTF with $(d\log n)$-lossyness for any $d \geq 1$, whereas Hofheinz's DCR based ABM-LTF is $((d-3)\log n)$-lossy.

Table 3. Comparison among ABM-LTFs

ABM-LTF	exp. factor	output-length	input-length	lossyness
ABM-LTF [18]	$5/3$	$(5(d+1)+1)\log n$	$3d\log n$	$(d-3)\log n$
ABM-LTF from Sect. 7	$5/3$	$(5(d+1)+1)\log n$	$3(d+1)\log n$	$d\log n$

Acknowledgments. We thank Kirill Morozov and his students for nice feedback in the early version of this work. We also thank Dennis Hofheinz for valuable discussion.

References

1. Bellare, M., Hofheinz, D., Yilek, S.: Possibility and impossibility results for encryption and commitment secure under selective opening. In: Joux, A. (ed.) EUROCRYPT 2009. LNCS, vol. 5479, pp. 1–35. Springer, Heidelberg (2009)

2. Bellare, M., Micali, S., Ostrovsky, R.: Perfect zero-knowledge in constant rounds. In: STOC 1990, pp. 482–493. ACM (1990)

3. Camenisch, J., Shoup, V.: Practical verifiable encryption and decryption of discrete logarithms. In: Boneh, D. (ed.) CRYPTO 2003. LNCS, vol. 2729, pp. 126–144. Springer, Heidelberg (2003)

4. Canetti, R.: Universally composable security: A new paradigm for cryptographic protocols. In: 42nd Annual IEEE Symposium on Foundations of Computer Science (FOCS 2001), pp. 136–145. IEEE Computer Society (2001), The full version available at at Cryptology ePrint Archive, http://eprint.iacr.org/2000/067

5. Canetti, R., Fischlin, M.: Universally composable commitments. In: Kilian, J. (ed.) CRYPTO 2001. LNCS, vol. 2139, pp. 19–40. Springer, Heidelberg (2001)

6. Canetti, R., Lindell, Y., Ostrovsky, R., Sahai, A.: Universally composable two-party and multi-party secure computation. In: STOC 2002, pp. 494–503. ACM, New York (2002), The full version is available at http://eprint.iacr.org/2002/140

7. Cramer, R., Damgård, I., Schoenmakers, B.: Proofs of partial knowledge and simplified design of witness hiding protocols. In: Desmedt (ed.) [11], pp. 174–187.

8. Damgård, I., Groth, J.: Non-interactive and reusable non-malleable commitment schemes. In: STOC 2003, pp. 426–437. ACM (2003)

9. Damgård, I., Jurik, M.: A generalisation, a simplification and some applications of Paillier's probabilistic public-key system. In: Kim, K.-c. (ed.) PKC 2001. LNCS, vol. 1992, pp. 125–140. Springer, Heidelberg (2001)

10. Damgård, I., Nielsen, J.B.: Perfect hiding and perfect binding universally composable commitment schemes with constant expansion factor. In: Yung, M. (ed.) CRYPTO 2002. LNCS, vol. 2442, pp. 581–596. Springer, Heidelberg (2002), The full version is available at http://www.brics.dk/RS/01/41/

11. Desmedt, Y.G. (ed.): CRYPTO 1994. LNCS, vol. 839. Springer, Heidelberg (1994)

12. Fehr, S., Hofheinz, D., Kiltz, E., Wee, H.: Encryption schemes secure against chosen-ciphertext selective opening attacks. In: Gilbert, H. (ed.) EUROCRYPT 2010. LNCS, vol. 6110, pp. 381–402. Springer, Heidelberg (2010)

13. Fischlin, M., Libert, B., Manulis, M.: Non-interactive and re-usable universally composable string commitments with adaptive security. In: Lee, D.H., Wang, X. (eds.) ASIACRYPT 2011. LNCS, vol. 7073, pp. 468–485. Springer, Heidelberg (2011)

14. Fujisaki, E.: All-but-many encryption: A framework for efficient fully-equipped UC commitments. IACR Cryptology ePrint Archive, 2012:379 (2012)

15. Fujisaki, E.: New constructions of efficient simulation-sound commitments using encryption and their applications. In: Dunkelman, O. (ed.) CT-RSA 2012. LNCS, vol. 7178, pp. 136–155. Springer, Heidelberg (2012)

16. Garay, J.A., Mackenzie, P.P., Yang, K.: Strengthening zero-knowledge protocols using signatures. In: Biham, E. (ed.) EUROCRYPT 2003. LNCS, vol. 2656, pp. 177–194. Springer, Heidelberg (2003)

17. Gennaro, R.: Multi-trapdoor commitments and their applications to proofs of knowledge secure under concurrent man-in-the-middle attacks. In: Franklin, M. (ed.) CRYPTO 2004. LNCS, vol. 3152, pp. 220–236. Springer, Heidelberg (2004), The full version available at at Cryptology ePrint Archive, http://eprint.iacr.org/2003/214

18. Hofheinz, D.: All-but-many lossy trapdoor functions. In: Pointcheval, D., Johansson, T. (eds.) EUROCRYPT 2012. LNCS, vol. 7237, pp. 209–227. Springer, Heidelberg (2012), http://eprint.iacr.org/2011/230 (last revised March 18, 2013)

19. Itoh, T., Ohta, Y., Shizuya, H.: Language dependent secure bit commitment. In: Desmedt (ed.) [11], pp. 188–201

20. Lindell, Y.: Highly-efficient universally-composable commitments based on the DDH assumption. In: Paterson, K.G. (ed.) EUROCRYPT 2011. LNCS, vol. 6632, pp. 446–466. Springer, Heidelberg (2011), The full version available at Cryptology ePrint Archive, http://eprint.iacr.org/2011/180

21. MacKenzie, P., Yang, K.: On simulation-sound trapdoor commitments. In: Cachin, C., Camenisch, J.L. (eds.) EUROCRYPT 2004. LNCS, vol. 3027, pp. 382–400. Springer, Heidelberg (2004)

22. Nishimaki, R., Fujisaki, E., Tanaka, K.: An efficient non-interactive universally composable string-commitment scheme. IEICE Transactions 95-A(1), 167–175 (2012)

23. Paillier, P.: Public-key cryptosystems based on composite degree residuosity classes. In: Stern, J. (ed.) EUROCRYPT 1999. LNCS, vol. 1592, pp. 223–238. Springer, Heidelberg (1999)

24. Peikert, C., Vaikuntanathan, V., Waters, B.: A framework for efficient and composable oblivious transfer. In: Wagner, D. (ed.) CRYPTO 2008. LNCS, vol. 5157, pp. 554–571. Springer, Heidelberg (2008)

25. Peikert, C., Waters, B.: Lossy trapdoor functions and their applications. In: Ladner, R.E., Dwork, C. (eds.) STOC 2008, pp. 187–196. ACM (2008)

A   Ideal Multi-commitment Functionality

Functionality $\mathcal{F}_{\mathsf{MCOM}}$

$\mathcal{F}_{\mathsf{MCOM}}$ proceeds as follows, running with parties, $P_1, \ldots, P_n$, and an adversary $\mathcal{S}$:

- **Commit phase:** Upon receiving input $(\mathtt{commit}, \mathtt{sid}, \mathtt{ssid}, P_i, P_j, x)$ from P_i, proceed as follows: If a tuple $(\mathtt{commit}, \mathtt{sid}, \mathtt{ssid}, \ldots)$ with the same $(\mathtt{sid}, \mathtt{ssid})$ was previously recorded, does nothing. Otherwise, record the tuple $(\mathtt{sid}, \mathtt{ssid}, P_i, P_j, x)$ and send $(\mathtt{receipt}, \mathtt{sid}, \mathtt{ssid}, P_i, P_j)$ to P_j and $\mathcal{S}$.
- **Reveal phase:** Upon receiving input $(\mathtt{open}, \mathtt{sid}, \mathtt{ssid})$ from P_i, proceed as follows: If a tuple $(\mathtt{sid}, \mathtt{ssid}, P_i, P_j, x)$ was previously recorded, then send $(\mathtt{reveal}, \mathtt{sid}, \mathtt{ssid}, P_i, P_j, x)$ to P_j and $\mathcal{S}$. Otherwise, does nothing.

Fig. 4. The ideal multi-commitment functionality

B   All-But-Many Lossy Trapdoor Functions

We recall all-but-many lossy trapdoor functions (ABM-LTF) [18], by slightly modifying the notation to fit our purpose. All-but-many lossy trapdoor function ABM.LTF = (ABM.gen, ABM.spl, ABM.eval, ABM.inv) consists of the following algorithms:

- ABM.gen is a PPT algorithm that takes 1^κ and outputs $(pk, (sk, w))$, where pk defines a set U_{pk}. We let $U'_{pk} = \{0, 1\}^\kappa \times U_{pk}$. pk also determines two disjoint sets, L^{loss}_{pk} and L^{inj}_{pk}, such that $L^{\mathsf{loss}}_{pk} \cup L^{\mathsf{inj}}_{pk} \subset U'_{pk}$.

- ABM.spl is a PPT algorithm that takes (pk, w, t), where $t \in \{0, 1\}^\kappa$, picks up inner random coins $v \leftarrow \mathsf{COIN}^{\mathsf{spl}}$, and computes $u \in U_{pk}$. We write $L_{pk}^{\mathsf{loss}}(t)$ to denote the image of ABM.spl on t under pk, i.e.,

$$L_{pk}^{\mathsf{loss}}(t) := \{u \in U_{pk} \mid \exists w, \exists v : u = \mathsf{ABM.spl}(pk, w, t; v)\}.$$

We require $L_{pk}^{\mathsf{loss}} = \{(t, u) \mid t \in \{0, 1\}^\kappa \text{ and } u \in L_{pk}^{\mathsf{loss}}(t)\}$. We set $\widehat{L}_{pk}^{\mathsf{loss}} :=$
$U'_{pk} \backslash L_{pk}^{\mathsf{inj}}$. Since $L_{pk}^{\mathsf{loss}} \cap L_{pk}^{\mathsf{inj}} = \emptyset$, we have $L_{pk}^{\mathsf{loss}} \subseteq \widehat{L}_{pk}^{\mathsf{loss}} \subset U'_{pk}$.
- ABM.eval is a DPT algorithm that takes pk, (t, u), and message $x \in \mathsf{MSP}$ and computes $c = \mathsf{ABM.eval}^{(t,u)}(pk, x)$, where MSP denotes the message space uniquely determined by pk.
- ABM.inv is a DPT algorithm that takes sk, (t, u), and c, and computes $x = \mathsf{ABM.inv}^{(t,u)}(sk, c)$.

All-but-many encryption schemes require the following properties:

1. **Adaptive All-but-many property:** (ABM.gen, ABM.spl) is a probabilistic pseudo random function (PPRF), as defined in Sect. 3.1, with *strongly* unforgeability on $\widehat{L}_{pk}^{\mathsf{loss}} = U'_{pk} \backslash L_{pk}^{\mathsf{inj}}$. Strong unforgeability in this paper is called evasiveness in [18].

2. **Inversion** For every $\kappa \in \mathbb{N}$, every $(pk, (sk, w)) \in \mathsf{ABM.gen}(1^\kappa)$, every $(t, u) \in L_{pk}^{\mathsf{inj}}$, and every $x \in \mathsf{MSP}$, it always holds that

$$\mathsf{ABM.inv}^{(t,u)}(sk, \mathsf{ABM.eval}^{(t,u)}(pk, x)) = x.$$

3. **ℓ-Lossyness** For every $\kappa \in \mathbb{N}$, every $(pk, (sk, w)) \in \mathsf{ABM.gen}(1^\kappa)$, and every $(t, u) \in L_{pk}^{\mathsf{loss}}$, the image set $\mathsf{ABM.eval}^{(t,u)}(pk, \mathsf{MSP})$ is of size at most $|\mathsf{MSP}| \cdot 2^{-\ell}$.

Here L_{pk}^{loss} (resp. L_{pk}^{inj}) in ABM-LTFs corresponds to L_{pk}^{td} (resp. L_{pk}^{ext}) in ABMEs. We remark that ABM-LTFs [18] require that (ABM.gen, ABM.spl) should be *strongly* unforgeable, whereas ABMEs requires that (ABM.gen, ABM.spl) be just unforgeable.

C   Assumptions and Some Useful Lemmas

Let us write $\Pi^{(d)}$ to denote DJ PKE with parameter d.

Assumption 3. *We say that the DCR assumption holds if for every PPT A, there exists a key generation algorithm $\mathbf{K}$ such that $\mathsf{Adv}_A^{\mathsf{dcr}}(\kappa) =$*

$$\Pr[\mathsf{Expt}_A^{\mathsf{dcr}-0}(\kappa) = 1] - \Pr[\mathsf{Expt}_A^{\mathsf{dcr}-1}(\kappa) = 1]$$

is negligible in κ, where

$$
\begin{array}{l|l}
\mathsf{Expt}_A^{\mathsf{dcr}-0}(\kappa): & \mathsf{Expt}_{d,A}^{\mathsf{dcr}-1}(\kappa): \\
\quad n \leftarrow \mathbf{K}(1^\kappa); R \xleftarrow{\cup} \mathbb{Z}_{n^2}^\times & \quad n \leftarrow \mathbf{K}(1^\kappa); R \xleftarrow{\cup} \mathbb{Z}_{n^2}^\times \\
\quad c = R^n \bmod n^2 & \quad c = (1+n)R^n \bmod n^2 \\
\quad return\ A(n,c). & \quad return\ A(n,c).
\end{array}
$$

Assumption 4 ([18]). *We say that the non-trivial divisor assumption holds on* $\Pi^{(d)}$ *if for every PPT A,* $\mathsf{Adv}^{\mathsf{divisor}}_{A,\Pi^{(d)}}(\kappa) = \mathsf{negl}(\kappa)$ *where*

$$\mathsf{Adv}^{\mathsf{divisor}}_{A,\Pi^{(d)}}(\kappa) = \Pr[(pk, sk) \leftarrow \mathbf{K}(1^\kappa); A(pk) = c : 1 < \gcd(\mathbf{D}(c), n) < n].$$

This assumes that an adversary cannot compute an encryption of a non-trivial divisor of n, i.e., $\mathbf{E}(p)$, under given public-key pk_{dj} only. Since the adversary is only given pk_{dj}, the assumption is plausible.

Lemma 2. *If A is an adversary against* $\Pi^{(d)}$, *there is adversary* A' *against* $\Pi^{(1)}$ *such that*
$$\mathsf{Adv}^{\mathsf{divisor}}_{A,\Pi^{(d)}}(\kappa) \le \mathsf{Adv}^{\mathsf{divisor}}_{A',\Pi^{(1)}}(\kappa).$$

Assumption 5 ([18]). *We say that the non-multiplication assumption holds on DJ PKE* $\Pi^{(d)}$ *if for every PPT adversary A, the advantage of A,* $\mathsf{Adv}^{\mathsf{mult}}_{A,\Pi^{(d)}}(\kappa) = \mathsf{negl}(\kappa)$, *where* $\mathsf{Adv}^{\mathsf{mult}}_{A,\Pi^{(d)}}(\kappa) = \Pr[(pk, sk) \leftarrow \mathbf{K}(1^\kappa); c_1, c_2 \leftarrow \mathbb{Z}^\times_{n^{d+1}}; c^* \leftarrow A(pk, c_1, c_2) : \mathbf{D}_{sk}(c^*) = \mathbf{D}_{sk}(c_1) \cdot \mathbf{D}_{sk}(c_2)].$

This assumes that an adversary cannot compute $\mathbf{E}(x_1 \cdot x_2)$ for given $(pk_{\mathsf{dj}}, \mathbf{E}(x_1), \mathbf{E}(x_2))$. If the multiplicative operation is easy, DJ PKE turns out a fully-homomorphic encryption (FHE), which is unlikely. Although breaking the non-multiplication assumption does not mean that DJ PKE turns out a FHE, this connection gives us some feeling that this assumption is plausible.

Lemma 3. *If A is an adversary against DJ PKE* $\Pi^{(d)}$, *there is an adversary* A' *against* $\Pi^{(1)}$ *such that*

$$\mathsf{Adv}^{\mathsf{mult}}_{A,\Pi^{(d)}}(\kappa) \le \mathsf{Adv}^{\mathsf{mult}}_{A',\Pi^{(1)}}(\kappa).$$

Multi-valued Byzantine Broadcast: The $t < n$ Case

Martin Hirt and Pavel Raykov

ETH Zurich, Switzerland
{hirt,raykovp}@inf.ethz.ch

Abstract. Byzantine broadcast is a distributed primitive that allows a specific party to consistently distribute a message among n parties in the presence of potential misbehavior of up to t of the parties. All known protocols implementing broadcast of an ℓ-bit message from point-to-point channels tolerating any $t < n$ Byzantine corruptions have communication complexity at least $\Omega(\ell n^2)$. In this paper we give cryptographically secure and information-theoretically secure protocols for $t < n$ that communicate $\mathcal{O}(\ell n)$ bits when ℓ is sufficiently large. This matches the optimal communication complexity bound for any protocol allowing to broadcast ℓ-bit messages. While broadcast protocols with the optimal communication complexity exist for $t < n/2$, this paper is the first to present such protocols for $t < n$.

1 Introduction

1.1 Byzantine Broadcast

The Byzantine broadcast problem (aka Byzantine generals) is stated as follows [PSL80]: A specific party (the sender) wants to distribute a message among n parties in such a way that all correct parties obtain the same message, even when some of the parties are malicious. The malicious misbehavior is modeled by a central adversary who corrupts up to t parties and takes full control of their actions. Corrupted parties are called *Byzantine* and the remaining parties are called *correct*. Broadcast requires that all correct parties agree on the same value v, and if the sender is correct, then v is the value proposed by the sender. Broadcast is one of the most fundamental primitives in distributed computing. It is used to implement various protocols like voting, bidding, collective contract signing, etc. Basically, this list can be continued with all protocols for secure multi-party computation as defined by Yao [Yao82, GMW87].

There exist various implementations of Byzantine broadcast from synchronous point-to-point communication channels with different security guarantees. In the model without trusted setup, perfectly-secure Byzantine broadcast is achievable when $t < n/3$ [PSL80, BGP92, CW92]. In the model with trusted setup, cryptographically or information-theoretically secure Byzantine broadcast is achievable for any $t < n$ [DS83, PW96].

P. Sarkar and T. Iwata (Eds.): ASIACRYPT 2014, PART II, LNCS 8874, pp. 448–465, 2014.

Closely related to the broadcast problem is the consensus problem. In consensus each party holds a value as input, and then parties agree on a common value as output of consensus. In this paper we consider the case where any number of parties may be Byzantine. In this case the consensus problem is not well-defined, and hence we do not treat it here.

1.2   Efficiency of Byzantine Broadcast

In this paper we focus on the efficiency of broadcast protocols. In particular, we are interested in optimizing their *communication complexity*. The communication complexity of a protocol is defined by Yao [Yao79] to be the number of bits sent/received by correct parties during the protocol run.[1]

Historically, the broadcast problem was introduced for binary values [PSL80]. However, in various applications *long* values are broadcast rather than bits. Examples of such applications are general purpose multi-party computation protocols and specific tasks like voting. Such a broadcast of long values is called *multi-valued* broadcast. In this paper we study the communication complexity of multi-valued broadcast protocols.

Many known protocols for multi-valued broadcast [TC84, FH06, LV11, Pat11] are actually *constructions* from a broadcast of short messages and point-to-point channels. Communication complexity of such constructions is computed in terms of the point-to-point channels and the broadcast for short messages usage. The security of the protocol is based on the security of the construction and the security of the broadcast for short messages.

Let us denote the communication complexity of a short s-bit message broadcast with $\mathcal{B}(s)$. The most trivial construction is to broadcast the message bit by bit, which is perfectly secure for $t < n$ and has communication complexity $\ell\mathcal{B}(1)$. The construction by Turpin and Coan [TC84] is perfectly secure and tolerates $t < n/3$ while communicating $\mathcal{O}(\ell n^2 + n\mathcal{B}(1))$ bits. The construction by Fitzi and Hirt [FH06] is information-theoretically secure and tolerates $t < n/2$ while communicating $\mathcal{O}(\ell n + n^3\kappa + n\mathcal{B}(n + \kappa))$ bits, where κ denotes a security parameter. The construction by Liang and Vaidya [LV11] is perfectly secure and tolerates $t < n/3$ while communicating $\mathcal{O}(\ell n + \sqrt{\ell}n^2\mathcal{B}(1) + n^4\mathcal{B}(1))$ bits. This construction can even be extended to tolerate more than $n/3$ corruptions [LV11]. However, the extended protocol inherently requires $t < n/2$ (see Appendix A for the details). The construction by Patra [Pat11] is perfectly secure and tolerates $t < n/3$ while communicating $\mathcal{O}(\ell n + n^2\mathcal{B}(1))$ bits.

In this paper we consider the case where $t < n$. In this model existing protocols [DS83, PW96] were designed to broadcast bits, but they can be easily adopted to broadcast long messages. A simple modification of the protocol by Dolev and Strong [DS83] is cryptographically secure and has communication complexity $\Omega(\ell n^2 + n^3\kappa)$. Analogously, the protocol by Pfitzmann and Waidner [PW96] is information-theoretically secure and has communication complexity $\Omega(\ell n^2 + n^6\kappa)$

---

[1] When counting the number of bits received by correct players, we take into account only messages which were *actively* received by them, i.e., messages which should be received according to the protocol specification.

[Fit03]. Also the protocols of [HMR14] can be seen as multi-valued constructions for $t < n$. However, their resulting communication complexity is $\Omega(\ell n^3)$.

Another measure of protocol efficiency often considered is round complexity. There are two principal classes of protocols with respect to this measure: constant-round and non-constant round. In the model without trusted setup, constant-round binary Byzantine broadcast is achievable when $t < n/3$ [FM88]. In the model where public-key infrastructure (PKI) has been set up via a trusted party, constant-round binary Byzantine broadcast is achievable for $t < n/2$ [KK06], but is not achievable for $t < n$ [GKKO07].

1.3   Contributions

Consider any protocol for multi-valued broadcast. Since every correct player must learn the value proposed by the sender, the communication costs of the broadcast protocol must be at least $\mathcal{O}(\ell n)$. In this paper we give two generic constructions for a multi-valued broadcast which allow to achieve optimal communication complexity of $\mathcal{O}(\ell n)$ bits for $t < n$. The first construction is cryptographically secure and communicates $\mathcal{O}(\ell n + n(\mathcal{B}(\kappa) + n\mathcal{B}(1)))$ bits. The second construction is information-theoretically secure and communicates $\mathcal{O}(\ell n + n^3(\mathcal{B}(\kappa) + n\mathcal{B}(1)))$ bits. The constructions take $\mathcal{O}(n^2)$ and $\mathcal{O}(n^3)$ rounds, respectively. Table 1 summarizes the complexity costs of the existing constructions for multi-valued broadcast.[2]

Table 1. The overview of multi-valued broadcast constructions

Threshold	Security	Bits Communicated	Literature
		$\mathcal{O}(\ell n^2 + n\mathcal{B}(1))$	[TC84]
$t < n/3$	perfect	$\mathcal{O}(\ell n + (\sqrt{\ell}n^2 + n^4)\mathcal{B}(1))$	[LV11]
		$\mathcal{O}(\ell n + n^2\mathcal{B}(1))$	[Pat11]
$t < n/2$	inf.-theor.	$\mathcal{O}(\ell n + n^3\kappa + (n^2 + n\kappa)\mathcal{B}(1))$	[FH06]
	perfect	$\ell\mathcal{B}(1)$	Trivial
$t < n$	inf.-theor.	$\mathcal{O}(\ell n + (n^4 + n^3\kappa)\mathcal{B}(1))$	This paper
	cryptographical	$\mathcal{O}(\ell n + (n^2 + n\kappa)\mathcal{B}(1))$	This paper

In order to obtain a concrete protocol for multi-valued broadcast one takes the above constructions and composes them with the existing protocols for a bit broadcast (e.g., [BGP92, DS83, PW96]). The security of the composed protocol is then the "minimal" security provided by the construction and the bit broadcast protocol employed. For example, when composing information-theoretical

---

[2] In order to facilitate comparison we substitute $\mathcal{B}(s)$ with $s\mathcal{B}(1)$ in the communication complexity of the constructions, which is trivially possible since $\mathcal{B}(s) \leq s\mathcal{B}(1)$ for all s and such arguments appear as summands inside the big $\mathcal{O}$.

construction for $t < n/2$ [FH06] with cryptographically secure protocol for $t < n$ [DS83] we obtain multi-valued broadcast protocol with cryptographic security tolerating $t < n/2$ and communication complexity $\mathcal{O}(\ell n + n^4(n + \kappa))$. Further instantiations are described in Table 2.

Table 2. Instantiations of multi-valued broadcast constructions

Threshold	Security	Bits Communicated	Literature
$t < n/3$	perfect	$\mathcal{O}(\ell n^2)$	Trivial with [BGP92]
		$\mathcal{O}(\ell n + \sqrt{\ell} n^4 + n^6)$	[LV11] with [BGP92]
		$\mathcal{O}(\ell n + n^4)$	[Pat11] with [BGP92]
$t < n/2$	inf.-theor.	$\mathcal{O}(\ell n + n^7 \kappa)$	[FH06] with [PW96]
	cryptogr.	$\mathcal{O}(\ell n + n^4(n + \kappa))$	[FH06] with [DS83]
$t < n$	inf.-theor.	$\Omega(\ell n^2 + n^6 \kappa)$	[PW96]
		$\mathcal{O}(\ell n + n^{10} \kappa)$	This with [PW96]
	cryptogr.	$\Omega(\ell n^2 + n^3 \kappa)$	[DS83]
		$\mathcal{O}(\ell n + n^5 \kappa)$	This with [DS83]

We note that all multi-valued constructions are only *asymptotically* optimal in ℓ, i.e., they only outperform the trivial construction when relatively long messages are broadcast. Such long messages appear, for example, in voting protocols [CGS97] (where the set of authorities agree on the set of ballots), or in multi-party computation protocols [GMW87] (when all gates on a particular level of the circuit are evaluated in parallel). In particular, multi-party computation protocols for $t < n$ (e.g., [AJLA+12, GGHR14]) achieve better communication complexity when combined with the broadcast constructions presented in this paper.

Furthermore, we investigate the round complexity of constructions for multi-valued broadcast. While for the case of $t < n/2$ constant-round constructions exist (e.g., [FH06]), we prove that in the settings with $t < n$ constant-round constructions do not exist.[3] This is a generalization of the impossibility result given in [GKKO07], because the underlying broadcast procedure for small messages can be used to distribute PKI (by letting the parties broadcast their public keys) and hence PKI cannot be sufficient to implement broadcast in a constant number of rounds.

2 Model and Definitions

Parties. We consider a setting consisting of n parties (players) $\mathcal{P} = \{P_1, \ldots, P_n\}$ with some designated party called the sender, which we denote with P_s for some

---

[3] In the notation of [HMR14] this means that no non-trivial constant-round broadcast-amplification protocols tolerating $t < n$ exist.

$s \in \{1, \ldots, n\}$. For a set of parties $A \subseteq \mathcal{P}$ let $\overline{A}$ denote $\mathcal{P} \setminus A$. We assume that the parties are connected with a synchronous authentic point-to-point network. Synchronous means that all parties share a common clock and that the message delay in the network is bounded by a constant.

Broadcast definition. A broadcast protocol allows the sender P_s to distribute a value v_s among parties $\mathcal{P}$ such that:

TERMINATION: Every correct party $P_i \in \mathcal{P}$ terminates.

CONSISTENCY: All correct parties in $\mathcal{P}$ decide on the same value.

VALIDITY: If the sender P_s is correct, then every correct party $P_i \in \mathcal{P}$ decides on the value proposed by the sender $v_i = v_s$.

Adversary. The faultiness of parties is modeled in terms of a central adversary corrupting up to $t < n$ parties, making them deviate from the protocol in any desired manner. We distinguish two types of security in this paper: *cryptographic* and *information-theoretic*. Cryptographic security guarantees that the protocol is secure based on some computational assumptions (e.g., signatures and/or collision-resistant hash functions), while information-theoretical (also called statistical) security captures the fact that even a computationally unbounded adversary cannot violate the security of the protocol with a non-negligible probability.

3   Protocols Overview

We present cryptographically and information-theoretically secure constructions for multi-valued broadcast. Both constructions are built over point-to-point channels and an oracle for broadcasting short messages. When describing protocols we often say that players broadcast messages, while meaning that they actually use the given broadcast oracle.

On the highest level both constructions broadcast the long message block by block, where each block is broadcast using a special protocol for block broadcast. This block broadcast protocol achieves optimal communication complexity only in *good* executions, while in *bad* executions more bits need to be communicated. We select the number of blocks in such a way that good executions outnumber bad ones and the total communication complexity is optimal. Whether an execution is good or bad is determined using the *Dispute Control Framework* [BH06]. Dispute control is a technique which keeps track of disputes (also called conflicts) between players and ensures that occurred disputes cannot show up again. Intuitively, an execution is good if it is dispute-free, and bad otherwise.

We employ the dispute control framework as follows. We consider a set of unordered pairs of parties Δ, where $\{P_i, P_j\} \in \Delta$ represents the fact that parties P_i and P_j accuse each other of being Byzantine. Parties start a protocol by setting Δ to be the empty set. Then during the protocol run they add new disputes to Δ when they learn about new accusations. We ensure that Δ always remains *valid*, meaning that if $\{P_i, P_j\} \in \Delta$ then at least one of the players P_i, P_j is Byzantine.

4 Cryptographically Secure Construction

First, we present a protocol CryptoBlockBC for broadcasting blocks. The protocol CryptoBlockBC makes use of an external procedure for broadcasting short values and a set of disputes Δ. Then we plug CryptoBlockBC in the protocol CryptoBC, which broadcasts an ℓ-bit message block by block q times. In each invocation of CryptoBlockBC we will use the same global variable Δ with the disputes among the players. This means that if parties P_i and P_j conflict during some block broadcast, then they conflict in all later invocations of CryptoBlockBC. Then, we count the communication complexity of the resulting construction and select q which makes its optimal.

4.1 Block Broadcast Protocol CryptoBlockBC

The protocol CryptoBlockBC employs a collision-resistant hash function CRHash, i.e., no efficient algorithm can find two different inputs v, v' with $\mathsf{CRHash}(v) = \mathsf{CRHash}(v')$.[4] In the beginning of the protocol the sender broadcasts a hash $h = \mathsf{CRHash}(v_s)$ of the value it holds. The goal of the protocol is to ensure that all correct players learn v_s. All parties during the protocol run are divided into two sets: H and $\overline{H}$. The set H consists of happy players who have already learned v_s, and $\overline{H}$ who have not. At each iteration of CryptoBlockBC we try to move a player from $\overline{H}$ to H. We select a pair of players P_x, P_y such that $P_x \in H$ and $P_y \in \overline{H}$. Then P_x sends the value it holds to P_y. This procedure is meaningless if parties P_x, P_y are in the dispute, so the pair is chosen such that $\{P_x, P_y\} \notin \Delta$. Once P_y receives a value from P_x it verifies that its hash is h; in the positive case P_y is included in H and in the negative case a conflict between P_x and P_y is found. Hence at each iteration we either include one player into H or we discover a new conflict between a pair of players.

Protocol CryptoBlockBC(v_s):
1. Parties initialize happy set H to be $\{P_s\}$.
2. Sender P_s: Broadcast $h := \mathsf{CRHash}(v_s)$.
3. While $\exists\, P_x, P_y \in \mathcal{P}$ s.t. $P_x \in H$ and $P_y \in \overline{H}$ and $\{P_x, P_y\} \notin \Delta$ do
   r.1 P_x: Send v_x to player P_y. Denote received value by v_y.
   r.2 P_y: If $h = \mathsf{CRHash}(v_y)$ broadcast 1, else broadcast 0.
   r.3 If P_y broadcasted 1 then parties add P_y to H, otherwise they add $\{P_x, P_y\}$ to Δ.
4. $\forall P_i \in \mathcal{P}$: If $P_i \in H$ decide on v_i, otherwise decide on $\bot$.

Lemma 1. *Given that the initial dispute set Δ_s is valid and CRHash is a collision-resistant hash function, protocol CryptoBlockBC achieves broadcast*

---

[4] This is rather informal definition of collision resistance for unkeyed hash functions, for a more formal treatment see [Rog06].

(of v_s) and terminates with a valid dispute set Δ_e. Furthermore, the protocol terminates in $\mathcal{O}(n+d)$ rounds communicating at most $\mathcal{B}(|h|)+(n+d)(|v_s|+\mathcal{B}(1))$ bits, where $d = |\Delta_e| - |\Delta_s|$, $|h|$ is the output length of CRHash, and $|v_s|$ is the block length.

Proof. First, we prove that at each iteration of the while loop all correct players in H always hold the same value v such that $\mathsf{CRHash}(v) = h$. A player is included into H under condition that it broadcasts 1 at Step $r.2$, which he does only if it holds a value v with $\mathsf{CRHash}(v) = h$. Hence for any two correct players $P_i, P_j \in H$ it must hold that $\mathsf{CRHash}(v_i) = h$ and $\mathsf{CRHash}(v_j) = h$. Since CRHash is collision-resistant it implies that $v_i = v_j$.[5]

(Validity of Δ_e). We show that whenever P_x and P_y are correct then $\{P_x, P_y\}$ is not added to Δ at Step $r.3$. A correct $P_x \in H$ holds v_x with $\mathsf{CRHash}(v_x) = h$ and sends $v_x = v_y$ to P_y at Step $r.1$, who successfully verifies that $\mathsf{CRHash}(v_y) = h$ and broadcasts 1 at Step $r.2$, hence $\{P_x, P_y\}$ is not added to Δ at Step $r.3$.

(Termination). At each iteration of the while loop either the happy set H or the dispute set Δ grows. $|H|$ is limited by n and $|\Delta|$ is limited by n^2, hence the number of iterations is limited.

(Consistency). We prove that in the end of the protocol all correct players belong either to H (and decide on the same value v) or to $\overline{H}$ (and decide on $\perp$). As shown above Δ remains valid in all iterations, hence for correct players P_x and P_y the pair $\{P_x, P_y\} \notin \Delta$. Hence, if $P_x \in H$ and $P_y \in \overline{H}$ then the while loop does not terminate.

(Validity). The sender P_s is always in H. If P_s is correct then it decides on v_s and due to the consistency criterion all other correct players decide on v_s as well.

(Complexity Analysis). At each iteration of the while loop either H or Δ grows. Hence, the total number of iterations of the while loop is upper bounded by $n + d$ where d is $|\Delta_e| - |\Delta_s|$. This implies that the number of rounds the construction employs is $\mathcal{O}(n + d)$. Furthermore, the total communication costs of the protocol are upper bounded by $\mathcal{B}(|h|) + (n + d)(|v_s| + \mathcal{B}(1))$. ☐

4.2   Constructing Broadcast for Long Messages

Now we plug in `CryptoBlockBC` in the protocol `CryptoBC` which broadcasts a message block by block.

---

[5] More formally, when an adversary can provoke two correct players to hold colliding values for CRHash with non-negligible probability, then this adversary can be used to construct an efficient collision-finding algorithm for CRHash.

Protocol CryptoBC(v_s, q):
1. Parties initialize dispute set Δ with the empty set.
2. Sender P_s: Cut v_s in q pieces $v^1, \ldots, v^q$ (add padding if required).
3. For $r = 1, \ldots, q$ invoke CryptoBlockBC(v^r), denote the output of party P_i by v_i^r.
4. $\forall P_i \in \mathcal{P}$: If one of $v_i^r = \bot$ then output $\bot$, otherwise output $v_i^1 || \cdots || v_i^q$.

Since block broadcast is invoked q times, due to Lemma 1 the total communication complexity is at most

$$\sum_{i=1}^{q} \left[\mathcal{B}(|h|) + (n + d_i)(\ell/q + \mathcal{B}(1)) \right] = q\mathcal{B}(|h|) + (qn + \sum_{i=1}^{q} d_i)(\ell/q + \mathcal{B}(1))$$

bits. We know that the sum of d_i is upper bounded by the total number of possible disputes n^2. Hence we have that communication complexity is upper bounded by $q\mathcal{B}(|h|) + (qn + n^2)(\ell/q + \mathcal{B}(1))$. By setting $q = n$ we get that the total communication is at most $2\ell n + 2n^2\mathcal{B}(1) + n\mathcal{B}(|h|)$ which is $\mathcal{O}(\ell n + n(\mathcal{B}(\kappa) + n\mathcal{B}(1)))$.

The number of rounds the construction employs is $\sum_{i=1}^{q} r_i$, where each $r_i \in \mathcal{O}(n + d_i)$. Hence, for $q = n$ we have that the total number of rounds is $\mathcal{O}(n^2)$.

The following theorem summarizes the cryptographically secure construction presented in this section:

Theorem 1. *In the setting with $t < n$, the construction* CryptoBC *with $q = n$ achieves cryptographically secure broadcast of ℓ-bit messages in $\mathcal{O}(n^2)$ rounds by communicating $\mathcal{O}(\ell n + n(\mathcal{B}(\kappa) + n\mathcal{B}(1)))$ bits (where κ is a security parameter and $\mathcal{B}(s)$ is the complexity of the underlying broadcast for short s-bit messages).*

In order to obtain a concrete multi-valued broadcast protocol we instantiate CryptoBC with the protocol [DS83]:

Theorem 2. *Instantiating the construction* CryptoBC *with $q = n$ and [DS83] as underlying broadcast for short messages results in a cryptographically secure multi-valued broadcast protocol for $t < n$ with communication complexity $\mathcal{O}(\ell n + n^5\kappa)$ (where κ is a security parameter).*

5   Information-Theoretically Secure Construction

This section is organized similar to the cryptographic case. First, we present a protocol ITBlockBC for broadcasting blocks which is analogous to CryptoBlockBC, with the difference that it relies on a universal hash function instead of a collision-resistant one. As in the cryptographic case we then plug ITBlockBC in the ITBC protocol, which broadcasts a message block by block q times. Then, we count the communication complexity of the resulting protocol ITBC, and select the number of blocks q which makes it optimal.

5.1   Universal Hash Functions

Consider a family of functions $\mathcal{U} = \{U_k\}_{k \in \mathcal{K}}$ indexed with a key set $\mathcal{K}$, where each function U_k maps elements of some set $\mathcal{X}$ to a fixed set of bins $\mathcal{Y}$. The family $\mathcal{U}$ is called ε-universal if for any two distinct messages v_1 and v_2,

$$\frac{|\{k \in \mathcal{K} \mid U_k(v_1) = U_k(v_2)\}|}{|\mathcal{K}|} \leq \varepsilon.^6$$

A ε-universal hash function can for example be constructed as follows: Let $\mathcal{X} = \{0,1\}^\ell$, $\mathcal{K} = \mathcal{Y} = \mathrm{GF}(2^\nu)$, and any value $v \in \{0,1\}^\ell$ be interpreted as a polynomial f_v over $\mathrm{GF}(2^\nu)$ of degree $\lceil \ell/\nu \rceil - 1$. The hash function is defined as $U_k(v) = f_v(k)$. We know that two distinct polynomials of degree $\lceil \ell/\nu \rceil - 1$ can match in at most $\lceil \ell/\nu \rceil - 1$ points. Hence, for any two distinct $v_1, v_2 \in \{0,1\}^\ell$,

$$\frac{|\{k \in \{0,1\}^\nu \mid U_k(v_1) = U_k(v_2)\}|}{2^\nu} \leq \frac{\lceil \ell/\nu \rceil - 1}{2^\nu} \leq 2^{-\nu}\ell.$$

So, $\{U_k\}_{k \in \{0,1\}^\nu}$ is a family of $(2^{-\nu}\ell)$-universal hash functions.

We will denote a ε-universal hash function with ITHash.

5.2   Block Broadcast Protocol ITBlockBC

Similarly to the cryptographic case all parties during the run of the protocol ITBlockBC are divided into two sets: H and $\overline{H}$. The set H consists of happy players who have already learned v_s, and $\overline{H}$ who have not. The difference to the cryptographic case is that the set H is not monotonically growing—it may happen that the same player may be added/removed from H several times. At each iteration of ITBlockBC we try to move a player from $\overline{H}$ to H. We select a pair of players P_x, P_y such that $P_x \in H$, $P_y \in \overline{H}$ and $\{P_x, P_y\} \notin \Delta$. Then P_x sends the value it holds to P_y. Now player P_y needs to verify that the value received from P_x is the value that correct parties in H hold. In order to do so, P_y broadcasts a key k for ε-universal hash function ITHash, and then P_s broadcasts a hash h for this key. As long as P_y honestly chooses k uniformly at random, with overwhelming probability correct players will obtain different hashes if they hold different values. If a party in $H \cup \{P_y\} \setminus \{P_s\}$ holds a value with a hash h, then he broadcasts 1, and 0 otherwise (the sender P_s does not broadcast because if he is correct he can broadcast only 1). If every party broadcasts 1, then the iteration was successful and P_y is added to H. Otherwise, some of the parties in $H \cup \{P_y\}$ do not hold the right value and we search for new disputes.

An important difference from the cryptographic case is that disputes may occur not only between P_x and P_y, but between any two parties in H. In order to find such disputes, one must be able to reason about the history of how H was formed. We will keep a history set T which will contain pairs of players (P_x, P_y) such that P_y learned the value it holds from P_x.

---

[6] This is a combinatorial definition of a universal hash function, usually the last condition is written probabilistically as $\Pr[k \xleftarrow{\$} \mathcal{K} : U_k(v_1) = U_k(v_2)] \leq \varepsilon$.

Protocol ITBlockBC(v_s):
1. Parties initialize happy set H to be $\{P_s\}$ and history set T to be $\emptyset$.
2. While $\exists\, P_x, P_y \in \mathcal{P}$ s.t. $P_x \in H$ and $P_y \in \overline{H}$ and $\{P_x, P_y\} \notin \Delta$ do
   - $r.1$ P_x: Send v_x to player P_y. Denote received value by v_y. Add (P_x, P_y) to T.
   - $r.2$ P_y: Generate random $k \in \mathcal{K}$ and broadcast it.
     Sender P_s: Broadcast $h := \mathsf{ITHash}_k(v_s)$.
   - $r.3$ $\forall P_i \in H \cup \{P_y\} \setminus \{P_s\}$: If $h = \mathsf{ITHash}_k(v_i)$ then broadcast 1, otherwise 0.
   - $r.4$ If all parties broadcasted 1
        - Add P_y to H.

     else
        - For all $(P_i, P_j) \in T$ s.t. P_i broadcasted 1 (resp. $P_i = P_s$) and P_j broadcasted 0, add $\{P_i, P_j\}$ to Δ.
        - Set H to $\{P_s\}$, T to $\emptyset$.
3. $\forall P_i \in \mathcal{P}$: If $P_i \in H$ decide on v_i, otherwise decide on $\perp$.

Lemma 2. *Given that the initial dispute set Δ_s is valid and ITHash is a universal hash function, protocol* ITBlockBC *achieves broadcast (of v_s) and terminates with a valid dispute set Δ_e (except with negligible probability). Furthermore, the protocol terminates in $\mathcal{O}(n + nd)$ rounds communicating at most $(n + nd)(|v_s| + \mathcal{B}(|h|) + \mathcal{B}(|k|) + n\mathcal{B}(1))$ bits, where $d = |\Delta_e| - |\Delta_s|$, $|h|$ is the output length of ITHash, $|k|$ is the key length of ITHash, and $|v_s|$ is the block length.*

Proof. First, we prove that at each iteration of the while loop all correct players in H always hold the same value v. More precisely, we need to show that if a correct player P_y is added to H, then, given that all correct players in H hold the same value v, it holds that $v_y = v$. We have that all parties in $H \cup \{P_y\} \setminus \{P_s\}$ broadcast 1 at Step $r.3$. This implies that P_y successfully verifies that $\mathsf{ITHash}_k(v_y) = h$, and all correct parties in H verify that $\mathsf{ITHash}_k(v) = h$. Due to the fact that P_y is correct, the key k is chosen uniformly at random, so given that $\mathsf{ITHash}_k(v_y) = \mathsf{ITHash}_k(v)$, it must hold with overwhelming probability $1 - \varepsilon$ that $v_y = v$.

Second, we show that if the condition at Step $r.4$ is false then at least one new conflict is found. We have that not all players in $H \cup \{P_y\} \setminus \{P_s\}$ broadcasted 1. Consider two possible cases:

(Exists $P_z \in H \setminus \{P_s\}$ which broadcasts 0 at step $r.3$) Since P_z is in H there must exist a sequence of players $P_{i_1}, P_{i_2}, \ldots, P_{i_k}$ in H such that $P_{i_1} = P_s, P_{i_k} = P_z$ and $(P_{i_j}, P_{i_{j+1}}) \in T$ for all $j = 1, \ldots, k - 1$ (see illustration in Figure 1). In the rth iteration some of the players in H stayed happy (P_s and those who broadcasted 1) and some become unhappy (broadcasted 0). We know that P_s stayed happy and P_z became unhappy. Hence in a row $P_{i_1}, P_{i_2}, \ldots, P_{i_k}$ there are players of both types. Then we have that exist two players $P_{i_u}, P_{i_{u+1}}$ such that P_{i_u} stays happy and $P_{i_{u+1}}$ becomes unhappy. By construction of T, $(P_{i_u}, P_{i_{u+1}}) \in T$ implies that $\{P_{i_u}, P_{i_{u+1}}\}$ is not yet in Δ.

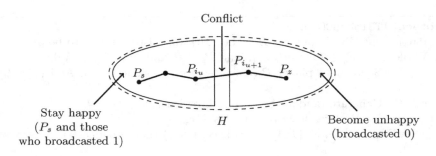

Fig. 1. Conflict finding in `ITBlockBC`

Consequently, the pair $\{P_{i_u}, P_{i_{u+1}}\}$ will be identified as having a conflict and will be added to Δ.

(Each $P_i \in H \setminus \{P_s\}$ broadcasts 1 at step r.3) It means that P_x broadcasts 1 (or $P_x = P_s$) and P_y broadcasts 0. Hence the new dispute $\{P_x, P_y\}$ will be added to Δ.

Now we proceed with the proof of the current lemma.

(Validity of Δ_e). We show that whenever P_i and P_j are correct then $\{P_i, P_j\}$ is never added to Δ. The pair $\{P_i, P_j\}$ is added to Δ only when P_i sent some v to P_j (i.e., $(P_i, P_j) \in T$), and they disagree for some key k whether $\mathsf{ITHash}_k(v)$ equals h. Hence, P_i or P_j is corrupted.

(Termination). There can be at most n successive iterations where the set H grows (condition at Step $r.4$ is true). As shown above whenever condition at Step $r.4$ is false a new conflict is found. The number of conflicts is limited and so must be the number of the while loop iterations.

(Consistency). We prove that in the end of the protocol all correct players belong either to H (and decide on the same value v) or to $\overline{H}$ (and decide on $\perp$). As shown above Δ remains valid in all iterations, hence for any two correct players P_x, P_y, the pair $\{P_x, P_y\} \notin \Delta$. Hence, if $P_x \in H$ and $P_y \in \overline{H}$ then the while loop does not terminate.

(Validity). The correct sender P_s is always in H. The sender P_s decides on v_s and due to the consistency criterion all other correct players decide on v_s as well.

(Complexity Analysis). There can be at most n consecutive iterations, where no conflict is found, hence the total number of iterations is at most $n + nd$, where $d = |\Delta_e| - |\Delta_s|$. This implies that the number of rounds the construction employs is $\mathcal{O}(n + nd)$. Furthermore, since the communication costs of each iteration are at most $|v_s| + \mathcal{B}(|h|) + \mathcal{B}(|k|) + n\mathcal{B}(1)$, we have that the total communication costs of the protocol are upper bounded by $(n + nd)(|v_s| + \mathcal{B}(|h|) + \mathcal{B}(|k|) + n\mathcal{B}(1))$. □

5.3   Constructing Broadcast for Long Messages

Similarly to the cryptographic case, we plug ITBlockBC in the protocol ITBC which simply broadcasts a message block by block. The protocol ITBC is a copy of the protocol CryptoBC with the only difference that CryptoBlockBC is substituted with ITBlockBC.

Due to Lemma 2 the total communication complexity of ITBC is at most

$$\sum_{i=1}^{q} \left[(n + d_i n)(\ell/q + \mathcal{B}(|h|) + \mathcal{B}(|k|) + n\mathcal{B}(1)) \right] =$$

$$n(q + \sum_{i=1}^{q} d_i)(\ell/q + \mathcal{B}(|h|) + \mathcal{B}(|k|) + n\mathcal{B}(1)).$$

This expression is bound by $n(q + n^2)(\ell/q + \mathcal{B}(|h|) + \mathcal{B}(|k|) + n\mathcal{B}(1))$. By setting $q = n^2$ we have that communication costs are at most $2\ell n + 2n^3(\mathcal{B}(|h|) + \mathcal{B}(|k|) + n\mathcal{B}(1)))$ which is $\mathcal{O}(\ell n + n^3(\mathcal{B}(\kappa) + n\mathcal{B}(1)))$.

The number of rounds the construction employs is $\sum_{i=1}^{q} r_i$, where each $r_i \in \mathcal{O}(n + nd_i)$. Hence, for $q = n^2$ we have that the total number of rounds is $\mathcal{O}(n^3)$.

The following theorem summarizes the information-theoretically secure construction presented in this section:

Theorem 3. *In the setting with $t < n$, the construction ITBC with $q = n^2$ achieves information-theoretically secure broadcast of ℓ-bit messages in $\mathcal{O}(n^3)$ rounds by communicating $\mathcal{O}(\ell n + n^3(\mathcal{B}(\kappa) + n\mathcal{B}(1)))$ bits (where κ is a security parameter and $\mathcal{B}(s)$ is the complexity of the underlying broadcast for short s-bit messages).*

In order to obtain a concrete multi-valued broadcast protocol we instantiate ITBC with the protocol [PW96]:

Theorem 4. *Instantiating the construction ITBC with $q = n^2$ and [PW96] as underlying broadcast for short messages results in an information-theoretically secure multi-valued broadcast protocol for $t < n$ with communication complexity $\mathcal{O}(\ell n + n^{10}\kappa)$ (where κ is a security parameter).*

6   On the Round Complexity of Multi-valued Constructions

While the primary goal of this paper is to build communication efficient protocols, one often optimizes the protocols with respect to another measure of the protocols' efficiency, number of rounds employed by a protocol. According to this measure there are two principal classes of the protocols: constant-round and non-constant round. In the following we investigate whether it is possible to obtain protocols optimal in both measures, that is, constant-round multi-valued broadcast protocols with optimal communication complexity for $t < n$.

The goal of this paper is to build protocols for efficient multi-valued constructions. We stress that by construction we understand a protocol for n players which realizes multi-valued broadcast on top of bilateral channels and a special procedure for broadcasting bits. We explicitly distinguish such constructions and *plain* multi-valued broadcast protocols (e.g., [DS83, PW96]) that directly implement broadcast from bilateral channels.

When $t < n/2$ both communication and round optimal multi-valued broadcast protocols can be built by combining constant-round construction [FH06] with a constant-round binary broadcast protocol (e.g., [KK06, GKKO07]). For the case of arbitrary $t < n$ it has been shown that no plain protocol can achieve broadcast in a constant number of rounds [GKKO07]. In the context of this paper this shows that no concrete instantiation of a multi-valued construction and a procedure for broadcasting bits can be constant-round. However, it is still interesting to understand whether a non-trivial constant-round construction for multi-valued broadcast exists separately. Next we show that this is not possible, i.e., there is a separation between $t < n/2$ and $t < n$ cases not only for broadcast protocols but between constructions for multi-valued broadcast as well.

A Construction's Failure Probability (Based on [GY89]). Consider any multi-valued construction protocol $\boldsymbol{\pi} = (\pi_1, \ldots, \pi_n)$. A scenario is a triple $(v, B, \mathcal{A})$ where $v \in \{0,1\}^\ell$ is a value that the sender broadcasts, $B \subseteq \mathcal{P}$ is a set of malicious players controlled with an adversarial strategy $\mathcal{A}$. We call an execution of the protocol $\boldsymbol{\pi}$ in a scenario *successful* if the outputs of honest parties $\mathcal{P} \setminus B$ satisfy broadcast properties (validity and consistency). We define the error $\varepsilon_{\boldsymbol{\pi}, v, B, \mathcal{A}}$ to be the probability of an unsuccessful execution over the randomness used by honest parties and the adversary in the corresponding scenario.[7] Then the failure probability of $\boldsymbol{\pi}$ is defined as $\max\limits_{v, B, \mathcal{A}} \varepsilon_{\boldsymbol{\pi}, v, B, \mathcal{A}}$, i.e., as the maximum failure among all scenarios.

Impossibility Framework. We employ a standard indistinguishability argument that is used to prove that certain security goals cannot be achieved by any protocol in the Byzantine environment [PSL80]. Such a proof goes by contradiction, i.e., by assuming that the security goals can be satisfied by means of some protocol $\boldsymbol{\pi} = (\pi_1, \ldots, \pi_n)$. Then the programs π_i are used to build a *configuration* with contradictory behavior. The configuration consists of (possibly) multiple copies of π_i connected with bilateral channels and given admissible inputs. Once the configuration is built, one simultaneously starts all the programs in the configuration and analyzes the outputs produced by the programs locally. By arguing that the view of some programs π_i and π_j in the configuration is indistinguishable from their view when run by the corresponding players P_i and P_j (while the adversary corrupts the remaining players in $\mathcal{P} \setminus \{P_i, P_j\}$) we can deduce consistency conditions on the outputs by π_i and π_j that lead to a contradiction. The main novelty in the following proof is that we consider an extended communication model where in addition to bilateral channels players are given

---

[7] In all executions we assume that the procedure to broadcast bits is perfectly secure, i.e., the values broadcast with it are consistently delivered to the parties.

access to a special procedure for broadcasting short messages. While following the path described above, we need to additionally describe how the calls to this procedure are handled.

Theorem 5. *Every non-trivial* [8] *multi-valued broadcast construction for* $t < n$ *which takes less than* $n - 1$ *rounds fails with probability at least* $1/(2n)$.

Proof. Take any non-trivial construction $\boldsymbol{\pi} = (\pi_1, \ldots, \pi_n)$ which requires $q < n - 1$ rounds and has error probability ε. Without loss of generality, assume as well that the sender is P_1, i.e., the sender's program is π_1. On the highest level our proof consists of three steps. (i) we define a configuration (inspired by [GKKO07]). (ii) we show that all programs in the configuration must output the same value v with probability $1 - n\varepsilon$. (iii) we use an information flow argument to prove that there is a program in the configuration that outputs v with probability at most $1/2$. Finally, we combine the probability inequalities given by (ii) and (iii) to conclude that $\varepsilon \geq 1/(2n)$.

(i) Consider a chain of n programs $\pi_1, \pi_2, \pi_3, \ldots, \pi_n$ connected with bilateral channels as shown in Figure 2. In this configuration only programs that are connected communicate, i.e., π_1 communicates only with π_2 and receives no messages from parties in $\mathcal{P} \setminus \{P_1, P_2\}$. Let π_1 be given as input a uniform random variable V chosen from the input domain $\{0,1\}^\ell$. Now we execute the programs. Whenever any program broadcasts any value using the broadcast procedure this value is delivered to all programs in the configuration.

(ii) First, we prove that any pair of connected programs (π_i, π_{i+1}) in the chain outputs the same value. One can view the configuration as the player P_i running the program π_i and P_{i+1} running π_{i+1} while the adversary corrupting $\mathcal{P} \setminus \{P_i, P_{i+1}\}$ is simulating the programs $\pi_1, \ldots, \pi_{i-1}$ and $\pi_{i+2}, \ldots, \pi_n$. Due to the consistency property, π_i and π_{i+1} must output the same value with probability at least $1 - \varepsilon$. Since every connected pair of programs in the chain outputs the same value with probability at least $1 - \varepsilon$, then all the programs in the configuration output the same value with probability at least $1 - (n-1)\varepsilon$. Moreover, the configuration can be viewed as P_1 executing π_1 while the adversary corrupts $\mathcal{P} \setminus \{P_1\}$ and simulates the remaining programs. Due to the validity property, π_1 must output V with probability at least $1 - \varepsilon$. Finally, all the programs in the chain output V with probability $1 - n\varepsilon$.

(iii) Let S_i^r be a random variable denoting the state of the program π_i in the chain after r rounds of the protocol execution. By state we understand the input that the program has, the set of all messages that the program received up to the rth round over point-to-point channels and via the underlying broadcast procedure together with the random coins it has used. Let B^r be a random

---

[8] By non-trivial we mean every construction which broadcasts strictly less bits with the broadcast procedure than the length of the message broadcast ℓ.

Fig. 2. The configuration to show the impossibility of non-trivial construction

variable denoting the list of the values that have been broadcast with the broadcast procedure up to the rth round.

After r rounds only programs $\pi_1, \pi_2, \ldots, \pi_{r+1}$ can receive full information about V. The remaining programs in the chain $\pi_{r+2}, \pi_{r+3}, \ldots, \pi_n$ can receive only the information that was distributed with the broadcast procedure, i.e., the information contained in B^r. That is, one can verify by induction that for any r and for all $i \geq r+2$ holds $I(V; S_i^r | B^r) = 0$. Hence, for the last program in the chain π_n after q rounds of computation it holds that $I(V; S_n^q | B^q) = 0$ and hence $I(V; S_n^q) \leq H(B^q)$. Because we assumed that the construction is non-trivial, at most $\ell - 1$ bits can be broadcast with the broadcast procedure. Hence, we have that $H(B^q) \leq \ell - 1$. Combining these facts we get that $I(V; S_n^q) \leq \ell - 1$. Hence, the last program π_n outputs V with probability at most $1/2$. However, we have shown above that all programs (including π_n) output V with probability at least $1 - n\varepsilon$. Hence, we have that $1/2 \geq 1 - n\varepsilon$ which implies that $\varepsilon \geq 1/(2n)$.   □

7   Conclusions

Existing multi-valued broadcast protocols achieve optimal communication complexity only for $t < n/3$ [LV11] or $t < n/2$ [FH06]. In this paper we proposed the first multi-valued broadcast protocols that tolerate any $t < n$ Byzantine corruptions and achieve optimal communication complexity $\mathcal{O}(\ell n)$ for sufficiently long messages of ℓ bits. One of the proposed protocols is cryptographically secure and the other one is information-theoretically secure. The cryptographically secure protocol is based on the security of the signature scheme and a collision-resistance of the hash function employed. It communicates $\mathcal{O}(\ell n + n^5 \kappa)$ bits. The information-theoretically secure protocol may fail with a negligible probability and needs to communicate $\mathcal{O}(\ell n + n^{10} \kappa)$ bits.

The presented constructions CryptoBC and ITBC require $\mathcal{O}(n^2)$ and $\mathcal{O}(n^3)$ rounds, respectively. While constant-round constructions are unachievable, it is still unresolved whether more round-efficient constructions exist. We leave round-complexity optimizations and proving stronger lower bounds as open questions.

References

[AJLA+12]   Asharov, G., Jain, A., López-Alt, A., Tromer, E., Vaikuntanathan, V., Wichs, D.: Multiparty computation with low communication, computation and interaction via threshold FHE. In: Pointcheval, D., Johansson, T. (eds.) EUROCRYPT 2012. LNCS, vol. 7237, pp. 483–501. Springer, Heidelberg (2012)

[BGP92]   Berman, P., Garay, J.A., Perry, K.J.: Bit optimal distributed consensus. In: Computer Science Research, pp. 313–322. Plenum Publishing Corporation, New York (1992); Preliminary version appeared in STOC 1989

[BH06]   Beerliová-Trubíniová, Z., Hirt, M.: Efficient multi-party computation with dispute control. In: Halevi, S., Rabin, T. (eds.) TCC 2006. LNCS, vol. 3876, pp. 305–328. Springer, Heidelberg (2006)

[CGS97]   Cramer, R., Gennaro, R., Schoenmakers, B.: A secure and optimally efficient multi-authority election scheme. In: Fumy, W. (ed.) EUROCRYPT 1997. LNCS, vol. 1233, pp. 103–118. Springer, Heidelberg (1997)

[CW92]   Coan, B.A., Welch, J.L.: Modular construction of a byzantine agreement protocol with optimal message bit complexity. Information and Computation 97, 61–85 (1992); Preliminary version appeared in PODC 1989

[DS83]   Dolev, D., Strong, H.R.: Authenticated algorithms for Byzantine agreement. SIAM Journal on Computing 12(4), 656–666 (1983); Preliminary version appeared in STOC 1982

[FH06]   Fitzi, M., Hirt, M.: Optimally efficient multi-valued Byzantine agreement. In: Proceedings of the 26th Annual ACM Symposium on Principles of Distributed Computing, PODC 2006, pp. 163–168. ACM, New York (2006)

[Fit03]   Fitzi, M.: Generalized Communication and Security Models in Byzantine Agreement. PhD thesis, ETH Zurich (March 2003), Reprint as vol. 4 of ETH Series in Information Security and Cryptography. Hartung-Gorre Verlag, Konstanz (2003) ISBN 3-89649-853-3

[FM88]   Feldman, P., Micali, S.: Optimal algorithms for byzantine agreement. In: Simon, J. (ed.) STOC, pp. 148–161. ACM (1988)

[GGHR14]   Garg, S., Gentry, C., Halevi, S., Raykova, M.: Two-round secure MPC from indistinguishability obfuscation. In: Lindell, Y. (ed.) TCC 2014. LNCS, vol. 8349, pp. 74–94. Springer, Heidelberg (2014)

[GKKO07]   Garay, J.A., Katz, J., Koo, C.-Y., Ostrovsky, R.: Round complexity of authenticated broadcast with a dishonest majority. In: Proceedings of the 48th Annual IEEE Symposium on Foundations of Computer Science, FOCS 2007, pp. 658–668. IEEE Computer Society, Washington, DC (2007)

[GMW87]   Goldreich, O., Micali, S., Wigderson, A.: How to play any mental game. In: Proceedings of the 19th Annual ACM Symposium on Theory of Computing, STOC 1987, pp. 218–229. ACM, New York (1987)

[GY89]   Graham, R.L., Yao, A.C.: On the improbability of reaching byzantine agreements. In: Proceedings of the twenty-first Annual ACM Symposium on Theory of Computing, STOC 1989, pp. 467–478. ACM, New York (1989)

[HMR14]   Hirt, M., Maurer, U., Raykov, P.: Broadcast amplification. In: Lindell, Y. (ed.) TCC 2014. LNCS, vol. 8349, pp. 419–439. Springer, Heidelberg (2014)

[KK06]   Katz, J., Koo, C.-Y.: On expected constant-round protocols for byzantine agreement. In: Dwork, C. (ed.) CRYPTO 2006. LNCS, vol. 4117, pp. 445–462. Springer, Heidelberg (2006)

[LV10a]   Liang, G., Vaidya, N.: Complexity of multi-value byzantine agreement. Technical report, University of Illinois at Urbana-Champaign (2010), http://www.crhc.illinois.edu/wireless/papers/ba_sum_capacity_0729.pdf

[LV10b]   Liang, G., Vaidya, N.: Short note on complexity of multi-value byzantine agreement. CoRR, abs/1007.4857 (2010)

[LV11]   Liang, G., Vaidya, N.: Error-free multi-valued consensus with Byzantine failures. In: Proceedings of the 30th Annual ACM Symposium on Principles of Distributed Computing, PODC 2011, pp. 11–20. ACM, New York (2011), The arxiv version is available at http://arxiv.org/abs/1101.3520

[LV14]   Liang, G., Vaidya, N.: Personal Communication (2014)

[Pat11]   Patra, A.: Error-free multi-valued broadcast and Byzantine agreement with optimal communication complexity. In: Fernàndez Anta, A., Lipari, G., Roy, M. (eds.) OPODIS 2011. LNCS, vol. 7109, pp. 34–49. Springer, Heidelberg (2011)

[PSL80]   Pease, M.C., Shostak, R.E., Lamport, L.: Reaching agreement in the presence of faults. Journal of the ACM 27(2), 228–234 (1980)

[PW96]   Pfitzmann, B., Waidner, M.: Information-theoretic pseudosignatures and Byzantine agreement for t ≥ n/3. Technical report, IBM Research (1996)

[Rog06]   Rogaway, P.: Formalizing human ignorance. In: Nguyên, P.Q. (ed.) VIETCRYPT 2006. LNCS, vol. 4341, pp. 211–228. Springer, Heidelberg (2006)

[TC84]   Turpin, R., Coan, B.A.: Extending binary Byzantine agreement to multivalued Byzantine agreement. Information Processing Letters 18(2), 73–76 (1984)

[Yao79]   Yao, A.C.: Some complexity questions related to distributive computing (preliminary report). In: Proceedings of the Eleventh Annual ACM Symposium on Theory of Computing, STOC 1979, pp. 209–213. ACM, New York (1979)

[Yao82]   Yao, A.C.: Protocols for secure computations. In: Proceedings of the 23rd Annual Symposium on Foundations of Computer Science, SFCS 1982, pp. 160–164. IEEE Computer Society, Washington, DC (1982)

A   On the Constructions of Liang and Vaidya [LV11, LV10a, LV10b]

In [LV11] it is stated that the broadcast constructions presented there can be extended to tolerate $t \geq n/3$. We contacted the authors and they said that this statement is misleading and it should have been "$t < n/2$" instead of "$t \geq n/3$" to be more clear [LV14]. Below we detail why [LV11] inherently requires $t < n/2$ and cannot be extended beyond this bound (this reasoning applies to the related constructions [LV10a, LV10b]).

Essentially, the construction relies on a player set S such that all players in S have the same value v and S is guaranteed to contain at least one correct player. The value v is the value that should be agreed on. This technique requires that such S is unique. Uniqueness of S can be guaranteed only when $t < n/2$. When $t \geq n/2$, even if all correct players do share the same value v, the Byzantine players can always pretend to have a different value v' and create a larger player set S' just among themselves to prevent protocol from reaching agreement.

Fairness versus Guaranteed Output Delivery in Secure Multiparty Computation*

Ran Cohen and Yehuda Lindell

Department of Computer Science, Bar-Ilan University, Israel
cohenrb@cs.biu.ac.il, lindell@biu.ac.il

Abstract. In the setting of secure multiparty computation, a set of parties wish to compute a joint function of their private inputs. The computation should preserve security properties such as privacy, correctness, independence of inputs, fairness and guaranteed output delivery. In the case of no honest majority, fairness and guaranteed output delivery cannot always be obtained. Thus, protocols for secure multiparty computation are typically of two disparate types: protocols that assume an honest majority (and achieve all properties *including* fairness and guaranteed output delivery), and protocols that do not assume an honest majority (and achieve all properties *except for* fairness and guaranteed output delivery). In addition, in the two-party case, fairness and guaranteed output delivery are equivalent. As a result, the properties of fairness (which means that if corrupted parties receive output then so do the honest parties) and guaranteed output delivery (which means that corrupted parties cannot prevent the honest parties from receiving output in any case) have typically been considered to be the same.

In this paper, we initiate a study of the relation between fairness and guaranteed output delivery in secure multiparty computation. We show that in the multiparty setting these properties are distinct and proceed to study under what conditions fairness implies guaranteed output delivery (the opposite direction always holds). We also show the existence of non-trivial functions for which complete fairness is achievable (without an honest majority) but guaranteed output delivery is not, and the existence of non-trivial functions for which complete fairness and guaranteed output delivery are achievable. Our study sheds light on the role of broadcast in fairness and guaranteed output delivery, and shows that these properties should sometimes be considered separately.

1 Introduction

1.1 Background

In the setting of secure multiparty computation, a set of mutually distrusting parties wish to jointly and securely compute a function of their inputs. This computation should be such that each party receives its correct output, and none

---

* This research was supported by THE ISRAEL SCIENCE FOUNDATION (grant No. 189/11). The first author was also supported by the Ministry of Science, Technology and Space and by the National Cyber Bureau of Israel.

P. Sarkar and T. Iwata (Eds.): ASIACRYPT 2014, PART II, LNCS 8874, pp. 466–485, 2014.

of the parties learn anything beyond their prescribed output. In more detail, the most important security properties that we wish to capture are: privacy (no party should learn anything more than its prescribed output), correctness (each party is guaranteed that the output that it receives is correct), independence of inputs (the corrupted parties must choose their inputs independently of the honest parties' inputs), fairness[1] (corrupted parties should receive their output if and only if honest parties do), and guaranteed output delivery (corrupted parties should not be able to prevent honest parties from receiving their output). The standard definition today, [3,7] formalizes the above requirements (and others) in the following general way. Consider an ideal world in which an external trusted party is willing to help the parties carry out their computation. An ideal computation takes place in this ideal world by having the parties simply send their inputs to the trusted party, who then computes the desired function and passes each party its prescribed output. The security of a real protocol is established by comparing the outcome of the protocol to the outcome of an ideal computation. Specifically, a real protocol that is run by the parties (without any trusted party) is secure, if an adversary controlling a coalition of corrupted parties can do no more harm in a real execution than in the above ideal execution.

The above informal description is "overly ideal" in the following sense. It is a known fact that unless an honest majority is assumed, it is impossible to obtain generic protocols for secure multi-party computation that guarantee output delivery and fairness [4]. The definition is therefore typically *relaxed* when no honest majority is assumed. In particular, under certain circumstances, honest parties may not receive any output, and fairness is not always guaranteed. Recently, it was shown that it is actually possible to securely compute some (in fact, many) two-party functionalities fairly [11,1]. In addition, it is possible to even compute some multiparty functionalities fairly, for any number of corrupted parties; in particular, the *majority* function may be securely computed fairly with 3 parties, and the *Boolean OR* function may be securely computed for any number of parties [10]. This has promoted interest in the question of fairness in the setting of no honest majority.

1.2   Fairness versus Guaranteed Output Delivery

The two notions of fairness and of guaranteed output delivery are quite similar and are often interchanged. However, there is a fundamental difference between them. If a protocol guarantees output delivery, then the parties always obtain output and cannot abort. In contrast, if a protocol is fair, then it is only guaranteed that *if* one party receives output then all parties receive output. Thus, it is possible that all parties abort. In order to emphasize the difference between the notions, we note that every protocol that provides guaranteed output delivery can be transformed into a protocol that provides fairness but *not* guaranteed

---

[1] Throughout this paper, whenever we say "fair" we mean "completely fair", and so if any party learns anything then all parties receive their entire output. This is in contrast to notions of partial fairness that have been studied in the past.

output delivery, as follows. At the beginning every party broadcasts OK; if one of the parties did not send OK then all the parties output ⊥; otherwise the parties execute the original protocol (that ensures guaranteed output delivery). Clearly every party can cause the protocol to abort. However, it can only do so before any information has been obtained. Thus, the resulting protocol is fair, but does not guarantee output delivery.

It is immediate to see that guaranteed output delivery implies fairness, since if all parties must receive output then it is not possible for the corrupted parties to receive output while the honest do not. However, the opposite direction is not clear. In the two-party case, guaranteed output delivery is indeed implied by fairness since upon receiving abort the honest party can just compute the function on its own input and a default input for the other party. However, when there are many parties involved, it is not possible to replace inputs with default inputs since the honest parties do not necessarily know who is corrupted (and security mandates that honest parties' inputs cannot be changed; otherwise, this could be disastrous in an election-type setting). This leads us to the following fundamental questions, which until now have not been considered at all (indeed, fairness and guaranteed output delivery are typically used synonymously):

> *Does fairness imply guaranteed output delivery? Do there exist function-alities that can be securely computed with fairness but not with guaranteed output delivery? Are there conditions on the function/network model for which fairness implies guaranteed output delivery?*

The starting point of our work is the observation that the *broadcast functionality* does actually separate guaranteed output delivery and fairness. Specifically, let n denote the overall number of parties, and let t denote an upper bound on the number of corrupted parties. Then, it is well known that secure broadcast can be achieved if and only if $t < n/3$ [14,13]. However, it is also possible to achieve *weak* broadcast (which means that either all parties abort and no one receives output, or all parties receive and agree upon the broadcasted value) for any $t < n$ [6]. In our terms, this is a secure computation of the broadcast functionality with *fairness* but *no guaranteed output delivery*. Thus, we see that for $t \geq n/3$ there exist functionalities that can be securely computed with fairness but not with guaranteed output delivery (the fact that broadcast cannot be securely computed with guaranteed output delivery for $t \geq n/3$ follows directly from the bounds on Byzantine Generals [14,13]). Although broadcast does provide a separation, it is an atypical function. Specifically, there is no notion of privacy, and the functionality can be computed information theoretically for any $t < n$ given a secure setup phase [15]. Thus, broadcast is a trivial functionality.[2] This leaves the question of whether fairness and guaranteed output delivery are distinct still holds for more "standard" secure computation tasks.

It is well known that for $t < n/2$ any multiparty functionality can be securely computed with guaranteed output delivery given a broadcast channel [8,16].

---

[2] We stress that "trivial" does not mean easy to achieve or uninteresting. Rather, it means that cryptographic hardness is not needed to achieve it in the setting of no honest majority [12].

Thus, using the weak broadcast of [6] in the protocols of [8,16] we have that *any* functionality can be securely computed with fairness for $t < n/2$. This leaves open the question as to whether there exist functionalities (apart from broadcast) that *cannot* be securely computed with guaranteed output delivery for $n/3 \leq t < n/2$.

In [10], they showed that the 3-party majority function and multiparty Boolean OR function can be securely computed with guaranteed output delivery for *any* number of corrupted parties (in particular, with an honest minority). However, the constructions of [10] use a broadcast channel. This leads us to the following questions for the range of $t \geq n/3$:

1. Can the 3-party majority function and multiparty Boolean OR function be securely computed with guaranteed output delivery without broadcast?
2. Can the 3-party majority function and multiparty Boolean OR function be securely computed with fairness without a broadcast channel?
3. Does the existence of broadcast make a difference with respect to fairness and/or guaranteed output delivery *in general*?

We remark that conceptually guaranteed output delivery is a stronger notion of security and that it is what is required in some applications. Consider the application of "mental poker"; if guaranteed output delivery is not achieved, then a corrupted party can cause the execution to abort in case it is dealt a bad hand. This is clearly undesirable.

1.3   Our Results

Separating Fairness and Guaranteed Output Delivery. We show that the 3-party majority function that can be securely computed with fairness [10] *cannot* be securely computed with guaranteed output delivery. Thus, there exist non-trivial functionalities (i.e., functionalities that cannot be securely computed in the information theoretic setting without an honest majority) for which fairness can be achieved but guaranteed output delivery cannot. Technically, we show this by proving that the 3-party majority function can be used to achieve broadcast, implying that it cannot be securely computed with guaranteed output delivery.

Theorem 1. *Consider a model without a broadcast channel and consider any $t \geq n/3$. Then, there exist non-trivial functionalities f (e.g., the majority function) such that f can be securely computed with fairness but f cannot be securely computed with guaranteed output delivery.*

This proves that fairness and guaranteed output delivery are distinct, at least in a model without a broadcast channel.

Feasibility of Guaranteed Output Delivery without Broadcast. The protocols of [10] for majority and Boolean OR both use a broadcast channel to achieve guaranteed output delivery. As we have seen in Theorem 1 this is essential for achieving their result for the majority function. However, is this also the case for the Boolean OR function? In general, do there exist non-trivial functionalities for which guaranteed output delivery is achievable without a broadcast channel and for any number of corrupted parties?

Theorem 2. *Consider a model without a broadcast channel and consider any number of corruptions. Then, there exist non-trivial functionalities f (e.g., the Boolean OR function) such that f can be securely computed with guaranteed output delivery.*

On the Role of Broadcast. We show that the existence or non-existence of broadcast is meaningless with respect to fairness, but of great significance with respect to guaranteed output delivery. Specifically, we show the following:

Theorem 3. *Let f be a multiparty functionality. Then:*

1. *There exists a protocol for securely computing f with fairness with a broadcast channel if and only if there exists a protocol for securely computing f with fairness without a broadcast channel.*
2. *If there exists a protocol for securely computing f with fairness (with or without a broadcast channel), then there exists a protocol for securely computing f with guaranteed output delivery with a broadcast channel.*

Thus, fairness and guaranteed output delivery are *equivalent* in a model with a broadcast channel, and *distinct* without a broadcast channel. In contrast, by Theorem 1 we already know that without broadcast it does not hold that fairness implies guaranteed output delivery (otherwise, the separation in Theorem 1 would not be possible). We also show that under black-box reductions, fairness *never* helps achieve guaranteed output delivery. That is:

Theorem 4. *Let f be a multiparty functionality and consider a hybrid model where a trusted party computes f fairly for the parties (i.e., either all parties receive output or none do). Then, there exists a protocol for securely computing f with guaranteed output delivery in the hybrid model if and only if there exists a protocol for securely computing f with guaranteed output delivery in the real model with no trusted party.*

Intuitively, Theorem 4 follows from the fact that an adversary can always cause the result of calls to f to be abort in which case they are of no help. This does not contradict item (2) of Theorem 3 since given a broadcast channel and nonblack-box access to the protocol that computes f with fairness, it is possible to apply a variant of the GMW compiler [8] and detect which party cheated and caused the abort to occur.

Conditions under Which Fairness Implies Guaranteed Output Delivery. We have already seen that fairness implies guaranteed output delivery given broadcast. We also consider additional scenarios in which fairness implies guaranteed output delivery. We prove that if a functionality can be securely computed with fairness and *identified abort* (meaning that the identity of the cheating party is detected) then the functionality can be securely computed with guaranteed output delivery. Finally, we show that in the fail-stop model (where the only thing an adversary can do is instruct a corrupted party to halt prematurely), fairness is always equivalent to guaranteed output delivery. This follows from the fact that broadcast is trivial in the fail-stop model.

Identified Abort and Broadcast. In the model of identified abort, the identity of the cheating party is revealed to the honest parties. This definition was explicitly used by [2], who remarked that it is met by most protocols (e.g., [8]), but not all (e.g., [9]). This model has the advantage that a cheating adversary who runs a "denial of service" attack and causes the protocol to abort cannot go undetected. Thus, it cannot repeatedly prevent the parties from obtaining output. An interesting corollary that comes out of our work—albeit not related to fairness and guaranteed output delivery—is that security with identified abort *cannot* be achieved in general for $t \geq n/3$ without broadcast. This follows from the fact that if identified abort can be achieved in general (even without fairness), then it is possible to achieve broadcast. Thus, we conclude:

Corollary 1. *Consider a model without a broadcast channel and consider any $t \geq n/3$. Then, there exist functionalities f that cannot be securely computed with identified abort.*

Summary of Feasibility. The table below summarizes the state of affairs regarding feasibility for secure computation with fairness and guaranteed output delivery, for different ranges regarding the number of corrupted parties.

Num. of Corrupted	With Broadcast	Without Broadcast
$t < n/3$	All f can be securely computed with guaranteed output delivery	
$n/3 \leq t < n/2$	All f can be computed with guaranteed output delivery	OR can be computed with guaranteed output delivery
$t \geq n/2$	Fairness implies guaranteed output delivery	MAJ *cannot* be computed with guaranteed output delivery
-	If f can be securely computed fairly with broadcast then it can be securely computed fairly without broadcast	

Preliminaries. Full definitions can be found in the full version [5]. We consider a number of different ideal models: security with guaranteed output delivery, with fairness, with abort, with identified abort (meaning that in the case of abort one of the corrupted parties is identified by the honest parties), and fairness with identified abort. The ideal models for these models are respectively denoted IDEAL$^{\text{g.d.}}$, IDEAL^{fair}, IDEAL^{abort}, IDEAL$^{\text{id-abort}}$, IDEAL$^{\text{fair,id-abort}}$. We also consider hybrid model protocols where the parties send regular messages to each other, and also have access to a trusted party who computes some function f for them. The trusted party may compute according to any of the specified ideal model. Letting type $\in \{\text{g.d.}, \text{fair}, \text{abort}, \text{id-abort}, (\text{fair}, \text{id-abort})\}$, we call this the (f, type)-hybrid model, and denote it HYBRID$^{f,\text{type}}$. The security parameter is denoted by κ, and the set of corrupted parties by $\mathcal{I}$. Unless stated otherwise, all adversaries considered are *malicious*.

2 Separating Fairness from Guaranteed Output Delivery

In this section we prove Theorem 1. As we have mentioned in the Introduction, it is known that secure broadcast can be t-securely computed with guaranteed output delivery if and only if $t < n/3$. In addition, secure broadcast can be computed with fairness, for any $t \leq n$, using the protocol of [6]. Thus, broadcast already constitutes a separation of fairness from guaranteed output delivery; however, since broadcast can be information theoretically computed (and is trivial in the technical sense; see Footnote 2), we ask whether or not such a separation also exists for more standard secure computation tasks.

In order to show a separation, we need to take a function for which fairness in the multiparty setting is feasible. Very few such functions are known, and the focus of this paper is not the construction of new protocols. Fortunately, in [10], it was shown that the 3-party majority function can be securely computed with fairness. (In [10] they use a broadcast channel. However, as we show in Section 4.1, this implies the result also without a broadcast channel.) We stress that the 3-party majority function is not trivial, and in fact the ability to securely compute it with any number of corruptions implies the existence of oblivious transfer (this is shown by reducing the 2-party greater-than functionality to it and applying [12]).

We show that the 3-party majority function f_{maj} *cannot* be securely computed with guaranteed output delivery and any number of corrupted parties in the point-to-point network model by showing that it actually implies broadcast. The key observation is that there exists an input $(1,1,1)$ for which the output of f_{maj} will be 1, even if a single corrupted party changes its input to 0. Similarly, there exists an input $(0,0,0)$ for which the output of f_{maj} will be 0, even if a single corrupt party changes its input to 1. Using this property, we show that if f_{maj} can be computed with guaranteed output delivery, then there exists a broadcast protocol for 3 parties that is secure against a single corruption. Given an input bit β, the sender sends β to each other party, and all parties compute f_{maj} on the input they received. This works since a corrupted dealer cannot make two honest parties output inconsistent values, since f_{maj} provides the same output to all parties. Likewise, if there is one corrupted receiver, then it cannot change the majority value (as described above). Finally, if there are two corrupted receivers, then it makes no difference what they output anyway.

Theorem 5. *Let t be a parameter and let $f_{\mathsf{maj}} : \{0,1\}^3 \to \{0,1\}^3$ be the majority functionality for 3 parties $f_{\mathsf{maj}}(x_1, x_2, x_3) = (y, y, y)$ where $y = (x_1 \wedge x_2) \vee (x_3 \wedge (x_1 \oplus x_2))$. If f_{maj} can be t-securely computed with guaranteed output delivery in a point-to-point network, then there exists a protocol that t-securely computes the 3-party broadcast functionality for any t.*

Proof: We construct a protocol π for securely computing the 3-party broadcast functionality $f_{\mathsf{bc}}(x, \lambda, \lambda) = (x, x, x)$ in the $(f_{\mathsf{maj}}, \mathsf{g.d.})$-hybrid model (i.e., in a hybrid model where a trusted party computes the f_{maj} functionality with guaranteed output delivery). Protocol π works as follows:

1. The sender P_1 with input $x \in \{0,1\}$ sends x to P_2 and P_3.
2. Party P_1 sends x to the trusted party computing f_{maj}. Each party P_i ($i \in \{2,3\}$) sends the value it received from P_1 to f_{maj}.
3. Party P_1 always outputs x. The parties P_2 and P_3 output whatever they receive from the trusted party computing f_{maj}.

Let $\mathcal{A}$ be an adversary attacking the execution of π in the $(f_{\mathsf{maj}}, \mathsf{g.d.})$-hybrid model; we construct an ideal model adversary $\mathcal{S}$ in the ideal model for f_{bc} with guaranteed output delivery. $\mathcal{S}$ invokes $\mathcal{A}$ and simulates the interaction of $\mathcal{A}$ with the honest parties and with the trusted party computing f_{maj}. $\mathcal{S}$ proceeds based on the following corruption cases:

- P_1 *alone is corrupted:* $\mathcal{S}$ receives from $\mathcal{A}$ the values $x_2, x_3 \in \{0,1\}$ that it sends to parties P_2 and P_3, respectively. Next, $\mathcal{S}$ receives the value $x_1 \in \{0,1\}$ that $\mathcal{A}$ sends to f_{maj}. $\mathcal{S}$ computes $x = f_{\mathsf{maj}}(x_1, x_2, x_3)$ and sends x to the trusted party computing f_{bc}. $\mathcal{S}$ simulates $\mathcal{A}$ receiving x back from f_{maj}, and outputs whatever $\mathcal{A}$ outputs.

- P_1 *and one of P_2 or P_3 are corrupted:* the simulation is the same as in the previous case except that if P_2 is corrupted then the value x_2 is taken from what $\mathcal{A}$ sends in the name of P_2 to f_{maj} (and not the value that $\mathcal{A}$ sends first to P_2); likewise for P_3. Everything else is the same.

- P_1 *is honest:* $\mathcal{S}$ sends an empty input λ to the trusted party for every corrupted party, and receives back some $x \in \{0,1\}$. Next, $\mathcal{S}$ simulates P_1 sending x to both P_2 and P_3. If both P_2 and P_3 are corrupted, then $\mathcal{S}$ obtains from $\mathcal{A}$ the values x_2 and x_3 that they send to f_{maj}, computes $x' = f_{\mathsf{maj}}(x, x_2, x_3)$ and simulates the trusted party sending x' back to all parties. If only one of P_2 and P_3 are corrupted, then $\mathcal{S}$ simulates the trusted party sending x back to all parties. Finally, $\mathcal{S}$ outputs whatever $\mathcal{A}$ outputs.

The fact that the simulation is good is straightforward. If P_1 is corrupted, then only consistency is important, and $\mathcal{S}$ ensures that the value sent to f_{bc} is the one that the honest party/parties would output. If P_1 is not corrupted, and both P_2 and P_3 are corrupted, then P_1 always outputs the correct x as required, and the outputs of P_2 and P_3 are not important. Finally, if P_1 and P_2 are corrupted, then $\mathcal{S}$ sends f_{bc} the value that P_3 would output in the real protocol as required; likewise for P_1 and P_3 corrupted. ∎

Theorem 5 implies that f_{maj} cannot be securely computed with guaranteed output delivery for any $t < 3$ in a point-to-point network; this follows immediately from the fact that the broadcast function can be securely computed if and only if $t < n/3$. Furthermore, by [10], f_{maj} *can* be securely computed fairly given oblivious transfer (and as shown in Section 4.1 this also holds in a point-to-point network). Thus, we have:

Corollary 2. *Assume that oblivious transfer exists. Then, there exist non-trivial functionalities f such that f can be securely computed with fairness but cannot be securely computed with guaranteed output delivery, in a point-to-point network and with $t \geq n/3$.*

Three-Party Functionalities That Imply Broadcast. It is possible to generalize the property that we used to show that f_{maj} implies broadcast. Specifically, consider a functionality f with the property that there exist inputs (x_1, x_2, x_3) and (x_1', x_2', x_3') such that $f(x_1, x_2, x_3) = 0$ and $f(x_1', x_2', x_3') = 1$, and such that if either of x_2 or x_3 (resp., x_2' or x_3') are changed arbitrarily, then the output of f remains the same. Then, this function can be used to achieve broadcast. We describe the required property formally inside the proof of the theorem below. We show that out of the 256 functions over 3-bit inputs, there are 110 of them with this property. It follows that none of these can be securely computed with guaranteed output delivery in the presence of one or two corrupted parties. We prove the following:

Theorem 6. *There are 110 functions from the family of all 3-party Boolean functions* $\{f : \{0,1\} \times \{0,1\} \times \{0,1\} \to \{0,1\}\}$ *that cannot be securely computed with guaranteed output delivery in a point-to-point network with* $t = 1$ *or* $t = 2$.

Proof: We provide a combinatorial proof of the theorem, by counting how many functions have the property that arbitrarily changing one of the inputs does not effect the output, and there are inputs that yield output 0 and inputs that yield output 1. As we have seen in the proof of Theorem 5, it is possible to securely realize the broadcast functionality given a protocol that securely computes any such functionality with guaranteed output delivery.

We prove that there are 110 functions $f : \{0,1\}^3 \to \{0,1\}$ in the union of the following sets F_1, F_2, F_3:

1. Let F_1 be the set of all functions for which there exist $(a, b, c), (a', b', c') \in \{0,1\}^3$ such that $f(a, b, \cdot) = f(a, \cdot, c) = 1$ and $f(a', b', \cdot) = f(a', \cdot, c') = 0$.
2. Let F_2 be the set of all functions for which there exist $(a, b, c), (a', b', c') \in \{0,1\}^3$ such that $f(a, b, \cdot) = f(\cdot, b, c) = 1$ and $f(a', b', \cdot) = f(\cdot, b', c') = 0$.
3. Let F_3 be the set of all functions for which there exist $(a, b, c), (a', b', c') \in \{0,1\}^3$ such that $f(\cdot, b, c) = f(a, \cdot, c) = 1$ and $f(\cdot, b', c') = f(a', \cdot, c') = 0$.

Observe that any function in one of these sets can be used to achieve broadcast, as described above. Based on the inclusion-exclusion principle and using Lemma 2 proven below, it follows that:

$$|F_1 \cup F_2 \cup F_3| = 3 \cdot 50 - 3 \cdot 16 + 8 = 110,$$

as required. We first prove the following lemma:

Lemma 1. *If* $f \in F_1$, *then* $a \neq a'$, *if* $f \in F_2$ *then* $b \neq b'$ *and if* $f \in F_3$ *then* $c \neq c'$.

Proof: Let $f \in F_1$ and let $a, a', b, b', c, c' \in \{0,1\}$ be inputs fulfilling the condition for set F_1. Assume by contradiction that $a = a'$. Thus,

$$f(a, b, c) = f(a, \bar{b}, c) = f(a, b, \bar{c}) = 1 \text{ and } f(a, b', c') = f(a, \bar{b}', c') = f(a, b', \bar{c}') = 0.$$

If $b = b'$ then $f(a, b, c') = f(a, b', c') = 0$. However, $f(a, b, c) = f(a, b, \bar{c}) = 1$ and so $f(a, b, c') = 1$ for any c', in contradiction. Thus $b \neq b'$. Similarly, $c \neq c'$. Therefore, $b' = \bar{b}$ and $c' = \bar{c}$ and by the condition, $f(a, b, c) = 1$ and $f(a, \bar{b}, \bar{c}) = 0$.

Consider $f(a, \bar{b}, c)$. From the condition, $f(a, b, c) = f(a, \bar{b}, c) = 1$. However, changing the c coordinate to $\bar{c}$ gives us $f(a, \bar{b}, \bar{c})$ which by the condition equals 0 (because $b' = \bar{b}$ and $c' = \bar{c}$). We therefore derive a contradiction, and so conclude that $a' = \bar{a}$. ∎

It remains to prove the following lemma, to derive the theorem.

Lemma 2. *The following hold:*

1. $|F_1| = |F_2| = |F_3| = 50.$
2. $|F_1 \cap F_2| = |F_1 \cap F_3| = |F_2 \cap F_3| = 16.$
3. $|F_1 \cap F_2 \cap F_3| = 8.$

Proof: Let $f : \{0,1\}^3 \to \{0,1\}$ be a function represented by the Boolean string $(\beta_0\beta_1\beta_2\beta_3\beta_4\beta_5\beta_6\beta_7)$ as shown in Table 1:

Table 1. Representation of a Boolean function $\{0,1\}^3 \to \{0,1\}$

0	0	0	β_0
0	0	1	β_1
0	1	0	β_2
0	1	1	β_3
1	0	0	β_4
1	0	1	β_5
1	1	0	β_6
1	1	1	β_7

1. Assume $f \in F_1$ (the proof for F_2, F_3 is similar). The first quadruple $(\beta_0\beta_1\beta_2\beta_3)$ corresponds to $a = 0$ and the second quadruple $(\beta_4\beta_5\beta_6\beta_7)$ corresponds to $a = 1$. There exists b, c such that $f(a, b, c) = f(a, \bar{b}, c) = f(a, b, \bar{c})$ and b', c' such that $f(\bar{a}, b', c') = f(\bar{a}, \bar{b}', c') = f(\bar{a}, b', \bar{c}')$, in addition, $f(a, b, c) \neq f(\bar{a}, b', c')$. Therefore, in each such quadruple there must be a triplet of 3 identical bits, and the two triplets have opposite values. Denote $\beta = f(a, b, c)$, there are 5 options for $(\beta_0\beta_1\beta_2\beta_3)$ in which 3 of the bits equal β:
$$(\beta\beta\beta\beta), (\beta\beta\beta\bar{\beta}), (\beta\beta\bar{\beta}\beta), (\beta\bar{\beta}\beta\beta), (\bar{\beta}\beta\beta\beta).$$

For each such option, there are 5 options for $(\beta_4\beta_5\beta_6\beta_7)$ in which 3 of the bits equal $\bar{\beta}$:
$$(\bar{\beta}\bar{\beta}\bar{\beta}\bar{\beta}), (\bar{\beta}\bar{\beta}\bar{\beta}\beta), (\bar{\beta}\bar{\beta}\beta\bar{\beta}), (\bar{\beta}\beta\bar{\beta}\bar{\beta}), (\beta\bar{\beta}\bar{\beta}\bar{\beta}).$$

There are 2 options for the value of β, so in total $|F_1| = 2 \cdot 5 \cdot 5 = 50$.
2. Assume $f \in F_1 \cap F_2$ (the proof for $F_1 \cap F_3, F_2 \cap F_3$ is similar). In this case $a' = \bar{a}$ and $b' = \bar{b}$ and the constraints are
$$f(a, b, c) = f(\bar{a}, b, c) = f(a, \bar{b}, c) = f(a, b, \bar{c})$$
$$\neq f(\bar{a}, \bar{b}, c') = f(a, \bar{b}, c') = f(\bar{a}, b, c') = f(\bar{a}, \bar{b}, \bar{c}').$$

Therefore, the string is balanced (there are 4 zeros and 4 ones), where 3 of the bits $(\beta_0\beta_1\beta_2\beta_3)$ are equal to β and one to $\bar{\beta}$, and 3 of the bits $(\beta_4\beta_5\beta_6\beta_7)$ are equal to $\bar{\beta}$ and one to β.
There are 4 options to select 3 bits in $(\beta_0\beta_1\beta_2\beta_3)$, and 2 options to select one bit in $(\beta_4\beta_5\beta_6\beta_7)$. These two options correspond either to $(\bar{a}, b, c)$ or $(\bar{a}, \bar{b}, \bar{c})$. Hence, $|F_1 \cap F_2| = 2 \cdot 4 \cdot 2 = 16$.
3. Assume $f \in F_1 \cap F_2 \cap F_3$. In this case $a' = \bar{a}$, $b' = \bar{b}$ and $c' = \bar{c}$ and the constraints are
$$f(a, b, c) = f(\bar{a}, b, c) = f(a, \bar{b}, c) = f(a, b, \bar{c})$$
$$\neq f(\bar{a}, \bar{b}, \bar{c}) = f(a, \bar{b}, \bar{c}) = f(\bar{a}, b, \bar{c}) = f(\bar{a}, \bar{b}, c).$$

Therefore, the string is of the form $(\beta_0\beta_1\beta_2\beta_3\bar{\beta}_0\bar{\beta}_1\bar{\beta}_2\bar{\beta}_3)$, where 3 of the bits $(\beta_0\beta_1\beta_2\beta_3)$ are equal to β and one to $\bar{\beta}$.

There are 4 options to select 3 bits in $(\beta_0\beta_1\beta_2\beta_3)$, and setting them to the same value determines the rest of the string. Hence, $|F_1 \cap F_2 \cap F_3| = 2 \cdot 4 = 8$.

∎

This completes the proof of Theorem 6.   ∎

As we have mentioned in the Introduction, in the case that $t = 1$ (i.e., when there is an honest majority), all functions can be securely computed with fairness in a point-to-point network. Thus, we have that all 110 functions of Theorem 6 constitute a *separation* of fairness from guaranteed output delivery. That is, in the case of $n/3 \leq t < n/2$, we have that *many functions* can be securely computed with fairness but not with guaranteed output delivery. In addition, 8 out of these 110 functions reduce to 3-majority and so can be computed fairly for any $t \leq n$. Thus, these 8 functions form a separation for the range of $t \geq n/2$.

3   Fairness Implies Guaranteed Output Delivery for Default-Output Functionalities

In this section we prove Theorem 2. In fact, we prove a stronger theorem, stating that fairness implies guaranteed output delivery for functions with the property that there exists a "default value" such that any single party can fully determine the output to that value. For example, the multiparty Boolean AND and OR functionalities both have this property (for the AND functionality any party can always force the output to be 0, and for the OR functionality any party can always force the output to be 1). We call such a function a default-output functionality. Intuitively, such a function can be securely computed with guaranteed output delivery if it can be securely computed fairly, since the parties can first try to compute it fairly. If they succeed, then they are done. Otherwise, they all received abort and can just output their respective output in the default value for the functionality. This can be simulated since any single corrupted party in the ideal model can choose an input that results in the default output value.

Definition 1. *Let $f : (\{0,1\}^*)^n \to (\{0,1\}^*)^n$ be an n-ary functionality. f is called a* default-output functionality *with default output $(\tilde{y}_1, \ldots, \tilde{y}_n)$, if for every $i \in \{1, \ldots, n\}$ there exists a special input $\tilde{x}_i$ such that for every x_j with $j \neq i$ it holds that $f(x_1, \ldots, \tilde{x}_i, \ldots, x_n) = (\tilde{y}_1, \ldots, \tilde{y}_n)$.*

Observe that $(0, \ldots, 0)$ is a default output for the Boolean AND function, and $(1, \ldots, 1)$ is a default output for the Boolean OR function. We now prove that if a functionality f has a default output value, then the existence of a fair protocol for f implies a protocol with guaranteed output delivery for f.

Theorem 7. *Let $f : (\{0,1\}^*)^n \to (\{0,1\}^*)^n$ be a default-output functionality. If f can be t-securely computed with fairness (with or without a broadcast channel), then f can be t-securely computed with guaranteed output delivery, in a point-to-point network.*

Proof: Let f be as in the theorem statement, and let the default output be $(\tilde{y}_1, \ldots, \tilde{y}_n)$. Assume that f can be securely computed with fairness with or without a broadcast channel. By Theorem 9, f can be securely computed with fairness without a broadcast channel. We now construct a protocol π that securely computes f with guaranteed output delivery in the (f, fair)-hybrid model:

1. Each P_i sends its input x_i to the trusted party computing f.
2. Denote by y_i the value received by P_i from the trusted party.
3. If $y_i \neq \bot$, P_i outputs y_i, otherwise P_i outputs $\tilde{y}_i$.

Let $\mathcal{A}$ be an adversary attacking the execution of π in the (f, fair)-hybrid model. We construct an ideal model adversary $\mathcal{S}$ in the ideal model with guaranteed output delivery. Let $\mathcal{I}$ be the set of corrupted parties, let $i \in \mathcal{I}$ be one of the corrupted parties (if no parties are corrupted then there is nothing to simulate), and let $\tilde{x}_i$ be the input guaranteed to exist by Definition 1. Then, $\mathcal{S}$ invokes $\mathcal{A}$ and simulates the interaction of $\mathcal{A}$ with the trusted party computing f (note that there is no interaction between $\mathcal{A}$ and honest parties). $\mathcal{S}$ receives the inputs that $\mathcal{A}$ sends to f. If any of the inputs equal abort then $\mathcal{S}$ sends $\tilde{x}_i$ as P_i's input to its own trusted party computing f (with guaranteed output delivery), and arbitrary inputs for the other parties. Then, $\mathcal{S}$ simulates the corrupted parties receiving $\bot$ as output from the trusted party in π, and outputs whatever $\mathcal{A}$ outputs. Else, if none of the inputs equal abort, then $\mathcal{S}$ sends its trusted party the inputs that $\mathcal{A}$ sent. $\mathcal{S}$ then receives the outputs of the corrupted parties from its trusted party, and internally sends these to $\mathcal{A}$ as the corrupted parties' outputs from the trusted party computing f in π. Finally, $\mathcal{S}$ outputs whatever $\mathcal{A}$ outputs.

If $\mathcal{A}$ sends abort, then in the real execution every honest party P_j outputs $\tilde{y}_j$. However, since $\mathcal{S}$ sends the input $\tilde{x}_i$ to the trusted party computing f, by Definition 1 we have that the output of every honest party P_j in the ideal execution is also $\tilde{y}_j$. Furthermore, if $\mathcal{A}$ does not send abort, then $\mathcal{S}$ just uses exactly the same inputs that $\mathcal{A}$ sent. It is clear that the view of $\mathcal{A}$ is identical in the execution of π and the simulation with $\mathcal{S}$. We therefore conclude that π securely computes f with guaranteed output delivery, as required. ∎

We have proven that fairness implies guaranteed output delivery for default-output functionalities; it remains to show the existence of fair protocols for some default-output functionalities. Fortunately, this was already proven in [10]. The only difference is that [10] uses a broadcast channel. Noting that the multiparty Boolean OR functionality is non-trivial (in the sense of Footnote 2), and that it has default output $(1, \ldots, 1)$ as mentioned above, we have the following corollary.

Corollary 3. *Assume that oblivious transfer exists. Then, there exist non-trivial functionalities f that can be securely computed with guaranteed output delivery in a point-to-point network, for any $t < n$.*

Feasibility of Guaranteed Output Delivery. In Theorem 8, we prove that 16 non-trivial functionalities can be securely computed with guaranteed output delivery in a point-to-point network (by showing that they are default-output functionalities). Thus, guaranteed output delivery can be achieved for a significant number of functions.

Theorem 8. *There are 16 non-trivial functions from the family of all 3-party Boolean functions $\{f : \{0,1\} \times \{0,1\} \times \{0,1\} \to \{0,1\}\}$ that can be securely computed with guaranteed output delivery in a point-to-point network for any number of corrupted parties.*

Proof: When represented using its truth table as a Binary string (see Table 1), the 3-party Boolean OR function is (01111111), similarly, the Boolean AND function is (00000001). Every function $(\beta_0\beta_1\beta_2\beta_3\beta_4\beta_5\beta_6\beta_7)$ such that there exists i for which $\beta_i = \beta$ and for every $j \neq i$ $\beta_j = \bar{\beta}$ can be reduced to computing Boolean OR. Since there are 8 ways to choose i and 2 ways to choose β, we conclude that there are 16 such functions. ∎

4   The Role of Broadcast

In this section, we prove Theorem 3, and show that a functionality can be securely computed fairly with broadcast if and only if it can be securely computed fairly without broadcast. In addition, we show that if a functionality can be securely computed with fairness, then with a broadcast channel it can be securely computed with guaranteed output delivery.

4.1   Fairness Is Invariant to Broadcast

Gordon and Katz construct two fair multiparty protocols in [10], both of them require a broadcast channel. In this section we show that fairness holds for both even without a broadcast channel. More generally, fairness can be achieved with a broadcast channel if and only if it can be achieved without a broadcast channel.

It is immediate that fairness without broadcast implies fairness with broadcast. The other direction follows by using the protocol of [6] for detectable broadcast. In the first stage, the parties execute a protocol that establishes a public key infrastructure. This protocol is independent of the parties' inputs and is computed with abort. If the adversary aborts during this phase, it learns nothing about the output and fairness is retained. If the adversary does not abort, the parties can use the public key infrastructure and execute multiple (sequential) instances of authenticated broadcast, and so can run the original protocol with broadcast that is fair.

One subtlety arises since the composition theorem replaces every ideal call to the broadcast functionality with a protocol computing broadcast. However, in this case, each authenticated broadcast protocol relies on the same public key infrastructure that is generated using a protocol with abort. We therefore define a reactive ideal functionality which allows abort only in the first "setup" call. If no abort was sent in this call, then the functionality provides a fully secure broadcast (with guaranteed output delivery) from there on. The protocol of [6] securely computes this functionality with guaranteed output delivery, and thus constitutes a sound replacement of the broadcast channel (unless an abort took place).

Theorem 9. *Let f be an n-ary functionality and let $t \leq n$. Then, f can be t-securely computed with fairness assuming a broadcast channel if and only if f can be t-securely computed with fairness in a point-to-point network.*

Proof Sketch: If f can be t-securely computed with fairness in a point-to-point network, then it can be t-securely computed with fairness with a broadcast channel by just having parties broadcast messages and stating who the intended recipient is. (Recall that in the point-to-point network we assume authenticated but not private channels.)

Next, assume that f can be t-securely computed with fairness assuming a broadcast channel. We now show that it can be t-securely computed with fairness in a point-to-point network. We define the reactive functionality for *conditional broadcast* f_{condbc}. In the first call to f_{condbc}, the functionality computes the AND function, i.e., each party has an input bit b_i and the functionality returns $b = b_1 \wedge \ldots \wedge b_n$ to each party. In addition, the functionality stores the bit b as its internal state for all future calls. In all future calls to f_{condbc}, if $b = 1$ it behaves exactly like f_{bc}, whereas if $b = 0$ it returns $\perp$ to all the parties in the first call and halts. By inspection, it is immediate that the protocol of [6] securely computes f_{condbc} with guaranteed output delivery, for any $t \leq n$ in a point-to-point network.

Let π be the protocol that t-securely computes f assuming a broadcast channel; stated differently, π t-securely computes f in the $(f_{\mathsf{bc}}, \mathsf{g.d.})$-hybrid model. We construct a protocol π' for t-securely computing f in the $(f_{\mathsf{condbc}}, \mathsf{fair})$-hybrid model. π' begins by all parties sending the bit 1 to f_{condbc} and receiving back output. If a party receives back $b = 0$, it aborts and outputs $\perp$. Else, it runs π with the only difference that all broadcast messages are sent to f_{condbc} instead of to f_{bc}. Since f_{condbc} behaves exactly like f_{bc} as long $b = 1$ is returned from the first call, we have that in this case the output of π and π' is identical. Furthermore, π' is easily simulated by first invoking the adversary $\mathcal{A}'$ for π' and obtaining the corrupted parties' inputs to f_{condbc} in the first call. If any 0 bit is sent, then the simulator $\mathcal{S}'$ for π' sends abort to the trusted party, outputs whatever $\mathcal{A}'$ outputs and halts. Otherwise, it invokes the simulator $\mathcal{S}$ that is guaranteed to exist for π on the residual adversary $\mathcal{A}$ that is obtained by running $\mathcal{A}'$ until the end of the first call to f_{condbc} (including $\mathcal{A}'$ receiving the corrupted parties' output bits from this call). Then, $\mathcal{S}'$ sends whatever $\mathcal{S}$ wishes to send to the trusted party, and outputs whatever $\mathcal{S}$ outputs. Since f_{condbc} behaves exactly like f_{bc} when $b = 1$ in the first phase, we have that the output distribution generated by $\mathcal{S}'$ is identical to that of $\mathcal{S}$ when $b = 1$. Furthermore, when $b = 0$, it is clear that the simulation is perfect. ∎

4.2 Fairness with Identified Abort Implies Guaranteed Output Delivery

Before proceeding to prove that fairness implies guaranteed output delivery in a model with a broadcast channel, we first show that fairness with identified abort implies guaranteed output delivery. Recall that a protocol securely computes a functionality f with identified abort, if when the adversary causes an abort all honest parties receive $\perp$ as output along with the identity of a corrupted party. If a protocol securely computes f with fairness and identified abort, then it is guaranteed that if the adversary aborts, it learns nothing about the output and all honest parties learn an identity of a corrupted party. In this situation, the parties can eliminate the identified corrupted party and execute the protocol again, where an arbitrary party emulates the operations of the eliminated party using a default input. Since nothing was learned by the adversary when an abort occurs, the parties can rerun the protocol from scratch (without the identified corrupted party) and nothing more than a single output will be revealed to the adversary. Specifically, given a protocol π that computes f with fairness and identified abort, we can construct a new protocol π' that computes f with guaranteed output delivery. In the protocol π', the parties iteratively execute π, where in each iteration, either the adversary does not abort and all honest parties receive consistent output, or the adversary aborts without learning anything and the parties identify a corrupted party, who is eliminated from the next iteration.

Theorem 10. *Let f be an n-ary functionality and let $t \leq n$. If f can be t-securely computed with fairness and identified abort, then f can be t-securely computed with guaranteed output delivery.*

Proof: We prove the theorem by constructing a protocol π that t-securely computes f with guaranteed output delivery in the $(f, \mathsf{fair\text{-}id\text{-}abort})$-hybrid model. For every party P_i, we assign a default input value $\tilde{x}_i$ and construct the protocol π as follows:

1. Let $\mathcal{P}_1 = \{1, \ldots, n\}$ denote the set of indices of all participating parties.
2. For $i = 1, \ldots, t+1$
   (a) All parties in $\mathcal{P}_i$ send their inputs to the trusted party computing f, where the party with the lowest index in $\mathcal{P}_i$ simulates all parties in $\mathcal{P}_1 \setminus \mathcal{P}_i$, using their predetermined default input values.
   For each $j \in \mathcal{P}_i$, denote the output of P_j from f by y_j.
   (b) For every $j \in \mathcal{P}_i$, party P_j checks if y_j is a valid output, if so P_j outputs y_j and halts. Otherwise all parties receive $(\perp, i^*)$ as output, where i^* is an index of a corrupted party. If $i^* \notin \mathcal{P}_i$ (and so i^* is a previously identified corrupted party), then all parties set i^* to be the party with the lowest index in $\mathcal{P}_i$.
   (c) Set $\mathcal{P}_{i+1} = \mathcal{P}_i \setminus \{i^*\}$.

First note that there are at most $t+1$ iterations; therefore π terminates in polynomial time. Let $\mathcal{A}$ be an adversary attacking π and let $\mathcal{I}$ be the set of corrupted parties. We construct a simulator $\mathcal{S}$ for the ideal model with f and guaranteed output delivery, as follows. $\mathcal{S}$ invokes $\mathcal{A}$ and receives its inputs to f in every iteration. If an iteration contains an abort, then $\mathcal{S}$ simulates sending the response $(\perp, i^*)$ to all parties, and proceeds to the next iteration. In the first iteration in which no abort is sent (and such an iteration must exist since there are $t+1$ iterations and in every iteration except for the last one corrupted party is removed), $\mathcal{S}$ sends the inputs of the corrupted parties that $\mathcal{A}$ sent to the trusted party computing f. In addition, S sends the values for any corrupted parties that were identified in previous iterations: if the lowest index remaining is honest, then S sets these values to be the default values; else, it sets these values to be the values sent by A for these parties. Upon receiving the output from its trusted party, S hands it to A as if it were the output of the corrupted parties in the iteration of π, and outputs whatever A outputs.

The simulation in the $(f, \text{fair-id-abort})$-hybrid model is perfect since S can perfectly simulate the trusted party for all iterations in which an abort is sent. Furthermore, in the first iteration for which an abort is not sent, S sends f the exact inputs upon which the function f is computed in the protocol. Thus, the view of A and the output of the honest parties in the simulation with S are identical to their view and output in an execution of π in the $(f, \text{fair-id-abort})$-hybrid model. ∎

4.3   Fairness with Broadcast Implies Guaranteed Output Delivery

In Section 4.2, we saw that if a functionality can be securely computed with fairness and identified abort, then it can be securely computed with guaranteed output delivery. In this section, we show that assuming the existence of a broadcast channel, there is a protocol compiler that given a protocol computing a functionality f with fairness, outputs a protocol computing f with fairness and identified abort. Therefore, assuming broadcast, fairness implies guaranteed output delivery.

The protocol compiler we present is a modification of the GMW compiler, which relies on the code of the underlying fair protocol and requires non-black-box access to the protocol. (Therefore, this result does not contradict the proof in Section 5 that black box access to an ideal functionality that computes f with fairness does not help to achieve guaranteed output delivery.) The underlying idea is to use the GMW compiler [8,7]. However, instead of enforcing semi-honest behaviour, the compiler is used in order to achieve security with identified abort. This is accomplished by tweaking the GMW compiler so that first only public-coin zero-knowledge proofs are used, and second if an honest party detects dishonest behaviour—i.e., if some party does not send a message or fails to provide a zero knowledge proof for a message it sent—the honest parties record the identity i^* of the cheating party. We stress that the parties do *not* abort the protocol at this point, but rather continue until the end to see if they received $\perp$ or not. If they received $\perp$, then they output $(\perp, i^*)$ and halt. Else, if they received proper output, then they output it. Note that if the parties were to halt as soon as they detected a cheating party, then this would not be secure since it is possible that some of the corrupted parties already received output by that point. Thus, they conclude the protocol to determine whether they should abort or not.

The soundness of this method holds because in the GMW compiler with public-coin zero-knowledge proofs, a corrupted party cannot make an honest party fail, and all parties can verify if the zero-knowledge proof was successful or not. A brief description of the GMW compiler appears in full version [5]. We prove the following:

Theorem 11. *Assume the existence of one way functions and let $t \leq n$. If a functionality f can be t-securely computed with fairness assuming a broadcast channel, then f can be t-securely computed with guaranteed output delivery.*

Proof: We begin by proving that fairness with a broadcast channel implies fairness with identified abort.

Lemma 3. *Assume the existence of one way functions and let $t \leq n$. Then, there exists a polynomial-time protocol compiler that receives any protocol π, running over a broadcast channel, and outputs a protocol π', such that if π t-securely computes a functionality f with fairness then π' t-securely computes f with fairness and identified abort.*

Proof Sketch: Since the protocol is run over a single broadcasts channel, if at any point a party does not broadcast a message when it is suppose to, then all the parties detect it and can identify this party as corrupted, in case the protocol outputs $\perp$. Therefore, we can assume that no party halts the protocol by not sending messages.

We consider a tweaked version of the GMW compiler. The input commitment phase and the coin generation phase are kept the same. In the protocol emulation phase, when a sender transmits a message to a receiver, they execute a strong zero knowledge proof of knowledge with perfect completeness, in which the sender acts as the prover and the receiver as the verifier. The statement is that the message was constructed by the next message function, based on the sender's input, random coins and the history of all the messages the sender received in the protocol. However, if the prover fails to prove the statement, unlike in the GMW compiler, the verifier does not immediately broadcast the verification coins, but stores the verification coins along with the identity of the sender in memory, and resumes the protocol.

At the end of the protocol emulation, each party checks if it received an output, if so it outputs it and halts. If a party did not receive an output and it received a message for which the corresponding zero knowledge proof failed, it broadcasts the verification coins it used during the zero knowledge proof. In this case, the other parties verify if this is a justified reject, and if so they output $\perp$ along with the identity of the prover. If the reject is not justified, the parties output $\perp$ along with the identity party that sent the false verification coins.

Since the zero knowledge proof has perfect completeness, a corrupted party cannot produce verification coins that will falsely reject an honest party. Hence, only parties that deviate from the protocol can be identified as corrupted.

It case each honest party finishes the execution of the compiled protocol with some output, the compiled protocol remains secure, based on the security of the underlying protocol and of the zero knowledge proof.

In case one of the honest parties did not get an output, there must be at least one message that does not meet the protocol's specification, hence at least one honest party received a message without a valid proof. Therefore, all the honest parties output $\perp$ along with an identity of a corrupted party. However, in this situation, the adversary does not learn anything about the output, since otherwise there exists an attack violating the fairness of the underlying protocol π. Hence, the compiled protocol retains fairness. ∎

Applying Theorem 10 to Lemma 3 we have that f can be t-securely computed with guaranteed output delivery, completing the proof of the theorem. ∎

5 Black-Box Fairness Does Not Help for Guaranteed Output Delivery

In this section we show that the ability to securely compute a functionality with complete fairness does not assist in computing the functionality with guaranteed output delivery, at least in a black box manner. More precisely, a functionality f can be securely computed with guaranteed output delivery in the (f, fair)-hybrid model if and only if f can be securely computed with guaranteed output delivery in the plain model.

The idea is simply that any protocol that provides guaranteed output delivery in the (f, fair)-hybrid model has to work even if the output of every call to the trusted party computing f fairly concludes with an abort. This is because a corrupted party can always send abort to the trusted party in every such call.

Proposition 1. *Let f be an n-ary functionality and let $t \leq n$. Then, f can be t-securely computed in the (f, fair)-hybrid model with guaranteed output delivery if and only if f can be t-securely computed in the real model with guaranteed output delivery.*

Proof Sketch: If f can be t-securely computes f in the real model with guaranteed output delivery, then clearly it can be t-securely computed in the (f, fair)-hybrid model with guaranteed output delivery by simply not sending anything to the trusted party.

For the other direction, let π be a protocol that t-securely computes f in the (f, fair)-hybrid model with guaranteed output delivery. We construct a protocol π' in the real model which operates exactly like π, except that whenever there is a call in π to the ideal functionality f, the parties in π' emulate receiving $\perp$ as output. It is immediate that for every adversary $\mathcal{A}'$ for π', there exists an adversary $\mathcal{A}$ for π so that the output distributions of the two executions are identical ($\mathcal{A}$ just sends abort to every ideal call in π, and otherwise sends the same messages that $\mathcal{A}'$ sends). By the assumption that π is secure, there exists a simulator $\mathcal{S}$ for the ideal model for f with guaranteed output delivery. This implies that $\mathcal{S}$ is also a good simulator for $\mathcal{A}'$ in π', and so π' t-securely computes f with guaranteed output delivery in the real model. ∎

6 Additional Results

In this section we prove two additional results. First, there exist functionalities for which identified abort cannot be achieved (irrespective of fairness), and fairness and guaranteed output delivery are equivalent for fail-stop adversaries.

6.1 Identified Abort Cannot Be Achieved without Broadcast

We show that security with identified abort cannot be achieved in general without assuming a broadcast channel.

Proposition 2. *Assume the existence of one-way functions. There exist functionalities that cannot be securely computed with identified abort, in the point-to-point network model and with $t \geq n/3$.*

Proof Sketch: Assume by contradiction that the PKI setup functionality defined by

$$f_{\mathsf{PKI}}(\lambda, \ldots, \lambda) = ((\boldsymbol{pk}, sk_1), \ldots, (\boldsymbol{pk}, sk_n)),$$

can be t-securely computed with identified abort for some $t = n/3$, where $\boldsymbol{pk} = (pk_1, \ldots, pk_n)$ and each (pk_i, sk_i) are a public/private key pair for secure digital signature scheme (that exists if one-way function exists). Then, we can t-securely compute f_{bc} by running the protocol π that is assumed to exist for f_{PKI}, where π is t-secure with identified abort. As in the proof of Theorem 10, if π ends with abort then the party who is identified as corrupted is removed (unless the dealer is identified as corrupted, in which case all parties just output 0 and halt). This continues iteratively until the π terminates without abort, in which case a valid PKI is established between all remaining parties. Given this PKI, the parties can run authenticated broadcast in order to securely compute f_{bc}. Since f_{bc} cannot be securely computed for $t = t/3$, we have a contradiction. ∎

6.2 Fairness Implies Guaranteed Output Delivery for Fail Stop Adversaries

In the presence of malicious adversaries, fairness and guaranteed output delivery are different notions, since there exist functionalities that can be computed with complete fairness but cannot be computed with guaranteed output delivery. In the presence of semi-honest adversaries, it is immediate that both notions are equivalent, since the adversary cannot abort. In this section, we show that in the presence of the fail-stop adversaries, i.e., when the corrupted parties follow the protocol with the exception that the adversary is allowed to abort, fairness implies guaranteed output delivery.

The underlying idea is that if a corrupted party does not send a message to an honest party during the execution of a fair protocol, the honest party can inform all parties that it identified a corrupted party. Since the adversary is fail-stop, corrupted parties cannot lie and falsely incriminate an honest party. Similarly to the proof of Theorem 11, the parties do not halt if a party is detected cheating (i.e., halting early). Rather, the parties continue to the end of the protocol: if the protocol ended with output then they take the output and halt; otherwise, they remove the cheating party and begin again. Since the original protocol is fair, this guarantees that nothing is learned by any party if anyone receives abort; thus, they can safely run the protocol again. As in the proof of Theorem 10, this process is repeated iteratively until no abort is received. We conclude that:

Theorem 12. *Let f be a, n-ary functionality and let $t \leq n$. Then, f can be t-securely computed with fairness in the presence of fail-stop adversaries, if and only if f can be t-securely computed with guaranteed output delivery in the presence of fail-stop adversaries.*

References

1. Asharov, G.: Towards characterizing complete fairness in secure two-party computation. In: Lindell, Y. (ed.) TCC 2014. LNCS, vol. 8349, pp. 291–316. Springer, Heidelberg (2014)
2. Aumann, Y., Lindell, Y.: Security against covert adversaries: Efficient protocols for realistic adversaries. In: Vadhan, S.P. (ed.) TCC 2007. LNCS, vol. 4392, pp. 137–156. Springer, Heidelberg (2007)
3. Canetti, R.: Security and composition of multiparty cryptographic protocols. J. Cryptology 13(1), 143–202 (2000)
4. Cleve, R.: Limits on the security of coin flips when half the processors are faulty (extended abstract). In: STOC, pp. 364–369 (1986)
5. Cohen, R., Lindell, Y.: Fairness versus guaranteed output delivery in secure multiparty computation. Cryptology ePrint Archive, Report 2014/668 (2014), http://eprint.iacr.org/
6. Fitzi, M., Gottesman, D., Hirt, M., Holenstein, T., Smith, A.: Detectable byzantine agreement secure against faulty majorities. In: PODC, pp. 118–126 (2002)
7. Goldreich, O.: The Foundations of Cryptography. Basic Applications, vol. 2. Cambridge University Press (2004)
8. Goldreich, O., Micali, S., Wigderson, A.: How to play any mental game or a completeness theorem for protocols with honest majority. In: STOC, pp. 218–229 (1987)
9. Goldwasser, S., Lindell, Y.: Secure multi-party computation without agreement. Journal of Cryptology 18(3), 247–287 (2005)
10. Gordon, S.D., Katz, J.: Complete fairness in multi-party computation without an honest majority. In: Reingold, O. (ed.) TCC 2009. LNCS, vol. 5444, pp. 19–35. Springer, Heidelberg (2009)
11. Gordon, S.D., Hazay, C., Katz, J., Lindell, Y.: Complete fairness in secure two-party computation. In: STOC, pp. 413–422 (2008)
12. Kilian, J.: A general completeness theorem for two-party games. In: STOC, pp. 553–560 (1991)
13. Lamport, L., Shostak, R.E., Pease, M.C.: The byzantine generals problem. ACM Trans. Program. Lang. Syst. 4(3), 382–401 (1982)
14. Pease, M.C., Shostak, R.E., Lamport, L.: Reaching agreement in the presence of faults. J. ACM 27(2), 228–234 (1980)
15. Pfitzmann, B., Waidner, M.: Unconditional byzantine agreement for any number of faulty processors. In: Finkel, A., Jantzen, M. (eds.) STACS 1992. LNCS, vol. 577, pp. 339–350. Springer, Heidelberg (1992)
16. Rabin, T., Ben-Or, M.: Verifiable secret sharing and multiparty protocols with honest majority (extended abstract). In: STOC, pp. 73–85 (1989)

Actively Secure Private Function Evaluation

Payman Mohassel[1,2], Saeed Sadeghian[1], and Nigel P. Smart[3]

[1] Dept. Computer Science, University of Calgary
{pmohasse,sadeghis}@ucalgary.ca
[2] Yahoo Labs
pmohassel@yahoo-inc.com
[3] Dept. Computer Science, University of Bristol
nigel@cs.bris.ac.uk

Abstract. We propose the first general framework for designing actively secure private function evaluation (PFE), not based on universal circuits. Our framework is naturally divided into pre-processing and online stages and can be instantiated using any generic actively secure multiparty computation (MPC) protocol.

Our framework helps address the main open questions about efficiency of actively secure PFE. On the theoretical side, our framework yields the first actively secure PFE with linear complexity in the circuit size. On the practical side, we obtain the first actively secure PFE for arithmetic circuits with $O(g \cdot \log g)$ complexity where g is the circuit size. The best previous construction (of practical interest) is based on an arithmetic universal circuit and has complexity $O(g^5)$.

We also introduce the first linear Zero-Knowledge proof of correctness of "extended permutation" of ciphertexts (a generalization of ZK proof of correct shuffles) which maybe of independent interest.

Keywords: Secure Multi-Party Computation, Private Function Evaluation, Malicious Adversary, Zero-Knowledge Proof of Shuffle.

1 Introduction

Private Function Evaluation (PFE) is a special case of Multi-Party Computation (MPC), where the parties compute a function which is a private input of one of the parties, say party P_1. The key additional security requirement is that all that should leak about the function to an adversary, who does not control P_1, is the size of the circuit (i.e. the number of gates and distinct wires within the circuit). Clearly, PFE follows immediately from MPC by designing an MPC functionality which implements a universal machine/circuit; thus the only open questions in PFE research are those of efficiency. Using universal circuits one can achieve complexity of $O(g^5)$ in case of arithmetic circuits [23] and $O(g \cdot \log g)$ for boolean circuits [26]. For ease of exposition we ignore the factors depending on the number of parties and the security parameters as they depend on the particular underlying MPC being used. We still provide some numbers for the specific SPDZ instantiation in section 5.

P. Sarkar and T. Iwata (Eds.): ASIACRYPT 2014, PART II, LNCS 8874, pp. 486–505, 2014.

A number of previous work [1,2,4,12,14,15,16,17,22,24] have considered the design and implementation of more efficient general- and special-purpose private function evaluation. A major motivation behind these solutions (and PFE in general) is to hide the function being computed since it is proprietary, private or contains sensitive information. Some applications of interest considered in the literature are software diagnostic [4], medical applications [2], and intrusion detection systems [20].

But all prior solutions are in the semi-honest model and fail in the presence of an active adversary who does not follow the steps of the protocol (with the exception of the generic approach of applying an actively secure MPC to universal circuits). For example, a malicious party who does not own the function can cheat to learn the proprietary function or modify the outcome of computation without the function-holders' knowledge. Or a malicious function-holder, can learn information about honest parties' inputs.

One may question the need for actively secure PFE as the function-holder can cheat and use a malicious function, which reveals information about the other party's input. While we consider the general scenario in our protocols, there are common practical scenarios where the function-holder has no output in the computation, and therefore maliciously changing the function still does not let him learn anything even if he is actively cheating.

1.1 Our Contribution

In this work, we present the first general framework for designing actively secure PFE, not based on universal circuits. Our framework can be instantiated upon a generic actively secure MPC protocol satisfying quite general properties; namely that they are secret sharing based, actively secure (either robust or with aborts), can implement reactive functionalities, and have an ability to open various sharings securely, as well as generate (efficiently) sharings of random values. Suitable actively secure MPC protocols include BDOZ [3] and SPDZ [8] (for the case of arithmetic circuits and an arbitrary number of players with a dishonest majority), Tiny-OT [19] (for binary circuits and two players), or protocols such as that implemented in VIFF [7] utilizing Shamir secret sharing with a threshold of $t < n/3$.

Our framework helps address the main open questions about efficiency of actively secure PFE. On a theoretical note, we use it to show that actively secure PFE with linear complexity (in circuit size) is indeed feasible while avoiding strong primitives such as fully-homomorphic encryption (FHE).[1] On a practical note, we obtain a practical actively secure PFE for arithmetic circuit with $O(g \cdot \log g)$ complexity (a significant reduction from $O(g^5)$ [23]), and the first actively secure PFE in the information-theoretic setting.

---

[1] Note that with the use of the right circuit-private FHE scheme [21], and appropriate ZK proofs for correctness of the computation on encrypted data, it is likely possible to achieve linear PFE based on FHE, but we are interested in the use of much weaker primitives such as singly homomorphic encryption.

Our Framework. Our framework can be seen as an extension of the new framework of [17] which is only secure against passive adversaries. The key idea in [17] is to divide the problem into two sub-problems, the problem of hiding the topology of the wiring between individual gates (topology hiding), and the problem of hiding exactly what gate is evaluated (gate hiding), i.e. an addition or a multiplication (or AND/OR/XOR in case of boolean circuits).

This framework yields better asymptotic and practical efficiency for passively secure PFE compared to the universal circuit approach (see [17] for a detailed efficiency comparison). An important open question is then how to extend their solution to the case of active adversaries efficiently. In this paper we do exactly that by providing a recipe for turning any actively secure MPC protocol that satisfies our general requirements into an actively secure PFE protocol.

Our framework operates in two phases, an offline phase and an online phase. As in the case of standard MPC in the pre-processing model, our offline phase is input independent but it depends on the function. The offline phase is use-once, in the sense that the data produced cannot be reused for multiple invocations of the online phase. We note that a similar function-dependent pre-processing model (referred to as *dedicated pre-processing*) was recently considered in [9]. Dedicated pre-processing is particularly natural in PFE applications where the sensitive/proprietary function stays fixed for a period of time and is used in multiple executions (clearly in the latter case we need to execute the pre-processing multiple times, but this can be done in advance). Of course, if one is not willing to count a function-dependent offline phase as valid, then our complexities would be the combination of the two phases. It maybe the case that our underlying MPC protocol is itself in the pre-processing model (e.g. [3,8,19]), in which case that pre-processing will be essentially independent of the input and function being evaluated. Our framework shows the feasibility of offline computation independent of inputs, which was not the case in [17]. We elaborate on the two phases next:

Offline Phase. Roughly speaking, our offline phase generates two vectors of random values, *maps* the second to a new vector using a mapping that captures the topology of the circuit (referred to as extended permutation in [17]), and subtracts the result from the first. The result of the subtraction (difference vector) is opened while the two original vectors are shared among the parties. The two random vectors are used as one-time pads of all the intermediate values in the circuit, while the "difference vector" is used by the function-holder to connect the output of one gate to the input of another without learning the values or revealing the circuit topology. The offline phase also generates one-time MACs of all the components of the "difference vector" computed above, using a fixed global MAC key. These MACs are used to check the function-holder's work in the online phase of the protocol. These steps commit P_1 privately to the topology of the circuit. We also privately commit P_1 to gate types, hence fully committing him to the function being computed.

Online Phase. Our online, or circuit evaluation, phase is very distinct from that deployed in the underlying MPC protocol we use. In existing instantiations of our underlying MPC protocol, parties evaluate gates on values whose secrecy is maintained due to the fact that one is working on secret shared values only. In our protocol the parties have public one-time pad encryptions of the values being computed on, but the encryption keys, which are the random values generated in the offline phase, remain secret-shared. Party P_1 (the function holder) then uses the random vectors computed in the offline phase to transform the encrypted output of one gate to the encrypted input of the upcoming gate while maintaining one-time MACs of all the values he computes. These MACs allow all other parties to check P_1's work without learning the circuit topology. These operations are carried out securely using the underlying MPC protocol.

In both the online and the offline phase, all parties check P_1's work by checking the MACs of the values he computes locally. If any of the MACs fail, in case of security with abort, parties can simply end the protocol. But in case of robust MPC (e.g. $t < n/3$ for robust information theoretically secure protocols) the protocol needs to continue without P_1. To achieve this, honest parties jointly recover P_1's function and play his role in the remainder of the protocol.

In our protocols, if any adversary deviates from the protocol then, except with negligible probability, the honest parties will either abort, or be able to recover from the introduced error. The exact response depends on the underlying MPC protocol on which our PFE protocol is built. In all cases the privacy of the honest players inputs is preserved, bar what can be obtained from the output of the private function chosen by player P_1. Note that P_1 may or may not be a recipient of output, but many application of PFE are concerned with scenarios where the function-holder has no output.

Efficient Instantiations. One can efficiently instantiate our online phase with a linear complexity, using any actively secure MPC satisfying our requirements. The main challenge, therefore, lies in efficient instantiation of the offline phase. It is possible to implement our offline phase using any actively secure MPC sub-protocol as well (by securely computing a circuit that performs the above mentioned task) but the resulting constructions would neither be linear nor constant-round.

- We introduce a instantiation with $O(g)$ complexity, proving the feasibility of linear actively secure PFE for the first time. Our main new technical ingredient is a linear zero-knowledge (ZK) proof of "correct extended permutation" of ElGamal ciphertexts. While linear ZK proofs of shuffles are well-studied, it is not clear how to extend the techniques to extended permutation (see our incomplete attempt in the full version [18]) Instead, we propose a generic and linear solution that uses ZK proof of a correct shuffle in a black-box manner, and may be of independent interest. Our solution is based on the switching network construction of EP [17]. This construction consists of three components, two of which are permutation networks. Instead of evaluating switches, we use singly homomorphic encryption to evaluate each component, and then re-randomize. We use existing ZK proofs of shuffle to prove the correctness of first and third components which perform permutation. The middle component requires a separate compilation of ZK protocols. Note that generically applying ZK proofs to UC circuit evaluation does not provide a linear solution, and applying ZK proofs for the EP component also does not work. Our customized linear $\mathcal{ZK}_{EP}$ gets around these problems.

– We introduce a *constant-round* instantiation with $O(g \cdot \log g)$ complexity (contrast with $O(g^5)$ complexity for universal arithmetic circuits) that is also of practical interest. Our technique is itself an extension of ideas from [17]. In particular the basic algorithm is that of [17] for oblivious evaluation of a switching network, but some care needs to be taken to make sure the protocol is actively secure. This is done by applying MACs to the data being computed on. However, instead of having the MAC values being secret shared (as in SPDZ) or kept secret (as in BDOZ and Tiny-OT), the MAC values are public with the keys remaining secret shared. Nevertheless, the MACs used are very similar to those used in the BDOZ and Tiny-OT protocols [3,19], since they are two-key MACs in which one key is a per message key and one is a global key. While using MAC's is quite standard for ensuring consistency of data, our efficient deployment in the framework is non-trivial and novel. For example, while addition of MACs in the offline phase is done using a generic MPC, the circuit evaluation (online phase) does not use an MPC. This is different from [17]'s approach and previous MPC work. General active security techniques can not be directly employed in this context. It is not clear how to use cut-and-choose in case of PFE, e.g. it is not clear how not to reveal the function in the opening, and there are additional components (i.e. EP) in a PFE protocol which cut-and-choose does not seem to resolve.

Efficiency Discussion. We emphasize that our linear complexity solution is a feasibility result at it was an open question whether active PFE with linear complexity in circuit size is possible given simple crypto primitive such as singly homomorphic encryption (as opposed FHE). Our "efficient" arithmetic PFE only requires $O(g \log g)$ multiplication gates and it is a significant improvement in comparison with applying of arithmetic MPC to universal arithmetic circuit of size $O(g^5)$ [23]. If we apply active secure MPC for arithmetic circuits to this universal circuit the complexity cannot get better than $O(g^5)$. One can turn an arithmetic circuit into a boolean circuit and use Valiant's boolean UC [26] to obtain a PFE. But this is highly inefficient, and therefore we do not discuss this in detail.

2   Notation and the Underlying MPC Protocol

We assume our function f to be evaluated will eventually be given by player P_1 as an arithmetic circuit over a finite field $\mathbf{F}_p$; note p may not necessarily be prime. We let $\mathbf{g}(f)$ denote the number of gates in the circuit representing f.

For gates with fan-out greater than one, we count each seperate output wire as a different wire. We also select a value k such that $p^k > 2^{\mathsf{sec}}$, where sec is the security parameter; this is to ensure security of our MAC checking procedure in the online phase.

We assume n parties $P_1, \ldots, P_n$, of which an adversary may corrupt (statically) up to t of them; the value of t being dependent on the specific underlying MPC protocol. The corrupted adversaries could include party P_1. The MPC protocol should implement the functionality described in Figure 1. This functionality is slightly different from standard MPC functionalities in that we try to capture both the honest majority and the dishonest majority setting; and in the latter setting the adversary can force the functionality to abort at any stage of the computation and not just the output. We also introduce another operation called **Cheat** which will be useful in what follows.

It is clear that modern actively secure MPC protocols such as [7,8,19], implement this functionality in different settings. Thus various different settings (i.e. different values of n, p and t) will be able to be dealt with in our resulting PFE protocol by simply plugging in a different underlying MPC protocol. To ease exposition later we express our MPC protocol as evaluating functions in the finite field $\mathbf{F}_{p^k}$. Clearly such an MPC protocol can be built out of one which evaluates functions over the base finite field $\mathbf{F}_p$.

To ease notation in what follows we shall let $[varid]$ denote the value stored by the functionality under $(varid, a)$; and will write $[z] = [x] + [y]$ as a shorthand for calling **Add** and $[z] = [x] \cdot [y]$ as a shorthand for calling **Multiply**. And by abuse of notation we will let $varid$ denote the value, x, of the data item held in location $(varid, x)$.

3   Our Active PFE Framework

In this section we describe our active PFE framework in detail. We start by describing the offline functionality which pre-processes the function/circuit the parties want to compute (Section 3.1). Then, in Section 3.2, we show that given a secure implementation of $\mathcal{F}_{\mathrm{OFFLINE}}$, one can efficiently (linear complexity) construct an actively secure PFE based on any actively secure MPC. We postpone efficient instantiations of $\mathcal{F}_{\mathrm{OFFLINE}}$ to later sections.

3.1   The Function Pre-processing (Offline) Phase

In this section we detail the requirements of our pre-processing step once player P_1 has decided on the function f to be evaluated. P_1 is only required to enter a valid circuit, equivalent to his function f into the protocol. Each non-output wire w in the circuit is connected at one end (which we shall call the *outgoing wire or left point*) to a source, this is either the output of a (non-output) gate or an input wire. Conversely each non-output wire is connected at the other end (which we shall call the *incoming wire or right point*) to a destination point which is always an input to a gate. We denote the number of distinct Incoming

---

Functionality $\mathcal{F}_{\text{MPC}}$

The functionality consists of seven externally exposed commands **Initialize**, **Cheat**, **Input Data**, **Random**, **Add**, **Multiply**, and **Output** and one internal subroutine **Wait**.

Initialize: On input $(init, p, k, flag)$ from all parties, the functionality activates and stores p and k; and a representation of $\mathbf{F}_{p^k}$. The value of $flag$ is assigned to the variable dhm, to signal whether the MPC functionality should operate in the dishonest majority setting. The set of "valid" players is initially set to all players. In what follows we denote the set of adversarial players by $\mathcal{A}$.

Cheat: This is a command which takes as input a player index i, it models the case of (most) robust MPC protocols in the honest majority case. On execution the functionality aborts if dhm is set to $true$. Otherwise the functionality waits for input from all players. If a majority of the players return OK then the functionality reveals all inputs made by player i, and player i is removed from the list of "valid" players (the functionality continues as if player i does not exist).

Wait: This does two things depending on the value of dhm.
  - If dhm is set to $true$ then it waits on the environment to return a $GO/NO\text{-}GO$ decision. If the environment returns $NO\text{-}GO$ then the functionality aborts.
  - If dhm is set to $false$ then it waits on the environment. The environment will either return GO, in which case it does nothing, or the environment returns a value $i \in \mathcal{A}$, in which case $\text{Cheat}(i)$ is called.

Input Data: On input $(input, P_i, varid, x)$ from P_i and $(input, P_i, varid, ?)$ from all other parties, with $varid$ a fresh identifier, the functionality stores $(varid, x)$. The functionality then calls **Wait**.

Random: On command $(random, varid)$ from all parties, with $varid$ a fresh identifier, the functionality selects a random value r in $\mathbf{F}_{p^k}$ and stores $(varid, r)$. The functionality then calls **Wait**.

Add: On command $(add, varid_1, varid_2, varid_3)$ from all parties (if $varid_1, varid_2$ are present in memory and $varid_3$ is not), the functionality retrieves $(varid_1, x)$, $(varid_2, y)$ and stores $(varid_3, x + y)$. The functionality then calls **Wait**.

Multiply: On input $(multiply, varid_1, varid_2, varid_3)$ from all parties (if $varid_1, varid_2$ are present in memory and $varid_3$ is not), the functionality retrieves $(varid_1, x)$, $(varid_2, y)$ and stores $(varid_3, x \cdot y)$. The functionality then calls **Wait**.

Output: On input $(output, varid)$ from all honest parties (if $varid$ is present in memory), the functionality retrieves $(varid, x)$ and outputs it to the environment. The functionality then calls **Wait**, and only if **Wait** does not abort then it outputs x to all players.

Fig. 1. The required ideal functionality for MPC

Wires on the right by $\mathsf{iw}(f)$. We let $\mathsf{ow}(f)$ denote the number of Outgoing Wires on the left. Note that $\mathsf{iw}(f) = 2g$ and $\mathsf{ow}(f) = n + g - o$ where o is the number of output gates in the circuit. Since we are dealing with arbitrary fan out we have that $\mathsf{ow}(f) \leq \mathsf{iw}(f)$.

Functionality $\mathcal{F}_{\text{OFFLINE}}$

Initialize: As for $\mathcal{F}_{\text{MPC}}$.
Wait: As for $\mathcal{F}_{\text{MPC}}$.
Input Data: As for $\mathcal{F}_{\text{MPC}}$.
Cheat: As for $\mathcal{F}_{\text{MPC}}$.
Random: As for $\mathcal{F}_{\text{MPC}}$.
Add: As for $\mathcal{F}_{\text{MPC}}$.
Multiply: As for $\mathcal{F}_{\text{MPC}}$.
Output: As for $\mathcal{F}_{\text{MPC}}$.
Input Function: On input $(inputfunction, \pi, f)$ from player P_1 the functionality performs the following operations
    – The functionality calls $(random, K)$.
    – If f is not a valid arithmetic circuit then the functionality aborts.
    – For $i \in \{1, \ldots, \mathsf{iw}(f)\}$ the functionality calls $(random, r_i)$ and $(random, s_i)$.
    – For $j \in \{1, \ldots, \mathsf{ow}(f)\}$ the functionality calls $(random, l_j)$ and $(random, t_j)$.
    – The functionality then computes, for all $i \in \{1, \ldots, \mathsf{iw}(f)\}$

$$[p_i] = [r_i] - [\ell_{\pi(i)}], \quad [q_i] = ([s_i] - [t_{\pi(i)}]) + ([r_i] - [\ell_{\pi(i)}]) \cdot [K]$$

    – The functionality then outputs (p_i, q_i) to all players, for $i \in \{1, \ldots, \mathsf{iw}(f)\}$, by calling $(output, p_i)$ and $(output, q_i)$.
    – For $i \in \{1, \ldots, g\}$ the functionality calls $(input, P_1, G_i, 0)$ if gate i in the description of f is an addition gate, and $(input, P_1, G_i, 1)$ if gate i is a multiplication gate.

Fig. 2. The required ideal functionality for the Offline Phase

To fully capture the topology of the circuit we give each outgoing wire and incoming wire in the circuit a unique label. The labels for the outgoing wires will be $\{1, \ldots, \mathsf{ow}(f)\}$ starting from the input wires and then moving to the output wires of each gate in a topological order decided by P_1, whilst the labels for the incoming wires will be $\{1, \ldots, \mathsf{iw}(f)\}$ labelling the input wires to each gate in the same topological order. The topology is then defined by a mapping from outgoing wires to incoming wires and is called an "extended permutation" in [17]. We denote the inverse of this mapping by a function π from $\{1, \ldots, \mathsf{iw}(f)\}$ onto $\{1, \ldots, \mathsf{ow}(f)\}$. If w is a wire in the circuit with incoming wire label i, then it's outgoing wire label is given by $j = \pi(i)$.

To execute the function pre-processing, player P_1 on input of f determines a mapping π corresponding to f. The offline phase functionality $\mathcal{F}_{\text{OFFLINE}}$ which is described in Figure 2, extends the $\mathcal{F}_{\text{MPC}}$ functionality of Figure 1 by adding an additional operation **Input Function**. The **Input Function** generates a vector of random (but correlated) values and their one-time MACs using a fixed global MAC key K. In particular, the functionality first stores a vector of random values (r_i) for each incoming wire and another vector of random values (ℓ_i) for the outgoing wires in the circuit. These random values will play the role of "pads" for one-time encryption of the computed wire values in the online phase. The functionality then computes p_i, the difference between each outgoing wire's value r_i and the corresponding incoming wires' value $\ell_{\pi(i)}$, and reveals p_i to all parties. This difference vector will allow P_1 to maintain one-time encryption of each wire value in the online phase without revealing the circuit topology. Additional random values (s_i, t_i) and the global MAC key K are used to compute one-time MACs of each p_i, namely q_i. These MACs will be used to check P_1's actions in the online phase. The **Input Function** also commits P_1 to the function of each gate in his circuit by storing a bit (0 for addition and 1 for multiplication) for each gate.

3.2   The Function Evaluation (Online) Phase

We can now present our framework for actively secure PFE. We wish to implement the functionality in Figure 3. We express the functionality as evaluating a function f provided by P_1 which takes as input n inputs in $\mathbf{F}_{p^k}$, one from each player. Again we present the functionality in both the honest majority and the dishonest majority settings.

Functionality $\mathcal{F}_{\text{ONLINE}}$

Initialize: On input $(init, p, k, flag)$ from all players, the functionality activates and stores p and k; and a representation of $\mathbf{F}_{p^k}$. The value of $flag$ is assigned to the variable dhm, to signal whether the underlying MPC functionality should operate in the dishonest majority setting.

Wait: If dhm is set to $false$ then this does nothing. Otherwise it waits on the environment to return a $GO/NO\text{-}GO$ decision. If the environment returns $NO\text{-}GO$ then the functionality aborts.

Input Function: On input $(inputfunction, f)$ from player P_1 the functionality stores $(function, f)$. The functionality now calls **Wait**.

Input Data: On input $(input, P_i, x_i)$ from player P_i the functionality stores $(input, i, x_i)$. The functionality now calls **Wait**.

Output: On input $(output)$ from all honest players the functionality retrieves the data x_i stored in $(input, i, x_i)$ for $i \in \{1, \ldots, n\}$ (if all do not exist then the functionality aborts). The functionality then retrieves f from $(function, f)$ and computes $y = f(x_1, \ldots, x_n)$ and outputs it to the environment (or aborts if $(function, f)$ has not been stored). The functionality now calls **Wait**. Only on a successful return from **Wait** will the functionality output y to all players.

Fig. 3. The required ideal functionality for PFE

Realizing $\mathcal{F}_{\text{Online}}$ Given $\mathcal{F}_{\text{Offline}}$ and $\mathcal{F}_{\text{MPC}}$. A generic instantiation of $\mathcal{F}_{\text{OFFLINE}}$ based on any MPC is give in Figure 5. The idea is to work with *one-time pad* encryptions of the values for all intermediate wires and the corresponding one-time MACs. Here, the pads (r, ℓ, s, t values), as well as the MAC Key K are generated by the offline functionality, and shared among the parties so no party can learn intermediate values or forge MACs on his own.

In more detail, the protocol proceeds as follows. Initially, parties compute one-time encryption of the input values to the circuit (pads are the corresponding ℓ values). Then, the following process is repeated for every gate in the circuit until every gate is processed. Parties then open the outcome of the output gates as their final result.

For each gate, party P_1 uses the "difference vectors" (p_i values) from the offline phase to transform the one-time encryption of output of the previous gate to the one-time encryption of input of the current gate (the result is denoted by d_{i_0}, d_{i_1} for the i-th gate.), without revealing the topology or learning the actual wire values. This is diagrammatically presented in Figure 4 to aid the reader. A similar transformation is done on MACs of the wire values (using q_i values) in order to keep P_1 honest in his computation (denoted by m_{i_0}, m_{i_1}).

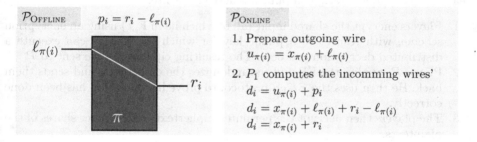

Fig. 4. Transformation of one-time encryption of an outgoing wire to the one-time encryption of an incoming wire using the values computes in $\mathcal{P}_{\text{OFFLINE}}$ protocol

Then, the protocol proceeds by jointly removing the one-time pads for the two inputs of the current gate and evaluating it together in order to compute a shared output z_i. Note that in this gate evaluation the gate type G_i is secret and shared among the players. This step can be performed using the $\mathcal{F}_{\text{MPC}}$ operations. Then, parties compute a one-time encryption of z_i using the corresponding ℓ value as the pad, and denote the result by u_j, just a relabeling where j is the outgoing wire's label of the output wire of the gate (note that $j = n + i$ since the outgoing wires are labeled starting with the n input wires and then the output wire of each gate).

Note, that if P_1 tries to deviate from the protocol in his local computation (i.e. when he connects outgoing wires to incoming wires) the generated MACs will not pass the jointly performed verifications and he will be caught. In that case, either the protocol aborts (in the case of dishonest majority) or his input (i.e. the function) is revealed (in the case of honest majority).

This leads to the following theorem, whose proof is given in full version [18].

Theorem 1. *In the $\mathcal{F}_{\text{OFFLINE}}$-hybrid model the protocol in Figure 5 securely implements the PFE functionality in Figure 3, with complexity $O(g)$.*

4   Implementing $\mathcal{F}_{\text{Offline}}$ with Linear Complexity

In this section we give a linear instantiation of the offline phase of the framework. Since our online phase has linear complexity, a linear offline phase implementation leads to a linear actively secure PFE. The main challenge in obtaining a linear solution is to design a linear method for applying the extended permutation π to values $\{[\ell_i]\}$ and $\{[t_i]\}$ to produce shared values $\{[\ell_{\pi(i)}]\}$ and $\{[t_{\pi(i)}]\}$. In the semi-honest case [17], linear complexity solution for this problem is achieved by employing a singly homomorphic encryption. The shared values are jointly encrypted; P_1 applies the extended permutation to the resulting ciphertexts and re-randomizes them in order to hide π; parties jointly decrypt in order to obtain the shares of the resulting plaintexts. To obtain active security, we need to make each step of the following computation actively secure:

1. Players encrypt the shared input (all of which lie in $\mathbf{F}_{p^k}$) using an encryption scheme, with respect to a public key for which the players can execute a distributed decryption protocol. The resulting ciphertexts are sent to P_1.
2. Player P_1 applies the EP and re-randomizes the ciphertexts and sends them back. He then uses the $\mathcal{ZK}_{\text{EP}}$ protocol to prove his operation has been done correctly.
3. The players then decrypt the permuted ciphertexts and recover shares of the plaintexts.

To implement the first and last steps we use an an instantiation based on El-Gamal encryption, see full version [18]. The middle step is more tricky, and we devote the rest of this section to describing this. For the middle step we need a linear zero-knowledge protocol to prove that P_1 applied a valid EP to the ciphertexts. Proof of a correct shuffle is a well studied problem in the context of Mix-Nets, and linear solutions for it exist [11]. As discussed in full version[18], however, extending these linear proofs to the case of extended permutations faces some subtle difficulties which we leave as an open question. Instead we aim for a more general construction that uses the currently available proofs of shuffling, in a black-box way.

4.1   Linear $\mathcal{ZK}_{\text{EP}}$ Protocol

After players compute the encryption of the shared inputs, P_1 knowing the circuit topology, applies the corresponding extended permutation to the ciphertexts. He then re-randomizes the ciphertexts and then "opens" the ciphertexts.

Protocol $\mathcal{P}_{\text{ONLINE}}$

The protocol is described in the $\mathcal{F}_{\text{OFFLINE}}$-hybrid model.

Input Function: Player P_1 given f selects the switching network mapping π and then calls $(inputfunction, \pi, f)$ on the functionality $\mathcal{F}_{\text{OFFLINE}}$.

Input Data: On input $(input, P_i, x_i)$ from player P_i the protocol executes the $(input, i, x_i)$ operation of the functionality $\mathcal{F}_{\text{OFFLINE}}$.

Output: The evaluation of the function proceeds as follows; where for ease of exposition we set $x_{\pi(h)} = y_h$ for all h, i.e. if a wire has input x_i on the left (as outgoing wire) then it has the same value y_h on the right (as incoming wire) where $i = \pi(h)$

- **Preparing Inputs to the Circuit:**
    - For each input wire i ($1 \le i \le n$) the players execute $[u_i] = [x_i] + [\ell_i]$, where i is the outgoing wire's label corresponding to that input wire, and $[v_i] = [t_i] + ([x_i] + [\ell_i]) \cdot [K]$ using the $\mathcal{F}_{\text{MPC}}$ functionality available via $\mathcal{F}_{\text{OFFLINE}}$.
    - Parties then call $(output, u_i)$ and $(output, v_i)$ to open $[u_i]$ and $[v_i]$.
- **Evaluating the Circuit:** For every gate $1 \le i \le g$ in the circuit players execute the following (here we assume that the gates are indexed in the same topological order P_1 chose to determine π):
    - P_1 **Prepares the Two Inputs for Gate i.**
        * Note that the two input wires for gate i have incoming wire labels $i_0 = 2i - 1$ and $i_1 = 2i$, and the (u, v) value for their corresponding outgoing wire labels are already determined, i.e. $u_{\pi(i_j)}$ and $v_{\pi(i_j)}$ are already opened for $j \in \{0, 1\}$.
        * Player P_1 computes, for $j = 0, 1$,
        $$d_{i_j} = u_{\pi(i_j)} + p_{i_j} \doteq (y_{i_j} + \ell_{\pi(i_j)}) + (r_{i_j} - \ell_{\pi(i_j)})$$
        $$\doteq y_{i_j} + r_{i_j},$$
        $$m_{i_j} = v_{\pi(i_j)} + q_{i_j} \doteq (t_{\pi(i_j)} + (y_{i_j} + \ell_{\pi(i_j)}) \cdot K)$$
        $$+ ((s_{i_j} - t_{\pi(i_j)}) + (r_{i_j} - \ell_{\pi(i_j)})) \cdot K)$$
        $$\doteq s_{i_j} + (y_{i_j} + r_{i_j}) \cdot K.$$
        * Player P_1 then broadcasts the values d_{i_j} and m_{i_j} to all players.
    - **Players Check P_1's Input Preparation.**
        * All players then use the $\mathcal{F}_{\text{MPC}}$ operations available (via the interface to the $\mathcal{F}_{\text{OFFLINE}}$ functionality) so as to store in the $\mathcal{F}_{\text{MPC}}$ functionality the values $[n_{i_j}] = [s_{i_j}] + (y_{i_j} + r_{i_j}) \cdot [K]$. The value is then opened to all players by calling $(Output, n_{i_j})$.
        * If $n_{i_j} \ne m_{i_j}$ then the players call **Cheat**(1) on the $\mathcal{F}_{\text{MPC}}$ functionality. This will either abort, or return the input of P_1 (and hence the function), in the latter case the players can now proceed with evaluating the function using standard MPC and without the need for P_1 to be involved.
    - **Players Jointly Evaluate Gate i.**
        * The players store the value $[y_{i_j}] = d_{i_j} - [r_{i_j}]$ in the $\mathcal{F}_{\text{MPC}}$ functionality.
        * The $\mathcal{F}_{\text{MPC}}$ functionality is then executed so as to compute the output of the gate as
        $$[z_i] = (1 - [G_i]) \cdot ([y_{i_0}] + [y_{i_1}]) + [G_i] \cdot [y_{i_0}] \cdot [y_{i_1}].$$
        * Note that the outgoing wire label corresponding to the output wire of the ith gate is $j = n + i$ so we just relabel $[z_i]$ to $[z_j]$.
        * If G_i is an output gate, players call $(Output, z_i)$ to obtain z_i, disregard next steps and continue to evaluate next gate.
        * The players compute via the MPC functionality $[u_j] = [z_j] + [\ell_j]$.
        * The players call $(Output, u_j)$ so as to obtain u_j.
        * The players then compute via the MPC functionality
        $$[v_j] = [t_j] + u_j \cdot [K] \doteq [t_j + (z_j + \ell_j) \cdot K].$$
        * The players call $(Output, v_j)$ so as to obtain v_j.

Fig. 5. The Protocol for implementing PFE

Next, we give a linear zero-knowledge protocol $\mathcal{ZK}_{EP}$, which enables P_1 to prove the correctness of his operation (i.e final ciphertexts are the result of P_1 applying a valid EP to the input ciphertexts). As our first attempt we considered the possibility of extending existing linear proofs of shuffle to get linear proofs of extended permutation. While plausible there are subtle difficulties that need to be addressed. For more details regarding our attempt on extending the method of Furukawa [11,10], refer to full version[18]. We leave this approach as an open problem. Instead we give a more general construction which makes black-box calls to proof of shuffle. This construction is inspired by the switching network construction of EP given in [17]. We first revisit the extended permutation construction of [17].

Assume the EP mapping represented by the function: $\pi : \{1...n\} \to \{1...m\}$ (Which maps m input wires to n output wires ($n \geq m$)). Note that in this section we use n and m to denote the size of EP. In a switching network, the number of inputs and outputs are the same, therefore, the construction takes m real inputs of the EP and $n - m$ additional *dummy inputs*. The construction is divided into three components. Each component takes the output of the previous one as input. Instead of applying the EP in one step, P_1 applies each component separately and uses a zero-knowledge protocol to prove its correctness. Figure 6 demonstrates the components. Next, we describe each component and identify the required ZK proof.

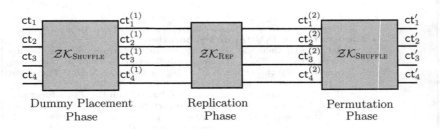

Fig. 6. EP construction. Components' names are written underneath. The zero-knowledge protocol for each component is written inside it's component box.

Table 1 lists the zero-knowledge protocols that we make a black-box use in our $\mathcal{ZK}_{EP}$ protocol. Note that we use P and Q for our EC instantiation instead of g and h.

- **Dummy-Value Placement Component:** This takes the real and dummy ciphertexts as input and for each ciphertexts of a real value that is mapped to k different outputs according to π, outputs the real ciphertexts followed by $k - 1$ dummy ciphertexts. This is repeated for each real ciphertext. The resulting output ciphertexts are all re-randomized. The dummy replacement step can be seen as a shuffling of the input ciphertexts. We use a proof of correct shuffle, $\mathcal{ZK}_{SHUFFLE}$, for correctness of this component.

Table 1. List of zero-knowledge protocols used in our $\mathcal{ZK}_{\text{EP}}$ protocol. Generator g and public key $h = g^{sk}$.

$\mathcal{ZK}$ Protocol	Relation/Language	Ref.
$\mathcal{ZK}_{\text{SHUFFLE}}(\{ct_i\}, \{ct'_i\})$	$\mathcal{R}_{\text{SHUFFLE}} = \{(G, g, h, \{ct_i\}, \{ct'_i\}) \mid \exists \pi, \text{st.}$ $C_1^{\prime (i)} = g^{r_i} C_1^{(\pi(i))} \wedge C_2^{\prime (i)} = h^{r_i} C_2^{(\pi(i))} \wedge \pi \text{ is perm.}\}$	[11]
$\mathcal{ZK}_{\text{EQ}}(ct_1, ct_2)$	$\mathcal{R}_{\text{EQ}} = \{(G, g, h, ct_i = \langle \alpha_i, \beta_i \rangle_{i \in \{1,2\}}) \mid \exists (m_1, m_2), \text{st.}$ $\alpha_i = g^{r_i} \wedge \beta_i = m_i h^{r_i} \wedge m_1 = m_2\}$	[5]
$\mathcal{ZK}_{\text{NO}}(ct)$	$\mathcal{L}_{\text{NO}} = \{(G, g, h, ct = \langle \alpha, \beta \rangle) \mid \exists (m_1 \neq 1), \text{st.}$ $\alpha = g^r \wedge \beta = m_1 h^r\}$	[13]

- **Replication Component:** This takes the output of the previous component as input. It directly outputs each real ciphertext but replaces each dummy ciphertext with an encryption of the real input that precedes it. At the end of this step, we have the necessary copies for each real input and the dummy inputs are eliminated. Naturally, all the ciphertexts are re-randomized. To prove correctness of this step, we need ZK proofs that the i-th output ciphertext has a plaintext equal to that of either the i-th input ciphertext or $(i - 1)$-th output ciphertext (these can be achieved using protocol $\mathcal{ZK}_{\text{EQ}}$ defined in Table 1 as a building block). But this is not sufficient to guarantee a correct EP, as we also have to make sure that after the replication component there are no dummy ciphertexts left. For this, we assume that all dummy ciphertexts are encryptions of one. Then for each output ciphertext in the replication component we use a protocol $\mathcal{ZK}_{\text{NO}}$, i.e. a ZK proof that the underlying plaintext is not one. The $\mathcal{ZK}_{\text{REP}}$ zero-knowledge protocol, is a compilation of three ZK protocols, two checking for equality of ciphertexts and one checking the inequality of plaintext to one.
- **Permutation Component:** This takes the output of the replication component as input and permutes each element to its final location as prescribed by π. We again use the proof of correct shuffle, $\mathcal{ZK}_{\text{SHUFFLE}}$. for this component.

$\mathcal{ZK}_{\text{EP}}$ *Protocol Description.* We assumed the inputs to the $\mathcal{ZK}_{\text{EP}}$, to be the outputs of our encryption functionality. Prover applies the extended permutation to the ciphertexts $(ct_1, \ldots, ct_n)$, where $ct_i = (C_1^{(i)}, C_2^{(i)})$. The prover obtains a re-randomized $(ct'_1, \ldots, ct'_n)$, where $ct'_i = (C_1^{\prime (i)}, C_2^{\prime (i)})$. We employ the techniques of Cramer et al. [6], to combine HVZK proof systems corresponding to each component, at no extra cost, into HVZK proof systems of the same class for any (monotonic) disjunctive and/or conjunctive formula over statements proved in the component proof systems. Figure 7 shows the complete description of our $\mathcal{ZK}_{\text{EP}}$ protocol. Note that we can choose dummy values from any set of random values S_d and substitute the $\mathcal{ZK}_{\text{NO}}(x)$ with $\vee_{\forall y \in S_d}(\mathcal{ZK}_{\text{EQ}}(x, y))$.

Protocol $\mathcal{ZK}_{\mathrm{EP}}(\{ct_i\}, \{ct'_i\})$

Shared Input: Ciphertexts $(ct_1, \ldots, ct_n)$

P_1**'s Input:** Extended permutation π

P_1 **Evaluates the components.**

- Player P_1 finds the corresponding permutation π_1, and π_2 for Dummy-placement component and permutation components.
- P_1 applies the Dummy-placement component to $(ct_1, \ldots, ct_n)$, and re-randomizes to find $(ct_1^{(1)}, \ldots, ct_n^{(1)})$.
- P_1 applies the Replication component to $(ct_1^{(1)}, \ldots, ct_n^{(1)})$, and re-randomizes them to find $(ct_1^{(2)}, \ldots, ct_n^{(2)})$.
- P_1 applies the permutation component to $(ct_1^{(2)}, \ldots, ct_n^{(2)})$, and re-randomizes them to find $(ct'_1, \ldots, ct'_n)$.

P_1 **Computes the ZK proofs and sends everything**

- Player P_1 uses the $\mathcal{ZK}_{\mathrm{SHUFFLE}}(\{ct_i\}, \{ct_i^{(1)}\})$ and $\mathcal{ZK}_{\mathrm{SHUFFLE}}(\{ct_i^{(2)}\}, \{ct'_i\})$ protocols to produce proof of correctness for his evaluation of Dummy-placement component and permutation component.
- Player P_1 used the $\mathcal{ZK}_{\mathrm{REP}}(\{ct_i^{(1)}\}, \{ct_i^{(2)}\})$ to produce proof of correctness for his evaluation of Replication component as follows(using [6] for combination) (and $\mathcal{ZK}_{\mathrm{REP}}^1 = \mathcal{ZK}_{\mathrm{NO}}\left(ct_1^{(2)}\right) \wedge \mathcal{ZK}_{\mathrm{EQ}}(ct_1^{(1)}, ct_1^{(2)})$):

  • For $2 \leq i \leq n$:
  $$\mathcal{ZK}_{\mathrm{REP}}^i = \left(\mathcal{ZK}_{\mathrm{EQ}}(ct_i^{(1)}, ct_i^{(2)}) \vee \mathcal{ZK}_{\mathrm{EQ}}(ct_{i-1}^{(2)}, ct_i^{(2)})\right) \wedge \mathcal{ZK}_{\mathrm{NO}}\left(ct_i^{(2)}\right)$$

  • $\mathcal{ZK}_{\mathrm{REP}} = \wedge_{i=1,\ldots,n}(\mathcal{ZK}_{\mathrm{REP}}^i)$

- Player P_1 sends $(ct_1^{(1)}, \ldots, ct_n^{(1)})$, $(ct_1^{(2)}, \ldots, ct_n^{(2)})$, $(ct'_1, \ldots, ct'_n)$ and all proofs to other players.

Players verify P_1 operations

- Players verify P_1's operations by verifying the the proofs sent by P_1.

Fig. 7. The protocol for zero-knowledge proof of extended permutation

Theorem 2. *The protocol described in Figure 7 is HVZK proof of an extended permutation* π, $(\mathsf{ct}_1, \ldots, \mathsf{ct}_n)$ *and* $(\mathsf{ct}'_1, \ldots, \mathsf{ct}'_n)$ *in the* $\mathcal{ZK}_{\text{SHUFFLE}}$, $\mathcal{ZK}_{\text{EQ}}$, $\mathcal{ZK}_{\text{NO}}$ *hybrid model, for the following relation:*

$$\mathcal{R}_{\text{EP}} = \{(G, g, h, \{\mathsf{ct}_i\}, \{\mathsf{ct}'_i\}) | \exists \pi, st.$$
$$C_1'^{(i)} = g^{r_i} C_1^{(\pi(i))} \wedge C_2'^{(i)} = h^{r_i} C_2^{(\pi(i))} \wedge \pi \text{ is EP.}\}$$

Proof. Refer to the full version [18] for proof.

Offline Protocol. Having all the parts of the puzzle, we can give the complete $O(g)$ protocol for the offline phase. Figure 8 shows the description, with the proof of security given in full version [18].

Linear Implementation of Protocol $\mathcal{P}_{\text{OFFLINE}}$-Linear

The protocol is described in the $\mathcal{F}_{\text{MPC}}$-hybrid model, thus the only operation we need to specify is the **Input Function** one.

Input Function:

P_1 **Shares his Circuit/Function.**
- Player P_1 calls $(input, G_j)$ for all $j \in \{1, \ldots, g\}$.
- Players evaluate and open $[G_j] \cdot (1 - [G_j])$ for $j \in \{1, \ldots, g\}$. If any of them is not 0, players abort (since in this case P_1 has not entered a valid function).

Players Generate Randomness for inputs and outputs of EP.
- Players call $(random, \cdot)$ of $\mathcal{F}_{\text{MPC}}$ to generate shared random values for inputs $\boldsymbol{\ell} = ([\ell_1], \ldots, [\ell_{\text{ow}(f)}])$ and outputs $([r_1], \ldots, [r_{\text{iw}(f)}])$ of EP.
- Players call $(random, \cdot)$ of $\mathcal{F}_{\text{MPC}}$ to generate shared random values for the MAC value corresponding to inputs $\boldsymbol{t} = ([t_1], \ldots, [t_{\text{ow}(f)}])$ and outputs $([s_1], \ldots, [s_{\text{iw}(f)}])$ of EP.

P_1 **applies the EP to $\boldsymbol{\ell}$ and $\boldsymbol{t}$.**
- The players call KeyGen on the Enc_{Elg} functionality.
- The playes call Encrypt on the Enc_{Elg} functionality with the plaintexts $([\ell_1], \ldots, [\ell_{\text{ow}(f)}])$ and the plaintexts $([t_1], \ldots, [t_{\text{ow}(f)}])$, to obtain ciphertexts $\mathsf{ct}_1, \ldots, \mathsf{ct}_{\text{ow}(f)}$ and $\mathsf{ct}_1^{\dagger}, \ldots, \mathsf{ct}_{\text{ow}(f)}^{\dagger}$.
- Player P_1 applies the extended permutation to $(\mathsf{ct}_1, \ldots, \mathsf{ct}_{\text{ow}(f)})$ and re-randomize to get $(\mathsf{ct}'_1, \ldots, \mathsf{ct}'_{\text{ow}(f)})$, the same is done with $(\mathsf{ct}_1^{\dagger}, \ldots, \mathsf{ct}_{\text{ow}(f)}^{\dagger})$ to obtain $(\mathsf{ct}_1'^{\dagger}, \ldots, \mathsf{ct}_{\text{ow}(f)}'^{\dagger})$.
- Player P_1 uses the $\mathcal{ZK}_{\text{EP}}$ to prove that he has used a valid extended permutation.
- Players call the Decrypt on the Enc_{Elg} functionality (given in full version [18]) with ciphertexts $(\mathsf{ct}'_1, \ldots, \mathsf{ct}'_{\text{ow}(f)})$ and $(\mathsf{ct}_1'^{\dagger}, \ldots, \mathsf{ct}_{\text{ow}(f)}'^{\dagger})$ so as to obtain $([\ell_{\pi(1)}], \ldots, [\ell_{\pi(\text{ow}(f))}])$ and $([t_{\pi(1)}], \ldots, [t_{\pi(\text{ow}(f))}])$.

Players Compute p_i, q_i.
- For $i \in \{1, \ldots, \text{iw}(f)\}$ players call $\mathcal{F}_{\text{MPC}}$ to compute:

$$[p_i] = [r_i] - [\ell_{\pi(i)}] \doteq [r_i - \ell_{\pi(i)}], \quad [q_i] = [s_i] - [t_{\pi(i)}] + p_i \cdot [K] \doteq [s_i - t_{\pi(i)} + p_i \cdot K]$$

Fig. 8. The protocol for linear implementation of the Offline Phase

5   A Practical Implementation of $\mathcal{F}_{\text{Offline}}$ with $O(g \cdot \log g)$ Complexity

A $O(g \cdot \log g)$ protocol to implement $\mathcal{F}_{\text{OFFLINE}}$ is given in full version [18], and is in the $\mathcal{F}_{\text{MPC}}$-hybrid model. Following the ideas in [17], we implement the functionality via secure evaluation of a *switching network* corresponding to the mapping π_f.

Switching Networks. A switching network SN is a set of interconnected switches that takes N inputs and a set of selection bits, and outputs N values. Each *switch* in the network accepts two ℓ-bit strings as input and outputs two ℓ-bit strings. In this paper we need to use a switching network that contains two switch types. In the first type (*type 1*), if the selection bit is 0 the two inputs remain intact and are directly fed to the two outputs, but if the selection bit is 1, the two input values swap places. In the second type (*type 2*), if the selection bit is 0, as before, the inputs are directly fed to outputs but if it is 1, the value of the first input is used for both outputs. For ease of exposition, in our protocol description we assume that all switches are of type 1, but the protocol can be easily extended to work with both switch types.

The *mapping* $\pi : \{1 \ldots N\} \to \{1 \ldots N\}$ corresponding to a switching network SN is defined such that $\pi(j) = i$ if and only if after evaluation of SN on the N inputs, the value of the input wire i is assigned to the output wire j (assuming a standard numbering of the input/output wires). In [17] it is shown how to represent any mapping with a maximum of N inputs and outputs via a network with $O(N \cdot \log N)$ type 1 and 2 switches (We refer the reader to [17] for the details). This yields a switching network with $O(g \cdot \log g)$ switches to represent the mapping for a circuit with g gates.

High Level Description. It is possible to implement the $\mathcal{F}_{\text{OFFLINE}}$ by securely computing a circuit for the above switching network using the $\mathcal{F}_{\text{MPC}}$. But for all existing MPC that meet our requirements, this would require $O(\log g)$ rounds of interaction which is the depth of the circuit corresponding to the switching network. We show an alternative constant-round approach with similar computation and communication efficiency. It follows the same idea as the OT-based protocol of [17] where the OT is replaced with an equivalent functionality implemented using $\mathcal{F}_{\text{MPC}}$. The main challenge in our case is to achieve *active security* and in particular to ensure that P_1 cannot cheat in his local computation. We do so by checking P_1's actions using one-time MACs of the values he computes on, and allow the other parties to learn his input and proceed without him, if he is caught cheating (or aborting).

Next we give an overview of the protocol. The protocol has four main components (as described in full version [18]). In the first step, P_1 converts his mapping π to selection bits for the switching network (i.e. b_is) and shares them with all players. He also shares a bit G_i indicating the function of gate i, with other players. In the second step, players generate random values for every wire in the network. P_1, based on his selection bit for the switch, learns two of the four possible "subtractions" of the random values for two output wires from those of the input wires i.e. $u_0^{\ell,i}$ and $u_1^{\ell,i}$. A similar process is performed for the t values to obtain $u_0^{t,i}$ and $u_1^{t,i}$ (Figure 9 shows this process in a diagram). These subtractions enable P_1 to transform a pair of values blinded with the random values of input wires, to the same pair of values permuted (based on the selection bit) and blinded with the random values of the output wires. All of the above can be implemented using the operations provided by the $\mathcal{F}_{\text{MPC}}$.

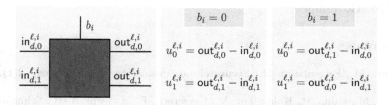

Fig. 9. The i-th switch. (superscripts: label of value subject to permute (ℓ or t), and switch index i) (subscripts: d refers to data, m refers to MAC, wire index 0 denotes the top wire in switch and 1 the bottom wire in switch).

In the third step, P_1 obtains the blinded ℓ and t values where the blinding for each is the random value for the corresponding input wire to the network (these are $h_d^{\ell,i}, h_d^{t,i}$, etc). Party P_1 can now process each switch as discussed above using the subtraction values in order to evaluate the entire network. At the end of this process, P_1 holds blinded values of the outputs of the switching network (blinded with randomness of the output wires).

In the final step, parties check that P_1 has not cheated during his evaluation, since he performed this step locally and not through the $\mathcal{F}_{\text{MPC}}$ operations. We use one-time MACs to achieve this goal. In particular, besides mapping blinded values through the network, P_1 also maps the corresponding one-time MACs (generated using the fixed-key K). This is done using a similar process described above and via the $v_j^{\ell,i}, v_j^{t,i}$ values. At the end of this process, P_1 holds one-time MACs for the blinded outputs of the switching network, in addition to the values themselves. Players then use the MPC functionality to jointly verify that the MACs indeed verify the values P_1 shared with them (i.e. $n^{\ell,i}$ and $m^{\ell,i}$ are the same, etc). As a result, P_1 can only cheat by forging the MACs which only happens with a negligible probability. If the MACs pass, parties compute and open the "difference vectors" by subtracting the mapped ℓ and t-value vectors from the r and s-value vectors. Refer to full version [18] for more details. If one instantiates the $\mathcal{F}_{\text{MPC}}$ by SPDZ [8], which has the $m.\log(p^k)$ complexity, then our complexity would be $m\left(10(2g\log 2g - 2g + 1) + 4g\right).\log(p^k)$. Refer to full version [18] for the proof of the following theorem.

Theorem 3. *In the $\mathcal{F}_{\text{MPC}}$-hybrid model the protocol $\mathcal{P}_{\text{OFFLINE}}$ in full version [18] securely implements the functionality in Figure 2, with complexity $O(g \cdot \log g)$.*

Acknowledgements. This work has been supported in part by ERC Advanced Grant ERC-2010-AdG-267188-CRIPTO, by EPSRC via grant EP/I03126X, and by Defense Advanced Research Projects Agency (DARPA) and the Air Force Research Laboratory (AFRL) under agreement number FA8750-11-2-0079. The US Government is authorized to reproduce and distribute reprints for Government purposes notwithstanding any copyright notation thereon. The views and conclusions contained herein are those of the authors and should not be interpreted as necessarily representing the official policies or endorsements, either expressed or implied, of Defense Advanced Research Projects Agency (DARPA) or the U.S. Government.

References

1. Abadi, M., Feigenbaum, J.: Secure circuit evaluation. J. Cryptology 2(1), 1–12 (1990)
2. Barni, M., Failla, P., Kolesnikov, V., Lazzeretti, R., Sadeghi, A.-R., Schneider, T.: Secure evaluation of private linear branching programs with medical applications. In: Backes, M., Ning, P. (eds.) ESORICS 2009. LNCS, vol. 5789, pp. 424–439. Springer, Heidelberg (2009)
3. Bendlin, R., Damgård, I., Orlandi, C., Zakarias, S.: Semi-homomorphic encryption and multiparty computation. In: Paterson, K.G. (ed.) EUROCRYPT 2011. LNCS, vol. 6632, pp. 169–188. Springer, Heidelberg (2011)
4. Brickell, J., Porter, D.E., Shmatikov, V., Witchel, E.: Privacy-preserving remote diagnostics. In: Ning, P., di Vimercati, S.D.C., Syverson, P.F. (eds.) ACM Conference on Computer and Communications Security, pp. 498–507. ACM (2007)
5. Chaum, D., Pedersen, T.P.: Wallet databases with observers. In: Brickell, E.F. (ed.) CRYPTO 1992. LNCS, vol. 740, pp. 89–105. Springer, Heidelberg (1993)
6. Cramer, R., Damgård, I., Schoenmakers, B.: Proof of partial knowledge and simplified design of witness hiding protocols. In: Desmedt, Y.G. (ed.) CRYPTO 1994. LNCS, vol. 839, pp. 174–187. Springer, Heidelberg (1994)
7. Damgård, I., Geisler, M., Krøigaard, M., Nielsen, J.B.: Asynchronous multiparty computation: Theory and implementation. In: Jarecki, S., Tsudik, G. (eds.) PKC 2009. LNCS, vol. 5443, pp. 160–179. Springer, Heidelberg (2009)
8. Damgård, I., Pastro, V., Smart, N.P., Zakarias, S.: Multiparty computation from somewhat homomorphic encryption. In: Safavi-Naini, Canetti (eds.) [25], pp. 643–662.
9. Damgård, I., Zakarias, S.: Constant-overhead secure computation of boolean circuits using preprocessing. In: Sahai, A. (ed.) TCC 2013. LNCS, vol. 7785, pp. 621–641. Springer, Heidelberg (2013)
10. Furukawa, J.: Efficient and verifiable shuffling and shuffle-decryption. IEICE Transactions 88-A(1), 172–188 (2005)
11. Furukawa, J., Sako, K.: An efficient scheme for proving a shuffle. In: Kilian, J. (ed.) CRYPTO 2001. LNCS, vol. 2139, pp. 368–387. Springer, Heidelberg (2001)
12. Gennaro, R., Hazay, C., Sorensen, J.S.: Text search protocols with simulation based security. In: Nguyen, P.Q., Pointcheval, D. (eds.) PKC 2010. LNCS, vol. 6056, pp. 332–350. Springer, Heidelberg (2010)
13. Hazay, C., Nissim, K.: Efficient set operations in the presence of malicious adversaries. In: Nguyen, P.Q., Pointcheval, D. (eds.) PKC 2010. LNCS, vol. 6056, pp. 312–331. Springer, Heidelberg (2010)

14. Ishai, Y., Paskin, A.: Evaluating branching programs on encrypted data. In: Vadhan, S.P. (ed.) TCC 2007. LNCS, vol. 4392, pp. 575–594. Springer, Heidelberg (2007)

15. Katz, J., Malka, L.: Constant-round private function evaluation with linear complexity. In: Lee, D.H., Wang, X. (eds.) ASIACRYPT 2011. LNCS, vol. 7073, pp. 556–571. Springer, Heidelberg (2011)

16. Kolesnikov, V., Schneider, T.: A practical universal circuit construction and secure evaluation of private functions. In: Tsudik, G. (ed.) FC 2008. LNCS, vol. 5143, pp. 83–97. Springer, Heidelberg (2008)

17. Mohassel, P., Sadeghian, S.: How to hide circuits in MPC an efficient framework for private function evaluation. In: Johansson, T., Nguyen, P.Q. (eds.) EUROCRYPT 2013. LNCS, vol. 7881, pp. 557–574. Springer, Heidelberg (2013)

18. Mohassel, P., Sadeghian, S., Smart, N.P.: Actively secure private function evaluation. Cryptology ePrint Archive, Report 2014/102 (2014), http://eprint.iacr.org/

19. Nielsen, J.B., Nordholt, P.S., Orlandi, C., Burra, S.S.: A new approach to practical active-secure two-party computation. In: Safavi-Naini, Canetti (eds.) [25], pp. 681–700

20. Niksefat, S., Sadeghiyan, B., Mohassel, P., Sadeghian, S.: Zids: A privacy-preserving intrusion detection system using secure two-party computation protocols. The Computer Journal (2013)

21. Ostrovsky, R., Paskin-Cherniavsky, A., Paskin-Cherniavsky, B.: Maliciously circuit-private fhe. Cryptology ePrint Archive, Report 2013/307 (2013), http://eprint.iacr.org/

22. Paus, A., Sadeghi, A.-R., Schneider, T.: Practical secure evaluation of semi-private functions. In: Abdalla, M., Pointcheval, D., Fouque, P.-A., Vergnaud, D. (eds.) ACNS 2009. LNCS, vol. 5536, pp. 89–106. Springer, Heidelberg (2009)

23. Raz, R.: Elusive functions and lower bounds for arithmetic circuits. In: Proceedings of the Fortieth Annual ACM Symposium on Theory of Computing, STOC 2008, pp. 711–720. ACM, New York (2008)

24. Sadeghi, A.-R., Schneider, T.: Generalized universal circuits for secure evaluation of private functions with application to data classification. In: Lee, P.J., Cheon, J.H. (eds.) ICISC 2008. LNCS, vol. 5461, pp. 336–353. Springer, Heidelberg (2009)

25. Safavi-Naini, R., Canetti, R. (eds.): CRYPTO 2012. LNCS, vol. 7417. Springer, Heidelberg (2012)

26. Valiant, L.: Universal circuits (preliminary report). In: Proceedings of the Eighth Annual ACM Symposium on Theory of Computing, pp. 196–203. ACM (1976)

Efficient, Oblivious Data Structures for MPC

Marcel Keller and Peter Scholl

Department of Computer Science, University of Bristol
{m.keller,peter.scholl}@bristol.ac.uk

Abstract. We present oblivious implementations of several data structures for secure multiparty computation (MPC) such as arrays, dictionaries, and priority queues. The resulting oblivious data structures have only polylogarithmic overhead compared with their classical counterparts. To achieve this, we give secure multiparty protocols for the ORAM of Shi et al. (Asiacrypt '11) and the Path ORAM scheme of Stefanov et al. (CCS '13), and we compare the resulting implementations. We subsequently use our oblivious priority queue for secure computation of Dijkstra's shortest path algorithm on general graphs, where the graph structure is secret. To the best of our knowledge, this is the first implementation of a non-trivial graph algorithm in multiparty computation with polylogarithmic overhead.

We implemented and benchmarked most of our protocols using the SPDZ protocol of Damgård et al. (Crypto '12), which works in the preprocessing model and ensures active security against an adversary corrupting all but one players. For two parties, the online access time for an oblivious array of size one million is under 100 ms.

Keywords: Multiparty computation, data structures, oblivious RAM, shortest path algorithm.

1 Introduction

In a secure multi-party computation (MPC) protocol, parties wish to perform some computation on their inputs without revealing the inputs to one another. The typical approach to securely implementing an algorithm for MPC is to rewrite the algorithm as a boolean circuit (or arithmetic circuit in some finite field) and then execute each gate of the circuit using addition or multiplication in the MPC protocol. For non-trivial functionalities, however, the resulting circuit can incur a large blow-up compared with the normal runtime. For example, algorithms that use a secret index as a lookup to an array search over the entire array when implemented naïvely in MPC, to avoid revealing which element was accessed. This means that the advantages of using complex data structures such as hash tables and binary trees cannot be translated directly to secure computation programs.

Oblivious RAM (ORAM) allows a client to remotely access their private data stored on a server, hiding the access pattern from the server. Ostrovsky and Shoup first proposed combining MPC and ORAM for a two-server writable PIR

P. Sarkar and T. Iwata (Eds.): ASIACRYPT 2014, PART II, LNCS 8874, pp. 506–525, 2014.

protocol [25], and Gordon et al. further explored this idea in a client-server setting, constructing a secure two-party computation protocol with amortized sublinear complexity in the size of the server's input using Yao's garbled circuits [16]. In the latter work, the state of an ORAM client is secret shared between two parties, whilst one party holds the server state (encrypted under the client's secret key). A secure computation protocol is then used to execute each ORAM instruction, which allows a secure lookup to an array of size N in $\mathsf{polylog}(N)$ time, in turn enabling secure computation of general RAM programs.

1.1 Our Contributions

Motivated by the problem of translating complex algorithmic problems to the setting of secure computation, we build on the work of Ostrovsky and Shoup [25] and Gordon et al. [16] by presenting new, efficient data structures for MPC, and applying this to the problem of efficient, secure computation of a shortest path on general graphs using Dijkstra's algorithm. Our contributions are outlined below.

Oblivious Array and Dictionary. In the context of MPC, we define an oblivious array as a secret shared array that can be accessed using a secret index, without revealing this index. Similarly, an oblivious dictionary can be accessed by a secret-shared key, which may be greater than the size of the dictionary.

We give efficient, polylogarithmic MPC protocols for oblivious array lookup based on two ORAM schemes, namely the SCSL ORAM of Shi et al. [27] (with an optimization of Gentry et al. [14]) and the Path ORAM scheme of Stefanov et al. [28], and evaluate the efficiency of both protocols. Our approach differs from that of Gordon et al., who consider only a client-server scenario where the server has a very large input. Instead, we use a method first briefly mentioned by Damgård et al. [10], where all inputs are secret shared across all parties, who also initially share the client state of the ORAM. The server's ORAM state is then constructed from the secret shared inputs using MPC and so is secret shared across all parties, but does not need to be encrypted since secret sharing ensures the underlying data is information theoretically hidden. This approach has two benefits: firstly, it scales naturally to any number of parties by simply using any secret-sharing based MPC protocol, and secondly it avoids costly decryption and re-encryption operations within MPC. Furthermore, Gordon et al. only achieve passive security, while our solution naturally provides active security when using an adequate MPC scheme such as SPDZ. Their approach of letting the server store the memory encrypted under a one-time pad for which the client generates the keys does not seem to extend to active security without losing efficiency.

Since the benefits of using ORAM only become significant for large input sizes (> 1000), we have also paid particular effort to creating what we call *Trivial ORAM* techniques for searching and accessing data on small input sizes with linear overhead, when full ORAM is less practical. The naive method for searching a list of size N involves performing a secure comparison between every

Data structure	Based on	Access complexity	Section
Oblivious array	Demultiplexing [21]	$O(N)$	3.1
Oblivious dictionary	{ Trivial ORAM	$O(N \cdot \ell)$	3.2
	{ Trivial ORAM	$O(N + \ell \cdot \log N)$	3.3
Oblivious array	{ SCSL ORAM	$O(\log^4 N)$	4.2
	{ Path ORAM	$O(\log^3 N)$	4.3
Oblivious priority queue	{ Oblivious array	$O(\log^4 N)$	5.1
	{ Modified Path ORAM	$O(\log^3 N)$	Full ver. [19]

Fig. 1. Overview of our oblivious data structures. For the dictionary, ℓ is the maximal size of keys.

element and the item being searched for, which takes time $O(N \cdot \ell)$ when comparing ℓ-bit integers. In Section 3 we present two $O(N)$ methods for oblivious array and dictionary lookup. These techniques come in useful for implementing the ORAM schemes for large inputs, but could also be used for applications on their own.

Figure 1 gives an overview of our algorithms. Note that the complexities may appear slightly higher than expected, due to the overhead of secure comparison in MPC. In a standard word model of computation, a $\log N$-bit comparison costs 1 operation, whereas in MPC this requires $O(\log N)$ operations, leading to the extra $O(\log N)$ factor seen in many of our data structures. As parameters for the complexity we use N for the size and ℓ for the maximal bit length of keys. Note that an oblivious dictionary implies an oblivious array with $\ell = \log N$. Furthermore, when choosing parameters for ORAM schemes we assume that the number of accesses to these data structures is in $O(N \log(N))$.

Secure Priority Queue and Dijkstra's Algorithm. In Section 5 we use our oblivious array to construct an oblivious priority queue, where secret shared items can be efficiently added and removed from the queue without revealing any information (even the type of operation being performed) and show how to use this to securely implement Dijkstra's shortest path algorithm in time $O(|E| \log^4 |E| + |V| \log^3 |V|)$, when only the number of vertices and edges in the graph is public. The previous best known algorithm [2] for this takes time in $O(|V|^3)$. In the full version [19], we also show how to modify the underlying ORAM to implement a priority queue directly, where each priority queue operation essentially takes just one ORAM access (instead of $\log |V|$ accesses), but we have not implemented this variant.

Secure Stable Matching. With our oblivious array, it is straightforward to implement the preference matrix used in the Gale-Shapley algorithm for stable matching. The resulting protocol again has polylogarithmic overhead compared to the complexity of Gale-Shapley, which is $O(N^2)$ for N agents of both kinds. Previous work by Franklin et al. [12] achieved $O(N^4)$ for two-party computation using semi-homomorphic encryption. We also implemented the case of every agent only having a constant number of preferences. In this case, secure stable matching takes quasi-linear time in the worst case.

Novel MPC and ORAM Techniques. We introduce several new techniques for MPC and ORAM that are used in our protocols, and may be of use in other applications. In Section 4.3 we describe a new method for *obliviously shuffling* a list of secret shared data points, actively secure against a dishonest majority of corrupted adversaries, using permutation networks. To the best of our knowledge, in the multi-party setting this has previously only been done for threshold adversaries. Section 4.4 describes a method for *batch initialization* of the SCSL ORAM in MPC using oblivious shuffling and sorting, which saves an $O(\log^3 N)$ factor compared with naively performing an ORAM access for each item, in practice giving a speedup of 10–100 times.

Implementation. We implemented and benchmarked the oblivious array and Dijkstra's algorithm using various ORAM protocols. Our implementation uses the SPDZ protocol of Damgård et al. [9,11], which is in the *preprocessing model*, separating the actual computation from a preprocessing phase where secret random multiplication triples and random bits are generated. The resulting online protocol is actively secure against a dishonest majority, so up to $n-1$ of n parties may be corrupted. We use the MPC compiler and framework of Keller et al. [20] to ensure that the minimal round complexity is achieved for each protocol.

1.2   Related Work

Other than the works already discussed [10,16,25], Gentry et al. [14] describe how to use homomorphic encryption combined with ORAM for reducing the communication cost of ORAM and also for secure computation in a client-server situation. These works are in a similar vein to ours, but our work is the first to thoroughly explore using ORAM for oblivious data structures, and applies to general MPC rather than a specific client-server scenario. We also expect that the access times from our interactive approach are much faster than what could be obtained using homomorphic encryption.

Gentry et al. [15] recently showed how to garble RAM programs for two party computation, which is an elegant theoretical result but does not seem to be practical at this time.

Toft described an oblivious secure priority queue with deterministic operations for use in MPC, without using ORAM [29]. The priority queue is based on a bucket heap, and supports insertion and removal in amortized $O(\log^2 N)$ time, but cannot support the decrease-key operation needed for Dijkstra's algorithm, unlike our ORAM-based priority queue.

In a recent, independent work, Wang et al. [31] consider oblivious data structures in the classical ORAM model. Their techniques are very similar to our oblivious priority queue shown in the full version [19], but the general method does not directly translate to the MPC setting due to the use of branching and a client-side cache.

Brickell and Shmatikov [4] were the first to consider graph algorithms in an MPC setting. Their solution only works for two parties and achieves only passive security because they rely on local computation by the two parties. Furthermore, the result is always public whereas our protocols allow for further secure computation without publishing the result.

Secure variants of shortest path and maximum flow algorithms were presented for use in MPC by Aly et al. [2]. They operate on a secret-shared adjacency matrix without using ORAM, which leads to an asymptotic complexity in $O(|V|^3)$ for Dijkstra's algorithm. More recently, Blanton et al. [3] presented an oblivious algorithm for finding shortest paths in a graph with complexity in $O(|V|^2)$. The same complexity is achieved in another recent work by Liu et al. [22]. All solutions do not come close to $O(|E| + |V|\log|V|)$ for the implementation on a single machine except for the case of dense graphs. Our solution incurs only an overhead in $O(\log^4|E| + \log^3|V|)$ over the classical algorithm for arbitrary graphs, improving upon previous solutions when the graph is relatively sparse.

Gentry et al. [14] also show how to modify their ORAM scheme to allow for lookup by key instead of index, by taking advantage of the tree-like structure in recursive ORAM. This was the inspiration for our second priority queue protocol given in the full version [19].

2 Preliminaries

The usual model of secure multiparty computation is the so-called arithmetic black box over a finite field. The parties have pointers to field elements inside the box, and they can order the box to add, multiply, or reveal elements. The box will follow the order if a sufficient number of parties (depending of the MPC scheme used) support it. In the case of the SPDZ protocol that we use for our experiments, only the full set of parties is sufficient. There is a large body of works in this model, some of which we refer to in the next section.

2.1 Building Blocks

In this section, we will refer to a few sub-protocols that we use later.

- $b \leftarrow$ EQZ($[a], \ell$) computes whether the ℓ-bit value a is zero. Catrina and de Hoogh [5] provide implementations for prime order fields that require either $O(\log \ell)$ rounds or a constant number of rounds. For fields of characteristic two, EQZ can be implemented by computing the logical OR of the bit decomposition of a. Generally, a simple protocol requires $O(\ell)$ invocations in $O(\log \ell)$ rounds, and PreMulC by Damgård et al. [7] allows for constant rounds. However, the constant-round version turns out to be slower in our implementation.
- $([b_0], \ldots, [b_{\ell-1}]) \leftarrow$ PreOR($[a_0], \ldots, [a_{\ell-1}]$) computes the prefix OR of the ℓ input bits. Catrina and de Hoogh [5] presented an implementation that requires $O(\ell)$ invocations in $O(\log \ell)$ rounds, while Damgård et al. [7] showed that a constant-round implementation is feasible. Again, the constant-round implementation is slower in our implementation.

- IfElse($[c], [a], [b]$) emulates branching by computing $[a] + [c] \cdot ([b] - [a])$. If c is zero, the result is a, if c is one, the result is b. Similarly, CondSwap swaps two secret-shared values depending on a secret-shared bit.
- $[b_0], \ldots, [b_{2^n-1}] \leftarrow$ Demux($[a_0], \ldots, [a_n]$) computes a vector of bits such that the a-th element is 1 whereas all others are 0 if $(a_0, \ldots, a_n)$ is the bit decomposition of a. We use this for our $O(N)$ oblivious array as well as the position map in Tree ORAM. Our implementation of Demux is due to Launchbury et al. [21]. It requires $O(2^\ell)$ invocations in $\lceil \log \ell \rceil$ rounds because one call of Demux with ℓ inputs induces 2^ℓ multiplications in one round and two parallel calls to Demux with at most $\lceil \ell/2 \rceil$ inputs.
- Sort($[x_0], \ldots, [x_{n-1}]$) sorts a list of n values. This can be done using Batcher's odd-even mergesort with $O(n \log^2 n)$ secure comparisons, or more efficiently using the method of Jónsson et al. [18] in $O(n \log n)$, with our oblivious shuffling protocol in Section 4.3. The latter is secure only if the sorted elements are unique.

2.2 Tree ORAM Overview

The ORAM schemes we use for our MPC protocols have the same underlying structure as the recursive Tree ORAM of Shi et al. [27] (SCSL), where N ORAM entries are stored in a complete binary tree with N leaves and depth $D = \lceil \log N \rceil$, encrypted under the client's secret key. Nodes in the tree are *buckets*, which each hold up to Z entries encrypted under the client's secret key, where the choice of Z affects statistical security. Each entry within a bucket consists of a triple (a, d_a, L_a), where a is an address in $\{0, \ldots, N-1\}$, d_a is the corresponding data, and L_a is a leaf node currently identified with a. The main invariant to ensure correctness is that at any given time, the entry (a, d_a, L_a) lies somewhere along the path from the root to the leaf node L_a.

The client stores a *position map*, which is a table mapping every address a to its corresponding leaf node L_a. To access the entry at address a, the client simply requests all buckets on the path to L_a, decrypts them and identifies the matching entry. If a write is being performed, the value is updated with a new value. Next, the client chooses a new leaf L'_a uniformly at random, updates the entry with L'_a and places the entry in the root node bucket of the tree. The path is then re-encrypted and sent back to the server.

Since a new leaf mapping is chosen at random every time an entry is accessed, the view of the server is identical at every access. Note that buckets are always padded to their full capacity, even if empty. To distribute the entries on the tree, an *eviction* procedure is used, which pushes entries further down the tree towards the leaves to spread out the capacity more evenly.

The original eviction method of Shi et al. is as follows:

- Client chooses at random two buckets from each level of the tree except the leaves, and requests these and each of their two children from server.

- For each chosen bucket, push an entry down into one of its child buckets, choosing the child based on the entry's corresponding leaf node.
- Re-encrypt the buckets and their children before sending them to the server.

Shi et al. showed that if the bucket size Z is set to k, the probability that any given bucket will overflow (during a single ORAM access) can be upper bounded by 2^{-k}.

Reducing Client Storage via Recursion. The basic ORAM described above requires a linear amount of client-side storage, due to the position map. To reduce this, the position map can itself be recursively implemented as an ORAM of size N/χ, by packing χ indices into a single entry. If, say, $\chi = 2$ then recursing this $\log_\chi N$ times results in a final position map of a single node. However, the communication and computation cost is increased by a factor of $O(\log N/\log \chi)$. Note that the entry size must be at least χ times as large as the index size, to be able to store χ indices in each entry.

Gentry et al. ORAM Optimizations. Gentry et al. [14] proposed two modifications to the SCSL ORAM to improve the parameter sizes. The first of these is to use a shallower tree, with only N/k instead of N leaves, for an ORAM of size N. The expected size of each leaf bucket is now k, so to avoid overflow it can be shown using a Chernoff bound that it suffices to increase the bucket size of the leaves to $2k$.

Gentry et al. also suggest k to be between 50 and 80, but do not justify their choices any further. By experimenting with probabilities calculated from the Chernoff bound, we found that the access complexity is actually minimized for bucket size $4k$, and choosing k between 12 and 24 (depending on the ORAM size) gives overflow probability at most 2^{-20} when the number of accesses is in $\tilde{O}(N)$.

The second optimization of Gentry et al. is to use higher degree trees to make the tree even shallower, which requires a more complex eviction procedure. We did not experiment with this variant, since working with higher degree trees and the new eviction method in MPC does not seem promising to us. When we refer to our implementation of the SCSL ORAM, we therefore mean the protocol with the first modification only.

Path ORAM. Path ORAM is another variant of tree ORAM proposed by Stefanov et al. [28]. It uses a small piece of client side storage, called the 'stash', and a new eviction method, which allows the bucket size Z to be made as small as 4.

For an ORAM query, the client first proceeds as usual by looking up the appropriate leaf node in the position map and requesting the path from the root to this leaf. The accessed entry is found on this path, reassigned another random leaf, and then every entry in the path and the stash is pushed as far down towards the leaf as possible, whilst maintaining the invariant that entries lie on the path towards their assigned leaves. In the case of bucket overflow, entries are placed in the client-stored stash. The proven overflow bound for Path ORAM is not currently very tight, so for our implementation we chose parameters based on simulation results instead of a formal analysis, again looking to achieve an overall overflow probability of 2^{-20} for our applications.

3   Oblivious Data Structures with Linear Overhead

The general model of our oblivious array and dictionary protocols is to secret share both the server and client state of an ORAM between all parties. This means that there is no distinction between client and server anymore because both roles are replaced by the same set of parties.

Since secret sharing hides all information from all parties, the server memory does not need to be encrypted; any requests from the client to the server that are decrypted in the original ORAM protocol are implemented by revealing the corresponding secret shared data. For server memory accesses, the address must be available in public, which allows the players to access the right shares. Computation on client memory is implemented as a circuit, as required for MPC. This means that any client memory access depending on secret data must be replaced by a scan of the whole client memory, similarly to a Trivial ORAM access. Note that we cannot use the ORAM scheme we are trying to implement for the client memory as this would introduce a circularity. Given the overhead for accessing the client memory, ORAM schemes with small client memory are most efficient in our model.

In this section, we outline our implementations based on *Trivial ORAM*, where all access operations are conducted by loading and storing the entire list of entries, and thus have linear overhead. In the context of MPC, the main cost factor is not memory accesses[1], but the actual computation on them. Nevertheless, the two figures are closely related here.

In our experiments (Section 6) we found that, for sizes up to a few thousand, the constructions in this section prove to be more efficient than the ones with better asymptotic complexity despite the linear overhead.

3.1   Oblivious Array

A possible way for obliviously searching an array of size N was proposed by Launchbury et al. [21], which we refer to as demultiplexing. It involves expanding the secret-shared index $i < N$ to a vector of size N that contains a one in the i-th position and zeroes elsewhere. The inner product of this index vector and the array of field elements gives the desired field element. The index vector can likewise be used to replace this field element by a new one while leaving the other field elements intact. The index expansion corresponds to a bit decomposition of the input followed by the demultiplexing operation. This procedure has cost in $O(N)$.

---

[1] In fact, just accessing the share of an entry comes at no cost because the shares are stored locally.

Storing an empty flag in addition to the array value proves useful for some applications. Thus, we require that the players store a list of tuples $([v_i], [e_i])_{i=0}^{N-1}$ containing the values and empty flags.

3.2   Oblivious Dictionary

We will use the dictionary in this section as a building block for implementing Tree ORAM in MPC. Previous work refers to the Trivial ORAM used in this section as non-contiguous ORAM, meaning that the index of an entry can be any number, in particular greater than the size of the array.

For simplification, we assume that the dictionary consists of index-value pairs (u, v) where both are field elements of the underlying field. The extension to several values is straightforward. In addition to one index-value pair per entry we store a bit e indicating whether the entry is empty. We will see shortly that this proves to be useful. In summary, the players store a list of tuples of secrets-shared elements $([u_i], [v_i], [e_i])_{i=0}^{N-1}$ for a dictionary of size N. Initially, e_i must be 1 for all i.

The Tree ORAM construction requires the dictionary to provide the following operations:

- ReadAndRemove($[u]$) returns and removes the value associated with index u if it is present in the dictionary, and a default entry otherwise.
- Add($[u], [v], [e]$) adds the entry (u, v, e) to the dictionary assuming that no entry with index v exists.
- Pop() returns and removes a non-empty entry $(u, v, 0)$ if the dictionary is not empty, and the default empty entry $(0, 0, 1)$ otherwise. Shi et al. [27] showed how to implement Pop using ReadAndRemove and Add, but for Trivial ORAM a dedicated implementation is more efficient.

In MPC, ReadAndRemove is the most expensive part because it contains a comparison for every entry. Essentially, it computes an index vector as in the previous section using comparison instead of demultiplexing. The complexity is dominated by the N calls to an equality test which cost $O(N \cdot \ell)$ invocations in $O(\log \ell)$ or $O(1)$ rounds for ℓ-bit keys, depending on the protocol used for equality testing.

Our implementations of Add and Pop make use of the bits indicating whether an entry is empty. This way, we avoid comparing every index in finding the first non-empty or empty entry, respectively. Both protocols require $O(N)$ invocations in $O(\log N)$ or $O(1)$ rounds, depending on the exact implementation.

Theorem 1. *An algorithm using arrays but no branching other than conditional writes to arrays can be securely implemented in MPC with linear overhead.*

Proof (Sketch). Implement arrays as described above and use a circuit for conditional writes. Since there is no branching otherwise, this effectively results in a circuit that can be securely executed using the arithmetic blackbox provided by MPC schemes.

3.3   Oblivious Dictionary in $O(N)$

The complexity of ReadAndRemove in the oblivious dictionary can be reduced to only $O(N + \ell \cdot \log N)$ instead of $O(N \cdot \ell)$, at the cost of increasing the round complexity from $O(1)$ to $O(\log N)$.

The protocol (see the full version [19]) starts by subtracting the item to be searched for from each input, and builds the *multiplication tree* from these values, which is a complete binary tree where each node is the product of its two children. The original values lie at the leaves, and at each level a single node will be zero, directing us towards the leaf where the item to be found lies. Now we can traverse the tree in a binary search-like manner, at each level obliviously selecting the next element to be compared until we reach the leaves. We do this by computing a bit vector indicating the position of a node which is equal to zero on the current level and then use this to determine the comparison to be performed next. Note that only a single call to an equality test is needed at each level of the tree, but a linear number of multiplications are needed to select the item on the next level.

The main caveat here is that, since the tree of multiplications can cause values to grow arbitrarily, we need to be able to perform an equality test on *any* field element. In $\mathsf{GF}(2^n)$ this works as usual, and in $\mathsf{GF}(p)$ this can be done by a constant-round protocol by Damgård et al. [7].

4   Oblivious Array with Polylogarithmic Overhead

In this section we give polylogarithmic oblivious array protocols based on the SCSL ORAM and Path ORAM schemes described in Section 2.2. These both have the same recursive Tree ORAM structure, so first we describe how to implement the position map, which is the same for both schemes.

4.1   Position Map in Tree ORAM

The most intricate part of implementing Tree ORAM in MPC is the position map. Since we implement an oblivious array, the positions can be stored as an ordered list without any overhead. Recall that for the ORAM recursion to work it is essential that at least two positions are stored per memory address of the lower-level ORAM. In an MPC setting, there are two ways of achieving this: storing several field elements in the same memory cell and packing several positions per field element. We use both approaches.

To simplify the indexing, we require that both the number of positions per field element and number of field elements per memory cell are powers of two. This allows to compute the memory address and index within a memory cell by bit-slicing. For example, if there are two positions per field element and eight field elements per memory cell, the least significant bit denotes the index within a field element, the next three bits denote the index within the memory cell, and the remaining bits denote the memory address. Because these parameters are publicly known, the bit-slicing is relatively easy to compute with MPC. In prime order fields, one can use Protocol 3.2 by Catrina and de Hoogh [5], which computes the remainder of the division by a public power of two; in fields of characteristic two one can simply extract the relevant bits by bit decomposition.

The bit-slicing used to extract a position from a field element is more intricate because the index is secret and must not be revealed. For prime-order fields, Aliasgari et al. [1] provide a protocol that allows to compute the modulo operation of a secret number and two raised to the power of a secret number. For fields of characteristic two, see the full version [19] for a similar protocol. The core idea of both protocols is to compute $[2^m]$ (or $[X^m]$) where m is the integer representation of secret-shared number. The bit decomposition of $[2^m]$ can then be used to mask the bit decomposition of another secret-shared number. Moreover, multiplying with $\mathsf{Inv}([2^m])$ allows to shift a number with the m least significant bits being zero by m positions to the right. The complexity of both versions of Mod2m is $O(\ell)$ invocations in $O(\log \ell)$ rounds or in a constant number of rounds if the protocols by Damgård et al. [7] are used. The latter proved to be less efficient in our experiments.

Finally, if the position map storage contains several field elements per memory cell, one also needs to extract the field element indexed by a secret number. This can be done using demultiplexing in the same way as for the oblivious array in Section 3.1.

4.2   SCSL ORAM

Many parts of Tree ORAM are straightforward to implement using the Trivial ORAM procedures from the last section.

For ReadAndRemove, the position is retrieved from the position map and revealed. All buckets on the path can then be read in parallel. Combining the results can be done similarly to the Trivial ORAM ReadAndRemove because the latter returns a value indicating whether the searched index was found in a particular bucket.

The Add procedure starts with adding the entry to the root bucket followed by the eviction algorithm. The randomness used for choosing the buckets to evict from does not have to be secret because the choice has to be made public. It is therefore sufficient to use a pseudorandom generator seeded by a securely generated value in order to reduce communication. However, it is crucial that the eviction procedure does not reveal which child of a bucket the evicted entry is added to. Therefore, we use a conditional swapping circuit. See the full version [19] for algorithmic descriptions of ReadAndRemove and Add.

Using ReadAndRemove and Add, it is straightforward to implement the universal access operation that we will later use in our implementation of Dijkstra's algorithm. Essentially, we start by ReadAndRemove and then write back the value just read or the new value depending on the write flag. Similarly, one can construct a read-only or write-only operation.

Complexity. Since the original algorithm by Shi et al. [27] does not involve branching, it can be implemented in MPC with asymptotically the same number of memory accesses. However, the complexity of an MPC protocol is determined by the amount of computation carried out. In ReadAndRemove, the index of every accessed element is subject to a comparison. These indices are $\log N$-bit numbers for an array of size N. Because the access complexity of the ORAM is in $O(\log^3 N)$, the complexity of all comparisons in ReadAndRemove is in $O(\log^4 N)$. It turns out that this dominates the complexity of the SCSL ORAM operations because Add and Pop do not involve comparisons. Furthermore, the algorithms in Section 4.1 have complexity in $O(\log N)$ and are only executed once per index structure and access. This leads to a minor contribution in $O(\log^2 N)$ per access of the oblivious array.

Theorem 2. *An algorithm using arrays but no branching other than conditional writes to arrays can be securely implemented in MPC with polylogarithmic overhead and negligible probability of an incorrect result.*

Proof (Sketch). Implement arrays as described above and use a circuit for conditional writes. Since there is no branching otherwise, this effectively results in a circuit that can be securely executed using the arithmetic blackbox provided by MPC schemes. The ORAM scheme accounts for the polylogarithmic overhead and the negligible failure probability.

For a complete simulation, we connect the ORAM and the MPC simulation in the following way: The ORAM simulation outputs the random paths used in ReadAndRemove. We input these values to the simulation of the arithmetic blackbox as they are revealed in the MPC protocol. Further random values are revealed by statistically secure algorithms. We sample those value according to the relevant distributions and input them to the MPC simulation as well. The indistinguishability of the resulting transcript follows using a hybrid argument with the two simulators.

4.3   Path ORAM

The Path ORAM access protocol is initially the same as in the Shi et al. and Gentry et al. schemes (with smaller bucket size), but differs in its eviction procedure. For eviction a leaf ℓ is chosen, either at random or in a deterministic bit-reversed order as in [14], and we consider the path from the root down to the leaf ℓ. For each entry $E = ([i], [\ell_i], [d_i])$ on the path we want to push E as far down the path as it can go, whilst still being on the path towards ℓ_i and avoiding bucket overflow. The high-level strategy for doing this in MPC is to first calculate the final position of each entry in the stash and the path to ℓ, then express this as a permutation and evaluate the resulting, secret permutation by oblivious shuffling. Since oblivious shuffling has only previously been studied in either the two party or threshold adversary setting, we describe a new protocol for oblivious shuffling with $m - 1$ out of m corrupted parties.

The eviction protocol for MPC first does the following for each entry $E = ([i], [\ell_i], [d_i])$ in the path to ℓ and the stash:

- Compute the *least common ancestor*, $\mathsf{LCA}(\ell, [\ell_i])$, of each entry's assigned leaf and the random leaf by XORing the bit-decomposed leaves together and identifying the first non-zero bit of this.
- For each level k in the path, compute a bit determining whether entry E goes to the bucket on level k or not, and bit indicating whether E ends up in the stash. To do this we maintain bits $u_{k,i}$ for $i = 0$ to $\lfloor \log_2 Z \rfloor$ that keep track of the size of bucket k, updating these as we go.

Now for each entry in the path, we have a list of bits, one of which is set to one to indicate the level we need to obliviously place the entry. At this point it might seem that we could just obliviously shuffle the entries and then reveal each level, but this is insecure. This is because in addition to the real entries in the path there are (an unknown number of) empty entries, which haven't been assigned a level. To be able to safely shuffle and reveal the real entries' levels, we first must assign a level to each empty entry, ensuring that every space in every bucket has had an entry assigned to it, so revealing the distribution of all levels gives no information.

To assign the empty entries with levels we use oblivious sorting: first the entries are sorted to separate the real and empty entries, and then the bucket positions are sorted to separate those that have already been used from the free ones. By aligning these two sorted lists, the free bucket levels can be assigned to the empty entries. Correctly aligning the two lists is actually slightly more involved than this due to the stash and some details have been omitted.

Parameters. In our implementation, we chose deterministic eviction with bucket size 2 and stash size 24. We estimate this gives an overflow probability of 2^{-75} for a single access, which should be comfortable for most applications. See the full version [19] for a complete description and extensive simulation results used to choose parameters.

Complexity. Computing the level that each entry ends up at requires $O(\log^2 N)$ multiplications. Our protocols for oblivious shuffling and bitwise sorting have complexity $O(n \log n)$ where here $n = |S| + Z \cdot \log N$, so the total complexity of the access protocol is $O(\log^2 N)$. Adding in the recursion gives an extra $O(\log N)$ factor, for an overall complexity of $O(\log^3 N)$.

For ReadAndRemove, the complexity is in $O(\log^3 N)$ instead of $O(\log^4 N)$ (for SCSL ORAM) because of the constant bucket size. This results in an overall access complexity in $O(\log^3 N)$.

Oblivious Shuffling with a Dishonest Majority. To carry out the oblivious shuffling above, we use a protocol based on *permutation networks*, which are circuits with N input and output wires, consisting only of CondSwap gates with the control bit for each gate hard-wired into the circuit. Given any permutation, there is an algorithm that generates the control bits such that the network applies this permutation to its inputs. The Waksman network [30] is an efficient permutation network with a recursive structure, with $O(N \log N)$ gates and depth $O(\log N)$. It has been used previously in cryptography for two-party oblivious shuffling [17], and also in fully homomorphic encryption [13] and private function evaluation [23].

Computing the control bits requires iterating through the orbits of the permutation, which seems expensive to do in MPC when the permutation is secret shared. Instead, for m players we use a composition of m permutations, as follows:

1. Each player generates a random permutation, and locally computes the control bits for the associated Waksman network [2]
2. Each player inputs (i.e. secret shares) their control bits.
3. For each value $[b]$ input by a player, open $[b] \cdot [b - 1]$ and check this equals 0, to ensure that b is a bit.
4. The resulting permutation is evaluated by composing together all the Waksman networks.

Step 3 ensures that all permutation network values input by the players are bits, even for malicious parties, so the composed networks will always induce a permutation on the data. As long as at least one player generates their input honestly, the resulting permutation will be random, as required. The only values that are revealed outside of the arithmetic black box are the bits in step 3, which do not reveal information other than identifying any cheaters. For m players, the complexity of the protocol for an input of size n is $O(mn \log n)$.

4.4   Batch Initialization of SCSL ORAM

The standard way of filling up an ORAM with N entries is to execute the access protocol once for each entry. This can be quite expensive, particularly for applications where the number of ORAM accesses is small relative to the size of the input. In the full version of this paper [19], we describe a new method for initializing the SCSL ORAM with N entries in time $O(N \log N)$, saving a factor of $O(\log^3 N)$ over the standard method. Note that this technique could also be used for standard ORAM (not in MPC), and then the complexity becomes $O(N)$, which seems optimal for this task. In practice our experiments indicate that this improves performance by at least 1-2 orders of magnitude, depending on the ORAM size.

5   Applications

In this section, we will use the oblivious array from the previous section to construct an oblivious priority queue, which we then will use in a secure multiparty computation of Dijkstra's shortest path algorithm.

---

[2] Note that generating the control bits randomly does not produce a random permutation, since for a circuit with m gates, there are 2^m possible paths and so any permutation will occur with probability $k2^{-m}$ for some integer k. However, for a uniform permutation we require $k2^{-m} = \frac{1}{n!}$.

5.1   Oblivious Priority Queue

A simple priority queue follows the design of a binary heap: a binary tree where every child has a higher key than its parent. The key-value pairs are put in an oblivious array, with the root at index 1, the left and right child of the root at indices 2 and 3, respectively, and so on. In order to be able to decrease the key of a value, we maintain a second array that links the keys to the respective indices in the first one.

The usage of ORAM requires that we set an upper limit on the size of the queue. We store the actual size of the heap in a secret-shared variable, which is used to access the first free position in the heap. Depending on the application, the size of the heap can be public and thus stored in clear. However, our implementation of Dijkstra's algorithm necessitates the size to be secret. This also means that the bubbling-up and -down procedures have to be executed on the tree of maximal size. The bubble-down procedure works similarly, additionally deciding whether to proceed with the left or the right child. Note that unlike the presentation in Sections 3 and 4, we allow tuples as array values here. This extension is straightforward to implement. Furthermore, we require a universal access operation for conditional writing. If such an operation is not available yet, it can implemented by reading first and then either writing the value just read or a new value depending on the write flag.

Our application also requires a procedure that combines insertion and decrease-key (depending on whether the value already is in the queue) and that can be controlled by a secret flag deciding whether the procedure actually should be executed or not. The combination is straightforward since both employ bubbling up, and the decision flag can be implemented using the universal access operation mentioned above. We refer to the full version [19] for exact descriptions of the procedures.

For a priority queue of maximal size N, both bubbling up and down run over $\log N$ levels accessing oblivious arrays of size in $O(N)$. Using Path ORAM therefore results in a complexity in $O(\log^4 N)$ per access operation. In the full version [19], we show how to modify the Tree ORAM structure to reduce the overhead to $O(\log^3 N)$ per access, similarly to the binary search ORAM of Gentry et al. [14] and, very recently, the oblivious priority queue of Wang et al. [31]. Unlike the latter, this priority queue supports decrease key so can be used for Dijkstra's algorithm, but we have not currently implemented this variant.

5.2   Secure Dijkstra's Algorithm

In this section we show how to apply the oblivious priority queue to a secure variant of Dijkstra's algorithm, where both the graph structure and the source vertex are initially secret shared across all parties. In our solution, the only information known by all participants is the number of vertices and edges. It is straightforward to have public upper bounds instead while keeping the actual figures secret. However, the upper bounds then determine the running time. This is inevitable in the setting of MPC because the parties are aware of the amount of computation carried out.

In the usual presentation [6], Dijkstra's algorithm contains a loop nested in another one. The outer loop runs over all vertices, and the inner loop runs over all neighbours of the current vertex. Directly implementing this in MPC would reveal the number of neighbours of the current vertex and thus some of the graph structure. On the other hand, executing the inner loop once for every other vertex incurs a performance punishment for graphs other than the complete graph. As a compromise, one could execute the inner loop as many times as the maximum degree of the graph, but even that would reveal some extra information about the graph structure. Therefore, we replaced the nested loops by a single loop that runs over all edges (twice in case of an undirected graph). Clearly, this has the same complexity as running over all neighbours of every vertex. The key idea of our variant of Dijkstra's algorithm is to execute the body of the outer loop for every edge but ignoring the effects unless the current vertex has changed. For this, conditional writing to an oblivious array plays a vital role.

In the following, we explain the data structure used by our implementation. Assume that the N vertices are numbered by 0 to $N - 1$. We use two oblivious arrays to store the graph structure. The first is a list of neighbours ordered by vertex: it starts with all neighbours of vertex 0 (in no particular order) followed by all neighbours of vertex 1 and so on. In addition, we store for every neighbour a bit indicating whether this neighbour is the last neighbour of the current vertex. The length of this list is the twice the number of edges for an undirected graph. In another array, we store for every vertex the index of its first neighbour in the first array. A third array is used to store the results, i.e., the distance from the source to every vertex. For the sake of simpler presentation, we omit storing the predecessor on the shortest path. Finally, we use a priority queue to store unprocessed vertices with the shortest encountered distance so far. See the full version [19] for our algorithm.

The main loop accesses a priority queue of size N_V and arrays of size N_E and N_V. Furthermore, we have to initialize a priority queue and an array of size N_V. Using the Path ORAM everywhere, this gives a complexity in $O(N_V \log^3 N_V + N_E(\log^4 N_V + \log^3 N_E))$. More concretely, for sparse graphs with $N_E \in O(N_V)$ (such as cycle graphs) the complexity is in $O(N_V \log^4 N_V)$, whereas for dense graphs with $N_E \in O(N_V^2)$ (such as complete graphs), the complexity is in $O(N_V^2 \log^4 N_V)$. The most efficient implementation of Dijkstra's algorithm on a single machine has complexity in $O(N_E + N_V \log N_V)$, and thus the penalty of using MPC is in $O(\log^4 N_V + \log^3 N_E)$.

6   Experiments

We implemented our protocols for oblivious arrays using the system described by Keller et al. [20], and ran experiments on two directly connected machines running Linux on an eight-core i7 CPU at 3.1 GHz. The online access times for different implementations of oblivious arrays containing one element of $GF(2^{40})$ per index are given in Figure 2. Most notably, the linear constructions are more efficient for sizes up to a few thousand. For these timings, we have also used the packing strategies described in Section 4.1 at the top level.

To estimate the offline time required to generate the necessary preprocessing data, we used the figures of Damgård et al. [8,9] for the finite fields $GF(2^n)$ and $GF(p)$, respectively. The offline time for accessing an oblivious array of size 2^{20} using Path ORAM would be 117 minutes in $GF(2^n)$ and 32 minutes in $GF(p)$, for active security with cheating probability 2^{-40}. Note that this could be easily improved using multiple cores, since the offline phase is trivially parallelizable, and also the $GF(2^n)$ times could be cut further by using the offline phase from TinyOT [24].

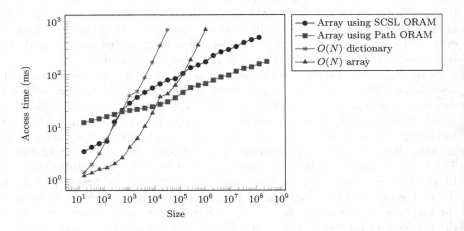

Fig. 2. Oblivious data structure online access times

6.1 Dijkstra's Algorithm

We have benchmarked Dijkstra's algorithm on cycle graphs of varying size. The edges of a cycle graph form a cycle passing every vertex exactly once. We chose it as a simple example of a graph with low degree. Figure 3 shows the timings of our implementation of the algorithm without ORAM techniques by Aly et al. [2] as well as our implementation using the oblivious array in Section 3.1 and the SCSL ORAM in Section 4.2. In all cases, we used MPC over a finite field modulo a 128-bit prime to allow for integer arithmetic. We generated our estimated figures using a fully functional secure protocol but running the main loop only a fraction of the times necessary.

For reference, we have also included timings for the offline phase, estimated using the costs given by Damgård et al. [9]. Note that these timings are to be understood as core seconds because the offline phase is highly parallelizable, unlike the online phase. It follows that the wall time of the offline phase can be brought down to the same order of magnitude of the online phase using ten thousand cores.

Our algorithms perform particularly well in comparison to the one by Aly et al. because cycle graphs have very low degree. For complete graphs, the full version [19] shows a different picture. The overhead for using ORAM is higher than the asymptotic advantage for all graph sizes that we managed to run our algorithms on.

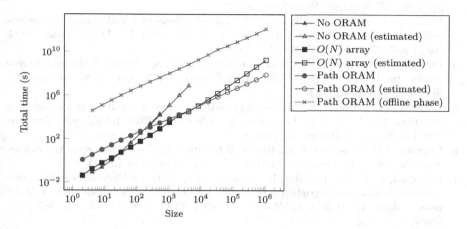

Fig. 3. Dijkstra's algorithm on cycle graphs

6.2 Gale-Shapley Algorithm for the Stable Marriage Problem

Using an oblivious array, it is straightforward to implement the Gale-Shapley algorithm in MPC. We have implemented a worst case example, where the main loop is executed $O(N^2)$ times. However, initialization of the preference matrices costs $O(N^2)$ in any case. With the overhead of the oblivious array, we get an overall complexity in $O(N^2 \log^3 N)$. Running a fully functional program for a limited time, we estimate that computing Gale-Shapley for 8192 pairs takes $9.1 \cdot 10^7$ seconds online and $1.5 \cdot 10^{12}$ seconds offline, using the figures by Damgård et al. [9] for the latter estimate. See the full version [19] for more timings.

Gale-Shapley with Limited Preferences. It does not always make sense to have full ranking of preferences. For example, it is hard to imagine that a human could ranking a thousand options. Therefore, we investigated the case of every agent only having a top twenty and feeling indifferent about the rest. In this case, both initialization and the worst-case main loop of a slightly modified algorithm become linear in the number of agents. Using the same techniques as above, we estimate that computing Gale-Shapley with 20 preferences for 2^{20} pairs takes $3.5 \cdot 10^7$ seconds online and $7.1 \cdot 10^{11}$ seconds offline.

Acknowledgements. We would like to thank Nigel Smart for various comments and suggestions. This work has been supported in part by EPSRC via grant EP/I03126X.

References

1. Aliasgari, M., Blanton, M., Zhang, Y., Steele, A.: Secure computation on floating point numbers. In: NDSS. The Internet Society (2013)
2. Aly, A., Cuvelier, E., Mawet, S., Pereira, O., Van Vyve, M.: Securely solving simple combinatorial graph problems. In: Sadeghi, A.-R. (ed.) FC 2013. LNCS, vol. 7859, pp. 239–257. Springer, Heidelberg (2013)
3. Blanton, M., Steele, A., Aliasgari, M.: Data-oblivious graph algorithms for secure computation and outsourcing. In: Chen, K., Xie, Q., Qiu, W., Li, N., Tzeng, W.G. (eds.) ASIACCS, pp. 207–218. ACM (2013)
4. Brickell, J., Shmatikov, V.: Privacy-preserving graph algorithms in the semi-honest model. In: Roy, B.K. (ed.) ASIACRYPT 2005. LNCS, vol. 3788, pp. 236–252. Springer, Heidelberg (2005)
5. Catrina, O., de Hoogh, S.: Improved primitives for secure multiparty integer computation. In: Garay, J.A., De Prisco, R. (eds.) SCN 2010. LNCS, vol. 6280, pp. 182–199. Springer, Heidelberg (2010)
6. Cormen, T.H., Leiserson, C.E., Rivest, R.L., Stein, C.: Introduction to Algorithms, 2nd edn. The MIT Press and McGraw-Hill Book Company (2001)
7. Damgård, I., Fitzi, M., Kiltz, E., Nielsen, J.B., Toft, T.: Unconditionally secure constant-rounds multi-party computation for equality, comparison, bits and exponentiation. In: Halevi, S., Rabin, T. (eds.) TCC 2006. LNCS, vol. 3876, pp. 285–304. Springer, Heidelberg (2006)
8. Damgård, I., Keller, M., Larraia, E., Miles, C., Smart, N.P.: Implementing AES via an actively/covertly secure dishonest-majority MPC protocol. In: Visconti, I., De Prisco, R. (eds.) SCN 2012. LNCS, vol. 7485, pp. 241–263. Springer, Heidelberg (2012)
9. Damgård, I., Keller, M., Larraia, E., Pastro, V., Scholl, P., Smart, N.P.: Practical covertly secure MPC for dishonest majority – or: Breaking the SPDZ limits. In: Crampton, J., Jajodia, S., Mayes, K. (eds.) ESORICS 2013. LNCS, vol. 8134, pp. 1–18. Springer, Heidelberg (2013)
10. Damgård, I., Meldgaard, S., Nielsen, J.B.: Perfectly secure oblivious RAM without random oracles. In: Ishai, Y. (ed.) TCC 2011. LNCS, vol. 6597, pp. 144–163. Springer, Heidelberg (2011)
11. Damgård, I., Pastro, V., Smart, N.P., Zakarias, S.: Multiparty computation from somewhat homomorphic encryption. In: Safavi-Naini, R., Canetti, R. (eds.) CRYPTO 2012. LNCS, vol. 7417, pp. 643–662. Springer, Heidelberg (2012)
12. Franklin, M.K., Gondree, M., Mohassel, P.: Improved efficiency for private stable matching. In: Abe, M. (ed.) CT-RSA 2007. LNCS, vol. 4377, pp. 163–177. Springer, Heidelberg (2006)
13. Gentry, C., Halevi, S., Smart, N.P.: Fully homomorphic encryption with polylog overhead. In: Pointcheval, D., Johansson, T. (eds.) EUROCRYPT 2012. LNCS, vol. 7237, pp. 465–482. Springer, Heidelberg (2012)
14. Gentry, C., Goldman, K.A., Halevi, S., Julta, C., Raykova, M., Wichs, D.: Optimizing ORAM and using it efficiently for secure computation. In: De Cristofaro, E., Wright, M. (eds.) PETS 2013. LNCS, vol. 7981, pp. 1–18. Springer, Heidelberg (2013)
15. Gentry, C., Halevi, S., Lu, S., Ostrovsky, R., Raykova, M., Wichs, D.: Garbled RAM revisited. In: Nguyen, P.Q., Oswald, E. (eds.) EUROCRYPT 2014. LNCS, vol. 8441, pp. 405–422. Springer, Heidelberg (2014)

16. Gordon, S.D., Katz, J., Kolesnikov, V., Krell, F., Malkin, T., Raykova, M., Vahlis, Y.: Secure two-party computation in sublinear (amortized) time. In: Yu, T., Danezis, G., Gligor, V.D. (eds.) ACM Conference on Computer and Communications Security, pp. 513–524. ACM (2012)

17. Huang, Y., Evans, D., Katz, J.: Private set intersection: Are garbled circuits better than custom protocols? In: NDSS. The Internet Society (2012)

18. Jónsson, K.V., Kreitz, G., Uddin, M.: Secure multi-party sorting and applications. IACR Cryptology ePrint Archive 2011, 122 (2011)

19. Keller, M., Scholl, P.: Efficient, oblivious data structures for MPC. Cryptology ePrint Archive, Report 2014/137 (2014), http://eprint.iacr.org/2014/137

20. Keller, M., Scholl, P., Smart, N.P.: An architecture for practical actively secure MPC with dishonest majority. In: Sadeghi, et al. (eds.) [26], pp. 549–560.

21. Launchbury, J., Diatchki, I.S., DuBuisson, T., Adams-Moran, A.: Efficient lookup-table protocol in secure multiparty computation. In: Thiemann, P., Findler, R.B. (eds.) ICFP, pp. 189–200. ACM (2012)

22. Liu, C., Huang, Y., Shi, E., Katz, J., Hicks, M.: Automating efficient RAM-model secure computation, http://www.cs.umd.edu/~liuchang/paper/oakland2014.pdf

23. Mohassel, P., Sadeghian, S.: How to hide circuits in MPC an efficient framework for private function evaluation. In: Johansson, T., Nguyen, P.Q. (eds.) EUROCRYPT 2013. LNCS, vol. 7881, pp. 557–574. Springer, Heidelberg (2013)

24. Nielsen, J.B., Nordholt, P.S., Orlandi, C., Burra, S.S.: A new approach to practical active-secure two-party computation. In: Safavi-Naini, R., Canetti, R. (eds.) CRYPTO 2012. LNCS, vol. 7417, pp. 681–700. Springer, Heidelberg (2012)

25. Ostrovsky, R., Shoup, V.: Private information storage. In: Proceedings of the Twenty-Ninth Annual ACM Symposium on Theory of Computing, pp. 294–303. ACM (1997)

26. Sadeghi, A.R., Gligor, V.D., Yung, M. (eds.): 2013 ACM SIGSAC Conference on Computer and Communications Security, CCS 2013, Berlin, Germany, November 4-8. ACM (2013)

27. Shi, E., Chan, T.-H.H., Stefanov, E., Li, M.: Oblivious RAM with $O((\log N)^3)$ worst-case cost. In: Lee, D.H., Wang, X. (eds.) ASIACRYPT 2011. LNCS, vol. 7073, pp. 197–214. Springer, Heidelberg (2011)

28. Stefanov, E., van Dijk, M., Shi, E., Fletcher, C.W., Ren, L., Yu, X., Devadas, S.: Path ORAM: an extremely simple oblivious RAM protocol. In: Sadeghi, et al. (eds.) [26], pp. 299–310.

29. Toft, T.: Secure datastructures based on multiparty computation. Cryptology ePrint Archive, Report 2011/081 (2011), http://eprint.iacr.org/2011/081

30. Waksman, A.: A permutation network. Journal of the ACM (JACM) 15(1), 159–163 (1968)

31. Wang, X., Nayak, K., Liu, C., Shi, E., Stefanov, E., Huang, Y.: Oblivious data structures. Cryptology ePrint Archive, Report 2014/185 (2014), http://eprint.iacr.org/2014/185

Author Index

Andreeva, Elena I-105
Applebaum, Benny II-162
Aranha, Diego F. I-262

Belaïd, Sonia II-306
Bellare, Mihir II-102
Benhamouda, Fabrice I-551
Bernstein, Daniel J. I-317
Bilgin, Begül II-326
Biryukov, Alex I-63
Bogdanov, Andrey I-105
Boneh, Dan I-42
Bos, Joppe W. I-358
Bouillaguet, Charles I-63
Boura, Christina I-179
Bruneau, Nicolas II-344
Brzuska, Christina II-122, II-142

Camenisch, Jan I-551
Catalano, Dario II-193
Chen, Yu II-366
Chuengsatiansup, Chitchanok I-317
Chung, Kai-Min II-62
Cohen, Ran II-466
Corrigan-Gibbs, Henry I-42
Costello, Craig I-338

Damgård, Ivan II-213
Danezis, George I-532
David, Bernardo II-213
de Portzamparc, Frédéric I-21
Dinur, Itai I-439
Doche, Christophe I-297
Ducas, Léo II-22
Dunjko, Vedran II-406
Dunkelman, Orr I-439

Emami, Sareh I-141

Faugère, Jean-Charles I-21
Fitzsimons, Joseph F. II-406
Fleischhacker, Nils I-512
Forler, Christian II-289
Fouque, Pierre-Alain I-262, I-420, II-306

Fournet, Cédric I-532
Fuchsbauer, Georg II-82
Fujisaki, Eiichiro II-426

Gérard, Benoît I-262, I-282, II-306
Giacomelli, Irene II-213
Gierlichs, Benedikt II-326
Gilbert, Henri I-200
Groth, Jens I-532
Guilley, Sylvain II-344
Guo, Jian I-458
Guo, Qian I-1
Guo, Yanfei II-366

Hanser, Christian I-491
Heuser, Annelie II-344
Hirt, Martin II-448
Hisil, Huseyin I-338
Hu, Lei I-158

Jager, Tibor I-512
Jarecki, Stanislaw II-233
Jean, Jérémy I-458, II-274
Johansson, Thomas I-1
Joo, Chihong II-173
Joux, Antoine I-378, I-420
Jovanovic, Philipp I-85
Joye, Marc II-1

Kammerer, Jean-Gabriel I-262
Keller, Marcel II-506
Keller, Nathan I-439
Khovratovich, Dmitry I-63
Khurana, Dakshita II-386
Kiayias, Aggelos II-233
Kleinjung, Thorsten I-358
Kohlweiss, Markulf I-532
Komargodski, Ilan II-254
Konstantinov, Momchil II-82
Krawczyk, Hugo II-233
Krenn, Stephan I-551

Lange, Tanja I-317
Lenstra, Arjen K. I-358
Libert, Benoît II-1

Lindell, Yehuda   II-466
Ling, San   I-141
Liu, Zhenming   II-62
Löndahl, Carl   I-1
Longo, Jake   I-223
Lucks, Stefan   II-289
Luykx, Atul   I-85, I-105
Lyubashevsky, Vadim   I-551, II-22

Ma, Xiaoshuang   I-158
Maji, Hemanta K.   II-386
Malkin, Tal   II-42
Marcedone, Antonio   II-193
Martin, Daniel P.   I-223
Mather, Luke   I-243
Mavromati, Chrysanthi   I-420
Mennink, Bart   I-85, I-105
Mittelbach, Arno   II-122, II-142
Mohassel, Payman   II-486
Mouha, Nicky   I-105

Nandi, Mridul   I-126, I-478
Naor, Moni   II-254
Naya-Plasencia, María   I-179
Neven, Gregory   I-551
Nielsen, Jesper Buus   II-213
Nikolić, Ivica   I-141, I-458, II-274
Nikov, Ventzislav   II-326
Nikova, Svetla   II-326

Oswald, Elisabeth   I-223, I-243

Page, Daniel   I-223
Pass, Rafael   II-62
Paterson, Kenneth G.   I-398
Perret, Ludovic   I-21
Peters, Thomas   II-1
Peyrin, Thomas   II-274
Pieprzyk, Josef   I-141
Pierrot, Cécile   I-378
Pietrzak, Krzysztof   II-82
Poettering, Bertram   I-398
Portmann, Christopher   II-406
Prest, Thomas   II-22
Puglisi, Orazio   II-193

Qiao, Kexin   I-158

Rao, Vanishree   II-82
Raykov, Pavel   II-448
Renner, Renato   II-406
Rijmen, Vincent   II-326
Rioul, Olivier   II-344

Sadeghian, Saeed   II-486
Sahai, Amit   II-386
Sasaki, Yu   I-458
Scholl, Peter   II-506
Schröder, Dominique   I-512
Schuldt, Jacob C.N.   I-398
Schwabe, Peter   I-317
Shamir, Adi   I-439
Slamanig, Daniel   I-491
Smart, Nigel P.   II-486
Song, Ling   I-158
Stam, Martijin   I-223
Standaert, François-Xavier   I-282
Stepanovs, Igors   II-102
Suder, Valentin   I-179
Sun, Siwei   I-158

Teranishi, Isamu   II-42
Tessaro, Stefano   II-102
Tibouchi, Mehdi   I-262
Tunstall, Michael J.   I-223

Veyrat-Charvillon, Nicolas   I-282

Wang, Huaxiong   I-141
Wang, Peng   I-158
Wenzel, Jakob   II-289
Whitnall, Carolyn   I-243

Yasuda, Kan   I-105
Yogev, Eylon   II-254
Yun, Aaram   II-173
Yung, Moti   II-1, II-42

Zapalowicz, Jean-Christophe   I-262
Zhang, Jiang   II-366
Zhang, Zhenfeng   II-366
Zhang, Zongyang   II-366